physica status solidi (b)

ISSN 0370-1972 * VOL. 125 · NO. 1 · SEPTEMBER 1984

AKADEMIE-VERLAG · BERLIN

THEORIA CUM PRAXI

Classification Scheme

1. Structure of Crystalline Solids
    1.1 Perfectly Periodic Structure
    1.2 Solid-State Phase Transformations
    1.3 Alloys. Metallurgy
    1.4 Microstructure (Magnetic Domains See 18; Ferroelectric Domains See 14.4.1)
    1.5 Films
    1.6 Surfaces
2. Non-Crystalline State
3. Crystal Growth
4. Bonding Properties
5. Mössbauer Spectroscopy
6. Lattice Dynamics. Phonons
7. Acoustic Properties
8. Thermal Properties
9. Diffusion
10. Defect Properties (Irradiation Defects See 11)
    10.1 Metals
    10.2 Non-Metals
11. Irradiation Effects (X-Ray Diffraction Investigations See 1 and 10)
12. Mechanical Properties (Plastic Deformations See 10)
    12.1 Metals
    12.2 Non-Metals
13. Electron States
    13.1 Band Structure
    13.2 Fermi Surfaces
    13.3 Surface and Interface States
    13.4 Impurity and Defect States
    13.5 Elementary Excitations (Phonons See 6)
        13.5.1 Excitons
        13.5.2 Plasmons
        13.5.3 Polarons
        13.5.4 Magnons
14. Electrical Properties. Transport Phenomena
    14.1 Metals. Semi-Metals
    14.2 Superconductivity. Superconducting Materials and Devices
    14.3 Semiconductors
        14.3.1 Films
        14.3.2 Surfaces and Interfaces
        14.3.3 Devices. Junctions (Contact Problems See 14.3.4)
        14.3.4 High-Field Phenomena, Space-Charge Effects, Inhomogeneities, Injected Carriers (Electroluminescence See 20.3; Junctions See 14.3.3)
    14.4 Dielectrics
        14.4.1 Ferroelectrics
15. Thermoelectric and Thermomagnetic Properties
16. Photoconductivity. Photovoltaic Effects
17. Emission of Electrons and Ions
    17.1 Field Emission Microscope Investigations
18. Magnetic Properties
    18.1 Paramagnetic Properties
    18.2 Ferromagnetic Properties
        18.2.1 Ferromagnetic Films
    18.3 Ferrimagnetic Properties
    18.4 Antiferromagnetic Properties

(Continued on cover three)

physica status solidi (b)

basic research

Volume 125 · Number 1 · Pages 1 to 430, K1 to K92, and A1 to A8

September 1, 1984

PSSB 125(1) 1—430, K1—K92, A1—A8 (1984)

ISSN 0370-1972

AKADEMIE-VERLAG · BERLIN

Subscriptions and orders for single copies should be directed
in the GDR:

to the Postzeitungsvertrieb or
to the Akademie-Verlag, DDR-1086 Berlin, Leipziger Str. 3 – 4;

in the other socialist countries:

to a book-shop for foreign language literature or
to the competent news-distributing agency;

in the FRG and BERLIN (WEST):

to a book-shop or
to the wholesale distributing agency Kunst und Wissen, Erich Bieber, OHG, D-7000 Stuttgart 1, Wilhelmstr. 4 – 6;

in the other Western European countries:

to Kunst und Wissen, Erich Bieber GmbH, CH-8008 Zürich, Dufourstr. 51;

in USA and CANADA:

to Verlag Chemie International Inc., 303 N.W. 12th Avenue, Deerfield Beach, FL 33441, USA;

in other countries:

to the international book- and journal-selling trade,
to Buchexport, Volkseigener Außenhandelsbetrieb der Deutschen Demokratischen Republik, DDR-7010 Leipzig, Postfach 160, or
to the Akademie-Verlag, DDR-1086 Berlin, Leipziger Str. 3 – 4.

Editorial Note:

"physica status solidi (b)" undertakes that an original paper *accepted for publication* before the 8th of any month will be published within 50 days of this date unless the author requests a postponement. In special cases there may be some delay between *receipt* and *acceptance* of a paper due to the review and, if necessary, revision of the paper.

Schriftleiter und verantwortlich für den Inhalt:
Professor Dr. Dr. h. c. P. Görlich, DDR-1086 Berlin, Leipziger Str. 3 – 4; Postfach 1233 bzw. DDR-6900 Jena, Schillbachstr. 24.

Verlag:
Akademie-Verlag, DDR-1086 Berlin, Leipziger Str. 3 – 4; Fernruf: 2236221 und 2236229; Telex-Nr.: 114420; Bank: Staatsbank der DDR, Berlin, Kto.-Nr.: 6836-26-20712.

Chefredakteur:
Dr. H.-J. Hänsch.

Redaktionskollegium:
Prof. Dr. E. Gutsche, Dr. H.-J. Hänsch, Dr. H. Lange, Dr. S. Oberländer.

Anschrift der Redaktion:
DDR-1086 Berlin, Leipziger Str. 3 – 4; Postfach 1233; Fernruf: 2236279.

Veröffentlicht unter der Lizenznummer 1310 des Presseamtes beim Vorsitzenden des Ministerrates der Deutschen Demokratischen Republik.

Gesamtherstellung:
VEB Druckerei „Thomas Müntzer", DDR-5820 Bad Langensalza.

Erscheinungsweise:
Die Zeitschrift „physica status solidi (b)" erscheint jeweils am 1. eines jeden Monats. Jährlich erscheinen 6 Bände zu je 2 Heften.

Bezugspreis:
Je Band 200,— M zuzüglich Versandspesen (Preis für die DDR: 130,— M).

Bestellnummer dieses Bandes: 1068/125.

Printed in the German Democratic Republic.

AN (EDV) 20831

Author Index

Contents

Review Article

Original Papers and Short Notes

Structure

Lattice Properties

Defects, atomistic aspects

Magnetism

Extended electronic states and transitions

Localized electronic states and transitions

Electric transport

Pre-Printed Titles

Systematic List

Subject classification: Corresponding papers begin on the following pages (pages given in italics refer to the principle subject classification):

Contents of Volume 125 Continued on Page 433

Review Article

phys. stat. sol. (b) **125**, 11 (1984)

Subject classification: 6 and 18.2; 13; 22.5

School of Physics, University of Hyderabad[1])

Origin of Superexchange Interactions in 3d Magnetic Solids (I) [2])

By

K. N. SHRIVASTAVA

Contents

1. Introduction

In 1926, Heisenberg discovered that repulsive Coulomb interactions between electrons could lead to an exchange effect strongly coupling their spins, which gives rise to the spin-parallel, ferromagnetic configuration of atoms [1]. Later, Dirac [2] showed that the exchange can be written as

$$\mathcal{H}' = -\sum_{ij} J_{ij}^0 \mathbf{s}_i \cdot \mathbf{s}_j \tag{1.1}$$

with

$$J_{ij}^0 = \langle \psi_i(r_1)\, \psi_j(r_2)|\, \frac{e^2}{r_{12}}\, |\psi_j(r_1)\, \psi_i(r_2)\rangle\,. \tag{1.2}$$

This exchange in terms of e^2/r_{12} which is correct only for orthogonal orbitals, is positive (ferromagnetic). However, in general, the exchange is determined from the total Hamiltonian, $\mathcal{H}$, so that

$$J_{ij} = \langle \psi_i(r_1)\, \psi_j(r_2)|\, \mathcal{H}\, |\psi_j(r_1)\, \psi_i(r_2)\rangle\,. \tag{1.3}$$

If two electrons are in an attractive potential $-V(r)$, the exchange is given by

$$\begin{aligned} J_{ij} &= \langle \psi_i(r_1)\, \psi_j(r_2)|\, \frac{e^2}{r_{12}}\, |\psi_j(r_1)\, \psi_i(r_2)\rangle - 2\,\langle \psi_i(r_1)\mid \psi_j(r_1)\rangle\, \langle \psi_j(r_2)|\, V\, |\psi_i(r_2)\rangle \\ &= J_{ij}^0 - 2T_{ij}V_{ij} \end{aligned} \tag{1.4}$$

---

[1]) P. O. Central University, Hyderabad 500134, India.
[2]) Part II see phys. stat. sol. (b) **125**, No. 2 (1984).

which can be negative if the exchange without the potential term, J_{ij}^0, is smaller than twice the product of the overlap integral $T_{ij} = \langle \psi_i \mid \psi_j \rangle$, and the matrix element of the attractive potential. In 1932 Néel [3] suggested that the negative sign of the interaction could lead to a state in which different subsets or sublattices of the spins in a crystal could align themselves antiparallel (antiferromagnetic). Landau [4] stressed the phenomenological similarity between ferromagnetism and antiferromagnetism and predicted the existence of the Néel point. The first correct ideas of superexchange via nonmagnetic atoms which lead to antiferromagnetism were given by Kramers [5] and developed by Anderson [6].

Kittel [7] first realized the importance of elastic forces and found that the competition between the exchange and the elastic forces gave rise to biquadratic superexchange interactions, $j(\boldsymbol{S}_1 \cdot \boldsymbol{S}_2)^2$ which were important for the understanding of phase transitions from ferro- to antiferromagnetic state and for the explanation of magnetostriction, i.e., a change in volume of the crystal near the Curie (or the Néel) point. The origin of the biquadratic superexchange has been traced by Shrivastava [8] in the interaction of the electron orbit with lattice vibrations. The strong dependence of static exchange interaction upon the interatomic separation has been predicted by Shrivastava and Jaccarino [9] in accord with the experimental behavior [10].

The phonon-induced superexchange interactions in magnetic insulators are discovered by Shrivastava leading to a strong temperature dependence in the effective exchange in both ferromagnetic [11] and antiferromagnetic [12] systems. These interactions arise because of the coupling of the electron orbit with lattice phonons in the solid.

In this article we discuss a formulation for the evaluation of eigenvalues of a four-electron system when the electrons are not necessarily on the same atom. This technique is used to calculate the singlet and triplet state energies leading to the exchange interaction in magnetic pairs of atoms. It is found that the exchange interaction is made up of various terms arising from the kinetic energy, the Coulomb repulsion, and the electron–nuclear attraction in which several terms arise as a result of rearrangement of charges due to transfer of electrons from the occupied orbits to the empty ones. Thus the superexchange interactions arising may lead to cancellations of the exchange integral owing to repulsive Coulomb interaction alone. The dominance of the repulsive Coulomb interactions between electrons in the 3d metal ions over the attractive electron–nuclear interaction integrals leads to ferromagnetism, otherwise, if the attractive forces dominate, there results antiferromagnetism. In the first-principles calculation the two types of interactions occur with different signs so that the phenomenon of "weak magnetism" is predicted and if a parameter is varied, there can be a phase transition from ferro- to antiferromagnetism. We have found the phonon-induced superexchange in which the temperature is a convenient parameter. It is indeed found that at a particular temperature characteristic of the material, called the "zero", the exchange interaction becomes zero. The exchange and superexchange depend strongly on the interatomic distance leading to the exchange Grüneisen parameter of the order of 12. The bond angle and interatomic distance dependent superexchange interactions also suggest the occurrence of both ferro- as well as antiferromagnetic compounds for the same magnetic atom such as Cr^{3+}. We have also found the deviations of the Heisenberg exchange Hamiltonian $\sum_{ij} J\boldsymbol{S}_i \cdot \boldsymbol{S}_j$ in terms of the dynamics of the lattice. The electron–phonon interaction leads to a biquadratic superexchange $\sum_{ij} j(\boldsymbol{S}_i \cdot \boldsymbol{S}_j)^2$ consistent with measured magnetostriction which corrects the Heisenberg exchange which is true in the zero-temperature limit of the proper dynamical theory.

2. Four-Electron Eigenvalue Technique

The eigenvalues of a many-electron system cannot be obtained easily as the method of determinantal wave functions is considerably involved and cannot be managed for a large number of electrons. The vector model of Dirac can be handled for a small number of electrons. Its development by Van Vleck [13] and Serber [14] permits the transfer of electrons from one atom to another so that a few configurations can be mixed and their eigenvalues can be obtained without evaluating the complete energy matrix. Mixing of a small number of configurations may be sufficient for a few atoms and for small molecules on the basis of energy arguments. The vector model followed here is sufficient to predict the order of magnitude of various interactions such as the hyperfine interaction, the transferred hyperfine interaction [15 to 18], the optical level separations, and the exchange interactions in ion pairs [19 to 27].

Consider a configuration A, specified by n orthogonal orbitals $a_1(r_1), a_2(r_2), \dots, a_n(r_n)$ which are all different. There are $n!$ wave functions, obtained by operating the permutation operators of the orbitals upon the product function

$$\psi_A = a_1(r_1)\, a_2(r_2) \dots a_n(r_n) \tag{2.1}$$

any one of which may be represented by

$$P_{ij\dots kl,\, st\dots pq\dots}\psi_A\,, \tag{2.2}$$

where P permutes the labels of the orbitals $a_i(r_i)$, i.e.,

$$P_{ij\dots kl,\, st\dots pq,\dots}\psi_A = \\ = a_1(r_1)\, a_2(r_2) \dots a_j(r_i) \dots a_l(r_k) \dots a_i(r_l) \dots a_t(r_s) \dots a_q(r_p) \dots a_s(r_q) \dots a_n(r_n)\,. \tag{2.3}$$

In addition to defining the permutation P of orbitals of ψ_A, we also define the inverse operator P_r^{-1} which permutes the coordinates r_i in ψ_A,

$$P\psi_A = P_r^{-1}\psi_A\,. \tag{2.4}$$

The $n!$ wave functions form an $n!n!$ energy matrix, the elements of which are of the form $\langle P_a\psi_A|\mathcal{H}|P_b\psi_A\rangle$, where P_a and P_b are two arbitrary elements of the permutation group for n orbits and $\mathcal{H}$ is the Hamiltonian of the system of n electrons. Using (2.4) and the invariance property of the Hamiltonian under the permutation group, we obtain

$$\mathcal{H} P_{b,r}^{-1} = P_{b,r}^{-1}\mathcal{H}\,, \qquad (P_{a,r}^{-1}\psi_A)^*\, P_{b,r}^{-1} = (P_{b,r}P_{a,r}^{-1}\psi_A)^*\,. \tag{2.5}$$

A particular matrix element can then be written as

$$\begin{aligned}\langle P_a\psi_A|\,\mathcal{H}\,|P_b\psi_A\rangle &= \\ &= \int (P_{a,r}^{-1}\psi_A)^*\,\mathcal{H} P_{b,r}^{-1}\psi_A\, \mathrm{d}r_1\, \mathrm{d}r_2 \dots \mathrm{d}r_n \\ &= \int [(P_{a,r}P_{b,r}^{-1})^{-1}\,\psi_A]^*\,\mathcal{H}\psi_A\, \mathrm{d}r_1\, \mathrm{d}r_2 \dots \mathrm{d}r_n \\ &= \int (P_aP_b^{-1}\psi_A)^*\,\mathcal{H}\psi_A\, \mathrm{d}r_1\, \mathrm{d}r_2 \dots \mathrm{d}r_n\,.\end{aligned} \tag{2.6}$$

We define

$$\mathcal{H}_P^{AA} = \langle P\psi_A|\,\mathcal{H}\,|\psi_A\rangle\,, \tag{2.7}$$

then it is seen from (2.6) that the coefficient of $\mathcal{H}_P^{AA}$ in the energy matrix is itself a matrix, whose elements are zero except when $P = P_aP_b^{-1}$, in which case they are unity. We define a product function, e.g.,

$$\psi_A = a_1(r_1)\, a_2(r_2)\, a_3(r_3)\,. \tag{2.8}$$

There are 3! permutations, out of which three are independent, the identity P_{I}, P_{13} and P_{23}, and three are dependent $P_{12} = P_{12}P_{\mathrm{I}}$, $P_{123} = P_{12}P_{23}$, $P_{132} = P_{12}P_{13}$. The inverse of these six permutation operators are

$$\begin{aligned} P_{\mathrm{I}}^{-1} &= P_{\mathrm{I}}\,, & P_{12}^{-1} &= P_{12}\,, & P_{13}^{-1} &= P_{13}\,, \\ P_{23}^{-1} &= P_{23}\,, & P_{123}^{-1} &= P_{132}\,, & P_{132}^{-1} &= P_{123}\,. \end{aligned} \tag{2.9}$$

For convenience, we write their products in Table 1, in which the inverse operator is applied first. Using (2.6) and Table 1 we construct the energy matrix given in Table 2 in which the elements $\langle P\psi_{\mathrm{A}}|\, \mathcal{H}\, |\psi_{\mathrm{A}}\rangle$ factor out to lead to a simpler form given in Table 3. The coefficient of $\mathcal{H}_P^{\mathrm{AA}}$ is a matrix whose elements are unity for $P_a P_b^{-1} = P$ and zero ortherwise. The matrix representing the permutation operator P itself consists of elements $\langle P_a\psi_{\mathrm{A}} \mid PP_b\psi_{\mathrm{A}}\rangle$ which are zero because of the orthogonality between the one-electron orbitals, except for $P = P_a P_b^{-1}$ in which case they are unity. Therefore, the coefficient of $\mathcal{H}_P^{\mathrm{AA}}$ is simply the matrix representation P of the permutation operator and the energy matrix arising from the configuration A becomes

$$\mathcal{H}^{\mathrm{AA}} = \sum_P \mathcal{H}_P^{\mathrm{AA}} P\,. \tag{2.10}$$

Similarly, for another configuration, the diagonal elements are

$$\mathcal{H}^{\mathrm{BB}} = \sum_P \langle P\psi_{\mathrm{B}}|\, \mathcal{H}\, |\psi_{\mathrm{B}}\rangle = \sum_P \mathcal{H}_P^{\mathrm{BB}} P \tag{2.11}$$

and the off-diagonal elements are

$$\mathcal{H}^{\mathrm{AB}} = \sum_P \langle P\psi_{\mathrm{A}}|\, \mathcal{H}\, |\psi_{\mathrm{B}}\rangle = \sum_P \mathcal{H}_P^{\mathrm{AB}} P\,. \tag{2.12}$$

So far we have discussed the problem of evaluating the permutation operator in which only two indices occur at a time. We have considered $n!$ wave functions in the product form rather than in the form of a Slater determinant. For a system of n electrons, there are 2^n linearly independent spin states. The orbital permutation operator P can be obtained from the spin representation P^σ by

$$P = (-1)^P\, P^\sigma\,. \tag{2.13}$$

The operator P_{ij}^σ which permutes the spins of electrons i and j can be expressed [2] in terms of the spins as

$$P_{ij}^\sigma = \tfrac{1}{2}\,(1 + 4\mathbf{s}_i \cdot \mathbf{s}_j) \tag{2.14}$$

so that we can write

$$P_{ij} = -\tfrac{1}{2}\,(1 + 4\mathbf{s}_i \cdot \mathbf{s}_j) \tag{2.15}$$

Table 1

Multiplication table for the permutation group for $n = 3$

	P_{I}^{-1}	P_{12}^{-1}	P_{13}^{-1}	P_{23}^{-1}	P_{123}^{-1}	P_{132}^{-1}
P_{I}	P_{I}	P_{12}	P_{13}	P_{23}	P_{132}	P_{123}
P_{12}	P_{12}	P_{I}	P_{132}	P_{123}	P_{13}	P_{23}
P_{13}	P_{13}	P_{123}	P_{I}	P_{132}	P_{23}	P_{12}
P_{23}	P_{23}	P_{132}	P_{123}	P_{I}	P_{12}	P_{13}
P_{123}	P_{123}	P_{13}	P_{23}	P_{12}	P_{I}	P_{132}
P_{132}	P_{132}	P_{23}	P_{12}	P_{13}	P_{123}	P_{I}

Table 2

The energy matrix for a product of three orbitals

	$P_I\psi_A$	$P_{12}\psi_A$	$P_{13}\psi_A$	$P_{23}\psi_A$	$P_{123}\psi_A$	$P_{132}\psi_A$
$P_I\psi_A$	$\langle P_I\psi_A \vert \mathcal{H} \vert \psi_A\rangle$	$\langle P_{12}\psi_A \vert \mathcal{H} \vert \psi_A\rangle$	$\langle P_{13}\psi_A \vert \mathcal{H} \vert \psi_A\rangle$	$\langle P_{23}\psi_A \vert \mathcal{H} \vert \psi_A\rangle$	$\langle P_{132}\psi_A \vert \mathcal{H} \vert \psi_A\rangle$	$\langle P_{123}\psi_A \vert \mathcal{H} \vert \psi_A\rangle$
$P_{12}\psi_A$	$\langle P_{12}\psi_A \vert \mathcal{H} \vert \psi_A\rangle$	$\langle P_I\psi_A \vert \mathcal{H} \vert \psi_A\rangle$	$\langle P_{132}\psi_A \vert \mathcal{H} \vert \psi_A\rangle$	$\langle P_{123}\psi_A \vert \mathcal{H} \vert \psi_A\rangle$	$\langle P_{13}\psi_A \vert \mathcal{H} \vert \psi_A\rangle$	$\langle P_{23}\psi_A \vert \mathcal{H} \vert \psi_A\rangle$
$P_{13}\psi_A$	$\langle P_{13}\psi_A \vert \mathcal{H} \vert \psi_A\rangle$	$\langle P_{123}\psi_A \vert \mathcal{H} \vert \psi_A\rangle$	$\langle P_I\psi_A \vert \mathcal{H} \vert \psi_A\rangle$	$\langle P_{132}\psi_A \vert \mathcal{H} \vert \psi_A\rangle$	$\langle P_{23}\psi_A \vert \mathcal{H} \vert \psi_A\rangle$	$\langle P_{12}\psi_A \vert \mathcal{H} \vert \psi_A\rangle$
$P_{23}\psi_A$	$\langle P_{23}\psi_A \vert \mathcal{H} \vert \psi_A\rangle$	$\langle P_{132}\psi_A \vert \mathcal{H} \vert \psi_A\rangle$	$\langle P_{123}\psi_A \vert \mathcal{H} \vert \psi_A\rangle$	$\langle P_I\psi_A \vert \mathcal{H} \vert \psi_A\rangle$	$\langle P_{12}\psi_A \vert \mathcal{H} \vert \psi_A\rangle$	$\langle P_{13}\psi_A \vert \mathcal{H} \vert \psi_A\rangle$
$P_{123}\psi_A$	$\langle P_{123}\psi_A \vert \mathcal{H} \vert \psi_A\rangle$	$\langle P_{13}\psi_A \vert \mathcal{H} \vert \psi_A\rangle$	$\langle P_{23}\psi_A \vert \mathcal{H} \vert \psi_A\rangle$	$\langle P_{12}\psi_A \vert \mathcal{H} \vert \psi_A\rangle$	$\langle P_I\psi_A \vert \mathcal{H} \vert \psi_A\rangle$	$\langle P_{132}\psi_A \vert \mathcal{H} \vert \psi_A\rangle$
$P_{132}\psi_A$	$\langle P_{132}\psi_A \vert \mathcal{H} \vert \psi_A\rangle$	$\langle P_{23}\psi_A \vert \mathcal{H} \vert \psi_A\rangle$	$\langle P_{12}\psi_A \vert \mathcal{H} \vert \psi_A\rangle$	$\langle P_{13}\psi_A \vert \mathcal{H} \vert \psi_A\rangle$	$\langle P_{123}\psi_A \vert \mathcal{H} \vert \psi_A\rangle$	$\langle P_I\psi_A \vert \mathcal{H} \vert \psi_A\rangle$

Table 3

The energy matrix of Table 2 in the factorized form with $\mathcal{H}_{123}^{AA} = \langle P_{123}\psi_A | \mathcal{H} | \psi_A \rangle$

$$\mathcal{H}_{I}^{AA}\begin{bmatrix}1&0&0&0&0&0\\0&1&0&0&0&0\\0&0&1&0&0&0\\0&0&0&1&0&0\\0&0&0&0&1&0\\0&0&0&0&0&1\end{bmatrix} + \mathcal{H}_{12}^{AA}\begin{bmatrix}0&1&0&0&0&0\\1&0&0&0&0&0\\0&0&0&0&0&1\\0&0&0&0&1&0\\0&0&0&1&0&0\\0&0&1&0&0&0\end{bmatrix} + \mathcal{H}_{13}^{AA}\begin{bmatrix}0&0&1&0&0&0\\0&0&0&0&1&0\\1&0&0&0&0&0\\0&0&0&0&0&1\\0&1&0&0&0&0\\0&0&0&1&0&0\end{bmatrix} +$$

$$+ \mathcal{H}_{23}^{AA}\begin{bmatrix}0&0&0&1&0&0\\0&0&0&0&0&1\\0&0&0&0&1&0\\1&0&0&0&0&0\\0&0&1&0&0&0\\0&1&0&0&0&0\end{bmatrix} + \mathcal{H}_{123}^{AA}\begin{bmatrix}0&0&0&0&0&1\\0&0&0&1&0&0\\0&1&0&0&0&0\\0&0&1&0&0&0\\1&0&0&0&0&0\\0&0&0&0&1&0\end{bmatrix} + \mathcal{H}_{132}^{AA}\begin{bmatrix}0&0&0&0&1&0\\0&0&1&0&0&0\\0&0&0&1&0&0\\0&1&0&0&0&0\\0&0&0&0&0&1\\1&0&0&0&0&0\end{bmatrix}$$

and any permutation P can be expressed in terms of the spins by writing it as a product of interchanges. The factors of the wave functions in a given configuration belonging to orbitals centered at different lattice sites may be nonorthogonal. Therefore, the secular equation is written as

$$\begin{vmatrix}(\mathcal{H}-W)^{AA} & (\mathcal{H}-W)^{AB} & . & . \\ (\mathcal{H}-W)^{BA} & (\mathcal{H}-W)^{BB} & . & . \\ . & . & . & .\end{vmatrix} = 0\,. \tag{2.16}$$

If two orbitals b_1 and b_2 of a given configuration $\psi_B = b_1b_2b_3$ are identical, but all the orbitals of the configuration $\psi_A = a_1a_2a_3$ are different, we choose a representation in which P_{12} is diagonal. Let g be the rank of the P^{AA}'s constructed for the configuration ψ_A and r be the number of eigenvalues of P_{12} which are $+1$, $(g - r)$ is then the number of eigenvalues which are -1. We observe that $(g - r)$ rows and columns arising from the configuration ψ_B for which P_{12} has the diagonal elements -1, must be removed because $b_1 = b_2$ will lead to zero contribution. Hence the matrices which enter in the expansion for $(\mathcal{H} - W)^{BB}$ are of the rank r. Then the secular equation contains block matrices where $(\mathcal{H} - W)^{AA}$ is $g \times g$, $(\mathcal{H} - W)^{BB}$ is $r \times r$, $(\mathcal{H} - W)^{AB}$ is $r \times g$, and $(\mathcal{H} - W)^{BA}$ is $g \times r$, where $(\mathcal{H} - W)^{BA} = \sum_P (\mathcal{H} - W)_P^{BA} P^{BA}$, etc. as

$$\begin{array}{r|c:c:c|l} & g \text{ columns} & r \text{ columns} & & \\ g \text{ rows} & \sum_P (\mathcal{H}-W)_P^{AA}\, P^{AA} & \sqrt{2}\,[\sum_P{}' (\mathcal{H}-W)_P^{BA}\, P^{BA}]^+ & \cdots & \\ \hdashline r \text{ rows} & \sqrt{2}\,\sum_P{}' (\mathcal{H}-W)_P^{BA}\, P^{BA} & \sum_P{}' (\mathcal{H}-W)_P^{BB}\, P^{BB} & \cdots & = 0\,. \\ & . & . & \cdots & \end{array}$$

Since b_1 and b_2 are identical orbitals for two permutations P' and P, $P' = P_{12}P$, we obtain

$$(\mathcal{H} - W)_{P'}^{BA} = (\mathcal{H} - W)_P^{BA}\,, \qquad (\mathcal{H} - W)_{P'}^{BB} = (\mathcal{H} - W)_P^{BB}\,. \tag{2.17}$$

Further,

$$P'_{ij} = (P_{12})_{ik}\, P_{kj} = \delta_{ik} P_{kj} = P_{ij} \tag{2.18}$$

so that

$$P'^{BA} = P^{BA}\,, \qquad P'^{BB} = P^{BB}\,. \tag{2.19}$$

Using (2.17) and (2.19) the contributions of P' and P to the energy matrix can be written as

$$\begin{aligned}(\mathcal{H} - W)^{BA}_{P'}\, P'^{BA} + (\mathcal{H} - W)^{BA}_{P}\, P^{BA} &= 2(\mathcal{H} - W)^{BA}_{P}\, P^{BA}\,,\\ (\mathcal{H} - W)^{BB}_{P'}\, P'^{BB} + (\mathcal{H} - W)^{BB}_{P}\, P^{BB} &= 2(\mathcal{H} - W)^{BB}_{P}\, P^{BB}\,.\end{aligned} \tag{2.20}$$

Therefore, we have to consider only $n!/2$ elements of the permutation group which are related to the $n!/2$ elements of a multiple of P_{12}. For example, for a system of four electrons in which two electrons 1 and 2 are on identical orbits, we form two sets from the four permutations,

$$\begin{aligned}&\mathcal{P}\colon P_{I}, P_{13}, P_{14}, P_{23}, P_{24}, P_{34}, P_{134}, P_{143}, P_{234}, P_{243}, P_{1324}, P_{1423}\,;\\ &\mathcal{P}_{12}\colon P_{12}, P_{132}, P_{142}, P_{123}, P_{124}, P_{1234}, P_{1342}, P_{1432}, P_{1234}, P_{1243}, P_{1324}, P_{1423}\,.\end{aligned} \tag{2.21}$$

In general there may be r_i pairs of identical orbitals in a given configuration ψ_{R_i}. We suppose that the first pair of filled orbitals is 1, 2, the second 3, 4, etc. The permutations $P_{12}, P_{34}, \ldots P_{2n-1,2n}, \ldots$ all commute so that we can find a representation in which they are simultaneously diagonal. The matrix element is then

$$(\mathcal{H} - W)^{R_iR_j} = 2^{(r_i-r_j)/2} \sum_P{}' (\mathcal{H} - W)^{R_iR_j}_{P}\, P^{R_iR_j}\,, \tag{2.22}$$

where $r_i > r_j$ and the summation is taken over $n!/2^{r_i}$ independent permutations. $P^{R_iR_j}$ are obtained from P by removing the rows for which an interchange of identical orbitals of R_i has the eigenvalue -1, and removing the columns for which an interchange of identical orbitals of R_j has the eigenvalue -1. The permutation operator P^{σ}_{ij} which interchanges the spins of electrons i and j is given by (2.14). It is invariant under a rotation of the spin axes and commutes with S^2. For the case of a four-electron system, there exist 2^4 linearly independent spin states. We show Van Vleck's branching diagram in Fig. 1. For one electron it shows one spin doublet, for two electrons there is one singlet and one triplet, for three electrons there are two doublets and one quartet, and for the case of a four-electron system there are two singlets, three triplets, and one quintet. Since P^{σ} is diagonal in S and M_S, its 16×16 dimensional matrix representation has the block form displayed in Fig. 2 in which Q, T, and σ are the reduced matrices describing the quintet, the triplet, and the singlet states, respectively.

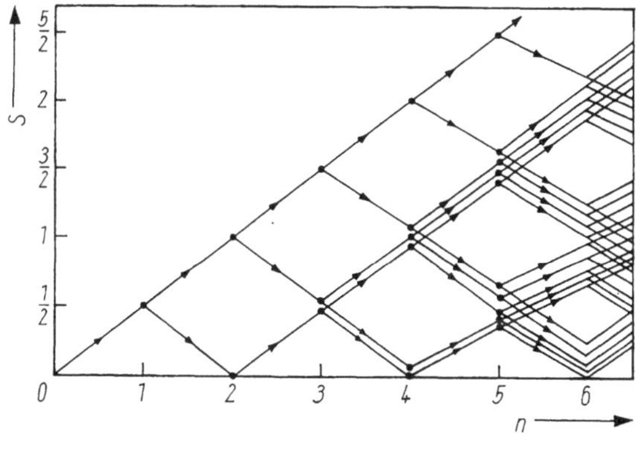

Fig. 1. Van Vleck's branching diagram for the eigenstates of S^2. For one-electron system, $n = 1$, there is one doublet. For two electrons, $n = 2$, there is one singlet and one triplet. For $n = 3$, there are two doublets and one quartet. For $n = 4$, there are two singlets, three triplets, and one quintet. For $n = 5$, there are five doublets, four quartets, and one sextet. For $n = 6$ there are five singlets, nine triplets, five quintets, and one septet

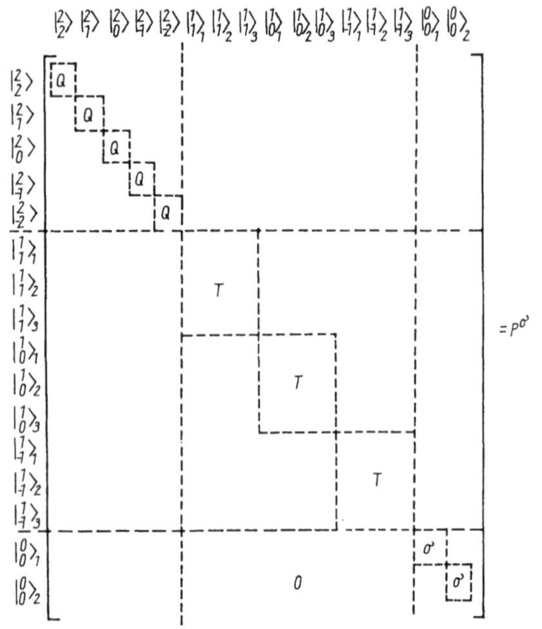

Fig. 2. Block form of P^σ showing two singlets, three triplets, and one quintet for a four-electron system

The states corresponding to $n = 4$ and $S = 0$ are found to be

$$\left|\begin{matrix}0\\0\end{matrix}\right\rangle_1 = \frac{1}{\sqrt{3}}\left[\alpha_1\alpha_2\beta_3\beta_4 - \frac{1}{2}(\alpha_1\beta_2 + \beta_1\alpha_2)(\alpha_3\beta_4 + \beta_3\alpha_4) + \beta_1\beta_2\alpha_3\alpha_4\right],$$
$$\left|\begin{matrix}0\\0\end{matrix}\right\rangle_2 = \frac{1}{2}(\alpha_1\beta_2 - \beta_1\alpha_2)(\beta_3\alpha_4 - \alpha_3\beta_4)\,. \tag{2.23}$$

Operating with the spin-permutation operators P^σ_{I}, P^σ_{12}, P^σ_{13}, and P^σ_{14}, we find

$$P^\sigma_{\mathrm{I}}\left|\begin{matrix}0\\0\end{matrix}\right\rangle_1 = \left|\begin{matrix}0\\0\end{matrix}\right\rangle_1, \quad P^\sigma_{12}\left|\begin{matrix}0\\0\end{matrix}\right\rangle_1 = \left|\begin{matrix}0\\0\end{matrix}\right\rangle_1, \quad P^\sigma_{\mathrm{I}}\left|\begin{matrix}0\\0\end{matrix}\right\rangle_2 = \left|\begin{matrix}0\\0\end{matrix}\right\rangle_2, \quad P^\sigma_{12}\left|\begin{matrix}0\\0\end{matrix}\right\rangle_2 = -\left|\begin{matrix}0\\0\end{matrix}\right\rangle_2, \tag{2.24}$$

$$P^\sigma_{13}\left|\begin{matrix}0\\0\end{matrix}\right\rangle_1 = \frac{1}{\sqrt{3}}\left[\beta_1\alpha_2\alpha_3\beta_4 - \frac{1}{2}(\alpha_3\beta_2 + \beta_3\alpha_2)(\alpha_1\beta_4 + \beta_1\alpha_4) + \alpha_1\beta_2\beta_3\alpha_4\right] =$$
$$= -\frac{1}{2}\left|\begin{matrix}0\\0\end{matrix}\right\rangle_1 - \sqrt{\frac{3}{4}}\left|\begin{matrix}0\\0\end{matrix}\right\rangle_2, \tag{2.25}$$

$$P^\sigma_{13}\left|\begin{matrix}0\\0\end{matrix}\right\rangle_2 = \frac{1}{2}(\alpha_3\beta_2 - \beta_3\alpha_2)(\alpha_1\beta_4 - \beta_1\alpha_4) = -\sqrt{\frac{3}{4}}\left|\begin{matrix}0\\0\end{matrix}\right\rangle_1 + \frac{1}{2}\left|\begin{matrix}0\\0\end{matrix}\right\rangle_2,$$

$$P^\sigma_{14}\left|\begin{matrix}0\\0\end{matrix}\right\rangle_1 = \sqrt{\frac{1}{3}}\left[\beta_1\alpha_2\beta_3\alpha_4 - \frac{1}{2}(\alpha_4\beta_2 + \beta_4\alpha_2)(\alpha_3\beta_1 + \beta_3\alpha_1) + \alpha_1\beta_2\alpha_3\beta_4\right] =$$
$$= -\frac{1}{2}\left|\begin{matrix}0\\0\end{matrix}\right\rangle_1 + \sqrt{\frac{3}{4}}\left|\begin{matrix}0\\0\end{matrix}\right\rangle_2, \tag{2.26}$$

$$P^\sigma_{14}\left|\begin{matrix}0\\0\end{matrix}\right\rangle_2 = \frac{1}{2}(\alpha_4\beta_2 - \beta_4\alpha_2)(\alpha_3\beta_1 - \beta_3\alpha_1) = \sqrt{\frac{3}{4}}\left|\begin{matrix}0\\0\end{matrix}\right\rangle_1 + \frac{1}{2}\left|\begin{matrix}0\\0\end{matrix}\right\rangle_2,$$

The matrix elements P^σ_{ij} of the representation of a spin-permutation operator P^σ are determined by

$$P^\sigma \left| \begin{matrix} S \\ M_S \end{matrix} \right\rangle_i = \sum_{j=i}^{m} P^\sigma_{ji} \left| \begin{matrix} S \\ M_S \end{matrix} \right\rangle_j , \tag{2.27}$$

where m is the number of a given spin multiplet. Consequently, the P^σ's are given by

$$P^\sigma_{\mathrm{I}} = \begin{pmatrix} 1 & 0 \\ 0 & 1 \end{pmatrix}, \qquad P^\sigma_{12} = \begin{pmatrix} 1 & 0 \\ 0 & -1 \end{pmatrix},$$
$$P^\sigma_{13} = \begin{pmatrix} -\frac{1}{2} & -\sqrt{\frac{3}{4}} \\ \sqrt{\frac{3}{4}} & \frac{1}{2} \end{pmatrix}, \qquad P^\sigma_{14} = \begin{pmatrix} -\frac{1}{2} & \sqrt{\frac{3}{4}} \\ \sqrt{\frac{3}{4}} & \frac{1}{2} \end{pmatrix}. \tag{2.28}$$

From the symmetry of (2.23) we obtain

$$P^\sigma_{34} = P^\sigma_{12} , \qquad P^\sigma_{24} = P^\sigma_{13} , \qquad P^\sigma_{23} = P^\sigma_{14} . \tag{2.29}$$

The P^σ's for higher-order permutations are given by the products of the above representations for the "interchange" operators.

The three triplet states for $M_S = 1$ are found to be

$$\left| \begin{matrix} 1 \\ 1 \end{matrix} \right\rangle_1 = \frac{1}{\sqrt{2}} \left[\alpha_1\alpha_2 \frac{1}{\sqrt{2}} (\alpha_3\beta_4 + \beta_3\alpha_4) - \frac{1}{\sqrt{2}} (\alpha_1\beta_2 + \beta_1\alpha_2)\, \alpha_3\alpha_4 \right],$$
$$\left| \begin{matrix} 1 \\ 1 \end{matrix} \right\rangle_2 = \alpha_1\alpha_2 \frac{1}{\sqrt{2}} (\alpha_3\beta_4 - \beta_3\alpha_4) , \tag{2.30}$$
$$\left| \begin{matrix} 1 \\ 1 \end{matrix} \right\rangle_3 = \frac{1}{\sqrt{2}} (\alpha_1\beta_2 - \beta_1\alpha_2)\, \alpha_3\alpha_4 .$$

The state $\left| \begin{matrix} 1 \\ 1 \end{matrix} \right\rangle_1$ couples $|S_{12} = 1\rangle$ with $|S_{34} = 1\rangle$, $\left| \begin{matrix} 1 \\ 1 \end{matrix} \right\rangle_2$ couples $|S_{12} = 1\rangle$ with $|S_{34} = 0\rangle$, and $\left| \begin{matrix} 1 \\ 1 \end{matrix} \right\rangle_3$ couples $|S_{12} = 0\rangle$ with $|S_{34} = 1\rangle$. Using (2.27) we find for the interchange operators,

$$P^\sigma_{12} = \begin{pmatrix} 1 & 0 & 0 \\ 0 & 1 & 0 \\ 0 & 0 & -1 \end{pmatrix}, \qquad P^\sigma_{34} = \begin{pmatrix} 1 & 0 & 0 \\ 0 & -1 & 0 \\ 0 & 0 & 1 \end{pmatrix},$$
$$P^\sigma_{13} = \begin{pmatrix} 0 & \sqrt{\frac{1}{2}} & -\sqrt{\frac{1}{2}} \\ \sqrt{\frac{1}{2}} & \frac{1}{2} & \frac{1}{2} \\ -\sqrt{\frac{1}{2}} & \frac{1}{2} & \frac{1}{2} \end{pmatrix}, \qquad P^\sigma_{14} = \begin{pmatrix} 0 & -\sqrt{\frac{1}{2}} & \sqrt{\frac{1}{2}} \\ -\sqrt{\frac{1}{2}} & \frac{1}{2} & \frac{1}{2} \\ -\sqrt{\frac{1}{2}} & -\frac{1}{2} & \frac{1}{2} \end{pmatrix}, \tag{2.31}$$

$$P_{23}^{\sigma} = \begin{pmatrix} 0 & \sqrt{\frac{1}{2}} & \sqrt{\frac{1}{2}} \\ \sqrt{\frac{1}{2}} & \frac{1}{2} & -\frac{1}{2} \\ \sqrt{\frac{1}{2}} & -\frac{1}{2} & \frac{1}{2} \end{pmatrix}, \qquad P_{24}^{\sigma} = \begin{pmatrix} 0 & -\sqrt{\frac{1}{2}} & \sqrt{\frac{1}{2}} \\ -\sqrt{\frac{1}{2}} & \frac{1}{2} & \frac{1}{2} \\ \sqrt{\frac{1}{2}} & \frac{1}{2} & \frac{1}{2} \end{pmatrix},$$

and the quintet state for $M_S = 2$ is given by

$$\left|\begin{matrix}2\\2\end{matrix}\right\rangle = |\alpha_1\alpha_2\alpha_3\alpha_4\rangle \tag{2.32}$$

for which the P^σ's are unity. The above representations for the P^σ's are valid when all the electron orbitals are different. If any two orbital wave functions a_1 and a_2 are identical, the spins s_1 and s_2 must be coupled into a singlet state. If the electron is transferred from the orbital $a_2\!\uparrow$ to $b_2\!\uparrow$ keeping the same spin state as without the transfer, the combinations formed by s_1 and s_2 cannot be changed so that only $\left|\begin{matrix}0\\0\end{matrix}\right\rangle_2$ in (2.23) and $\left|\begin{matrix}1\\1\end{matrix}\right\rangle_3$ in (2.30) are allowed.

Consider a four-electron system with $S = 0$. The one-electron orbitals are represented by a_1, a_2, a_3, and a_4 which are orthonormal. The secular equation is given by

$$|\sum_P (\mathcal{H} - W)_P\, P| = 0 \tag{2.33}$$

with

$$(\mathcal{H} - W)_P = \langle Pa_1a_2a_3a_4|\, \mathcal{H} - W\, |a_1a_2a_3a_4\rangle \tag{2.34}$$

and the Hamiltonian is

$$\mathcal{H} = -\sum_i \frac{\hbar^2}{2m} \nabla_i^2 + \sum_{i>j} \frac{e^2}{r_{ij}}. \tag{2.35}$$

Using (2.28) and (2.29) we find

$$\begin{vmatrix} A_{11} & A_{12} \\ A_{21} & A_{22} \end{vmatrix} = 0\,; \tag{2.36}$$

$$\begin{aligned} A_{11} &= \mathcal{H}_\mathrm{I} - \mathcal{H}_{12} - \mathcal{H}_{34} - \tfrac{1}{2}(\mathcal{H}_{13} + \mathcal{H}_{14} + \mathcal{H}_{23} + \mathcal{H}_{24}) - W\,, \\ A_{22} &= \mathcal{H}_\mathrm{I} + \mathcal{H}_{12} + \mathcal{H}_{34} - \tfrac{1}{2}(\mathcal{H}_{13} + \mathcal{H}_{14} + \mathcal{H}_{23} + \mathcal{H}_{24}) - W\,, \\ A_{12} &= A_{21} = (\tfrac{3}{4})^{1/2}(\mathcal{H}_{13} + \mathcal{H}_{24} - \mathcal{H}_{14} - \mathcal{H}_{23})\,, \end{aligned} \tag{2.37}$$

where

$$\mathcal{H}_{ij} = \langle a_j a_i|\, \frac{e^2}{r_{12}}\, |a_i a_j\rangle\,, \qquad \mathcal{H}_\mathrm{I} = \sum_i \langle a_i|\, \frac{\hbar^2}{2m} \nabla_i^2\, |a_i\rangle\,. \tag{2.38}$$

The solution of (2.36) is then

$$\begin{aligned} W = \mathcal{H}_\mathrm{I} \pm [&(\mathcal{H}_{12} + \mathcal{H}_{34})^2 + (\mathcal{H}_{14} + \mathcal{H}_{23})^2 + (\mathcal{H}_{13} + \mathcal{H}_{24})^2 - \\ &- (\mathcal{H}_{12} + \mathcal{H}_{34})(\mathcal{H}_{13} + \mathcal{H}_{24}) - (\mathcal{H}_{12} + \mathcal{H}_{34})(\mathcal{H}_{14} + \mathcal{H}_{23}) - \\ &- (\mathcal{H}_{13} + \mathcal{H}_{24})(\mathcal{H}_{14} + \mathcal{H}_{23})]^{1/2}\,. \end{aligned} \tag{2.39}$$

The same result can be obtained by Slater's determinantal wave functions.

For a configuration A consider an antisymmetrized wave function Φ_i^A which is the i-th eigenstate of S^2 and S_z. For the four electron system singlet state, one of the eigenfunctions is given by

$$\frac{1}{2}\left[\begin{pmatrix} a_1 & a_2 & a_3 & a_4 \\ \alpha & \beta & \alpha & \beta \end{pmatrix} - \begin{pmatrix} a_1 & a_2 & a_3 & a_4 \\ \beta & \alpha & \alpha & \beta \end{pmatrix} - \begin{pmatrix} a_1 & a_2 & a_3 & a_4 \\ \alpha & \beta & \beta & \alpha \end{pmatrix} + \begin{pmatrix} a_1 & a_2 & a_3 & a_4 \\ \beta & \alpha & \alpha & \beta \end{pmatrix}\right] =$$

$$= \frac{1}{(4!)^{1/2}} \sum_P (-1)^P PP^\sigma a_1 a_2 a_3 a_4 \left[\frac{1}{2}(\alpha_1\beta_2 - \beta_1\alpha_2)(\alpha_3\beta_4 - \beta_3\alpha_4)\right] =$$

$$= \frac{1}{(4!)^{1/2}} \sum_P (-1)^P PP^\sigma \psi_A |S = 0\rangle. \tag{2.40}$$

The i-th eigenstate is, therefore,

$$\Phi_i^A = \frac{1}{(4!)^{1/2}} \sum_P (-1)^P PP^\sigma \psi_A \theta_i , \tag{2.41}$$

where ψ_A is the product of the one-electron orbitals and θ_i is the i-th eigenstate of S^2 and S_z, P permutes the orbitals and P^σ permutes the spins. The matrix elements are then of the form

$$\langle \Phi_i^A | \mathcal{H} - W | \Phi_i^B \rangle = \langle \mathcal{A}\psi_A \theta_i | \mathcal{H} - W | \mathcal{A}\psi_B \theta_j \rangle =$$

$$= \frac{1}{n!} \langle \sum_P (-1)^P PP^\sigma \psi_A \theta_i | \mathcal{H} - W | \sum_{P'} (-1)^{P'} P'P'^\sigma \psi_B \theta_j \rangle . \tag{2.42}$$

Since all the permutations have been accounted for, a multiplication by the inverse operator $(-1)^{P'} (P'P'^\sigma)^{-1}$ leaves the wave function unchanged,

$$\sum_{P'} P'P'^\sigma (P'P'^\sigma)^{-1} \psi_B \theta_j = n! \psi_B \theta_j . \tag{2.43}$$

The matrix elements for a given value of S and M_S then become

$$\langle \Phi_i^A | \mathcal{H} - W | \Phi_i^B \rangle = \langle \sum_P (-1)^P PP^\sigma \psi_A \theta_i | \mathcal{H} - W | \psi_B \theta_j \rangle =$$

$$= \sum_P (-1)^P \langle P\psi_A | \mathcal{H} - W | \psi_B \rangle \langle \theta_i | P^\sigma | \theta_j \rangle =$$

$$= \sum_P (-1)^P \langle P\psi_A | \mathcal{H} - W | \psi_B \rangle P_{ij}^\sigma . \tag{2.44}$$

Thus the use of the permutation operator leads to the antisymmetrization of the wave functions for a constant spin.

3. The Heisenberg Hamiltonian

The exchange energy coupling two individual electrons of spins $s_1 = \pm\frac{1}{2}$ and $s_2 = \pm\frac{1}{2}$ is given by

$$W_{12} = -2J_{12}\, \mathbf{s}_1 \cdot \mathbf{s}_2 \tag{3.1}$$

for the singlet state (↑↓), we have $(s_1 + s_2)^2 = (s_1 + s_2)(s_1 + s_2 + 1) = 0$ for $s_1 = +\frac{1}{2}$, $s_2 = -\frac{1}{2}$, $S = 0$ or the dot product $\mathbf{s}_1 \cdot \mathbf{s}_2 = -3/4$. For the triplet state (↓↓) $S = 1$,

$(s_1 + s_2)^2 = S(S + 1) = 2$ or $\mathbf{s}_1 \cdot \mathbf{s}_2 = 1/4$. If W_0 is the spin-unperturbed part of the energy, the singlet W_s and the triplet W_t state energies are given by

$$W_s = W_0 + \tfrac{3}{2} J_{12}\,, \qquad W_t = W_0 - \tfrac{1}{2} J_{12}\,, \tag{3.2}$$

so that the exchange in terms of singlet and triplet state energies is

$$2J_{12} = W_s - W_t\,. \tag{3.3}$$

Thus if the energies of the singlet and the triplet states could be calculated from the fundamental Hamiltonian

$$\mathcal{H} = -\sum_i \frac{\hbar^2}{2m} \nabla_i^2 - \sum_{ig} \frac{Z_g e^2}{r_{ig}} + \sum_{i>j} \frac{e^2}{r_{ij}}, \tag{3.4}$$

one would have a first-principle calculation of the exchange J_{ij}. Let us consider a two-electron system. Then the first term in (3.4) is the kinetic energy of the two electrons and the second the potential due to attraction with nuclei. The distance r_{ig} is measured from the i-th electron to g-th nucleus. The third term is the Coulomb repulsion between two electrons. The exchange arising from the first and the second terms is called the "kinetic exchange" and the "potential exchange", respectively, by Anderson [6]. If the two electrons are orthogonal, such exchange interactions do not arise because of their equal contribution to the singlet and triplet states. The third term being repulsive (positive) gives rise to the Heisenberg exchange between orthogonal orbitals, which is positive. Indeed, *as long as the interaction is repulsive it gives only ferromagnetic coupling. The electron nuclear attractive (negative) potential term in the fundamental Hamiltonian* (3.4) *gives rise to the antiferromagnetic coupling.* For the two-electron problem,

$$\mathcal{H} = \mathcal{H}(1) + \mathcal{H}(2) + \mathcal{H}(1, 2) \tag{3.5}$$

with

$$\begin{aligned} \mathcal{H}(1) &= -\frac{\hbar^2}{2m} \nabla_1^2 - \frac{Z_1 e^2}{|\boldsymbol{r}_1 - \boldsymbol{R}_1|} - \frac{Z_2 e^2}{|\boldsymbol{r}_1 - \boldsymbol{R}_2|}, \\ \mathcal{H}(2) &= -\frac{\hbar^2}{2m} \nabla_2^2 - \frac{Z_1 e^2}{|\boldsymbol{r}_2 - \boldsymbol{R}_1|} - \frac{Z_2 e^2}{|\boldsymbol{r}_2 - \boldsymbol{R}_2|}, \\ \mathcal{H}(1, 2) &= \frac{e^2}{|\boldsymbol{r}_1 - \boldsymbol{r}_2|}. \end{aligned} \tag{3.6}$$

In general, we define the one-electron Hamiltonian

$$\mathcal{H}_i = -\frac{\hbar^2}{2m} \nabla_i^2 - \sum_{ig} \frac{Z_g e^2}{r_{ig}}. \tag{3.7}$$

The first-order energy W is found from the solution of

$$-\langle \psi_i(r_1)\, \psi_j(r_2)|\; \mathcal{H} - W\; |\psi_i(r_1)\, \psi_j(r_2)\rangle \pm \langle \psi_i(r_1)\, \psi_j(r_2)|\; \mathcal{H} - W\; |\psi_j(r_1)\, \psi_i(r_2)\rangle = 0\,, \tag{3.8}$$

where the upper sign leads to the energy of the triplet and the lower sign of the singlet. The first term being common to both the singlet and the triplet does not affect the exchange. However, the second term in (3.8) for orthogonal orbitals, $\langle \psi_i \mid \psi_j \rangle = 0$, reduces to

$$\pm \langle \psi_i(r_1)\, \psi_j(r_2)|\; \frac{e^2}{r_{12}}\; |\psi_j(r_1)\, \psi_i(r_2)\rangle\,, \tag{3.9}$$

which does not have kinetic and potential energy terms of the one-electron Hamiltonian and always leads to positive exchange corresponding to ferromagnetic coupling. When the orbitals ψ_i and ψ_j are nonorthogonal (3.9) are not true any more. Instead one obtains from (3.8) terms of the form

$$\langle\psi_i(1)\,\psi_j(2)|\,\mathcal{H}(1)+\mathcal{H}(2)\,|\psi_j(1)\,\psi_i(2)\rangle = 2T\,\langle\psi_i(1)|\,\mathcal{H}(1)\,|\psi_j(1)\rangle\,, \tag{3.10}$$

where

$$T = \langle\psi_j(2)\mid\psi_i(2)\rangle$$

is the overlap integral and the matrix of the one-electron operator $\mathcal{H}(1)$ is negative due to large attractive nuclear potential. The exchange is now determined by

$$\langle\psi_i(1)\,\psi_j(2)|\,\frac{e^2}{r_{12}}\,|\psi_j(1)\,\psi_i(2)\rangle + 2T\,\langle\psi_i(1)|\,\mathcal{H}(1)\,|\psi_j(1)\rangle -$$
$$- T^2\,\langle\psi_i(1)\,\psi_j(2)|\,\mathcal{H}\,|\psi_i(1)\,\psi_j(2)\rangle\,. \tag{3.11}$$

If the second term which is negative dominates, the first positive term, the exchange, can be antiferromagnetic. Introduction of nonmagnetic ions in between the magnetic ions normally introduces large negative potential terms that dominate the positive Coulomb terms and thus exchange is largely negative or antiferromagnetic. Another way of looking at the situation is that the nonmagnetic ions destroy the orthogonality of orbitals, which are not centered on the same atom anyway, so that (3.9) is not valid any more and the eigenvalue problem is more involved. This idea of introducing a large negative potential through intervening ions is called the superexchange (or indirect exchange) as opposed to direct exchange which is very likely dominated by the Coulomb interactions and hence ferromagnetic. The superexchange gives rise to antiferromagnetic contributions. Since the exchange and superexchange arise as a result of cancellations between the ferromagnetic Coulomb and the antiferromagnetic potential terms it is not easy to predict these interactions. Thus, for example, EuO is a ferromagnet, yet MnO is antiferromagnetic. Nevertheless, there have been several efforts to predict the exchange interactions in several systems. Shrivastava's phonon-induced exchange interactions appear to be dominantly antiferromagnetic indicating the possibility of a phase transition from a ferromagnetic to an antiferromagnetic state with increasing temperature. The f electrons in rare earths are much too extended so that the ferromagnetic contribution dominates. The first-principles calculations of the exchange interaction have been done by Yamashita and Kondo [28], Keffer and Oguchi [20], Huang and Orbach [21 to 25], and Rimmer [29]. Falkovskaya and Sapozhnikov [26] have included the smaller two-electron excitation terms which were found to be antiferromagnetic. These two-electron excitation terms were first suggested by Nesbet [30]. The only first-principles calculation of exchange as a function of interatomic distance using Hartree-Fock wave functions is done by Shrivastava and Jaccarino [9]. We shall adopt the notation and scheme of these works as far as possible.

4. The Exchange in Linear Bonding

4.1 Exchange between nonorthogonal orbitals

We consider exchange between two electrons localized on two magnetic atoms with a nonmagnetic ion in between. In order for the central orbit to be nonmagnetic it must be doubly occupied, one electron with spin up and the other with spin down. The minimum number of electrons to be considered in a superexchange calculation is therefore equal to four and the minimum number of atoms is three. Once a model exchange for these four electrons is established additional electrons can be easily incorporated by performing appropriate summation. We assume the highest symmetry, namely cubic, in which the three atoms are localized on a straight line in [001] direction. We designate the central electrons as 1 and 2 and the magnetic electron on the atom at $+R$ as 3 and on the atom at $-R$ as 4. The orbitals are designated as $a_i (i = 1$ to $4)$. Our four-electron three-atom model then appears as d(4) $(-R)$–p(1, 2) (0)–d(3) $(+R)$. In view of the role of orthogonality in determining the exchange, we emphasize that, for the time being, the orbitals of the four electrons a_i under consideration are mutually nonorthogonal as they are not centered on the same atom, e.g.

$$3d_{xz}(-R) - 2p_x(\uparrow\downarrow)\ (0) - 3d_{xz}(+R)\,.$$

The spin density on the $2p_x$ orbital brought about by the presence of the magnetic ion at $(-R)$ favors an antiferromagnetic coupling to the electron on the $3d_{xz}(+R)$ orbital because of the attractive nuclear potential which enters the exchange for nonorthogonal orbitals. Besides, the charge transfer between the nonorthogonal magnetic orbitals also gives rise to antiferromagnetic contributions to the exchange.

We estimate the contributions to the superexchange interaction arising from the following five configurations:

(i) the ionic configuration d(4) $(-R)$–p(1, 2) (0)–d(3) $(+R)$;

(ii) the configuration in which one electron is transferred from the central atom to the atom at $+R$;

(iii) the configuration in which one electron is transferred from the central atom to the atom at $-R$;

(iv) the configuration in which one electron is transferred from the central atom to the atom at $-R$, and at the same time an electron is transferred from the atom at $+R$ to the vacated orbit on the central atom;

(v) as (iv), but the atoms at $-R$ and $+R$ are interchanged.

Nesbet's [30] higher-order antiferromagnetic terms involving the transfer on both the p electrons to the empty orbits on the atoms at $+R$ and $-R$ by the Coulomb interaction are ignored. In the configuration (i), exchange occurs because of Coulomb forces and the overlap between the occupied orbitals. Configurations (ii) and (iii) are degenerate as are (iv) and (v). Therefore, it is convenient to take linear combinations of four-electron wave functions of (ii) with (iii) and (iv) with (v). We represent the state corresponding to the configuration (i) by A, (ii) and (iii) by B and C, and (iv) and (v) by D and E, and describe the matrix elements of the unperturbed static Hamiltonian $\mathcal{H}_0$ by superscripts of the reduced states in the notation of Serber [14].

The eigenvalue problem can normally be solved by applying the method of determinental wave functions and solving the secular equation. In our case this is a rather complicated problem. However, the quantum-mechanical vector model of Dirac as extended by Van Vleck and Serber can be applied with comparative ease. What it does is that the energy of any one of the selected configurations can be determined without calculating the energy of any other configuration. For example, the configuration d^3 has several doublet and quartet states in which Van Vleck manages to calculate the energies of the doublets without determining those of the quartets. This is not so in the Heisenberg matrix method where the entire eigenvalues and eigenstates must be calculated simultaneously.

The matrix element $(\mathcal{H} - W)^{ij}$ connecting configurations i and j is given by Serber,

$$(\mathcal{H} - W)^{ij} = 2^{(q_i - q_j)/2} \sum_P{}' \, (\mathcal{H} - W)^{ij}_P P^{ij}\,; \qquad (q_i \geqq q_j)\,,$$

where

$$(\mathcal{H} - W)_P^{ij} = \langle P\psi_i | \mathcal{H} - W | \psi_j \rangle = \mathcal{H}_P^{ij} - W\langle P\psi_i | \psi_j \rangle \tag{4.1}$$

and $\mathcal{H}$ is the sum of the total kinetic energy of the electrons, the electron–nuclear potentials, and the Coulomb interaction between electrons, W is the energy variable in the secular equation, q_i denotes the number of pairs of filled orbitals in configuration i and the sum is taken only over the $n!/2^{q_i}$ permutations which do not differ merely by interchanges of identical orbits of i, and

$$\psi_i = \varphi_1^i(r_1)\,\varphi_2^i(r_2)\,\dots\,\varphi_n^i(r_n) \tag{4.2}$$

is the product of the one-electron orbitals specifying configuration i. We define the overlap integrals between d and p orbitals by

$$S = \langle a_1 | a_4 \rangle = -\langle a_1 | a_3 \rangle \tag{4.3}$$

and the overlap integral between two $3d_{xz}$ orbitals centered at $+R$ and at $-R$ by

$$T = \langle a_3 | a_4 \rangle\,. \tag{4.4}$$

We note that in each configuration there is at least one pair of identical orbitals which are occupied by two electrons with opposite spin, i.e., s_1 and s_2 form a singlet in all our configurations. Following the description for the construction of the P^σ the representations of the interchange operators are calculated by Keffer and Oguchi, and Huang and Orbach from Serber's formulation. The P^σ for higher-order permutations are

$$P_{\rm I}^\sigma = -1\,, \qquad P_{13}^\sigma = P_{23}^\sigma = P_{14}^\sigma = P_{24}^\sigma = -\tfrac{1}{2}$$

for both the singlet and triplet states,

$$P_{34}^\sigma = +1\,, \qquad P_{134}^\sigma = P_{234}^\sigma = P_{143}^\sigma = P_{243}^\sigma = -\tfrac{1}{2}\,, \qquad P_{1324}^\sigma = P_{1423}^\sigma = 1$$

for the singlet state, and

$$P_{34}^\sigma = -1\,, \qquad P_{134}^\sigma = P_{234}^\sigma = P_{143}^\sigma = P_{243}^\sigma = \tfrac{1}{2}\,, \qquad P_{1324}^\sigma = P_{1423}^\sigma = 0$$

for the triplet state.

The configurations (i) to (v) all contribute to the energy of the singlet state, hence the secular determinant is five-dimensional. Since the ionic configuration is even, it is connected only with even combinations of the excited configurations. Therefore, we form the following two even states:

$$\psi_{\rm G} \approx \frac{1}{\sqrt{2}}\,(\psi_{\rm B} - \psi_{\rm C})\,, \qquad \psi_{\rm G'} \approx \frac{1}{\sqrt{2}}\,(\psi_{\rm D} + \psi_{\rm E})\,, \tag{4.5}$$

where $\psi_{\rm B}$ etc. are defined by the product of the one-electron orbitals

$$\begin{aligned}
\psi_{\rm A} &= |{\rm d}_{xz}(-R),\ {\rm p}_x(\uparrow),\ {\rm p}_x(\downarrow),\ {\rm d}_{xz}(+R)\rangle\,,\\
\psi_{\rm B} &= |{\rm d}'_{xz}(\downarrow)\,(-R),\ {\rm d}'_{xz}(\uparrow)\,(-R),\ {\rm p}_x(\downarrow),\ {\rm d}_{xz}(+R)\rangle\,,\\
\psi_{\rm C} &= |{\rm d}_{xz}(-R),\ {\rm p}_x,\ {\rm d}_{xz}(\uparrow)\,(+R),\ {\rm d}_{xz}(\downarrow)\,(+R)\rangle\,,\\
\psi_{\rm D} &= |{\rm d}_{xz}(\uparrow)\,(-R),\ {\rm d}_{xz}(\downarrow)\,(-R),\ {\rm p}_x(\uparrow),\ {\rm p}_x(\downarrow)\rangle\,,\\
\psi_{\rm E} &= |{\rm p}_x(\downarrow),\ {\rm p}_x(\uparrow),\ {\rm d}_{xz}(\downarrow)\,(+R),\ {\rm d}_{xz}(\uparrow)\,(+R)\rangle\,.
\end{aligned} \tag{4.6}$$

We have ignored the overlaps between ψ_B, ψ_C and ψ_D, ψ_E in evaluating the normalization constants, and the spin zero is not defined when both the orientations are allowed. Out of the four electrons, two electrons always make an antiparallel pair while the remaining two are true to be parallel or antiparallel giving rise to the singlet and triplet possibilities. The 3×3 determinant of ψ_A, ψ_G, and $\psi_{G'}$ is of interest to evaluate the perturbed energy of the configuration ψ_A,

$$\begin{vmatrix} (\mathcal{H} - W)^{AA} & (\mathcal{H} - W)^{AG} & (\mathcal{H} - W)^{AG'} \\ (\mathcal{H} - W)^{GA} & (\mathcal{H} - W)^{GG} & (\mathcal{H} - W)^{GG'} \\ (\mathcal{H} - W)^{G'A} & (\mathcal{H} - W)^{G'G} & (\mathcal{H} - W)^{G'G'} \end{vmatrix} = 0 . \tag{4.7}$$

The Hamiltonian is just

$$\mathcal{H} = \sum_{i=1}^{4} \mathcal{H}_i + \sum_{i>j} \frac{e^2}{r_{ij}}, \tag{4.8}$$

where the one-electron Hamiltonian is

$$\mathcal{H}_i = -\sum_i \frac{\hbar^2}{2m} \nabla_i^2 - \sum_{ig} \frac{Z_g e^2}{r_{ig}} . \tag{4.9}$$

In view of the symmetry of the states,

$$\begin{aligned} (\mathcal{H} - W)^{AB} &= -(\mathcal{H} - W)^{AC} , \qquad & (\mathcal{H} - W)^{AD} &= (\mathcal{H} - W)^{AE} , \\ (\mathcal{H} - W)^{BB} &= (\mathcal{H} - W)^{CC} , & (\mathcal{H} - W)^{DD} &= (\mathcal{H} - W)^{EE} , \\ (\mathcal{H} - W)^{CD} &= -(\mathcal{H} - W)^{BE} , & (\mathcal{H} - W)^{CE} &= -(\mathcal{H} - W)^{BE} . \end{aligned} \tag{4.10}$$

The matrix elements in the secular equation are given by

$$\begin{aligned} (\mathcal{H} - W)^{AA} &= (\mathcal{H}_I^{AA} - W) - (\mathcal{H}_{13}^{AA} - S^2 W) - (\mathcal{H}_{14}^{AA} - S^2 W) + (\mathcal{H}_{34}^{AA} - T^2 W) - \\ &\quad - (\mathcal{H}_{134}^{AA} + S^2 T W) - (\mathcal{H}_{143}^{AA} + S^2 T W) + 2(\mathcal{H}_{13,24}^{AA} - S^4 W) , \\ (\mathcal{H} - W)^{AG} &= \sqrt{2}\, (\mathcal{H} - W)^{AB} = \\ &= \sqrt{2}\, [(\mathcal{H}_I^{AB} - S^3 W) - (\mathcal{H}_{13}^{AB} + S^2 T W) - (\mathcal{H}_{14}^{AB} - S W) + \\ &\quad + (\mathcal{H}_{34}^{AB} + S^3 T W) - (\mathcal{H}_{134}^{AB} - S T^2 W) - (\mathcal{H}_{143}^{AB} - S^3 W) + \\ &\quad + 2(\mathcal{H}_{13,24}^{AB} + S T W)] , \\ (\mathcal{H} - W)^{AG'} &= \sqrt{2}\, (\mathcal{H} - W)^{AD} = \\ &= \sqrt{2}\, [(\mathcal{H}_I^{AD} - T W) - (\mathcal{H}_{13}^{AD} + S^2 W) - (\mathcal{H}_{14}^{AD} - S^2 T W) + \\ &\quad + (\mathcal{H}_{13,24}^{AD} + S^4 W)] , \\ (\mathcal{H} - W)^{GG} &= (\mathcal{H} - W)^{BB} - (\mathcal{H} - W)^{BC} = \\ &= (\mathcal{H}_I^{BB} - W) - (\mathcal{H}_{13}^{BB} - T^2 W) - (\mathcal{H}_{14}^{BB} - S^2 W) + (\mathcal{H}_{34}^{BB} - S^2 W) - \\ &\quad - (\mathcal{H}_{134}^{BB} + S^2 T W) - (\mathcal{H}_{143}^{BB} + S^2 T W) + 2(\mathcal{H}_{13,24}^{BB} - S^2 T^2 W) - \\ &\quad - (\mathcal{H}_I^{BC} + S^2 T^2 W) + (\mathcal{H}_{13}^{BC} - S^2 T W) + (\mathcal{H}_{14}^{BC} - S^2 T W) - \\ &\quad - (\mathcal{H}_{34}^{BC} - T^3 W) + (\mathcal{H}_{134}^{BC} - T W) + (\mathcal{H}_{143}^{BC} + S^2 T^2 W) - \\ &\quad - 2(\mathcal{H}_{13,24}^{BC} + S^2 W) , \end{aligned}$$

$$(\mathcal{H} - W)^{\mathrm{GG'}} = (\mathcal{H} - W)^{\mathrm{BD}} + (\mathcal{H} - W)^{\mathrm{BE}} =$$
$$= \sqrt{2}\,[(\mathcal{H}_{\mathrm{I}}^{\mathrm{BD}} - S^3TW) - (\mathcal{H}_{13}^{\mathrm{BD}} + S^3W) - (\mathcal{H}_{14}^{\mathrm{BD}} - STW) +$$
$$+ (\mathcal{H}_{13,24}^{\mathrm{BD}} + SW) + (\mathcal{H}_{\mathrm{I}}^{\mathrm{BE}} + S^3W) - (\mathcal{H}_{13}^{\mathrm{BE}} - S^3TW) -$$
$$- (\mathcal{H}_{14}^{\mathrm{BE}} - STW) + (\mathcal{H}_{13,24}^{\mathrm{BE}} + ST^2W)]\,,$$
$$(\mathcal{H} - W)^{\mathrm{G'G'}} = (\mathcal{H} - W)^{\mathrm{DD}} + (\mathcal{H} - W)^{\mathrm{DE}} =$$
$$= (\mathcal{H}_{\mathrm{I}}^{\mathrm{DD}} - W) - (\mathcal{H}_{13}^{\mathrm{DD}} - S^2W) - (\mathcal{H}_{14}^{\mathrm{DD}} - S^2W) +$$
$$+ (\mathcal{H}_{13,24}^{\mathrm{DD}} - S^4W) + (\mathcal{H}_{\mathrm{I}}^{\mathrm{DE}} - T^2W) - (\mathcal{H}_{13}^{\mathrm{DE}} + S^2TW) -$$
$$- (\mathcal{H}_{14}^{\mathrm{DE}} + S^2TW) + (\mathcal{H}_{13,24}^{\mathrm{DE}} - S^4W)\,. \tag{4.11}$$

We replace W by its unperturbed value $\mathcal{H}_{\mathrm{I}}^{\mathrm{AA}}$ in all terms except in $\mathcal{H}_{\mathrm{I}}^{\mathrm{AA}} - W$ as was done by Keffer and Oguchi [20].

This is equivalent to a perturbation approximation. The energy corrections to the configuration ψ_{A} should then be small. Since the matrix elements are of the order of $\mathcal{H}_{\mathrm{I}}^{ii}\langle P\psi_i \mid \psi_j\rangle$, our approximation is adequate. The energy of the singlet state, calculated in this way is given by

$$W_{\mathrm{s}} = \mathcal{H}_{\mathrm{I}}^{\mathrm{AA}} - (\mathcal{H}_{13}^{\mathrm{AA}} - S^2\mathcal{H}_{\mathrm{I}}^{\mathrm{AA}}) - (\mathcal{H}_{14}^{\mathrm{AA}} - S^2\mathcal{H}_{\mathrm{I}}^{\mathrm{AA}}) + (\mathcal{H}_{34}^{\mathrm{AA}} - T^2\mathcal{H}_{\mathrm{I}}^{\mathrm{AA}}) -$$
$$- (\mathcal{H}_{134}^{\mathrm{AA}} + S^2T\mathcal{H}_{\mathrm{I}}^{\mathrm{AA}}) - (\mathcal{H}_{143}^{\mathrm{AA}} + S^2T\mathcal{H}_{\mathrm{I}}^{\mathrm{AA}}) + 2(\mathcal{H}_{13,24}^{\mathrm{AA}} - S^4\mathcal{H}_{\mathrm{I}}^{\mathrm{AA}}) +$$
$$+ 2(\mathcal{H}_{14}^{\mathrm{AB}} - S\mathcal{H}_{\mathrm{I}}^{\mathrm{AA}})(\mathcal{H}_{\mathrm{I}}^{\mathrm{BB}} - \mathcal{H}_{\mathrm{I}}^{\mathrm{AA}})^{-1}[-(\mathcal{H}_{14}^{\mathrm{AB}} - S\mathcal{H}_{\mathrm{I}}^{\mathrm{AA}}) + 2(\mathcal{H}_{\mathrm{I}}^{\mathrm{AB}} - S^3\mathcal{H}_{\mathrm{I}}^{\mathrm{AA}}) -$$
$$- 2(\mathcal{H}_{143}^{\mathrm{AB}} - S^3\mathcal{H}_{\mathrm{I}}^{\mathrm{AA}}) + 4(\mathcal{H}_{13,24}^{\mathrm{AB}} + ST\mathcal{H}_{\mathrm{I}}^{\mathrm{AA}})] -$$
$$- 4[(\mathcal{H}_{\mathrm{I}}^{\mathrm{AD}} - T\mathcal{H}_{\mathrm{I}}^{\mathrm{AA}}) - (\mathcal{H}_{13}^{\mathrm{AD}} + S^2\mathcal{H}_{\mathrm{I}}^{\mathrm{AA}})]^2 (\mathcal{H}_{\mathrm{I}}^{\mathrm{DD}} - \mathcal{H}_{\mathrm{I}}^{\mathrm{AA}})^{-1} +$$
$$+ 8(\mathcal{H}_{14}^{\mathrm{AB}} - S\mathcal{H}_{\mathrm{I}}^{\mathrm{AA}})(\mathcal{H}_{13,24}^{\mathrm{BD}} + S\mathcal{H}_{\mathrm{I}}^{\mathrm{AA}})[(\mathcal{H}_{13}^{\mathrm{AD}} + S^2\mathcal{H}_{\mathrm{I}}^{\mathrm{AA}}) - (\mathcal{H}_{\mathrm{I}}^{\mathrm{AD}} - T\mathcal{H}_{\mathrm{I}}^{\mathrm{AA}})] \times$$
$$\times (\mathcal{H}_{\mathrm{I}}^{\mathrm{BB}} - \mathcal{H}_{\mathrm{I}}^{\mathrm{AA}})^{-1}(\mathcal{H}_{\mathrm{I}}^{\mathrm{DD}} - \mathcal{H}_{\mathrm{I}}^{\mathrm{AA}})^{-1} + 2(\mathcal{H}_{14}^{\mathrm{AB}} - S\mathcal{H}_{\mathrm{I}}^{\mathrm{AA}})^2(\mathcal{H}_{\mathrm{I}}^{\mathrm{BB}} - \mathcal{H}_{\mathrm{I}}^{\mathrm{AA}})^{-2} \times$$
$$\times [(\mathcal{H}_{34}^{\mathrm{BB}} - S^2\mathcal{H}_{\mathrm{I}}^{\mathrm{AA}}) - (\mathcal{H}_{14}^{\mathrm{BB}} - S^2\mathcal{H}_{\mathrm{I}}^{\mathrm{AA}}) - 2(\mathcal{H}_{13,24}^{\mathrm{BC}} + S^2\mathcal{H}_{\mathrm{I}}^{\mathrm{AA}}) +$$
$$+ (\mathcal{H}_{134}^{\mathrm{BC}} - T\mathcal{H}_{\mathrm{I}}^{\mathrm{AA}})] - 4(\mathcal{H}_{14}^{\mathrm{AB}} - S\mathcal{H}_{\mathrm{I}}^{\mathrm{AA}})^2(\mathcal{H}_{13,24}^{\mathrm{BD}} + S\mathcal{H}_{\mathrm{I}}^{\mathrm{AA}})^2 \times$$
$$\times (\mathcal{H}_{\mathrm{I}}^{\mathrm{BB}} - \mathcal{H}_{\mathrm{I}}^{\mathrm{AA}})^{-2}(\mathcal{H}_{\mathrm{I}}^{\mathrm{DD}} - \mathcal{H}_{\mathrm{I}}^{\mathrm{AA}})^{-1}\,. \tag{4.12}$$

We now calculate the energy of the triplet state. Since the configurations ψ_{D} and ψ_{E} being of the form $|\mathrm{d}(\uparrow)\,\mathrm{d}(\downarrow)\,\mathrm{p}(\uparrow)\,\mathrm{p}(\downarrow)\rangle$, do not contribute to the energy of this state, they need not be considered. We are then left with the configurations A, B, and C and since B and C are degenerate, we are left with only a 2×2 secular determinant

$$\begin{vmatrix} (\mathcal{H} - W)^{\mathrm{AA}} & (\mathcal{H} - W)^{\mathrm{AG}} \\ (\mathcal{H} - W)^{\mathrm{GA}} & (\mathcal{H} - W)^{\mathrm{GG}} \end{vmatrix} = 0\,. \tag{4.13}$$

Following the sampe procedure as used in the calculation of the energy of the singlet state, we have

$$(\mathcal{H} - W)^{AA} = (\mathcal{H}_{I}^{AA} - W) - (\mathcal{H}_{13}^{AA} - S^2W) - (\mathcal{H}_{14}^{AA} - S^2W) - (\mathcal{H}_{34}^{AA} - T^2W) + + (\mathcal{H}_{134}^{AA} + S^2TW) + (\mathcal{H}_{143}^{AA} + S^2TW),$$

$$(\mathcal{H} - W)^{AG} = \sqrt{2}\,(\mathcal{H} - W)^{AB} = = \sqrt{2}\,[(\mathcal{H}_{I}^{AB} - S^3W) - (\mathcal{H}_{13}^{AB} + S^3TW) - (\mathcal{H}_{14}^{AB} - SW) - - (\mathcal{H}_{34}^{AB} + S^3TW) + (\mathcal{H}_{134}^{AB} - ST^2W) + (\mathcal{H}_{143}^{AB} - S^3W)], \tag{4.14}$$

$$(\mathcal{H} - W)^{GG} = (\mathcal{H} - W)^{BB} - (\mathcal{H} - W)^{BC} = = (\mathcal{H}_{I}^{BB} - W) - (\mathcal{H}_{13}^{BB} - T^2W) - (\mathcal{H}_{14}^{BB} - S^2W) - (\mathcal{H}_{34}^{BB} - S^2W) + + (\mathcal{H}_{134}^{BB} + S^2TW) + (\mathcal{H}_{143}^{BB} + S^2TW) - (\mathcal{H}_{I}^{BC} + S^2T^2W) + + (\mathcal{H}_{13}^{BC} - S^2TW) + (\mathcal{H}_{14}^{BC} - S^2TW) + (\mathcal{H}_{34}^{BC} - T^3W) - - (\mathcal{H}_{134}^{BC} - TW) - (\mathcal{H}_{143}^{BC} - S^2T^2W).$$

The energy of the triplet state is

$$W_t = \mathcal{H}_{I}^{AA} - (\mathcal{H}_{13}^{AA} - S^2\mathcal{H}_{I}^{AA}) - (\mathcal{H}_{14}^{AA} - S^2\mathcal{H}_{I}^{AA}) - (\mathcal{H}_{34}^{AA} - T^2\mathcal{H}_{I}^{AA}) + + (\mathcal{H}_{134}^{AA} + S^2T\mathcal{H}_{I}^{AA}) + (\mathcal{H}_{143}^{AA} + S^2T\mathcal{H}_{I}^{AA}) + + 2(\mathcal{H}_{14}^{AB} - S\mathcal{H}_{I}^{AA})\,(\mathcal{H}_{I}^{BB} - \mathcal{H}_{I}^{AA})^{-1} \times \times [-(\mathcal{H}_{14}^{AB} - S\mathcal{H}_{I}^{AA}) + 2(\mathcal{H}_{I}^{AB} - S^3\mathcal{H}_{I}^{AA}) + 2(\mathcal{H}_{143}^{AB} - S^3\mathcal{H}_{I}^{AA})] - - 2(\mathcal{H}_{14}^{AB} - S\mathcal{H}_{I}^{AA})^2\,(\mathcal{H}_{I}^{BB} - \mathcal{H}_{I}^{AA})^{-2} \times \times [(\mathcal{H}_{14}^{BB} - S^2\mathcal{H}_{I}^{AA}) + (\mathcal{H}_{34}^{BB} - S^2\mathcal{H}_{I}^{AA}) + (\mathcal{H}_{134}^{BC} - T\mathcal{H}_{I}^{AA})]. \tag{4.15}$$

The exchange energy coupling two individual d electrons each belonging to the $3d_{xz}$ orbital of the magnetic ions, is given by

$$W_{34} = -2J_{34}\,\mathbf{s}_3 \cdot \mathbf{s}_4\,, \tag{4.16}$$

in which the exchange is defined by $2J_{34} = W_s - W_t$. Therefore, from (4.12) and (4.15), the exchange is easily calculated to be

$$J_{34} = J_{io} + J_{ac} + J_{cc} + J_{acc}\,, \tag{4.17}$$

where

$$J_{io} = (\mathcal{H}_{34}^{AA} - T^2\mathcal{H}_{I}^{AA}) - (\mathcal{H}_{134}^{AA} + S^2T\mathcal{H}_{I}^{AA}) - (\mathcal{H}_{143}^{AA} + S^2T\mathcal{H}_{I}^{AA}) + + (\mathcal{H}_{13,24}^{AA} - S^4\mathcal{H}_{I}^{AA}), \tag{4.18}$$

$$J_{ac} = 4\gamma_{14}[-(\mathcal{H}_{143}^{AB} - S^3\mathcal{H}_{I}^{AA}) + (\mathcal{H}_{13,24}^{AB} + ST\mathcal{H}_{I}^{AA})] + + 2\gamma_{14}^2[(\mathcal{H}_{34}^{BB} - S^2\mathcal{H}_{I}^{AA}) - (\mathcal{H}_{13,24}^{BC} + S^2\mathcal{H}_{I}^{AA}) + (\mathcal{H}_{134}^{BC} - T\mathcal{H}_{I}^{AA})], \tag{4.19}$$

$$J_{cc} = -2\Delta_D^{-1}[(\mathcal{H}_{I}^{AD} - T\mathcal{H}_{I}^{AA}) - (\mathcal{H}_{13}^{AD} + S^2\mathcal{H}_{I}^{AA})]^2, \tag{4.20}$$

$$J_{\text{acc}} = 4\gamma_{14}\,\varDelta_{\text{D}}^{-1}(\mathcal{H}_{13,24}^{\text{BD}} + S\mathcal{H}_{\text{I}}^{\text{AA}})\,[(\mathcal{H}_{13}^{\text{AD}} + S^2\mathcal{H}_{\text{I}}^{\text{AA}}) - (\mathcal{H}_{\text{I}}^{\text{AD}} - T\mathcal{H}_{\text{I}}^{\text{AA}})] - - 2\gamma_{14}^2\varDelta_{\text{D}}^{-1}(\mathcal{H}_{13,24}^{\text{BD}} + S\mathcal{H}_{\text{I}}^{\text{AA}})^2\,, \tag{4.21}$$

and

$$\gamma_{14} = \varDelta_{\text{B}}^{-1}(\mathcal{H}_{14}^{\text{AB}} - S\mathcal{H}_{\text{I}}^{\text{AA}})\,, \qquad \varDelta_{\text{B}} = \mathcal{H}_{\text{I}}^{\text{BB}} - \mathcal{H}_{\text{I}}^{\text{AA}}\,, \qquad \varDelta_{\text{D}} = \mathcal{H}_{\text{I}}^{\text{DD}} - \mathcal{H}_{\text{I}}^{\text{AA}}\,. \tag{4.22}$$

Here J_{io} is the contribution to the exchange integral due to overlap and Coulomb effects within the ionic configuration (i). The second term J_{ac} gives the effects resulting from the admixture of (ii), i.e. anion–cation charge transfer, J_{cc} comes from the d–d transfer, and J_{acc} is a result of the cross terms among ionic, the anion–cation, and cation–cation transfer configurations. The matrix elements in (4.19) to (4.21) are given in terms of the one-electron and the two-electron integrals, in the notation

$$\mathcal{H}_i = -\frac{\hbar^2}{2m}\nabla_i^2 - \sum_j \frac{Z_j e^2}{r_{ij}}\,, \tag{4.23}$$

$$\mathcal{H} = \sum_i \mathcal{H}_i + \sum_{i>j} \frac{e^2}{r_{ij}}\,, \tag{4.24}$$

$$\langle a_i|\,\mathcal{H}_1\,|a_j\rangle = \langle ij\rangle\,, \tag{4.25}$$

$$\langle a_1 a_2|\,\frac{e^2}{r_{ij}}\,|a_3 a_4\rangle = \langle a_1 a_2|\,|a_3 a_4\rangle = \langle 1234\rangle\,, \quad \text{etc.}\,; \tag{4.26}$$

$$a_1 = a_2 = \text{p}_x\,, \qquad a_3 = \text{d}_{xz}(+R)\,, \qquad a_4 = \text{d}_{xz}(-R)\,, \tag{4.27}$$

$$\mathcal{H}_{34}^{\text{AA}} - \mathcal{H}_{\text{I}}^{\text{AA}}T^2 = 2\langle 34\rangle\,T - (\langle 33\rangle + \langle 44\rangle)\,T^2 + + 4\langle 1314\rangle\,T + \langle 3443\rangle - 2\langle 1313\rangle\,T^2 - - (\langle 1414\rangle + \langle 3434\rangle)\,T^2\,, \tag{4.28}$$

$$\mathcal{H}_{134}^{\text{AA}} + S^2T\mathcal{H}_{\text{I}}^{\text{AA}} = \mathcal{H}_{143}^{\text{AA}} + S^2T\mathcal{H}_{\text{I}}^{\text{AA}} = = (\langle 31\rangle - \langle 14\rangle)\,ST - \langle 34\rangle\,S^2 + + (\langle 11\rangle + \langle 33\rangle + \langle 44\rangle)\,S^2T + (\langle 1113\rangle - \langle 1114\rangle)\,ST + + \langle 3114\rangle\,T + (\langle 3413\rangle - \langle 4134\rangle)\,S - - \langle 1413\rangle\,S^2 + (\langle 1111\rangle + 2\langle 1313\rangle + 2\langle 1414\rangle + + \langle 3434\rangle)\,S^2T\,, \tag{4.29}$$

$$\mathcal{H}_{1324}^{\text{AA}} - \mathcal{H}_{\text{I}}^{\text{AA}}S^4 = 2(\langle 41\rangle - \langle 31\rangle)\,S^3 - 2(\langle 11\rangle + \langle 33\rangle + \langle 44\rangle)\,S^4 + + (\langle 3113\rangle + \langle 4114\rangle - 2\langle 3411\rangle - 2\langle 3114\rangle)\,S^2 - - (\langle 1111\rangle + 2\langle 1313\rangle + 2\langle 1414\rangle + \langle 3434\rangle)\,S^4\,, \tag{4.30}$$

$$\mathcal{H}_{14}^{\text{AB}} - \mathcal{H}_{\text{I}}^{\text{AA}}S = \langle 14\rangle - \langle 11\rangle\,S + \langle 4144\rangle + \langle 1343\rangle + \langle 1114\rangle - - \langle 1111\rangle\,S - (\langle 1313\rangle + \langle 1414\rangle + \langle 3434\rangle)\,S\,, \tag{4.31}$$

$$\mathcal{H}_{143}^{AB} - \mathcal{H}_{I}^{AA} S^3 = (\langle 14\rangle - 2\langle 31\rangle)\, S^2 - 2\langle 11\rangle\, S^3 + \langle 33\rangle\, S^3 + + (\langle 4144\rangle - 2\langle 4143\rangle)\, S^2 + + (\langle 1331\rangle - \langle 1143\rangle - \langle 1341\rangle)\, S - - (\langle 1111\rangle + 2\langle 1313\rangle + 2\langle 1414\rangle + \langle 3434\rangle)\, S^3\,, \tag{4.32}$$

$$\mathcal{H}_{1324}^{AB} + \mathcal{H}_{I}^{AA} ST = -\langle 34\rangle\, S + \langle 13\rangle\, T + \langle 11\rangle\, ST + \langle 33\rangle\, ST + + \langle 3143\rangle - \langle 3444\rangle\, S - \langle 3141\rangle\, S + + (\langle 4143\rangle - \langle 1131\rangle)\, T + + (\langle 1111\rangle + 2\langle 1313\rangle + \langle 1414\rangle + \langle 3434\rangle)\, ST\,, \tag{4.33}$$

$$\mathcal{H}_{34}^{BB} - \mathcal{H}_{I}^{AA} S^2 = (\langle 44\rangle + \langle 33\rangle - 2\langle 11\rangle)\, S^2 - 2\langle 13\rangle\, S + + \langle 4444\rangle\, S^2 - 4\langle 4143\rangle\, S + \langle 1331\rangle - - (\langle 1111\rangle + \langle 1313\rangle)\, S^2 + (2\langle 1414\rangle + \langle 3434\rangle)\, S^2\,, \tag{4.34}$$

$$\mathcal{H}_{1324}^{BC} + \mathcal{H}_{I}^{AA} S^2 = (\langle 13\rangle - \langle 14\rangle)\, S + 2\langle 11\rangle\, S^2 + (\langle 3133\rangle - \langle 3431\rangle - - \langle 1434\rangle - \langle 4414\rangle)\, S + (\langle 1111\rangle + 2\langle 1313\rangle + + 2\langle 1414\rangle)\, S^2\,, \tag{4.35}$$

$$\mathcal{H}_{134}^{BD} - \mathcal{H}_{I}^{AA} T = \langle 43\rangle - \langle 11\rangle\, T + \langle 3433\rangle + \langle 4131\rangle + \langle 4434\rangle - - (\langle 1111\rangle + \langle 1313\rangle + \langle 1414\rangle)\, T\,, \tag{4.36}$$

$$\mathcal{H}_{I}^{AD} - \mathcal{H}_{I}^{AA} T = \langle 43\rangle - \langle 33\rangle\, T + 2\langle 1413\rangle + \langle 4434\rangle - - (2\langle 1313\rangle + \langle 3434\rangle)\, T\,, \tag{4.37}$$

$$\mathcal{H}_{13}^{AD} + \mathcal{H}_{I}^{AA} S^2 = (\langle 13\rangle - \langle 41\rangle)\, S + \langle 11\rangle\, S^2 + \langle 33\rangle\, S^2 + + (\langle 1113\rangle + \langle 1434\rangle)\, S - (\langle 4111\rangle + \langle 4414\rangle)\, S + + \langle 4113\rangle + (\langle 1111\rangle + 2\langle 1313\rangle + \langle 1414\rangle + + \langle 3434\rangle)\, S^2\,, \tag{4.38}$$

$$\mathcal{H}_{13,24}^{BD} + \mathcal{H}_{I}^{AA} S = \langle 13\rangle + (\langle 11\rangle + \langle 33\rangle)\, S - \langle 44\rangle\, S + + 2\langle 4143\rangle + \langle 1131\rangle - \langle 4444\rangle\, S + + (\langle 1111\rangle + 2\langle 1313\rangle + \langle 3434\rangle)\, S\,. \tag{4.39}$$

The problem of calculating the exchange is now reduced to calculating the one-electron integrals of the kinetic energy and the potential energy and the two-electron integrals of Coulomb interaction and a small number of three- and four-electron charge distribution integrals.

4.2 *Effect of other electrons*

The effect of electrons other than those considered in the three-center four-electron model can be taken into account by considering the three ions in an effective potential. Thus our four electrons are now in a uniform background. The effect of other electrons therefore occurs in the potential but not in the interaction. In view of the strong dependence of exchange on distance this is not a serious approximation. The modified one-electron Hamiltonian can be written as

$$\mathcal{H}_i = -\frac{\hbar^2\nabla_i^2}{2m} - \sum_j \frac{Z_j e^2}{r_{ij}} + V(r) . \tag{4.40}$$

In the linear pair of vanadium ions, $V^{2+}(-R)$–F^-–$V^{2+}(+R)$ the effective potential may be written as $V_{(-R)}(V^{3+}) + V_{(+R)}(V^{3+}) + V(F^+)$, i.e.,

$$\mathcal{H}_1 = -\frac{\hbar^2\nabla_1^2}{2m} + V_{(-R)}(V^{3+}) + V_{(+R)}(V^{3+}) + V(F^+) , \tag{4.41}$$

where $V_{(-R)}(V^{3+})$, $V_{(+R)}(V^{3+})$, and $V(F^+)$ are now the effective Hartree one-electron potential energies arising from V^{3+} ions at $+R$ and at $-R$ and the F^+ ion in the center. The ionicities are chosen so as to exclude the potential already taken into account. For example, consider the one-electron integral

$$\begin{aligned}\langle 13\rangle &= \langle a_1|\, \mathcal{H}_1\, |a_3\rangle = \\ &= \langle a_1|\, -\frac{\hbar^2\nabla^2}{2m}\, |a_3\rangle + \langle a_1|\, V_{(-R)}(V^{3+})\, |a_3\rangle + \\ &\quad + \langle a_1|\, V_{(+R)}(V^{3+})\, |a_3\rangle + \langle a_1|\, V(F^+)\ a_3\rangle .\end{aligned} \tag{4.42}$$

The Hartree equation for the orbital a_1 of F^- is

$$\left[-\frac{\hbar^2\nabla^2}{2m} + V(F^+) + \langle a_2(r_2)|\, \frac{e^2}{r_{12}}\, |a_2(r_2)\rangle\right] = \varepsilon_H(F^-)\, a_1(r_1) \tag{4.43}$$

and that of a_3 is

$$\left[-\frac{\hbar^2}{2m}\nabla^2 + V_{(+R)}(V^{3+})\right] a_3 = \varepsilon_H(V^{2+})\, a_3(r_3) . \tag{4.44}$$

Multiplying (4.43) by a_3 and (4.44) by a_1 and integrating,

$$\begin{aligned}\langle 13\rangle &= -\varepsilon_H(F^-)\, S - \varepsilon_H(V^{2+})\, S - \langle 2321\rangle + \langle a_1|\, \frac{\hbar^2\nabla^2}{2m}\, |a_3\rangle + \\ &\quad + \langle a_1|\, V_{(-R)}\, |a_3\rangle .\end{aligned} \tag{4.45}$$

The Hartree energies $\varepsilon_H(F^-)$ and $\varepsilon_H(V^{2+})$ appearing in the above equation can be approximated by the one-electron energies in the Hartree-Fock approximation calculated by Watson [31].

Shrivastava and Jaccarino have performed the detailed computations of various integrals using the ionic Hartree-Fock wave functions [31] for the linear V^{2+}–F^-–V^{2+} pair which is experimentally found by Smith and Owen [32] in $KMgF_3$:V^{2+}. The overlaps, the one-electron, and two-electron integrals are computed with sufficient accuracy, the normalization integrals are correct up to six decimals. The computer program evaluates seven types of integrals given in Tables 4 to 7. The estimation of

Table 4

Various types of integrals which are calculated accurately

1	$\int \psi^*(i, r)\, \psi(j, r')\, d\tau$	overlap
2	$\int \psi^*(i, r)\, \psi(j, r')\, \frac{d\tau}{r}$	split nuclear attraction
3	$\int \psi(j, r')\, [\frac{1}{2}\, \nabla_r^2 \psi^*(i, r)]\, d\tau$	kinetic energy
4	$\int \psi^*(i, r')\, \psi(j, r')\, \frac{d\tau}{r}$	nuclear attraction
5	$\int\int \psi^*(i, r_1)\, \psi(j, r_1)\, \psi^*(k, r_2')\, \psi(l, r_2')\, \frac{d\tau_1\, d\tau_2}{r_{12}}$	Coulomb
6	$\int\int \psi^*(i, r_1)\, \psi(j, r_1)\, \psi^*(k, r_2)\, \psi(l, r_2')\, \frac{d\tau_1\, d\tau_2}{r_{12}}$	hybrid
7	$\int\int \psi^*(i, r_1)\, \psi(j, r_1')\, \psi^*(k, r_2)\, \psi(l, r_2')\, \frac{d\tau_1\, d\tau_2}{r_{12}}$	exchange

Table 5a

The Hartree-Fock wave functions calculated by Watson [31] for F^-(2p) and V^{2+}(3d) are identified with the seven types of integrals calculated by Shrivastava as defined in Table 4

atomic orbitals on center A, F^-(2p)

ORB.	L	M	CMP.	N	KAPPA	WEIGHT
101	1	1	1	2	4.696799278	0.208194971
			2	2	2.432199478	0.463508964
			3	2	1.572199821	0.238420963
			4	2	0.926099956	0.232508957

atomic orbitals on center B, V^{2+}(3d)

ORB.	L	M	CMP.	N	KAPPA	WEIGHT
201	2	1	1	3	1.828899384	0.498819649
			2	3	3.610199928	0.524881542
			3	3	6.801999092	0.109604120
			4	3	12.432199478	0.006752737

Here ORB. indicates a nomenclature for an orbital wave function, for example $\psi[F^-(2p)] = 101$ on the first center and $\psi[V^{2+}(3d)] = 201$ on the second center. The two centers are at a distance apart called the INTERNUCLEAR DISTANCE. L is the orbital quantum number equal to 1 for p and 2 for d electrons and M is the projection of L varying from L to $-L$, N is the principal quantum number, CMP is a composition number which designates the Slater orbitals within a Hartree-Fock wave function. KAPPA is the orbital exponent and WEIGHT gives the weightage of the Slater orbital in the construction of the Hartree-Fock wave function. For example $\psi[F^-(2p)] = 0.2081949\, \chi_1 + 0.4635089\, \chi_2 + 0.2384209\, \chi_3 + 0.2325089\, \chi_4$ with $\chi_1 = N_1 r^{n-1} \times \times \exp(-4.6967992r)\, Y_1^1$; $\chi_2 = N_2 r^{n-1} \exp(-2.4321994r)\, Y_1^1$, $\chi_3 = N_3 r^{n-1} \exp(-1.5721998r)\, Y_1^1$; $\chi_4 = N_4 r^{n-1} \exp(-0.9260999r)\, Y_1^1$. The normalization constants, N(CMP.) are given by (4.50) and the angular part of the wave function is given by the spherical harmonics, Y_1^1 for $L = 1$, $M = 1$. The value of n for F^-(2p) is 2.

Table 5b

Seven types of integrals of the wave functions continued from Table 5a

	INTERNUCLEAR DISTANCE				3.40150833	3.59048080	3.78071975	3.96842575	4.15739822	4.34637165
1	101	101	0	0	0.99999475	0.99999475	0.99999475	0.99999428	0.99999428	0.99999475
1	101	201	0	0	0.10535914	0.08942807	0.07568049	0.06410241	0.05416194	0.04572399
1	201	201	0	0	0.99999368	0.99999321	0.99999321	0.99999321	0.99999368	0.99999321
2	101	101	0	0	1.16128635	1.16128635	1.16128635	1.16128635	1.16128540	1.16128635
2	101	201	0	0	0.05917401	0.04798973	0.03881466	0.03145087	0.02542318	0.02053695
2	201	201	0	0	1.01924229	1.01924133	1.01924229	1.01924229	1.01924229	1.01924229
2	201	101	0	0	0.04323365	0.03559754	0.02929204	0.02418646	0.01996465	0.01649769
3	101	101	0	0	−2.86059570	−2.86059952	−2.86059761	−2.86059761	−2.86059570	−2.86059380
3	101	201	0	0	−0.03564832	−0.02554283	−0.01804748	−0.01262207	−0.00863859	−0.00575784
3	201	201	0	0	−4.91632652	−4.91632080	−4.91632462	−4.91632271	−4.91632652	−4.91631889
4	101	101	0	0	0.28290188	0.26906723	0.25634533	0.24495721	0.23442024	0.22473550
4	201	201	0	0	0.29082245	0.29303467	0.27232641	0.27462667	0.24887657	0.26115662
5	101	101	101	101	0.77438754	0.77438700	0.77438754	0.77438647	0.77438700	0.77438647
5	101	−101	−101	101	0.07828730	0.07828730	0.07828736	0.07828724	0.07828712	0.07828730
5	201	201	201	201	0.73504484	0.73504436	0.73504514	0.73504466	0.73504514	0.73504514
5	201	−201	−201	201	0.05989688	0.05989685	0.05989686	0.05989680	0.05989691	0.05989686
5	101	101	201	201	0.27984422	0.28344864	0.26434988	0.26800317	0.24265188	0.25619304
5	101	−101	−201	201	0.02634291	0.02429964	0.02083505	0.01922201	0.01609168	0.01551707
6	101	101	−101	−201	0.05272835	0.04307594	0.03507879	0.02860295	0.02325844	0.01889317
6	101	101	101	201	0.05272835	0.04307594	0.03507879	0.02860295	0.02325844	0.01889317
6	101	−101	−101	201	0.00359667	0.00281674	0.00219542	0.00171186	0.00132872	0.00102855
6	201	201	−201	−101	0.04218340	0.03480291	0.02868460	0.02371513	0.01959567	0.01620596
6	201	201	201	101	0.04218340	0.03480291	0.02868460	0.02371513	0.01959567	0.01620596
6	201	−201	−201	101	0.00188351	0.00150663	0.00120757	0.00097425	0.00078778	0.00063935
7	101	201	−101	−201	0.00576852	0.00402622	0.00279706	0.00194967	0.00135400	0.00093978
7	101	201	101	201	0.00576852	0.00402622	0.00279706	0.00194967	0.00135400	0.00093978
7	101	−201	−101	201	0.00063431	0.00043965	0.00030342	0.00021017	0.00014509	0.00010011

Table 6

Seven types of integrals as a function of interatomic distance when the same orbital is on both atoms

atomic orbitals on center A, V^{2+}3d (Watson [31])

ORB.	L	M	CMP.	N	KAPPA	WEIGHT			
101	2	1	1	3	1.828899384	0.498819649			
			2	3	3.610199928	0.524881542			
			3	3	6.801999092	0.109604120			
			4	3	12.432199478	0.006752737			
	INTERNUCLEAR DISTANCE				7.41339778	6.98254013	6.28522968	5.34414959	3.77999973
1	101	101	0	0	0.99999368	0.99999321	0.99999368	0.99999321	0.99999321
1	101	201	0	0	0.00210977	0.00355040	0.00799818	0.02231645	0.09656584
2	101	101	0	0	1.01924229	1.01924229	1.01924133	1.01924229	1.01924229
2	101	201	0	0	0.00055824	0.00098864	0.00243355	0.00776355	0.04441216
3	101	101	0	0	−4.91632462	−4.91632652	−4.91632462	−4.91632271	−4.91632652
3	101	201	0	0	0.00064019	0.00085306	0.00099547	−0.00150701	−0.05363765
4	101	101	0	0	0.58556044	0.41841060	0.29711223	0.23087388	0.27213937
5	101	101	101	101	0.73504567	0.73504466	0.73504466	0.73504466	0.73504531
5	101	−101	−101	101	0.05989691	0.05989687	0.05989690	0.05989685	0.05989688
5	101	101	201	201	0.58600813	0.41961765	0.29757321	0.23188645	0.27497530
5	101	−101	−201	201	0.00760892	0.00628889	0.00562818	0.00621983	0.01627427
6	101	101	−101	−201	0.00055737	0.00098576	0.00241943	0.00767222	0.04299658
6	101	101	101	201	0.00055737	0.00098576	0.00241943	0.00767222	0.04299658
6	101	−101	−101	201	0.00001302	0.00002511	0.00007110	0.00027492	0.00217710
7	101	201	−101	−201	0.00000163	0.00000480	0.00002600	0.00022484	0.00542199
7	101	201	101	201	0.00000163	0.00000480	0.00002600	0.00022484	0.00542199
7	101	−201	−101	201	0.00000016	0.00000049	0.00000272	0.00002436	0.00064837

Table 7

Seven types of integrals when V^{2+}(3d) is on both atoms

atomic orbitals on center A

ORB.	*L*	*M*	CMP.	*N*	KAPPA	WEIGHT				
101	2	1	1	3	1.828899384	0.498819649				
			2	3	3.610199928	0.524881542				
			3	3	6.801999092	0.109604120				
			4	3	12.432199478	0.006752737				
	INTERNUCLEAR DISTANCE				8.6927	8.31479	7.93685	7.5589	7.18096	6.80529
1	101	101	0	0	0.99999321	0.99999321	0.99999368	0.99999321	0.99999368	0.99999321
1	101	201	0	0	0.00041904	0.00068233	0.00110219	0.00176440	0.00279837	0.00439230
2	101	101	0	0	1.01924229	1.01924229	1.01924229	1.01924229	1.01924229	1.01924229
2	101	201	0	0	0.00009661	0.00016353	0.00027500	0.00045916	0.00076087	0.00125039
3	101	101	0	0	−4.91632462	−4.91632843	−4.91632652	−4.91633034	−4.91632652	−4.91632843
3	101	201	0	0	0.00019369	0.00028602	0.00041126	0.00057078	0.00075579	0.00093189
4	101	101	0	0	0.60530716	1.03734779	0.52735621	0.42154998	0.68888563	0.17069775
5	101	101	101	101	0.73504585	0.73504484	0.73504514	0.73504466	0.73504466	0.73504466
5	101	−101	−101	101	0.05989693	0.05989688	0.05989696	0.05989688	0.05989687	0.05989687
5	101	101	201	201	0.60455263	1.03856659	0.52765173	0.42111689	0.69116837	0.17091757
5	101	−101	−201	201	0.00586468	0.01091239	0.00591706	0.00518122	0.00971030	0.00271853
6	101	101	−101	−201	0.00009671	0.00016360	0.00027492	0.00045861	0.00075915	0.00124593
6	101	101	101	201	0.00009671	0.00016360	0.00027492	0.00045861	0.00075915	0.00124593
6	101	−101	−101	201	0.00000176	0.00000321	0.00000579	0.00001041	0.00001858	0.00003291
7	101	201	−101	−201	0.00000006	0.00000016	0.00000043	0.00000113	0.00000293	0.00000747
7	101	201	101	201	0.00000006	0.00000016	0.00000043	0.00000113	0.00000293	0.00000747
7	101	−201	−101	201	0.00000001	0.00000002	0.00000004	0.00000011	0.00000030	0.00000077

three- and four-electron distributions is not so good. However, the number of such integrals required is small and therefore the error is not too serious [33]. The major difficulty arises in estimating the energy denominators. What it means is that the energies of the unperturbed configurations (i), (ii), and (iv) are very uncertain and thus the numerical values of $\mathcal{H}_{\mathrm{I}}^{\mathrm{AA}}$, $\mathcal{H}_{\mathrm{I}}^{\mathrm{BB}}$, and $\mathcal{H}_{\mathrm{I}}^{\mathrm{DD}}$ which define the energy differences Δ_{B} and Δ_{D} are not found to the desirable accuracy. Since the higher-order terms are small and occur with both positive and negative signs, it may appear that the error is small. The arbitrariness in the calculation should however be emphasized. Since the electron transfers only to the nearest neighbor and there is considerable overlap, the transfer energy is not related with the ionization potential. We consider the evaluation of the one-electron transfer energy, $\mathcal{H}_{\mathrm{I}}^{\mathrm{BB}} - \mathcal{H}_{\mathrm{I}}^{\mathrm{AA}} = \Delta_{\mathrm{B}}$. According to Huang and Orbach [21] this energy is given by the sum of the following terms: a) the difference between the energy of an electron in $V^+(t_{2g}^3\uparrow, t_{2g}\downarrow)$ configuration and the ionization energy of the F^- ion, b) the difference between the Madelung energies at the V site and the F site, c) the electronic polarization energy of the surrounding ions subsequent to transferring an electron from F^- to V^{2+} site, and d) the interaction energy between the extra electron at the V site and the hole at the F site. We estimate a) by taking the difference between the unoccupied site energy of the V^{2+} ion and the ionization energy of F^- ion which is ≈ -4.6 eV. The increase in the Madelung energy together with the electron–hole interaction energy subsequent to charge transfer is approximately 26.9 eV as obtained by Hubbard et al. [33]. The electronic polarization energy is estimated by using the static expression $-\sum_i \alpha_i E_i^2$ where α_i is the electronic polarizability and E_i the electric field at the i-th ion site. We reduce this value by a factor of 3 to include the many-body effects. Considering only the nearest neighbors of the V^{2+}–F^- pair and taking the dielectric constant of 2, the polarization energy is estimated to be about -1.1 eV. The sum of all contributions gives $\Delta_{\mathrm{B}} = 0.7$ at. units. Similarly, $\Delta_{\mathrm{D}} = \mathcal{H}_{\mathrm{I}}^{\mathrm{DD}} - \mathcal{H}_{\mathrm{I}}^{\mathrm{AA}} = 0.57$ at. units. The arbitrariness in these values is obvious. Particularly, the many-body effects which are discussed by Marshall and Stuart [34] in terms of expansion of cation orbits, and hence contraction of anion orbits, cannot be estimated easily. Marshall and Stuart have discussed them in terms of simple scaling of distances whereas Hubbard et al. have suggested the use of renormalized charged states and accordingly readjusted Hartree-Fock wave functions.

The potential terms occurring in (4.42) are estimated from a point-charge model and are necessarily incomplete. Considerable errors are expected to be introduced by these potential terms. We compute the elements in (4.17) using the Hartree-Fock wave

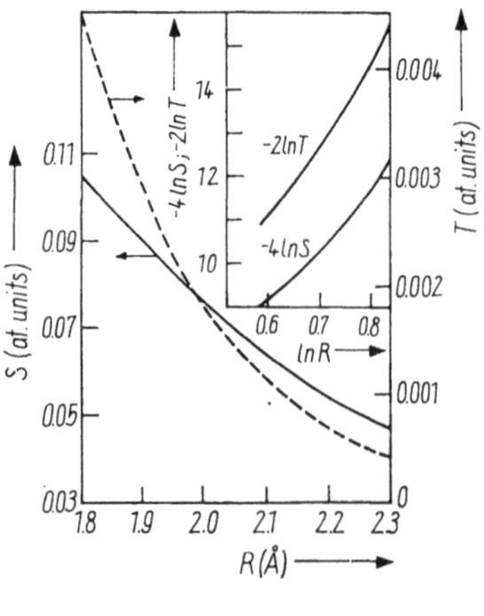

Fig. 3. Variation of overlap integrals $S = \langle d_x(V^{2+}) \mid p_x(F^-) \rangle$ and $T = \langle d_x V^{2+}(+R) \mid d_x V^{2+}(-R) \rangle$ as a function of nearest-neighbor distance R

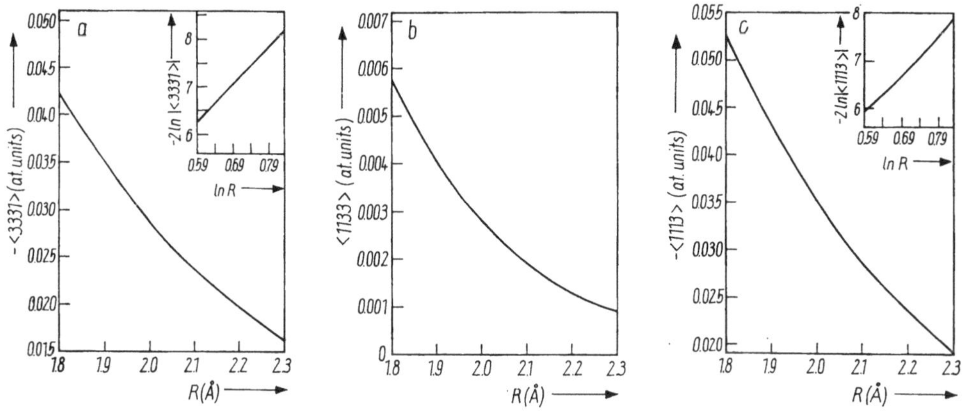

Fig. 4. Variation of Coulomb integral a) $\langle d_x(+R)\, d_x(+R)|\, |d_x(+R)\, p_x(F^-)\rangle$ called $\langle 3331\rangle$, b) $\langle p_x(F^-)\, p_x(F^-)|\, |d_x(+R)\, d_x(+R)\rangle$ called $\langle 1133\rangle$, and c) $\langle p_x(F^-)\, p_x(F^-)|\, |p_x(F^-)\, d_x(+R)\rangle$ called $\langle 1113\rangle$ as a function of V^{2+}–F^- distance

functions of V^{2+} and of F^-, considering the V^{2+}–F^- distance as a parameter. In order to get a feeling of how various terms vary with interatomic distances, some of the integrals are shown in Fig. 3 to 5. Some composite terms of ionic contribution (4.18) are plotted in Fig. 6 and all the terms of (4.17) are given in Fig. 7. The exchange interaction between two d electrons, defined by $-2J_{34}\mathbf{S}_3 \cdot \mathbf{S}_4$ with $2J_{34} = W_s - W_t$, consists of several ferro- and several antiferromagnetic terms. The superexchange for the atomic case $-2J_{AB}\mathbf{S}_A \cdot \mathbf{S}_B$ is obtained from the electronic case $-2J_{34}\mathbf{s}_3 \cdot \mathbf{s}_4$ by performing the sum over all the unpaired electrons. The exchange energy for the ion–ion coupling

$$W_{AB} = -2 \sum_{i=1}^{n_A} \sum_{j=1}^{n_B} J_{ij}\mathbf{S}_i\mathbf{S}_j = -\frac{2}{n_A n_B}\, \mathbf{S}_A \cdot \mathbf{S}_B \sum_{i>j} J_{ij} \tag{4.46}$$

is just the energy between the electron–electron coupling summed over the number of unpaired electrons on the atom A, n_A, and the number of electrons on the atom B,

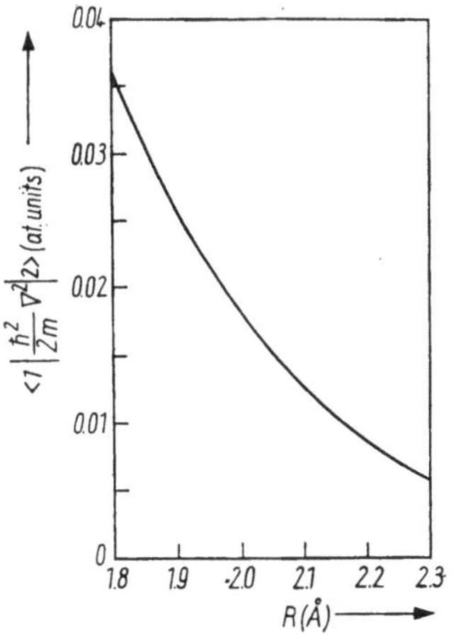

Fig. 5. Variation of split kinetic energy integral between $p_x(F^-)$ electron and $d_x(V^{2+})$ electron as a function of V^{2+}–F^- distance

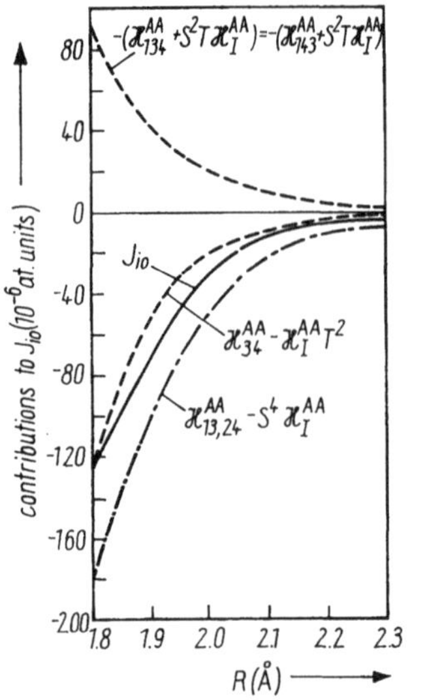

Fig. 6

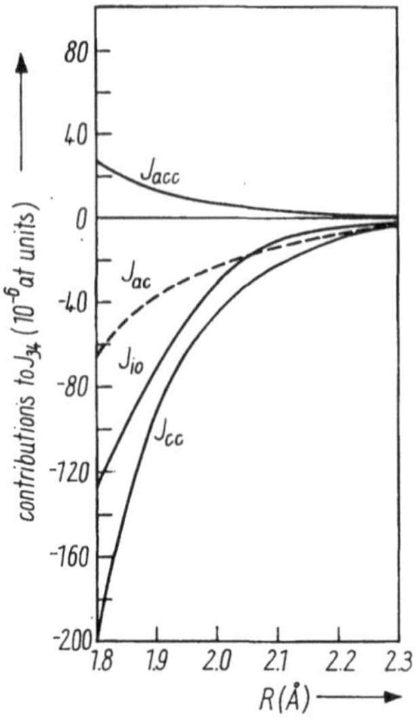

Fig. 7

Fig. 6. Various contributions of ionic exchange (4.18) as a function of V^{2+}–F^- distance

Fig. 7. Collective terms of (4.17) as a function of V^{2+}–F^- distance

n_B, so that

$$J_{AB} = \frac{1}{n_A n_B} \sum_{i=1}^{n_A} \sum_{j=1}^{n_B} J_{ij} \,. \tag{4.47}$$

In the case of V^{2+}–F^-–V^{2+} this gives rise to a factor of $\frac{2}{9}$ in going from the electronic to the atomic exchange since the atomic spin is $\frac{3}{2}$ for the configuration d^3. The calculated value of exchange in linear V^{2+}–F^-–V^{2+} as a function of the V^{2+}–F^- distance

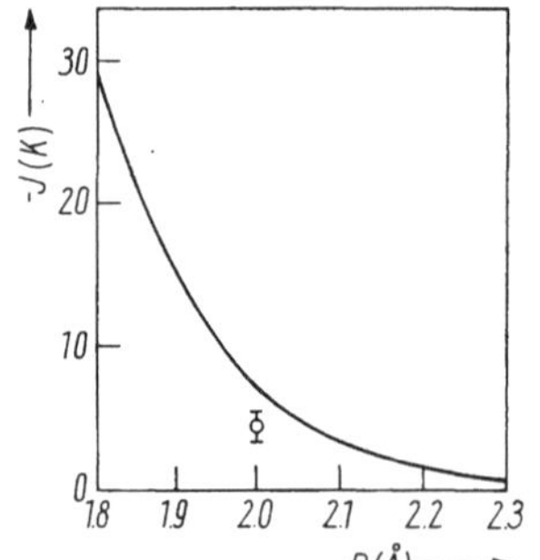

Fig. 8

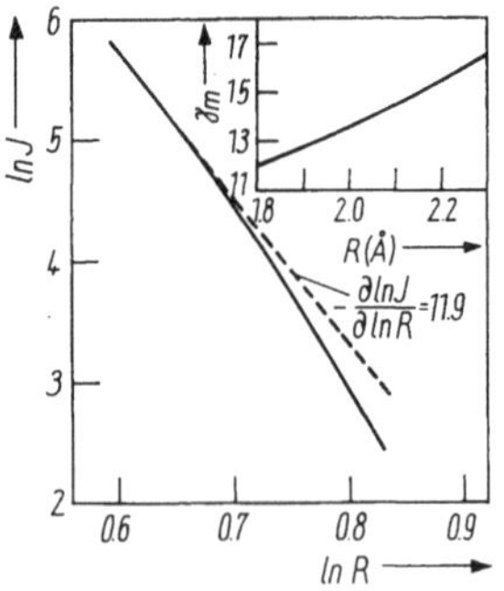

Fig. 9

Fig. 8. Calculated exchange in a linear pair V^{2+}–F^-–V^{2+} as a function of V^{2+}–F^- distance. The experimental value of Smith and Owen is shown at $R \approx 2$ Å. This distance in V^{2+}-doped $KMgF_3$ is not known because of local readjustment of distances near the impurity

Fig. 9. The magnetic Grüneisen parameter, $\gamma_m = -\partial \ln J/\partial \ln R$, for linear V^{2+}–F^-–V^{2+} exchange coupled pair from the calculation of Fig. 8. The value of γ_m appears to be a function of interatomic separation, about 12 for $R \approx 1.8$ Å and ≈ 17 for $R \approx 2.3$ Å

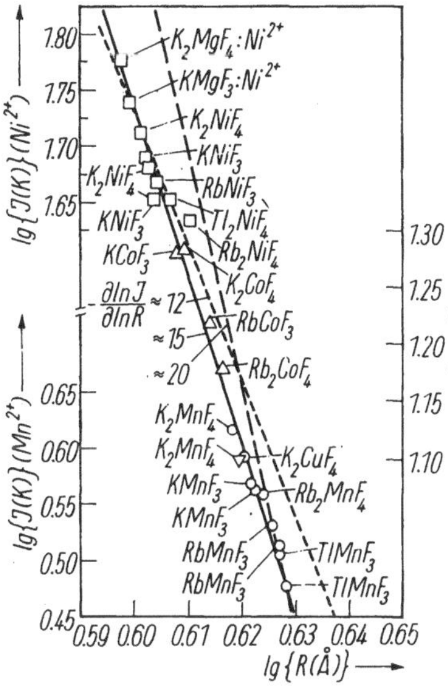

Fig. 10. Double-logarithmic plot of exchange with cation–cation distance in several compounds of Mn^{2+}, Cu^{2+}, Co^{2+}, and Ni^{2+} indicates the value of exchange Grüneisen parameter varying from about 12 to 20. The data are taken from Table 8

is given in Fig. 8, and a logarithmic plot of the same is displayed in Fig. 9. It is found that $\gamma_m = -\partial \ln J/\partial \ln R$ is a function of interatomic distance as $\gamma_m(R) \approx 12$ for the V^{2+}–F^- distance of 0.18 nm and ≈ 17 for 0.23 nm. In Fig. 10, we show a plot of lg J versus the logarithm of cation–cation distance for several systems of Ni^{2+}, Co^{2+}, Cu^{2+}, and Mn^{2+} from Table 8, as discussed by Shrivastava [10] from the data of various workers [35 to 64]. It is seen that Ni compounds have shorter bonds and larger exchange compared with isomorphous compounds of Mn^{2+} which have longer bonds and smaller exchange. The exchange Grüneisen parameter $\gamma_m(R)$ for Ni^{2+} is about

Table 8

Experimental results of exchange and cation–cation separation including the approximate temperature of measurement

material	T (K)	J (K)	R (Å)	No. in Fig. 12	ref.
$RbMnF_3$	295	3.26	4.238	3	[35, 36, 37, 55]
	$<$ 50	3.40	4.225	4	[36, 37, 38]
$TlMnF_3$	295	3.0	4.250	2	[39, 40, 41]
	$<$ 50	3.2	4.238	1	[39, 40, 41]
$KMnF_3$	295	3.65	4.191	6	[37, 42, 43]
	200	3.70	4.183	7	[37, 42, 43]
Rb_2MnF_4	$<$ 30	3.62	4.207	5	[44, 45, 46, 54, 56]
K_2MnF_4	295	3.90	4.172	8	[44, 47, 37, 57]
	4.2	4.13	4.152	9	[47, 48]
$RbCoF_3$	$\approx$100	16.6	4.114	11	[49, 50, 44]
$KCoF_3$	$<$100	19.2	4.058	13	[42, 44]
Rb_2CoF_4	$<$120	15.1	4.135	10	[44, 51, 58]
K_2CoF_4	$<$120	19.3	4.068	12	[44, 58]
K_2CuF_4	4.2	10.0	4.155		[63]
$RbNiF_3$	4.2	46.6	4.020	16	[54]
$KNiF_3$	$\approx$300	44.0	4.014	17	[52, 59, 60]
	4.2	49.0	4.000	19	[42]
Rb_2NiF_4	$<$200	43.0	4.078	14	[53, 54, 56]
Tl_2NiF_4	$\approx$200	45.0	4.043	15	[53, 54, 56]
K_2NiF_4	$\approx$300	48.0	4.006	18	[52, 54, 56]
	4.2	51.5	3.992	20	[46, 60, 62]
$KMgF_3:Ni^{2+}$	$<$100	55.0	3.975	22	[54]
$K_2MgF_4:Ni^{2+}$	$<$100	60.0	3.960	21	[54, 60, 61, 64]

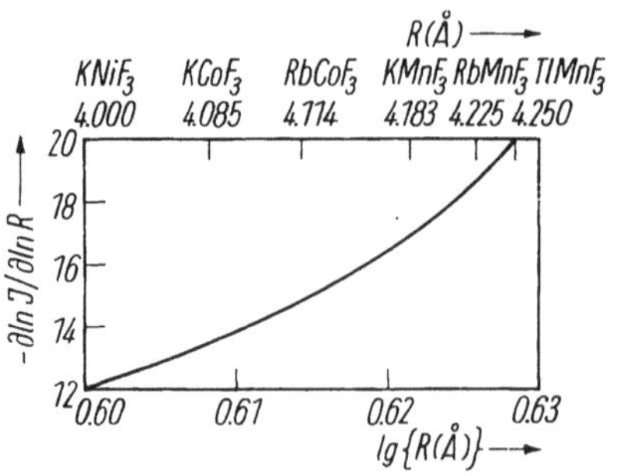

Fig. 11

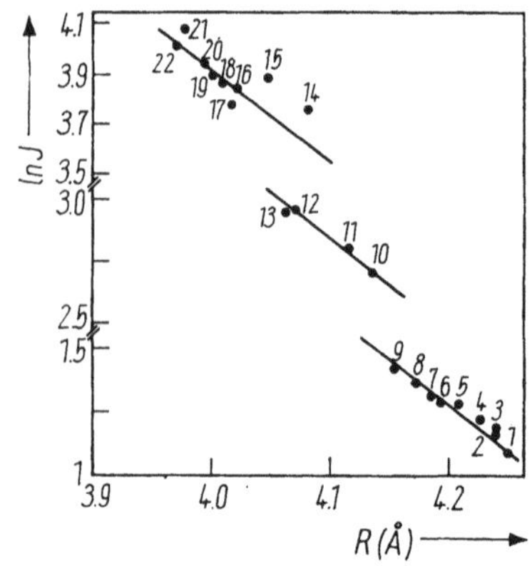

Fig. 12

Fig. 11. The apparent universality of $\gamma_m(R)$ deduced from Fig. 10. The exchange Grüneisen parameter appears to depend on the interatomic separation but not on the magnetic ion. This experimental value may be compared with that predicted for V^{2+} ion wave functions discussed in the text

Fig. 12. Plot of ln J with R, the distance between two magnetic ions. The data are given in Table 8. The exchange is given by $J = J_0 \exp(-\beta R)$ with $\beta = 3.55$ for all 22 points

12 whereas at larger cation–cation distances appropriate to the compounds of Mn^{2+}, the value of γ_m is about 20. An essentially similar result is predicted by our computation of V^{2+}–F^-–V^{2+} and therefore it appears that γ_m is *independent of the magnetic ion*. The variation which we find is perhaps already inherent in the nature of Hartree-Fock wave functions which are linear combinations of functions like [65]

$$\Phi_{i\lambda\alpha} = \sum_p \chi_{p\lambda\alpha} C_{i\lambda p}\,, \tag{4.48}$$

where

$$\chi_{p\lambda\alpha}(r, \theta, \varphi) = R_{\lambda p}(r)\, Y_{\lambda\alpha}(\theta, \varphi) \tag{4.49}$$

and

$$R_{\lambda p}(r) = [(2n_{\lambda p})!]^{-1/2} (2\zeta_{p\lambda})^{n_{\lambda p}+1/2} r^{n_{\lambda p}-1} e^{-\zeta_{\lambda p} r}\,. \tag{4.50}$$

Here the orbitals are characterized by an index λ indicating the orbital angular momentum, by an index α which indicates the subspecies labeling the individual members of the orbitally degenerate set that transforms according to the representation λ, and by an index i which refers to the i-th orbital of symmetry λ. The orbital $\Phi_{i\lambda\alpha}$ is expanded in terms of basis functions with p as a particular basis function of symmetry λ. The expansion coefficient depends on i, λ, and p but not on the projection α. The basis functions χ are Slater-type orbitals with integer quantum numbers.

The universality of $\gamma_m(R)$ is therefore understandable but is not immediately obvious on the basis of theoretical considerations. The natural value is shown in Fig. 11. Instead of a power law, if we assume an exponential behaviour, then in Fig. 12 we find that $J = J_0 \exp(-\beta R)$ with $\beta = 3.55$.

4.3 Exchange between orthogonal orbitals

Consider the configuration (i) of four electrons on three orbitals $d_{xz}(-R)$-$p_x(\uparrow\downarrow)$-$-d_{xz}(+R)$. In the configurations (ii) and (iii) one of the electrons of the p_x is transferred either to the empty orbit on the atom at $(-R)$ or to the atom at $(+R)$. The resulting states are like $d_{xz}(+R)$ $(\uparrow\downarrow)$-$p_z(\uparrow)$-$d_{xz}(+R)$ in which the orientation of the $d_{xz}(+R)$ is still arbitrary so that although $d_{xz}(+R)$ $(\uparrow\downarrow)$ is a singlet, $d_{xz}(+R)$ is still free to form either a singlet or a triplet with $p_z(\uparrow)$. We can obtain the energies of these states separately. However, it is simpler to construct the projection operators P^σ by demanding $\mathbf{s}_1$ to be antiparallel to $\mathbf{s}_2$ and leaving the orientations $\mathbf{s}_3$ and $\mathbf{s}_4$ unspecified. The solution to the secular equation then occurs in the form of the inner product $\mathbf{s}_3 \cdot \mathbf{s}_4$ which is exactly what is desired. Realizing that

$$P^\sigma_{ij} = \tfrac{1}{2}(1 + 4\mathbf{s}_i \cdot \mathbf{s}_j)$$

with

$$P^\sigma |S, M_S\rangle_i = \sum_{j=1}^{m} P^\sigma_{ji} |S, M_S\rangle_j ,$$

one obtains

$$\begin{aligned} &P_{\mathrm{I}} = 1 , \qquad P_{13} = P_{23} = P_{14} = P_{24} = -\tfrac{1}{2} , \\ &P_{134} = P_{234} = P_{143} = P_{243} = -\tfrac{1}{2} P_{34} , \\ &P_{13,24} = P_{1423} = \tfrac{1}{2}(1 + P_{34}) \end{aligned} \tag{4.51}$$

for the twelve permutations which differ by more than an interchange of the orbitals a_1 and a_2. The secular equation is

$$\begin{vmatrix} (\mathcal{H} - W)^{\mathrm{AA}} & (\mathcal{H} - W)^{\mathrm{CA}} \\ (\mathcal{H} - W)^{\mathrm{AC}} & (\mathcal{H} - W)^{\mathrm{CC}} \end{vmatrix} = 0 \tag{4.52}$$

in which the matrix elements, as before are

$$\begin{aligned} (\mathcal{H} - W)^{\mathrm{AA}} &= (\mathcal{H}^{\mathrm{AA}}_{\mathrm{I}} - W) - (\mathcal{H}^{\mathrm{AA}}_{13} - WS^2) - \mathcal{H}^{\mathrm{AA}}_{14} + \mathcal{H}^{\mathrm{AA}}_{13,24} + \\ &\quad + P_{34}(\mathcal{H}^{\mathrm{AA}}_{34} - \mathcal{H}^{\mathrm{AA}}_{134} - \mathcal{H}^{\mathrm{AA}}_{143} + \mathcal{H}^{\mathrm{AA}}_{13,24}) , \\ (\mathcal{H} - W)^{\mathrm{CC}} &= (\mathcal{H}^{\mathrm{CC}}_{\mathrm{I}} - W) - (\mathcal{H}^{\mathrm{CC}}_{13} - WS^2) - \mathcal{H}^{\mathrm{CC}}_{14} + \mathcal{H}^{\mathrm{CC}}_{13,24} + \\ &\quad + P_{34}(\mathcal{H}^{\mathrm{CC}}_{34} - \mathcal{H}^{\mathrm{CC}}_{134} - \mathcal{H}^{\mathrm{CC}}_{143} + \mathcal{H}^{\mathrm{CC}}_{13,24}) , \\ (\mathcal{H} - W)^{\mathrm{AC}} &= (\mathcal{H}^{\mathrm{AC}}_{\mathrm{I}} - WS^3) - (\mathcal{H}^{\mathrm{AC}}_{13} - WS) - \mathcal{H}^{\mathrm{AC}}_{14} + \mathcal{H}^{\mathrm{AC}}_{10,24} + \\ &\quad + P_{34}(\mathcal{H}^{\mathrm{AC}}_{34} - \mathcal{H}^{\mathrm{AC}}_{134} - \mathcal{H}^{\mathrm{AC}}_{143} + \mathcal{H}^{\mathrm{AC}}_{13,24}) . \end{aligned} \tag{4.53}$$

The dominant terms, bilinear in spin, are calculated to be

$$\begin{aligned} W &= -2\mathbf{s}_3 \cdot \mathbf{s}_4 [\mathcal{H}^{\mathrm{AA}}_{1324} - 2\gamma_{13}\mathcal{H}^{\mathrm{AC}}_{134} + \gamma^2_{13}\mathcal{H}^{\mathrm{CC}}_{34} + \mathcal{H}^{\mathrm{AA}}_{34} - 2\mathcal{H}^{\mathrm{AA}}_{134} + 2\gamma_{13}\mathcal{H}^{\mathrm{AC}}_{1324}] = \\ &= -2J'_{34}\mathbf{s}_3 \cdot \mathbf{s}_4 . \end{aligned} \tag{4.54}$$

The matrix elements in the above are

$$\begin{aligned} &\mathcal{H}^{\mathrm{AA}}_{1324} = \langle 4114\rangle S^2 , \\ &\mathcal{H}^{\mathrm{AC}}_{13} - \mathcal{H}^{\mathrm{AA}}_{\mathrm{I}} S = \langle 13\rangle - \langle 11\rangle S + \langle 3133\rangle + \langle 1131\rangle + \langle 1434\rangle - \\ &\qquad\qquad - (\langle 1111\rangle + \langle 1313\rangle + \langle 1414\rangle) S , \\ &\mathcal{H}^{\mathrm{AC}}_{134} = \langle 4114\rangle S , \\ &\mathcal{H}^{\mathrm{CC}}_{34} = \langle 4114\rangle , \end{aligned}$$

$$\mathcal{H}_{34}^{AA} = \langle 3443 \rangle ,$$
$$\mathcal{H}_{1324}^{AC} = \langle 4134 \rangle ,$$
$$\mathcal{H}_{134}^{AA} = \langle 4134 \rangle S ,$$
$$\gamma_{13} = -\Delta_C^{-1}(\mathcal{H}_{13}^{AC} - \mathcal{H}_I^{AA} S) , \quad (4.55)$$

so that (4.54) reduces to

$$J'_{34} = [(S + \gamma_{13})^2 \langle 4114 \rangle - 2\langle 4134 \rangle (S + \gamma_{13}) + \langle 3443 \rangle] . \quad (4.56)$$

If the magnetic electron wave functions are written as

$$\alpha_3 = a_3 - (S + \gamma_{13}) a_1 , \quad (4.57)$$

then the exchange (4.56) is just the integral of the Coulomb interaction

$$J'_{34} \approx \langle \alpha_3 \alpha_4 | \frac{e^2}{r_{34}} | \alpha_4 \alpha_3 \rangle , \quad (4.58)$$

which in the case of linear V^{2+}–F^-–V^{2+} is ferromagnetic, ≈ -0.7 K, a little too small to be calculated reliably. However, this ferromagnetic contribution becomes very large if the angle is reduced from 180°. For a bond angle of 90° this contribution makes the pair ferromagnetic.

References

[1] W. HEISENBERG, Z. Phys. **38**, 411 (1926); **49**, 619 (1926).
[2] P. A. M. DIRAC, Proc. Roy. Soc. **A123**, 714 (1929); The Principles of Quantum Mechanics, Oxford University Press 1958.
[3] L. NÉEL, Ann. Phys. (France) **18**, 5 (1932).
[4] L. D. LANDAU, Phys. Z. SU **4**, 675 (1933).
[5] H. A. KRAMERS, Physica (Utrecht) **1**, 182 (1934).
[6] P. W. ANDERSON, Phys. Rev. **79**, 350 (1950).
[7] C. KITTEL, Phys. Rev. **120**, 335 (1960).
[8] K. N. SHRIVASTAVA, Phys. Letters A **56**, 399 (1976).
[9] K. N. SHRIVASTAVA and V. JACCARINO, Phys. Rev. B **13**, 299 (1976).
[10] K. N. SHRIVASTAVA, phys. stat. sol. (b) **78**, 749 (1976).
[11] K. N. SHRIVASTAVA, Phys. Letters A **54**, 441 (1975).
[12] K. N. SHRIVASTAVA, J. Phys. C **8**, L554 (1975).
A. SUGUNA and K. N. SHRIVASTAVA, Phys. Letters A **71**, 121 (1979); phys. stat. sol. (b) **99**, 305 (1980).
[13] J. H. VAN VLECK, Phys. Rev. **45**, 405 (1934).
[14] R. SERBER, Phys. Rev. **45**, 461 (1934).
[15] K. N. SHRIVASTAVA, Phys. Rev. B **20**, 2634 (1979).
[16] K. N. SHRIVASTAVA, Phys. Rev. B **20**, 5375 (1979).
[17] K. N. SHRIVASTAVA, Phys. Rev. B **21**, 2702 (1980).
[18] K. N. SHRIVASTAVA, Physica (Utrecht) **B100**, 67 (1980).
[19] C. HERRING, in: Magnetism, Vol. IV, Ed. G. T. RADO and H. SUHL, Academic Press, New York 1966.
[20] F. KEFFER and T. OGUCHI, Phys. Rev. **109**, 730 (1959).
[21] N. L. HUANG and R. ORBACH, Phys. Rev. **154**, 487 (1967).
[22] N. L. HUANG and R. ORBACH, AIP Conf. Proc. **10**, 1238 (1972).
[23] N. L. HUANG, Phys. Rev. **157**, 378 (1967).
[24] N. L. HUANG, Phys. Rev. **164**, 636 (1967).
[25] N. L. HUANG, Ph. D. Thesis, University of California, Los Angeles 1966.

[26] L. D. Falkovskaya and V. A. Sapozhnikov, phys. stat. sol. (b) 71, 469 (1975).
[27] N. Fuchikami, J. Phys. Soc. Japan 28, 871 (1970).
[28] J. Yamashita and J. Kondo, Phys. Rev. 109, 730 (1958).
[29] D. E. Rimmer, J. Phys. C 2, 329 (1969).
[30] R. K. Nesbet, Phys. Rev. 119, 658 (1960).
[31] R. E. Watson, MIT Solid State and Molecular Theory Group, Quarterly Progr. Rep. No. 12, 1958.
[32] S. R. P. Smith and J. Owen, J. Phys. C 4, 1399 (1971).
[33] J. Hubbard, D. E. Rimmer, and F. R. A. Hopegood, Proc. Phys. Soc. 88, 13 (1966).
[34] W. Marshall and R. Stuart, Phys. Rev. 123, 2048 (1961).
[35] Y. P. Simanov, L. P. Batsanova, and L. M. Kovba, Soviet Phys. – inorg. Chem. 2, 207 (1957).
[36] C. G. Windsor and R. W. H. Stevenson, Proc. Phys. Soc. 87, 501 (1966).
[37] D. J. Breed, Thesis, University of Amsterdam, 1969.
[38] D. T. Teaney, V. L. Moruzzi, and B. E. Argyle, J. appl. Phys. 37, 1122 (1966).
[39] S. A. Kizhaev, A. G. Tutov, and V. A. Bokov, Soviet Phys. – Solid State 7, 2325 (1966).
[40] L. M. Rao and N. S. Satya Murthy, J. Physique 32, Suppl. C-1, 617 (1971).
[41] D. E. Eastman and M. W. Shafer, J. appl. Phys. 38, 1274 (1967).
[42] A. Okazaki and Y. Suemune, J. Phys. Soc. Japan 16, 671 (1961).
[43] O. Beckman and K. Knox, Phys. Rev. 121, 376 (1961).
[44] D. J. Breed, K. Gilijamse, and A. R. Miedema, Physica (Utrecht) 45, 205 (1969).
[45] A. H. M. Schrama, Physica (Utrecht) 68, 279 (1973).
[46] R. J. Birgeneau, H. J. Guggenheim, and G. Shirane, Phys. Rev. B 1, 2211 (1970).
[47] B. O. Loopstra, B. van Laar, and D. J. Breed, Phys. Letters A 26, 526 (1968); 27, 188 (1968).
[48] R. J. Birgeneau, H. J. Guggenheim, and G. Shirane, Phys. Rev. B 6, 304 (1973).
[49] M. P. Petrov and G. M. Nedlin, J. appl. Phys. 39, 1012 (1968).
[50] Y. Allain, J. Denis, A. Hirpin, J. Lecomte, P. Meriel, J. Nouet, F. Plique, and A. Zarembovitch, J. Physique 32, Suppl. C-1, 611 (1971).
[51] E. J. Samuelson, J. Phys. Chem. Solids 35, 785 (1974).
[52] E. P. Maarschall, A. C. Botterman, S. Vega, and A. R. Miedema, Physica (Utrecht) 41, 473 (1969).
[53] W. Rudorff, J. Kandler, and D. Babel, Z. anorg. allg. Chem. 317, 261 (1962).
[54] L. J. de Jong and A. R. Miedema, Adv. Phys. 23, 1 (1974).
[55] L. J. de Jong and D. J. Breed, Solid State Commun. 15, 1061 (1974).
[56] L. J. de Jong, AIP Conf. Proc. 10, 561 (1973).
[57] D. J. Breed, Physica (Utrecht) 37, 35 (1967).
[58] W. J. L. Buyers, T. M. Holden, E. C. Svensson, R. A. Cowley, and M. T. Hutchings, J. Phys. C 4, 2139 (1971).
[59] M. E. Lines, Phys. Rev. 164, 736 (1967).
[60] S. R. Chinn, H. J. Zeiger, and J. R. O'Connor, Phys. Rev. B 3, 1709 (1971).
[61] J. Skalyo, Jr., G. Shirane, R. J. Birgeneau, and H. J. Guggenheim, Phys. Rev. Letters 23, 1394 (1969).
[62] I. Yamada and M. Ikebe, J. Phys. Soc. Japan 33, 1334 (1972).
[63] Y. Yamaguchi and N. Sakamoto, J. Phys. Soc. Japan 27, 1444 (1969).
[64] H. W. de Wijn, L. R. Walker, and R. E. Walstedt, Phys. Rev. B 8, 285 (1973).
[65] E. Clementi and C. Roetti, Atomic Data nuclear Data Tables 14, 177 (1974).

(Received July 21, 1983)

Original Papers

phys. stat. sol. (b) **125**, 45 (1984)

Subject classification: 1.2 and 20.1; 1.5

Department of Physics, University of Essex, Colchester[1]) *(a)*
Departamento de Fisica, Universidade Federal da Bahia, Salvador[2]) *(b)*

Thick Films near Phase Transitions

Green Functions, Fluctuation Spectra, and Light Scattering Theory

By

A. Ribeiro Filho (a), (b) and D. R. Tilley (a)

A film of thickness L with a one-component order parameter η is considered. It is assumed that the free energy is given by a simple Landau-type expansion in η, that the boundary condition is linear in η and $\nabla\eta$, and that fluctuations in the dielectric function depend linearly on fluctuations in η. With the use of phenomenological equations of motion the order-parameter Green function is derived. Expressions are given, and illustrated numerically, for the power spectrum of order-parameter fluctuations and the cross section for inelastic light scattering.

Eine Schicht der Dicke L, mit einem einkomponentigen Ordnungsparameter η wird untersucht. Es wird angenommen, daß die Dichte der freien Energie durch eine einfache Landauentwicklung in η gegeben ist, daß die Randbedingung in η und $\nabla\eta$ linear ist, und daß die Schwankungen der dielektrischen Funktion linear von den Schwankungen von η abhängig sind. Mit Hilfe von phänomenologischen Bewegungsgleichungen wird die Greensche Funktion des Ordnungsparameters abgeleitet. Ausdrücke für das Kraftspektrum von Ordnungsparameterschwankungen und der Streuquerschnitte der inelastischen Lichtstreuung werden angegeben und numerisch illustriert.

1. Introduction

In a previous paper [1], we showed how the phenomenological theory of light scattering from bulk solids in the vicinity of phase transitions [2] can in principle be extended to describe inelastic light scattering from a semi-infinite medium. The order-parameter Green function was derived for the high-temperature (prototype) phase, and fluctuation spectra were obtained. These fluctuation spectra are enhanced as the temperature is lowered towards the critical value. It was then shown how the theoretical methods developed to describe Brillouin scattering from surfaces [3 to 5] can be applied to find expressions for the cross-section.

In this paper, we extend our calculations to films. The free energy F is written in terms of a single-component order parameter η as

$$FS^{-1} = \int_{-L}^{0} \mathrm{d}z\, \frac{1}{2}\left[A\eta^2 + \frac{1}{2}B\eta^4 + C\left(\frac{\mathrm{d}\eta}{\mathrm{d}z}\right)^2\right] - \frac{1}{2}C\lambda^{-1}(\eta_-^2 + \eta_+^2)\,, \tag{1}$$

where S is the surface area of the film with plane surfaces at $z = 0$ and $z = -L$. We write $\eta_+ = \eta(0)$ and $\eta_- = \eta(-L)$. It is assumed that $B > 0$, $C > 0$, $A(T \gtrless T_c^0) \gtrless 0$, so that T_c^0 is the bulk critical temperature. The Euler-Lagrange equations derived from (1) give the equation for the equilibrium value of η

$$C\frac{\mathrm{d}^2\eta}{\mathrm{d}z^2} - A\eta - B\eta^3 = 0 \tag{2}$$

[1]) Wivenhoe Park, Colchester C04 3SQ, Great Britain.
[2]) Salvador (BA), Brazil.

with boundary conditions

$$\frac{d\eta}{dz} - \lambda^{-1}\eta = 0 \quad \text{at} \quad z = 0\,, \tag{3}$$

$$\frac{d\eta}{dz} + \lambda^{-1}\eta = 0 \quad \text{at} \quad z = -L\,. \tag{4}$$

It has been shown elsewhere [6] that (2) to (4) can be solved at all temperatures for η. For $\lambda > 0$, it is seen from (3) and (4) that η increases at the film boundaries, so it is to be expected that the critical temperature T_c is greater than the bulk value T_c^0. In fact [6], T_c is given by

$$\tanh\left(\tfrac{1}{2}\,\bar{\alpha}L\right) = (\bar{\alpha}\lambda)^{-1}\,, \tag{5}$$

where

$$\bar{\alpha}^2 = \frac{|A|}{C}\,. \tag{6}$$

Conversely, for $\lambda < 0$, one expects $T_c < T_c^0$, and in fact T_c is given by

$$\tan\left(\tfrac{1}{2}\,\bar{\alpha}L\right) = -(\bar{\alpha}\lambda)^{-1}\,. \tag{7}$$

Numerical solutions of (5) and (7) are given in [6].

The plan of this paper is similar to that of [1]. In Section 2 we evaluate the order-parameter Green function in the high-temperature phase. In Section 3 we derive and discuss the expressions for thermal fluctuations of the order parameter, and in Section 4 we derive the cross-section. Finally, Section 5 contains some discussion.

2. Green Functions

We derive the Green function by linear response theory, as was done in [1]. We assume the equation of motion

$$m\ddot{\eta} + \gamma\dot{\eta} + \frac{\delta F}{\delta\eta} = 0 \tag{8}$$

and we augment F with a term ΔF involving a driving force $J(\boldsymbol{r})$:

$$\Delta F = -\frac{1}{S}\int dx\,dy \int_{-L}^{0} dz\, J(\boldsymbol{r})\,\eta^*(\boldsymbol{r})\,. \tag{9}$$

We take

$$J(\boldsymbol{r}) = J_0 \exp(-i\omega t)\,\delta(z - z') \tag{10}$$

so that the linear response of $\eta(z)$ gives the Green function $\langle\langle\eta(z);\eta^*(z')\rangle\rangle_\omega$.

The Green function for the bulk material is

$$\langle\langle\eta(z);\eta^*(z')\rangle\rangle_\omega = (2SC\alpha)^{-1}\exp(-\alpha\,|z - z'|)\,, \tag{11}$$

where

$$\alpha^2 = \frac{A - m\omega^2 - i\omega\gamma}{C}\,. \tag{12}$$

This function serves as the particular integral in the differential equation for $\eta(z)$. We add the complementary function $A_1\exp(\alpha z) + A_2\exp(-\alpha z)$, in which A_1 and A_2

are determined by the boundary conditions (3) and (4). We find

$$\begin{aligned}\langle\!\langle \eta(z); \eta(z')^* \rangle\!\rangle = (2SC\alpha)^{-1} \{&\exp(-\alpha |z - z'|) + \\ &+ D^{-1}[g_1 (\exp[\alpha(z + z')] + \exp[-\alpha(2L + z + z')]) + \\ &+ g_2(\exp[-\alpha(2L + z - z')] + \exp[-\alpha(2L - z + z')])]\} ,\end{aligned} \tag{13}$$

where

$$D = (\lambda\alpha - 1)^2 - g_2 \exp(-2\alpha L) , \tag{14}$$

$$g_1 = (\lambda\alpha + 1)(\lambda\alpha - 1) , \tag{15}$$

$$g_2 = (\lambda\alpha + 1)^2 . \tag{16}$$

In the limit $L \to \infty$, this film Green function reduces to the Green function for the semi-infinite medium [1]. This can be seen since $D \to (\lambda\alpha - 1)^2$, and as $L \to \infty$ the exponentials in which L appears explicitly vanish.

It can be seen from (13) that the Green function has a pole for $D = 0$, that is

$$(\lambda\alpha - 1)(\lambda\alpha + 1)^{-1} = \pm \exp(-\alpha L) . \tag{17}$$

For general ω this may be expected to give the soft-mode frequency, while for $\omega = 0$ and consequently $\alpha^2 = C^{-1}A = \bar{\alpha}^2$ it should give the critical temperature T_c. It is easily seen in fact that with the choice of the positive sign (17) reduces to (5) for $\lambda > 0$ and to (7) for $\lambda < 0$.

For calculation of the light-scattering cross-section it is necessary to know the Fourier transform of (13), defined by

$$\langle\!\langle \eta(Q); \eta^*(Q') \rangle\!\rangle_\omega = L^{-2} \int_{-L}^{0} \mathrm{d}z \int_{-L}^{0} \mathrm{d}z' \exp[i(Q'z' - Qz)] \langle\!\langle \eta(z); \eta^*(z') \rangle\!\rangle_\omega . \tag{18}$$

The integrals are elementary, and the result is

$$\begin{aligned}\langle\!\langle \eta(Q); \eta^*(Q') \rangle\!\rangle = (2SCL^2\alpha)^{-1} \{&(Q + i\alpha)^{-1} (Q' - Q)^{-1} \times \\ &\times (1 - \exp[-i(Q' - Q) L]) - \\ &- (Q + i\alpha)^{-1} (Q' + i\alpha)^{-1} [\exp\{i(Q + i\alpha) L\} - \\ &- \exp\{i(Q - Q') L\}] + \\ &+ (Q' - i\alpha)^{-1} (Q - i\alpha)^{-1} [1 - \exp\{-i(Q' - i\alpha) L\}] - \\ &- (Q' - Q)^{-1} (Q - i\alpha)^{-1} \times \\ &\times [1 - \exp\{-i(Q' - Q) L\}] + (\lambda\alpha + 1)(\lambda\alpha - 1) \times \\ &\times [(\lambda\alpha - 1)^2 - (\lambda\alpha + 1)^2 \exp(-2\alpha L)]^{-1} [(Q + i\alpha)^{-1} \times \\ &\times (Q' - i\alpha)^{-1} \times \\ &\times [1 - \exp\{i(Q + i\alpha) L\}] [1 - \exp\{-i(Q' - i\alpha) L\}] + \\ &+ (Q - i\alpha)^{-1} (Q' + i\alpha)^{-1} \times \\ &\times [1 - \exp\{i(Q - i\alpha) L\}] [1 - \exp\{-i(Q' + i\alpha) L\}] \times \\ &\times \exp(-2\alpha L)] + \\ &+ (\lambda\alpha + 1)^2 [(\lambda\alpha - 1)^2 - (\lambda\alpha + 1)^2 \exp(-2\alpha L)]^{-1} \times \\ &\times [(Q + i\alpha)^{-1} (Q' + i\alpha)^{-1} \times \\ &\times [1 - \exp\{i(Q + i\alpha) L\}] [1 - \exp\{-i(Q' + i\alpha) L\}] \times \\ &\times \exp(-2\alpha L) + (Q - i\alpha)^{-1} \times \\ &\times (Q' - i\alpha)^{-1} [1 - \exp\{i(Q - i\alpha) L\}] \times \\ &\times [1 - \exp\{i(i\alpha - Q') L\} \exp(-2\alpha L)]\} .\end{aligned} \tag{19}$$

3. Fluctuation Power-Spectrum

The fluctuation power-spectrum follows from (13) with the help of the fluctuation–dissipation theorem [1]. It is

$$\langle \eta(z)\, \eta^*(z') \rangle_\omega = (2\pi S C\omega)^{-1}\, k_B T\, \mathrm{Im}\, \{\alpha^{-1} [\exp(-\alpha |z - z'|) + D^{-1} g_1 [\exp\{\alpha(z + z')\} + \exp\{-\alpha(2L + z + z')\}] + D^{-1} g_2 [\exp\{-\alpha(2L - z + z')\} + \exp\{-\alpha(2L + z - z')\}]\}\,, \tag{20}$$

where D_1, g_1, and g_2 have been defined by (14) to (16). Equation (20) reduces to the semi-infinite or surface power spectrum expression, in the limit $L \to \infty$.

The power spectrum (20) depends on a number of different parameters, including frequency, temperature, thickness L and z and z' separately. In order to gain insight without undue complexity, we have performed illustrative numerical calculations for selected values of the parameters. As before [1], we first specialise to the case $z = z'$, so that we consider

$$\langle |\eta(z)|^2 \rangle = (2\pi S C\omega_{00}\bar{\omega})^{-1}\, k_B T \xi_0\, \mathrm{Im}\, \{\alpha_0^{-1} \times [1 + D_0^{-1} g_1^0 [\exp(2\alpha_0 z_0) + \exp\{-2\alpha_0(z_0 + L_0)\}] + 2D_0^{-1} g_2^0 \exp(-2\alpha_0 L_0)]\}\,, \tag{21}$$

where

$$D_0 = (\lambda_0\alpha_0 - 1)^2 - (\lambda_0\alpha_0 + 1)^2 \exp(-2\alpha_0 L_0)\,, \tag{22}$$

$$g_1^0 = (\lambda_0\alpha_0 + 1)\,(\lambda_0\alpha_0 - 1)\,, \tag{23}$$

$$g_2^0 = (\lambda_0\alpha_0 + 1)^2 \tag{24}$$

and we have introduced scaling factors and dimensionless variables by

$$\xi_0 = \left(\frac{C}{A_0}\right)^{1/2}, \qquad \omega_{00} = \left(\frac{C}{m}\right)^{1/2}, \qquad A = A_0\left(\frac{T}{T_c^0} - 1\right), \tag{25}$$

$$L_0 = \xi_0^{-1} L\,, \quad z_0 = \xi_0^{-1} z\,, \quad \bar{\omega} = \omega \left(\frac{m}{A_0}\right)^{1/2}, \quad \alpha_0 = \xi_0 \alpha\,, \quad \lambda_0 = \xi_0^{-1}\lambda\,. \tag{26}$$

Equation (21) reduces to the semi-infinite fluctuation power spectrum in the limit $L_0 \to \infty$ when $D_0 \to (\lambda_0\alpha_0 - 1)^2$. The first term $k_B T \xi_0 (2\pi S C\omega_{00}\bar{\omega})^{-1}\, \mathrm{Im}\, \alpha_0^{-1}$ represents the bulk power spectrum, while the others describe the effect of the surfaces at $z = 0$ and $z = -L$. It is also helpful to write explicitly the $z = 0$ form of (21)

$$\langle |\eta(0)|^2 \rangle = (2\pi S C\omega_{00}\bar{\omega})^{-1}\, k_B T \xi_0\, \mathrm{Im}\, \{\alpha_0^{-1} [1 + D_0^{-1} \times [g_1^0 (1 + \exp\{-2\alpha_0 L_0\}) + 2 g_2^0 \exp\{-2\alpha_0 L_0\}]]\}\,. \tag{27}$$

Some graphs illustrating (21) and (27) are given in Fig. 1 to 6. In these figures the extrapolation length $\lambda_0 = \xi_0^{-1}\lambda$ is taken as 1, while the damping parameter $\sigma_0 = \gamma/(A_0 m)^{1/2}$ is either 0.5 or 1; these are the same values as used previously [1] for the semi-infinite medium. All the figures are for a film of intermediate thickness $L_0 = 1$; corresponding figures for $L_0 = 10$ and $L_0 = 0.2$ have also been drawn [7]. The value $\lambda_0 = 1$ corresponds to a surface-ordering critical temperature T_c^* for the semi-infinite medium of $T_c^* = 2T_c^0$. Each of Fig. 1 to 6 has four curves, corresponding to temperatures $T = 1.5T_c^*$, $2T_c^*$, $3T_c^*$, and $4T_c^*$, or equivalently $3T_c^0$, $4T_c^0$, $6T_c^0$, and $8T_c^0$.

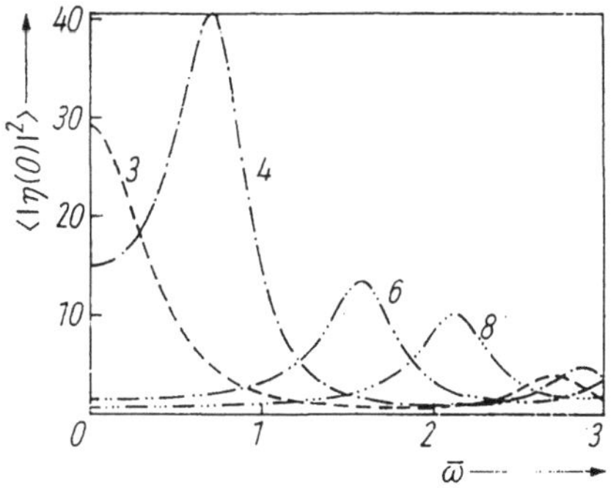

Fig. 1. Power spectrum $\langle|\eta(0)|^2\rangle$ vs. $\bar{\omega}$ for a film of thickness $L_0 = 1$ with $\sigma_0 = 0.5$, $\lambda_0 = 1.0$. The function plotted is $2\pi S\omega_{00}(k_B\xi_0)^{-1}\langle|\eta(0)|^2\rangle$. Values of T/T_c^0 are marked

Fig. 1 shows $\langle|\eta(0)|^2\rangle$ of (27), the boundary value of the fluctuation spectrum, as a function of reduced frequency $\bar{\omega}$ for $\sigma_0 = 0.5$. The corresponding graphs for $L_0 = 10$ [7] are essentially those obtained [1] for a semi-infinite medium; this result is as expected, although it serves as a useful check on the calculations. Those graphs show that the maximum magnitude of the frequency spectrum increases and the frequency at the maximum decreases as T decreases towards T_c^*, as is usual for phase transitions. A similar effect is seen in Fig. 1, except that the curve for the lowest value $T/T_c^0 = 3$ is somewhat suppressed. This reflects the influence of the boundary conditions combined with a longer coherence length. A stronger suppression is seen in the corresponding curves for $L_0 = 0.2$ [7].

Fig. 2 and 3 show the z dependence, (21), for the two frequency values $\bar{\omega} = 0.01$ and 2.10, the other parameters being the same as in Fig. 1. The two frequency values correspond to the maxima in Fig. 1 for $T/T_c^0 = 3$ and 8, respectively. As is easily seen from (21), the curves are all symmetrical about the mid-point $z = -L/2$. Fig. 2 and 3, like subsequent graphs in Fig. 5 and 6 for the z dependence, show a maximum in the fluctuation spectrum at the boundaries, as is to be expected for positive λ. The temperature dependent coherence length $\xi(T) = (C/A)^{1/2}$ varies only from $\xi_0/2^{1/2}$ to $\xi_0/7^{1/2}$ between the curves for $3T_c^0$ and $8T_c^0$. This is why all the curves show a length scale of order ξ_0 for changes in the magnitude of the fluctuation spectrum.

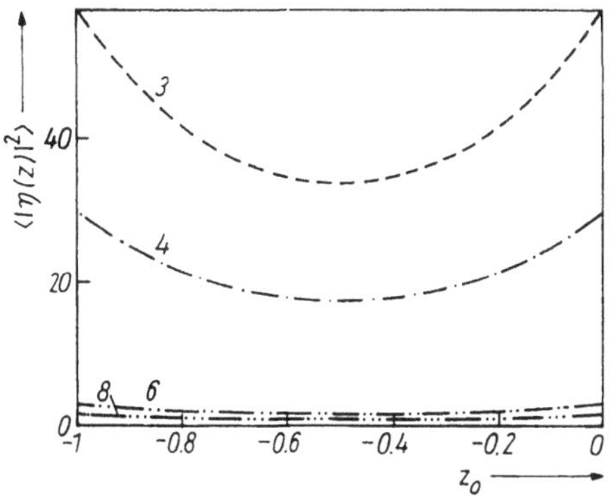

Fig. 2

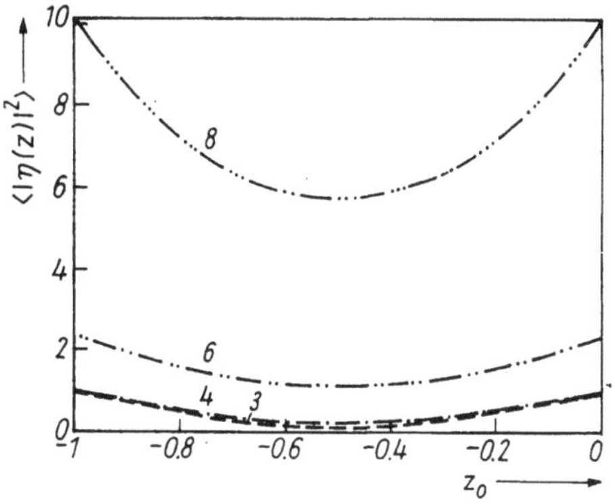

Fig. 3

Fig. 2. Power spectrum $\langle|\eta(z)|^2\rangle$ vs. $z_0 = z/\xi_0$ for a film of thickness $L_0 = 1$ at frequency $\bar{\omega} = 0.001$. Normalising factor and other parameters as in Fig. 1

Fig. 3. As Fig. 2 for $\bar{\omega} = 2.101$

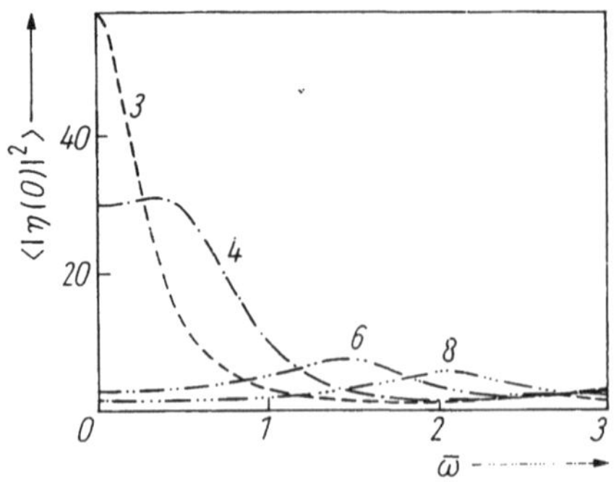

Fig. 4. As Fig. 1 for $\sigma_0 = 1.0$

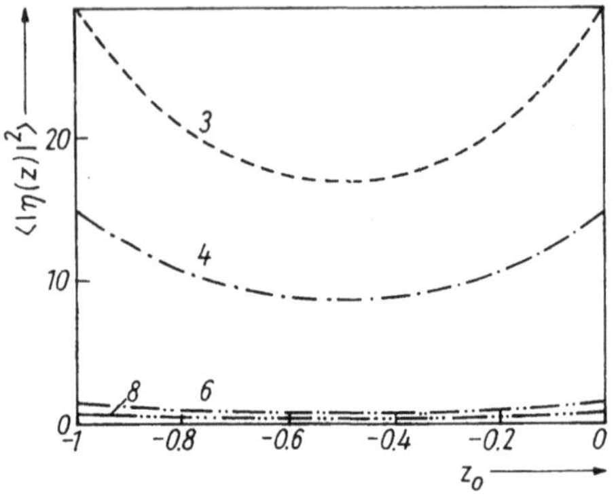

Fig. 5

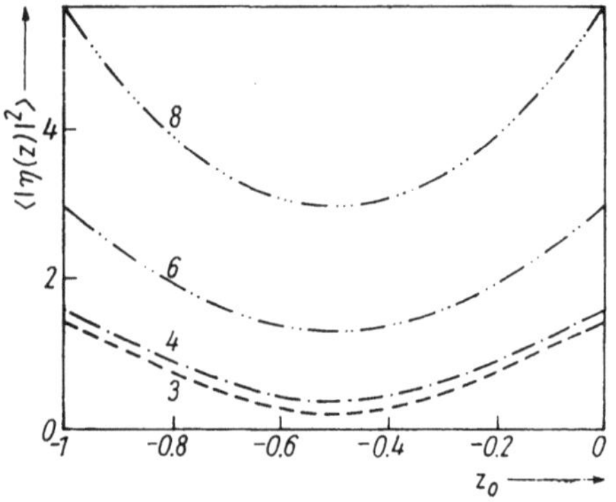

Fig. 6

Fig. 5. As Fig. 2 for $\sigma_0 = 1.0$

Fig. 6. As Fig. 3 for $\sigma_0 = 1.0$

In order to illustrate the effect of increased damping, we show in Fig. 4 to 6 graphs similar to those in Fig. 1 to 3 but for $\sigma_0 = 1$. Comparison of Figs 1 and 4 shows that as in the semi-infinite medium one effect of increased damping is to broaden the frequency dependence of $\langle|\eta(0)|^2\rangle$. Fig. 5 and 6 for z-dependence are similar to Fig. 2 and 3. The corresponding graphs to Fig. 2 and 3 for a thick film, $L_0 = 10$ [7], show some oscillations in the z dependence, and these oscillations are reduced by an increase in damping.

4. Cross-Section for Inelastic Light Scattering

A major application of the Green function of (19) is the evaluation of the cross-section for inelastic light scattering. The details of the calculation are nowadays well known [3, 8, 9], and we therefore only define the notation and state the results; details are given elsewhere [7].

We consider light of frequency ω_I normally incident on the film; we give the cross-sections for back scattering into the medium of incidence and for forward scattering into the substrate supporting the film. Notation for the optical wave vectors is defined in Fig. 7. As before [1], we assume that fluctuations in the dielectric function are linear in order-parameter fluctuations, $\Delta\varepsilon = a'\,\Delta\eta$; the implications of this have already been discussed [1]. As in Sections 2 and 3, we restrict attention to the high-temperature phase, $T > T_c^*$.

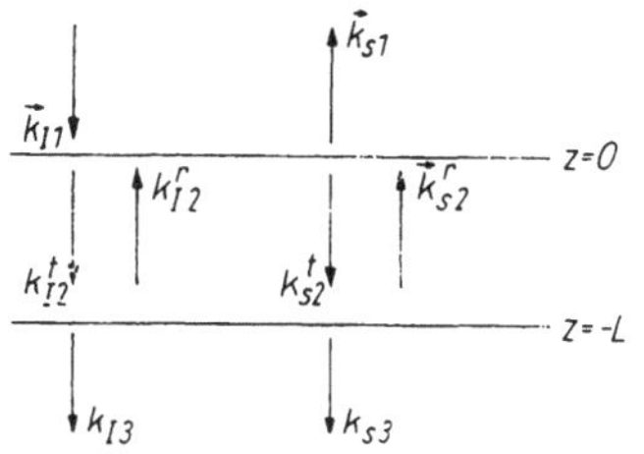

Fig. 7. Optical wave vectors in the light scattering process

The cross-section for backscattering is found to be

$$\begin{aligned}
\frac{d^2\sigma}{d\Omega\, d\omega_s} = {} & \frac{\varepsilon_1 \omega_I \omega_s S k_B T}{4\pi^3 c^2 \omega} \operatorname{Im} \{|\alpha_1|^2 [|k_+|^2 (|f_1|^2 \Lambda^1_2 \Lambda^{1*}_2 \langle\langle \eta(-K^*_+) ; \eta^*(-K_+)\rangle\rangle - \\
& - f_1^* f_2 \Lambda^1_4 \Lambda^{1*}_2 \langle\langle \eta(-K^*_+); \eta^*(K_-)\rangle\rangle + |f_2|^2 \Lambda^1_4 \Lambda^{1*}_4 \langle\langle \eta(K^*_-); \eta^*(K_-)\rangle\rangle - \\
& - f_1 f_2^* \Lambda^1_2 \Lambda^{1*}_4 \langle\langle \eta(K^*_-); \eta^*(-K_+)\rangle\rangle) + |k_-|^2 \times \\
& \times (|f_1|^2 \Lambda^1_1 \Lambda^{1*}_1 \langle\langle \eta(-K^*_-); \eta(-K_-)\rangle\rangle + \\
& + |f_2|^2 \Lambda^1_3 \Lambda^{1*}_3 \langle\langle \eta(K^*_+); \eta^*(K_+)\rangle\rangle - f_1^* f_2 \Lambda^1_3 \Lambda^{1*}_1 \langle\langle \eta(-K^*_-); \eta^*(K_+)\rangle\rangle - \\
& - f_1 f_2^* \Lambda^1_1 \Lambda^{1*}_3 \langle\langle \eta(K^*_+); \eta^*(-K_-)\rangle\rangle) + \\
& + k_+^* k_- (|f_1|^2 \Lambda^1_1 \Lambda^{1*}_2 \langle\langle \eta(-K^*_+); \eta(-K_-)\rangle\rangle + \\
& + |f_2|^2 \Lambda^1_3 \Lambda^{1*}_4 \langle\langle \eta(K^*_-); \eta^*(K_+)\rangle\rangle - f_1^* f_2 \Lambda^1_3 \Lambda^{1*}_2 \langle\langle \eta(-K^*_+); \eta^*(K_+)\rangle\rangle - \\
& - f_1 f_2^* \Lambda^1_1 \Lambda^{1*}_4 \langle\langle \eta(K^*_-); \eta^*(-K_-)\rangle\rangle) + \\
& + k_+ k_-^* (|f_1|^2 \Lambda^1_2 \Lambda^{1*}_1 \langle\langle \eta(-K^*_+); \eta^*(-K_+)\rangle\rangle + \\
& + |f_2|^2 \Lambda^1_4 \Lambda^{1*}_3 \langle\langle \eta(K^*_+); \eta(K_-)\rangle\rangle - f_1^* f_2 \Lambda^1_4 \Lambda^{1*}_1 \langle\langle \eta(-K^*_-); \eta^*(K_-)\rangle\rangle - \\
& - f_1 f_2^* \Lambda^1_2 \Lambda^{1*}_3 \langle\langle \eta(K^*_+); \eta^*(\;\; K_+)\rangle\rangle)]\} , \qquad (28)
\end{aligned}$$

where

$$f_1 = 2k_{I1}(k_{I2} + k_{I3}) \exp(-ik_{I2}L)\, \Delta_{rt}^{-1} , \qquad (29)$$

$$f_2 = 2k_{I1}(k_{I2} - k_{I3}) \exp(ik_{I2}L)\, \Delta_{rt}^{-1} , \qquad (30)$$

$$\begin{aligned}\Delta_{rt} = {} & (k_{I1} + k_{I2})(k_{I2} + k_{I3}) \exp(-ik_{I2}L) + (k_{I1} - k_{I2}) \times \\ & \times (k_{I2} - k_{I3}) \exp(ik_{I2}L) , \qquad (31)\end{aligned}$$

$$\alpha_1 = \omega_s^2 c^{-2} a' \Delta_0^{-1} , \qquad (32)$$

$$\begin{aligned}\Delta_0 = {} & (k_{s1} + k_{s2})(k_{s2} + k_{s3}) \exp(-ik_{s2}L) + (k_{s1} - k_{s2}) \times \\ & \times (k_{s2} - k_{s3}) \exp(ik_{s2}L) , \qquad (33)\end{aligned}$$

$$\Lambda^1_1 = \exp(ik_{s2}L) - \exp(-iK_{0-}L) , \qquad (34)$$

$$\Lambda^1_2 = \exp(-ik_{s2}L) - \exp(-iK_{0-}L) , \qquad (35)$$

$$\Lambda^1_3 = \exp(ik_{s2}L) - \exp(-iK_{0+}L) , \qquad (36)$$

$$\Lambda^1_4 = \exp(-ik_{s2}L) - \exp(-iK_{0+}L) , \qquad (37)$$

$$k_\pm = k_{s2} \pm k_{s3} , \qquad (38)$$

$$K_\pm = k_{I2} \pm k_{I2} . \qquad (39)$$

ε_1 is the dielectric constant in $z > 0$, $K_{0\pm}$ are the wave-numbers of the driving polarizations, ω_s is the frequency of the scattered light, and k_{s2} etc. are the positive scalar magnitudes of the corresponding wave vectors shown in Fig. 7.

The cross-section for forward scattering into the substrate is obtained from (28) with the replacements $\alpha_1 \to \alpha_1 \nu_3$, $k_- \Lambda_1^1 \to k'_+ \Lambda_1^3$, $k_+ \Lambda_2^1 \to -k'_- \Lambda_2^3$, $k_- \Lambda_3^1 \to k'_+ \Lambda_3^3$, $k_+ \Lambda_4^1 \to -k'_- \Lambda_4^3$, and $\varepsilon_1 \to \varepsilon_3$, where

$$\nu_3 = \exp(-ik_{s3}L)\,, \tag{40}$$

$$\Lambda_1^3 = 1 - \exp(-ik_{s2}L)\exp(-iK_{0-}L)\,, \tag{41}$$

$$\Lambda_2^3 = -1 + \exp(ik_{2s}L)\exp(-iK_{0-}L)\,, \tag{42}$$

$$\Lambda_3^3 = 1 - \exp(-ik_{2s}L)\exp(-iK_{0+}L)\,, \tag{43}$$

$$\Lambda_4^3 = -1 + \exp(ik_{s2}L)\exp(-iK_{0+}L)\,, \tag{44}$$

$$k'_\pm = k_{s2} \pm k_{s1}\,. \tag{45}$$

These general expressions are quite complicated, and we therefore specialise to limiting cases of physical interest. We first consider an optically opaque film in which the imaginary parts k''_{I2} and k''_{s2} are large, $k''_{I2}L \gg 1$ and $k''_{s2}L \gg 1$.

In this case it is seen that $f_2 \to 0$ and $f_1 \Lambda_1^1 \to 0$. Another similar simplification occurs for the case of transparent materials of fairly low refractive index, where

$$k_{In} + k_{I2} \gg |k_{In} - k_{I2}| \qquad (n = 1, 3) \tag{46}$$

with corresponding inequalities for the wave numbers of the scattered field. For both situations we find the same expression for the backscattering cross-section:

$$\begin{aligned}
\frac{d^2\sigma}{d\Omega\, d\omega_s} = {} & \frac{\varepsilon_1 \omega_I \omega_s S k_B T \xi_0^3}{8\pi^3 C c^2 \omega_{00} \overline{\omega}} |\alpha_1 k_+ f_1|^2 \,\mathrm{Im}\, \{\exp(-ik_{s20}L_0) - \\
& - \exp[i(k_{I20} - K_0) L_0])\, (\exp[ik_{s20}^* L_0] - \exp[i(K_0^* - k_{I20}^*) L_0]) \times \\
& \times \alpha_0^{-1} [(K_0^* - i\alpha_0)^{-1} (2i\, \mathrm{Im}\, K_0)^{-1} [1 - \exp(-2\mathrm{Im}\, K_0 L_0)] - \\
& - (K_0^* - i\alpha_0)^{-1} (K_0 - i\alpha_0)^{-1} [\exp\{-i(K_0^* - i\alpha_0) L_0\} - \\
& - \exp(-2\mathrm{Im}\, K_0 L_0)] + \\
& + (K_0^* + i\alpha_0)^{-1} (K_0 + i\alpha_0)^{-1} [1 - \exp\{i(K_0 + i\alpha_0) L_0\}] - \\
& - [2i\, \mathrm{Im}\, K_0 (K_0^* + i\alpha_0)]^{-1} [1 - \exp(-2\mathrm{Im}\, K_0 L_0)] + \\
& + (\lambda_0\alpha_0 + 1)(\lambda_0\alpha_0 - 1) [(\lambda_0\alpha_0 - 1)^2 - \\
& - (\lambda_0\alpha_0 + 1)^2 \exp(-2\alpha_0 L_0)]^{-1} \times \\
& \times [(K_0^* - i\alpha_0)^{-1} (K_0 + i\alpha_0)^{-1} (1 - \exp\{-i(K_0^* - i\alpha_0) L_0\}) \times \\
& \times (1 - \exp\{i(K_0 + i\alpha_0) L_0\}) + (K_0^* + i\alpha_0)^{-1} (K_0 - i\alpha_0)^{-1} \times \\
& \times (1 - \exp\{-i(K_0^* + i\alpha_0) L_0\}) \times \\
& \times (1 - \exp\{i(K_0 - i\alpha_0) L_0) \exp(-2\alpha_0 L_0)] + \\
& + (\lambda_0\alpha_0 + 1)^2 [(\lambda_0\alpha_0 - 1)^2 - (\lambda_0\alpha_0 + 1)^2 \exp(-2\alpha_0 L_0)]^{-1} \times \\
& \times [(K_0^* - i\alpha_0)^{-1} (K_0 - i\alpha_0)^{-1} (1 - \exp\{-i(K_0^* - i\alpha_0) L_0) \times \\
& \times (1 - \exp\{i(K_0 - i\alpha_0) L_0\}) \exp(-2\alpha_0 L_0) + (K_0^* + i\alpha_0)^{-1} \times \\
& \times (K_0 + i\alpha_0)^{-1} (1 - \exp\{-i(K_0^* + i\alpha_0) L_0\}) \times \\
& \times (1 - \exp\{i(K_0 + i\alpha_0) L_0\}) \exp(-2\alpha_0 L_0)]]\}\,,
\end{aligned} \tag{47}$$

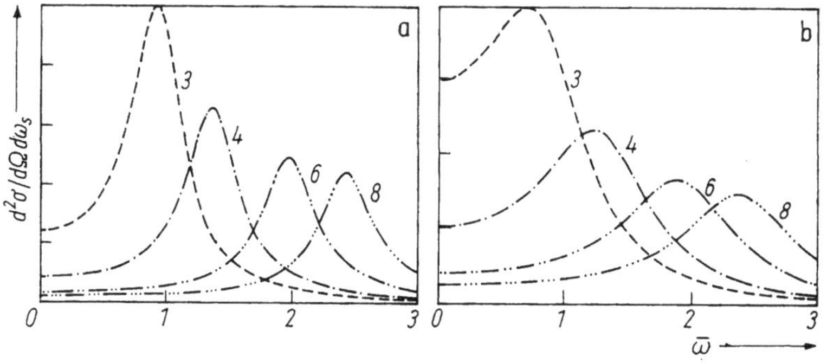

Fig. 8. Differential cross-section for backscattering, (48). Parameters are $L_0 = 1$, $\lambda_0 = 1$, $K_0 = 1 + i$, $k_{I20} = 1.1 + 5i$, $k_{s20} = 1.05 + 5i$ for a) $\sigma_0 = 0.5$ and b) 1.0. We plot $(8\pi^3 Cc^2\omega_{00}\bar{\omega}/\varepsilon_1\omega_1\omega_s S k_B \xi_0^3)\,|\alpha_1 k_+ f_1|^{-2}\,(d^2\sigma/d\Omega\,d\omega_s)$ (in a linear scale) versus frequency shift $\bar{\omega}$. Values of T/T_c^0 are marked

where we have used the reduced variables (25) and (26) and defined new ones by

$$K_0 = \xi_0 K_+\,, \qquad k_{I20} = \xi_0 k_{12}\,, \quad \text{and} \quad k_{s20} = \xi_0 k_{s2}\,. \tag{48}$$

We can also write a similar expression for the forward scattering cross section from a transparent medium satisfying (46). The light scattering cross-section in this case can be written similarly to (47), with $\xi_0 K_+$ being substituted by $\xi_0 K_-$, k_+ by k'_+ and Λ'_2 by $\nu_3 \Lambda_1^3$.

We have carried out numerical calculations to illustrate (47). The frequency dependence of the backscattering cross-section for $L_0 = 10$ [7] agrees with that found for a semi-infinite medium [1]; this fact seves as a useful check on both the algebra leading to (47) and the computer programming. We show here, in Fig. 8a and b, the numerical results for $L_0 = 1.0$ and damping $\sigma_0 = 0.5$ and 1.0, respectively. The curves are very similar to those for the underlying frequency spectra shown in Fig. 1 and 4, and similar observations may be made about them. As the temperature is lowered the cross-section increases and its peak shifts to lower temperature, except that the cross-section for $T/T_c^0 = 3$ in Fig. 8a is somewhat suppressed, presumably because of the boundary conditions. Comparison of Fig. 8a and b shows that an increase in the damping parameter σ_0 gives the expected broadening in the cross-section peak.

5. Conclusions

We have shown how the calculations for a semi-infinite medium [1] may be extended to films. As before [1], the calculations are somewhat schematic, and a number of extensions are possible and perhaps desirable. We wish to emphasize that light-scattering investigations of phase transitions in films are worth consideration, and that at least in mean-field theory the corresponding calculations are well-understood and tractable.

Acknowledgement

A.R.F. thanks CNPq for financial support.

References

[1] A. Ribeiro Filho, D. R. Tilley, E. L. Albuquerque, N. S. Almeida, and M. C. Oliveros, phys. stat. sol. (b) **124**, 457 (1984).
[2] V. L. Ginzburg, A. P. Levanyuk, and A. A. Sobyanin, Physics Rep. **57**, 151 (1980).
[3] R. Loudon, J. Phys. C **11**, 403 (1978).
[4] N. L. Rowell and G. I. Stegeman, Phys. Rev. B **18**, 2598 (1978).
[5] D. R. Tilley, J. Phys. C **13**, 781 (1980).
[6] D. R. Tilley and B. Žekš, Solid State Commun. **49**, 823 (1984).
[7] A. Ribeiro Filho, Ph. D. Thesis, University of Essex, 1984.
[8] E. L. Albuquerque, R. Loudon, and D. R. Tilley, J. Phys. C **13**, 1775 (1980).
[9] M. C. Oliveros and D. R. Tilley, J. Phys. C **15**, 4289 (1982).

(Received March 27, 1984)

phys. stat. sol. (b) **125**, 55 (1984)

Subject classification: 4; 12.1; 13.1; 21.1

Zentralinstitut für Physik der Erde, Institutsteil Jena[1]),
Akademie der Wissenschaften der DDR

Elastic and Bonding Properties of Transition Metals

By

U. WALZER

A theory linking the nearly-free-electron theory to d-band broadening and hybridization effects is further developed in such a way that a model, called MA, for the energy, the pressure p, the bulk modulus and its derivatives with respect to pressure as a function of the relative volume x is created. The theoretical values of MA are compared with all known experimental data for transition metals. For predictions of the relative volume as a function of pressure as well as for the predictions of the first derivative of the bulk modulus with respect to pressure at zero pressure significant improvements are obtained. The approximation of measured values in the p–x plot is very good for most transition metals, in particular for iron, too, and good for the other metals.

Eine Theorie, die die Theorie der fast freien Elektronen mit der der d-Bandspreizung und der Hybridisation verbindet, wird so weiterentwickelt, daß ein Modell, genannt MA, für die Energie, den Druck p, den Kompressionsmodul und seine Druckableitungen als Funktion des relativen Volumens x entsteht. Die theoretischen Werte von MA werden mit allen bekannten Meßdaten von Übergangsmetallen verglichen. Dabei ergeben sich sowohl für die Voraussagen des relativen Volumens als Funktion des Druckes als auch für die der ersten Druckableitung des Kompressionsmoduls bei Nulldruck erhebliche Verbesserungen. Für die meisten Übergangsmetalle werden die Meßwerte im p–x-Diagramm sehr gut approximiert. Das gilt insbesondere für Eisen. Auch für die anderen stellt MA eine gute Näherung dar.

1. Introduction

In principle, it is possible to compute the elastic and bonding properties of transition metals from a relativistic generalization of the Schrödinger equation. Since a solution of the Schrödinger equation is possible only for very simple cases, a number of approximation methods have been developed. However, for the zero-pressure bulk modulus, its first derivative with respect to pressure and for the equation of state, these methods yield theoretical values constituting a rather poor approximation of measured values. Therefore, empirical and semi-empirical equations of state are used to the present day in applied high-pressure research, in geophysics and astrophysics. It is the objective of this paper to modify the afore-mentioned methods such as to make the density of transition metals so well predictable as a function of pressure that the user may renounce empirical methods. Wigner and Seitz [1] studied the bonding in monovalent sodium. Studies on polyvalent metals were published only much later. The reason for this was, first, the necessity to self-consistently compute the potential in the Schrödinger equation not only in dependence on the ion core, but also on the Coulomb field of the valence electrons. The existence of spin follows from the relativistic quantum mechanics. Therefore, secondly, exchange and correlation terms had to be introduced. This was done by the local density functional formalism by Kohn et al. [2, 3]. Moruzzi et al. [4, 5] applied it to 3d and 4d metals. Using self-consistent

[1]) Burgweg 11, DDR-6900 Jena, GDR.

energy band calculations, they were able to compute very well the Wigner-Seitz radius, the cohesive energy, and the zero-pressure bulk modulus of the 4d transition metals.

Andersen [6, 7], basing his studies on the local density functional formalism, found the atomic sphere approximation and the related muffin-tin orbital theory. From the linear superposition of the muffin-tin orbitals, a solution of the Schrödinger equation is obtained. Pettifor [8, 9] used the atomic sphere approximation for calculating the band structure of the transition metals. Many details of this structure proved to be immaterial for the calculation of the bonding forces. The pseudopotential theory [10, 11] shows a possibility for reaching a separation between free-electron-like and d-like states. Harrison and Froyen [12] combined the pseudopotential theory with the muffin-tin orbital theory. As a result, in a similar way as Pettifor [8], they achieved a significant simplification of the theory of d-band structure of transition metals. Harrison and Wills, basing their work on this, developed a theory of the binding energy of simple metals [13] and of transition metals [14]. These theories are very promising. Yet, the agreement between the theoretical and experimental values is not sufficient for further practical application. Therefore, we shall further develop this theory, laying particular emphasis on the comparison with experimental values. For a further familiarization with this field, the following three review articles are recommended: [15, 16, 17].

2. Theory

For the purpose of simplification, we neglect the redistribution of electrons to the two external, not fully filled shells. The number Z of valence electrons per ion results from the sum of free s electrons and of the d electrons,

$$Z = Z_s + Z_d \, . \tag{1}$$

We treat the outer electron shell as a uniform electron gas under the influence of an empty-core pseudopotential of radius r_c (see [16]). The energy per ion for the free-electron-like states is

$$E_{fe} = \frac{3}{10} Z_s \frac{\hbar^2}{m} k_F^2 - \frac{3}{4\pi} Z_s e^2 k_F - \frac{Z_s^2 e^2 \alpha k_F}{(18\pi Z_s)^{1/3}} + \frac{2}{3\pi} Z_s e^2 r_c^2 k_F^3 \, . \tag{2}$$

For the radius r_0 of the atomic volume Ω_0 we have

$$r_0 = k_F^{-1} (9\pi Z_s/4)^{1/3} \quad \text{and} \quad \Omega_0 = 4\pi r_0^3/3 \, . \tag{3}$$

Thus, from (2) we get

$$E_{fe} = k_1 \Omega_0^{-2/3} - k_2 \Omega_0^{-1/3} + k_3 \Omega_0^{-1} \tag{4}$$

with

$$k_1 = 2.210 \frac{\hbar^2}{2m} Z_s^{5/3} \left(\frac{4\pi}{3}\right)^{2/3} , \tag{5}$$

$$k_2 = \frac{e^2}{2} [0.916 \, Z_s^{4/3} + \alpha Z_s^2] \left(\frac{4\pi}{3}\right)^{1/3} , \tag{6}$$

$$k_3 = 2\pi e^2 r_c^2 Z_s^2 \, . \tag{7}$$

The first term of (2) defines the mean kinetic energy of free electrons, the second one the exchange energy, the third term is the Madelung energy, the assumption we are making from now on being $\alpha = 1.8$. The fourth summand is a correction term for the Madelung energy which is introduced on account of the empty-core pseudopotential.

e is the electron charge, m the electron mass, and $\hbar$ Planck's quantum of action divided by 2π.

When dealing with the coupling which broadens the d states into bands, we calculate the energy E_b according to Friedel [18], the bandwidth W_d according to Harrison and coworkers [12, 14, 16], introducing, however, a first (differing) hypothesis.

$$E_b = -\tfrac{1}{2} Z_d(1 - Z_d/10)\, W_d\,, \tag{8}$$

$$W_d = \left[\frac{12}{N_a}\right]^{1/2} \Big(\sum_i \sum_{j \neq i} V_D^2(r_{ij})\Big)^{1/2} \quad \text{and} \quad V_D \equiv \left(\frac{1}{5} \sum_m V_{ddm}^2(r)\right)^{1/2}. \tag{9}$$

$$V_{ddm} = \sum_k \frac{\langle d'|\, \Delta\, |k\rangle \langle k|\, \Delta\, |d\rangle}{E_d - E_k} \tag{10}$$

is an expression for the coupling between the local nonoverlapping d states and the orthogonalized plane waves. Δ is a hybridization term in the Hamiltonian. The plane wave can be expressed as follows (Schiff [19]):

$$|k\rangle = \Omega^{-1/2} \sum_l (2l+1)\, i^l j_l(kr)\, [4\pi/(2l+1)]^{1/2}\, Y_l^0(\theta_2, \Phi_2)\, e^{ik\cdot r_j}\,.$$

r_j is the position vector. The atomic d states are expressed by $|d\rangle = R_2(r)\, Y_2^m(\theta_1, \Phi_1)$, where R_2 is the radial d-state wave function. It is known [12] that

$$V_D(r) = \frac{\hbar^2 r_d^3}{m r^5} [(\eta_{dd0}^2 + 2\eta_{dd1}^2 + 2\eta_{dd2}^2)/5]^{1/2} \tag{11}$$

with

$$\eta_{dd0} = -\frac{45}{\pi}\,, \qquad \eta_{dd1} = \frac{30}{\pi}\,, \qquad \eta_{dd2} = -\frac{15}{2\pi}. \tag{12}$$

r is the interionic distance, N_a the number of atoms, n the coordination number, d the nearest-neighbor separation. We now introduce the hypothesis that the ratio of the nearest-neighbor separation to the atomic radius remains a constant K also in the case of compression of the crystal. Moreover, we use only the interaction exerted by the nearest neighbors, thus obtaining

$$E_b = -Z_d \left(1 - \frac{Z_d}{10}\right) (30.9)\, n^{1/2} \frac{\hbar^2}{2m} \frac{r_d^3}{K^5} \left(\frac{4\pi}{3}\right)^{5/3} \Omega_0^{-5/3} \tag{13}$$

because

$$30.9 = (12)^{1/2} [(\eta_{dd0}^2 + 2\eta_{dd1}^2 + 2\eta_{dd2}^2)/5]^{1/2}\,.$$

We define a constant k_4,

$$E_b \equiv -k_4 \Omega_0^{-5/3}\,, \tag{14}$$

r_d is the d-state radius applicable to the free atom. We take the r_d from [12] which are adapted to the bandwidths given by Andersen and Jepsen [20]. For r_0 and r_c, we take the values from [14], with the core-radius r_c being so adapted that the correct atomic volume at zero pressure is obtained. In addition to the afore-mentioned energy contributions, a small contribution E_c is added which describes the shift in the d-band center. The procedure in which it has been derived is found in the paper by Wills and Harrison [14].

It is a consequence of the nonorthogonality of d-like states. In our hypothesis, it is

$$E_c = Z_d n (11.40) \frac{\hbar^2}{m} \frac{r_d^6}{K^8} \left(\frac{4\pi}{3}\right)^{8/3} \Omega_0^{-8/3} \equiv k_5 \Omega_0^{-8/3}\,. \tag{15}$$

Magnetic effects, making a contribution above all to the energy of the 3d metals [21, 22], as well as spin-orbit coupling are not directly taken into account by us. Instead, we use the following expression for the energy

$$E = k_1\Omega^{-2/3} - k_2\Omega^{-1/3} + k_3\Omega^{-1} - bk_4\Omega^{-5/3} + ck_5\Omega^{-8/3}\,. \tag{16}$$

From this, we obtain the new equation of state

$$p = k_1 \frac{2}{3}\Omega^{-5/3} - k_2\,\Omega^{-4/3} + k_3\,\Omega^{-2} - b\,k_4 \frac{5}{3}\,\Omega^{-8/3} + c\,k_5 \frac{8}{3}\,\Omega^{-11/3}\,. \tag{17}$$

From the pressure p, the bulk modulus B is calculated

$$B = k_1 \frac{10}{9}\,\Omega^{-5/3} - k_2 \frac{4}{9}\,\Omega^{-4/3} + k_3\, 2\,\Omega^{-2} - - b\,k_4 \frac{40}{9}\,\Omega^{-8/3} + c\,k_5 \frac{88}{9}\,\Omega^{-11/3}\,. \tag{18}$$

One point each of the curve is fixed in (17) and (18): We postulate that for the atomic volume $\Omega = \Omega_0$ at vanishing pressure $p = 0$ really follows from (17) and that the

Table 1

List of input parameters of the theoretical models

element	$B_{0,\mathrm{ex}}$ (100 MPa)	r_0 (10^{-10} m)	r_c (10^{-10} m)	r_d (10^{-10} m)	d (10^{-10} m)	n	Z_s	Z_d
Sc	435	1.81	0.95	1.24	3.25	12	1.5	1.5
Ti	1051	1.61	0.91	1.08	2.89	12	1.5	2.5
V	1619	1.49	0.87	0.98	2.62	8	1.5	3.5
Cr	1901	1.42	0.82	0.90	2.50	8	1.5	4.5
Mn	596	1.43	0.78	0.86	2.24	8	1.5	5.5
Fe	1683	1.41	0.71	0.80	2.48	8	1.5	6.5
Co	1914	1.39	0.62	0.76	2.50	12	1.5	7.5
Ni	1860	1.38	0.53	0.71	2.49	12	1.5	8.5
Cu	1370	1.41	0.46	0.67	2.56	12	1.5	9.5
Y	366	1.99	1.09	1.58	3.55	12	1.5	1.5
Zr	833	1.77	1.06	1.41	3.17	12	1.5	2.5
Nb	1702	1.62	1.01	1.28	2.86	8	1.5	3.5
Mo	2725	1.55	0.95	1.20	2.72	8	1.5	4.5
Tc	(2970)	1.50	0.84	1.11	2.71	12	1.5	5.5
Ru	3208	1.48	0.72	1.05	2.65	12	1.5	6.5
Rh	2704	1.49	0.62	0.99	2.69	12	1.5	7.5
Pd	1808	1.52	0.52	0.94	2.75	12	1.5	8.5
Ag	1007	1.59	0.45	0.89	2.89	12	1.5	9.5
Lu	411	1.92	1.06	1.58	3.43	12	1.5	1.5
Hf	1090	1.75	1.05	1.44	3.13	12	1.5	2.5
Ta	2000	1.62	1.02	1.34	2.86	8	1.5	3.5
W	3232	1.56	0.94	1.27	2.74	8	1.5	4.5
Re	3720	1.52	0.79	1.20	2.74	12	1.5	5.5
Os	(4180)	1.49	0.66	1.13	2.68	12	1.5	6.5
Ir	3550	1.50	0.51	1.08	2.71	12	1.5	7.5
Pt	2783	1.53	0.33	1.04	2.77	12	1.5	8.5
Au	1732	1.59	0.40	1.01	2.88	12	2	9

experimental zero-pressure bulk modulus $B_{0,\mathrm{ex}}$ equals the theoretical zero-pressure bulk modulus B_0. From these two assumptions, b and c have been calculated.

From

$$-\Omega \frac{\partial B}{\partial \Omega} = k_1 \frac{50}{27} \Omega^{-5/3} - k_2 \frac{16}{27} \Omega^{-4/3} + k_3 4\Omega^{-2} - - bk_4 \frac{320}{27} \Omega^{-8/3} + ck_5 \frac{968}{27} \Omega^{-11/3} \tag{19}$$

the pressure derivative of the bulk modulus at vanishing pressure is formed,

$$B_1 = \frac{1}{B(\Omega_0)} \left(-\Omega \frac{\partial B}{\partial \Omega}\right)_{\Omega=\Omega_0} . \tag{20}$$

The model proposed here for the behavior of transition metals under high pressure, i.e., (16) to (20) will be hereafter called model A (abbreviated MA). If we assume that b and c equal unity, we obtain a modified Wills-Harrison model (abbreviated WH).

3. Comparison of the Theory with Experimental Results

By comparing the curves $p(\Omega)$ and $B(\Omega)$ with measured values, the theory can be tested for its applicability to transition metals at $\Omega \neq \Omega_0$. Another test can be performed by comparing B_1 with observed data.

Table 1 shows the well known input parameters. There are eight for each of the metals, seven of which having a microphysical nature. The only macroscopic input parameter is $B_{0,\mathrm{ex}}$, the experimental isothermal bulk modulus at vanishing pressure according to [23], which parameter is given in the second column. The following columns contain the radius of the atomic volume r_0 and the pseudopotential core radius r_c, both taken from [14], the d-state radius r_d according to [12], the nearest neighbor separation d according to [23], n the number of nearest neighbors in equilibrium according to [24, 25], Z_s the number of electrons per ion being in the free-electron-like states, and Z_d which is the number of electrons in the d-band according to [14].

In Fig. 1 to 9, the theoretical curves obtained with the models MA (full curve) and WH (dashed curve) are compared with the measured points for all transition

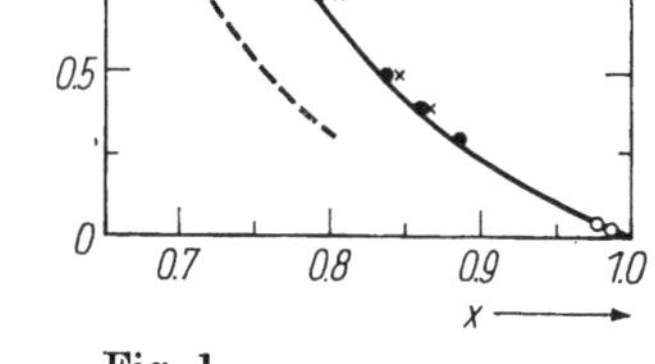

Fig. 1

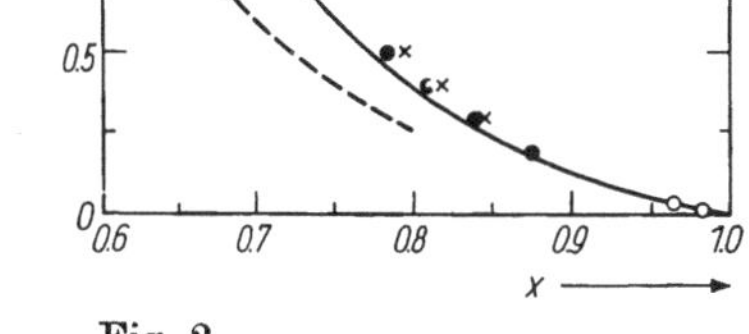

Fig. 2

Fig. 1. Gold. Comparison of equations of the state MA (full curve) and WH (dashed) with experimental data: ○ static data [27], ● room isotherm [28] calculated from Hugoniot data (crosses)
Fig. 2. Silver. For text, see Fig. 1

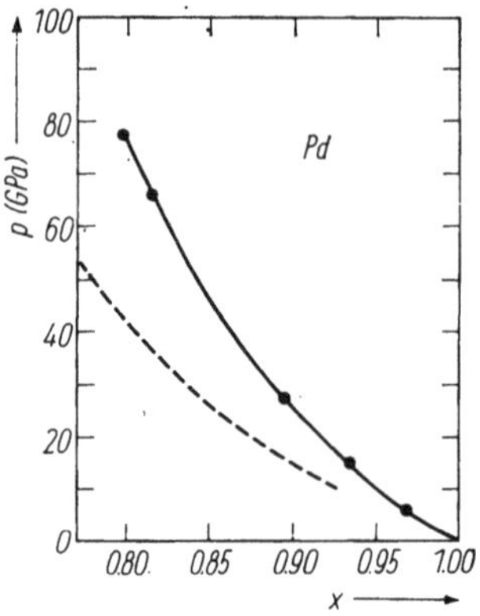

Fig. 3

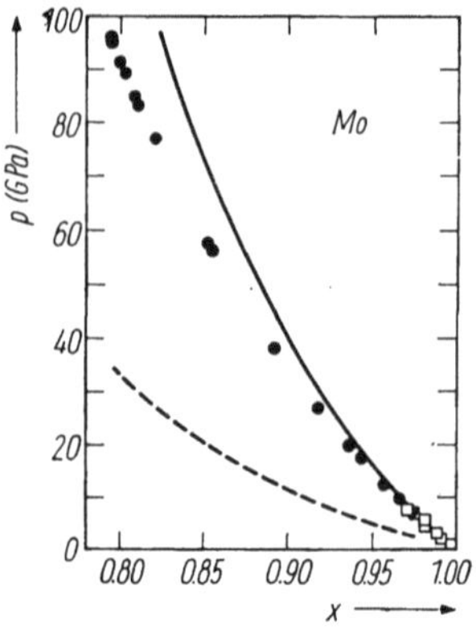

Fig. 4

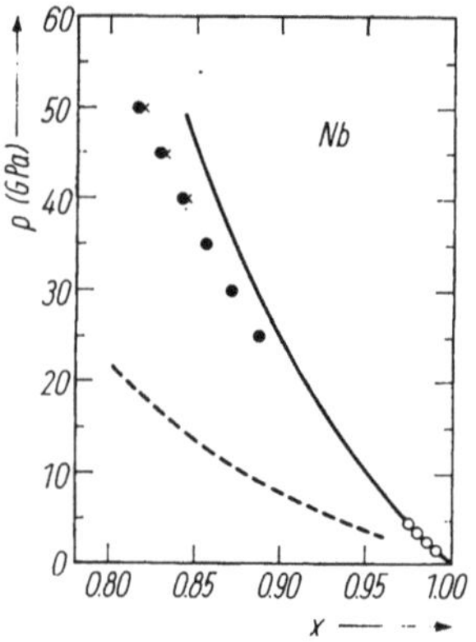

Fig. 5

Fig. 3. Palladium. For text, see Fig. 4

Fig. 4. Molybdenum. Comparison of the equations of state MA and WH with measured data. The experimental values are taken from Mao et al. [29] (293 K isothermal data, ●) and Li Chung Ming and Manghnani [30] (□)

Fig. 5. Niobium. Comparison of the equations of state MA and WH with measured data: ○ static data [31]; Hugoniot data and room isotherms calculated from them have been taken from [32]

metals. For this purpose, the pressure p is plotted versus the relative volume $x = \Omega/\Omega_0$. Crosses denote Hugoniot data, insignificant for the comparison, from which room isotherms, in some cases also zero isotherms, have been calculated in literature. The knowledge we have of the compression comes from shock-wave data, X-ray investigations, and volumetric measurements. The graphic representation of the compared experimental results in Fig. 1, 2, 5, 6, 7, 8a, and 9 is based on that used by

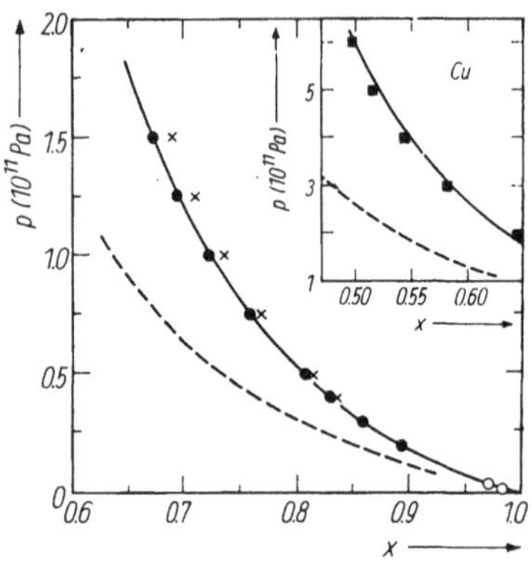

Fig. 6

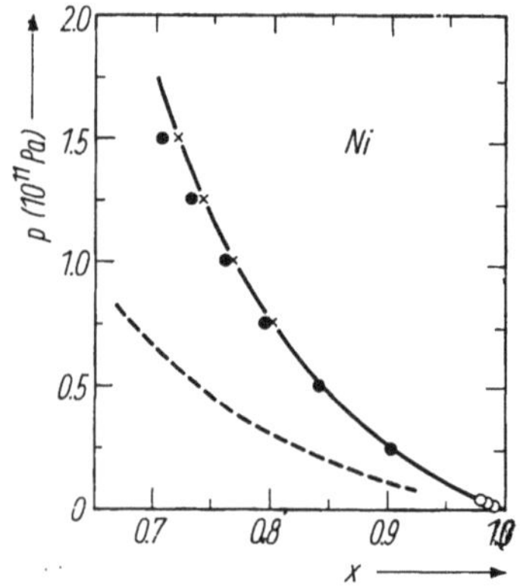

Fig. 7

Fig. 6. Copper. Comparison of the equations of state MA and WH with experimental data: ■ zero isotherm [33]. For further text, see Fig. 1

Fig. 7. Nickel. Comparison of the equations of state MA and WH with experimental data: × Hugoniot values, ● isotherms calculated from them according to [34], ○ static data according to [27]

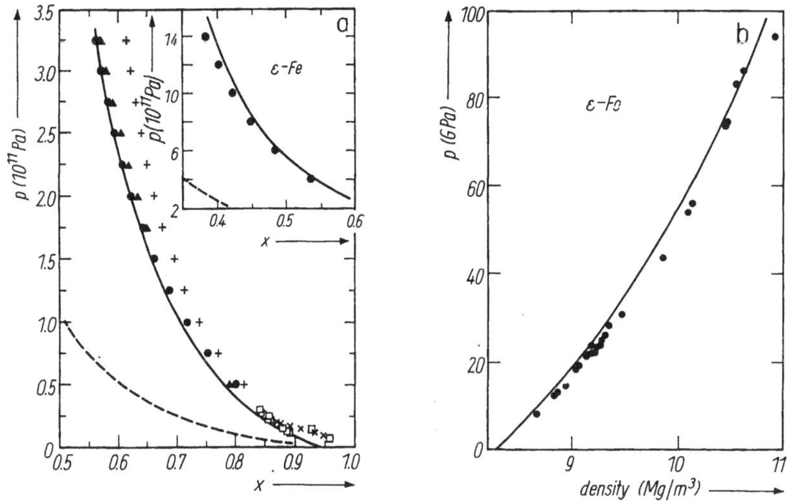

Fig. 8. Iron. Comparison of the equations of state MA and WH with experimental data. The curves are calculated for h.c.p. iron. The relative volume shown is $\varrho_0^{(\alpha)}/\varrho$ with $\varrho_0^{(\alpha)}$ being the zero-pressure density for b.c.c. iron. The experimental data in a) have been taken from the following references: ● isotherms from Hugoniot data [36], ▲ isotherms from [34, 37], □ X-ray data from [38], × shock-wave data from [39], + smoothed Hugoniot data from [36]. The representation of 300 K isothermal experimental values in b) has been taken from [40], one value has been corrected according to the original work [35]

Ullmann and Pankov [26] who have compiled the numerical values from literature. Since we have $\varrho_0^{(\alpha)}/\varrho_0^{(\varepsilon)} = \frac{7.87}{8.36} = 0.9414$ for the ratio of the zero-pressure densities of b.c.c. iron to h.c.p. iron, the p–x curve for ε-Fe begins only at $x = 0.9414$. When we look at Fig. 1 to 9, we see that MA yields a significant improvement in the approximation of the experimental data over WH. In particular, the excellent approximation obtained for iron shows good promise for an application in the study of the cores of the terrestrial planets. (Since the mantles of these planets are composed in a complex manner of silicates, oxides, and other materials, simpler equations of state will still have to be used for them. It is advisable to employ the model MZ [41], developed from [42] for this purpose.)

Table 2 shows a few further comparisons between the model WH and the new model MA, respectively, and observational results. In the first column, the element is

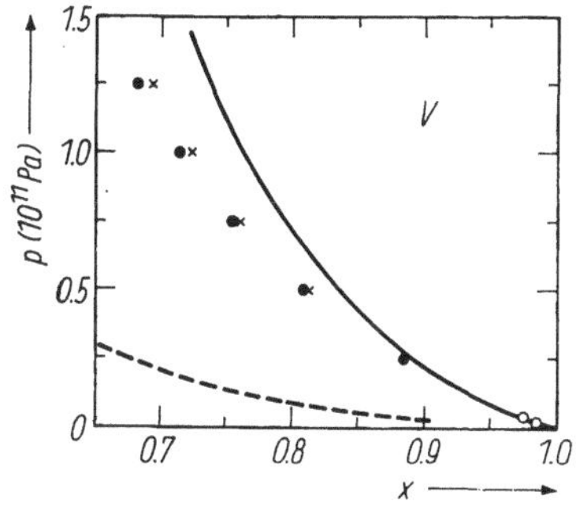

Fig. 9. Vanadium. Comparison of the equations of state MA and WH with experimental data: × and ● Hugoniot and room isotherm values calculated from them according to [34], ○ static data from [31]

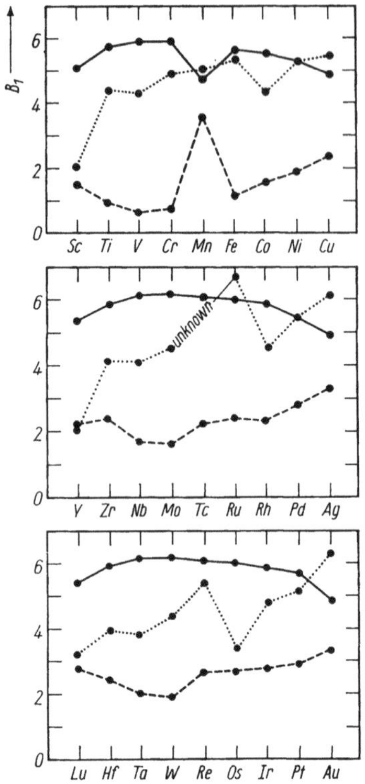

Fig. 10. The full line links the first derivatives of the bulk modulus with respect to pressure at vanishing pressure as predicted by MA, the dashed line the corresponding values as predicted by WH. The dotted line links the experimental values $B_{1,\mathrm{ex}}$. An interesting paper on the systematics shown here has been written by Vohra et al. [46]

given. Unless something else is specified, it is in the α-phase. The second column shows by how many times the experimental zero-pressure bulk modulus is greater than the theoretical one according to WH. For MA, this ratio equals unity for all transition metals according to the construction of the model and, thus, is ideal. The experimentalinitial first derivative of the bulk modulus with respect to pressure, called $B_{1,\mathrm{ex}}$, is given in the third column. The corresponding theoretical values B_1 in the next two columns prove that MA gives a somewhat better approximation of the measured values than WH (see also Fig. 10). It is a large advantage of MA that, in contrast to the usual equations of state, it does not require $B_{1,\mathrm{ex}}$ as an input parameter, since this quantity is only insufficiently known for most materials. Unless specified otherwise, most $B_{1,\mathrm{ex}}$ values have been taken from Steinberg [43]. As Fig. 1 to 9 have shown that MA yields a realistic equation of state, the sixth column of Table 2 gives a view as to how the pressure values in the Periodic Table that are necessary for compressing the substance to $x = 0.75$ change.

4. Conclusions

A combination [12, 14] of the muffin-tin orbital theory and the transition-metal pseudopotential theory was used for formulating the d-band structure of transition metals. This permitted to write the total energy as a sum of the kinetic energy of the free electrons, the exchange energy, a Madelung term, and its pseudopotential correction as well as two terms representing band broadening and hybridization effects. Three assumptions were introduced:

(i) The nearest-neighbor separation to atomic radius ratio is constant.

(ii) At zero-pressure volume, the pressure has to vanish exactly.

(iii) At zero-pressure volume, the experimental zero-pressure bulk modulus must be exactly obtained.

Table 2

Supplementation of the comparison of the models MA and WH through experimental data. For an explanation, see text

element	$B_{0,ex}/B_0$ defined by WH	$B_{1,ex}$	B_1 defined by WH	B_1 defined by MA	p (100 MPa) at $x = 0.75$, defined by MA	b	c
Sc	2.427	2.0*)	1.432	5.111	269.92	1.846	4.190
Ti	5.030	4.37	0.814	5.736	707.07	2.071	4.844
V	6.173	4.26	0.609	5.913	1113.60	2.217	5.530
Cr	5.513	4.89*)	0.709	5.916	1308.37	2.148	5.041
Mn	−5.526	5.0**)	3.502	4.724	355.79	0.657	0.632
Fe	3.348	5.29	1.137	5.606	1112.61	2.274	4.962
Fe(ε)					657.93		
Co	2.764	4.26*)	1.569	5.511	1246.95	2.409	4.118
Ni	2.175	5.26	1.876	5.311	1178.45	3.150	4.393
Cu	1.615	5.48	2.411	4.879	816.82	5.573	4.073
Y	2.143	2.0*)	2.182	5.354	234.33	1.567	2.399
Zr	2.290	4.11	2.340	5.858	568.82	1.590	2.240
Nb	3.326	4.06	1.628	6.093	1194.34	1.975	3.105
Mo	3.507	4.50	1.612	6.152	1922.75	2.197	3.293
Tc	2.480	unknown	2.232	6.070	2073.75	2.030	2.634
Ru	2.312	6.61*)	2.402	5.978	2212.19	2.113	2.536
Rh	2.262	4.50*)	2.314	5.822	1827.48	2.461	2.800
Pd	1.716	5.42	2.826	5.495	1171.02	2.550	2.415
Ag	1.369	6.12	3.313	4.920	602.45	3.178	1.951
Lu	1.784	3.2***)	2.762	5.372	263.95	1.396	1.881
Hf	2.314	3.95	2.399	5.962	752.75	1.670	2.269
Ta	2.742	3.79	1.994	6.140	1410.41	1.954	2.773
W	3.070	4.33	1.901	6.169	2282.98	2.185	2.962
Re	2.149	5.41	2.675	6.085	2598.37	1.986	2.286
Os	2.031	3.4**)	2.727	6.026	2896.00	2.207	2.357
Ir	1.948	4.83*)	2.790	5.886	2415.42	2.402	2.350
Pt	1.768	5.18*)	2.946	5.705	1849.06	2.906	2.318
Au	1.572	6.29	3.331	4.865	1027.17	1.428	1.498

*) According to [44], **) according to [45], ***) estimated.

The resulting model was tested using experimental compression data from nine transition metals, a significant numerical improvement of prediction being obtained.

Acknowledgements

The author wishes to express his sincere gratitude to W. A. Harrison, D. G. Pettifor, and J. M. Wills for communicating their results [14, 17] prior to publication.

References

[1] E. P. Wigner and F. Seitz, Phys. Rev. **43**, 804 (1933).
[2] P. Hohenberg and W. Kohn, Phys. Rev. B **136**, 864 (1964).
[3] W. Kohn and L. J. Sham, Phys. Rev. A **140**, 1133 (1965).
[4] V. L. Moruzzi, A. R. Williams, and J. F. Janak, Phys. Rev. B **15**, 2854 (1977).

[5] V. L. MORUZZI, J. F. JANAK, and A. R. WILLIAMS, Calculated Electronic Properties of Metals, Pergamon Press, New York 1978.
[6] O. K. ANDERSEN, Solid State Commun. **13**, 133 (1973).
[7] O. K. ANDERSEN, Phys. Rev. B **12**, 3060 (1975).
[8] D. G. PETTIFOR, J. Phys. F 7, 613 (1977).
[9] D. G. PETTIFOR, J. Phys. F 8, 219 (1978).
[10] W. A. HARRISON, Phys. Rev. **181**, 1036 (1969).
[11] J. A. MORIARTY, Phys. Rev. B **5**, 2066 (1972).
[12] W. A. HARRISON and S. FROYEN, Phys. Rev. B **21**, 3214 (1980).
[13] W. A. HARRISON and J. M. WILLS, Phys. Rev. B **25**, 5007 (1982).
[14] J. M. WILLS and W. A. HARRISON, Phys. Rev. B **28**, 4363 (1983).
[15] A. R. MACKINTOSH and O. K. ANDERSEN, The Electronic Structure of Transition Metals, in: Electrons at the Fermi Surface, Ed. M. SPRINGFORD, Cambridge University Press, Cambridge 1980.
[16] W. A. HARRISON, Electronic Structure and the Properties of Solids, Freeman, San Francisco 1980.
[17] D. G. PETTIFOR, Electron Theory of Metals, in: Physical Metallurgy, Ed. R. W. CAHN and P. HAASEN, 3rd ed., North-Holland Publ. Co., Amsterdam 1984.
[18] J. FRIEDEL, in: The Physics of Metals, Ed. J. M. ZIMAN, Cambridge University Press, Cambridge 1969.
[19] L. I. SCHIFF, Quantum Mechanics, McGraw-Hill Publ. Co., New York 1968.
[20] O. K. ANDERSEN and O. JEPSEN, Physica (Utrecht) **91** B, 317 (1977).
[21] J. F. JANAK and A. R. WILLIAMS, Phys. Rev. B **14**, 4199 (1976).
[22] A. M. OLEŚ, Phys. Rev. B **28**, 327 (1983).
[23] C. KITTEL, Einführung in die Festkörperphysik, Akademische Verlagsgesellschaft Geest & Portig, Leipzig 1973.
[24] G. LEIBFRIED and N. BREUER, Point Defects in Metals, Vol. 1., Introduction to the Theory, Springer-Verlag, Berlin/Göttingen/Heidelberg/New York 1978.
[25] G. E. R. SCHULZE, Metallphysik, Akademie-Verlag, Berlin 1967.
[26] W. ULLMANN and V. L. PANKOV, Veröffentlichungen des Zentralinstituts für Physik der Erde **41**, 1 (1976).
[27] S. N. VAIDYA and G. C. KENNEDY, J. Phys. Chem. Solids **31**, 2329 (1970).
[28] W. J. CARTER, S. P. MARSH, and J. N. FRITZ, in: Accurate Characterization of the High Pressure Environment, Ed. E. C. LLOYD, Nat. Bur. Standards spec. Publ. **326**, 147 (1971).
[29] H. K. MAO, P. M. BELL, J. W. SHANER, and D. J. STEINBERG, J. appl. Phys. **49**, 3276 (1978).
[30] LI CHUNG MING and M. H. MANGHNANI, J. appl. Phys. **49**, 208 (1978).
[31] S. N. VAIDYA and G. C. KENNEDY, J. Phys. Chem. Solids **33**, 1377 (1972).
[32] M. H. RICE, R. G. MCQUEEN, and J. M. WALSH, Solid State Phys. **6**, 1 (1958).
[33] L. V. ALTSHULER, A. A. BAKANOVA, and R. F. TRUNIN, Zh. eksper. teor. Fiz. **42**, 91 (1962).
[34] R. G. MCQUEEN and S. P. MARSH, J. appl. Phys. **31**, 1253 (1960).
[35] H. K. MAO and P. M. BELL, Carnegie Institution Washington, Yearboock, Washington 1978 (p. 519).
[36] H. TAKEUCHI and H. KANAMORI, J. geophys. Res. **71**, 3985 (1966).
[37] V. L. PANKOV and V. A. KALININ, Fiz. Zemli **3**, 3 (1974).
[38] T. TAKAHASHI, W. A. BASSETT, and H. MAO, J. geophys. Res. **73**, 4717 (1968).
[39] D. BANCROFT, E. L. PETERSON, and S. MINSHALL, J. appl. Phys. **27**, 291 (1956).
[40] J. M. BROWN and R. G. MCQUEEN, Adv. Earth Planet. Sci. **12**, 611 (1982); High-Pressure Research in Geophysics, Ed. S. AKIMOTO and M. H. MANGHNANI, Center Acad. Publ. Tokyo
[41] U. WALZER, Two Equations of State for Metals, Halides, and Oxides, Gerlands Beitr. Geophysik, **93**, 37 (1984).
[42] U. WALZER, Phys. Earth Planet. Interiors **30**, 62 (1982).
[43] D. J. STEINBERG, J. Phys. Chem. Solids **43**, 1173 (1982).
[44] M. W. GUINAN and D. J. STEINBERG, J. Phys. Chem. Solids **35**, 1501 (1974).
[45] K. A. GSCHNEIDER, Jr., Solid State Phys. **16**, 275 (1964).
[46] Y. K. VOHRA, S. K. SIKKA, and W. B. HOLZAPFEL, J. Phys. F **13**, L 107 (1983).

(Received April 26, 1984)

phys. stat. sol. (b) 125, 65 (1984)

Subject classification: 6 and 8; 10.2; 22; 22.1.1

Department of Physics, Garhwal University, Srinagar Garhwal[1])

Low-Temperature Lattice Thermal Conductivity of Nonmetallic Solids with Isotopic Impurities

By

R. P. GAIROLA

Using thermodynamic Green's function technique, it is shown that the inclusion of anharmonic terms along with the mass and force constant change terms arising due to isotopic impurities in the crystal Hamiltonian, leads to a cross term Γ^{AD} of defect with anharmonic parameters in the expression for the phonon width. This interaction term Γ^{AD}, which varies as $D\omega^3T$, contributes significantly in the vicinity of thermal conductivity maximum, thereby giving a good agreement between theory and experiment, as seen in the analysis for germanium and magnesium stannide solids. A plausible estimate is given for the interaction parameter D taking a linear chain model as example. Due to the occurrence of this interaction term Γ^{AD}, Mathiessen's inverse summation rule no longer remains valid. The present analysis differs from those using Callaway's phenomenological model or extensions thereof, in the sense that it does not explicitly assume the validity of the inverse summation rule in the conductivity calculations, as done by earlier workers.

Unter Benutzung der Technik Greenscher Funktionen wird gezeigt, daß die Einbeziehung anharmonischer Terme zusammen mit den Termen der Massen- und Kraftkonstantenänderung, die von Isotopenstörstellen im Kristall-Hamiltonian herrühren, zu einem Cross-Term Γ^{AD} des Defekts mit anharmonischen Parametern im Ausdruck für die Phononenbreite führen. Dieser Wechselwirkungsterm Γ^{AD}, der sich wie $D\omega^3T$ ändert, ergibt in der Nähe des thermischen Leitfähigkeitsmaximums einen wesentlichen Beitrag und führt zu einer guten Übereinstimmung zwischen Theorie und Experiment, wie aus der Analyse von Germanium- und Magnesiumstannid entnommen werden kann. Eine plausible Berechnung für den Wechselwirkungsparameter D wird angegeben, wobei ein lineares Kettenmodell als Beispiel gewählt wird. Infolge des Auftretens dieses Wechselwirkungsterms Γ^{AD} bleibt die inverse Mathiessensche Summenregel nicht länger gültig. Die vorgestellte Analyse unterscheidet sich von dem phänomenologischen Callawayschen Modell oder dessen Erweiterungen, in dem Sinne, daß die Gültigkeit der inversen Summationsregel bei den Leitfähigkeitsberechnungen, die in früheren Arbeiten benutzt wurden, nicht explizit angenommen wird.

1. Introduction

The lattice thermal conductivity of nonmetallic solids, at low temperatures, has been widely studied by many workers [1 to 5] in the past, employing the Callaway model [6] or extensions of it. In all these studies, it has been implicitly assumed that the relaxation rates due to isotopic impurities ($\tau_D^{-1} \sim A\omega^4$) and anharmonic interactions ($\tau_A^{-1} \sim B\omega^2T^3$) are independent of each other which are then added in accordance with Mathiessen's rule [7] and that there is no interference or cross term [8] of defect and anharmonic interactions appearing in the total relaxation time. However, the inclusion of these cross terms (Γ^{AD}) cannot be ignored in the exact calculation of thermal conductivity of doped crystals, particularly near the low-temperature maximum where the conventional domination of τ_D^{-1} ends and that of τ_A^{-1} starts. The present study is aimed at investigating the role of these cross terms in explaining the low-

[1]) Post Box 25, Srinagar Garhwal 246174, India.

temperature lattice thermal conductivity of nonmetallic solids doped with isotopic impurities.

It is now revealed [8 to 10] that the additivity of inverse relaxation time (i.e. Mathiessen's rule) is only approximately valid. Altukhov and Zavt [9] have studied the validity of the inverse relaxation time summation approximation but with the assumption that one should add the matrix elements of the various scattering processes rather than the squares of the respective elements. It is however seen [11] that following Altukhov and Zavt, the analyzed thermal conductivity curves for germanium lie much below the experimental curves. In another attempt to analyze the low-temperature thermal conductivity of germanium in view of the interference terms arising due to the invalidity of the inverse relaxation time summation method, Srivastava and Kumar [11] have taken into account the nonlinearity in the phonon dispersion relation, as also suggested by Walton [12]. Although the agreement between theoretical and experimental conductivity curves is shown to be satisfactory, the results cannot be very faithfully relied upon, since the interference parameter ξ, which they have taken constant in the analysis, depends on temperature considerably and no longer remains constant. This shall be discussed further in the subsequent analysis.

It has been remarked earlier [8, 13] that the cross term (Γ^{AD}) in general is not obtained in thermal conductivity expressions, because whenever isotopic impurities are accounted for in the Hamiltonian, the anharmonicity is not considered [14 to 16]. This leads us to the usual Rayleigh scattering of phonons by the isotopic impurities with a relaxation rate given by $\tau_{\text{D}}^{-1} = A\omega^4$. But if we retain anharmonic terms in the Hamiltonian along with the defect terms, then the cross term τ_{AD}^{-1} appears in the expression for the total relaxation time, indicating that Mathiessen's inverse summation rule no longer remains valid, consequently the anharmonic and defect-induced phonon scattering relaxation rates are not quite independent of each other as considered in thermal conductivity analysis up to now. In the present paper, this cross term has been analyzed in detail, for a particular model crystal and the low-temperature lattice thermal conductivity of germanium and magnesium stannide solids has been investigated in view of the role exhibited by this cross term.

2. Formulation

We shall now formulate the necessary theory to evaluate the interaction (cross) term, and the thermal conductivity of doped nonmetallic solids.

2.1 *Green's functions and the interaction term*

The Hamiltonian for our system, which includes anharmonicity along with the mass and force constant change terms, arising due to the substitution of isotopic impurities, can be conveniently written as [8]

$$H = \sum_k \hbar\omega_k(a_k^\dagger a_k + \tfrac{1}{2}) - \sum_{k_1k_2} \hbar C(k_1, k_2)\, B_{k_1}B_{k_2} + \sum_{k_1k_2} \hbar D(k_1, k_2) \times$$
$$\times A_{k_1}A_{k_2} + \sum_{S>3} \sum_{k_1} \dots \sum_{k_s} \hbar V_S(k_1, k_2, \dots, k_s)\, A_{k_1} \dots A_{k_s}\,, \tag{1}$$

where the symbols have their usual meaning as defined elsewhere [8]. Now considering the evaluation of the following retarded double-time Green's function,

$$G_{kq}(t - t') = \langle\langle A_k(t), A_q^+(t')\rangle\rangle = -i\theta(t - t')\,\langle[A_k(t), A_q^+(t')]\rangle\,, \tag{2}$$

with the aid of Hamiltonian (1), using the equation-of-motion technique [17], one obtains [8]

$$G_{kq}(\omega + i\varepsilon) = \omega_k \xi_{kq} \pi^{-1} [\omega^2 - \varepsilon_k^2 + 2i\omega_k \Gamma_k(\omega)] , \tag{3}$$

where

$$\Gamma_k(\omega) = \Gamma_k^A(\omega) + 8\pi \sum_{k_1} \alpha(-k, k_1)\, \alpha^+(q, k_1) \left(\frac{\omega_k}{\tilde{\omega}_{k_1}}\right) [\delta(\omega - \tilde{\omega}_{k_1}) - - \delta(\omega + \tilde{\omega}_{k_1})] + \sum_{k_1} \left[4 \, \frac{C(-k, k_1)}{\omega_k}\right]^2 \Gamma_{k_1}^A(\omega) , \tag{4}$$

and other symbols have the same meaning as defined earlier [8]. Here $\Gamma_k^A(\omega)$ represents the width in the phonon spectrum arising due to anharmonic interactions only (excepting the renormalization effects). If we retain only the dominant third- and fourth-order anharmonic terms, then $\Gamma_k^A(\omega)$ (i.e. τ_A^{-1}) varies [8] in the classical high-temperature limit as $\sim \omega_k^2 T + \omega_k^2 T^2$, where the $\omega_k^2 T$ dependence is due to three-phonon scattering processes while the $\omega_k^2 T^2$ variation arises due to four-phonon scattering. The second term in (4) gives the scattering rate (τ_D^{-1}) due to the impurity–phonon scattering. If the force constant change is ignored, then τ_D^{-1} can be shown [8] to vary as $A\omega^4$. The last term in (4) is the cross term (Γ^{AD}) of defect and anharmonic parameters.

Let us now evaluate this cross term, for a linear chain model [18, 19], in a closed form, ignoring quartic anharmonicity for instance. It is shown in Appendix A that its value turns out to be

$$\Gamma^{AD} = \frac{\lambda V k_B T}{64\pi^4 \hbar N v^3} \, \frac{\omega_k}{\omega_L^2} \, \frac{M_0^2}{\mu^2} \, f(1 - f)\, \omega_k^4 \theta(\omega_L - \omega_k) , \tag{5}$$

where λ is a dimensionless parameter, v the sound velocity, $\omega_L = (4\gamma/M)^{1/2}$, γ being the harmonic force constant, V is the crystal volume which contains a total number of N atoms whose $(N - n)$ lattice sites are occupied by atoms of mass M and n sites are occupied by randomly distributed substitutional impurities each of mass M'; $\mu = MM'/(M' - M)$ and M_0 is the weighted harmonic mean of masses of all atoms defined by $M_0^{-1} = fM'^{-1} + (1 - f)\, M^{-1}$, where $f = n/N$. Equation (5) shows that $\Gamma^{AD} \sim D\omega_k^3 T$, where D is the interaction parameter which can be explicitly expressed in terms of the defect parameter A and the dimensionless cubic anharmonic parameter λ, if we recall [8], $A = (V/12N\pi v^3)\,(M_0/\mu)^2 f(1 - f)$ and $\theta(0) = \frac{1}{2}$. One then obtains

$$D = \frac{3k_B}{32\pi^3 \hbar} \lambda A = 3.96472 \times 10^8 \lambda A \text{ s}^2 \text{ K}^{-1} . \tag{6}$$

In our opinion, (6) is a very good estimate for the interaction parameter.

2.2 *Thermal conductivity*

Starting with the Kubo formula [20] and evaluating the required correlation functions in the energy flux operator with the aid of Green's function (3), the thermal conductivity expression can be obtained [8] as

$$K(T) = \frac{\hbar^2 k_B \beta^2}{3V} \sum_k v_k^2 \omega_k^2 \, e\beta\hbar\omega_k (e^{\beta k \omega_k} - 1)^{-2}\, \Gamma_k^{-1}(\omega_k) , \tag{7}$$

where the renormalization effect on phonon frequencies has been ignored. Carruthers [21] and Klemens [1] have obtained a similar expression for the thermal conductivity, using Boltzmann's transport equation. In the above expression $\Gamma_k(\omega)$ is given by (4), which we have already discussed in the preceding section. It is, however, difficult to estimate thermal conductivity directly from (7). For making numerical estimates, one has to reduce (7) into a simpler form. Replacing the summation over k by integration as usual and averaging the various terms over all possible angles, (7) in the Debye approximation $\omega_k = vk$ becomes

$$K(T) = \frac{k_B}{2\pi^2 v} \int_0^{\omega_D} \tau(\omega_k) \left(\frac{\hbar\omega_k}{k_B T}\right)^2 (e^{\beta\hbar\omega_k} - 1)^{-2} \omega_k^2 e^{\beta\hbar\omega_k} d\omega_k , \tag{8}$$

where we have written $\Gamma_k^{-1}(\omega) = \tau(\omega)$ and ω_D is the Debye frequency. Equation (8) looks, at the first instance, similar to Callaway's expression [6] for the thermal conductivity (which, however, assumes the additivity of the inverse relaxation times for various phonon scattering processes) but, in fact, it differs from Callaway's expression in the sense that $\tau^{-1}(\omega_k)$, which is now given by (4), no longer follows the inverse relaxation time summation rule. This is explained in Appendix B. We shall use (8) to analyse the phonon conductivity of Ge and Mg_2Sn solids.

3. Calculations and Results

Expression (4) for $\tau^{-1}(\omega_k)$ is still not sufficient to analyze the thermal conductivity at low temperatures because it includes the effects of scattering of phonons from each other (phonon-phonon scattering, given by $\Gamma_A(\omega)$ and scattering of phonons from the defect atoms (phonon–defect scattering given by $\Gamma^D(\omega) + \Gamma^{AD}(\omega)$), but does not include the scattering of phonons from crystal boundaries [22] which becomes important at low temperatures. This is due to the fact that heat transfer across the crystal boundary was not included in the formulation while deriving (7). Therefore, we have to include this boundary scattering term in (4), in order to analyze the lattice thermal conductivity at low temperatures via (8). The expression for the effective relaxation time can, therefore, be written as

$$\tau^{-1}(\omega) = \Gamma_k(\omega) = \frac{v}{FL} + A\omega^4 + (B_1 + B_2)\,\omega^2 T^3 + D\omega^3 T , \tag{9}$$

where L is the Casimir length [22], F is a constant due to acoustic mismatch at the boundary or it may be some internal boundary parameter [23], A is the constant due to isotopic scattering, B_1 and B_2 are the constants for normal and umklapp phonon–phonon scattering, and D is the interaction parameter. In (9), v/FL is the boundary scattering term and we have chosen $\Gamma_k^A(\omega) \sim (B_1 + B_2)\,\omega^2 T^3$ for the normal and umklapp three-phonon scattering processes, in analogy with Callaway's phenomenological model. A few comments about (9) are worth mentioning:

(i) One can directly include the boundary scattering v/FL (say, Γ^B) as an additive term to the total phonon width, because the detailed calculations of Casimir [22] clearly indicate that the phonon boundary scattering is not due to any phonon transitions by phonon boundary scattering and hence its transition matrix elements can be neglected [11]. Stated otherwise, there will be no interference or cross term between boundary and defect scattering rates, and between the boundary and anharmonic (i.e. phonon–phonon) scattering rates, the result had been expressed [11] in the form

$$\tau^{-1} = \tau_B^{-1} + \tau_D^{-1} + \tau_A^{-1} ;$$

(ii) Equation (9) can be expressed in the form $\Gamma = \Gamma^{B} + \Gamma^{D} + \Gamma^{A} + \Gamma^{AD}$. However, we cannot express it as $\tau^{-1} = \tau_B^{-1} + \tau_D^{-1} + \tau_A^{-1}$ because of the presence of the cross- or interaction term Γ^{AD}. So the individuality of relaxation times becomes less meaningful.

(iii) Only, if the term Γ^{AD} can be neglected or when it does not appear for some solids (such as chemically pure solids or solids with defects other than isotopic impurities), the additivity of inverse relaxation time is valid as an approximation and the individuality of relaxation times regains its significance.

Since in the present work we are interested in examining the role of the cross- or interaction term in the temperature dependence of phonon conductivity curves, we shall use (8) (analogous to Callaway's phenomenological model) for the estimation of thermal conductivity rather than other modified models incorporating the longitudinal and transverse phonon modes separately (such as Holland's model [24]).

3.1 Phonon conductivity of germanium

The thermal conductivity of germanium [24] has been analyzed in the range 2 to 50 K by Srivastav and Kumar [11] where it is shown that by using Callaway's model ($\tau^{-1} = v/FL + A\omega^4 + (B_1 + B_2)\,\omega^2T^3$), one often observes a deviation near $K_m(T)$. Further it is also shown there that following Altukhov and Zavt [9], if one adds the matrix elements rather than the squares of the matrix elements, in an attempt to account for the interference term, the theoretically analyzed curves using $\tau^{-1} = v/FL + A\omega^4 + (B_1 + B_2)\,\omega^2T^3 + A''\omega^3T^{3/2}$, fall much below the experimental points near $K_m(T)$. These deviations near $K_m(T)$ have been explained by them and a good theoretical fit was obtained by taking into account the nonlinearity in the phonon dispersion which modifies τ^{-1} as $v/FL + A(1 + \xi)^4\,\omega^4 + (B_1 + B_2)\,\omega^2T^3$, where ξ is referred to as the interference factor and has been taken as a constant ($\xi = 0.15$) in the analysis. As shown in Appendix A, one notes that ξ depends on T considerably and hence cannot be taken as constant throughout the temperature range of analysis. However, even if we ignore the temperature dependence of ξ and assume it constant, one can still analyze the thermal conductivity of germanium using Callaway's model but with a different value of A say $A' = (1 + \xi)^4\,A$, i.e. taking $A' = 4.197615 \times 10^{-44}\ s^3$ instead of $2.40 \times 10^{-44}\ s^3$, other parameters remaining the same.

In the present work we have analyzed the thermal conductivity of germanium as given by (8), using (9) for the effective relaxation time which includes the term $D\omega^3T$ as the interaction term between defect and anharmonic parameters. This term has been obtained in the present work by using a nonperturbative technique (i.e. Green's function technique) and clearly differs from Altukhov and Zavt's term, $A''\omega^3T^{3/2}$ (as regards the temperature dependence), which has been obtained using a perturbation method [11] and enforcing a trivial assumption of adding linear terms of

Table 1

Values of different parameters used in the analysis for germanium

v	3.5×10^5 cm/s
L	0.24 cm
F	0.8
θ_D	376 K
A	$2.4 \times 10^{-44}\ s^3$
$B_1 + B_2$	2.77×10^{-23} s K^{-3}
D	$1.203 \times 10^{-33}\ s^2\ K^{-1}$

matrix elements rather than squares, as mentioned earlier. The different parameters used in the present analysis are listed in Table 1, and Fig. 1 shows the theoretically analyzed curves for germanium along with the experimental points of Holland [24]. If one neglects the interaction term $D\omega^3T$, the thermal conductivity curve reaches a higher value in the vicinity of the low-temperature maximum $K_m(T)$ and the deviation between experimental and theoretical points is maximum near $K_m(T)$. This is shown by the dashed curve in Fig. 1. However, if the interaction term $D\omega^3T$ is included in the analysis, a good agreement between theory and experiment is found, as shown by the continuous line in Fig. 1. We have used $D = 1.203 \times 10^{-33}\ s^2\ K^{-1}$, or $\delta = (k_B^3/\hbar^3)\ D = 2.7$, to get a best fit.

3.2 Phonon conductivity of magnesium stannide

As another example to investigate the importance of the interaction term $D\omega^3T$, we analyze the thermal conductivity of Mg_2Sn, as measured by Martin and Danielson [25]. It is evident from the thermal conductivity curves for Ge as shown in Fig. 1 that the interaction term contributes dominantly near the conductivity maximum. Therefore, we analyze the experimental conductivity data of Mg_2Sn in the range 5 to 40 K and set the different parameters in the analysis so as to get a best fit near the conductivity maximum. It has been a usual practice to generally neglect the exponential temperature dependence of the parameter B_2 in (9), representing umklapp phonon scattering, for simplicity of calculations and B_2 is taken as a parameter independent of temperature as we have also considered in the preceding section. This is because at low temperatures B_2 does not change appreciably with temperature but at higher temperatures this change is significant. However, as we are interested in examining the role of the interaction term, in the temperature dependence of thermal conductivity let us also consider in this section the explicit temperatu redependence of the parameter B_2, as considered elsewhere [25], to take a better account of the thermal conduc-

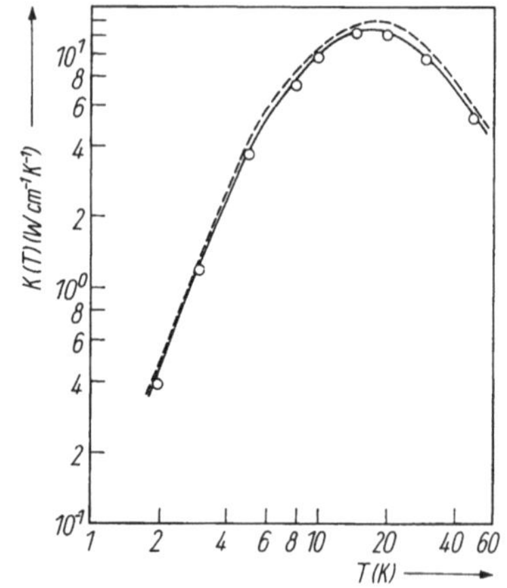

Fig. 1

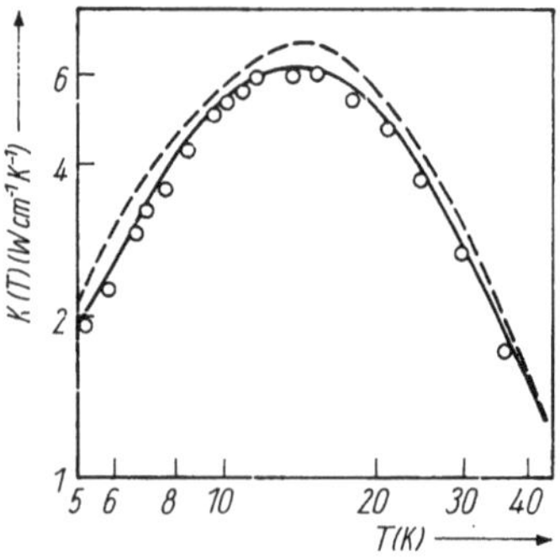

Fig. 2

Fig. 1. Thermal conductivity curves for germanium. Open circles are the experimental points from [24], the dashed curve shows the conductivity without $D\omega^3T$ term and the continuous curve is the present analysis

Fig. 2. Thermal conductivity curves for magnesium stannide. Experimental points are taken from [25], the continuous curve is the present analysis while the dashed curve represents the conductivity calculated without the interaction term $D\omega^3T$

Table 2

Values of parameters used in the calculation of $K(T)$ for Mg_2Sn

v	3.59×10^5 cm/s
L	0.22 cm
F	0.54
θ_D	154 K
A	5.6×10^{-44} s^3
B_1	9×10^{-23} K^{-3}
B_2	5×10^{-22} K^{-3}
a	3
D	1.203×10^{-33} s^2 K^{-1}

tivity of Mg_2Sn. We therefore consider the following expression for the relaxation time:

$$\tau^{-1}(\omega) = \frac{v}{FL} + A\omega^4 + \left[B_1 + B_2 \exp\left(\frac{-\theta}{aT}\right)\right]\omega^2 T^3 + D\omega^3 T . \tag{10}$$

The different parameters used in the analysis are listed in Table 2. Martin and Danielson [25] arbitrarily increased the value$A$ of to 6.3×10^{-44} s^3 instead of 5.6×10^{-44} s^3, a value estimated from Klemen's expression, (cf.) [26] and decreased the value of L from 2.2 to 1 mm, in order to improve the fit. This decrease in L, as computed from Casimir's [22] formula, has been explained by us, due to the creation of internal boundaries, owing to the microscale fluctuations in the crystal composition [23] or it may decrease due to a factor F [27], that is responsible for the acoustic mismatch at the boundary. We have, however, neglected the electron–phonon [7] and four-phonon scattering [28] that may be important at high temperatures. Since our goal is to investigate the significance of the interaction term $D\omega^3T$, in determining thermal conductivity as a function of temperature and also, since the contribution of this term is seen to be most significant near the thermal conductivity maximum $K_m(T)$, as evident from the preceding section, we have chosen the values of different parameters as listed in Table 2, so as to give a best fit near $K_m(T)$. The continuous curve in Fig. 2 shows such a fit, while the dashed curve is obtained when the term $D\omega^3T$ is ignored. We can conclude from Fig. 2 that the interaction term contributes significantly near the thermal conductivity maximum, so as to render a satisfactory fit which otherwise is not possible when this term is ignored. We have, in the present analysis, chosen the same value of the interaction parameter D as considered in the analysis of Ge, i.e. 1.203×10^{-33} s^2 K^{-1}. However, to obtain a good fit in the entire temperature region, a computer programme, with least-squares fitting method, would be useful.

4. Discussion and Conclusions

In the present work we have analyzed thermal conductivity of nonmetallic solids doped with isotopic impurities in the low-temperature region. It is seen that the inclusion of anharmonic terms, along with the mass and force constant change terms arising due to isotopic impurities, in the crystal Hamiltonian leads us to a cross term of defect with anharmonic parameters in the expression for the phonon width. This interaction term Γ^{AD} varies as $D\omega^3T$ and is much responsible for the thermal conductivity of isotopically disordered crystals near the low-temperature maximum.

If the anharmonic terms are not retained in the Hamiltonian, as done by a series of authors [14 to 16], then this cross term is not obtained and so using Callaway's phenomenological model, the thermal conductivity curve lies much above the experimental points, particularly in the vicinity of the thermal conductivity maximum. Although the agreement is then usually brought about by adjusting the various parameters but by doing so we rely on the additivity of relaxation rates (so the presence of Γ^{AD} is ignored) and hence increase the actual phonon width by some factor of four or so (as shown in Appendix B). In actual situations, we cannot ignore this interaction term, when dealing with thermal conductivity of doped crystals and hence Mathiessen's inverse summation rule no longer remains valid. However, this term shall not arise when dealing with a chemically pure crystal or with crystals having defects other than isotopic impurities, then Mathiessen's rule will be approximately valid (still within a factor stated in (B3), in the latter case). Our interaction term $D\omega^3T$ can be compared with Altukhov and Zavt's term [9, 11] $A''\omega^3T^{3/2}$, which has been obtained by using a perturbation method and assuming a trivial assumption of adding single powers of scattering matrix elements rather than squares. This is why the theoretically analyzed conductivity curve with $A''\omega^3T^{3/2}$ term lies [11] much below the experimental points for Ge. The inclusion of the $D\omega^3T$ term which we have presently obtained by using a nonperturbative technique, removes the discrepancy between theory and experiment as we have seen for germanium and magnesium stannide solids. Our analysis is also valid to other nonmetallic solids doped with isotopic impurities. However, it still remains to see whether the interaction term Γ^{AD} arises for the crystals having other types of defects such as dislocations, plastic deformations, etc. This is left for future communications.

Appendix A

Here we wish to obtain an expression for the interaction term Γ^{AD}, which is given by the last term of (4), i.e.

$$\Gamma^{AD} = \sum_{k_1} \left[\frac{4C(-k, k_1)}{\omega_k}\right]^2 \Gamma^{A}_{k_1}(\omega)\,, \tag{A1}$$

where $\Gamma^{A}_{k_1}(\omega)$ is given by (14b) of our earlier study [8], except for the replacement of k by k_1. If we ignore the four-phonon interactions for instance, which may be important at high temperatures, and choose the third-order anharmonic coefficient [18], $|V_3(-k_1, k_2, k_3)|^2 = (\lambda/288\omega N_L)\,\omega_1\omega_2\omega_3\,\Delta(-k_1 + k_2 + k_3)$, where λ is a dimensionless parameter, then following Pathak [18] one obtains

$$\Gamma^{A}_{k_1} = \frac{3\lambda}{8\pi^2\,\hbar\beta}\left(\frac{\omega_k}{\omega_L}\right)^2 \theta(\omega_L - \omega_k)\,, \tag{A2}$$

where $\beta = (k_BT)^{-1}$, $\theta(\omega)$ is the Heaviside step function and the dispersion relation $\omega_k = \omega_L \sin(\pi k a_0/2)$ has been used, a_0 being the lattice constant. In Pathak's result, the square in ω_L is missing, which is dimensionally incorrect and its substitution has led us to an erroneous result (24) of [8] giving $\Gamma^{AD} \sim \omega^4T$. Substituting (A2) into (A1), replacing summation over k_1 by integration as usual, and taking the directional average over all possible atomic configurations, one gets in the Debye approximation

$$\Gamma^{AD} = \frac{\lambda V}{64\pi^4\hbar v^3N}\,\frac{\omega_k}{\omega_L^2}\left(\frac{M_0}{\mu}\right)^2 f(1-f)\,\omega_{k_1}^4\theta(\omega_L - \omega_k)\,, \tag{A3}$$

which shows that $\Gamma^{AD} \sim D\omega_k^3T$, where $D = 3.96472A\,\lambda \times 10^8\ \mathrm{s^2\,K^{-1}}$, as in (6). A rough estimate for D is possible for a linear chain model of lead [18] if we arbitrarily assume $A = 8.0 \times 10^{-44}\ \mathrm{s^3}$ and $\lambda = 5$, then $D = 0.1586 \times 10^{-33}\ \mathrm{s^{20}\,K^{-1}}$ for the one-dimensional linear-chain model. In the present analysis we have taken $D = 1.203\ \times$

$\times\ 10^{-33}\ s^2\ K^{-1}$ (for three-dimensional solids) which can be compared with Altukhov and Zavt's term [2, 11], $\xi_{AZ}\omega^3T^{3/2}$ with $\xi_{AZ} = 2\,[A(B_1 + B_2)]^{1/2}$, which comes out to be 1.63×10^{-33} units for Ge.

The interference parameter ξ used in a previous analysis [11], should be given, when the ψ term is ignored, by

$$\xi = \frac{\hbar\omega_L}{4}\left(\frac{\varepsilon}{\gamma^2} - \frac{\delta^2}{\gamma^3}\right)\left(1 + \frac{\pi^2 k_B^2 T^2}{3\hbar^2\omega_L^2}\right), \tag{A4}$$

which is a dimensionally correct form, instead of that given by (20) (in units of $\hbar = 1$) of [11], since ξ must be dimensionless. Again if we consider a linear chain model of Pb as an example, then substituting the numerical values of ε, δ, γ, and ω_L, as suggested elsewhere [18, 29], it is easy to see from (A4) that ξ varies from -0.006 at 4 K to a value -0.0067 at 30 K for lead. So ξ, as given by (A4), depends significantly on T and hence cannot be taken constant as assumed earlier [11].

Appendix B

A few comments about the validity of Mathiessen's inverse summation rule will now be mentioned. According to this rule, when two (or more) distinct scattering processes, say, isotopic and phonon–phonon scattering, for example, operate simultaneously, having relaxation times τ_D and τ_A, respectively, for a particular mode, it is assumed that the scattering rates are additive, so that

$$\tau^{-1} = \tau_D^{-1} + \tau_A^{-1}\,, \tag{B1}$$

Expressing this result in terms of phonon widths Γ^D and Γ^A, we have

$$\frac{1}{\Gamma} = \frac{1}{\Gamma^D} + \frac{1}{\Gamma^A}\,. \tag{B2}$$

which can be maipulated as

$$\frac{1}{\Gamma} = \frac{1}{\Gamma^D + \Gamma^A} + \left\{\frac{(\Gamma^A - \Gamma^D)^2}{\Gamma^A\Gamma^D(\Gamma^A + \Gamma^D)} + \frac{3}{\Gamma^A + \Gamma^D}\right\}. \tag{B3}$$

Now the actual phonon width Γ will be the sum of phonon widths Γ^D and Γ^A, so that we can write

$$\frac{1}{\Gamma} = \frac{1}{\Gamma^A + \Gamma^D}\,. \tag{B4}$$

Comparing (B3) and (B4), it is obvious that resorting ourselves to the inverse summation rule, one does so approximately within a factor given by curly brackets in (B3). In the vicinity of the low-temperature maximum, if we assume $\Gamma^A \approx \Gamma^D = \gamma$, this factor is $3/(2\gamma)$ and so we increase $1/\Gamma$ by a factor of four from its actual value. In actual calculations this factor may be balanced by adjusting different parameters to get a good agreement between theory and experiment, using Callaway's phenomenological model or its modifications. However, all these models assume that Γ_D^{-1} and Γ_A^{-1} are independent of each other, but as evident from (4) we have $\Gamma = \Gamma^D + \Gamma^A + \Gamma^{AD}$, indicating that the processes represented by Γ^A and Γ^D are not independent but there is an interference term given by Γ^{AD}, which in the present calculations has been shown to vary as $D\omega^3T$.

Acknowledgement

The author is thankful to Prof. S. K. Joshi for bringing to his notice the importance of the validity of the "individuality of relaxation times".

References

[1] P. G. KLEMENS, Solid State Phys. 7, 1 (1958).
[2] P. CARRUTHERS, Rev. mod. Phys. **33**, 92 (1961).
[3] J. J. MARTIN, J. Phys. Chem. Solids **33**, 1139 (1972).
[4] J. J. MARTIN and H. R. SHANKS, J. appl. Phys. **45**, 2428 (1974).
[5] P. ERDOS and S. B. HALEY, Phys. Rev. **184**, 951 (1969).
[6] J. CALLAWAY, Phys. Rev. **113**, 1046 (1959).
[7] J. M. ZIMAN, Electrons and Phonons, Oxford University Press, 1960.
[8] R. P. GAIROLA, Acta phys. Polon. **A63**, 121 (1983).
[9] V. I. ALTUKHOV and G. S. ZAVT, phys. stat. sol. (b) **64**, 403 (1974); (b) **65**, 83 (1974).
[10] A. KUMAR, Indian J. pure appl. Phys. **15**, 304 (1977); Phys. Rev. B **24**, 4493 (1981).
[11] G. C. SRIVASTAV and A. KUMAR, Phys. Rev. B **25**, 2560 (1982).
[12] D. WALTON, Phys. Rev. B **7**, 3925 (1973).
[13] R. P. GAIROLA and M. K. GUPTA, Phys. Rev. B **28**, 2775 (1983).
[14] A. A. MARADUDIN, J. Amer. Chem. Soc. **B6**, 3405 (1964).
[15] B. S. SEMWAL and P. K. SHARMA, J. math. Phys. (New York) **15**, 648 (1974).
[16] P. K. SHARMA and R. BAHADUR, Phys. Rev. B **12**, 1522 (1975).
[17] D. N. ZUBAREV, Soviet Phys. – Uspekhi **3**, 320 (1960).
[18] K. N. PATHAK, Phys. Rev. **139**, A 1569 (1965).
[19] A. A. MARADUDIN, P. A. FLINN, and R. A. COLDWELL-HORSFALL, Ann. Phys. (U.S.A.) **15**, 337, 360 (1961).
[20] R. KUBO, M. YOKOTA, and S. NAKAJIMA, J. Phys. Soc. Japan **12**, 1203 (1957).
[21] P. CARRUTHERS, Rev. mod. Phys. **33**, 92 (1961).
[22] H. B. G. CASIMIR, Physica (Utrecht) **5**, 495 (1938).
[23] R. P. GAIROLA and B. D. INDU, Nuovo Cimento **58B**, 338 (1980).
[24] M. G. HOLLAND, Phys. Rev. **132**, 2461 (1963).
[25] J. J. MARTIN and G. C. DANIELSON, Phys. Rev. **166**, 879 (1968).
[26] W. D. LAWSON, S. NIELSON, E. H. PUTLEY, and V. ROBERTS, J. Electronics **1**, 203 (1955).
[27] A. E. KAZAKOV and E. L. NAGEOV, Fiz. tverd. Tela **9**, 1255 (1967).
[28] I. POMERANCHUK, J. Phys. (USSR) **4**, 259 (1941); **6**, 237 (1942).
[29] A. A. MARADUDIN and A. E. FEIN, Phys. Rev. **128**, 2589 (1962).

(Received March 13, 1984)

phys. stat. sol. (b) **125**, 75 (1984)

Subject classification: 6 and 20.1; 22.8.1

Azerbaidzhan State University, Baku[1]) *(a) and Institute of Spectroscopy, Troitsk (b)*

Anharmonicity Effects in IR-Reflection Spectra of Sheelites

By

SH. M. EFENDIEV (a), V. M. NAGIEV (a), and V. M. BURLAKOV (b)

Reflection spectra of $PbWO_4$ crystals vs. the temperature are studied over the 700 to 1000 cm^{-1} frequency region. On the basis of an invertibility criterion of the oscillator obtained in the present paper, the anomaly in the reststrahlen band in the reflection spectrum is found to be connected with a strong frequency dependence of the damping $\Gamma(\omega)$ of the transverse optical vibration. A possible decay channel for an intramolecular excitation corresponding to the peak of the function $\Gamma(\omega)$ is proposed.

Проведены температурные исследования спектра отражения кристалла $PbWO_4$ в области частот 700 до 1000 cm^{-1}. На основании критерия инвертированности осциллятора, полученного в работе, установлено, что аномалия в полосе остаточных лучей спектра отражения связана с сильной частотной зависимостью затухания $\Gamma(\omega)$ поперечного оптического колебания. Предложен возможный канал распада внутримолекулярного возбуждения, соответствующий пику функции $\Gamma(\omega)$.

1. Introduction and Experimental Results

In the vibrational spectrum of some crystals with the sheelite structure ($PbWO_4$, $PbMoO_4$, $CaWO_4$, $CaMoO_4$) five dipole-active vibrations are observed instead of four predicted by the group-theoretical analysis. One of the vibrations in the 800 to 900 cm^{-1} frequency region (Fig. 1) is supposed to be "spare" and its occurrence in the IR spectrum is caused by the selection rule due to internal stress in the crystal [1, 2] or to multiphonon processes [3]. The investigations carried out up to now do not allow an unambiguous determination of the mechanism of this vibration.

The reflection spectrum region of $PbWO_4$ shown in Fig. 1 is satisfactorily described within the framework of the dispersion oscillator model (Fig. 2). Oscillator parameters are presented in Table 1.

It should be noted that the numeration of TO and LO vibration frequencies is purely conventional in terms of the dielectric function $\varepsilon(\omega)$. However, it proved to be of principal importance in this case, since the present paper is aimed at the elucidation of the nature of the "spare" oscillator.

The above numeration of TO and LO frequencies has been performed on the basis of a comparison of the intensities and the half-widths of the bands in the $\mathrm{Im}(\varepsilon)$ and $\mathrm{Im}(-1/\varepsilon)$ spectra (Fig. 2). The numeration indicates that the spectrum discussed consists of an external oscillator No. 2 with a large longitudinal–transverse splitting and an internal, inverted oscillator with a small difference, $\omega_{LO}^2 - \omega_{TO}^2$, and a large damping of TO and LO vibrations. Before discussing the mechanism of the "spare" oscillator (which apparently appears in case of the oscillator No. 1), the problem of its affiliation to the first-order spectrum should be considered.

[1]) Ul. P. Lumumba 23, Baku 370602, USSR.

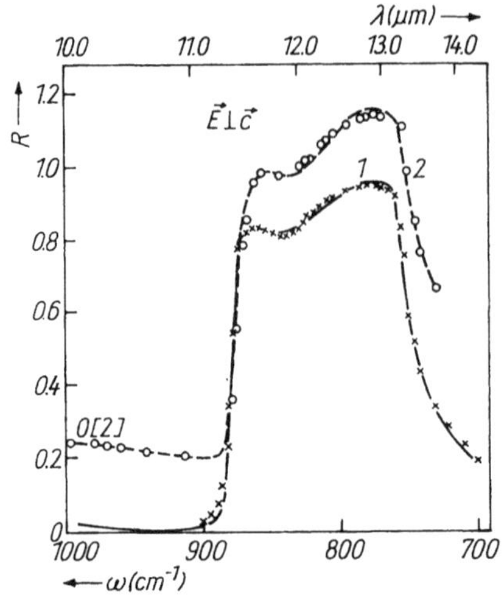

Fig. 1

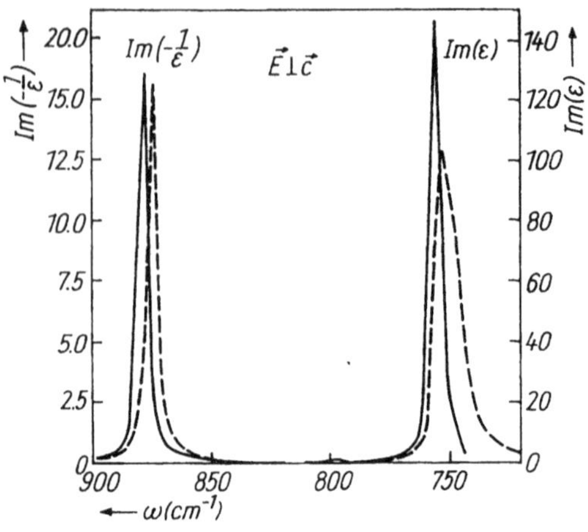

Fig. 2

Fig. 1. Reflection spectra of a $PbWO_4$ single crystal in $\boldsymbol{E} \perp \boldsymbol{c}$ polarization (solid curves are experimental at $T = 77$ K; the dashed lines at $T = 473$ K; ×, ○ are the values calculated by the dispersion oscillator method, 0[2] means the origin for curve (2))

Fig. 2. Imaginary and real parts of $\varepsilon(\omega)$ calculated from the $PbWO_4$ reflection spectrum (Fig. 1). ——— $T = 77$ and – – – 473 K

2. Discussion of the Results

Inversion of any oscillator in the region of the longitudinal–transverse splitting of another oscillator occurs due to interaction of TO and LO vibrations through the polarization field. The equation of motion for the i-th mode is

$$\ddot{Q}_i^x = -\omega_{0i}^2 Q_i^x + A_x e_{\mathrm{S}}^i \sum_{j \neq i} e_x^j Q_j^x \,. \tag{1}$$

Here x denotes TO or LO modes, e_{S}^i is the Szigeti mode charge, e_x^j the transverse (longitudinal) mode charge [4]

$$e_{\mathrm{T}}^j = \varepsilon_\infty e_{\mathrm{L}}^j = e_{\mathrm{S}}^j \left[1 + \frac{(\varepsilon_\infty - 1)\, L}{4\pi}\right], \qquad A_{\mathrm{T}} = \frac{L}{v}\,, \qquad A_{\mathrm{L}} = \frac{4\pi - L}{v}\,,$$

where L is the Lorentz factor, v the unit cell volume, ω_{0i} the mode frequency determined by a short-range force matrix.

In the case of two oscillators the solution of the system of (1) is

$$\begin{aligned}
&\tilde{\omega}_{1(2)x}^2 = \frac{\omega_{1x}^2 + \omega_{2x}^2}{2} + \sqrt{\left(\frac{\omega_{1x}^2 - \omega_{2x}^2}{2}\right)^2 + A_x' \,\Delta\omega_1^2\, \Delta\omega_2^2}\;; \\
&\Delta\omega_i^2 = \omega_{\mathrm{L}i}^2 - \omega_{\mathrm{T}i}^2 = \frac{4\pi + L(\varepsilon_\infty - 1)}{\varepsilon_\infty}\, e_{\mathrm{S}}^i e_{\mathrm{T}}^i\;; \qquad i = 1, 2, \\
&A_{\mathrm{T}}' = \left[\frac{L\varepsilon_\infty}{4\pi + L(\varepsilon_\infty - 1)}\right]^2, \qquad A_{\mathrm{L}}' = \left[\frac{4\pi - L}{4\pi + L(\varepsilon_\infty - 1)}\right]^2, \\
&\omega_{ix}^2 = \omega_{0i}^2 + A_x e_{\mathrm{S}}^i e_x^i\,.
\end{aligned} \tag{2}$$

Substituting into (2) the expression

$$\omega_{ix}^2 = \tilde{\omega}_{ix}^2 - \delta_{ix}\;; \qquad i = 1, 2\,, \qquad \delta_{1x} = -\,\delta_{2x}\,,$$

Table 1

Dispersion oscillator parameters obtained upon processing the $PbWO_4$ reflection spectrum, $\boldsymbol{E} \perp \boldsymbol{c}$ (frequencies and damping are given in cm^{-1})

T (K)	ω_{TO1}	γ_{TO1}	ω_{LO1}	γ_{LO1}	ω_{TO2}	γ_{TO2}	ω_{LO2}	γ_{LO2}	ε_∞
80	848	46	844	50	756	3.1	878	3.7	3.6
293	848	54	841	53	751	3.0	878	4.4	4.3
473	851	54	846	65	752	3.1	876	6.0	4.5

where $\tilde{\omega}_{ix}^2$ are the observed frequency values, we obtain a system of nonlinear equations with respect to the unknown values of δ_x. For the ratio δ_T/δ_L two solutions are obtained,

$$\Phi_{1,2} = \frac{\delta_T}{\delta_L} = \sqrt{K} - \frac{B}{2C} \pm \sqrt{\left(\sqrt{K} - \frac{B}{2C}\right)^2 - K}\,, \tag{3}$$

where

$$K = \frac{A'_L}{A'_T}\,, \qquad B = A'_T(M - \sqrt{K}N)^2\,,$$

$$C = A'_T(\Delta\tilde{\omega}_1^2\,\Delta\tilde{\omega}_2^2 - MN)\,, \qquad M = \tilde{\omega}_{1L}^2 - \tilde{\omega}_{2L}^2\,, \qquad N = \tilde{\omega}_{1T}^2 - \tilde{\omega}_{2T}^2\,.$$

Here and below we assume that $\tilde{\omega}_{1T}^2 > \tilde{\omega}_{2T}^2$, then $N > 0$, $M > 0$ in the case of successive oscillators ($\tilde{\omega}_{2T}^2 < \tilde{\omega}_{2L}^2 < \tilde{\omega}_{1T}^2 < \tilde{\omega}_{1L}^2$) and $N > 0$, $M < 0$ in the case of inserted oscillators ($\tilde{\omega}_{2T}^2 < \tilde{\omega}_{1L}^2 < \tilde{\omega}_{1T}^2 < \tilde{\omega}_{2L}^2$).

The obtained solutions (3) have to satisfy certain relationships depending on that of the values M and N and on the type of oscillators chosen, i.e. successive or inserted ones. These relationships arise from limitations on the values of δ_T and δ_L,

$$0 < \delta_T < N/2\,, \qquad |\delta_L| = M/2\,, \tag{4}$$

and they are as follows in the case of inserted oscillators:

$$\frac{KN}{M} - \sqrt{\left(\frac{KN}{M}\right)^2 - K} < \Phi_1 < \frac{KN}{M}\,; \quad -M < \sqrt{K}N\,, \tag{5a}$$

$$\frac{M}{N} + \sqrt{\left(\frac{M}{N}\right)^2 - K} > \Phi_2 > \frac{KN}{M}\,; \quad -M > \sqrt{K}\,N\,. \tag{5b}$$

Substituting (3) into (5a) and (5b) we obtain conditions which must be fulfilled for the inverted oscillator,

$$\tilde{\omega}_{1T}^2 - \tilde{\omega}_{1L}^2 < \frac{N + |M|}{2} - \sqrt{\left(\frac{N + |M|}{2}\right)^2 - \frac{N}{2}(|M| - \sqrt{K}N)} \tag{6a}$$

if $\quad |M| > \sqrt{K}N\,,$

$$\tilde{\omega}_{1T}^2 - \tilde{\omega}_{1L}^2 < \frac{N + |M|}{2} - \sqrt{\left(\frac{N + |M|}{2}\right)^2 - \frac{|M|}{2\sqrt{K}}(\sqrt{K}N - |M|)} \tag{6b}$$

if $\quad |M| < \sqrt{K}N\,.$

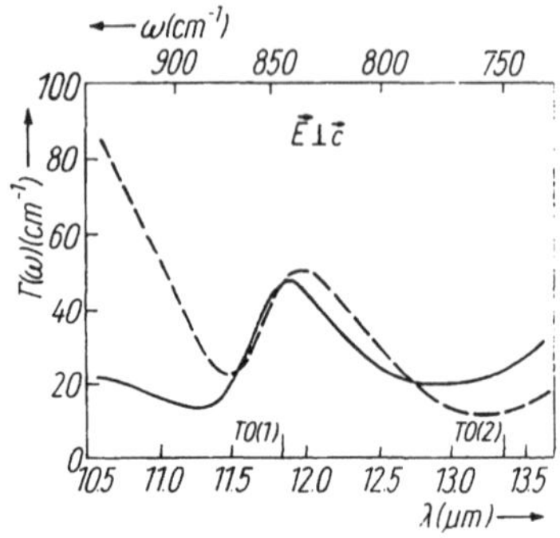

Fig. 3. Damping function $\Gamma(\omega)$ of a high-frequency, transverse phonon mode in $PbWO_4$ single crystal. ——— $T = 77$ and – – – 473 K

Analogous expressions, but with opposite sign in the inequalities, are obtained for successive oscillators. It may be shown that the calculation performed gives evaluation from above for the magnitude of the longitudinal–transverse splitting, $\Delta\tilde{\omega}_1^2$, of the inverted oscillator. It may easily be seen that by introducing damping, the absolute values of δ_T and δ_L become less than those corresponding to harmonic approximation. It means that inequalities (4) become stronger when anharmonicity is taken into account and consequently inequalities (6) will do the same. The validity of the obtained invertibility criterion has been checked by the results of [5] where two oscillators in the $Cd_{0.2}Zn_{0.8}S$ reflection spectrum were observed,

$$\omega_{1T} = 331\,, \qquad \omega_{1L} = 330\,, \qquad \omega_{2T} = 268\,, \qquad \omega_{2L} = 348\ \text{cm}^{-1}\,.$$

Invertibility of the first oscillator was reliably determined by the authors of [5] on the basis of studying the character of the phonon spectra readjustment depending on concentration of Cd atoms. Inequality (6) is fulfilled for the first oscillator at $L = 4\pi/3$ and $\varepsilon_\infty = 5.1$ that corresponds to their experimental values.

In the case of the $PbWO_4$ crystal the invertibility criterion for oscillator No. 1 is not fulfilled at all temperatures. At room temperature the result does not change for the values of parameters L and ε_∞ over the ranges $4\pi/3 \pm \pi/3$ and 4.3 ± 0.3, respectively; this is surely in excess of the reliable interval of variation of these values. Thus, the inverted oscillator model that describes the gap in the reststrahlen band (Fig. 1) cannot correspond to the real oscillator of the first-order spectrum. It also cannot correspond to a higher-order spectrum caused by electro-optical anharmonicity since in this case the formal fulfilment of one of the inequalities (6) is necessary. Consequently, the above-mentioned gap is most probably due to the anomaly in the damping frequency dependence of the transverse phonon, i.e. by mechanical anharmonicity [6]. As the reflection spectra are measured in S-polarized light, surface vibration excitations are found not to occur due to the surface roughness. Such vibrations can be observed only in P-polarized light.

The dielectric permittivity of the crystal with respect to mechanical anharmonicity is

$$\varepsilon(\omega) = \varepsilon_\infty + \frac{\omega_T^2(\varepsilon_0 - \varepsilon_\infty)}{(\omega_T^h)^2 - \omega^2 - 2i\omega_T^h\Gamma(\omega)}\,, \tag{7}$$

where ω_T^h is the frequency of the transverse phonon in harmonic approximation. The $\Gamma(\omega)$ function is expressed in terms of the $\varepsilon'' = \text{Im}\,\varepsilon(\omega)$ imaginary and $\varepsilon' = \text{Re}\,\varepsilon(\omega)$ real parts [7],

$$\Gamma(\omega) = \frac{(\varepsilon_0 - \varepsilon_\infty)\,\omega_T^h\,\varepsilon''(\omega)}{[\varepsilon'(\omega) - \varepsilon_\infty]^2 + [\varepsilon''(\omega)]^2}\,. \tag{8}$$

In Fig. 3 the function $\Gamma(\omega)$ is shown graphically at two temperatures. As seen in the figure, the peak intensity at the frequency 840 cm^{-1} (11.9 μm) is practically independent of temperature and this is not typical for anharmonicity effects. It should be noted, however, that the above peak according to [2] appears due to the distortion of the crystal surface. The region and the magnitude of deformation arising due to surface stresses may increase as the temperature decreases (the elastic modulus increases) and thus compensate a decrease of occupation numbers of the phonon states which determine the temperature dependence of $\Gamma(\omega)$ at invariable anharmonicity constant. Consider, for example, anharmonicity of the type $\gamma_{12} \langle Q_2(E_g)\rangle \times Q_2(E_g)\, Q_1^2(E_u)$, where $Q_2(E_g) \neq 0$ is the deformation that increases with decreasing temperature. Such anharmonicity always occurs if $Q_1(E_u)$ is the internal dipole oscillation of the molecular ion WO_4^{2-}, and $Q_2(E_g)$ is a translational oscillation of these ions with respect to one another. At the phased oscillation of intramolecular dipole moments, due to interaction of the latter with one another, there appears a periodic force which tends to change mutual positions of dipole so that it would become advantageous from the viewpoint of dipole–dipole interaction. In other words, this force excites a translational oscillation of WO_4^{2-} ions and this means a decay of the intramolecular excitation into intramolecular and intermolecular ones. Comparison of the sums of frequencies of the appropriate oscillations $\omega(E_u) = 751$ cm^{-1} and $\omega(E_g) = 86.5$ cm^{-1} [8] with the frequency of the peak maximum position $\Gamma(\omega)$, $\omega(\Gamma) = 840$ cm^{-1} (Fig. 3) confirms this conclusion.

Note that in an ideal crystal $Q_2(E_g) > 0$, and such a decay does not occur since the intramolecular dynamic dipoles form a stable configuration.

A similar explanation can be given for the anomaly in the reflection spectrum of another polarization, i.e. $\boldsymbol{E} \parallel \boldsymbol{c}$. In this case, the appropriate intermolecular oscillation is that of B_g symmetry. The comparison of frequencies analogous to that mentioned above also gives a satisfactory results: $\omega(A_u) = 758$ cm^{-1}, $\omega(B_g) = 53.7$ cm^{-1} [8], $\omega(\Gamma) = 822$ cm^{-1}.

Thus, the investigation carried out showed that the decay processes induced by surface stresses and arising apparently upon crystal polishing are the most probable reasons for the anomaly in the reflection spectrum of sheelites.

Acknowledgement

The authors are thankful to E. A. Vinogradov for useful remarks in the discussion of the results.

References

[1] J. F. Scott, J. chem. Phys. 48, 874 (1968).
[2] P. J. Miller, R. K. Khanna, and E. R. Lippincott, J. Phys. Chem. Solids **34**, 533 (1973).
[3] J. M. Stencel, E. Silberman, and J. Springer, Phys. Rev. B **14**, 5435 (1976).
[4] V. M. Burlakov, Preprint Inst. Spektrosk. Akad. Nauk SSSR, No. 7, Troitsk 1981 (p. 31).
[5] E. A. Vinogradov and Yu. A. Mityagin, Fiz. tverd. Tela **20**, 3162 (1978).
[6] E. R. Cowley, J. Phys. C **5**, 345 (1972).
[7] K. F. Pai, A. Rastogi, N. E. Tornberg, J. J. Parker, and R. P. Lowndes, J. Opt. Soc. Amer. **67**, 314 (1977).
[8] V. M. Nagiev, Sh. M. Efendiev, N. G. Darvishov, and N. N. Melnik, Fiz. tverd. Tela **24**, 2872 (1982).

(Received March 8, 1984)

phys. stat. sol. (b) 125, 81 (1984)

Subject classification: 2 and 4; 21.2; 21.7

Institute of Inorganic and Physical Chemistry, Academy of Sciences of the Kirgizian, SSR, Frunze[1]) *(a) and Polytechnic Institute, Frunze (b)*

On Correlation between Structural Features of Liquid Metals and the Effective Ion-Ion Potential

By

B. R. Gelchinskii (a) and R. M. Ryskulbekova (b)

A comparison of ion-ion interaction potentials obtained by various methods is performed. A certain similarity of potentials using empirical information and their distinction from priori potentials are noted. A correlation is found of the short-range part of the effective ion-ion potential obtained by the Percus-Yevick equation with radial distribution function form. It is suggested to use this correlation both, for developing methods of a theoretical estimation of ion-ion potentials and for revealing some specific features in the structure of melts.

Проведено сопоставление потенциалов межионного взаимодействия, найденных различными методами. Отмечается сходство потенциалов, использующих эмпирическую информацию и их отличие от априорных потенциалов. Выявлена корреляция короткодействующей части эффективного межионного потенциала, найденного по уравнению Перкуса-Йевика с видом функции радильного распределения. Предлагается использовать отмеченную корреляцию как для развития методов теоретической оценки межионных потенциалов, так и для выявления особенностей в структуре расплавов.

1. Introduction

The problem of the atomic structure of liquid or amorphous metals is of great interest for our further insight into the randomly condensed state. The qualitative similarity of liquid and crystalline states in atomic and electron structures, revealed as early as in the thirties, makes it possible to study liquid metals on the basis of their quasicrystalline structure. However, quasicrystal models of melts postulating their atomic structure cannot serve as a basis for developing their microscopic theory. Furthermore, a great number of investigations carried out have proved by various methods that liquid metals have their specific structure, which does not amount to a random crystalline lattice. These are, for example, the works by Bernal and King [1] on mechanical models of random systems, studies by molecular dynamics methods [2, 3], and interpretation of diffraction data [4, 5] termed two-structure model.

2. Comparison of Effective Ion-Ion Potentials Obtained by Various Methods

In the statistic theory of simple liquids and liquid metals in particular, the Bogolubov-Born-Green (BBG) and the Percus-Yevick (PY) equations and hypernetted chain (HC) are widely used to determine both the radial distribution function (RDF) by the given potential and the potential of ion-ion interaction $F(r)$. The question was discussed in detail in literature, for example, for liquid metals in [6] an analysis treating certain limitations of the approach for a qualitative estimation of ion-ion inter-

[1]) Leninskii Prospekt 267, 720071 Frunze-71, USSR.

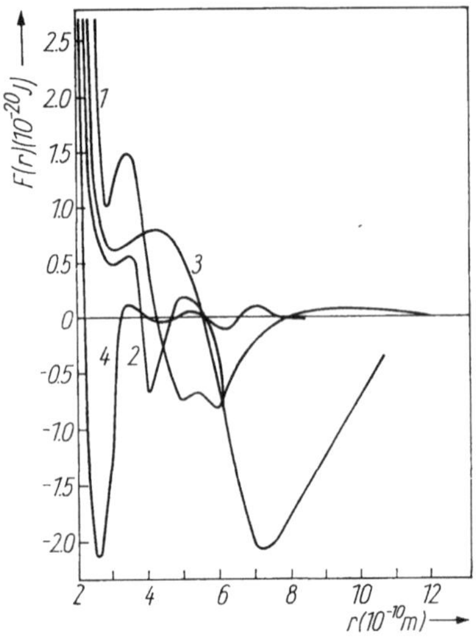

Fig. 1. The potentials of ion–ion interaction for aluminium. (1) in terms of the PY equation ($T = 1023$ K); (2) using molecular dynamics method by fitting to the experimental structure factor [3], (3) by fitting to the experimental phonon spectra of a crystal [7], (4) theoretical calculation by the pseudopotential method. ($F(r)$ here and in the other figures in 10^{-20} J)

action potentials is given. However, the equations mentioned above give qualitatively reliable information on potentials, which can be confirmed by comparing $F(r)$ obtained by various methods utilizing empirical information for some of their properties. Thus, Fig. 1 shows $F(r)$ Al obtained by various methods: from diffraction data in terms of the PY equation (curve 1[2])), by self-consistent molecular dynamics (curve 2), and pseudopotential fit to experimental phonon spectra of crystalline aluminium (curve 3).

Though there is no agreement between the long-range "tails" of these potentials, one can observe their obvious correlation in the short repulsive range and the first minimum, which, as is known, are principal for atomic structure formation. Note that all the curves obtained by the use of empirical information of a metal, are qualitatively different from the theoretical $F(r)$ (curve 4), found by pseudopotential method (Heine-Abarenkov pseudopotential, screening according to Singwi et al. [11]). The theoretical $F(r)$ can be essentially changed, if in the process of their calculation short-range interaction contributions as those of Born-Mayer and Van der Waals are additively taken into account [9, 10]. In particular, a much better agreement between theoretical and empirical potentials is achieved, as observed for Rb [10]. These contributions may be effectively taken into account in the initial data and they influence the results in semiempirical calculations.

3. Correlations between Effective Potentials and Structures of Melts

One can expect that short-range correlation on the $F(r)$ curve, which defines the atomic structure, will have definite correlations with the latter. It should be an analysis of some features of potentials from the point of view of the two-structure model [4, 5]. As is known, the conclusions of the model are made on the basis of diffraction data interpretation. It appears that the best agreement between the experimental intensity curves and theoretical ones is achieved, if one assumes, while calculating them, that the atomic structure of a melt is the combination of several (usually of two) substruc-

[2]) In the present paper the experimental diffraction data of [8] was used.

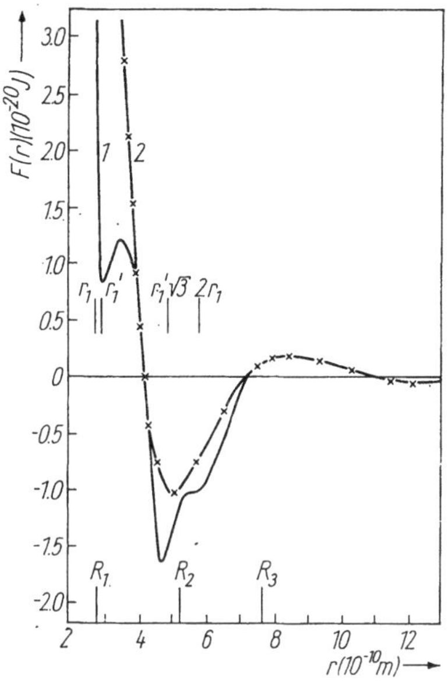

Fig. 2

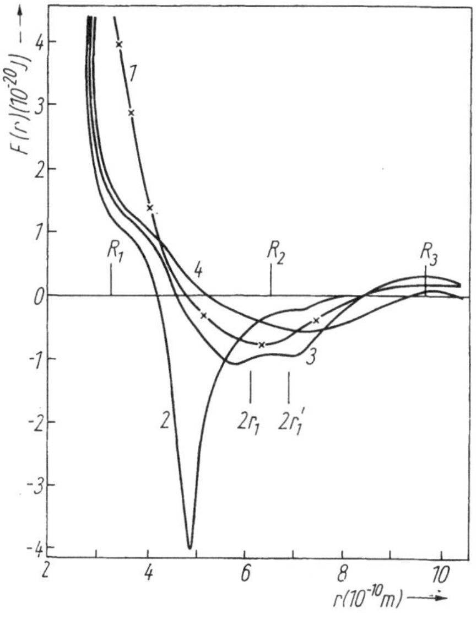

Fig. 3

Fig. 2. The potentials of ion–ion interaction for aluminium ($T = 943$ K). (1) in terms of the PY equation, (2) in terms of the HC equation; R_1, R_2, R_3 positions of the experimental RDF peaks, r_1 and r_1' distances found in two-structure model [4, 5] typical of chained and spherical packing, respectively

Fig. 3. The potentials of ion–ion interaction in liquid Bi. (1) in terms of the HC equation ($T =$ $= 1073$ K), (2) to (4) in terms of the PY equation ($T' = 823, 1073, 1273$ K); R_1, R_2, R_3, r_1, and r_1' are the same as in Fig. 2

tures. One of them is close-packed in the form of three-dimensional chains, having the length of 1.5 to 2.0 nm and the other is a spherically symmetrical packing of atoms with a little number of coordinational spheres. Sometimes it is considered that the remainders of the most close-packed crystal palnes form a chain structure, for example (111) for f.c.c. or (110) for b.c.c. structures. The basis of the conclusion is the proximity of typical distances for chained structures and these planes, which is displayed in interpreting diffraction data and computer simulation of melt structures (see e.g. [2, 3]).

In Fig. 2 positions of three RDF peaks taken from diffraction data [8] as well as characteristic distances, revealed in two-structure models alongside with $F(r)$ potentials are pointed out. It is clearly observed that the first $F(r)$ minimum (in the positive energy region) is close to the position of the first RDF peak, though the second minimum is split. Comparing the coordinates of its minima with the typical distances, obtained from the two-structure model for liquid aluminium [22] one can observe their distinct correlation.

With temperature increase the $F(r)$ changes also correlate with the conclusion of the two-structure model, which shows that the close-packed chain structures become shorter with increasing temperature and their contribution in scattering decreases. A hump on the second $F(r)$ minimum, that is at the distance, equal to the close-packed chain structure decreases by nearly a factor two with temperature (see Fig. 2, curve 1 and Fig. 1, curve 1). At the same time the depth and position of another hump of

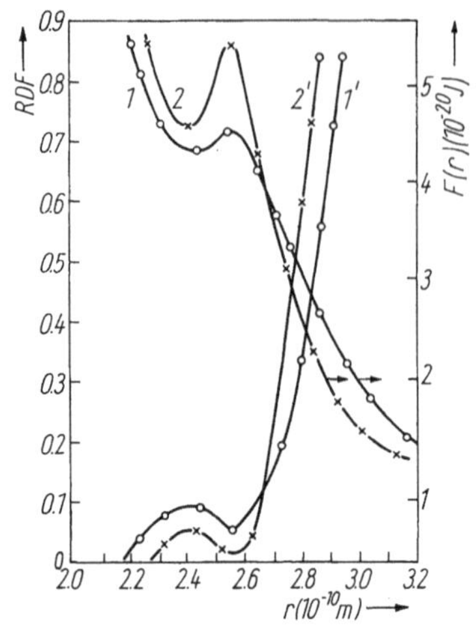

Fig. 4. The ion–ion potentials in terms of the PY equation (curves 1, 2) and RDF (curves 1′, 2′) in the region of small ion–ion distances. (1) and (1′) for Bi ($T = 603$ K), (2) and (2′) for Sn ($T = = 523$ K)

the second minimum at the distance characteristic of the spherical half-closed packing actually does not change. It is interesting to note that these correlations are observed for curves calculated in terms of the PY equation. The $F(r)$ calculations based on the same initial data in the same computer program in terms of the HC equation give notably different results (see Fig. 2, curve 2). There are no specific features in the co-ordination sphere region of this curve, and there is no minimum at the distance corresponding to the position of the first RDF peak.

Now we consider liquid Bi, having a complex structure and more half-closed packings in the crystallin state. Metals can be divided into two groups in the two-structure model: a) with basically metallic binding, b) with an essential part of directed homopolar binding. The present classification puts aluminium in the first group and bismuth in the second one.

Fig 3 represents $F(r)$ calculated for liquid Bi at different temperatures. Here we also can observe the correlation of specific features of $F(r)$ in the region corresponding to the second co-ordination sphere, with distances characteristic of two different structures. The temperature dependence of the potential changes is much stronger than for aluminium. Evidently, homopolar binding in Bi is strong enough at the melting point which can be seen in the deep, first minimum at 500 °C. There is a characteristic extremum in the region of the first co-ordination sphere. In terms of the two-structure model the decay of the close-packed structures leads to the appearance of symmetric, spherical packing with temperature due to a decrease of nonmetallic binding. These conclusions also agree with the change of the form for Bi depending on temperature. It can be seen that the depth of the potential well, the position of which correlates with the peak of the close-packed chain structure, vitally decreases, and at higher temperature (950 °C) the potential form is smooth, without any special features. The HC equation is insensitive to the feature of the structure obtained by diffraction data both for Bi and Al.

It should be noted that $F(r)$ is calculated in terms of these two equations based on the same initial data used for RDF calculations and the Ornstein-Zernike correlation function ($C(r)$). However, potentials calculated in terms of the HC equation are completely defined evidently by $C(r)$, then according to the PY equation for small r they are essentially dependent on the RDF form and $C(r)$ defines long-range oscillations only.

Therefore, some features should be noted of the $F(r)$ curves in terms of the PY equation, which correlate with those of RDF. They display only at temperatures near the melting point and disappear at insignificant overheating, moreover, mainly for the metals of the second group in accordance with the two-structure classification, such as Bi, In, Tl, and others. In Fig. 4 the short-range $F(r)$ branches versus RDF are reproduced. It is seen that these features are displayed in the form of additional minima for small distances, which correspond to small peaks of RDF for distances shorter than the position of the first principal maximum.

The distances characteristic of these features displayed in the two-structure model are much less, and their values somewhat larger than the double ion radius. One can assume that there is one more close-packed substructure with characteristic distances of the order of the ion diameter in the melt at the melting point for some time (at slight overheating), moreover, this substructure is not extended, since the maxima being multiples of these distances are not found. The features noted on the experimental RDF may be consequences of the influence of the short-range non-Coulombic forces, the account of which in [9] during the calculation of theoretical structure factor by the Monte-Carlo method produced asymmetrical structure factors typical of Ga and also the additional RDF peak at small r. It should be emphasized that the metals having simple crystalline structure in solid state may have notable features in RDF and $F(r)$. In particular an obvious, small peak somewhat left of the principal one in RDF of liquid K can be seen. As is quite noticeable in Fig. 11 of [12], however, $F(r)$ curves there in terms of the PY and HC equations are not brought to corresponding distances.

RDF and $F(r)$ calculations made in the work with the same initial data showed. that the peak in RDF correlates with the local minimum on the descending $F(r)$ branch (cf. Fig. 4 for Sn and Bi).

Note that for liquid Na these features are observed neither in RDF nor in $F(r)$. The difference in the $F(r)$ behaviour of metals seemingly similar in their properties can be accounted for by the distinctions in electrons in the electronic structure of their ion cores. The polarization coefficient of K is essentially (4.5 times) larger than that of Na. Hence, one can think that in K the forces of the short-range Van-der-Waals interaction dependent on the ion polarization coefficient are more significant, which in its turn leads to specific features in $F(r)$ and RDF.

4. Conclusions

Considering the data obtained in the present work, one can conclude that the PY equation reveals some characteristic features of metallic structures and the use of the latter for interpreting diffraction data yields additional information support the idea of a specific atomic structure of melts, which does not amount to simple crystalline analogues. Furthermore, one can expect that this equation would be successfully used for a theoretical estimation of structure factors in complex, polystructural melts, if adequate ion–ion potentials with several forms of short-range interaction taking place in metals, i.e. Coulombic, Born-Mayer, and Van-der-Waals (polarizational) potentials are used as initial data.

References

[1] J. D. Bernal and S. Y. King, in: Physics of Simple Liquids, Ed. H. N. V. Temperley, J. S. Rowlinson, and G. S. Rushbrooke, Nord-Holland Publ. Co., Amsterdam 1968 (p. 116).
[2] V. P. Manov, V. V. Pavlov, S. G. Popel, and L. A. Popova, Zh. fiz. Khim. 12, 3093 (1977).
[3] V. A. Poluchin, V. F. Uchov, and M. M. Dzugutov, Computernoe modelirovanie dinamiki i structury zhidkich metallov, Izd. Nauka, Moskva 1981.

[4] H. RICHTER, Proc. II. Internat. Conf. Properties of Liquid Metals, Tokyo, Sept. 1972, Tayler & Francis Ltd. London 1973 (p. 77).
[5] V. P. MANOV and S. J. POPEL, Izv. Akad. Nauk SSSR, Ser. Metally 2, 68 (1978).
[6] V. F. UCHOV, N. A. VATOLIN, B. R. GELCHINSKII, V. P. BESKACHKO, and O. A. ESIN, Mezhchastichnoe vzaimodeistvie v zhidkich metallakh, Izd. Nauka, Moskva 1979.
[7] H. C. GUPTA and B. B. TRIPATHI, Physica (Utrecht) 58, 136 (1972).
[8] Y. WASEDA and K. SUZUKI, Sci. Rep. Res. Inst. Tôhoku Univ. **A24**, 149 (1973).
[9] K. K. MON, N. W. ASHCROFT, and G. V. CHESTER, Phys. Rev. B 19, 5103 (1979).
[10] W. Z. SCHOMMERS, Z. Phys. **B24**, 171 (1976).
[11] K. S. SINGVI, A. SJOLANDER, M. P. TOSI, and R. H. LAND, Phys. Rev. B **1**, 1044 (1970).
[12] M. J. HASEGAWA, J. Phys. F. **6**, 649 (1976).

(Received December 1, 1983)

phys. stat. sol. (b) **125**, 87 (1984)

Subject classification: 6 and 20.1; 22.9

Ecole Nationale d'Ingénieurs Sfax[1])

Raman Study of Mono-, Di-, and Trimethyl Ammonium Hexachlorostannates

By

M. H. BEN GHOZLEN, Y. MLIK, and A. DAOUD

The low-frequency Raman spectra (0 to 350 cm^{-1}) of $(CH_3NH_3)_2SnCl_6$: 1 MHS, $[(CH_3)_2NH_2]_2 \cdot SnCl_6$: 2 MHS and $[(CH_3)_3NH]_2SnCl_6$: 3 MHS are recorded on single crystals at room temperature. The spectra are assigned on the basis of group theory analysis of vibrations and in view of the previously reported structural data of these hexachlorostannate compounds. The effect of site symmetry on the internal vibrations of the $SnCl_6^{2-}$ octahedra are discussed; the distortions of the $SnCl_6^{2-}$ octahedra are taken into account in the expressions of the Raman tensors relative to their internal modes.

Les spectres Raman de basse fréquence (0 à 350 cm^{-1}) de monocristaux de $(CH_3NH_3)_2SnCl_6$: (1 MHS), $[(CH_3)_2NH_2]_2SnCl_6$: (2 MHS) et $[(CH_3)_3NH]_2SnCl_6$: (3 MHS) ont été enregistrés à la temperature ambiante. Une attribution des vibrations observées sur les spectres est presentée en se basant sur une analyse des vibrations par la théorie des groupes ainsi que sur les données structurales concernant ces composés. L'effet du groupe de site sur les modes internes des octaèdres $SnCl_6^{2-}$ est discuté; les distorsions des entités octaédriques $SnCl_6^{2-}$ ont été mise en évidence dans les expressions des tenseurs Raman associés à leurs modes internes.

1. Introduction

The hexahalometallates R_2MX_6 where R is an organic cation, M is a metallic cation ($M = Sn^{4+}, Pt^{4+}$), and $X = Cl^-, Br^-$, have been the subject of several studies: X-ray, Raman, and infrared spectroscopy etc. [1 to 12]. Most of the spectroscopic studies are performed on powder samples. The structure of the family of compounds R_2MX_6 is characterized by the existence of slightly distorted MX_6^{2-} octahedra, on the other hand, the reported space groups are subgroups of Fm3m-O_h^5 which is the structure of K_2SnCl_6 at room temperature [7].

Table 1
Summary of structural data of 1 MHS, 2 MHS, and 3 MHS

	1 MHS [1]	2 MHS [2]	3 MHS [3]
space group	R$\bar{3}$m-D_{3d}^5	Pnnm-D_{2h}^{12}	Pa3-T_h^6
Z formula units in the primitive cell	1	2	4
octahedron site symmetry	D_{3d} (000)	C_{2h} (000, ...)	C_{3i} (000, ...)
cation site symmetry	C_{3v} (XXX, ...)	C_s ($XY0$, ...)	C_3 (XXX, ...)

[1]) 3038 Sfax, Tunisia.

In this study we report Raman spectra (0 to 350 cm^{-1}) recorded at room temperature on single crystals of three compounds: the mono-(1 MHS), di-(2MHS), and trimethylammonium hexachlorostannates (3MHS). These salts undergo phase transitions at 156 K for 1 MHS [1], 100 K for 2 MHS [4], and at 170 K for 3 MHS [13]. The structure of these compounds is summarized in Table 1 [1 to 4]. We notice that there is no disorder in these structures and that all these structures can be derived from Fm3m-O_h^5 by some distortions [14]. The distortions of Fm3m which lead to R$\bar{3}$m, Pnnm, and Pa3 affect the Raman tensors of the internal modes of the $SnCl_6^{2-}$ octahedra. Some tensor coefficients which are zero in the cubic Fm3m structure become different from zero for the low-symmetry structures; these coefficients will characterize the distortions.

Section 2 will be devoted to a group-theoretical analysis of the external vibrations of R_2MX_6 including the internal modes of the octahedra. In Section 3, the Raman spectra of these salts will be presented and an assignment of the different modes will be given. The distortions of the cubic Fm3m structure will be discussed in Section 4.

2. External Vibrations and Group Theory

In a previous work [14] we have concluded that the space groups R$\bar{3}$m (1 MHS), Pnnm (2 MHS), and Pa3 (3 MHS) are subgroups of Fm3m; the order parameters which describe the symmetry change do transform as $F_{2g}^{(\Gamma)}$ for R$\bar{3}$m and $E_g^{(x)}$ for Pnnm and Pa3, where Γ and X are high-symmetry points of the cubic face-centered Brillouin zone [15]. This means that the vibrational modes in the case of R$\bar{3}$m are derived from Γ point modes of Fm3m, for the two other cases the vibrational modes can be divided into two sets, the first comes from Γ point vibrations and the second from vibrations at zone boundary X. Consequently an enumeration of the vibrations of R_2MX_6 in Fm3m at Γ and X points will be useful for the correlation diagrams between the vibrations in Fm3m and in the different subgroups.

The enumaration of the external vibrations of R_2MX_6 is [16, 17]:

$$\Gamma_{vib}(\Gamma, Fm3m) = A_{1g} + E_g + 2F_{1g} + 2F_{2g} + 4F_{1u} + 2F_{2u}\,,$$

$$\Gamma_{vib}(X, Fm3m) = 3A_{1g} + A_{2g} + 2B_{1g} + B_{2g} + 4E_g + A_{1u} + 3A_{2u} + B_{1u} + B_{2u} + 6E_u\,,$$

$$\Gamma_{vib}(R\bar{3}m) = \underline{3A_{1g}} + 2A_{2g} + \underline{5E_g} + 2A_{1u} + 4A_{2u} + 6E_u\,,$$

$$\Gamma_{vib}(Pnnm) = \underline{8A_g} + \underline{8B_{1g}} + \underline{7B_{2g}} + \underline{7B_{3g}} + 8A_u + 8B_{1u} + 10B_{2u} + 10B_{3u}\,,$$

$$\Gamma_{vib}(Pa3) = \underline{5A_g} + \underline{5E_g} + \underline{15F_g} + 6A_u + 6E_u + 18F_u\,.$$

The underlined modes are Raman active [16]. The enumerated modes of R_2SnCl_6 include rotatory (R_c, R_0) and translatory (T_c, T_0) vibrations of the two entites: the organic cations and the octahedra and also the internal modes of $SnCl_6^{2-}$ noted as ν_1 to ν_6 [18]. Note that the translatory modes of the octahedra T_0 are inactive in Raman scattering since the local symmetry of the octahedra is centrosymmetric. All these enumerations are in agreement as it can be seen on the correlation diagrams (Tables 2, 3, and 4).

In a recent work Knop et al. [4] have mentioned that the structures of 2 MHS (Pnmn) and 3 MHS (Pa3) are not subgroups of the Fm3m antifluorite structure. The previous discussion shows that Pnnm and Pa3 are obviously Fm3m subgroups. The correlation diagrams (Tables 3, 4) show that the low-symmetry structures are ob-

Table 2

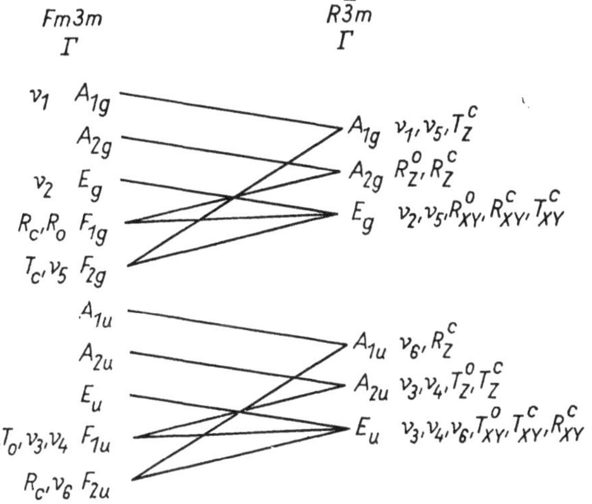

tained from Fm3m by distortions which can be described by the displacements (T_c, R_c, R_0); these motions belong to the irreducible representation E_g at the X point of Fm3m and become totally symmetric in Pnnm and Pa3 subgroups.

3. Raman Spectra

The Raman spectra were recorded on a triple monochromator Dilor RT30 coupled to an argon ion laser Spectra-Physics model 164 operating at 514.5 nm with output powers up to 600 mW.

Colourless single crystals of R_2SnCl_5 were obtained by slow evaporation of a supersaturated acid solution containing stoichiometric amounts of RCl and $SnCl_4$.

Prismatic crystals of about $3 \times 3 \times 1.5$ mm³ are selected for the experiments. The obtained samples are previously oriented on an X-ray camera. The wave vectors

Table 3

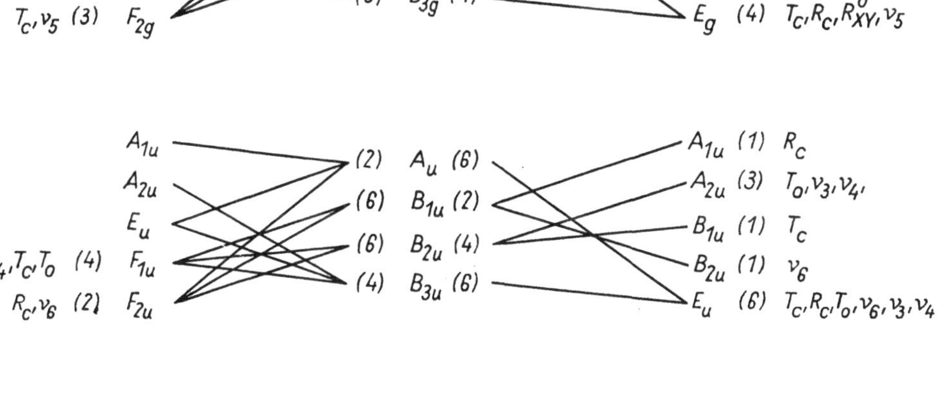

Table 4

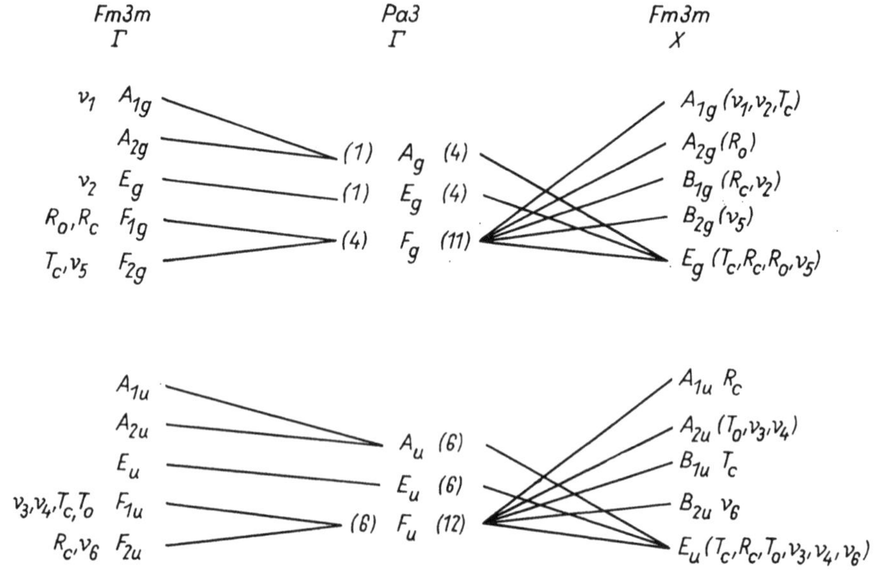

of incident and scattered light $\boldsymbol{k}_i$ and $\boldsymbol{k}_s$ are always along the crystallographic axes exept for the cubic compound (3 MHS) where these wave vectors are along the crystallographic directions [$\bar{1}$10] and [110] this geometry allows to record a pure E_g spectrum [16, 19]. It is noted that in this case the A_g spectrum cannot be obtained separately.

The Raman spectra of the three compounds are reproduced on Fig. 1 for 1 MHS, on Fig. 2 for 2 MHS, and on Fig. 3 for 3 MHS. It can be seen that:

(i) the Raman spectra confirm that the structures of this compounds are ordered; the observed band-widths did not reveal any disorder;

(ii) the internal modes of the octahedra ν_1, ε_2, and ν_5 are located approximately at the same frequencies as for the free octahedron, i.e. at 310, 235, and 160 cm^{-1}, respectively [18]. This means that the interaction forces between the octahedra and the organic cations are very weak. One can therefore neglect coupling between the external modes of the two entities;

(iii) the observed Raman spectra are compatible with the group-theoretical analyis.

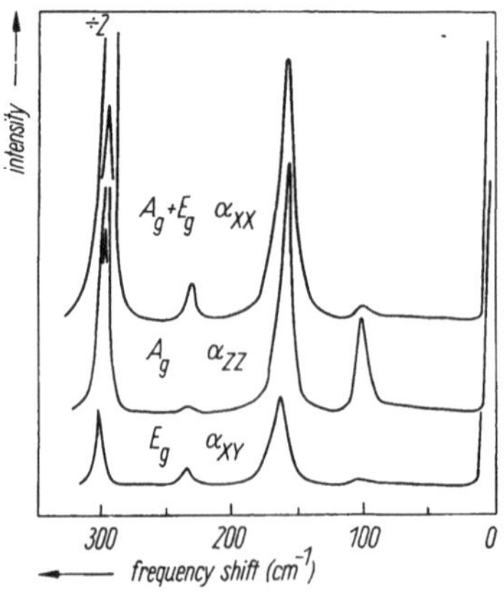

Fig. 1. Low-frequency Raman spectra of 1 MHS at room temperature. The X, Y, and Z laboratory axes coincide with the a, b, and c crystallographic axes, respectively

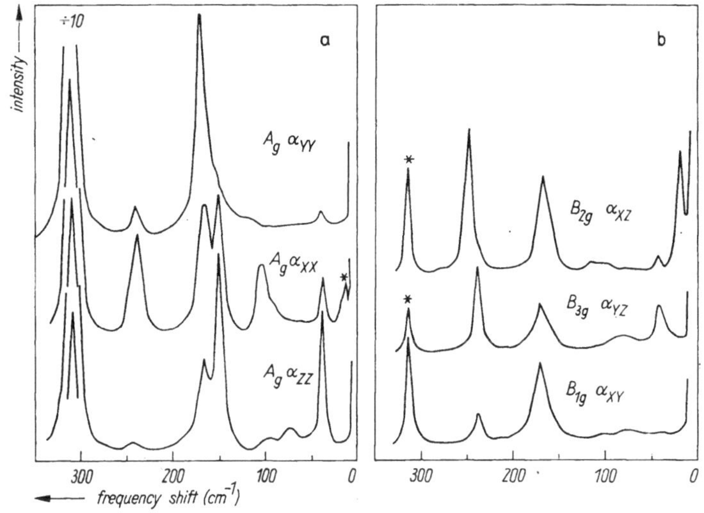

Fig. 2. Low-frequency Raman spectra of 2 MHS at room temperature. a) A_g modes, b) B_{1g}, B_{2g}, and B_{3g} modes. The asterisk indicate orientational spill-over of scattering from phonons active in other scattering geometry

For the assignment of the observed Raman lines, these spectra are compared and analysed on the basis of group theory, for instance in the case of 1 MHS the assignment is obvious and gives information about the frequency of the translatory vibrations of the cations. Also, previous results on compounds of the same family gives information about external mode frequencies of the $SnCl_6^{2-}$ octahedra [5 to 12].

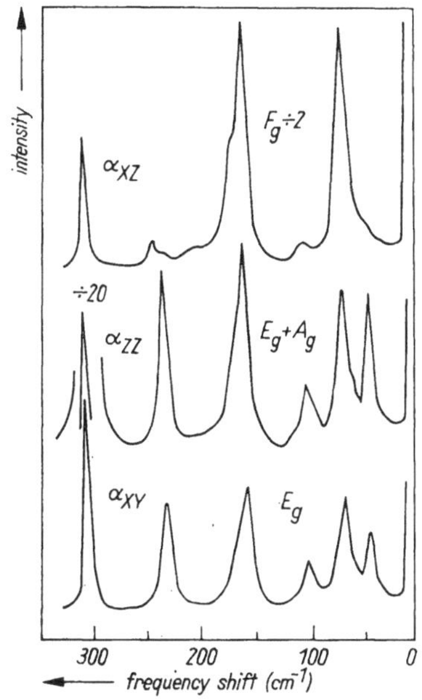

Fig. 3. Low-frequency Raman spectra of 3 MHS at room temperature. The X, Y, and Z laboratory axes coincide with the $a + b$, $a - b$, and c crystallographic directions, respectively. $\boldsymbol{k}_i = (110)$, $\boldsymbol{k}_s = (\bar{1}10)$

Table 5

attribution	1 MHS R$\bar{3}$m, $Z = 1$		2 MHS Pnnm, $Z = 2$				3 MHS Pa3, $Z = 4$			observation
	A_{1g}	E_g	A_g	B_{1g}	B_{2g}	B_{3g}	A_g	E_g	F_g	
R_0^{Γ}	—	(1) R_{0XY}	—	(1) R_{0Z}	(1) R_{XY} 40	(1) R_{XY} 40	—	—	(1)	the Z axis contain Sn^{4+} and 2 Cl^-
R_0^X	—	—	(1) R_{0XY} 40	—	(1) R_{0Z} 16	(1) R_{0XY} 40	(1) 42	(1) 42	(2) 48	
R_c^{Γ}	—	(1) R_{cXY}	—	(1) R_{cZ} 73	(1) ≈ 72	(1) 75	—	—	(1) 61	Z is the axis of symmetry of the molecule R^+
R_c^X	—	—	(1) R_{cXY} 73	—	(1) R_{cXY} ≈ 72	(1) R_{cXY} 75	(1) 69	(1) 69	(2) 70	
T_c^{Γ}	(1) T_{cZ} 105	(1) T_{cXY} 105	(1) 108	—	(1) ≈ 100	—	—	—	(1) 105	
T_c^X	—	—	(1) 90	(1) 95	—	(1) 95	(1) 100	(1) 100	(2) ≈ 100	
$SnCl_6^{=}$ internal modes										
ν_5^{Γ}	(1) 167	(1) 167	(1) 169	(1) 169	(1) 166	—	—	—	(1) 171	
ν_5^X	—	—	(1) 153	(1)	—	(1) 166	(1) 163	(1) 163	(1) 163	
ν_5^X	—	—	—	—	—	—	—	—	(1)	209 t_{CH_3}
ν_2^{Γ}	—	(1) 238	(1) 240	—	—	(1) 235	—	(1) 230	—	
ν_2^X	—	—	—	(1) 235	(1) 245	—	—	(2) 235 242	—	
ν_1^{Γ}	(1) 310	—	(1) 310	—	—	—	(1) 305	—	—	
ν_1^X	—	—	—	(1) 310	—	—	—	—	(1) 306	

The weak F_g band of 3MHS at 209 cm^{-1} is probably not the third F_g component of ν_5 since its frequency deviates significantly from the other two components observed at 163 and 171 cm^{-1}. Therefore, we assign this band to the torsional mode of the methyl groups expected in this spectrum [20].

The assignment summarized in Table 5 can be improved by total or partial deuteration of these compounds.

4. Distortion Effects on the $SnCl_6^{2-}$ Internal Modes

As it has been mentioned above, all these compounds 1 MHS, 2 MHS, and 3 MHS contain almost free $SnCl_6^{2-}$ octahedra. Consequently the distortion effects on $SnCl_6^{2-}$ octahedra can be seen through the induced change of the Raman tensors associated to the internal modes ν_1, ν_2, and ν_5. This is possible since the site symmetries of the octahedra in the three compounds are subgroups of O_h, which is the symmetry of a free octahedron.

The method consists in the calculation of linear combinations of Raman tensors adapted to the factor group symmetry.

The first step consists in writing the Raman tensors associated to $\nu_1(A_{1g})$, $\nu_2(E_g)$, and $\nu_5(F_{2g})$ in a coordinate diagram relative to the site group. A comparison between the obtained tensors and those tabulated for the corresponding site group allow to evaluate all tensor coefficients in terms of O_h tensor factors (a_0, b_0 and d_0). At this level some new coefficients are introduced and have been labelled ε_k^{ij}; where the indices i, j, and k are associated to the crystal number (i MHS), the internal mode ν_j, and the position in the corresponding Raman tensor, respectively. These coefficients which are equal to zero when the octahedron is fully symmetric take into account the symmetry lowering from O_h to the site group.

The projection operator technique provides the symmetry-adapted Raman tensors from the previously obtained generator tensors. The set of tensors associated to ν_1, ν_2 and ν_5 for each compound are given in Table 6. The results allow to compare the relative intensities of a given mode for the different diffusion geometries; for example in the case of 2 MHS one notes

$$I_{XY}(\nu_5, B_{1g}) \sim 2d_0^2 ,$$

$$I_{XZ}(\nu_5, B_{2g}) \sim 2d_0^2 \Rightarrow I_{XY} = I_{XZ} = \frac{I_{YZ}}{(\varepsilon_6^{25})^2},$$

$$I_{YZ}(\nu_5, B_{3g}) \sim 2(d_0\varepsilon_6^{25})^2 .$$

This type of relations can be checked when the spectra are recorded under the same experimental conditions. In the case of 2 MHS(ν_5) Fig. 2b, the observed Raman intensities are in agreement with the previous relations and allow to conclude that ε_6^{25} is approximately equal to one. Note that the total intensity for a given mode is proportional to Z times the intensity produced by an octahedron, where Z is the number of octahedra, $SnCl_6^{2-}$ in the primitive cell. The intensity of ν_1, ν_2, and ν_5 produced by a free octahedron are

$$I(\nu_1) \approx 3a_0^2 , \qquad I(\nu_2) \approx 6b_0^2 , \quad \text{and} \quad I(\nu_5) \approx 6d_0^2 .$$

For example in the case of 3 MHS the total expected intensity of the different components of ν_5 is equal to

$$I_{tot}(\nu_5) = I(A_g) + I(E_g) + I(F_g^{(1)}) + I(F_g^{(2)}) + I(F_g^{(3)}) .$$

This type of relations allows to check the different calculation steps. The observed Raman intensities of the spectra when compared to the calculated ones from the established tensors, show that the introduced coefficients ε_k^{ij} in the case of 2MHS and 3 MHS are relatively large. This means that the Raman intensities of the internal octahedron modes are very sensitive to distortions.

Table 6a

Raman tensors associated to the internal modes ν_1, ν_2 in the case of 1 MHS ($R\bar{3}m$), 2 MHS (Pnnm), and 3 MHS (Pa3)

	ν_1		ν_2			
1 MHS	$A_{1g}:a_0$	$\begin{pmatrix} 1 & & \\ & 1 & \\ & & 1+\varepsilon_3^{11} \end{pmatrix}$	$E_{g,1}:b_0$	$\begin{pmatrix} 1 & & \\ & -1 & \sqrt{2} \\ & \sqrt{2} & \end{pmatrix}$	$E_{g,2}:b_0$	$\begin{pmatrix} & -1 & -\sqrt{2} \\ -1 & & \\ -\sqrt{2} & & \end{pmatrix}$
2 MHS	$A_g:a_0\sqrt{2}$	$\begin{pmatrix} 1 & & \\ & 1 & \\ & & 1 \end{pmatrix}$	$A_g:b_0\sqrt{2}$	$\begin{pmatrix} 2 & & \\ & -1 & \\ & & -1 \end{pmatrix}$	$B_{1g}:b_0\sqrt{2}$	$\begin{pmatrix} & \varepsilon_4^{22} & \\ \varepsilon_4^{22} & & \\ & & \end{pmatrix}$
	$B_{1g}:a_0\sqrt{2}$	$\begin{pmatrix} & \varepsilon_4^{21} & \\ \varepsilon_4^{21} & & \\ & & \end{pmatrix}$	$B_{2g}:b_0\sqrt{2}$	$\begin{pmatrix} & & \varepsilon_5^{22} \\ & & \\ \varepsilon_5^{22} & & \end{pmatrix}$	$B_{3g}:b_0\sqrt{2}$	$\begin{pmatrix} & & \\ & & -\sqrt{3} \\ & -\sqrt{3} & \end{pmatrix}$
3 MHS	$A_g:2a_0$	$\begin{pmatrix} 1+\varepsilon_3^{31} & & \\ & 1+\varepsilon_3^{31} & \\ & & 1+\varepsilon_3^{31} \end{pmatrix}$	$E_{g,1}:\frac{2b_0}{3}$	$\begin{pmatrix} 3-\sqrt{3}\,\alpha' & & \\ & 3+\sqrt{3}\,\alpha' & \\ & & -6 \end{pmatrix}$	$E_{g,2}:\frac{2b_0}{3}$	$\begin{pmatrix} -3\sqrt{3}-\alpha' & & \\ & 3\sqrt{3}-\alpha' & \\ & & 2\alpha' \end{pmatrix}$
	$F_{g,1}:2a_0$	$\begin{pmatrix} & & \\ & & \varepsilon_3^{31} \\ & \varepsilon_3^{31} & \end{pmatrix}$	$F_{g,1}^{(1)}:\frac{2b_0}{3}$	$\begin{pmatrix} & & \\ & & \sqrt{3}\,\beta' \\ & \sqrt{3}\,\beta' & \end{pmatrix}$	$F_{g,1}^{(2)}:\frac{2b_0}{3}$	$\begin{pmatrix} & & \\ & & \beta' \\ & \beta' & \end{pmatrix}$
	$F_{g,2}:2a_0$	$\begin{pmatrix} & & \varepsilon_3^{31} \\ & & \\ \varepsilon_3^{31} & & \end{pmatrix}$	$F_{g,2}^{(1)}:\frac{2b_0}{3}$	$\begin{pmatrix} & & \sqrt{3}\,\beta' \\ & & \\ \sqrt{3}\,\beta' & & \end{pmatrix}$	$F_{g,2}^{(2)}:\frac{2b_0}{3}$	$\begin{pmatrix} & & \beta' \\ & & \\ \beta' & & \end{pmatrix}$
	$F_{g,3}:2a_0$	$\begin{pmatrix} & \varepsilon_3^{31} & \\ \varepsilon_3^{31} & & \\ & & \end{pmatrix}$	$F_{g,3}^{(1)}:\frac{2b_0}{3}$	$\begin{pmatrix} & \sqrt{3}\,\beta' & \\ \sqrt{3}\,\beta' & & \\ & & \end{pmatrix}$	$F_{g,3}^{(2)}:\frac{2b_0}{3}$	$\begin{pmatrix} & \beta' & \\ \beta' & & \\ & & \end{pmatrix}$

$\alpha' = \varepsilon_6^{32}\sqrt{2} + \varepsilon_4^{32}$; $\beta' = \varepsilon_4^{32}\sqrt{2} - \varepsilon_4^{32}$.

Table 6b

Raman tensors associated to the internal mode ν_5 in the case of 1 MS ($R\bar{3}m$), 2 MHS (Pnnm), and 3 MHS (Pa3)

	ν_5		
1 MHS	$A_{1g}: \frac{d_0}{\sqrt{3}} \begin{pmatrix} -1 & & \\ & -1 & \\ & & 2 \end{pmatrix}$	$E_{g,1}: \frac{d_0}{\sqrt{3}} \begin{pmatrix} \sqrt{2} & & \\ & -\sqrt{2} & -1 \\ & -1 & \end{pmatrix}$	$E_{g,2}: \frac{d_0}{\sqrt{3}} \begin{pmatrix} & -\sqrt{2} & 1 \\ -\sqrt{2} & & \\ 1 & & \end{pmatrix}$
2 MHS	$A_g^{(1)}: d_0\sqrt{2} \begin{pmatrix} \varepsilon_1^{25} & & \\ & -1 & \\ & & 1 \end{pmatrix}$	$A_g^{(2)}: d_0\sqrt{2} \begin{pmatrix} \varepsilon_1^{25} & & \\ & \varepsilon_2^{25} & \\ & & \varepsilon_3^{25} \end{pmatrix}$	$B_{1g}^{(1)}: d_0\sqrt{2} \begin{pmatrix} & 1 & \\ 1 & & \\ & & \end{pmatrix}$
	$B_{1g}^{(2)}: d_0\sqrt{2} \begin{pmatrix} & \varepsilon_4^{25} & \\ \varepsilon_4^{25} & & \\ & & \end{pmatrix}$	$B_{2g}: d_0\sqrt{2} \begin{pmatrix} & & 1 \\ & & \\ 1 & & \end{pmatrix}$	$B_{3g}: d_0\sqrt{2} \begin{pmatrix} & & \\ & & \varepsilon_6^{25} \\ & \varepsilon_6^{25} & \end{pmatrix}$
3 MHS	$A_g: \frac{2d_0}{3\sqrt{3}} \begin{pmatrix} \varepsilon_3^{35} & & \\ & \varepsilon_3^{35} & \\ & & \varepsilon_3^{35} \end{pmatrix}$	$E_{g,1}: \frac{2d_0}{3\sqrt{6}} \begin{pmatrix} -\alpha\sqrt{3} & & \\ & \alpha\sqrt{3} & \\ & & \end{pmatrix}$	$E_{g,2}: \frac{2d_0}{3\sqrt{6}} \begin{pmatrix} -\alpha & & \\ & -\alpha & \\ & & 2\alpha \end{pmatrix}$
	$F_{g,1}^{(1)}: \frac{2d_0}{3\sqrt{3}} \begin{pmatrix} & & \\ & & 3+\varepsilon_3^{35} \\ & 3+\varepsilon_3^{35} & \end{pmatrix}$	$F_{g,1}^{(2)}: \frac{d_0}{3\sqrt{3}} \begin{pmatrix} & & \\ & & -3\sqrt{6}+2\beta \\ & -3\sqrt{6}+2\beta & \end{pmatrix}$	$F_{g,1}^{(3)}: \frac{2d_0}{3\sqrt{3}} \begin{pmatrix} & & \\ & & 3\sqrt{2}+2\sqrt{3}\,\beta \\ & 3\sqrt{2}+2\sqrt{3}\,\beta & \end{pmatrix}$
	$F_{g,2}^{(1)}: \frac{2d_0}{3\sqrt{3}} \begin{pmatrix} & & 3+\varepsilon_3^{35} \\ & & \\ 3+\varepsilon_3^{35} & & \end{pmatrix}$	$F_{g,2}^{(2)}: \frac{d_0}{3\sqrt{3}} \begin{pmatrix} & & -3\sqrt{6}+2\beta \\ & & \\ -3\sqrt{6}+2\beta & & \end{pmatrix}$	$F_{g,2}^{(3)}: \frac{d_0}{3\sqrt{3}} \begin{pmatrix} & & 3\sqrt{2}+2\sqrt{3}\,\beta \\ & & \\ 3\sqrt{2}+2\sqrt{3}\,\beta & & \end{pmatrix}$
	$F_{g,3}^{(1)}: \frac{2d_0}{3\sqrt{3}} \begin{pmatrix} & 3+\varepsilon_3^{35} & \\ 3+\varepsilon_3^{35} & & \\ & & \end{pmatrix}$	$F_{g,3}^{(2)}: \frac{d_0}{3\sqrt{3}} \begin{pmatrix} & -3\sqrt{6}+2\beta & \\ -3\sqrt{6}+2\beta & & \\ & & \end{pmatrix}$	$F_{g,3}^{(3)}: \frac{d_0}{3\sqrt{3}} \begin{pmatrix} & 3\sqrt{2}+2\sqrt{3}\,\beta & \\ 3\sqrt{2}+2\sqrt{3}\,\beta & & \\ & & \end{pmatrix}$

$\alpha = \varepsilon_6^{35}\sqrt{2} + \varepsilon_4^{35}$, $\beta = \varepsilon_6^{35}\sqrt{2} - \varepsilon_4^{35}$.

5. Conclusion

The assignment of low-frequency Raman spectra of 1 MHS, 2 MHS, and 3 MHS is useful for further studies of phase transitions observed in these compounds. This assignment has been made easier by the group to subgroup relations and by simultaneous examination of the corresponding Raman spectra. The distortions of the $SnCl_6^{2-}$ octahedra seem difficult to quantify from the calculated Raman tensors. We can also conclude that the observed spectra are in agreement with the previously resolved structures.

Acknowledgements

The authors wish to thank Dr. R. M. Pick and Dr. H. Poulet (Université de Paris) for valuable discussions.

References

[1] K. Kitahama, H. Kiriyama, and Y. Baba, Bull. Chem. Soc. Japan **52**, 324 (1979).
[2] M. H. Ben Ghozlen, A. Daoud, and J. W. Bats, Acta cryst. **B37**, 1415 (1981).
[3] M. H. Ben Ghozlen and J. W. Bats, Acta cryst. **B38**, 1308 (1982).
[4] O. Knop, T. S. Cameron, and M. A. James, Canad. J. Chem. **61**, 1620 (1983).
[5] G. P. O'Leary and R. G. Wheeler, Phys. Rev. B **1**, 4409 (1970).
[6] R. W. Berg, F. W. Pulsen, and N. J. Bjerrum, J. chem. Phys. **67**, 1829 (1977).
[7] S. L. Chodos and R. W. Berg, J. chem. Phys. **70**, 4864 (1979).
[8] R. W. Berg, J. chem. Phys. **69**, 1325 (1978).
[9] R. W. Berg, J. chem. Phys. **71**, 2531 (1979).
[10] W. Vonder Ohe, J. chem. Phys. **62**, 3923 (1975).
[11] K. Negita, N. Nakamura, and H. Chihara, Solid State Commun. **34**, 949 (1980).
[12] J. Pelzl, P. Engels, and R. Florian, phys. stat. sol. (b) **82**, 145 (1977).
[13] M. H. Ben Ghozlen, unpublished result.
[14] M. H. Ben Ghozlen and Y. Mlik, J. Phys. C **16**, 4365 (1983).
[15] C. J. Bradley and A. P. Cracknell, The Mathematical Theory of Symmetry in Solids, Clarendon Press, Oxford 1972.
[16] H. Poulet and J. P. Mathieu, Spectres de vibration et symetrie des cristaux, Gordon & Breach, Paris 1970.
[17] W. Fateley, F. Dollish, N. McDevitt, and F. Bentley, Infrared and Raman Selection Rules for Molecular and Lattice Vibrations, Wiley, New York 1972.
[18] K. Nakamoto, Infrared Spectra of Inorganic and Coordination Compounds, Wiley, New York 1970.
[19] M. Krauzmann, Thèse d'état, Paris 1969.
[20] M. Schlaak, M. Couzi, and P. V. Huong, Ber. Bunsenges. phys. Chem. **80**, 881 (1976).

(Received April 2, 1984)

phys. stat. sol. (b) **125**, 97 (1984)

Subject classification: 4 and 13.1; 21.1; 21.6

Department of Physics, R. K. Mission Residential College, Narendrapur (a),
Department of Physics, Jadavpur University (b),
and Instrumentation Centre, University of Kalyani[1]) *(c)*

Charge Transfer and Energy of Formation in Ordered Binary Alloys

By

S. Chakraborty (a), A. Manna (b), and A. K. Ghosh (c)

The study of charge transfer between constituents plays a central role in the study of alloy formation. So far it has been studied by the pseudopotential, self-consistent KKR-Xα, Thomas-Fermi, and Thomas-Fermi-Dirac methods. Recently it is shown that the α-expansion method of Löwdin, as adopted by Matheiss, can be successfully employed to derive electron density distribution in a solid from free atomic statefunctions, and hence charge transfer and energy of formation of ordered binary alloys can be calculated via Hodges and Stott mechanism as proposed by Alonso and Girifalco. Here a still simpler direct procedure is presented which yields equally good, if not better, results.

Die Untersuchung von Ladungstransfer zwischen den Konstituenten spielt eine zentrale Rolle bei der Diskussion der Legierungsbildung. Bisher wurde er mittels Pseudopotential-, selbstkonsistenten KKR-Xα-, Thomas-Fermi- und Thomas-Fermi-Dirac-Methoden untersucht. Kürzlich wurde gezeigt, daß die α-Entwicklungsmethode von Löwdin, wie sie von Matheiss entwickelt wurde, erfolgreich angewendet werden kann, um die Verteilung der Elektronendichte in einem Festkörper aus den Zustandsfunktionen freier Atome abzuleiten, somit lassen sich Ladungstransfer und Bildungsenergie von geordneten binären Legierungen über den Hodges-Stott-Mechanismus, wie er von Alonso und Girifalco vorgeschlagen wird, berechnen. Es wird nun eine noch einfachere direkte Methode vorgestellt, die mindestens gleich gute, wenn nicht bessere Ergebnisse liefert.

1. Introduction

To study the binding in an alloy, it is convenient to consider it as an equivalent lattice of equivolume components. An alloy being composed of dissimilar metals, equivolume components will be charged, the charge having been transferred from one component to another. As a result of net charges on equivolume components a Madelung-type attraction originates. This attraction favours ordering in a binary alloy.

Of course, this simple description of binding in a transition metal alloy has been contested by Pettifor and co-worker and others [1] who showed that the energy of formation does not depend *explicitly* on the charge transfer, the binding, as in a pure metal, originating from strong bonding of the d valence electrons.

In any case, different workers have studied charge transfer by different methods and many have successfully calculated the energy of formation even for transition metal alloys from charge transfer values. Inglesfield [2] followed the pseudopotential method to study alloys of Mg, Cd, and Hg. Moruzzi et al. [3] utilized a self-consistent KKR-Xα method to study six intermetallic compounds of Cu, Zn, Ni, Al, Co, and Be. The energy density functional of Hohenberg and Kohn [4] was utilized by Hodges and Stott [5] to study alloys of simple metals. The Thomas-Fermi method was used

[1]) Kalyani 741235, W. Bengal, India.

by Ratti and Ziman [6] and Cragg and Fletcher [7], and the Thomas-Fermi-Dirac method by Alonso and Girifalco [8] to study a number of alloys and intermetallic compounds.

We have shown ([9], hereafter referred to as I) that the charge transfer and energy of formation of equiatomic inter-transition-metal compounds can be satisfactorily estimated by a combination of the α-expansion technique of Löwdin [10] and the Hodges and Stott [5] mechanism. However, in I we followed a rather round-about procedure, used by Alonso and Girifalco [8] to calculate the amount of charge transfer. In this paper we present a direct procedure for this purpose and show that the present procedure is much simpler and it produces equally good, if not better, results for the energy of formation of such compounds.

2. Procedure

In I we needed the R_s versus ϱ_s curve for each constituent metal, R_s and ϱ_s being Wigner-Seitz (WS) radius and cell-boundary charge density, respectively. Each R_s and ϱ_s, in turn, had to be calculated via the α-expansion method corresponding to an arbitrary lattice parameter value. Several such R_s- and ϱ_s-values were necessary to draw only one curve. The charge transfer between elements could be calculated with the help of these curves (see Fig. 4 in I) by an appeal to the Hodges and Stott mechanism [5] which may be briefly described as follows:

Step 1: Compress or expand parent metals, preserving their original structure, to a common atomic volume equal to the volume per atom of the alloy.

Step 2: Transform parent metal structures to that of the alloy.

Step 3: 'Cut' them up into their constituent cells and fit them together to form the alloy. While doing this, one notices that the charge density is discontinuous at the boundary and chemical potentials in the two cells are different.

Step 4: Allow the charge density to relax to its equilibrium distribution so that it is continuous across cell boundaries and hence chemical potentials equalize.

In the present procedure we skip all these four steps. We just place three shells of free atoms in their respective positions in accordance with the observed structure of the alloy and carry out the α-expansion as described in I or earlier [11] to find the equilibrium charge density distribution in the alloy with the help of the relation

$$\Psi(a \mid r) = \frac{1}{2ar} \int\limits_{|a-r|}^{a+r} R\Psi(R)\, \mathrm{d}R\,, \tag{1}$$

where $\Psi(R)$ is a function centred around a certain origin and $\Psi(a \mid r)$ is its transform around another origin, the two being separated by a distance a. From this equilibrium charge density distribution, the equal-volume postulate of alloy components helps to find out the amount of charge transfer between components. The radii of neutral cells in the alloy are calculated by an appropriate volume integration of the charge density. It is obvious that there will be no discontinuity of the charge density across the cell boundary in this procedure.

Amounts of charge transfer, ΔQ, radii of neutral cells in the alloy, R_A^0 and R_B^0, and cell boundary charge densities, ϱ_s, thus obtained for eleven alloys, are presented in Table 1 along with the same values reported in I.

The energy of formation of the alloy is calculated by following the method as outlined in I. The justification for doing so is that although we have skipped the four steps to find out ΔQ here, indeed all the four steps are tacitly involved in the forma-

Table 1

Charge transfer, ΔQ (in fractions of e), radii of neutral cells in the alloy, R_A^0 and R_B^0 (in at. units) and cell boundary charge densities, ϱ_s (in e/(at. units)^3) in the alloy. Charge is transferred from A to B elements

compound (AB)	ΔQ (A → B)		R_A^0		R_B^0		ϱ_s	
	I	present	I	present	I	present	I	present
ZnCu	0.16	0.19	2.820	2.808	2.670	2.688	0.0232	0.0247
ScCo	0.22	0.30	3.012	3.040	2.855	2.812	0.0240	0.0240
ScNi	0.445	0.34	3.140	3.070	2.730	2.830	0.0200	0.0240
ScCu	0.45	0.68	3.220	3.296	2.810	2.762	0.0186	0.0193
ScRh	0.0064	0.14	2.988	3.080	2.970	2.886	0.0242	0.0250
ScAg	0.23	0.40	3.280	3.340	3.065	3.010	0.0171	0.0180
TiFe	0.085	0.20	2.799	2.834	2.731	2.704	0.0315	0.0315
TiNi	0.46	0.31	2.978	2.910	2.620	2.700	0.0245	0.0290
TiCo	0.21	0.25	2.852	2.870	2.730	2.696	0.0297	0.0300
VFe	0.07	0.16	2.735	2.750	2.680	2.546	0.0340	0.0340
FeCo	0.03	0.026	2.671	2.660	2.640	2.656	0.0345	0.0350

tion of the alloy from the pure metals. Therefore, the energy of alloy formation, ΔH, comprises three contributions: elastic, E_{el}, electrochemical, E_{ec}, and Madelung, E_M. E_{el} was estimated via the Morse function representation of the crystal energy as

$$E_{el} = \frac{1}{2} \sum_k \Delta E_k; \qquad k = \mathrm{A, B}\,, \tag{2}$$

where

$$\Delta E_k = \frac{1}{2} D_k \{1 + [e^{-2\alpha_k(R_k - R_k^0)} - 2\,e^{-\alpha_k(R_k - R_k^0)}]\}\,, \tag{3}$$

R_k, α_k, and D_k are WS radius, repulsion parameter, and cohesive energy per atom, and R_k^0 is the equilibrium neutral cell radius of the k-th component of the alloy AB. Values of R_k, α_k, and D_k were taken from Girifalco and Alonso [12].

E_{ec} was calculated from the following relation of Hodges and Stott [5]:

$$E_{ec} = -\frac{1}{4} \Delta\mu\, \Delta Q\,, \tag{4}$$

where $\Delta\mu$ is the difference in chemical potential between A and B components. $\Delta\mu$'s were estimated from μ-values given by Alonso and Girifalco [13].

E_M was calculated from

$$E_M = -\alpha_M \frac{|\Delta Q|^2}{a}\,, \tag{5}$$

where the Madelung constant, α_M, equals 2.0353618 for the CsCl structure to which these compounds belong and a is the lattice constant of the compound. a-values were taken from Nevit [14].

Our calculated values of ΔH, along with those of others and experimental values, have been presented in Table 2. It may be seen that our present values compare favourably with experimental ones where they are available or with semiempirical values obtained from Miedema's [15] relation. Although it is not expected that this type of calculation will produce quantitative results, one notices that the present method generates all the values with correct sign. This is rather gratifying because other calculations failed even to predict the correct sign for ΔH for a few systems.

Table 2
Energy of formation (in Ryd/atom)

compound	Alonso and Girifalco [8]	Girifalco and Alonso [12]	Miedema [15]	experiment [16]	I [9]	this work
ZnCu	−0.0100	+0.0002		−0.0092	−0.0122	−0.0162
ScCo	−0.0069	−0.0490	−0.0320		−0.0168	−0.0435
ScNi	−0.0118		−0.0418		−0.0707	−0.0525
ScCu	−0.01555	−0.0527	−0.0328		−0.0917	−0.1945
ScRh	+0.0015		−0.0642		+0.0138	−0.0174
ScAg	+0.0056		−0.0380		−0.0336	−0.0831
TiFe	−0.0017	−0.0166	−0.0214	−0.0154	−0.0014	−0.0204
TiNi	−0.0156		−0.0459	−0.0258	−0.1102	−0.0518
TiCo	−0.0083		−0.0379		−0.0247	−0.0365
VFe	−0.0231		−0.0086	−0.0064	−0.0035	−0.0149
FeCo	+0.0004		−0.0007		−0.0007	−0.0005

3. Concluding Remarks

From the results presented in this paper, as well as those presented in I, we may conclude that the α-expansion technique as applied here or in the earlier work is a satisfactory, simple procedure to study the formation of an alloy. Of course, of the two procedures proposed by us, the present one is a lot simpler and involves much less computation time.

References

[1] D. G. Pettifor, Phys. Rev. Letters **42**, 846 (1979).
D. G. Pettifor and C. M. Verma, J. Phys. C **12**, L253 (1979).
J. van der Rest, F. Gautier, and F. Brouers, J. Phys. F **5**, 2283 (1975).
[2] J. E. Inglesfield, J. Phys. C **2**, 1285, 1293 (1969); Acta metall. **17**, 1395 (1969), J. Phys. F **2**, 878 (1972).
[3] V. Moruzzi, A. Williams, and J. Janak, Phys. Rev. B **10**, 4856 (1974).
[4] P. Hohenberg and W. Kohn, Phys. Rev. **136**, B 864 (1964).
[5] C. H. Hodges and M. J. Stott, Phil. Mag. **26**, 375 (1972).
[6] V. K. Ratti and J. M. Ziman, J. Phys. F **4**, 1684 (1974).
[7] D. M. Cragg and G. C. Fletcher, J. Phys. F **7**, 87 (1977).
[8] J. A. Alonso and L. A. Girifalco, J. Phys. Chem. Solids **38**, 869 (1977).
[9] S. Chakraborty, A. Manna, and A. K. Ghosh, phys. stat. sol. (b) **122**, 469 (1984).
[10] P. O. Löwdin, Adv. Phys. **5**, 1 (1956).
[11] A. K. Ghosh, S. Chakraborty, and A. Manna, phys. stat. sol. (b) **118**, 373 (1983).
[12] L. A. Girifalco, and J. A. Alonso, Theory of Alloy Phase Formation, Ed. L. H. Benett, The Metallurgical Society of AIME, New York 1979.
[13] J. A. Alonso and L. A. Girifalco, Phys. Rev. B **19**, 3889 (1979).
[14] M. Nevit, Electronic Structure and Alloy Chemistry of Transition Elements, Ed. P. Beck, Wiley, New York 1963.
[15] A. Miedema, J. less-common Metals **46**, 67 (1976).
[16] R. Hultgren, P. Desai, D. Hawkins, M. Gleiser, and K. Kelly, Selected Values of Thermodynamic Properties of Binary Alloys, A.S.M., New York 1973.

(Received March 26, 1984)

phys. stat. sol. (b) **125**, 101 (1984)

Subject classification: 4 and 13.1; 21.1; 21.4

Laboratoire de Thermodynamique et Physico-Chimie Métallurgiques, Ecole Nationale Supérieure d'Electrochimie et d'Electrométallurgie de Grenoble, Saint Martin d'Heres[1]) *(a)*
and Chimie Métallurgique des Terres Rares, C.N.R.S., Meudon[2]) *(b)*

A Theoretical and Experimental Study of the Enthalpies of Formation of $LaNi_5$-Type Compounds

By

A. PASTUREL (a), C. COLINET (a), C. ALLIBERT (a), P. HICTER (a), A. PERCHERON-GUEGAN (b), and J. C. ACHARD (b)

The heats of formation of $LaNi_5$-type compounds are computed within a tight-binding scheme for the d-band. It is shown that the relatively large difference between the energy levels of the two elements is a dominant factor in the determination of the formation energy. The evolution of the energy level of the rare-earth metal can explain the experimental behaviour of the heats of formation of the RNi_5 compounds along the rare-earth series. In order to discuss the calculated results the enthalpies of formation of $LaNi_5$, $PrNi_5$, $SmNi_5$, and YNi_5 are determined by a calorimetric method.

Die Bildungswärmen von $LaNi_5$-Verbindungen werden numerisch innerhalb eines "tight-binding"-Modells für das d-Band berechnet. Es wird gezeigt, daß die relativ große Differenz zwischen den Energieniveaus der beiden Elemente ein dominierender Faktor bei der Bestimmung der Bildungsenergie ist. Die Entwicklung des Energieniveaus des Seltenerd-Metalls kann das experimentelle Verhalten der Bildungswärmen der RNi_5-Verbindungen in der Serie der Seltenen Erden erklären. Um die berechneten Ergebnisse zu diskutieren, werden die Bildungsenthalpien von $LaNi_5$, $PrNi_5$, $SmNi_5$ und YNi_5 mittels kalorimetrischen Methoden bestimmt.

1. Introduction

It is now well established that the cohesive energy of the pure transition metal constituents arises from the strong bonding of the valence d-electrons which are well described by the tight-binding approximation [1]. With an identical scheme, predictions of the heats of formation of transition alloy have been the object of considerable attention recently; van der Rest et al. [2] used the CPA with off-diagonal disorder to compute the alloy density of states while Cyrot and Cyrot-Lackmann [3] used a simple model for this density, based on an expansion in its moments. More recently, Pettifor [4] and Varma [5] have applied Friedel's rectangular d-band density-of-states scheme [1] to the alloy problem and almost identical ΔH to those obtained from the more detailed CPA calculations were found. Pasturel et al. [6] have shown that charge transfer and heats of formation in these alloys can be easily calculated with moments technics and with the method of the continued fraction developed to the first step.

It is therefore tempting to apply this kind of tight-binding approach to calculate the heats of formation of transition–rare-earth metal alloys. The problem is how the rare-earth metal can be treated similarly to a transition metal in the case of a calcula-

[1]) B.P. 75, 38402 Saint Martin d'Hères Cédex, France.
[2]) 1 Place A. Briand, 92190 Meudon, France.

tion of formation energy. On the one hand, many physical properties of rare-earth metals are similar to those of transition metals. Band calculations have shown that the band structures of rare-earth metals consist of a wide s–p band together with a narrow d-band. Dimmock et al. [7] have evaluated the width of the 5d band of gadolinium and generally, for pure rare-earth metals, the density of states is much like that of transition metals. The results of Johansson [8] and Duthie and Pettifor [9] indicate that between 1.5 and 2.5 electrons should be attributed to rare-earth metal. On the other hand, the effect of the additional d–f interaction between f- and d-electron spins is only to split the up- and down-spin bands; for the calculation of cohesive energy, this effect is negligible since we use an integral value of the density of states.

After recalling and developing some theoretical points, we present our experimental data and will discuss our calculated values in comparison with these data.

2. Theoretical Aspects

We have seen that the cohesive energy of transition metals is essentially due to the band energy of d-electrons; similarly, it is possible to predict the general behaviour of the heats of formation of transitional alloys from the change in energy associated with the deformation of the d-band upon alloying.

Neglecting the contributions to ΔH arising from changes in magnetic interactions, the heat of formation can be written in the Hartree-Fock approximation as

$$\Delta H = \Delta H_{\text{H.F.}} - \Delta H_{\text{el-el}} , \tag{1}$$

where $\Delta H_{\text{H.F.}}$ is the change of the Hartree-Fock energy and $\Delta H_{\text{el-el}}$ the change of the electron–electron interaction energy which has been counted twice in $\Delta H_{\text{H.F.}}$. For concentrated disordered alloys we neglect the change in the interatomic electron–electron interactions which is much smaller than the change of the intraatomic one,

$$\Delta_{\text{el-el}} H = \Delta H_{\text{el-el}}^{\text{intra}} . \tag{2}$$

In the presence of charge transfer, we have to define new effective atomic energy levels E_i,

$$E_i = E_i^0 + U\,\Delta N_i , \tag{3}$$

where $\Delta N_i = N_i - N_i^0$ is the charge transfer (N_i^0 and N_i are the number of d-electrons in the pure metal i and in the alloy, respectively) and E_i^0 the atomic energy level in pure metal i. U is some effective intraatomic Coulomb integral [6], assumed to be identical for the two metals, it will be taken to be equal to 4 eV.

If x_i is the i-composition in the alloy, the change of the electron–electron intraatomic energy is given by

$$\Delta H_{\text{el-el}}^{\text{intra}} = \sum_{i=\text{A,B}} x_i U (N_i^2 - N_i^{02}) \tag{4}$$

and the change of one-electron energies by

$$\Delta H_{\text{H.F.}} = \int^{E_{\text{F}}} E n(E)\,\mathrm{d}E - \sum_{i=\text{A,B}} x_i \int^{E_{\text{F}i}} E n_i^0(E)\,\mathrm{d}E , \tag{5}$$

where $n_i^0(E)$ and $E_{\text{F}i}$ are the density of states per atom and the Fermi level in the pure i metal, respectively, $n(E)$ and E_{F} the same quantities for the alloy.

With the approximation of the continued fraction developed to the first step, the density of states and the d-band contribution to the cohesive energy of the i metal can be written as

$$n_i^0(E) = \frac{1}{\pi \sqrt{b_i^0}} \left[1 - \left(\frac{E - E_i^0}{2\sqrt{b_i^0}}\right)^2\right]^{1/2}, \tag{6}$$

$$E_{B,i} = \int^{E_{Fi}} (E - E_i^0)\, n_i^0(E)\, dE = -\frac{40\sqrt{b_i^0}}{3\pi} [1 - (A_i^0)^2]^{3/2}. \tag{7}$$

b_i^0 is defined from the first two moments of the density of states of i metal according the following relation:

$$b_i^0 = \mu_{2,i}^0 - (\mu_{1,i}^0)^2. \tag{8}$$

A_i^0 is related to the number of d-electrons of i metal by the following expression:

$$N_i^0 = 10\left\{\frac{1}{2} + \frac{1}{\pi}\sin^{-1} A_i^0 + \frac{A_i^0}{\pi}[1 - (A_i^0)^2]^{1/2}\right\} \tag{9}$$

with

$$A_i^0 = \frac{E_{Fi} - E_i^0}{2\sqrt{b_i^0}}. \tag{10}$$

The pure i metal is thus characterized by its number of d-electrons, N_i^0, the atomic energy level, E_i^0, and by b_i^0 which is proportional to its bandwidth, W_i^0 [3].

For the alloy, the total density of d-states is deduced from the local density of d-states of i metal, which can be also determined as previously by knowing the parameters A_i and b_i. An equation identical to (9) gives N_i as a function of A_i. For the determination of b_i, we use also a relation identical to (8); the two moments of the local density of i metal are

$$\mu_{1,i} = E_i, \qquad \mu_{2,i} = E_i^2 + Zx_i t_{ii}^2 + Z(1 - x_i)\, t_{ij}^2, \tag{11}$$

where Z is the coordination number and t_{ii}, t_{ij} are the transfer integrals between two identical atoms or two different atoms.

In Shiba's approximation [10] t_{ij} is given by

$$t_{ij} = \sqrt{t_{ii} t_{jj}}, \tag{12}$$

where t_{ii} and t_{ij} are related to the bandwidths of i and j metals [3].

The charge transfer calculated from equations identical to (10) can be different from the one deduced from the effective level (see (3)) and it is necessary to have a self-consistent calculation to equalize these two transfers; the two self-consistency equations necessary to this calculation are the following:

$$E_i - E_j = E_i^0 - E_j^0 + U\,\Delta N_i - U\,\Delta N_j, \tag{13}$$

$$E_i - E_j = 2\sqrt{b_j}\, A_j - 2\sqrt{b_i}\, A_i. \tag{14}$$

Table 1

Electronic characteristics of the elements

	N_i^0	E_i^0 (eV)	W_i^0 (eV)
Ni	9.3	−15.2	4.7
La	1.9	− 6.2	7
Pr	1.9	− 6.1	7
Sm	1.9	− 6	7
Y	1.9	− 5.6	7

As an input to calculate the enthalpy of formation of a disordered binary alloy according (1), (4), and (5), the values of E_i^0, W_i as well as the number of d-electrons are required. The atomic energy levels E_i^0 are obtained from Hermann and Skillman's tables [11]. The nickel bandwidth is obtained from the renormalized atom method [12]. For rare-earth metals, we will assume an average number of 1.9 d-electrons and the same bandwidth as given by Dimmock et al.'s calculations [7]. All these parameters are collected in Table 1.

3. Experimental

The crystalline samples used in this investigation were prepared by means of induction melting of the appropriate amounts of pure elements. The samples were checked by metallographic examination, micropobe analysis, and X-ray diffraction. The nominal composition of the different compounds and their lattice parameters are reported in Table 2.

The enthalpy of formation of the RNi_5 compound corresponds to the reaction

$$\langle R\rangle_{T_0} + 5\langle Ni\rangle_{T_0} \rightarrow \langle RNi_5\rangle_{T_0} .$$

Its value is deduced from the dissolution enthalpies of the compound and its components in liquid aluminium by using the relation [13]

$$\Delta H_f^{T_0}\langle RNi_5\rangle = Q_R^\infty + 5Q_{Ni}^\infty - Q_{RNi_5}^\infty ,$$

where Q_X^∞ (X = R, Ni, RNi_5) is the heat of dissolution at infinite dilution of the elements or of the compound. The enthalpies of dissolution of RNi_5 in liquid aluminium are obtained using the previously described calorimetric method [13, 14]. The values of the formation enthalpies ΔH_f^{298} obtained in this way are collected in Table 3. Note that the values are expressed per mol of atoms.

4. Discussion

As seen in Section 2, the determination of the heats of formation from (1) requires the knowledge of the bandwidths of the pure metals, W_i^0, the relative position of the atomic energy levels, E_i^0, and the number of d-electrons, N_i^0. Using the parameters given in Table 1, the values of the heats of formation of the RNi_5 compounds are

Table 2
Structural characteristics of RNi_5 compounds

compound	lattice parameter (nm)
$La_{0.987}Ni_{5.026}$	$a = 0.5016$ $c = 0.3987$
$Pr_{1.004}Ni_{4.992}$	$a = 0.4964$ $c = 0.3983$
$Sm_{1.04}Ni_{4.972}$	$a = 0.4926$ $c = 0.3980$
$Y_{0.99}Ni_{5.01}$	$a = 0.4882$ $c = 0.3959$

Table 3

Experimental enthalpies of formation of RNi_5 compounds

compound	T (K)	number of runs	Q (kJ/mol)	ΔH_f (kJ/mol of atoms)	ref.
Ni	948	10	−119.6		[15]
Y	1000	7	−138.5		[16]
La	931	10	−164.4		[15]
Pr*)			−155.9		[16]
Sm	966	10	−145.2		[16]
$LaNi_5$	987	8	−605.3	−26.5	
$PrNi_5$	970	9	−589	−26.8	
$SmNi_5$	987	9	−548	−30.3	
YNi_5	995	9	−525	−34.1	

*) Obtained by interpolation.

calculated. They are compared with the experimental results in Table 4. We can see from this comparison that the agreement is satisfying between experimental and calculated values. However, we must point out that the use of only two moments of the density of states does not allow us to take into account the crystalline structure of these compounds strictly. On this point two remarks must be made: in these alloys the general feature of the density of states is given by the chemical order and is not due to the topological disorder [17]; the second point is that we compare compounds which have the same structure. The evolution of the energy level of the rare-earth metal, as it is shown on Fig. 1, seems to reproduce well the evolution of the enthalpies of formation of RNi_5 compounds along the rare-earth series, i.e. that the enthalpies of formation of these compounds are less exothermic at the beginning of the series. More generally, we can predict a similar behaviour for the enthalpies of formation of the rare earth 3d alloys (for a given 3d metal) since the energy level of the rare-earth metal is a parameter independent of the composition of the alloy.

Although our model is too crude to study quantitatively either valence state of the rare-earth metal or magnetism in these compounds, it renders an account of their general physical properties. The rare-earth (or Y) atomic energy level is always higher than the Ni one and the electron transfer occurs from the rare-earth to the transition metal. In these Ni-rich alloys, the local density of states on the transition atom is much like that in pure metal and on the contrary the local density of states on the rare-earth atom is much lower and wider than in pure metals. This remark allows us to study qualitatively two facts:

Table 4

Comparison between calculated and experimental enthalpies of formation

compound	ΔH_f (exp.) (kJ/mol of atoms)	ΔH_f (calc.)
$LaNi_5$	−26.5	−26.1
$PrNi_5$	−26.8	−26.8
$SmNi_5$	−30.3	−32.3
YNi_5	−34.1	−35.2

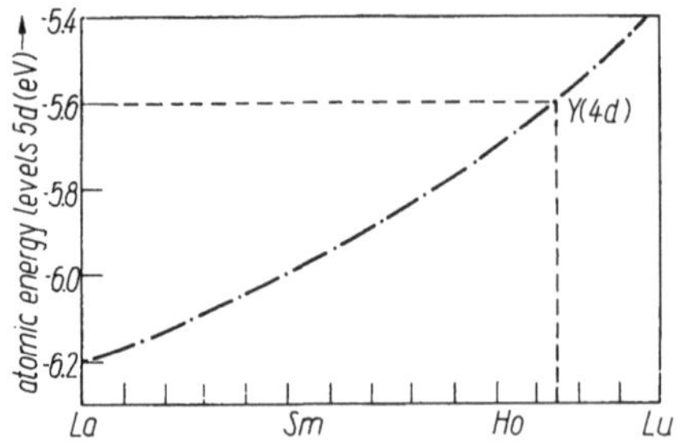

Fig. 1. Evolution of the atomic energy level of the rare-earth metals

For RNi_5 compounds, the charge transfer is large enough such that the Ni d-band is nearly full and the Fermi level lies in the part of the density of states mainly due to the rare-earth atom. The density of states at the Fermi level is small and Ni is never magnetic. In the same way, we can also explain the valence state of cerium in the $CeNi_5$ compound. We show experimentally [18] that the enthalpy of formation of $CeNi_5$ compound has an abnormal, highly negative value compared with the other ones; considering always the possibility of a charge transfer, 5d electrons of the cerium are transferred to the 3d band of Ni, lowering the Fermi level E_F of the compound compared to the Fermi level of the rare-earth metal. If the $4f^1$ level is higher than E_F, Ce becomes tetravalent.

References

[1] J. FRIEDEL, The Physics of Metals, Chap. 8, Ed. J. M. ZIMAN, Cambridge University Press, London 1969.
[2] J. VAN DER REST, F. GAUTIER, and F. BROUERS, J. Phys. F **5**, 2283 (1975).
[3] M. CYROT and F. CYROT-LACKMANN, J. Phys. F **6**, 2257 (1976).
[4] D. G. PETTIFOR, Solid State Commun. **28**, 621 (1978).
[5] C. M. VARMA, Solid State Commun. **31**, 295 (1979).
[6] A. PASTUREL, P. HICTER, and F. CYROT-LACKMANN, Solid State Commun. **48**, 561 (1983).
[7] J. O. DIMMOCK, A. J. FREEMAN, and R. E. WATSON, Phys. Rev. **167**, 497 (1968).
[8] B. JOHANSSON, Inst. Phys. Conf. Ser. No. 37, 39 (1978).
[9] J. C. DUTHIE and D. G. PETTIFOR, Phys. Rev. Letters **38**, 564 (1977).
[10] H. SHIBA, Progr. theor. Phys. (Kyoto) **46**, 77 (1971).
[11] F. HERMAN and S. SKILLMAN, Atomic Structure Calculations, Prentice-Hall, Inc., New York 1963.
[12] L. HODGES, R. E. WATSON, and H. EHRENREICH, Phys. Rev. B **5**, 3953 (1972).
[13] C. CHATILLON-COLINET and J. C. MATHIEU, Rapp. LTPCM, ENSEEG, Saint Martin d'Hères (France) 1979 (TMO1).
[14] M. JEYMOND, D. LANDAUD, M. LEGARDEUR, and A. PASTUREL, Thermochim. Acta **55**, 301 (1981).
[15] H. DIAZ, A. PERCHERON-GUEGAN, J. C. ACHARD, C. CHATILLON-COLINET, and J. C. MATHIEU, Internat. J. Hydrogen Energy **4**, 445 (1979).
[16] A. PASTUREL, C. CHATILLON-COLINET, A. PERCHERON-GUEGAN, and J. C. ACHARD, J. less-common Metals **90**, 21 (1983).
[17] F. CYROT-LACKMANN, J. non-crystall. Solids **61/62**, 1030 (1984).
[18] C. COLINET and A. PASTUREL, phys. stat. sol. (a) **80**, K75 (1983).

(Received May 23, 1984)

phys. stat. sol. (b) 125, 107 (1984)

Subject classification: 6; 12.1; 21

Department of Applied Physics and Mathematics, Mining College, Akita University[1])

Pressure Effect on Phonon Frequencies and Elastic Stiffness Constants of Al

By

T. SOMA, T. ITOH, and H.-MATSUO KAGAYA

The lattice dynamics of Al under pressure is investigated using the local Heine-Abarenkov model potential with volume-dependent parameters in the perturbational method. Then, using the homogeneous deformation method, the bulk modulus and the shear elastic constants under compression are studied theoretically, and the mode Grüneisen parameters and the pressure-derivatives of the elastic constants for Al are obtained.

Die Gitterdynamik des Aluminium unter Druck wird mit einem lokalen Heine-Abarenkov-Modell-potential mit volumen-abhängigen Parametern in Störungsrechnung behandelt. Die Volumen-Elastizitätsmoduli und die Schermoduli bei Kompression werden dann mittels der Methode der homogenen Deformation untersucht, und die Grüneisenparameter und die Druck-Differential-koeffizienten der Elastizitätsmoduln für Aluminium berechnet.

1. Introduction

The phonon frequencies of Al have been experimentally determined using the technique of coherent inelastic neutron scattering [1]. Recently, we [2] presented the local Heine-Abarenkov model potential with volume-dependent parameters for Al, and our model potential reproduced the experimental data of the static properties such as the cohesive energy, the equation of state and, the bulk modulus [3]. Then, two parameters of our model potential were determined by satisfying the zero-pressure condition and the first zero of the empirical pseudopotential interpolated from band calculations, but we have no fitting parameter to the observed phonon dispersion curves and elastic constants themselves. Recently we [4] applied our treatment to the study of the lattice dynamics of Al. The experimentally observed dispersion curves were reproduced using second-order perturbational treatment and our local Heine-Abarenkov model potential. Then, the effect of different forms of exchange and correlation on the phonon frequency, the compressibility problem for long-wave phonons and the contribution from the three-body interaction were also discussed in detail.

In the present work, using the dynamical treatment from first principles, we study the pressure effect on the phonon frequency of Al and obtain also the mode Grüneisen parameters in Section 2. Then, using the homogeneous deformation method, we investigate the compression effect on the elastic stiffness constants, and obtain also the pressure derivatives of the elastic constants and the mode Grüneisen parameters for long-wave phonons in Section 3.

2. Lattice Dynamics under Pressure and Mode Grüneisen Parameter

We adopt the same formulations as our previous work [4, 5] to calculate the phonon curves. The phonon dispersion curves $\nu(\boldsymbol{q})$ are calculated numerically in dependence

[1]) Akita 010, Japan.

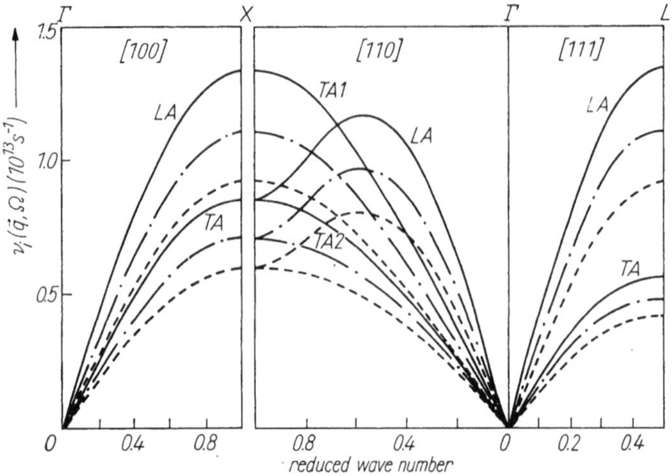

Fig. 1. Phonon dispersion curves of Al along the symmetry directions under pressure. Full and dashed-dotted curves correspond to the compressed volume $\Omega/\Omega_0 = 0.8$ and 0.9. The dashed lines show the previous ones at atmospheric pressure $\Omega/\Omega_0 = 1$ [4]

on the compressed volume Ω/Ω_0, where Ω_0 is the crystal volume at atmospheric pressure. The calculated dispersion curves for Al against the reduced wave number along the symmetry directions [100], [110], and [111] using our local Heine-Abarenkov model potential with Vashishta-Singwi screening are shown in Fig. 1. In this figure, full and dash-dotted curves represent those at the representative compressed volume $\Omega/\Omega_0 = 0.8$ and 0.9, and the dashed curves the previous ones at atmospheric pressure $\Omega/\Omega_0 = 1$ [4]. Then we show the compression effect on the characteristic frequencies at the representative points X and L in Fig. 2. In this figure, full, dashed, dash-dotted, and double dash-dotted curves represent ν_{LA}, ν_{TA} at X and ν_{LA}, ν_{TA} at L, respectively. Then, the phonon frequency $\nu_i(\Omega)$ normalised to the corresponding ones under atmospheric pressure $\nu_i(\Omega_0)$ are given. From Fig. 1 and 2, we see that the large shift of phonon frequencies for both the longitudinal and the transverse modes is notable for Al. The variation width of phonon frequency using different forms [2, 4] of exchange and correlation in Fig. 1 and 2 are within 3% at the compressed volume $\Omega/\Omega_0 = 0.8$.

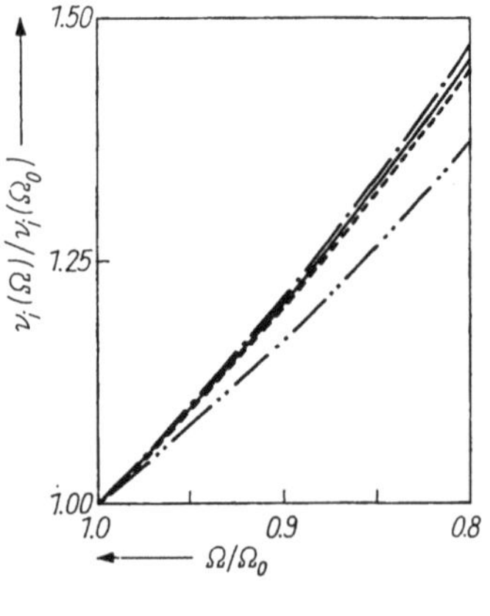

Fig. 2. Volume dependence of characteristic phonon frequencies for Al. Full, dashed, dash-dotted, and double dash-dotted curves correspond to ν_{LA}, ν_{TA} at X and ν_{LA}, ν_{TA} at L, respectively

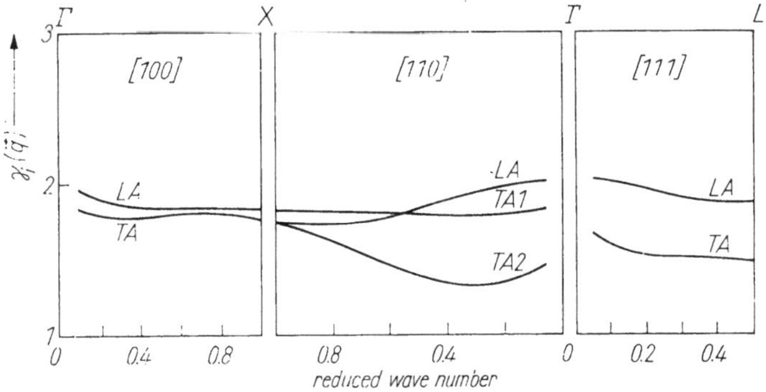

Fig. 3. Mode Grüneisen parameters of Al along the symmetry directions

The mode Grüneisen parameter $\gamma_i(\boldsymbol{q})$ for the i-th mode with phonon frequency $\nu_i(\boldsymbol{q})$ is a measure of the volume (Ω) dependence and is defined by

$$\gamma_i(\boldsymbol{q}) = -\frac{\mathrm{d}[\ln \nu_i(\boldsymbol{q})]}{\mathrm{d}[\ln \Omega]}. \tag{1}$$

We show the obtained results of $\gamma_i(\boldsymbol{q})$ along the symmetry directions [100], [110], and [111] in Fig. 3, where the numerical results with Vashishta-Singwi screening function are given. Then, we show in Table 1 the numerical results of $\gamma_i(\boldsymbol{q})$ using other screening functions ([2], [4]) at the symmetry points X and L. In the long-wavelength region in Fig. 3, we cannot determine the corresponding mode Grüneisen parameter because of the very small shift of phonon frequency.

3. Elastic Constants under Pressure and Long-Wave Mode Grüneisen Parameter

Elastic constants can be determined by two methods, such as the static treatment (the homogeneous deformation method) and the dynamical treatment (the long-wave phonon-velocity method). Brovman and Kagan [6] solved clearly that the former was more favorable as compared with the latter. We [4] obtained the influence of this compressibility problem by Brovman and Kagan to be important for Al, and adopt the static treatment in order to study the compression effect on the elastic constants of Al.

Table 1

Screening effect referred to as H, K-L, S-S-T-L, and V-S [2 to 4] on the mode Grüneisen parameter of Al at symmetry points X and L

		H	K-L	S-S-T-L	V-S
X	γ_{LA}	1.596	1.717	1.816	1.828
	γ_{TA}	1.534	1.693	1.748	1.751
L	γ_{LA}	1.629	1.737	1.850	1.866
	γ_{TA}	1.362	1.468	1.483	1.478

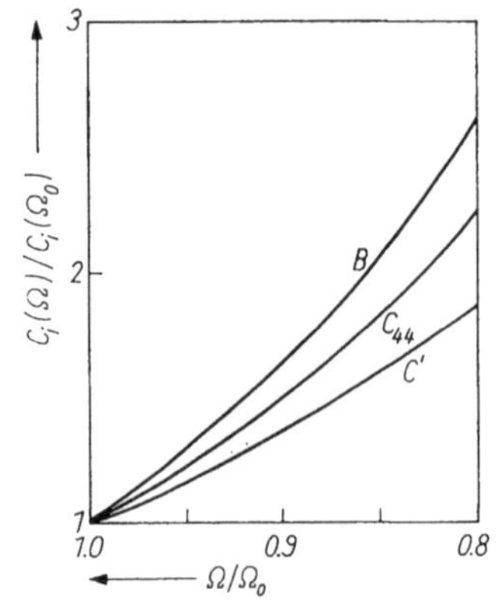

Fig. 4. The normalized elastic constants $C_i(\Omega)/C_i(\Omega_0)$ of Al as a function of the volume strain

Three independent elastic stiffness constants of cubic crystals are given by considering the following homogeneous deformations. First, considering uniform volume expansion denoted by $R_x' = v^{1/3}R_x$, $R_y' = v^{1/3}R_y$, and $R_z' = v^{1/3}R_z$, the bulk modulus B is given by

$$B = \frac{1}{\Omega}\frac{\mathrm{d}^2E}{\mathrm{d}v^2} = \Omega\frac{\mathrm{d}^2E}{\mathrm{d}\Omega^2}. \tag{2}$$

In (2), E is the crystal energy of Al [2] and the partial variations of the crystal energy E with respect to crystal volume Ω, reciprocal lattice vectors $\boldsymbol{G}$, and electron density n is given by

$$\frac{\mathrm{d}}{\mathrm{d}\Omega} = \left(\frac{\partial}{\partial\Omega}\right)_{\boldsymbol{G},n} - \frac{G_\alpha}{3\Omega}\left(\frac{\partial}{\partial G_\alpha}\right)_{\Omega,n} - \frac{n}{\Omega}\left(\frac{\partial}{\partial n}\right)_{\Omega,\boldsymbol{G}}. \tag{3}$$

Next, considering shear deformation in one plane, denoted by $R_x' = R_x + \gamma_1 R_y$, $R_y' = R_y$, and $R_z' = R_z$, where the crystal volume is taken to remain constant, the shear modulus C_{44} is obtained by

$$C_{44} = \frac{1}{\Omega}\frac{\mathrm{d}^2E}{\mathrm{d}\gamma_1^2}. \tag{4}$$

Thirdly, considering compression in one direction, denoted by $R_x' = (1 + \varepsilon_1)\, R_x'$, $R_y' = R_y/(1 + \varepsilon_1)$, and $R_z' = R_z$, where the crystal volume is also maintained constant, the shear modulus C' is obtained by

$$C' = \frac{1}{4\Omega}\frac{\mathrm{d}^2E}{\mathrm{d}\varepsilon_1^2}. \tag{5}$$

Table 2

The pressure derivatives of elastic stiffness constants of Al near zero pressure

	H	K-L	S-S-T-L	V-S	obs.*)
$\mathrm{d}B/\mathrm{d}P$	4.85	5.11	5.14	5.12	4.74
$\mathrm{d}C_{44}/\mathrm{d}P$	2.56	2.73	3.02	2.83	2.25
$\mathrm{d}C'/\mathrm{d}P$	1.18	1.27	1.43	1.38	1.65

*) Ho and Ruoff [7].

Table 3

The mode Grüneisen parameters $\gamma_i(\boldsymbol{q} \to \boldsymbol{0})$ of Al for long-wave phonons along [100], [110], and [111] directions

		H	K-L	S-S-T-L	V-S	obs.*), **)
[100]	LA	2.14	2.18	2.28	2.21	2.24
	TA	1.75	1.94	2.16	1.97	2.66
[110]	LA	2.04	2.17	2.29	2.19	2.33
	TA1	1.75	1.94	2.16	1.97	2.66
	TA2	1.84	1.71	1.98	1.82	2.33
[111]	LA	2.02	2.18	2.29	2.19	2.36
	TA	1.80	1.82	2.07	1.91	2.46

*) Ho and Ruoff [7].
**) Kamm and Alers [8].

The numerical results with Vashishta-Singwi screening for the compression effect on the elastic constants of Al are shown in Fig. 4. In this figure, we introduce the normalized elastic constants $C_i(\Omega)/C_i(\Omega_0)$, where $C_i(\Omega_0)$ is the value at atmospheric pressure, in order to isolate the compression effect. The variation width of the obtained results using different forms of exchange and correlation in Fig. 4 are within 5% at the compressed volume $\Omega/\Omega_0 = 0.8$. From Fig. 4, we see that all of B, C_{44}, and C' increase as the crystal is compressed.

The pressure derivatives of the elastic stiffness constants $\mathrm{d}C_i/\mathrm{d}P$ ($C_i = B, C_{44}, C'$) at low temperatures are given by

$$\frac{\mathrm{d}C_i}{\mathrm{d}P} = -\frac{\Omega}{B}\frac{\mathrm{d}C_i}{\mathrm{d}\Omega}. \tag{6}$$

In the long-wavelength limit, we obtain the corresponding mode Grüneisen parameters $\gamma_i(\boldsymbol{q} \to \boldsymbol{0})$ from this pressure derivatives of the elastic constants $\mathrm{d}C_i/\mathrm{d}P$. For instance, in the main symmetry directions such as [100], [110], and [111], we obtain

$$\gamma_i(\boldsymbol{q} \to \boldsymbol{0}) = -\frac{1}{6} + B \sum_i b_i \frac{\mathrm{d}C_i}{\mathrm{d}P} \left(2 \sum_i b_i C_i\right)^{-1}, \tag{7}$$

where b_i, C_i are the linear coefficients and elastic constants for the i-th mode in question. The calculated results for the pressure derivatives of the elastic constants are shown in Table 2, and the good agreement with the observed data [7] is obtained. Then, using (7) and data in Table 2, the mode Grüneisen parameters $\gamma_i(\boldsymbol{q} \to \boldsymbol{0})$ for long-wave phonons along [100], [110], and [111] directions are calculated and the obtained results are given in Table 3. The agreement between the obtained results of $\gamma_i(\boldsymbol{q} \to \boldsymbol{0})$ in Table 3 and the gradual approach from $\gamma_i(\boldsymbol{q})$ in Section 2 (for example, see Fig. 3) is good for TA modes, and poor for LA modes, because the so-called compressibility problem is not applied to the shear modulus, i.e. TA modes for long-wave phonons.

In conclusion, we studied theoretically the compression effect on the lattice dynamics and elastic constants of Al, and our obtained data for the mode Grüneisen parameters are useful in studies of the anharmonicity of polyvalent metal Al.

References

[1] R. STEDMAN and G. NILSON, Phys. Rev. **145**, 492 (1966).
[2] T. SOMA, Y. KONNO, and H.-M. KAGAYA, phys. stat. sol. (b) **117**, 743 (1983).
[3] T. SOMA, Y. KONNO, and H.-M. KAGAYA, phys. stat. sol. (b) **118**, K 155 (1983).
[4] T. SOMA, T. ITOH, and H.-MATSUO KAGAYA, phys. stat. sol. (b) **123**, 463 (1984).
[5] T. SOMA, phys. stat. sol. (b) **99**, 195 (1980).
[6] E. G. BROVMAN and YU. KAGAN, Soviet Phys. – J. exper. theor. Phys. **30**, 721 (1970).
[7] P. S. HO and A. L. RUOFF, J. appl. Phys. **40**, 3151 (1969).
[8] G .N. KAMM and G. ALERS, J. appl. Phys. **35**, 327 (1964).

(Received May 14, 1984)

phys. stat. sol. (b) **125**, 113 (1984)

Subject classification: 4; 6; 12.2; 22; 22.5.3

Condensed Matter Research Centres, Department of Postgraduate Studies and Research in Physics, Rani Durgawati University, Jabalpur[1])

Interionic Forces and Elastic Properties of Na_2S

By

NEETA AGNIHOTRI, R. K. SINGH, and S. P. SANYAL[2])

An interionic potential model is proposed for crystals with antifluorite structure which consists of long-range Coulomb and three-body interactions (TBI), and short-range dispersive and repulsive interactions up to second-neighbour ions. This potential is used to calculate the cohesive energy, second-order elastic constants, and their pressure derivatives of Na_2S. The calculated values of elastic constants and cohesive energy agree well with their measured data which shows that the TBI contribute significantly to these properties.

Es wird ein Interionenpotentialmodell für Kristalle mit Antifluoritstruktur vorgeschlagen, das aus weitreichenden Coulomb- und Dreikörper-Wechselwirkungen (TBI) und kurzreichweitigen dispersiven und repulsiven Wechselwirkungen bis zu den zweiten Ionennachbarn besteht. Dieses Potential wird benutzt, um die Kohäsionsenergie, die elastischen Konstanten zweiter Ordnung und deren Druckableitungen für Na_2S zu berechnen. Die berechneten Werte der elastischen Konstanten und der Kohäsionsenergie stimmen gut mit ihren gemessenen Werten überein, was zeigt, daß die TBI beträchtlich zu diesen Eigenschaften beitragen.

1. Introduction

There are several fluorite and anti-fluorite structure materials with fast ionic conduction which have attracted considerable attention due to their many useful technological applications. One of such materials is sodium sulphide (Na_2S) [1] with antifluorite structure (Fig. 1). This solid remained relatively unexplored until recently when its phonon frequencies and elastic constants were measured by Buhrer and Bill [2] using the neutron scattering technique. As far as the chemical properties are concerned, Na_2S has a strong resemblance with calcium fluorite. However, its physical properties, namely the phonon dispersion curves and interatomic forces show considerable difference [2]. This is so mainly because of the less compact crystal structure of Na_2S (Fig. 1) in comparison to that of CaF_2. Thus, due to the availability of large cavities in the host Na_2S lattice, the formation of Frenkel defects is easier in it than in CaF_2. This leads to significantly different physical properties of Na_2S.

An inspection of the existing literature reveals that very scanty attention has been paid to understand the nature of the interionic forces and elastic properties of solids with antifluorite structure. Buhrer and Bill [2] have, recently, tried to interpret the measured phonon dispersion frequencies and fast ion conduction phenomena in Na_2S using a shell model. The values of their model parameters have been derived by fitting the experimental frequencies at zone boundaries and as such they do not carry any physical significance [3] as remarked by them. Due to this fact the elastic constants derived from this model do not agree with the experimental values. In view of this limitation it appears that their study on lattice dynamics [2] does not provide any useful

[1]) Jabalpur 482001, India.

[2]) Present address: International Centre for Theoretical Physics, P.O. Box No. 586, I-34100, Trieste, Italy.

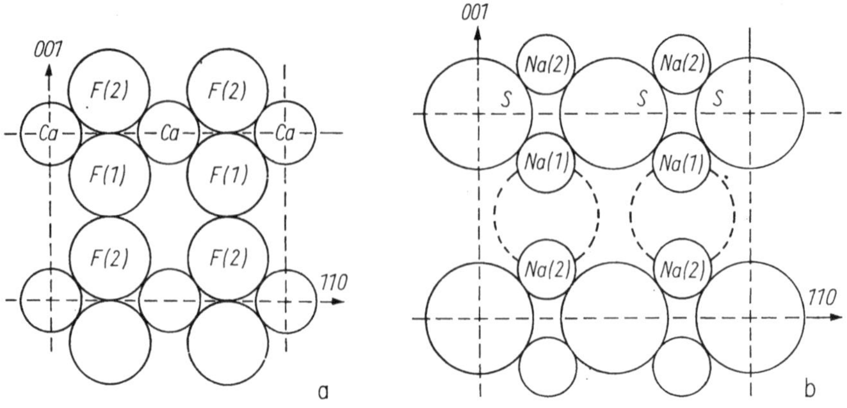

Fig. 1. a) Fluorite and b) antifluorite structures

information about the nature of the interionic forces, bonding characters, etc. of Na_2S. Besides, their model does not take account of the dispersive and three-body forces in their model. While the three-body forces have been found responsible for Cauchy violation ($C_{12} \neq C_{44}$) and various other phenomena in ionic crystals [4, 5] the dispersive forces contribute significantly to the lattice energy.

In some recent papers, Singh and co-workers [6 to 9] have already emphasised the necessity of inclusion of three-body interaction (TBI) effects in crystals with fluorite structure. These TBI arise from the electron shell deformation [4, 5, 9] of an ion and their strength depends on the overlap of the nearest-neighbour ions. These TBI have been found to influence the elastic, dynamical, and allied properties considerably [6, 9] of alkaline earth fluorites. Since the fluorites and antifluorities are isomorphous, it is expected that these TBI will also have considerable effects on the elastic, cohesive, thermodynamical properties of antifluorite crystals. Moreover, the dispersive forces arising from the dipole–dipole and dipole–quadrupole interactions in this solid are also expected to have important effect in the static and cohesive properties because of the high cation polarizability.

In the present paper we have reported an interionic potential for the antifluorite crystals, consisting of the long-range Coulomb and three-body interactions (TBI) and short-range dispersive and repulsive interactions effective up to the second-neighbour ions. This potential has been used to calculate the harmonic (second-order) and anharmonic elastic constants and their pressure derivatives and the cohesive energy of Na_2S. A comparison of the results on these properties have been made with those of CaF_2 to find a discrimination. A brief outline of the present model is described in Section 2. The results have been presented and discussed in Section 3.

2. Theory

2.1 The interionic potential

In the present analysis, we have considered that four types of interionic forces are responsible for cohesion in crystals with antifluorite structure, viz. long-range Coulomb and three-body forces and short-range forces due to dispersive and repulsive interactions. Accordingly, the crystal potential energy can be expressed as

$$W(r) = \sum_{ij}' \frac{Z_i Z_j e^2}{r_{ij}} + \sum_{ijk}' \frac{Z_i Z_j e^2}{r_{ij}} f(r_{ik}) + \sum_{ij} c_{ij} r_{ij}^{-6} + \sum_{ij} d_{ij} r_{ij}^{-8} + W_R(r_{ij}) , \tag{1}$$

where the first term represents the usual Coulomb energy with r_{ij} as the interionic separation. The second term corresponds to the three-body interaction energy with $f(r_{ik})$ as a parameter, which depends on the amount of overlap of two adjacent ions, i and k. Third and fourth terms in (1) represent the van der Waals dispersive forces due to dipole–dipole and dipole–quadrupole interactions, with c_{ij} and d_{ij} as coefficients. The last term is the repulsive interaction energy, expressed as

$$W_R(r) = b[n_1\beta_{ij} \exp\left((r_i + r_j - r_{ij})/p\right) + n_2\beta_{ii} \exp((2r_i - K_1 r_{ij})/p) + \\ + n_2'\beta_{jj} \exp\left((2r_j - K_2 r_{ij})/p\right)] \tag{2}$$

in terms of the parameters b and p and ionic radii r_i and r_j. β_{ij} are the Pauling coefficients; $n_1(= 8)$ and $n_2(= 6)$, $n_2'(= 12)$ are the numbers of nearest unlike (n_1) and like (n_2 and n_2') neighbours. $K_1(= 1.1547)$ and $K_2(= 1.633)$ are the ratio of shortest cation–cation and anion–anion distances to the interionic separation, respectively.

To the best of our knowledge no reliable values of dispersive coefficients (c_{ij} and d_{ij}) for Na_2S are available in the literature, except those obtained by Agarwal et al. [10], based on London's perturbation approach. Their values are, however, inaccurate, due to considerable uncertainties in estimating the excitation energy. We have, therefore, calculated them from the variational method [11, 12] according to which these coefficients are expressed as

$$c_{ij} = \frac{3}{2} \frac{e\hbar}{m^{1/2}} \alpha_i \alpha_j \left[\left(\frac{\alpha_i}{N_i}\right)^{1/2} + \left(\frac{\alpha_j}{N_j}\right)^{1/2}\right]^{-1}, \tag{3a}$$

$$d_{ij} = \frac{27}{8} \frac{\hbar^2}{m} \alpha_i \alpha_j \left[\left(\frac{\alpha_i}{N_i}\right)^{1/2} + \left(\frac{\alpha_j}{N_j}\right)^{1/2}\right]^2 \left[\frac{\alpha_i}{N_i} + \frac{20}{3}\left(\frac{\alpha_i \alpha_j}{N_i N_j}\right)^{1/2} + \left(\frac{\alpha_j}{N_i}\right)\right]^{-1}. \tag{3b}$$

These expressions contain only the ionic polarizabilities (α_i, α_j) and the effective number of electrons of ions (N_i, N_j), accurate values of which can be obtained in literature. The overall values of dispersive interaction coefficients (C and D) can be obtained from

$$C = 2c_{ij}S_R^{(6)}(r) + c_{ii}S_R^{(6)}(0) + c_{jj}S_R^{(6)}(\tfrac{1}{2}a, \tfrac{1}{2}a, \tfrac{1}{2}a) + \tfrac{1}{2}c_{jj}S_R^{(6)}(0) , \tag{4a}$$

$$D = 2d_{ij}T_R^{(8)}(r) + d_{ii}T_R^{(8)}(0) + d_{jj}T_R^{(8)}(\tfrac{1}{2}a, \tfrac{1}{2}a, \tfrac{1}{2}a) + \tfrac{1}{2}d_{jj}T_R^{(8)}(0) \tag{4b}$$

Table 1

Values of van der Waals coefficients, input data, and model parameters of Na_2S

van der Waals coefficients				input data		model parameters	
c_{ij} (10^{-55} Jm^6)		d_{ij} (10^{-67} Jm^8)		properties*	values	parameters	values
c_{+-}	65.4	d_{+-}	44.4	r_0 (nm^{-1})	2.82	p (nm^{-1})	0.311
c_{++}	9.0	d_{++}	1.0	B_T	4.94	b (10^{-17} J)	0.541
c_{--}	640.2	d_{--}	546.6	C_{44}	2.32**)	f_0	−35.720
C	845.9	D	436.0				

*) The values of r and C_{44} and B_T (in units of 10^{10} Nm^{-2}) have been taken from [2].
**) Value used in calculation.

with S and T as the lattice sums [13]. The calculated values of the dispersive coefficients are listed in Table 1.

2.2 The elastic constants

In order to obtain the expressions for the second- and third-order elastic constants for the crystals with antifluorite structure, we have made use of the interionic potential (1) and the method of homogeneous deformation [14] as followed by Bedi and Verma [9]. The expressions for the second-order elastic constants (SOEC) are

$$C_{11} = L[3.0511g + \tfrac{1}{3}(A_{13} + 2B_{13}) + (A_{12} + A_{33}) + 8.9558g'] , \quad (5)$$

$$C_{12} = L[-3.4651g + \tfrac{1}{3}(A_{13} - B_{13}) + 8.9558g'] , \quad (6)$$

$$C_{44} = L[-1.5225g + \tfrac{1}{3}(A_{13} + 2B_{13}) + B_{12} + B_{33} + \tfrac{1}{3} R^2/Q] , \quad (7)$$

where $L = e^2/4a^4$, $g = Z(Z' + 8f(r))$, $g' = Zr_0(\mathrm{d}f(r)/\mathrm{d}r)$, and

$$R = A_{13} - B_{13} - 15.086g , \quad (8)$$

$$Q = A_{13} + 2B_{13} + 3(A_{12} + 2B_{12}) + 3(A_{33} + 2B_{33}) . \quad (9)$$

In the above expressions, A_{ij} and B_{ij} are the short-range force constants for pairwise interactions, where the suffixes $(i, j = 1, 2, 3)$ with A and B stand for the interaction between Na(1), Na(2), and S(3) ions, respectively. These force constants in terms of the interionic potentials are written as

$$A_{ij} = \frac{2V}{e^2} \frac{\mathrm{d}^2 W_{ij}^{\mathrm{SR}}(r)}{\mathrm{d}r^2} , \qquad B_{ij} = \frac{2V}{e^2} \frac{1}{r} \frac{\mathrm{d} W_{ij}^{\mathrm{SR}}(r)}{\mathrm{d}r} , \quad (10)$$

where $W_{ij}^{\mathrm{SR}}(r)$ consists of repulsive and dispersive interaction potentials, defined in (1) and (2).

The expressions for third-order elastic constants for the crystals with antifluorite structures have been derived using the method mentioned by Bedi and Verma [9] and using (1). These expressions are

$$C_{111} = L[-17.74g + M - 3(B_{13} + B_{12} + B_{33}) + (C_{12} - 3A_{12} + 3B_{12}) + + C_{33} - 3A_{33} + 3B_{33} + 7.76g'' + 33.33g'] , \quad (11)$$

$$C_{112} = L[3.18g + M + 7.76g'' - 26.87g''] , \quad (12)$$

$$C_{123} = L[10.96g + M + 7.76g'' - 24.96g'] , \quad (13)$$

$$C_{144} = L[10.96g + M + 2.56g'' - 1.07g' + U(-6.70g - \tfrac{2}{9}C_{13} - - 5.17g'' + 29.20g') + U^2(-18.16g + M' + C_{12} + C_{33} + + 2.58g'' + 5.97g')] , \quad (14)$$

$$C_{166} = L[3.18g + M + 2.58g'' - 1.07g' + U(-16.76g - 2M - - 5.17g'' + 29.20g') + U^2(9.08g + \tfrac{2}{9}C_{13} + A_{12} + A_{33} - - B_{12} - B_{33} + 2.58g'' + 5.97g')] , \quad (15)$$

$$C_{456} = L[10.96g + M + U(-10.06g - \tfrac{1}{9}C_{13} + 3(A_{13} + A_{33}) - - 3(B_{12} + B_{33})) + U^3(-9.60g - M)] , \quad (16)$$

where $U = R^2/Q$, $M = (C_{13} - 3A_{13} + 3B_{13})/9$, $M' = (6A_{13} - 6B_{13} + C_{13})/9$. The pressure derivatives of the SOE constants can be obtained from the knowledge of the above-mentioned SOE and TOE constants and the method mentioned by Bedi and Verma [9].

3. Results and Discussion

The model described above has only three parameters, namely, b, p, and $f(r)$ which completely describe the interionic forces and elastic properties of solids with antifluorite structure. These parameters have been determined from bulk modulus (B_T) and equilibrium condition and one elastic constant. The higher-order derivatives of $f(r)$ have been evaluated by assuming $f(r)$ to vary exponentially, such that

$$f(r) = f_0 \exp(-r_0/p) , \tag{17}$$

where f_0 is a constant. The input data and derived model parameters for Na_2S are listed in Table 1.

Table 2

Cohesive energy (kJ/mol) of Na_2S

$W_{Coulomb}$	W_{TBI}	W_{vdW}	W_R	W_{total}	ref.
−2482	82	−108	256	−2248	present
				−2203	exp. [15]
				−2261	[10]

It is seen from Table 2 that our calculated values of cohesive energy for Na_2S agree well with the experimental data. The contributions from both van der Waals and three-body interactions are quite large and is about 5% each. A little higher value of cohesive energy (in comparison to that of experimental) is obtained because in the present case we neglect the repulsive interactions beyond second neighbours and also the thermal energy contributions. A similar agreement, but a little higher value of cohesive energy of Na_2S, has also been obtained by Agarwal et al. [10] who used a two-body potential model for such analysis.

The values of elastic constants, B_T, ionic charge (Z), and zone-boundary frequencies (W_{LO} and W_{TO}) obtained from the present model are listed in Table 3 and compared with experimental and theoretical values obtained by Buhrer and Bill [2]. A fairly good agreement between experimental and present results reveals that TBI contribute significantly to these elastic constants, and as is seen from (5) to (7) the TBI are responsible for the Cauchy violation in them ($C_{12} \neq C_{44}$). It is worth mentioning that in no way the present analysis aims to achieve a numerical fit to the elastic constants, and

Table 3

Elastic constants (10^{10} Nm^{-2}) and phonon frequencies (meV) of Na_2S

	C_{11}	C_{12}	C_{44}	B_T	Z	W_{LO}	W_{TO}
present	8.55	2.86	2.32	4.94	0.96	33.0	23.8
exp. [2]	8.14	3.34	2.13	4.94	—	33.5	24.1
others [2]	6.44	2.17	1.79	3.59	0.72	—	—

Table 4
Third-order elastic constants (10^{11} Nm^{-2}) and first pressure derivatives of Na_2S

properties	Na_2S	CaF_2*)
C_{111}	−6.62	−12.45
C_{112}	−2.21	− 4.00
C_{123}	−1.52	− 2.54
C_{144}	−1.28	− 1.24
C_{166}	−0.83	− 2.14
C_{456}	−1.12	− 0.75
dC_{11}/dP	5.45	5.76
dC_{12}/dP	4.97	4.62
dC_{44}/dP	0.89	1.17

*) Experimental value [16].

thus the model parameters have their own physical significance. Further, the present model predicts a little higher value of ionic charge ($Z = 0.96$) in comparison to that obtained by Buhrer and Bill ($Z = 0.72$) [2]. This is mainly because the present theory is suitable for ionic solids and thus neglects effects due to covalency.

In addition to the SOE constants, we have also calculated the third-order elastic constants (TOEC) and pressure derivatives of SOEC of Na_2S, as they are indispensible for understanding the anharmonic and thermal behaviour. The calculated values of TOEC and pressure derivatives of SOEC are listed in Table 4 and compared with those of CaF_2 in the absence of any measured data on them. It is seen from this Table that the qualitative features of the TOEC are identical in Na_2S and CaF_2 though they differ in magnitude. The larger values of C_{111}, C_{123}, and C_{112} in CaF_2 might be because the first-neighbour force constants are quite larger in CaF_2 in comparison to those of Na_2S [12]. The similarity in the magnitudes of C_{166}, C_{144}, and C_{456} in both the solids indicates that the short-range interactions are not so effective in these constants. The calculated values of pressure derivatives of SOEC of Na_2S are surprisingly quite similar to those of CaF_2. Thus, it is expected that the high-pressure behaviour of these two solids may be similar to some respects. At present, in the absence of any measured data on these properties, no definite conclusion can be drawn. Further measurements on these properties will, however, lead to a better understanding of various thermal properties of Na_2S. It is further interesting to note that the calculated values of zone-boundary frequencies from the present model agree fairly (see Table 3) with their measured data. This good agreement is mainly because the TBI modify the dielectric properties of ionic solids [5] and hence the polarizabilities of ions. The shell model parameters were obtained in the usual way. Finally it may be concluded that the present model explains satisfactorily the interionic forces and other crystal properties of Na_2S and is expected to explain its dynamical behaviour.

Acknowledgements

The authors are thankful to University Grants Commission, New Delhi, India for financial support. One of us (N.A.) is thankful to UGC, New Delhi for a fellowship. S.P.S. is thankful to Department of Science and Technology, New Delhi, India for an associateship.

References

[1] H. H. MOBIUS, H. WITZMAN, and R. HARTUNG, Z. phys. Chem. **227**, 40 (1964).
[2] W. BUHRER and H. BILL, J. Phys. C **13**, 5495 (1980).
[3] H. BILZ and W. KRESS, Solid State Sci. **10**, (1979).
[4] S. O. LUNDQVIST, Ark. Fys. (Sweden) **12**, 365 (1957).
[5] R. K. SINGH, Physics Rep. **85**, 259 (1982).
[6] R. K. SINGH, S. P. SANYAL, and N. K. PANDEY, J. chem. Phys. **67**, 2596 (1982).
[7] R. K. SINGH and N. K. PANDEY, phys. stat. sol. (b) 115, 555 (1983).
[8] R. K. SINGH and N. K. PANDEY, to be published.
[9] S. S. BEDI and M. P. VERMA, phys. stat. sol (b) **94**, 297 (1979); **97**, 203 (1980).
[10] G. G. AGARWAL, O. P. SHARMA, J. C. SHARMA, H. P. SHARMA, and J. SHANKAR, J. inorg. nuclear Chem. **39**, 2131 (1977).
[11] R. NARAYAN, J. Phys. Chem. Solids **38**, 1097 (1977).
[12] R. K. SINGH, N. AGNIHOTRI, S. V. SURYANARAYANA, and S. P. SANYAL, phys. stat. sol. (b) **123**, 453 (1984).
[13] M. P. TOSI, Solid State Phys. **16**, 1 (1964).
[14] F. BIRCH, Phys. Rev. 71, 809 (1947).
[15] R. C. WEAST. (Ed.), C.R.C. Handbook of Chemistry and Physics, CRC Press Inc., Florida 1981 (D-98).
[16] S. ALTERVITZ and D. GERLICH, Phys. Rev. **184**, 999 (1969).

(Received February 7, 1984)

phys. stat. sol. (b) **125**, 121 (1984)

Subject classification: 6; 22.1; 22.1.1 22.2.2; 22.5.2; 22.8

Department of Semiconductor Physics, Charles University, Prague[1]) *(a), and Sektion Physik der Bergakademie Freiberg*[2]) *(b)*

A Frequency Sum Rule and Spectral Moments in Lattice Dynamics

By

V. FREI (a) and P. DEUS (b)

A sum rule of powers of phonon frequencies is derived generalizing a sum rule of squared frequencies given previously by Blackman, Brout, and Rosenstock. The connection of these sums with the moments of the frequency spectrum of the crystal is investigated by choosing gradually enlarged unit cells (GEUC) in the lattice and calculating the sums of the powers of phonon frequencies in suitable points of the correspondingly smaller Brillouin zones. The convergence of these sums in the GEUC-process is shown using published dispersion curves of cubic crystals and group-theoretical formulae for the high symmetry points in the Brillouin zones of f.c.c. lattices. Applications of the generalized sum rule are outlined.

Eine Summenregel für die Potenzen der Phononenfrequenzen wird durch Verallgemeinerung einer bereits früher von Blackman, Brout und Rosenstock angegebenen Summenregel für die Frequenzquadrate hergeleitet. Der Zusammenhang der entsprechenden Summen mit den Momenten des Gitterspektrums wird mit Hilfe von stufenweise vergrößerten Einheitszellen im Gitter und der Berechnung der Summen der Potenzen der Phononenfrequenzen für geeignete Punkte der entsprechend kleineren Brillouinzonen untersucht. Die Konvergenz dieser Summen im Laufe der fortschreitenden Vergrößerung der Einheitszellen wird anhand veröffentlichter Dispersionskurven kubischer Kristalle mittels gruppentheoretischer Formeln für die hochsymmetrischen Punkte der Brillouinzonen von k.f.z.-Kristallen gezeigt. Anwendungsmöglichkeiten der verallgemeinerten Summenregel werden angegeben.

1. Introduction

A frequency sum rule in lattice dynamics has been formulated by Blackman [1], by Brout [2], and in a more detail by Rosenstock [3 to 5.] The conclusions of Rosenstock may be summarized in our notation as follows. The sum $B_2(\boldsymbol{q})$ of the squared angular frequencies $\omega_j^2(\boldsymbol{q})$ over all the branches of the vibrational spectrum of a crystal consists of a constant and a $\boldsymbol{q}$-dependent term,

$$B_2(\boldsymbol{q}) = \sum_{j=1}^{3s} \omega_j^2(\boldsymbol{q}) = B_2' + B_2''(\boldsymbol{q}) \,. \tag{1}$$

Here $\boldsymbol{q}$ is the wave vector, s the number of atoms in the primitive cell of the crystal. As a rule, $B_2' \gg B_2''$ is valid or, in other words, B_2 is almost $\boldsymbol{q}$-independent.

In contrast to [1] and [2] Rosenstock did not deal with the value and the physical meaning of B_2', but he found that B_2' results (i) from the electrostatic interactions among all the atoms and (ii) from the non-electrostatic or short-range interactions of the crystallographically inequivalent atoms in the crystal lattice; $B_2''(\boldsymbol{q})$ results from the non-electrostatic interactions of the crystallographically equivalent atoms (i.e. of atoms the position vectors of which differ by a translation vector of the lattice).

[1]) Ke Karlovu 5, 121 16 Prague 2, Czechoslovakia.

[2]) Silbermannstrasse 1, DDR-9200 Freiberg, GDR.

From $B_2''(\boldsymbol{q})$ Rosenstock obtained qualitative results concerning some near-neighbour interactions in various crystals. He also pointed out that the crystallographical equivalence of the atoms may be changed by choosing a larger unit cell in the crystal under consideration.

Recently, following a suggestion by Leibfried [6], the connection between sums of the type (1) and the spectral moments of the frequency spectrum has been employed [7] to estimate the force constants of some ternary semiconducting compounds; the spectral moments were taken from thermal measurements. In [7] the Rosenstock sum rule has in fact been generalized in some sense, but neither [6] nor [7] give a proof of this generalized sum rule. In the present communication an even more general frequency sum rule than that used in [7] is shown to be valid and its possible applications are outlined.

2. The Frequency Sum Rule. Gradually Enlarged Unit Cells

Let us consider a crystal lattice with a primitive cell (Ω^a) which contains s^a atoms and has the volume Ω^a. The upper index a refers to the underlying geometrical lattice denoted by (a). We shall use the Born-von Karman (BvK) boundary conditions writing the crystal volume as $\Omega = N_1N_2N_3\Omega^a = N\Omega^a$. In the corresponding Brillouin zone BZ^a there are N equidistant $\boldsymbol{q}$-points. To each of them corresponds the volume $(2\pi)^3/\Omega$ in the reciprocal space. The centre of BZ^a is Γ^a.

We shall now define the non-negative scalar quantities

$$B_n(\boldsymbol{q}^a) = \sum_{j=1}^{3s^a} \omega_j^n(\boldsymbol{q}) \,, \qquad j = j^a \,, \qquad \boldsymbol{q} \in \mathrm{BZ}^a \,, \qquad n \neq 0 \,. \tag{2a}$$

In particular, $B_2(\boldsymbol{q}^a) = B_2(\boldsymbol{q})$ is the Blackman or Brout sum occurring in (1). Note that there are $3Ns^a$ (not necessarily different) eigenvalues $\omega_j(\boldsymbol{q})$ corresponding to N values of $\boldsymbol{q} \in \mathrm{BZ}^a$ and to $3s^a$ branches of the vibrational spectrum. Thus, $B_n(\boldsymbol{q}^a)$ involves $3s^a$ values of ω.

In the lattice studied larger unit cells can be chosen, based on integral linear combinations of the primitive translations $\boldsymbol{a}_i^a$, $i = 1, 2, 3$. For instance, we may take $\boldsymbol{a}_i^b = \lambda_i \boldsymbol{a}_i^a$, λ_i integers, $\lambda_1\lambda_2\lambda_3 = \lambda$. We thus obtain a larger unit cell (Ω^b) with the volume $\Omega^b = \lambda\Omega^a$. It contains $\lambda s^a = s^b$ atoms. To (Ω^b) it corresponds a λ-times smaller Brillouin zone BZ^b [3]), which contains N/λ equally spaced $\boldsymbol{q}$-vectors. The centre of BZ^b is Γ^b. The corresponding frequency sum is defined by

$$B_n(\boldsymbol{q}^b) = \lambda^{-1} \sum_{j'=1}^{3s^b} \omega_{j'}^n(\boldsymbol{q}) \,, \qquad j' = j^b \,, \qquad \boldsymbol{q} \in \mathrm{BZ}^b \,. \tag{2b}$$

Each sum $B_n(\boldsymbol{q}^b)$ involves $3s^b$ values of ω.

By repeating this step for $\lambda < \lambda' < \lambda'' < \ldots < \lambda^\infty = N$ we obtain a sequence of gradually enlarged unit cells (GEUC) and, correspondingly, a sequence of gradually smaller Brillouin zones. Of course, the values of N_1, N_2, and N_3 are to be chosen accordingly. The "infinity" value $\lambda^\infty = N \gg 1$ finally leads to the Brillouin zone BZ$^\infty$ which has the volume $(2\pi)^3/\Omega$ and contains, therefore, only the point Γ^∞. Similarly as in (2b) we then write

$$B_n(\boldsymbol{q}^\infty) = N^{-1} \sum_{j'=1}^{3s^\infty} \omega_{j'}^n(\boldsymbol{q}) \,, \qquad j' = j^\infty \,, \qquad \boldsymbol{q} \in \mathrm{BZ}^\infty \,. \tag{3}$$

This sum now involves all the $3s^\infty = 3Ns^a$ values of ω. Otherwise expressed, it holds,

[3]) In this point Rosenstock [4] followed another way.

with $\boldsymbol{q} \in \mathrm{BZ}^{\mathrm{a}}$ and $j = j^{\mathrm{a}}$,

$$B_n(\boldsymbol{q}^\infty) = N^{-1} \sum_{j=1}^{3s^{\mathrm{a}}} \sum_{\boldsymbol{q}} \omega_j^n(\boldsymbol{q}) = N^{-1} \int_0^{\omega_{\max}} \omega^n g(\omega)\, \mathrm{d}\omega \,, \tag{4}$$

where $\omega_{\max}$ is the maximum frequency and the density of states is normalized to $3Ns^{\mathrm{a}}$,

$$\int_0^{\omega_{\max}} g(\omega)\, \mathrm{d}\omega = 3Ns^{\mathrm{a}} \,.$$

Using the well-known expression for the n-th spectral moment of the lattice

$$\overline{\omega^n} = (3Ns^{\mathrm{a}})^{-1} \int_0^{\omega_{\max}} \omega^n g(\omega)\, \mathrm{d}\omega \tag{5}$$

we see that the sequence of the GEUC implies the convergence of the sequence of the frequency sums. One thus arrives at the frequency sum rule (FSR) in the limit form

$$\lim_{\mathrm{p}\to\infty} B_n(\boldsymbol{q}^{\mathrm{p}}) = 3s^{\mathrm{a}}\overline{\omega^n} \,, \qquad \mathrm{p} = \mathrm{a, b, c, \dots}; \qquad n > 0 \,. \tag{6a}$$

In principle, the points $\boldsymbol{q}^{\mathrm{p}} \in \mathrm{BZ}^{\mathrm{p}}$ can be chosen arbitrarily, but the choice may not be irrelevant (see Sections 3.1 and 5). Especially, if we avoid the zero values of the acoustic frequencies at $\boldsymbol{q} = (0, 0, 0)$, we can write

$$B_n(\boldsymbol{q}^{\mathrm{p}}) \to 3s^{\mathrm{a}}\overline{\omega^n} \,, \qquad \boldsymbol{q} \neq (0, 0, 0) \,, \qquad n \neq 0 \tag{6b}$$

with $\mathrm{p} = \mathrm{a, b, c, \dots}, N_0 < N$. Here the negative spectral moments are included, too.

3. Lattice-Dynamical and Group-Theoretical Treatment of the Frequency Sum Rule

3.1 The frequency sums and the dynamical matrix

An insight into the physical meaning of (6 a, b) may be obtained within the framework of the harmonic approximation. Let us consider here an ideal non-ionic crystal lattice with the primitive cell (Ω^{a}).

In the harmonic approximation the squared frequencies $\omega_j^2(\boldsymbol{q})$, $j = j^{\mathrm{a}} = 1, 2, \dots, 3s^{\mathrm{a}}$, $\boldsymbol{q} \in \mathrm{BZ}^{\mathrm{a}}$, are the eigenvalues of the Fourier-transformed, mass-adjusted dynamical matrix $D(\boldsymbol{q})$ (see, e.g., [8 to 10]). To define the notation, let us write

$$|D(\boldsymbol{q}) - \omega^2(\boldsymbol{q})\, 1| = 0 \,, \tag{7}$$

where 1 is the $3s^{\mathrm{a}} \times 3s^{\mathrm{a}}$ unit matrix and

$$D^{kk'}(\boldsymbol{q}) = (M_k M_{k'})^{-1/2}\, D_0^{kk'}(\boldsymbol{q}) \tag{8a}$$

with

$$D_0^{kk'}(\boldsymbol{q}) = \sum_l \Phi^{kk'}(\boldsymbol{X}_{lk'}^{\mathrm{a}} - \boldsymbol{X}_{0k}^{\mathrm{a}}) \exp\,[i\boldsymbol{q}(\boldsymbol{X}_{lk'}^{\mathrm{a}} - \boldsymbol{X}_{0k}^{\mathrm{a}})] \tag{8b}$$

is a 3×3 submatrix of $D(\boldsymbol{q})$ representing the interactions of the k-th atoms with the k'-th atoms in the lattice; M_k, $M_{k'}$ are the respective atomic masses. The index l labels the cell, the index k labels the atomic position in the cell (Ω^{a}), so that

$$\boldsymbol{X}_{lk}^{\mathrm{a}} = \boldsymbol{X}_l^{\mathrm{a}} + \boldsymbol{x}_{k'}^{\mathrm{a}} \qquad \boldsymbol{X}_l^{\mathrm{a}} = \sum_{i=1}^{3} l_i \mathrm{a}_i^{\mathrm{a}} \,, \qquad l_i = l_i^{\mathrm{a}} \text{ integers} \,, \qquad k = k^{\mathrm{a}} = 1, \dots, s^{\mathrm{a}} \tag{9}$$

is the equilibrium position vector of the lk-th atom. It is convenient to put $\boldsymbol{x}_1^{\mathrm{a}} = 0$.

The elements of the 3×3 matrices $\Phi^{kk'}(\boldsymbol{X}^{\rm a}_{lk'} - \boldsymbol{X}^{\rm a}_{0k})$ are the force constants in the BvK formalism. We do not impose any restrictions on l, i.e. on the range of the interactions.

From the properties of the Hermitian matrix $D(\boldsymbol{q})$ it follows for $\boldsymbol{q} \in {\rm BZ}^{\rm a}$:

$$B_n(\boldsymbol{q}^{\rm a}) = \sum_{j=1}^{3s^{\rm a}} \omega_j^n(\boldsymbol{q}) = {\rm tr}\,[D(\boldsymbol{q})]^{n/2}\,, \qquad j = j^{\rm a}\,, \tag{10}$$

where $n \neq 0$ may be any rational nubmer, due to the Born theorem (see [8], equation (2.2.15) and (2.2.23)). It is known, however, that $n \geqq -3$ applies for the harmonic crystal (see, e.g., [11]).

We shall consider the Blackman sum B_2 first.

Although ${\rm tr}\, D(\boldsymbol{q})$ involves only the diagonal submatrices $D^{kk}(\boldsymbol{q})$ of $D(\boldsymbol{q})$, it also contains the traces of the force-constant matrices $\Phi^{kk'}$, $k' \neq k$, which appear in the expression for $\Phi^{kk}(0)$ [10],

$$\Phi^{kk}(0) = -\sum_{l\neq 0} \Phi^{kk}(\boldsymbol{X}^{\rm a}_l) - \sum_{k'\neq k} D_0^{kk'}(0)\,. \tag{11}$$

Using (8), (10), and (11) we obtain

$$B_2(\boldsymbol{q}^{\rm a}) = -\sum_{k=1}^{s^{\rm a}} \left\{\frac{1}{M_k} \sum_{l\neq 0} [1 - \exp(i\boldsymbol{q}\boldsymbol{X}^{\rm a}_l)]\, {\rm tr}\, \Phi^{kk}(\boldsymbol{X}^{\rm a}_l) - \right.$$
$$\left. - \sum_{k'\neq k} \sum_l {\rm tr}\, \Phi^{kk'}(\boldsymbol{X}^{\rm a}_{lk'} - \boldsymbol{X}^{\rm a}_{0k})\right\}; \qquad \boldsymbol{q} \in {\rm BZ}^{\rm a}\,. \tag{12a}$$

From this equation important lattice-dynamical conclusions concerning the FSR can be drawn; here, only a brief outline will be given.

(i) By comparing (12a) and (1) we see that

$$B_2'(\boldsymbol{q}^{\rm a}) = \sum_{k=1}^{s^{\rm a}} \frac{1}{M_k} \sum_{k'\neq k} {\rm tr}\, D_0^{kk'}(0) = B_2(\Gamma^{\rm a})\,, \tag{13a}$$

$$B_2''(\boldsymbol{q}^{\rm a}) = -\sum_{k=1}^{s^{\rm a}} \frac{1}{M_k} \sum_{l\neq 0} [1 - \exp(i\boldsymbol{q}\boldsymbol{X}^{\rm a}_l)]\, {\rm tr}\, \Phi^{kk}(\boldsymbol{X}^{\rm a}_l); \qquad q \in {\rm BZ}^{\rm a}\,. \tag{14a}$$

This proves the Rosenstock's assertions concerning the crystallographical equivalence of atoms, see Section 1. Of course, the equivalence is taken with respect to (9), i.e. to the lattice (a).

(ii) If we choose a λ-times greater cell ($\Omega^{\rm p}$), $\lambda > 1$, the atomic positions read

$$\boldsymbol{X}^{\rm p}_{l'k'} = \boldsymbol{X}^{\rm p}_{l'} + \boldsymbol{x}^{\rm p}_{k'}\,, \qquad \boldsymbol{X}^{\rm p}_{l'} = \sum_{i=1}^{3} l_i \boldsymbol{a}^{\rm p}_i\,, \qquad l_i' = l_i^{\rm p} \text{ integers}\,, \tag{15}$$
$$k' = 1, \dots, s^{\rm p} = \lambda s^{\rm a};$$

$\boldsymbol{x}^{\rm a}_1 = 0$ implics $\boldsymbol{x}^{\rm p}_1 = 0$. In terms of the vectors $\boldsymbol{a}^{\rm p}_i$ fewer atoms have integral coordinates and there are more fractional coordinates if compared to (9). The sum

$$B_2(\boldsymbol{q}^{\rm p}) = -\frac{1}{\lambda} \sum_{k'=1}^{s^{\rm p}} \frac{1}{M_{k'}} \sum_{l'\neq 0} \left\{[1 - \exp(i\boldsymbol{q}\boldsymbol{X}^{\rm p}_{l'})]\, {\rm tr}\, \Phi^{k'k'}(\boldsymbol{X}^{\rm p}_{l'}) - \right.$$
$$\left. - \sum_{k''\neq k'} {\rm tr}\, D_0^{k'k''}(0)\right\}; \qquad \boldsymbol{q} \in {\rm BZ}^{\rm p} \tag{12b}$$

(cf. (2b)) can be divided, again, into two terms

$$B_2'(\boldsymbol{q}^{\rm p}) = \frac{1}{\lambda} \sum_{k'=1}^{s^{\rm p}} \frac{1}{M_k} \sum_{k''\neq k'} {\rm tr}\, D_0^{k'k''}(0) = B_2(\Gamma^{\rm p})\,, \tag{13b}$$

$$B_2''(\boldsymbol{q}^{\mathrm{p}}) = -\frac{1}{\lambda} \sum_{k'=1}^{s^{\mathrm{p}}} \frac{1}{M_{k'}} \sum_{l'\neq 0} [1 - \exp(i\boldsymbol{q}\boldsymbol{X}_{l'}^{\mathrm{p}})] \operatorname{tr} \Phi^{k'k'}(\boldsymbol{X}_{l'}^{\mathrm{p}}); \qquad \boldsymbol{q} \in \mathrm{BZ}^{\mathrm{p}} . \tag{14b}$$

We see that more pair interactions are involved in (13b) then in (13a) whereas there are fewer in (14b) than in (14a).

(iii) Due to the square brackets in (14 a, b) some pair interaction contributions to $B_2''(\boldsymbol{q}^{\mathrm{p}})$ vanish at Γ^{p}, p = a, b, ... The number of the contributions with this property decreases if the cell volume Ω^{p} is increased.

(iv) As a consequence of (iii), the term $B_2(\Gamma^{\mathrm{p}})$ contains — among other contributions — all the pair interaction contributions of the atoms which lie in the symmetrical or Wigner-Seitz cell $(\Omega^{\mathrm{p}})_{\mathrm{WS}}$. With $\Omega^{\mathrm{p}} > \Omega^{\mathrm{a}}$ the sum $B_2(\Gamma^{\mathrm{p}})$ involves more distant coordination sphere(s) of atoms, as compared with $B_2(\Gamma^{\mathrm{a}})$. The sequence $B_2(\Gamma^{\mathrm{p}})$, p = a, b, c, ..., converges because of the decreasing interaction of more distant atoms, but this convergence may not be monotonic (see the figures in Section 4).

A more detailed lattice-dynamical treatment of $B_2(\Gamma^{\mathrm{p}})$ with applications will be given in subsequent papers or short notes (under preparation for phys. stat. sol. and Czech. J. Phys.).

As for $B_n(\boldsymbol{q}^{\mathrm{p}})$, $n \neq 2$, let us only mention that explicit expressions become clumsy with increasing, negative or fractional values of n. Nevertheless, for a simple model, the expressions for $B_4(\boldsymbol{q}^{\mathrm{a}})$ and $B_6(\boldsymbol{q}^{\mathrm{a}})$ have been used successfully [7]. Let us also remark that B_2 depends only on the traces of the matrices $\Phi^{kk'}$ and so it does not contain all the force constants in general. The opposite is true for $B_4 = \operatorname{tr} D^2$, because for each Hermitian matrix H the following relation holds:

$$\operatorname{tr} H^2 = \sum_i \sum_k |H_{ik}|^2 . \tag{16}$$

Of course, it is always possible to obtain B_n numerically after diagonalizing the matrix $D(\boldsymbol{q})$.

It would be straightforward, but inconvenient, to solve repeatedly the secular equations of the form (7) for s^{b}, s^{c}, ... atoms in the enlarged unit cells. Fortunately, many of the quantities $B_n(\boldsymbol{q}^{\mathrm{p}})$ can be obtained easily from $\omega_j(\boldsymbol{q})$, $\boldsymbol{q} \in \mathrm{BZ}^{\mathrm{a}}$, by means of group theory. This will be the subject of another communication [12]. The next section gives only a brief outline, which is necessary to show the verification and possible applications of the sum rule.

3.2 Group-theoretical formulae concerning the frequency sums for cubic crystals

From now on, the symbols (p) will denote specific lattices with full cubic point symmetry $\mathrm{O_h}$-m3m. It is convenient to choose a sequence of lattices defined by their respective primitive translations $\boldsymbol{a}_i^{\mathrm{p}}$, $i = 1, 2, 3$. The first three of them are the Bravais lattices: (a) face-centred cubic; (b) body-centred cubic (not of direct importance here); (c) simple cubic. The cell volumes are $\Omega^{\mathrm{a}} = a^3/4$, $\Omega^{\mathrm{b}} = a^3/2$, $\Omega^{\mathrm{c}} = a^3$, where a is the lattice constant. Putting

$$\boldsymbol{a}_i^{\mathrm{d}} = 2\boldsymbol{a}_i^{\mathrm{a}} , \quad \boldsymbol{a}_i^{\mathrm{e}} = 2\boldsymbol{a}_i^{\mathrm{b}} , \quad \boldsymbol{a}_i^{\mathrm{f}} = 2\boldsymbol{a}_i^{\mathrm{c}}; \quad \boldsymbol{a}_i^{\mathrm{g}} = 4\boldsymbol{a}_i^{\mathrm{a}}; \quad \boldsymbol{a}_i^{\mathrm{t}} = 3\boldsymbol{a}_i^{\mathrm{a}} ,$$
$$i = 1, 2, 3 , \tag{17}$$

we obtain lattices with the respective cell volumes $\Omega^{\mathrm{d}} = 2a^3$, $\Omega^{\mathrm{e}} = 4a^3$, $\Omega^{\mathrm{f}} = 8a^3$;

$\Omega^g = 16a^3$; $\Omega^t = (27/4)\, a^3$. If we further write the ratio of the cell volumes as

$$\lambda(\mathrm{r, p}) = \frac{\Omega^\mathrm{r}}{\Omega^\mathrm{p}}, \tag{18}$$

we see that the integral quotients λ are now $\lambda(\mathrm{c, a}) = 4$ to $\lambda(\mathrm{g, a}) = 64$; $\lambda(\mathrm{t, a}) = 27$.

The upper indices p = a, b, ... make it possible to retain the commonly used notation of symmetry points in the pertinent Brillouin zones $\mathrm{BZ^p}$. The cartesian coordinates of the $\boldsymbol{q}$-points are now commonly expressed by the reduced wave vector coordinates $\zeta_1, \zeta_2, \zeta_3$ in units of $2\pi/a$. In accordance with [13] we use the following symbols:

$$\left.\begin{array}{llll} \Gamma^\mathrm{a} : (000), & \mathrm{X^a} : (010), & \mathrm{L^a} : (\frac{1}{2}\,\frac{1}{2}\,\frac{1}{2}), & \mathrm{W^a} : (\frac{1}{2}\,1\,0), \\ \Delta^\mathrm{a}_\zeta : (0\zeta 0), \ \Sigma^\mathrm{a}_\zeta : (\zeta\zeta 0), \ \Lambda^\mathrm{a}_\zeta : (\zeta\zeta\zeta), & & & \end{array}\right\} \tag{19a}$$

$$\mathrm{H^b} : (010), \tag{19b}$$

$$\mathrm{X^c} : (0\,\tfrac{1}{2}\,0), \qquad \mathrm{M^c} : (\tfrac{1}{2}\,\tfrac{1}{2}\,0), \qquad \mathrm{R^c} : (\tfrac{1}{2}\,\tfrac{1}{2}\,\tfrac{1}{2}). \tag{19c}$$

Since all the lengths in the zone $\mathrm{BZ^d}$ are twice shorter than those in $\mathrm{BZ^a}$, due to (17), and so on, one easily finds further reduced wave vector coordinate expressions as

$$\mathrm{X^d, H^e} \text{ or } \Delta^\mathrm{a}_{1/2} : (0\,\tfrac{1}{2}\,0), \qquad \mathrm{R^f} \text{ or } \Lambda^\mathrm{a}_{1/4} : (\tfrac{1}{4}\,\tfrac{1}{4}\,\tfrac{1}{4}) \tag{20}$$

Now, by mapping the Brillouin zone $\mathrm{BZ^a}$ into the smaller zones $\mathrm{BZ^p}$ and using the well-known properties of $\omega_j(\boldsymbol{q})$ we obtain, at least for some high symmetry $\boldsymbol{q}$-points in some Brillouin zones $\mathrm{BZ^p}$, simple expressions for the sums $B_n(\boldsymbol{q}^\mathrm{p})$ in terms of $\omega_j(\boldsymbol{q})$, $\boldsymbol{q} \in \mathrm{BZ^a}$ [12]. Important results may be summarized here in the following notation:

$$\left.\begin{array}{l} \Gamma^\mathrm{c} \ldots \frac{1}{4}\,(\Gamma^\mathrm{a}, 3\mathrm{X^a}), \qquad \Gamma^\mathrm{d} \ldots \frac{1}{8}\,(\Gamma^\mathrm{a}, 3\mathrm{X^a}, 4\mathrm{L^a}), \\ \Gamma^\mathrm{e} \ldots \frac{1}{16}\,(\Gamma^\mathrm{a}, 3\mathrm{X^a}, 12\Sigma^\mathrm{a}_{1/2}), \qquad \Gamma^\mathrm{f} \ldots \frac{1}{32}\,(\Gamma^\mathrm{a}, 3\mathrm{X^a}, 4\mathrm{L^a}, 6\mathrm{W^a}, 6\Delta^\mathrm{a}_{1/2}, 12\Sigma^\mathrm{a}_{1/2}); \\ \Gamma^\mathrm{t} \ldots \frac{1}{27}\,(\Gamma^\mathrm{a}, 8\Lambda^\mathrm{a}_{1/3}, 6\Delta^\mathrm{a}_{2/3}, 12\Sigma^\mathrm{a}_{2/3}), \end{array}\right\} \tag{21a}$$

$$\left.\begin{array}{l} \mathrm{X^c} \ldots \frac{1}{2}\,(\mathrm{W^a}, \Delta^\mathrm{a}_{1/2}), \qquad \mathrm{X^d} \ldots \frac{1}{4}\,(\mathrm{W^a}, \Delta^\mathrm{a}_{1/2}, 2\Sigma^\mathrm{a}_{1/2}), \\ \mathrm{H^e} \ldots \frac{1}{8}\,(2\mathrm{L^a}, 3\mathrm{W^a}, 3\Delta^\mathrm{a}_{1/2}). \end{array}\right\} \tag{21b}$$

The first of the formulae (21a) means

$$B_n(\Gamma^\mathrm{c}) = \tfrac{1}{4}\,[B_n(\Gamma^\mathrm{a}) + 3B_n(\mathrm{X^a})], \qquad n \neq 0 \tag{22}$$

and so on. It is seen that in the above cases the knowledge of $\omega_j(\boldsymbol{q})$ along some symmetry lines in $\mathrm{BZ^a}$ is sufficient. On the other hand, already in the formulae of the type (21 a, b) for the points Γ^g and $\mathrm{X^f}$ general $\boldsymbol{q}$-points of $\mathrm{BZ^a}$ emerge.

4. Verification of the Frequency Sum Rule

By means of the published dispersion curves (e.g., in [14]) and of (21 a, b) we have verified the FSR for some integral values of n. A quite rapid convergence was found, at least for $n = 2, 4, 6$. Examples are given in Fig. 1.

Further we treated some materials for which the spectral moments were available to us. The values of $B_n(\Gamma^\mathrm{f})$ or $B_n(\Gamma^\mathrm{t})$, $n = 2, 4, 6$, differ from $3s^\mathrm{a}\overline{\omega^n}$ by less than 10% in many cases. Again, examples are shown (Fig. 2). The disagreement between the sums and the moments can result from three different sources: (i) the sequence obtained by us is not sufficiently long, (ii) inexact values of ω were taken (e.g., due to plot reading), (iii) the values of the moments are not precise.

Fig. 3 gives a more natural representation of the converging sequences $B_n(\Gamma^\mathrm{p})$. Increasing values of B_2 and decreasing values of B_4, B_6 are typical, at least for $s = 2$.

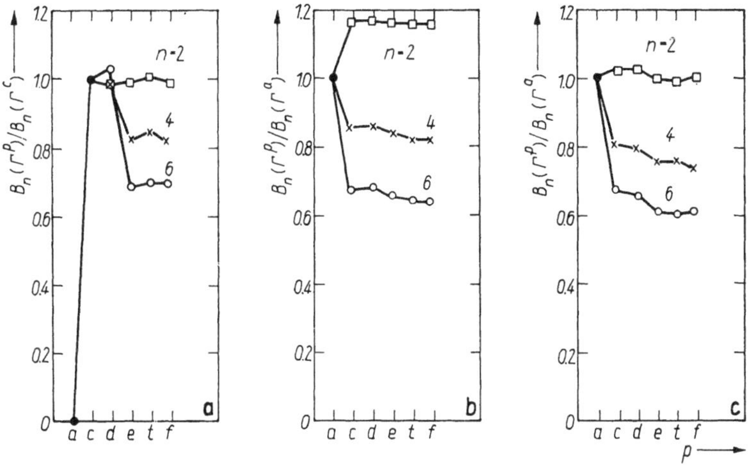

Fig. 1. Variation of the frequency sums $B_n(\Gamma^p)$ related to the value of $B_n(\Gamma^c)$ and $B_n(\Gamma^a)$, respectively, in the sequence of enlarged cells with $\lambda = 1, 4, 8, 16, 27, 32$ (corresponding to the labelling of the abscissa by p = a, c, d, e, t, f, respectively): a) xenon [15], b) diamond [16], c) germanium [17, 18]

Due to a very careful study of GaP by Pödör [26] probable error limits in determining the sums B_n could be displayed in Fig. 3 c to e.

As for details, with Xe ($s = 1$) a rapid convergence is to be expected due to the absence of long-range interactions. The variation of B_2 for $\lambda > 15$ may be caused by inexact values of ω, see point (ii) above. A comparison of $B_2(\Gamma^p)$ for diamond and for Ge (Fig. 1) supports the Rosenstock's conjecture that the near-neighbour interactions are stronger in C than in Ge [3]. But the comparable variation of B_4 and B_6 for both the substances suggests that off-diagonal elements of the respective $\Phi^{kk'}$ matrices

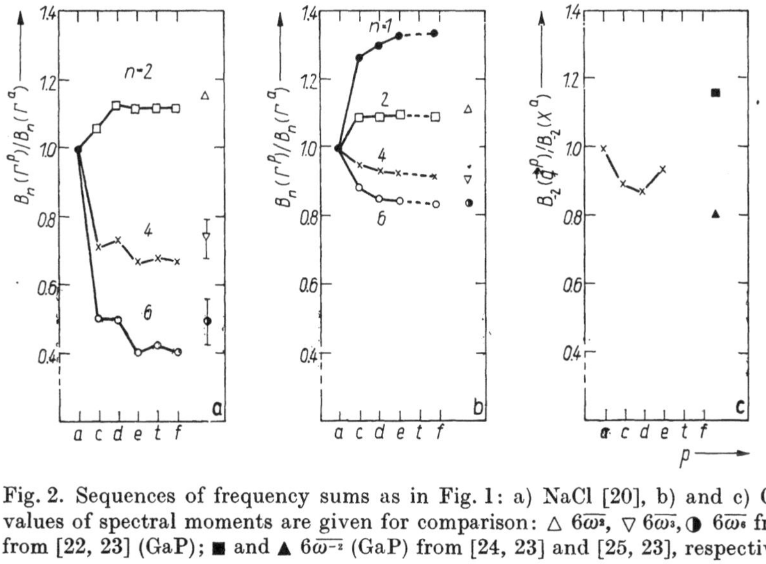

Fig. 2. Sequences of frequency sums as in Fig. 1: a) NaCl [20], b) and c) GaP [21]. Published values of spectral moments are given for comparison: △ $6\overline{\omega^2}$, ▽ $6\overline{\omega^4}$, ◑ $6\overline{\omega^6}$ from [11] (NaCl) and from [22, 23] (GaP); ■ and ▲ $6\overline{\omega^{-2}}$ (GaP) from [24, 23] and [25, 23], respectively

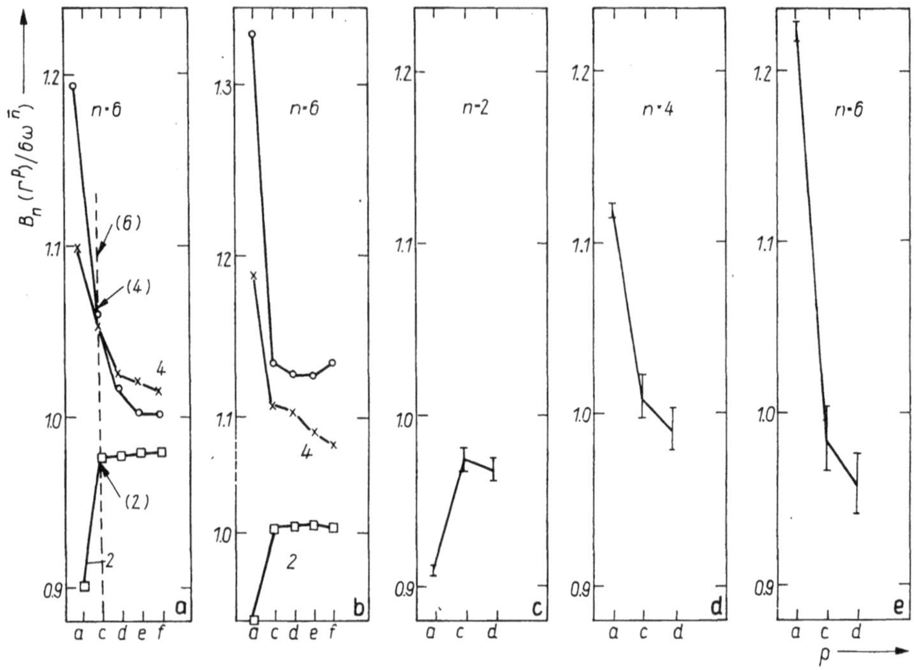

Fig. 3. Frequency sums $B_n(\Gamma^p)$ for GaP related to the values of $6\overline{\omega^n}$, $n = 2, 4, 6$, according to [23]. In a), b) the ω-values are taken from [21] ("model B" in the first case), c) to e) from [26], with error limits. As for the arrows in a), see Appendix

may not greatly differ, see (16). The strong variation of $B_4(\Gamma^p)$ and $B_6(\Gamma^p)$ for the ionic crystal NaCl may result from long-range electrostatic interactions by which, on the other hand, the values of $B_2(\Gamma^p)$ are not affected (see Rosenstock's conclusion (i), Section 1).

For GaP the values of $\overline{\omega^2}$, $\overline{\omega^4}$, $\overline{\omega^6}$ determined in [23] from the experimental data given in [22] are surprisingly near to $B_n(\Gamma^f)$, but the error limits are not known. If these values of spectral moments are reliable, then from the two lattice-dynamical models, treated in [21], the second one is less realistic, cf. Fig. 3 a, b. For the moment $\overline{\omega^1}$ no value was available, but the points in Fig. 2b suggest a good convergence. With B_{-2} (Fig. 2c) the $\boldsymbol{q}$-point (000) had to be avoided, see (6b); we chose X^a, X^c, X^d, and H^e and used (21b). Here the convergence is not satisfactory ([24] and [25] give the lowest values of $\overline{\omega^{-2}}$ known to us [23]). Nevertheless, the convergence $B_{-2}(\boldsymbol{q}^p) \to 6\overline{\omega^{-2}}$ beyond the very limited sequence given by (21b) remains possible.

The $\omega_j(\boldsymbol{q})$ values given in [26] limit the use of (21a) to p = a, c, d only. The behaviour of $B_4(\Gamma^p)$ and $B_6(\Gamma^p)$ suggests that the values of $\overline{\omega^4}$, $\overline{\omega^6}$ used for reference in Fig. 3 are too high. The error limits according to [26] are the maximum error limits, supposing simple addition of errors as estimated for individual branches. It is hoped that the sums $B_n(\Gamma^p)$ may, in fact, be obtained with a higher precision, as required for some applications.

5. Applications of the Frequency Sum Rule

The frequency sum rule interconnects measurable quantities, namely the frequencies ω and the spectral moments $\overline{\omega^n}$, with lattice-dynamical quantities $D(\boldsymbol{q})$, $\Phi^{kk'}$, and

$g(\omega)$. In applying the FSR different approaches are possible, for instance:

(i) Fairly accurate approximations to $3s^a\overline{\omega^n}$ may be obtained, if sufficiently large unit cells are used, see (6 a, b). This has common features with the LUC (large unit cell) approach concerning the special points in the Brillouin zone, see, e.g., [27].

(ii) Any term in the sequence $B_n(\boldsymbol{q}^p)$, p = a, b, ..., may be considered as an approximation to $3s^a\overline{\omega^n}$. There are, e.g., promisingly simple, rough estimates θ^p_∞, p = = a, b, c, ..., of the high-temperature value of the Debye temperature, $\theta_\infty = (\hbar/k)\,[(5/3)\,\overline{\omega^2}]^{1/2}$. So θ^a_∞ based on $B_2(\Gamma^a)$ gives 1750 K for diamond and 270 K for NaCl[4]).

(iii) The variations of B_n within BZ^a yield some information on the force constants in the crystal. This has been partly used by Rosenstock [3 to 5], who, however, dealt with the very near interactions only.

(iv) From variations in the sequence of $B_n(\Gamma^p)$ (or, more generally, of $B_n(\boldsymbol{q}^p)$), p = a, b, c, ..., far interactions may be studied. E.g., the drop from $B_2(\Gamma^d)$ to $B_2(\Gamma^e)$ for Ge (Fig. 2c) is in accordance with the negative tr Φ for the eighth neighbours as found in [28]. Of course, the question of accuracy in determining the values of $B_n(\boldsymbol{q}^p)$ is especially important here.

Problems concerning various applications are under study.

Acknowledgements

The authors are grateful to Prof. Dr. J. Monecke and Dr. W. Cordts for their interest in our work and for valuable discussions. One of us (V.F.) wishes to thank the Department of Physics of the Mining Academy Freiberg for their hospitality and the possibility to accomplish this work.

Appendix

In the previous paper [7] the force constants α_A, α_B, and β of a Keating model for two chalcopyrite ternary semiconducting compounds $CuInS_2$ and $AgInTe_2$ ($s = 8$) have been estimated using the FSR in the form

$$B_n(\Gamma^a) \approx 3s^a\overline{\omega^n}\,, \qquad n = 2, 4, 6\,. \tag{A1}$$

The failure of the same method applied to similar diamond- or sphalerite-type crystals ($s = 2$) was tentatively explained in [7] by assuming that (A1) is not well fulfilled for lattices having smaller primitive cells. This assumption has in fact been confirmed by the results of the present communication. Since the chalcopyrite cell $(\Omega^a)_{ch}$ is comparable with the diamond or sphalerite cell $(\Omega^c)_d$ and the bonding conditions in ternary semiconductors and in GaP are not very different, it may be expected that $B_n(\Gamma^a)_{ch}$ is a comparably good approximation to $3s^a\overline{\omega^n}$ as $B_n(\Gamma^c)_d$, see Fig. 3 a. In [7] we supposed the error limits in the values of $\overline{\omega^2}$, $\overline{\omega^4}$, and $\overline{\omega^6}$ to be $\pm 3\%$, $\pm 6\%$, and $\pm 9\%$, respectively. The corresponding arrows in Fig. 3a suggest that $B_2(\Gamma^a)_{ch}$ may not exceed these error limits, in contrast to $B_n(\Gamma^a)_d$. This explains the failure of the method with sphalerites. Moreover, we now know that

$$B_2(\Gamma^a)_{ch} < 24\overline{\omega^2}\,, \qquad B_n(\Gamma^a)_{ch} > 24\overline{\omega^n} \quad \text{for} \quad n = 4, 6 \tag{A2}$$

(cf. the figures) which enables us to make a more specific study of the error limits than done in the Appendix of [7]. As a result, the force constant β has a lower value than that estimated in [7]; this is in accordance with results of other authors quoted in [7].

---

[4]) See "Note added in proof".

For the above reasons it may be supposed that a more detailed formulation of the Keating model for the chalcopyrite crystals, even within the unit cell $(\Omega^a)_{ch}$ [29], is physically meaningful.

References

[1] M. BLACKMAN, Proc. Roy. Soc. A **181**, 58 (1942).
[2] R. BROUT, Phys. Rev. **113**, 43 (1959).
[3] H. B. ROSENSTOCK, Phys. Rev. **129**, 1959 (1963).
[4] H. B. ROSENSTOCK, in: Proc. Internat. Conf. Lattice Dynamics, Copenhagen (Denmark), August 1963, Ed. R. P. WALLIS, Pergamon Press, 1965.
[5] H. B. ROSENSTOCK and G. BLANKEN, Phys. Rev. **145**, 546 (1966).
[6] G. LEIBFRIED, in: Handbuch der Physik VII/l, Ed. S. FLÜGGE, Springer-Verlag, Berlin/Göttingen/Heidelberg 1955.
[7] P. DEUS, V. FREI, O. BILEK, and W. CORDTS, phys. stat. sol. (b) **112**, 269 (1982).
[8] A. A. MARADUDIN, E. W. MONTROLL, G. H. WEISS, and I. P. IPATOVA, Theory of Lattice Dynamics in the Harmonic Approximation, 42nd ed. , Academic Press, New York/London 1971.
[9] B. T. M. WILLIS and A. W. PRYOR, Thermal Vibrations in Crystallography, Cambridge University Press, 1975.
[10] M. LAX, Symmetry Principles in Solid State and Molecular Physics, Wiley, New York/London/Sydney/Toronto 1974.
[11] T. H. K. BARRON, W. T. BERG, and J. A. MORRISON, Proc. Roy. Soc. A **242**, 478 (1957).
[12] V. FREI, Czech. J. Phys., to be published.
[13] C. J. BRADLEY and A. P. CRACKNELL, The Mathematical Theory of Symmetry in Solids, Clarendon Press, Oxford 1972.
[14] H. BILZ and W. KRESS, Phonon Dispersion Relations in Insulators, Springer-Verlag, Berlin/Heidelberg/New York 1979.
[15] N. A. LURIE, G. SHIRANE, and J. SKALYO, JR., Phys. Rev. B **9**, 5300 (1974), cited in [14].
[16] J. L. WARREN, J. L. YARNELL, G. DOLLING, and R. A. COWLEY, Phys. Rev. **158**, 805 (1967), cited in [14].
[17] G. NILSSON and G. NELIN, Phys. Rev. B **3**, 364 (1971), cited in [19].
[18] G. NELIN and G. NILSSON, Phys. Rev. B **5**, 3151 (1972), cited in [19].
[19] W. WEBER, Phys. Rev. B **15**, 4789 (1977).
[20] G. RAUNIO and S. ROLANDSON, Phys. Rev. B **2**, 2098 (1970).
[21] P. H. BORCHERDS, K. KUNZ, G. F. ALFREY, and R. L. HALL, J. Phys. C **12**, 4699 (1979).
[22] R. W. HILL and D. H. PARKINSON, Phil. Mag. **43**, 309 (1952).
[23] P. DEUS, Dissertation B, Bergakademie Freiberg, DDR (1982).
[24] M. G. SHUMSKII, V. T. BUBLIK, S. S. GORELIK, and M. A. GUREVICH, Kristallografiya **16**, 779 (1971).
[25] G. PEPE, P. MASRI, M. BIENFAIT, and L. DOBRZYNSKI, Acta cryst. A **30**, 290 (1974).
[26] B. PÖDÖR, phys. stat. sol. (b) **120**, 207 (1983).
[27] R. A. EVARESTOV and V. P. SMIRNOV, phys. stat. sol. (b) **119**, 9 (1983).
[28] A. D. ZDETSIS and C. S. WANG, Phys. Rev. B **19**, 2999 (1979).
[29] P. DEUS and V. FREI, Czech. J. Phys. B **34**, 680 (1984).

(Received April 10, 1984)

Note added in proof:

More examples of the values of θ_∞^a and θ_∞^c whith a brief discussion see V. FREI and M. POLČÍK, Czech. J. Phys, B **34**, 883 (1984).

phys. stat. sol. (b) **125**, 131 (1984)

Subject classification: 4; 1.1; 6; 20.1; 22.7

Physikalisches Institut der Universität Würzburg[1] *(a),*
I. Physikalisches Institut der Rheinisch-Westfälischen Technischen Hochschule Aachen (b),
and Institut Max von Laue-Paul Langevin, Grenoble (c)

Effect of Hydrostatic and Uniaxial Pressure on Structural Properties and Raman Active Lattice Vibrations in Bi_2Te_3

By

W. KULLMANN (a), J. GEURTS (b), W. RICHTER (b), N. LEHNER[2] (c),
H. RAUH (a), U. STEIGENBERGER (c), G. EICHHORN (a), and R. GEICK (a)

The lattice parameters of the hexagonal unit cell, the atomic parameters, and the shift of the Raman active vibrations are measured of p-Bi_2Te_3 under uniaxial and hydrostatic pressure. The structural investigations are performed by neutron diffraction, and the lattice vibrations are studied by Raman scattering. The results of the diffraction experiments constitute a direct experimental confirmation of the weak bonding between the sandwiches in this narrow gap semiconductor with sandwich structure. As a consequence, some Raman active phonon modes show a rather strong dependence upon uniaxial pressure up to $\Delta\tilde{\nu}/\Delta p = 1.9 \times 10^{-8}\ cm^{-1}/Pa$.

Die Gitterparameter der hexagonalen Einheitszelle, die Atompositionen und die Verschiebung der Raman-aktiven Gitterschwingungen von p-Bi_2Te_3 unter uniaxialem und hydrostatischem Druck werden gemessen. Die strukturellen Untersuchungen werden mittels Neutronenbeugung durchgeführt, die Gitterschwingungen mit Ramanstreuung untersucht. Die Ergebnisse der Diffraktionsexperimente konstatieren einen direkten experimentellen Beweis für die schwachen Bindungen zwischen den Sandwiches in diesem Schmalband-Halbleiter mit Sandwichstruktur. Als Konsequenz zeigen einige Raman-aktive Gitterschwingungen eine relativ starke Druckabhängigkeit der Anregungsfrequenzen, unter uniaxialem Druck bis zu $\Delta\tilde{\nu}/\Delta p = 1{,}9 \times 10^{-8}\ cm^{-1}/Pa$.

1. Introduction

Bi_2Te_3 is a narrow-gap semiconductor with an energy gap of 150 meV. Beside its thermoelectric properties [1 to 4] the bonding characteristics are of great interest because of the sandwich character of its structure and of the highly anisotropic electronic and other properties. In $Bi_{14}Te_{11}S_{10}$, a compound closely related to Bi_2Te_3, two-dimensional conduction has been observed [5]. The crystal structure of Bi_2Te_3 is isomorph with that of other B^V–B^{VI} compounds like Sb_2Te_3 and Bi_2Se_3. All these substances crystallize in a rhombohedral unit cell of space group D_{3d}^5 ($R\bar{3}m$), containing five atoms. This crystal structure consists of layers formed by equal atoms in hexagonal arrangement. The sequence of layers parallel to the c-axis is as follows:

–Bi–Te(1)]–[Te(1)–Bi–Te(2)–Bi–Te(1)]–[Te(1)–Bi–

That means five of these layers form a structural unit called a sandwich. The bonding within a sandwich is assumed to be of covalent type [6] with additional ionic contribution between Bi and Te(1). The hexagonal superstructure unit cell and the sandwich character are shown schematically in Fig. 1.

---

[1]) Röntgenring 8, D-8700 Würzburg, FRG.

[2]) Present address: Siemens AG, Zentralbereich Technik, D-8000 München 83.

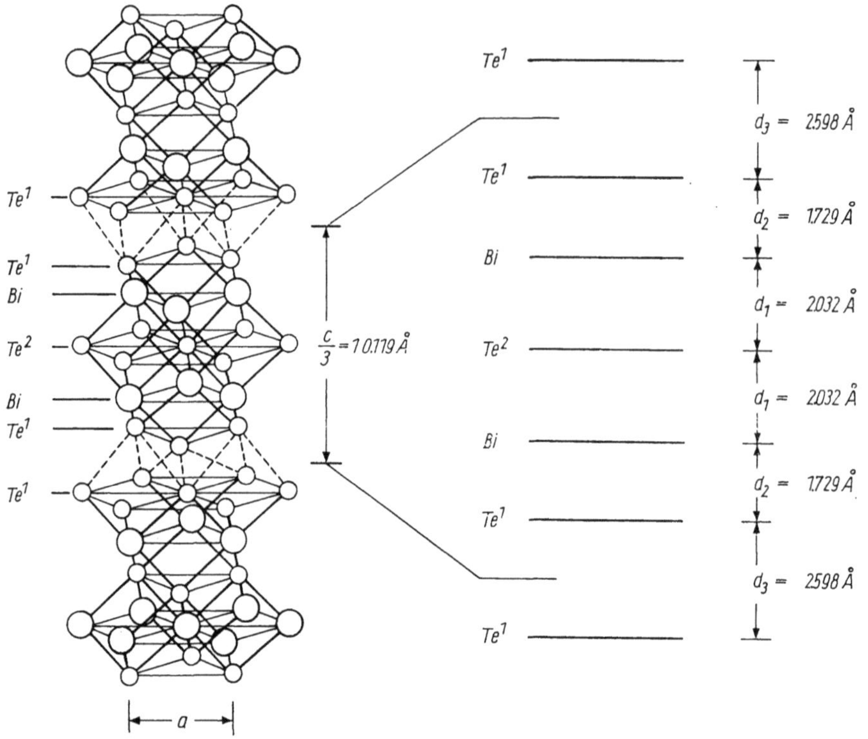

Fig. 1. Hexagonal unit cell and schematic illustration of the sandwich structure (data for $T = 77$ K)

For a long time the bonding between the sandwiches (Te(1)–Te(1)) has been thought of purely van-der-Waals type because of the easy cleavage behaviour and the results of LEED experiments [7] on the cleavage planes perpendicular to the c-axis. But in view of the assumed electron transfer from Te(1) to Bi [6] and of the high polarizability of Te, long-range Coulomb forces cannot be excluded and may play an important part in the bonding properties. Consequently, the results of a number of other experiments, like ultrasonic measurements [8], far infrared studies [9], inelastic neutron scattering [10, 11], inelastic tunnel spectroscopy [12], and measurements of the mechano-caloric effect [13, 14], have been found not to be consistent with the assumption of van-der-Waals forces between sandwiches. The experimental data indicate that Coulomb interactions cannot be neglected. An increased intersandwich bonding may alternatively arise in the case of lattice imperfections like Te(2)-vacancies [13, 14].

In order to elucidate the bonding properties, especially the Te(1)–Te(1) bondings, we studied the changes of the lattice spacings by neutron diffraction and the shift of Raman active phonon frequencies both under uniaxial and hydrostatic pressure.

2. Experimental

The elastic neutron scattering experiments have been performed at the D15 diffractometer at the ILL Grenoble. All measurements were done in normal beam geometry with lifting counter. A Cu(331) monochromator and an incoming wavelength of $\lambda =$

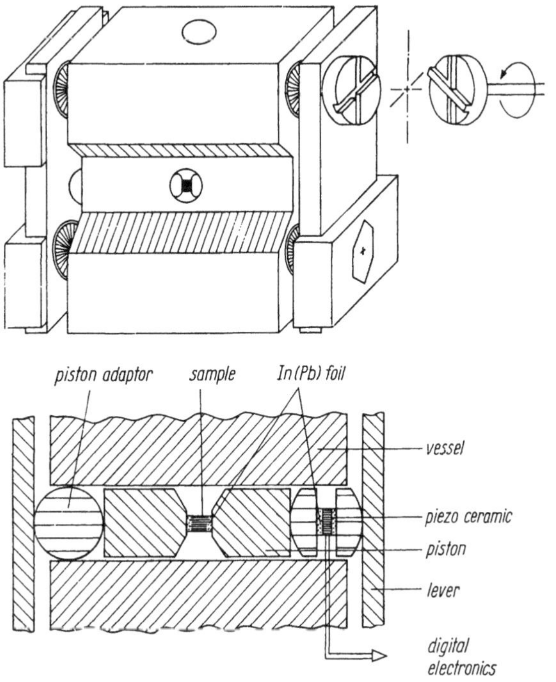

Fig. 2. Schematic view and cross section of the uniaxial pressure cell

= 1.174 Å have been used. The diffraction experiments under hydrostatic pressure were performed with an ILL standard pressure cell [15, 16] with helium gas as the pressure medium.

The pressure vessel for uniaxial pressure is made of a special aluminum alloy with sufficiently high mechanical stability. There is a hole of 12 mm diameter at the sample position, which allows the use of the pressure cell both for neutron experiments and optical measurements. The sample is pressurized by two guided movable aluminum pistons. In order to overcome surface roughness and to establish pressure equalization lead foils (for n-diffraction) or indium foils (for Raman scattering) are put between the sample and the piston faces. Spherical adapter pieces prevent the pistons from canting. The pressure cell can be mounted in an optical cryostat or cryostats for neutron scattering. Force is produced by compressing Belleville spring washers by turns of mechanical pivots from outside the cryostat at low temperatures. The pressure is detected by a piezo ceramic and analysed with a digital integration technique. Fig. 2 shows a schematic view and a cross section of the pressure cell.

The Raman experiments have been performed with an argon ion laser of wavelengths $\lambda = 514.5$ and 496.5 nm. For reasons of the small energy gap all measurements employed the backscattering geometry. The scattered light was analysed by a SPEX double monochromator with spatial filter for stray light reduction and was detected with a cooled GaAs photomultiplier, photon counting electronics, and a multichannel analyser. For uniaxial pressure, we have used the same pressure equipment as for neutron diffraction and as described above, apart from the cryostat. Optical hydrostatic pressure experiments were carried out with the high pressure equipment system PS-100 100000 PSI pumping system (High Pressure Equipment Comp. Inc, Linden, Pennsylvania) with oil as hydrostatic pressure medium.

Samples were grown from stoichiometric melt by Czochralski method. They all showed p-type behaviour. The crystals were cut by an acid string saw (Honeywell etch) and the noncleavage surfaces polished with Honeywell etch for the optical experiments.

3. Results of the Structural Investigations under Pressure

Structural investigations under pressure are a very direct method to study the bonding properties in solids. Therefore, we have measured the lattice parameters of the hexagonal unit cell and the atomic parameters under uniaxial pressure up to 6×10^7 Pa and hydrostatic pressure up to 4×10^8 Pa with elastic neutron scattering at $T = 77$ K. Unfortunately, uniaxial pressure above 6×10^7 Pa caused plastic deformation and destruction of the samples. Under hydrostatic conditions however the crystals are stable. In the low pressure region, where our experiments were performed, no phase transition is expected. At room temperature and for hydrostatic pressure about $p = 2.3 \times 10^9$ Pa, a phase transition occurs accompanied by a change of the electronic parameters [17]. A phase transformation to metallic behaviour is observed at $p = 6.7 \times 10^9$ Pa [17, 18].

For the determination of the lattice constants and atomic parameters 215 Bragg reflections have been measured for each pressure. The data evaluation was performed with an ILL refinement program of the PROMETHEUS SYSTEM [19]. Extinction effects were taken into consideration by the method of Becker and Coppens [20 to 22]. The structure refinement suffered mainly from two facts. There are, on the one hand, the geometrical restrictions resulting from the construction of the cell for uniaxial pressure (cf. Fig. 2) and, therefore, limiting the number of measureable reflections. There was encountered, on the other hand, very strong anisotropic extinction which required important corrections in the refinement. Thus we only present here the pressure dependence of the lattice parameters and the positional parameters $d_1 = (u - \frac{1}{3})\,c$, $d_2 = (v - u - \frac{1}{3})\,c$, and $d_3 = (c/3 - 2d_1 - 2d_2)$ (cf. Fig. 1). u, v are the atomic positions for Bi and Te, respectively, in the rhombohedral unit cell in Wyckoff notation [26].

Fig. 3 shows the decrease of the c-lattice constant under pressure. Uniaxial pressure parallel [001] influences the c-lattice parameter about five times stronger than hydrostatic pressure. While in the uniaxial case the variation is nearly linear, a pronounced nonlinear behaviour of concave form is observed with increasing hydrostatic pressure. This nonlinearity indicates the strong anharmonicity of the lattice. It could be a hint that repulsion forces increase under pressure and strengthen the intersandwich bonds. A similar reversible nonlinear dependence of c under hydrostatic pressure has been found in the homologous compound Sb_2Te_3 [23]. The a-lattice constant does not show a significant variation under uniaxial pressure within experimental errors but decreases slightly by about 0.3% in the hydrostatic case at 4×10^8 Pa. In Fig. 3 we also show the variation of the c-lattice parameter under pressure as calculated from the elastic constants [8]. Our experimental data show a much stronger pressure dependence. The calculation by means of the elastic constants yields a value of 2.5, which is too small for the ratio of the variation of c under uniaxial pressure to that

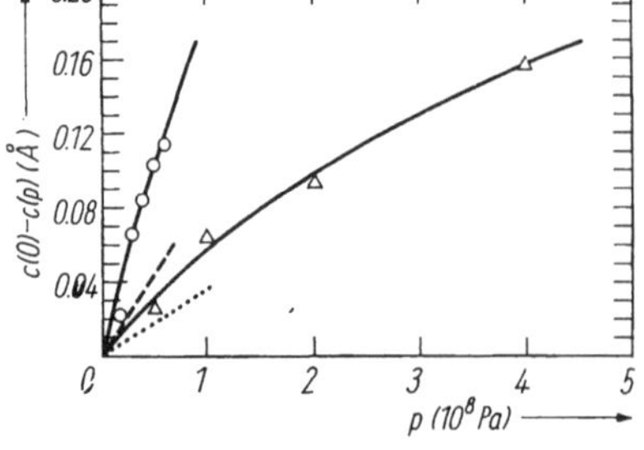

Fig. 3. Variation of the c-lattice parameter as determined by neutron diffraction under uniaxial (○) and hydrostatic pressure (△) ($T = 77$ K). The solid lines are guide for the eye. Also indicated are the predictions from the elastic constants for $c(0) - c(p)$ in case of uniaxial (— — —) and hydroststatic (. . . . . .) pressure, respectively

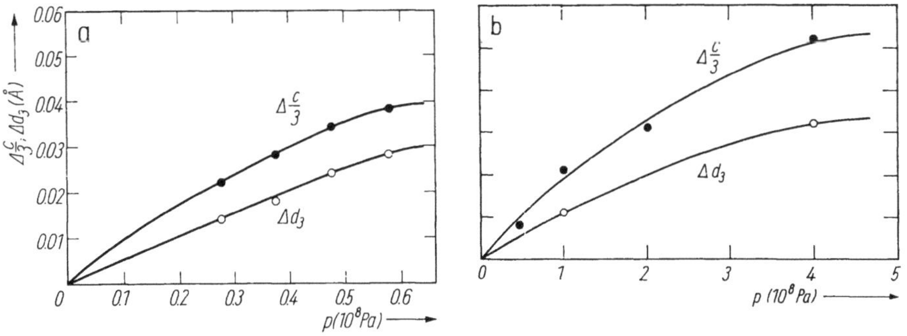

Fig. 4. Pressure dependence of the sandwich dimension $\Delta c/3 = (c(0) - c(p))/3$ and that of the intersandwich distance $\Delta d_3 = d_3(0) - d_3(p)$ for a) uniaxial pressure and b) hydrostatic pressure

under hydrostatic pressure. Results similar to ours were also found in the measurements of the mechano-caloric effect [13]. Qualitatively the strong anharmonic properties in Bi_2Te_3 may be the reason for these discrepancies between pressure experiments and the ultrasonic investigations.

In the previous considerations only the mean effect of pressure on the whole c-axis is considered. But when looking at Fig. 1 the different spacings between the atomic sheets indicate a different reaction on external stress for the sandwich and intersandwich regions. Fig. 4 compares the relation between the variation with pressure of the whole sandwich dimension $\Delta c/3$ in c-direction with that of the intersandwich distance Δd_3 (as determined from the positional parameters). In both cases, under uniaxial and hydrostatic pressure, nearly $\frac{2}{3}$ of the changes in the c-lattice constant are caused by that of the intersandwich distances. Only one third of the change can be attributed to the intrasandwich distances. The ratio $\Delta c/(3\,\Delta d_3) = 1.4$ of the changes of the sandwich dimension $c/3$ and of the intersandwich distance d_3 must be seen in comparison with the ratio of the corresponding total dimensions $c(0)/(3d_3(0)) = 3.9$. While the sandwich extension in c-direction is around four times longer than the intersandwich distance, external stress changes the whole sandwich dimensions only up to 0.5%, but the intersandwich distance up to 1.2% at $p = 4 \times 10^8$ Pa. Furthermore the small change of the intrasandwich layer seperation is nearly equally spread over the four distances between the layers within the sandwich. This behaviour is a clear and direct experimental evidence that the intersandwich bonding is much weaker than the intrasandwich (covalent) bond.

The difference in the variation of c and d_3 under uniaxial and hydrostatic pressure is well understandable under these circumstances. Uniaxial pressure parallel to the c-axis should cause a rather strong decrease of the c-lattice parameter and additionally a transverse dilatation of the a-lattice parameter. However, strong interatomic potentials in the basal plane perpendicular to the external uniaxial stress will hardly be affected at the low pressures used in our experiments. Under hydrostatic conditions also in a-direction a compression is expected which counteracts the compression parallel to c. The alternating strong and weak bondings in Bi_2Te_3 along the c-axis may well be the reason for the great differences obtained under hydrostatic and uniaxial pressure upon c and for the anharmonicities observed in our results (deviation from linearity in Fig. 4).

4. Results of the Raman Experiments

Further insight into the bonding properties can be gained by the study of the dynamical properties of the crystal. Group theoretical considerations for the Bi_2Te_3 structure yields 15 lattice dynamical modes at the zone centre,

$$\Gamma = 2\Gamma_1^+(A_{1g}) + 2\Gamma_3^+(E_g) + 3\Gamma_2^-(A_{2u}) + 3\Gamma_3^-(E_u)\,,$$

where $1\Gamma_2^-(A_{2u})$ and $1\Gamma_3^-(E_u)$ represent the acoustic modes. The Raman active vibrations Γ_1^+ $A_{1g})$ and $\Gamma_3^+(E_g)$ are of special interest for the intersandwich bondings because in these modes neighbouring Te(1) atoms are vibrating out of phase. Rigid layer modes and Davydov splittings encountered in other layer structures are not possible in this case because of the identical symmetry of sandwich and unit cell.

Fig. 5 shows the four expected Raman active lattice vibrations for $\boldsymbol{k} \parallel \boldsymbol{c}$ at $T = 300$ K with polarized incident light but without using an analyser. The frequencies are: 36.5 cm^{-1} (E_g^1), 62.0 cm^{-1} (A_{1g}^1), 102.3 cm^{-1} (E_g^2), 134.0 cm^{-1} (A_{1g}^2). The results are in good agreement with earlier experiments of Richter et al. [9] and inelastic neutron experiments [10, 11] but disagree with the results of Raman investigations of Nikolić et al. [24]. It should be noted that our results include the low-frequency mode E_g^1, which has not been observed in the previous experiments [9], for reasons of its low intensity. At low temperatures this vibration could not be seen because of the low signal-to-noise ratio inside the cryostat and the pressure cell.

Fig. 6 shows the dependence of the Raman active modes A_{1g}^1, E_g^2, A_{1g}^2 under uniaxial pressure up to 6×10^7 Pa parallel to the c-axis at $T = 77$ K. Fig. 6 also illustrates schematically the displacements of the five basis atoms inside a sandwich calculated by factor group analysis [9]. The frequencies of the A_{1g}^1 and E_g^2 modes increase linearly with increasing pressure. The high-frequency A_{1g}^2 vibration is independent of pressure within experimental accuracy. Linewidths and intensities of all modes do not change in the whole pressure region. The strong pressure dependence of the A_{1g}^1 vibration in comparison to the A_{1g}^2 mode becomes evident when considering the symmetry coordinates [9] which are, to a good approximation, the actual eigenvectors as can be seen from lattice dynamical model calculations [25]. In the A_{1g}^1 mode Bi and Te(1) are vibrating in phase while in the A_{1g}^2 vibration Bi and Te(1) are out of phase. Te(1) atoms in adjacent sandwiches are always vibrating out of phase. So in the low-frequency A_{1g}^1 vibration the pressure variation of frequency especially illustrates the anharmonicities of the intersandwich Te(1)–Te(1) bondings. In the case of A_{1g}^2 the covalent ionic bondings between Bi and Te(1) cover these effects of the intersandwich forces.

In the hydrostatic Raman investigation up to 6×10^8 Pa only the modes A_{1g}^1 and E_g^2 could be observed because of the high scattering background in the pressure cell. The low signal-to-noise ratio, which even decreased with pressure, made it impossible to localize the corresponding frequencies above $p = 4.5 \times 10^8$ Pa. Fig. 7 shows the dependence of the A_{1g}^1 and E_g^2 modes under hydrostatic pressure at $T = 300$ K. Also in this case the observed mode frequencies increase linearly with pressure.

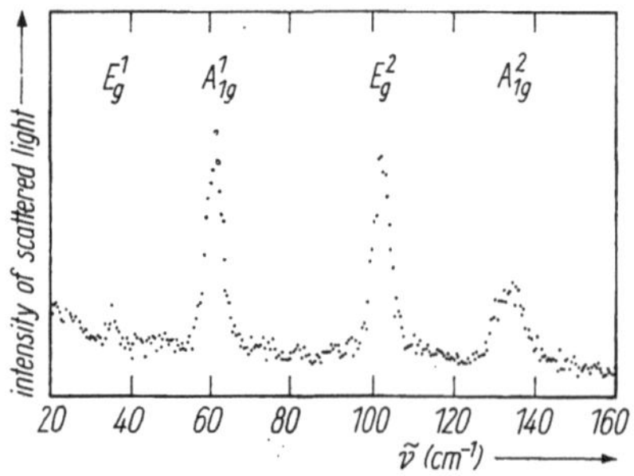

Fig. 5. Raman spectrum of Bi_2Te_3 at $T = 300$ K for $\boldsymbol{k} \parallel \boldsymbol{c}$ (backscattering geometry) without using analyser

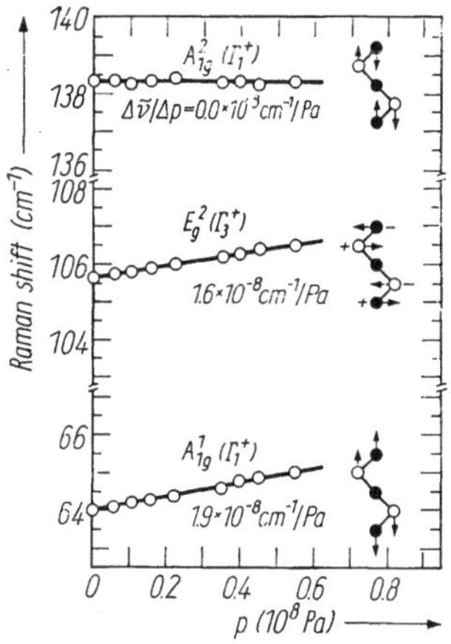

Fig. 6

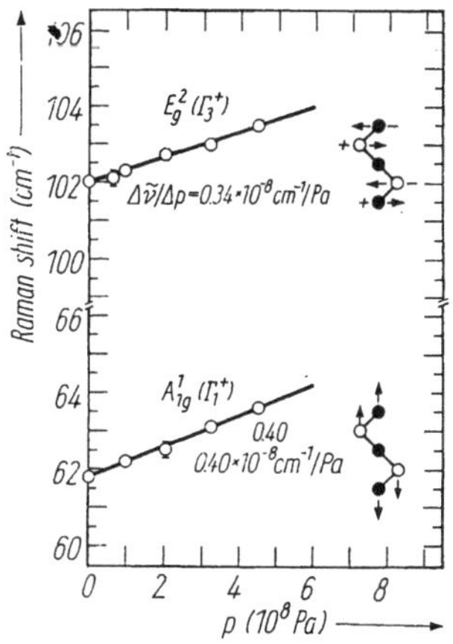

Fig. 7

Fig. 6. Pressure dependence of the Raman modes under uniaxial pressure ($T = 77$ K). For each mode, the symmetry species is indicated and the value is given for the pressure coefficient $d\tilde{\nu}/dp$ At the right, there are shown the atomic displacements (○ Bi, ● Te) according to a factor. group analysis [9]

Fig. 7. Pressure dependence of the Raman modes under hydrostatic pressure ($T = 300$ K). For each mode, the symmetry species is indicated and the value is given for the pressure coefficient $d\tilde{\nu}/dp$. At the right, there are shown the atomic displacements (○ Bi, ● Te) according to a factor group analysis [9]

A comparison of the frequency shifts observed under uniaxial and hydrostatic pressure are qualitatively very similar in both cases. The results again can be related to each other by a scaling factor $(d\tilde{\nu}/dp_{uniaxial})/(d\tilde{\nu}/dp_{hydro})$ of order 4.8. Under both conditions the pressure variation of the E_g^2 mode preponderates that of the A_{1g}^1 vibration by an amount of 18 to 19%.

5. Conclusion

The structural investigations demonstrate the existence of weak bondings between the sandwiches. The main pressure effect is the shortening of the intersandwich distances and the increase of the forces between the sandwiches (anharmonicity), in contrast to the properties inside a sandwich. Electrostatic repulsion, resulting from the partly ionic character of the Te(1) atoms, may be the origin for our findings. In addition to pure van-der-Waals forces, stronger bondings probably resulting from the long-range Coulomb interactions must be taken into account.

In comparison, the linear dependence of the Raman active lattice vibrations under pressure results from the influence of the strong intrasandwich bondings on the Raman modes which becomes evident by considering the corresponding eigenvectors (cf. Fig. 6, 7). The relatively strong pressure dependence of the A_{1g}^1 and E_g^2 modes again confirm the weak character of the Te(1)–Te(1) bonds and its strong anharmonicity.

Both the structural and dynamical investigations show the principal equivalence of uniaxial and hydrostatic pressure for Bi_2Te_3. Sandwich dimensions and interlayer distances as well as the Raman active lattice vibrations show the same qualitative behaviour under external stress, however with a much stronger dependence upon uniaxial pressure. For both experimental studies, i.e. neutron diffraction and Raman scattering, we obtain a scaling factor of about 4.5 to 5 connecting the results for the two kinds of pressure. This unique scaling factor means that, in future, we need not perform investigations under uniaxial pressure with the risk of damaging the sample but are allowed to use hydrostatic pressure instead.

Acknowledgements

We are much indebted to H. Koubek for growing the Bi_2Te_3 single crystals. We also would like to thank Ch. Vettier for his advises concerning the pressure cells and W. F. Kuhs for placing the refinement computer program for structural data reduction to our disposal. We are grateful to Fa. Rosenthal Technik AG, Werksgruppe III, Lauf, for providing the piezo ceramics for the uniaxial pressure measurements. Financial support of the Deutsche Forschungsgemeinschaft for the Raman experiments and of the Bundesministerium für Forschung und Technologie for the neutron diffraction experiments is gratefully acknowledged.

References

[1] H. J. GOLDSMID and R. W. DOUGLAS, Brit. J. appl. Phys. 5, 386 (1954).
[2] A. D. GOLETSKAYA, V. V. SOLOGUB, and S. S. SHALYT, Soviet Phys. — Semicond. 5, 416 (1971).
[3] W. M. YIM and F. D. ROSI, Solid State Electronics **15**, 1121 (1972).
[4] D. R. LOVETT, Semimetals and Narrow-Bandgap Semiconductors, Pion Ldt., London 1977.
[5] H. H. SOONPAA and W. A. SCHWALM, Solid State Commun. **42**, 201 (1982).
[6] J. R. DRABBLE and C. H. L. GOODMAN, J. Phys. Chem. Solids **5**, 142 (1958).
[7] D. HANEMAN, Phys. Rev. **119**, 567 (1960).
[8] J. O. JENKINS, J. A. RAYNE, and R. W. URE, JR., Phys. Rev. B **5**, 3171 (1972).
[9] W. RICHTER, H. KÖHLER, and C. R. BECKER, phys. stat. sol. (b) **84**, 619 (1977).
[10] V. WAGNER, G. DOLLING, B. M. POWELL, and G. LANDWEHR, phys. stat. sol. (b) **85**, 311 (1978).
[11] W. KULLMANN, G. EICHHORN, R. GEICK, N. LEER, and U. STEIGENBERGER, ILL Experimental Reports and Theory College Activities 1983, Grenoble, Exp.-Nr. 4-01-245, (p. 34).
[12] A. MORITANI, M. FUKJIOKA, and J. NAKAI, Japan. J. appl. Phys. **15**, 1523 (1976).
[13] P. DATO, Diploma Thesis, Univ. of Würzburg, 1979.
[14] P. DATO and H. KÖHLER, to be published.
[15] J. PAUREAU and C. VETTIER, Rev. sci. Instrum. **46**, 1484 (1975).
[16] J. PAUREAU and C. VETTIER, High Temp.-High Press. **7**, 529 (1975).
[17] L. G. KHVOSTANTSEV, A. I. ORLOV, N. KH. ABRIKOSOV, T. E. SVECHNIKOVA, and S. N. CHIZHEVSKAYA, phys. stat. sol. (a) **71**, 49 (1982).
[18] L. F. VERESHCHAGIN, E. YA. ATABAEVA, and N. A. BENDELIANI, Soviet Phys. — Solid State **13**, 2051 (1972).
[19] U. H. ZUCKER, E. PERENTHALER, W. F. KUHS, R. BACHMANN, and H. SCHULTZ, J. appl. Cryst. **16**, 358 (1983).
[20] P. J. BECKER and P. COPPENS, Acta cryst. **A30**, 129 (1974).
[21] P. J. BECKER and P. COPPENS, Acta cryst. **A30**, 148 (1974).
[22] P. J. BECKER and P. COPPENS, Acta cryst. **A31**, 417 (1975).
[23] N. SAKAI, T. KAJIWARA, K. TAKEMURA, S. MINOMURA, and Y. FUJII, Solid State Commun. **40**, 1045 (1981).
[24] P. M. NIKOLIĆ, L. J. A. MILJKOVIĆ, and B. LAVRENČIĆ, Fizika, Suppl. **10**, 354 (1978).
[25] G. EICHHORN, H. RAUH, W. KULLMANN, R. GEICK, and N. LEHNER, Verh. DPG (VI) **18**, 569 (1983).
[26] R. W. G. WYCKOFF, Crystal Structures, Vol. 2, Interscience Publ., New York/London/Sydney 1964.

(Received May 7, 1984)

phys. stat. sol. (b) **125**, 139 (1984)

Subject classification: 10.2 and 20.3; 22.5.2

Institute of Physics, Academy of Sciences of the Estonian SSR, Tartu[1])

Photostimulated Reorientation of $H_z^{[001]}(Pb^{2+})$ Centres in a KCl:$PbCl_2$ Crystal

By

V. NAGIRNYI, T. SOOVIK, and S. ZAZUBOVICH

Temperature dependences of polarization and kinetic characteristics of luminescence are investigated for Pb^{2+} ions perturbed by a (nnn) cation vacancy and an interstitial chlorine atom ($H_z^{[001]}(Pb^{2+})$ centres) in a KCl:$PbCl_2$ crystal. At $T > 110$ K the reorientation of these centres in the triplet excited state of the Pb^{2+} ion stimulated by polarized UV radiation becomes evident in the polarization spectra and decay kinetics of the A_T emission. It results neither in the depolarization of the emission nor in the "alignment" of the centres. Most probably the effect can be explained by a strong covalent interaction between the excited Pb^{2+} ion and an interstitial, and it is caused by the dissociation of the two-component defect near the Pb^{2+} ion and the following jump of the interstitial chlorine atom into one of the ⟨100⟩ directions perpendicular to the $\{Pb^{2+}-v_c^-\}$ axis.

Изучены температурные зависимости поляризационных и кинетических характеристик люминесценции ионов Pb^{2+}, возмущенных далекой катионной вакансией и междоузельным атомом хлора ($H_z^{[001]}(Pb^{2+})$-центров) в KCl–$PbCl_2$. При $T > 110$ K в поляризационных спектрах и кинетике затухания A_T-излучения проявляются переориентации этих центров в триплетном возбужденном состоянии, стимулируемые поляризованной УФ-радиацией. Они не приводят ни к деполяризации излучения, ни к "выстраиванию" центров в определенном направлении. Наиболее вероятно, переориентации объясняются сильным ковалентным взаимодействием возбужденного иона Pb^{2+} с междоузельным атомом хлора и происходят в результате диссоциации двухкомпонентного дефекта рядом с ионом Pb^{2+} и перескока междоузельного атома на одно из направлений ⟨100⟩, перпендикулярное оси $\{Pb^{2+}-v_c^-\}$.

1. Introduction

In X-irradiated KCl:$PbCl_2$ crystals associates of an interstitial chlorine atom (H centre) with $Pb^{2+}v_c^-$ dipoles of two types are found and investigated by optical [1, 2] and ESR [3] methods. They differ by the position of the cation vacancy v_c^- (in [110] or [001] direction) with respect to the Pb^{2+} ion and are labelled in [2] as $H_z^{[110]}(Pb^{2+})$ and $H_z^{[001]}(Pb^{2+})$, respectively. At 80 K $H_z^{[110]}(Pb^{2+})$ centres with 3.35 eV emission are mainly created. $H_z^{[001]}(Pb^{2+})$ centres with 1.8 eV emission can be formed thermally from $H_z^{[110]}(Pb^{2+})$ owing to the jumping of v_c^- from (nn) to (nnn) position with respect to Pb^{2+}. The luminescence of these associates is caused by electronic transitions from the triplet excited state of the Pb^{2+} ion perturbed by v_c^- and an interstitial chlorine atom [2]. The latter, in the form of the Cl_2^- molecular ion, occupies an anion (a) site and a cation (c) vacancy [3]. Thus, the perturbing defect near Pb^{2+} can be labelled as $[Cl_2^-)]_{ac}$.

In this paper, several interesting peculiarities are observed in the polarization and decay kinetics of $H_z^{[001]}(Pb^{2+})$ emission at $T > 110$ K, which may be interpreted as a result of the photostimulated reorientation of $H_z^{[001]}(Pb^{2+})$ centres in the excited state.

[1]) 142 Riia Street, 202400 Tartu, USSR.

2. Experimental Procedure

Single $KCl:PbCl_2$ crystals and the methods of their preparation for experiments were described in [2]. The samples X-irradiated for 3 to 4 h at 80 K, were heated up to the temperature of V_k delocalization. Then, at 80 K F, Pb^+, and Pb^{3+} centres were optically destroyed to obtain the maximum amount of $H_z^{[110]}$ (Pb^{2+}) centres. On heating up to 230 K the $H_z^{[110]}$ (Pb^{2+}) centres were transformed into $H_z^{[001]}$ (Pb^{2+}) centres.

The spectral characteristics of the emission were measured by the method described in [2]. The polarization properties of the stationary emission were investigated in the temperature region from 4.2 to 300 K at various excitation and observation conditions schematically shown in Fig. 1. The emission was excited by monochromatic, linearly polarized light produced by a Glan prism. The electric vector $\boldsymbol{E}_{exc}$ was parallel to the z-axis ([001] direction of the crystal) or to the y-axis ([010] direction). The polarized emission was observed through red optical filters (ZhS-4 + KS-11) and an Ahrens prism either in the direction of the exciting light propagation (along the x-axis) or in the perpendicular direction (along y-axis). The polarization degree

$$P = \frac{I_{\parallel} - I_{\perp}}{I_{\parallel} + I_{\perp}},$$

where $I_{\parallel}$ and $I_{\perp}$ are the intensities of the emission polarized parallel and perpendicular either to the z-axis (if the observation was along y-axis) or to $\boldsymbol{E}_{exc}$ (if the observation was along x-axis). The crystal orientation was the same in all experiments: the [001] axis parallel to z-axis.

The decay of the polarized emission was investigated under excitation by about 45 ns spark flashes in air through a monochromator ZMR-3 and a polaroid ($\boldsymbol{E}_{exc} \parallel z$-axis). The emission was observed in y-direction through another polaroid and ZhS + + KC-11 filters and detected by a cooled photomultiplier of type 9816B, EMI, England working in the photon counting regime. The decay curves $I_{\parallel}(t)$ and $I_{\perp}(t)$ were successively registered with a multichannel time analyzer (NTA 1024, EMG, Hungary, with the minimum channel time $t = 5$ μs or LP 4900, Nokia, Finland, with $t = 40$ ns). Each decay curve was separated into (two or three) exponential components taking into account the corresponding computer-calculated decay times.

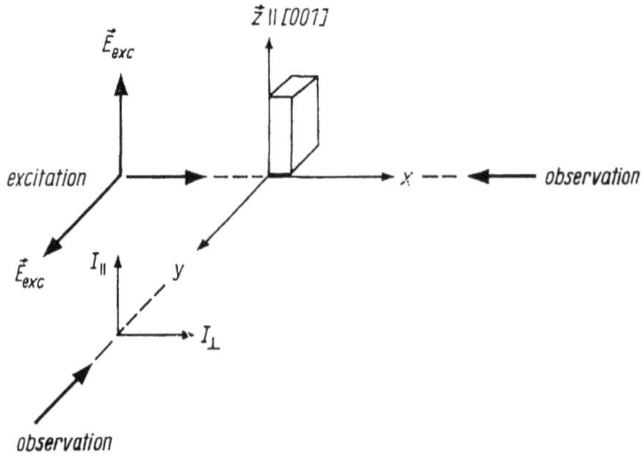

Fig. 1. Excitation and observation conditions of the study of polarized luminescence

3. Experimental Results

3.1 Spectral and polarization characteristics of the stationary luminescence

The 1.8 eV emission of $H_z^{[001]}$ (Pb^{2+}) centres is caused by electronic transitions from tetragonal minima of the triplet excited state of the Pb^{2+} ion and, thus, it is analogous to the A_T emission of $Pb^{2+}v_c^-$ centres. The 4.38 eV absorption band is analogous to the A band of $Pb^{2+}v_c^-$ [2]. The Stokes shift is about 2.6 eV.

Spectral and polarization characteristics of the A_T emission of $H_z^{[001]}$ (Pb^{2+}) centres are shown in Fig. 2. As the temperature grows from 80 to 210 K, the A_T emission band (curve 1) shifts slightly to the high-energy side (by about 0.03 eV), and the excitation band (curve 2), to the low-energy side (by about 0.01 eV). The degree of polarization observed on excitation in the A absorption band region (curve 3) changes from −33 to +18% at 80 K and from +50 to −27% at 210 K.

Thus, the heating of the crystal from 80 to 210 K results in a sharp inversion of the polarization spectra as well as in an increase (by about two times) of the absolute values of the polarization degree, but it practically does not influence the spectral characteristics of the $H_z^{[001]}$ (Pb^{2+}) emission.

3.2 Temperature dependences of the polarization of stationary luminescence

Fig. 3 demonstrates the temperature dependences of the polarization degree (P) of the $H_z^{[001]}$ (Pb^{2+}) emission on excitation in the high-energy (P_h) and low-energy (P_l)

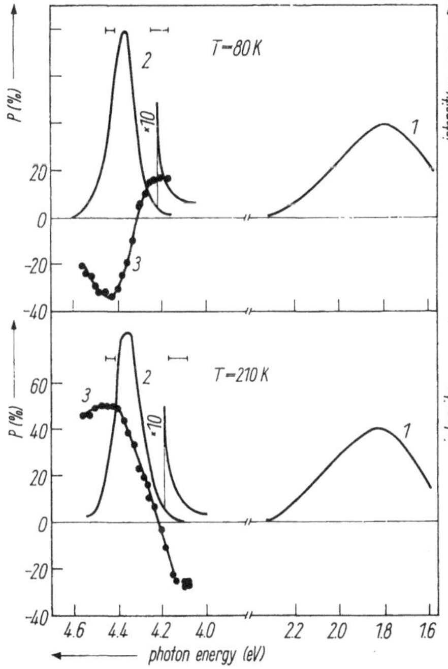

Fig. 2

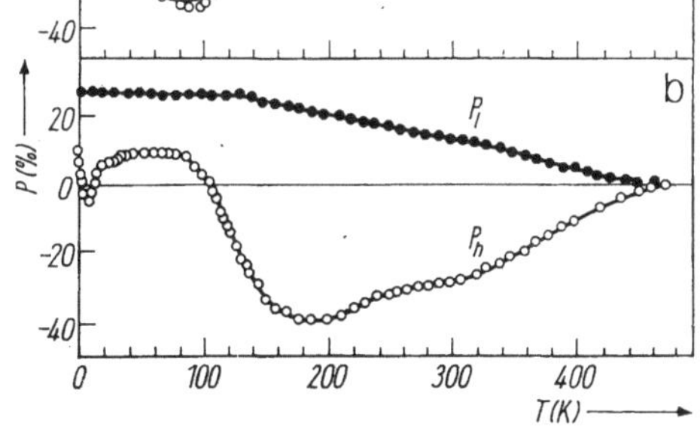

Fig. 3

Fig. 2. (1) Emission, (2) excitation, and (3) polarization spectra of the A_T emission of $H_z^{[001]}(Pb^{2+})$ centres at 80 and 210 K

Fig. 3. Temperature dependences of the polarization degrees (P) of the A_T emission of a) $H_z^{[001]}(Pb^{2+})$ and b) $Sn^{2+}v_c^-$ centres in KCl crystal measured on excitation in the high-energy (P_h) and low-energy (P_l) regions of the A absorption band indicated in Fig. 2

tails of the A absorption band. For a comparison, the analogous dependences $P_h(T)$ and $P_l(T)$ are represented in Fig. 3b for the A_T emission of $Sn^{2+}v_c^-$ centres in a KCl: $SnCl_2$ crystal, which have been discussed in detail earlier [4]. Up to 100 K $P(T)$ of $H_z^{[001]}$ (Pb^{2+}) and $Sn^{2+}v_c^-$ centres are qualitatively similar. At $T < 25$ K the emission of $H_z^{[001]}$ (Pb^{2+}) occurs mainly from the initially populated minima of the triplet excited state, and $P_h > 0$, $P_l > 0$. As the temperature rises, P_h becomes negative, since the lower minima are now more populated due to thermally stimulated transitions between the upper and lower minima. At $T \gtrapprox 70$ K in $H_z^{[001]}$ (Pb^{2+}) (at $T \gtrapprox 170$ K in $Sn^{2+}v_c^-$) thermal equilibrium is established between the X, Y, and Z minima of the excited state.[2]) Under these conditions the absolute value of the negative polarization degree (P_h) in both centres is approximately two times larger than that of the positive one (P_l). It means [5, 6] [3]) that in both cases the Z minimum is the lowest one (in the case of the $H_z^{[001]}$ (Pb^{2+}) centre z is directed from Pb^{2+} to v_c^-). On high-energy excitation the X_y and Y_x states, and on low-energy excitation the X_z, Z_x, Y_z, Z_y states are populated preliminarily. As in the $H_z^{[001]}$ (Pb^{2+}) centre the perturbing defect is located in the [001] direction with respect to the Pb^{2+} ion [3], the signs and absolute values of P_h and P_l point to the dominating role of non-electrostatic interactions of Pb^{2+} optical electrons with the surroundings [6]. On further heating of the KCl:$SnCl_2$ crystal, the P_h and P_l values decrease gradually owing to Z $\rightleftarrows$ X, Y transitions. Above 300 K the diffusion jumps of v_c^- around Sn^{2+} start influencing the polarization properties of the emission and lead to its complete depolarization at 460 K.

In contrast to $Sn^{2+}v_c^-$, the P_h and P_l values of the $H_z^{[001]}$ (Pb^{2+}) centre undergo a drastic change at $T < 110$ K. Near 140 K the signs of P_h and P_l are inverted. As the temperature rises their absolute values increase quickly and become saturated at 210 K. Above 230 K $H_z^{[001]}$ (Pb^{2+}) centres become thermally destroyed but P_h and P_l values keep constant up to 300 K. At 210 K as well as at 80 K the polarization degrees, if $\boldsymbol{E}_{exc} \parallel z$-direction, are the same when observed either in x- or y-direction (see Fig. 1). If $\boldsymbol{E}_{exc} \parallel y$-direction, $P = 0$ when observed along y-direction and it is the same as in the previous case when observed along x-direction.

The $P_h(T)$ and $P_l(T)$ dependences measured on cooling and heating the crystal coincide completely. This fact as well as a very weak dependence of the emission and excitation spectra on temperature indicate that the inversion of the polarization spectrum in the temperature region from 110 to 210 K is not connected with any nonreversible change in the geometrical structure of the $H_z^{[001]}$ (Pb^{2+}) centre.

3.3 Decay kinetics of the emission. Characteristics of separate decay components

At 80 K the decay of the $H_z^{[001]}$ (Pb^{2+}) emission consists of two — slow (s) and fast (f) — components. The decay time of the fast component τ_f is independent of temperature. It is shorter than 45 ns on high-energy excitation and equal to 55 ns on low-energy excitation (Fig. 4a). The fast component is polarized by about +50%. Its light sum is about several per cent of that of the slow component and decreases as the temperature rises (Fig. 4c).

The decay time of the slow component (τ_s) at 80 K is equal to 40 μs on high-energy excitation and 27 μs on low-energy excitation and becomes shorter as the temperature increases (Fig. 4a). The slow component is polarized only up to 140 K. At 80 K its

---

[2]) Due to the large energy distance between the radiative and metastable minima of lead centres ($\delta \approx 10^{-2}$ eV) mainly the metastable X_x, Y_y, Z_z minima are populated at 70 K.

[3]) The model [5] may be applied to $H_z^{[001]}$(Pb^{2+}) centres as their Stokes shift is larger (about 2.6 eV) than the spin–orbit interaction constant (about 1 eV for the Pb^{2+} ion).

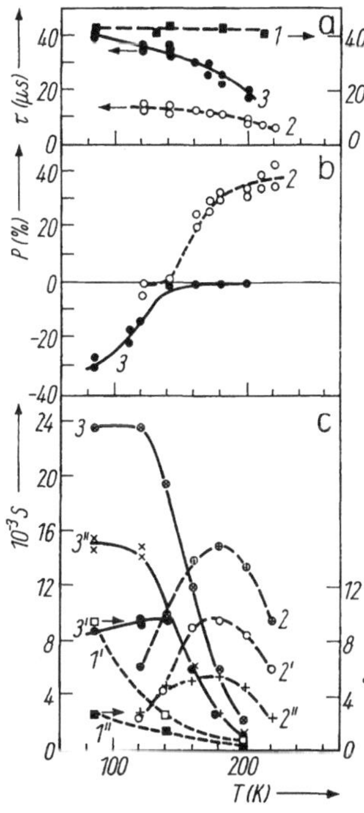

Fig. 4. Temperature dependences of a) decay time τ; b) degrees of polarizaton P; c) light sums S of the fast ((1), (1′), (1″)), "intermediate" ((2), (2′), (2″)), and slow ((3), (3′), (3″)) components of the A_T emission decay measured on high-energy excitation ($h\nu_{exc} = 4.42$ eV). c) (1′), (2′), (3′) $S_{||}$; (1″), 2″), (3″) $S_{\perp}$; (2), (3), $S_{||}+S_{\perp}$

polarization degrees are $P_h = -30\%$ (Fig. 4b) and $P_l = +18\%$. Two similar components have been observed in the decay of the A_T emission of $Pb^{2+}v_c^-$ centres and studied in [7, 8].

At $T > 120$ K a new component, here labelled as "intermediate", appears in the decay of the $H_z^{[001]}$ (Pb^{2+}) emission. Its decay time τ_i is about two to three times shorter than τ_s, and also decreases as the temperature grows (Fig. 4a). At 210 K τ_i is about 7.5 μs on high-energy and 4 μs on low-energy excitation, and the corresponding polarization degrees are +40% and −20%.[4]) At $T > 120$ K the light sum of the slow component sharply decreases but that of the "intermediate" component increases (Fig. 4c). At 220 K the latter is by at least an order larger than the former. Thus, the high polarization degree P_h of the stationary emission at 220 K is caused by the polarization of the "intermediate" component.

Its degree exceeds the limit value (+1/3) expected at perfect thermal equilibrium between all radiative and metastable minima of the triplet excited state [5].

4. Discussion

The temperature dependences of the polarization degrees, light sums, and decay times observed here for the $H_z^{[001]}$ (Pb^{2+}) emission can be explained by the photostimulated reorientation of these centres in the excited state which becomes possible at $T > 110$ K.

Photostimulated reorientations are thoroughly studied for anisotropic F_A, M, M_A, V_K colour centres in alkali halides (see, e.g., [9 to 12]). It is established that the activation energy for their reorientation in the excited state (E_{reor}^{exc}) is by an order smaller (near 0.1 eV) than that in the ground state (E_{reor}^{g}) (about 1 eV). Therefore, a statistical reorientation process tends to populate more and more orientations, where the centres are no longer excited by the incident polarized light, and the centres become "aligned". In [13] the photostimulated reorientations were detected by the ITC method also for $Pb^{2+}v_c^-$ dipoles in KCl:PbCl$_2$. In [4] the depolarization of the A_T emission has been observed in KCl:SnCl$_2$ at $T < 320$ K owing to the reorientations of the excited

[4]) Here smaller absolute values of P are obtained than in the stationary case (Fig. 2 and 3), due to the imperfection of samples used.

$Sn^{2+}v_c^-$ centres. However, the "optical alignment" of $Sn^{2+}v_c^-$ dipoles does not occur possibly due to the small difference between E_{reor}^{exc} and E_{reor}^{g}.

In contrast to all the investigated cases, the reorientation of $H_z^{[001]}$ (Pb^{2+}) centres under excitation by polarized light does not lead to the depolarization of their emission. The specially performed experiments show that the "optical alignment" of $H_z^{[001]}$ (Pb^{2+}) centres is also impossible. Most probably, such peculiarities of $H_z^{[001]}$ (Pb^{2+}) centres are caused by the complicated (two-component) structure of the defect perturbing the Pb^{2+} ion as well as by a strong covalent interaction between the excited Pb^{2+} ion and an interstitial. The processes in the excited state of these centres may be explained by using the following assumptions: (1) orientation of the p-function of the excited Pb^{2+} perpendicular to the $\{Pb^{2+}–[Cl_2^-)_{ac}\}$ axis is most preferable; (2) strong covalent interaction makes the interstitial chlorine atom (probably in the form of the $[Cl_2^-]_a$ molecular ion) orient along the p-function axis, and the excited state minimum in the $\{Pb^{2+}–[Cl_2^-]_a\}$ direction (i.e. perpendicular to the $\{Pb^{2+}–v_c^-\}$ axis) is now the lowest one. Therefore, the dissociation of the two-component defect occurs in the Pb^{2+} excited state, and the interstitial chlorine atom jumps from v_c^- onto one of the $\langle 100\rangle$ directions perpendicular to the $\{Pb^{2+}–v_c^-\}$ axis. The cation vacancy remains in its site that guarantees the memorizing of the original situation. This process needs activation energy, therefore its probability increases as the temperature grows. In the totolly symmetric ground state of Pb^{2+} the interstitial atom more preferentially occupies the cation vacancy: $[Cl_2^-]_{ac}$. Therefore, after the radiative transition it returns into its "own" v_c^-, and the original, isotropic distribution of $H_z^{[001]}$ (Pb^{2+}) centres between various $\langle 100\rangle$ directions is retrieved in the ground state. Thus, the "aligned" $H_z^{[100]}$ (Pb^{2+}) centres exist only for about 10^{-6} s after the absorption of the polarized light

Let us try to explain in more detail the polarization and kinetic properties of the $H_z^{[001]}$ (Pb^{2+}) emission. As has been mentioned above, z-axis is directed from Pb^{2+} to v_c^-, and the excited state minimum oriented in the direction from Pb^{2+} to the interstitial atom is the lowest one. On high-energy excitation by polarized light with $\boldsymbol{E}_{exc} \parallel z$-axis the X_y and Y_x states of the centres with the z-axis parallel to $\boldsymbol{E}_{exc}$ are initially populated. On low-energy excitation the X_z, Z_x, Y_z, and Z_y minima of the centres with z-axis $\perp \boldsymbol{E}_{exc}$ are initially populated. At $T \gtrapprox 70$ K the thermal transitions between X, Y, and Z minima result in the preferable population of the lowest Z minima on both excitations. At $T < 110$ K, when the reorientations of $H_z^{[001]}$ (Pb^{2+}) cannot occur, the emission from the Z minima dominates (in x- or y-directions), and $P_h < 0$, $P_l > 0$. The fast component is caused by the transitions from the preliminarily excited states, and the slow component occurs after the thermal equilibrium between the radiative and metastable minima is achieved.

At high temperatures, when the interstitial atom jumps from v_c^- into x- or y-directions, the X or Y states become the lowest ones. They are rapidly populated due to thermal transitions from the Z minimum, which is the upper one now, but it is the most populated one at the initial moment. As the most probable transitions are the thermal transitions without spin reorientation (e.g. $Z_y \rightleftarrows X_y$, $Z_x \rightleftarrows Y_x$), the X_y and Y_x minima (emitting along z-axis) become preferentially populated due to that process. The appearance of the "intermediate" component in the emission decay connected with the reorientation of $H_z^{[001]}$ (Pb^{2+}) centres, and its dominating role at $T > 170$ K points to the fact that after reorientation the thermal equilibrium is not established between the radiative and metastable minima. Therefore, the emission occurs mainly from the X_y and Y_x minima and has a high positive polarization degree on high-energy excitation and a negative one, on low-energy excitation. At temperatures near 140 K, when $P_h \approx 0$, and $P_l \approx 0$, the contributions of the transitions occurring from the Z minimum and from the X or Y minima are comparable, i.e. either before and after reorientation or after reorientation but before or after the new thermal equilibrium is established in the excited state due to the $Z \rightleftarrows X, Y$ transitions.

Thus, it may be concluded that owing to the complicated (two-component) structure of the defect perturbing the Pb^{2+} ion, the process of $H_z^{[001]}$ (Pb^{2+}) centre reorientation differs drastically from that of the centres investigated earlier. The reorientation of the excited $H_z^{[001]}$ (Pb^{2+}) centre results neither in the depolarization of emission nor in the "alignment" of the centres, as the system "remembers" its original position and returns into it after radiative transition.

Acknowledgement

The authors would like to thank Prof. Dr. V. Hizhnyakov for useful discussions.

References

[1] Zh. Egemberdiev, Trudy Inst. Fiz. Akad. Nauk EstSSR **51**, 163 (1980).
[2] Zh. Egemberdiev, S. Zazubovich, and V. Nagirnyi, phys. stat. sol. (b) **109**, 173 (1982).
[3] W. van Puymbroeck and D. Schoemaker, Phys. Rev. B **23**, 167 (1981).
[4] V. Hizhnyakov, S. Zazubovich, and T. Soovik, phys. stat. sol. (b) **66**, 727 (1974).
[5] V. Hizhnyakov, Preprint FI-36, Tartu 1975.
[6] V. Hizhnyakov and S. Zazubovich, phys. stat. sol. (b) **86**, 733 (1978).
[7] M. F. Trinkler, Optika i Spektroskopiya **16**, 854 (1964).
[8] S. Benci, R. Capelletti, F. Fermi, M. Manfredi, J. Z. Damm, and E. Mugenski, phys. stat. sol. (b) **90**, 657 (1978).
R. Capelletti, M. Manfredi, and E. Zecchi, Solid State Commun. **43**, 221 (1982).
[9] F. Lüty, Physics of Color Centers, Ed. W. Beall Fowler, Academic Press, Inc., 1968 (p. 209).
[10] R. W. Dreyfus, Phys. Rev. B **1**, 4826 (1970).
[11] I. Schneider, Appl. Optics **6**, 2197 (1967).
I. Schneider, M. Marrone, and M. N. Kabler, Appl. Optics **9**, 1163 (1970).
[12] R. B. Murray and P. G. Bethers, Phys. Rev. **177**, 1269 (1969).
P. G. Bethers and R. B. Murray, Phys. Letters A **28**, 9 (1969).
[13] R. Capelletti, R. Fieschi, and E. Okuno, Internat. Conf. Color Centers Ionic Cryst., Sendai (Japan) 1974 (p. G 136).

(Received February 15, 1984)

phys. stat. sol. (b) **125**, 147 (1984)

Subject classification: 9 and 10

Institute for Electrical Investigations, Cuernavaca[1]) *(a) and*
Metallurgical Engineering Department, The Ohio State University, Columbus[2]) *(b)*

Excess Solute Atmosphere Carried by a Potential Well Characteristic of a Moving Dislocation Core

By

R. FUENTES-SAMANIEGO (a) and J. P. HIRTH (b)

An expression is obtained for the excess solute atmosphere carried by a potential well characteristic of a dislocation core and moving at constant velocity in a solid solution. The results agree with the predictions of the dissipation theorem at limiting low values of the velocity. Results for all velocities are presented which can be interrelated with the drag force exerted by the excess solute on the well through the Einstein mobility relation.

Für die Atmosphäre eines Überschußstoffes, die von einer Potentialmulde getragen wird, wie sie für eine Versetzung charakterisisch ist, und mit konstanter Geschwindigkeit in einer Festkörperlösung wandert, wird ein Ausdruck abgeleitet. Die Ergebnisse stimmen mit Vorhersagen des Dissipationstheorems im Grenzfall niedriger Geschwindigkeiten überein. Ergebnisse für alle Geschwindigkeiten werden angegeben und lassen sich mit der Schleppkraft verknüpfen, die vom Überschußstoff an der Potentialmulde über die Einsteinsche Beweglichkeitsbeziehung ausgeübt wird.

1. Introduction

A solution has recently been presented [1] for the diffusion of interstitial solute in or near a moving potential well characteristic of a dislocation core. This result is a generalization of the one-dimensional problem, characteristic of a moving grain boundary or interface, that has been treated by several groups [2 to 6]. In the low-velocity limit the interaction forces for such problems can be deduced from a simple dissipation theorem [6], a variation of which also has been specialized [5] to a one-dimensional case. If the excess solute associated with the moving well is known, the forces can also be determined by the Einstein relation [6].

In the present work we first briefly review the one-dimensional solution and the dissipation theorem. Then the previous work on the two-dimensional case [1] is extended to calculate the excess solute carried by the dislocation and its relation to the drag force.

2. One-Dimensional Case

The solute distribution in a frame of reference (x, t) moving at constant velocity v has been presented for the repulsive-well case of Fig. 1a [6]. Continuity at the interface $x = 0$ yields the condition

$$\Delta C = \frac{vl}{D^*}(c_1^0 - c_2^0) = c_1^0 - c_3 \,, \tag{1}$$

[1]) Interior Internado Palmera, A Postal 475, Cuernavaca Morelas 62000, Mexico.
[2]) 116 West 19th Avenue, Columbus, Ohio 43210, USA.

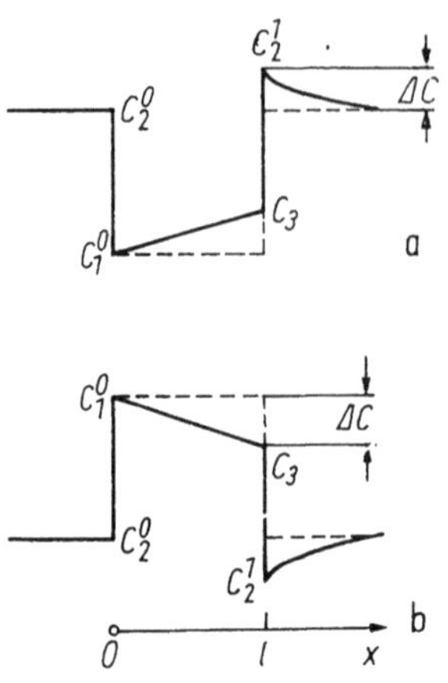

Fig. 1. Solute concentration in the vicinity of a square potential well of extent l moving with constant velocity in the x-direction for a) a repulsive well and b) an attractive well

where D^* is the constant solute diffusivity as defined in [1], and the solute concentration terms are defined in Fig. 1. The concentration in the well differs from that in the matrix because the solute has a binding energy W to the well. The solution for the solute distribution is

$$c = c_2^0; \qquad x < 0 ,$$

$$c = c_1^0 + \frac{x\,\Delta c}{l}; \qquad 0 < x < l ,$$

$$c = -\frac{c_2^0\,\Delta c}{c_1^0} \exp - \frac{v(x - l)}{D^*} + c_2^0: \qquad x > l . \tag{2}$$

The solute deficit in the well is

$$N_1 = (c_1^0 - c_2^0)\, l + \frac{l\,\Delta c}{2} . \tag{3}$$

The solute excess in the region $x > l$ is

$$N_2 = -\frac{l c_2^0}{c_1^0} (c_1^0 - c_2^0); \tag{4}$$

The total excess solute is the sum of (3) and (4),

$$N = \frac{l(c_1^0 - c_2^0)^2}{c_1^0} + \frac{1}{2}\, l\,\Delta c . \tag{5}$$

According to the Einstein mobility relation, the well velocity is related to a mobility (D^*/kT), the number of atoms exerting a drag force N, and the drag force F_x where k is Boltzmann's constant and T is absolute temperature. Thus,

$$N = \frac{F_x D^*}{vkT} . \tag{6}$$

For velocities so low that $v \ll D^*/l$, the second term on the right side of (5) is negligible and F_x is simply related to c_1^0 and c_2^0. The same result can be obtained from the dissipation theorem, where free-energy dissipation is related to power expended by F_x. Generalized to three-dimensional diffusion fluxes $\boldsymbol{J}$ in a volume V, the dissipation theorem [6] reads

$$\boldsymbol{F} \cdot \boldsymbol{v} = \frac{kT}{D^*} \int_v \frac{\boldsymbol{J} \cdot \boldsymbol{J}}{c} \, \mathrm{d}V \,. \tag{7}$$

Substituting in the *static* values of $\boldsymbol{J}$ and c, one recovers the simple equality of (5) and (6), without the term $l\,\Delta c/2$, in the low-velocity limit.

For the attractive well of Fig. 1 b, $(c_1^0 - c_2^0)$ changes sign. With these changes the same results (1) to (7) are valid, but (3) gives an excess and (4) a deficit. The value N in (5) is still an excess in the low velocity limit where the term $l\,\Delta c/2$ is negligible. The fact that N is an excess for the well moving at constant velocity for either an attractive or a repulsive well has received little attention. It has important obvious implications in the transport of impurities by moving grain boundaries or interphase interfaces where a square well interaction for solutes is a fair approximation.

3. Two-Dimensional Example

As a physical analog of the dislocation core interaction of solutes, solute is considered to have a binding energy W to a cylindrical core of radius a, Fig. 2, but zero interaction otherwise [1]. For an explicit example let us consider the attractive well case with initial conditions $c_1^0(r < a) > c_2^0(r > a)$. The excess in the well per unit length is composed of the static excess $\pi a^2(c_1^0 - c_2^0)$ and the incremental deficit in the well because it is moving $\pi a^2(c_1^0 Q_{\mathrm{ss}} - c_1^0)$, giving

$$\frac{N_1}{L} = \pi a^2(c_1^0 Q_{\mathrm{ss}} - c_2^0) \,, \tag{8}$$

where Q_{ss} is the ratio of the total solute in the well in steady state motion at velocity v in the x direction to the total solute in the static well: Q_{ss} is numerically evaluated in Fig. 5 of [1]. The excess solute outside the well is given by

$$\frac{N_2}{L} = \int_0^{2\pi} \mathrm{d}\theta \int_a^{\infty} [c_2(r, \theta) - c_2^0] \, r \, \mathrm{d}r \,, \tag{9}$$

where the solution for $c_2(r, \theta)$ is given in (58) and (61) of [1] as

$$c_2(r, \theta) = c_2^0 - c_2^0 \exp(-v^* r \cos \theta) \sum_{n=0}^{\infty} g_n(z) \, \mathrm{K}_n\left(\frac{zr}{a}\right) \cos(n\theta) \tag{10}$$

with the reduced parameters

$$z = \frac{va}{2D^*} \,, \qquad v^* = \frac{v}{2D^*} \,, \qquad \text{and} \qquad g_n(z) = \frac{A_n \mathrm{I}_n(z)}{\mathrm{K}_n(z)} \,, \tag{11}$$

where I_n and K_n are modified Bessel functions of order n of the first and second kind, respectively. As derived in (62) and (63) of [1], the coefficients A_n are found from the recursion formula

$$\sum_{n=0}^{\infty} A_n \mathrm{I}_n(z) \, \{\cos[(n+1)\,\theta] + \cos[(n-1)\,\theta]\} =$$

$$= \sum_{n=0}^{\infty} \left\{(4\varepsilon_n - 2A_n)\, \mathrm{I}_n'(z) - \frac{2A_n}{(S-1)\, z\mathrm{K}_n(z)}\right\} \cos(n\theta) \tag{12}$$

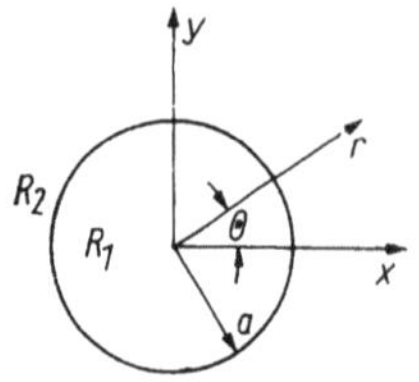

Fig. 2. Cylindrical well of radius a with constant solute interaction energy in region R

with the added condition

$$\sum_{n=0}^{\infty} (-1)^n g_n(z) = 0 . \tag{13}$$

Here $S = c_1^0/c_2^0$ and ε_n (= 1 when $n = 0$, = 2 otherwise) is Neumann's factor. Substituting (10) into (9) we find

$$\frac{N_2}{L} = -c_2^0 \int_a^{\infty} r\,\mathrm{d}r \sum_{n=0}^{\infty} g_n(z)\, \mathrm{K}_n\left(\frac{zr}{a}\right) \int_0^{2\pi} \exp\,(-v^* r \cos\theta) \cos\,(n\theta)\, \mathrm{d}\theta . \tag{14}$$

The use of Jacobi's expansion

$$\exp\,(R \cos\theta) = \sum_{n=0}^{\infty} \varepsilon_n \mathrm{I}_n(R) \cos\,(n\theta) \tag{15}$$

in (14) gives

$$\frac{N_2}{L} = -2\pi c_0^2 \int_a^{\infty} r\,\mathrm{d}r \sum_{n=0}^{\infty} (-1)^n g_n(z)\, \mathrm{K}_n\left(\frac{zr}{a}\right) \mathrm{I}_n(v^* r) =$$

$$= -2\pi c_2^0 \frac{a^2}{z^2} \int_z^{\infty} \mathrm{d}u \sum_{n=0}^{\infty} (-1)^n g_n(z)\, u\mathrm{K}_n(u)\, \mathrm{I}_n(u) \tag{16}$$

with the introduction of the variable $u = zr/a = v^* r$. Because of the property (13), the above relation becomes

$$\frac{N_2}{L} = -2\pi c_2^0 \frac{a^2}{z^2} \int_z^{\infty} \mathrm{d}u \sum_{n=0}^{\infty} (-1)^n g_n(z)\, [u\mathrm{I}_n(u)\, \mathrm{K}_n(u) - 0.5] . \tag{17}$$

The asymptotic form [7]

$$u\mathrm{I}_n(u)\, \mathrm{K}_n(u) - 0.5 = -\frac{4n^2 - 1}{16x^2} + \mathrm{O}\left(\frac{1}{x^3}\right); \qquad u \gg 1$$

shows that the integral in (17) can be introduced into the summation

$$\frac{N_2}{L} = -2\pi c_2^0 \frac{a^2}{z^2} \sum_{n=0}^{\infty} (-1)^n g_n(z) \int_z^{\infty} [u\mathrm{I}_n(u)\, \mathrm{K}_n(u) - 0.5]\, \mathrm{d}u . \tag{18}$$

Use of (A.1) from Appendix A gives an expression for the integral in (18) which is of the order of $(1/u)$ so the upper limit gives zero contribution and the lower limit yields

$$\frac{N_2}{L} = \frac{\pi}{2} a^2 c_2^0 \sum_{n=0}^{\infty} (-1)^n g_n(z) \Big[2\mathrm{I}_n(z)\,\mathrm{K}_n(z) + $$

$$+ \mathrm{I}_{n-1}(z)\,\mathrm{K}_{n+1}(z) + \mathrm{I}_{n+1}(z)\,\mathrm{K}_{n-1}(z) - \frac{2}{z} \Big] . \tag{19}$$

Substitution into (19) of (11), the explicit form for Q_{ss},

$$Q_{ss} = 1 - \sum_{n=0}^{\infty} (-1)^n A_n [\mathrm{I}_n^2(z) - \mathrm{I}_{n+1}(z)\,\mathrm{I}_{n-1}(z)] , \tag{20}$$

and the Wronskian of $\{\mathrm{K}_n(z), \mathrm{I}_n(z)\}$ [7],

$$\mathrm{I}_n(z)\,\mathrm{K}_{n+1}(z) + \mathrm{I}_{n+1}(z)\,\mathrm{K}_n(z) = \frac{1}{z} , \tag{21}$$

gives

$$\frac{N_2}{L} = \pi a^2 c_2^0 (1 - Q_{ss}) + \frac{\pi a^2 c_2^0}{2z} \sum_{n=0}^{\infty} (-1)^n A_n \frac{[\mathrm{I}_{n+1}(z) + \mathrm{I}_{n-1}(z)]}{\mathrm{K}_n(z)} . \tag{22}$$

With the relation $2\mathrm{I}_n'(z) = \mathrm{I}_{n+1}(z) + \mathrm{I}_{n-1}(z)$, the sum in (22) can be evaluated in terms of the quantity

$$P = \frac{1}{2(S-1)\,z} \sum_{n=0}^{\infty} (-1)^n \frac{A_n}{\mathrm{K}_n(z)} [\mathrm{I}_{n+1}(z) + \mathrm{I}_{n-1}(z)]$$

$$= -\frac{1}{2} \sum_{n=0}^{\infty} (-1)^n \mathrm{I}_n'(z) [\beta_n - 4\varepsilon_n \mathrm{I}_n'(z)] , \tag{23}$$

where

$$\beta_n = 4\varepsilon_n \mathrm{I}_n'(z) - \frac{2A_n}{(S-1)\,z\mathrm{K}_n(z)} . \tag{24}$$

With the use of (B.5) from Appendix B, (23) reduces to

$$P = -1 - \tfrac{1}{2} \sum_{n=0}^{\infty} (-1)^n \beta_n \mathrm{I}_n'(z) . \tag{25}$$

From (12), it follows that

$$\beta_0 = (A_1 + 2A_0)\,\mathrm{I}_1(z); \qquad n = 0 ,$$

$$\beta_n = 2A_n \mathrm{I}_n'(z) + A_{n-1}\mathrm{I}_{n-1}(z) + A_{n+1}\mathrm{I}_{n+1}; \qquad n \geqq 1 , \tag{26}$$

therefore

$$P = -1 - \tfrac{1}{2} \{\beta_0 \mathrm{I}_1'(z) - \beta_1 \mathrm{I}_1'(z) + \sum_{n=2}^{\infty} (-1)^n \mathrm{I}_n'(z) [2A_n \mathrm{I}_n'(z) + $$

$$+ A_{n-1}\mathrm{I}_{n-1}(z) + A_{n+1}\mathrm{I}_{n+1}(z)] . \tag{27}$$

Rearranging the indices, we obtain

$$P = -1 - \tfrac{1}{2} \sum_{n=0}^{\infty} (-1)^n A_n \{2[I'_n(z)]^2 - I_n(z) [I'_{n+1}(z) + I'_{n-1}(z)]\}$$

$$= -1 + \sum_{n=0}^{\infty} (-1)^n A_n \{I_n(z) I''_n(z) - [I'_n(z)]^2\}$$

$$= -1 + \sum_{n=0}^{\infty} (-1)^n A_n [I_n^2(z) - I_{n+1}(z) I_{n-1}(z) - \frac{1}{z} I_n(z) I'_n(z)] . \quad (28)$$

The diffusion solution [1] contains the expression for the drag force per unit length

$$F_x = -2\pi kTac_1^0 \left(1 - \frac{1}{S}\right) \sum_{n=0}^{\infty} (-1)^n A_n I_n(z) I'_n(z) \quad (29)$$

which, together with (20) and (28) yields

$$P = -Q_{ss} + \frac{SF_x}{2\pi kTac_1^0(S - 1) z} . \quad (30)$$

The use of (23) and (30) in (22) gives the expression for the excess outside the well

$$\frac{N_2}{L} = \pi a^2(c_2^0 - c_1^0 Q_{ss}) + \frac{aF_x}{2kTz} . \quad (31)$$

Together, (8) and (31) give for the total excess solute carried by the moving well, relative to the uniform bulk,

$$\frac{N}{L} = \frac{aF_x}{2kTz} = \frac{D^* F_x}{kTv} . \quad (32)$$

Thus, the more complicated two-dimensional problem gives the same result, as expected from the Einstein mobility relation, as does the simple one-dimensional analysis of (6). Moreover, the result in (32), together with the sums carried out for F_x in [1], enables us to determine N/L as a function of the reduced velocity parameter z as shown in Fig. 3.

According to the dissipation theorem, J is given by $v(c_1^0 - c_2)$ in region R_1 of Fig. 2 and is zero elsewhere, so (7) becomes

$$F_x = \frac{kT}{D^*} \pi a^2 v \frac{(c_1^0 - c_2^0)^2}{c_1^0} \quad (33)$$

which should be valid in the low-velocity limit $v \ll D^*/a$, or $z \ll 1$. In this limit, (32)

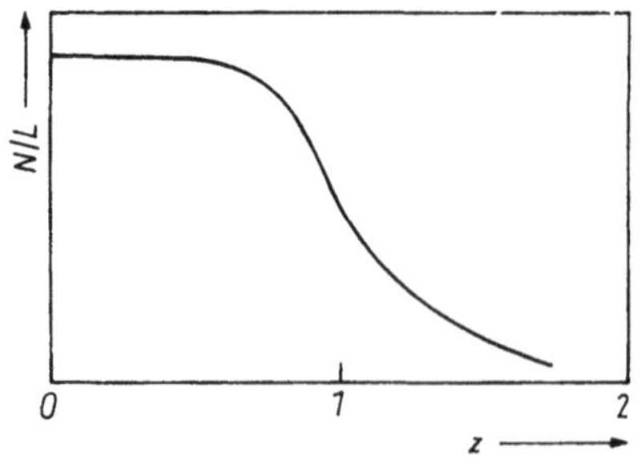

Fig. 3. Plot of total solute excess N/L as a function of the reduced velocity parameter z

should approach the asymptotic limit

$$\frac{N}{L} = \frac{\pi a^2(c_1^0 - c_2^0)^2}{c_1^0} . \tag{34}$$

As shown in Fig. 3, the limiting value for small z indeed approaches (34).

As for the one-dimensional analysis, the analysis holds throughout for the repulsive well case where W is negative and $c_1^0 < c_2^0$. In particular, (32) shows that the steady-state atmosphere being dragged by the moving repulsive well is a solute excess arising from the contribution from region R_2 in Fig. 2.

An attractive well is the usual physical model assumed as an analog for the dislocation core interaction for interstitial solutes where the strain energy contribution to the heat of solution of the solute is large. For substitutional solutes, where the "chemical" terms are the dominant contribution to the heat of solution, W could plausibly be of either sign and either the attractive or the repulsive well analog would be physically reasonable.

4. Summary

Previous work [1] contained results for the drag force acting on and solute concentration influenced by a cylindrical potential well, characteristic of a dislocation core moving at constant velocity. This work is extended here to calculate the excess solute concentration dragged by the well. Although the solution is considerably more complex than the simple one-dimensional equivalent [6], the results agree with the latter in several respects. First, the relation between the drag force, F_x, and the total solute excess, N/L, agrees with expectation from the Einstein mobility relation. Second, in the limit of low velocity, $v \ll D/a$, the results for N/L and F_x agree with the predictions of the dissipation theorem [6]. The present results together with the previous ones for F_x [1] give predictions of N/L for all velocities. The sign of N/L for the moving well is such that the solute atmosphere is an excess relative to the bulk for either an attractive or a repulsive solute interaction with the well. The results should be of importance in the analysis both of solute transport by dislocations and of the solute atmospheres around static dislocations which had previously been moving.

Acknowledgement

The authors are grateful for support of this research by the National Science Foundation under Grant No. NSF-DMR-83-11620.

Appendix A

With the use of standard forms for differentials of Bessel functions [7], one finds

$$\frac{\mathrm{d}}{\mathrm{d}u}\{u^2[\mathrm{I}_{n-1}(u)\,\mathrm{K}_{n+1}(u) + \mathrm{I}_{n+1}(u)\,\mathrm{K}_{n-1}(u)]\} =$$

$$= \frac{\mathrm{d}}{\mathrm{d}u}[u^{-n+1}\mathrm{I}_{n-1}(u)\,u^{n+1}\mathrm{K}_{n+1}(u) + u^{n+1}\mathrm{I}_{n+1}(u)\,u^{-n+1}\mathrm{K}_{n-1}(u)] =$$

$$= u^{-n+1}\mathrm{I}_{n-1}(u)\,[-u^{n+1}\mathrm{K}_n(u)] + [u^{-n+1}\mathrm{I}_n(u)]\,u^{n+1}\mathrm{K}_{n+1}(u) +$$
$$+ u^{n+1}\mathrm{I}_{n+1}(u)\,[-u^{-n+1}\mathrm{K}_n(u)] + [u^{n+1}\mathrm{I}_n(u)]\,u^{-n+1}\mathrm{K}_{n-1}(u) =$$
$$= -u^2\mathrm{K}_n(u)\,[\mathrm{I}_{n-1}(u) + \mathrm{I}_{n+1}(u)] + u^2\mathrm{I}_n(u)\,[\mathrm{K}_{n+1}(u) + \mathrm{K}_{n-1}(u)] =$$
$$= -2u^2[\mathrm{K}_n(u)\,\mathrm{I}_n'(u) + \mathrm{I}_n(u)\,\mathrm{K}_n'(u)] = -2u^2\frac{\mathrm{d}}{\mathrm{d}u}[\mathrm{I}_n(u)\,\mathrm{K}_n(u)] =$$

$$= \frac{\mathrm{d}}{\mathrm{d}u}[-2u^2\mathrm{I}_n(u)\,\mathrm{K}_n(u)] + 4u\mathrm{I}_n(u)\,\mathrm{K}_n(u) .$$

Therefore

$$u\mathrm{I}_n(u)\,\mathrm{K}_n(u) = \frac{1}{4}\frac{\mathrm{d}}{\mathrm{d}u}\{u^2[2\mathrm{I}_n(u)\,\mathrm{K}_n(u) + \mathrm{I}_{n-1}(u)\,\mathrm{K}_{n+1}(u) + \mathrm{I}_{n+1}(u)\,\mathrm{K}_{n-1}(u)]\}\,. \tag{A.1}$$

Appendix B

Beginning with the identity

$$-\cos^2\theta = \frac{\partial}{\partial z}[\exp(-z\cos\theta)]\frac{\partial}{\partial z}[\exp(z\cos\theta)]$$

we find that

$$\frac{1}{\pi}\int_0^{2\pi}\mathrm{d}\theta\left\{\frac{\partial}{\partial z}[\exp(-z\cos\theta)]\frac{\partial}{\partial z}[\exp(z\cos\theta)]\right\} = \frac{1}{\pi}\int_0^{2\pi}\cos^2\theta\,\mathrm{d}\theta = -1\,. \tag{B.1}$$

The substitution of (15) into (B.1) gives

$$-1 = \frac{1}{\pi}\int_0^{2\pi}\mathrm{d}\theta\,\frac{\partial}{\partial z}\left[\sum_{n=0}^{\infty}(-1)^n\,\varepsilon_n\mathrm{I}_n(z)\cos(n\theta)\right]\frac{\partial}{\partial z}\left[\sum_{m=0}^{\infty}\varepsilon_m\mathrm{I}_m(z)\cos(m\theta)\right]. \tag{B.2}$$

Rearranging (B.2) we see that

$$-1 = \frac{1}{\pi}\sum_{n=0}^{\infty}\sum_{m=0}^{\infty}(-1)^n\,\varepsilon_n\varepsilon_m\mathrm{I}_n'(z)\,\mathrm{I}_m'(z)\int_0^{2\pi}\cos(n\theta)\cos(m\theta)\,\mathrm{d}\theta\,. \tag{B.3}$$

But

$$\frac{1}{\pi}\int_0^{2\pi}\cos(n\theta)\cos(m\theta)\,\mathrm{d}\theta = \frac{2\delta_{mn}}{\varepsilon_n} = \begin{cases}2; & m = n = 0\\ 1; & m = n \neq 0\\ 0; & m \neq n\end{cases} \tag{B.4}$$

using the properties of the Kronecker delta function δ_{mn}. Therefore

$$-1 = 2\sum_{n=0}^{\infty}\sum_{m=0}^{\infty}(-1)^n\,\varepsilon_m\mathrm{I}_n'(z)\,\mathrm{I}_m'(z)\,\delta_{mn} = 2\sum_{n=0}^{\infty}(-1)^n\,\varepsilon_n[\mathrm{I}_n'(z)]^2\,. \tag{B.5}$$

References

[1] R. FUENTES-SAMANIEGO and J. P. HIRTH, phys. stat. sol. (b) **106**, 359 (1981).
[2] J. W. CAHN, Acta metall. **10**, 789 (1962).
[3] P. GORDON and R. A. VANDERMEER, Trans. MS AIME **224**, 917 (1962).
[4] K. LÜCKE and H. STÜWE, Acta metall. **19**, 1087 (1971).
[5] M. HILLERT and B. SUNDMAN, Acta metall. **24**, 731 (1976).
[6] J. P. HIRTH and J. LOTHE, Theory of Dislocations, Chap. 18, McGraw-Hill Publ. Co., New York 1968.
[7] M. ABRAMOWITZ and I. A. STEGUN, Handbook of Mathematical Functions, Dover Publ., Inc., New York 1965 (p. 374).

(Received March 27, 1984)

phys. stat. sol. (b) 125, 155 (1984)

Subject classification: 10.2; 22.5.2

Institute of Physics, Academy of Sciences of the Georgian SSR, Tbilisi[1])

A Theoretical Study of Simple Complexes of Point Defects in Alkali Halides by the Lattice Static Method

By

G. R. AUGST and YU. I. VLADISLAVSKII

Expressions are obtained for the energy of alkali halide crystals with two kinds of simple complexes of point defects — divalent impurity with vacancy and bivacancy — using the lattice static method. The heat of solution of Mg^{2+}, Ca^{2+}, Sr^{2+}, and Ba^{2+} in NaCl, KCl, and KBr is calculated. Numerical calculations show that in some cases the nearest-neighbour complex is less stable than the next-nearest-neighbour complex. The formation energy of a bivacancy is calculated in the same crystals. The results are consistent with those obtained experimentally. It is shown that in the latter case the configuration of the nearest-neighbour complex is the most stable one.

С использованием метода статики решетки получены выражения для энергии двух типов простых комплексов точечных дефектов-двухвалентная примесь с компенсирующей вакансией и бивакансия в щелочно-галоидном кристалле. Вычислена теплота растворения Mg^{2+}, Ca^{2+}, Sr^{2+} и Ba^{2+} в NaCl, KCl и KBr. Показано, что в ряде случаев энергетически более выгодным для вакансии является не положение ближайшего соседа двухвалентной примеси, а следующее за ним. Расчитана энергия образования бивакансии в тех же кристаллах. Результаты вычислений хорошо согласуются с экспериментом. Показано, что для бивакансии во всех случаях должна реализовыватьсья конфигурация ближайших соседей.

1. Introduction

The majority of theoretical studies of the point defect properties in ionic crystals are based on the Mott-Littleton method [1 to 4]. There is an alternative, consecutively microscopic approach which was used in [5, 6] in order to calculate the heat of solution of an isovalent, substitutional impurity and the defect energy of hydrogen impurity in alkali halide crystals. The advantage of this method consists in the fact that all calculations are carried out by "first principles" and allow to obtain analytical expressions for the displacements of lattice ions and for the relaxation energy.

In the present paper we show that this method may be used for complexes of point defects, too. In particular, we investigate the systems representing NaCl, KCl, and KBr crystals with Mg^{2+}, Ca^{2+}, Sr^{2+}, and Ba^{2+} impurity ions. The calculations are performed for different distances between the impurity and the associated vacancy. Our results are in agreement with the model calculations [2] mentioned above. This comparison is even more interesting, since we use the same interionic interaction potentials, although restrict ourselves to the polarizable point dipole models. The heat of solution of these impurites is calculated, too. In the present paper, the energy of bivacancy formation has been calculated for NaCl, KCl, and KBr crystals. The energies of the complexes are compared for different distances between the vacancy pairs.

[1]) Tbilisi 77, USSR.

2. Divalent Impurity with Vacancy

Let us consider an ionic crystal containing a divalent impurity in the origin and an associated vacancy. The charge density of such a system in the point $\boldsymbol{r}$ is

$$\varrho(\boldsymbol{r}) = e \sum_{n\nu} [(1 + \delta_{n0} - \delta_{n v})\, \delta_{\nu 1} - \delta_{\nu 2}]\, \delta(\boldsymbol{r} - \boldsymbol{R}_{n\nu}) , \tag{1}$$

where $\boldsymbol{R}_{n\nu}$ is the radius vector of the ion in the site $n\nu$, and $\nu = 1, 2$ corresponds to positive ($\nu = 1$) and negative ($\nu = 2$) ions with charges $\pm e$, respectively. The vacancy is located in the site $\boldsymbol{R}_v(\boldsymbol{n} = \boldsymbol{v}, \nu = 1)$.

The energy is calculated using the Ewald method, as suggested in [7].

The adiabatic part of the defect energy can be represented as follows:

$$E^{(0)} = -\frac{e^2}{R_v} + \sum_{n\nu} [\tilde{\varphi}_{1\nu}(\boldsymbol{R}_{n\nu}) - 2\varphi_{1\nu}(\boldsymbol{R}_{n\nu})] - \tilde{\varphi}_{11}(\boldsymbol{R}_v) + \varphi_{11}(\boldsymbol{R}_v) , \tag{2}$$

where $\varphi_{\nu\mu}(\boldsymbol{R})$ and $\tilde{\varphi}_{\nu\mu}(\boldsymbol{R})$ are the non-Coulomb energy of interaction between the host matrix ions and between the impurity ion and matrix ions, respectively. In (2) and further $\boldsymbol{R}_{n\nu}$ represent the positions of ions of the regular lattice. Using the lattice static method and according to [7], the relaxation energy can be represented as an iteration series, the two first terms of the latter being

$$E_{\mathrm{R}}^{(1)} = \frac{1}{2N} \sum_{\boldsymbol{Q}} P_\nu^\alpha(\boldsymbol{Q})\, G_{\nu\mu}^{\alpha\beta}(\boldsymbol{Q})\, P_\mu^\beta(-\boldsymbol{Q}) , \tag{3}$$

$$E_{\mathrm{R}}^{(2)} = \frac{1}{2N^2} \sum_{\boldsymbol{Q},\boldsymbol{Q}'} P_\nu^\alpha(\boldsymbol{Q})\, G_{\nu\mu}^{\alpha\beta}(\boldsymbol{Q})\, \Gamma_\mu^{\beta\zeta}(\boldsymbol{Q}, -\boldsymbol{Q}')\, G_{\mu\lambda}^{\zeta\gamma}(\boldsymbol{Q}')\, P_\lambda^\gamma(-\boldsymbol{Q}') . \tag{4}$$

The summation over $\boldsymbol{Q}$ is performed over the first Brillouin zone. The upper and lower indices in (3), (4) specify the Cartesian components and the number of the sublattice, respectively. The Fourier transforms of the forces $P_\nu^\alpha(\boldsymbol{Q})$ and the function $\Gamma_\nu^{\alpha\beta}(\boldsymbol{Q}, \boldsymbol{Q}')$ for asymmetrical defects are complex and given as

$$P_\nu^\alpha(\boldsymbol{Q}) = \frac{4\pi e^2}{\Omega_0} [1 - \exp(i\boldsymbol{Q}\boldsymbol{R}_v)] \sum_{\boldsymbol{k}} [\delta_{\nu 1} - \cos(\boldsymbol{k}\boldsymbol{\varrho})\, \delta_{\nu 2}] \frac{(\boldsymbol{k} + \boldsymbol{Q})^\alpha}{(\boldsymbol{k} + \boldsymbol{Q})^2} + \\ + \sum_{\boldsymbol{n}}{}' \left[\frac{\partial \tilde{\varphi}_{1\nu}(\boldsymbol{R}_{n\nu})}{\partial R_{n\nu}^\alpha} - \big(1 + \exp(i\boldsymbol{Q}\boldsymbol{R}_v)\big) \frac{\partial \varphi_{1\nu}(\boldsymbol{R}_{n\nu})}{\partial R_{n\nu}^\alpha}\right] \exp(i\boldsymbol{Q}\boldsymbol{R}_{n\nu}) , \tag{5}$$

$$\Gamma_\nu^{\alpha\beta}(\boldsymbol{Q}, \boldsymbol{Q}') = \frac{4\pi e^2}{\Omega_0} [1 - \exp\big(i(\boldsymbol{Q} + \boldsymbol{Q}')\, \boldsymbol{R}_v\big)] \times \\ \times \sum_{\boldsymbol{k}} [\delta_{\nu 1} - \cos(\boldsymbol{k}\boldsymbol{\varrho})\, \delta_{\nu 2}] \frac{(\boldsymbol{k} + \boldsymbol{Q} + \boldsymbol{Q}')^\alpha\, (\boldsymbol{k} + \boldsymbol{Q} + \boldsymbol{Q}')^\beta}{(\boldsymbol{k} + \boldsymbol{Q} + \boldsymbol{Q}')^2} + \\ + \sum_{\boldsymbol{n}}{}' \left[\frac{\partial^2 \tilde{\varphi}_{1\nu}(\boldsymbol{R}_{n\nu})}{\partial R_{n\nu}^\alpha\, \partial R_{n\nu}^\beta} - \big(1 + \exp\big(i(\boldsymbol{Q} + \boldsymbol{Q}')\, \boldsymbol{R}_v\big)\big) \frac{\partial^2 \varphi_{1\nu}(\boldsymbol{R}_{n\nu})}{\partial R_{n\nu}^\alpha\, \partial R_{n\nu}^\beta}\right] \times \\ \times \exp\big(i(\boldsymbol{Q} + \boldsymbol{Q}')\, \boldsymbol{R}_{n\nu}\big) ; \qquad \boldsymbol{\varrho} = \boldsymbol{R}_{n2} - \boldsymbol{R}_{n1} , \tag{6}$$

$G_{\nu\mu}^{\alpha\beta}(\boldsymbol{Q})$ is the zero Green function defined in [7]. For the numerical calculations the Ewald method has been used. However, in the present paper, for the sake of simplicity, the expressions are represented in the limit $\eta \to \infty$.

The non-Coulomb part of the interaction between the host matrix ions was approximated by set II of potentials [8], and the interaction between the impurity and matrix ions by the potentials with parameters taken from [2]. In the calculations of integrals and double integrals over the Brillouin zone in $E_{\mathrm{R}}^{(1)}$ and $E_{\mathrm{R}}^{(2)}$, 576 and 5184 points, respectively, were taken.

Table 1

Calculated defect energies and solution energies for crystals containing an alkali earth cation impurity (in eV)

system	$-E^{(0)}$	$-E_R^{(1)}$	$-E_R^{(2)}$	$-E$	h	h_{exp}
NaCl:Mg^{2+}	4.87	3.30	0.60	8.77	1.36	—
NaCl:Ca^{2+}	3.96	2.48	0.05	6.49	0.79	0.64—0.99 [9, 10]
NaCl:Sr^{2+}	3.10	2.02	0.02	5.14	1.18	0.78—1.82 [11, 10] 0.97 [12]
NaCl:Ba^{2+}	2.17	1.93	0.02	4.12	0.99	0.99—2.00 [10] 1.41 [12]
KCl:Mg^{2+}	4.64	3.58	1.65	9.87	1.52	—
KCl:Ca^{2+}	4.28	3.04	0.94	8.26	0.28	—
KCl:Sr^{2+}	3.94	2.65	0.48	7.07	0.50	1.02—1.30 [11]
KCl:Ba^{2+}	3.35	2.29	0.21	6.05	0.32	1.00 [12]
KBr:Mg^{2+}	4.38	3.37	1.55	9.30	1.44	—
KBr:Ca^{2+}	4.06	2.92	0.91	7.89	0.51	0.73 [10]
KBr:Sr^{2+}	3.78	2.57	0.48	6.83	0.64	1.13 [10]
KBr:Ba^{2+}	3.49	2.25	0.30	6.04	0.31	1.86 [10]

The results of the calculations of the energies of divalent impurities in alkali halides $E^{(0)}$, $E_R^{(1)}$, $E_R^{(2)}$, and

$$E = E^{(0)} + E_R$$

for the case of the associated vacancy in the nearest-neighbour position (110) are given in Table 1. The calculated energies E for the impurities in KCl and KBr slightly exceed (by less than 3%) the results of [2], and at the same time, for the case of NaCl this difference reaches 5%. However, for the case of Mg^{2+} in all the crystals our results practically coincide with [2].

The results of the calculations of the solution energy of divalent impurities is given in Table 1, too. The heat of solution can be determined as

$$h = E + 2E_{AX} - E_{BX_2},$$

where E_{AX} and E_{BX_2} represent lattice energies of the host lattice AX and the impurity lattice BX_2, respectively.

The E_{AX} lattice energy is easily calculated, and the values obtained using the parameters [8] are close to the experimental ones. Since the calculation of E_{BX_2} is less

Table 2

Calculated defect energies for (110), (200), and (211) dipolar complexes (in eV)

system	R_v		
	(110)	(200)	(211)
KCl:Mg^{2+}	−9.87	−10.01	−9.74
KCl:Ca^{2+}	−8.26	−8.39	−8.03
KCl:Sr^{2+}	−7.07	−6.98	−6.83
KCl:Ba^{2+}	−6.05	−5.83	−5.74

reliable, the experimental E_{BX_2} values from [3] are used. As can be seen, the calculated values of h listed in Table 1 are all positive, and the comparison with the solution energies of the impurities found mainly from the measurements of low-temperature ionic conductivity [10 to 12] shows a satisfactory agreement.

Calculated values of the defect energies E of vacancy–impurity complexes formed when substituting Mg^{2+}, Ca^{2+}, Sr^{2+}, and Ba^{2+} into KCl as a function of the distance between the vacancy and the impurity are given in Table 2. The calculations were made for the nearest-neighbour complex (110) and for the next-nearest-neighbour complexes (200) and (211). For larger values of $\boldsymbol{R}_v$ the calculations have not been carried out, since in this case, due to oscillations of the function $\exp(i\boldsymbol{Q}\boldsymbol{R}_v)$, the number of points in the Brillouin zone should be increased.

Similarly to the result of [2], our calculations show that for the impurities with small radii, Mg^{2+} and Ca^{2+}, the next-nearest-neighbour position (200) is energetically most favourable, while for the impurities with large radii, Sr^{2+} and Ba^{2+}, the nearest-neighbour configuration is found to be more stable. In our calculations the level difference is more substantial than in [2].

3. Bivacancy

Let us consider the case of vacancy pair formation with the positive ion vacancy placed in the origin, and the negative ion vacancy in the site $\boldsymbol{R}_v(\boldsymbol{n} = \boldsymbol{v}, \nu = 2)$. The formation energy of such a pair in the adiabatic approximation is

$$E^{(0)} = \frac{e^2}{\varrho}\alpha_M - \frac{e^2}{R_v} - 6\,[\varphi_{12}(\varrho) + \varphi_{11}(\sqrt{2}\varrho) + \varphi_{22}(\sqrt{2}\varrho)] + \varphi_{12}(\boldsymbol{R}_v)\,. \tag{7}$$

To obtain the relaxation energy we use (3) and (4), where in the case of a bivacancy

$$P_\nu^\alpha(\boldsymbol{Q}) = -\frac{4\pi e^2}{\Omega_0}\sum_{\boldsymbol{k}}\,[\delta_{\nu 1} - \cos(\boldsymbol{k}\varrho)\,\delta_{\nu 2}]\,[1 - \exp\left(i(\boldsymbol{k}+\boldsymbol{Q})\,\boldsymbol{R}_v\right)]\frac{(\boldsymbol{k}+\boldsymbol{Q})^\alpha}{(\boldsymbol{k}+\boldsymbol{Q})^2} -$$
$$- i\sum_{\boldsymbol{n}}{}'\left[\frac{\partial\varphi_{1\nu}(\boldsymbol{R}_{n\nu})}{\partial R_{n\nu}^\alpha}\exp(i\boldsymbol{Q}\boldsymbol{R}_{n\nu}) + \frac{\partial\varphi_{2\nu}(\boldsymbol{R}_{n\mu})}{\partial R_{n\mu}^\alpha}\exp\left(i\boldsymbol{Q}(\boldsymbol{R}_{n\mu} + \boldsymbol{R}_v)\right)\right], \tag{8}$$

$$\Gamma_\nu^{\alpha\beta}(\boldsymbol{Q}, \boldsymbol{Q}') = -\frac{4\pi e^2}{\Omega_0}\sum_{\boldsymbol{k}}\,[\delta_{\nu 1} - \cos(\boldsymbol{k}\varrho)\,\delta_{\nu 2}]\,[1 - \exp\left(i(\boldsymbol{k}+\boldsymbol{Q}+\boldsymbol{Q}')\,\boldsymbol{R}_v\right)]\times$$
$$\times\frac{(\boldsymbol{k}+\boldsymbol{Q}+\boldsymbol{Q}')^\alpha\,(\boldsymbol{k}+\boldsymbol{Q}+\boldsymbol{Q}')^\beta}{(\boldsymbol{k}+\boldsymbol{Q}+\boldsymbol{Q}')^2} -$$
$$-\sum_{\boldsymbol{n}}{}'\left[\frac{\partial^2\varphi_{1\nu}(\boldsymbol{R}_{n\nu})}{\partial R_{n\nu}^\alpha\,\partial R_{n\nu}^\beta}\exp\left(i(\boldsymbol{Q}+\boldsymbol{Q}')\,\boldsymbol{R}_{n\nu}\right) +\right.$$
$$\left.+\frac{\partial^2\varphi_{2\nu}(\boldsymbol{R}_{n\mu})}{\partial R_{n\mu}^\alpha\,\partial R_{n\mu}^\beta}\exp\left(i(\boldsymbol{Q}+\boldsymbol{Q}')\,(\boldsymbol{R}_{n\mu} + \boldsymbol{R}_v)\right)\right], \tag{9}$$

in (8) and (9) we imply that $\nu \neq \mu$.

Naturally, the Coulomb parts of $\boldsymbol{P}(\boldsymbol{Q})$ and $\Gamma(\boldsymbol{Q}, \boldsymbol{Q}')$ both of the bivacancy and of the divalent impurity with vacancy have identical structures, since both are due to the dipoles in the lattice.

In contrast to the bivacancy, in the case of a divalent impurity with a vacancy the function $\exp(i\boldsymbol{k}\boldsymbol{R}_v) \equiv 1$, and except for the sign (since the directions of moments are opposite), the Coulomb parts of (5), (6) and (8), (9) coincide. As in both cases $\boldsymbol{P}_\nu(\boldsymbol{Q})$ represents the superposition of the Fourier transforms of forces acting on sublattice

Table 3

Calculated formation energies of bivacancy (in eV)

crystal	R_v	$E^{(0)}$	$-E_R^{(1)}$	$-E_R^{(2)}$	U_0	U	U_{exp}	E
NaCl	(100)	3.08	1.38	0.03	1.67	1.45	1.30 [13] 1.38 [14]	—
	(111)	5.06	2.72	0.18	—	—	—	0.49
	(210)	5.73	3.14	0.20	—	—	—	0.72
KCl	(100)	2.89	1.17	0.08	1.64	1.48	1.34 [13] 1.45 [14]	—
	(111)	4.74	2.28	0.24	—	—	—	0.58
	(210)	5.34	2.57	0.26	—	—	—	0.87
KBr	(100)	2.63	1.06	0.03	1.54	1.43	1.41 [13]	—
	(111)	4.49	2.03	0.11	—	—	—	0.81
	(210)	5.07	2.32	0.26	—	—	—	0.95

ν from every component of the complex defect, these expressions can be easily generalized to the case of more complicated defects. An analogous situation takes place for the function $\Gamma_\nu(\mathbf{Q}, \mathbf{Q}')$, too.

The calculations at $R_v = \varrho$ give the formation energies of the bivacancy

$$U_0 = E^{(0)} + E_R$$

which are given in Table 3 for NaCl, KCl, and KBr crystals. The energies U_0 are obtained for the following lattice parameters: for NaCl $\varrho = 0.2789$ nm, for KCl $\varrho = 0.3116$ nm, and for KBr $\varrho = 0.3262$ nm, which corresponds to absolute zero.

These U_0 values exceed the experimental ones [13, 14] obtained at high temperatures. We have also calculated how the formation energies of a bivacancy are affected when the lattice expands by 3%; this is a rough measure of the linear thermal expansion of an alkali halide between absolute zero and the melting point. These results are in good agreement with the experimental values [13, 14] and close to model calculations [15].

The energies calculated for the next two values of R_v (at absolute zero), allow us to follow the change of the bivacancy energy as a function of the distance between the vacancies. Table 3 contains the results for

$$\Delta E = E^{(0)} + E_R - U_0 .$$

In contrast to the divalent impurity–vacancy complex, for which the nearest-neighbour configuration is not always energetically favourable, in the case of the binvacancy, the energy sharply increases with increasing distance between the vacancies.

Acknowledgement

The authors are grateful to E. L. Andronikashvili for his interest in the investigation.

References

[1] N. F. Mott and M. I. Littleton, Trans. Faradey Soc. **34**, 485 (1939).
[2] C. R. A. Catlow, J. Corish, J. M. Quigley, and P. W. M. Jacobs, J. Phys. Chem. Solids **41**, 231 (1980).
[3] V. S. Urusov, V. B. Dudnikova, and A. V. Garanin, phys. stat. sol. (b) **102**, 695 (1980).

[4] Y. V. G. S. Murti and V. Usha, Physica (Utrecht) **BC83**, 275 (1976).
[5] G. R. Augst and L. G. Bakhshetsyan, phys. stat. sol. (b) **104**, K85 (1981).
[6] G. R. Augst and L. G. Bakhshetsyan, phys. stat. sol. (b) **107**, 497 (1981).
[7] G. R. Augst, phys. stat. sol. (b) **83**, 55 (1977).
[8] C. R. A. Catlow, K. M. Diller, and M. J. Norgett, J. Phys. C **10**, 1395 (1977).
[9] A. Lidiard, Hdb. Phys., Vol. 20, Springer-Verlag, Berlin 1957 (p. 263).
[10] N. B. Hannay (Ed.), Defects in Crystals, Plenum Press, New York 1975.
[11] R. C. Bowman, J. chem. Phys. **59**, 2215 (1973).
[12] V. S. Urysov and V. B. Dudnikova, Izv. Akad. Nauk SSSR, Ser. neorg. Mater. **15**, 278 (1979).
[13] N. A. Economou and P. V. Sastry, phys. stat. sol. **6**, 135 (1964).
[14] F. Beniere, M. Beniere, and M. Chemla, Z. Naturf. **26a**, 915 (1971).
[15] C. R. A. Catlow, J. Corish, K. M. Diller, P. W. M. Jacobs, and M. J. Norgett, J. Phys. C **12**, 451 (1979).

(Received April 12, 1984)

phys. stat. sol. (b) **125**, 161 (1984)

Subject classification: 11

Department of Physics, Tokyo Gakugei University[1])

The Recoil Effect on the Electronic Stopping Power of Solids for Slowly Moving Atoms

By

S. SHINDO

The velocity dependence of the electronic stopping power of solids for low velocity atoms is discussed. It is shown that the stopping power contains not only the velocity-proportional term (friction term) but also the velocity anti-proportional term (recoil term). Calculations of the stopping power using the Born approximation, phase shift, and the Anderson model are presented.

Die Geschwindigkeitsabhängigkeit der elektronischen Bremsleistung von Festkörpern wird für niedrige Geschwindigkeiten diskutiert. Es wird gezeigt, daß die Bremsleistung nicht nur einen geschwindigkeitsproportionalen Term (Reibungsterm), sondern auch einen der Geschwindigkeit anti-proportionalen Term (Abstoßungsterm) enthält. Berechnungen der Bremsleistung unter Benutzung der Born-Näherung, Phasenverschiebung und des Anderson-Modells werden mitgeteilt.

1. Introduction

The stopping power of solids for slowly moving atoms is an important kinetic coefficient to discuss the phenomena originating from the atomic motion in solids, such as the radiation effects, sputtering, and the ion implantation in solids. Since Fermi and Teller [1] presented the pioneering theory of the stopping power for slow atoms, various [1 to 6] formulas of the stopping power as a function of the atomic number of the projectile Z_1, that of the target Z_2, and the velocity of the projectile v are derived. These theories predict that the electronic stopping power is proportional to the velocity of the moving atom v when the low-velocity condition

$$v \ll v_0 Z_2^{2/3} \tag{1}$$

is satisfied, where $v_0 = e^2/\hbar$.

Recently the interest in the theory of the electronic stopping power for slow atoms has been renewed in the field of surface physics [7 to 11], where the knowledge of the energy dissipation (stopping power) of the atom moving with thermal energy is necessary to make clear the dynamics of adsorbed atoms and catalysis. In this field, an important advance on the stopping power is represented using the Anderson model with which one may take into account the effect of Coulomb integral. However, the velocity dependence of the stopping power has not been discussed sufficiently.

In this paper, we show that the low-velocity stopping power (rate of the energy loss of moving atom per travelling path length) due to the electronic excitation contains the term inversely-proportional to velocity in addition to the previous velocity-proportional stopping power. The recoil effect which has not been taken into account previously is responsible for the velocity anti-proportional term.

In Section 2 the general theory of the electronic stopping power is presented. Both the velocity-proportional friction force and the recoil term inversely-proportional to velocity are derived. In Section 3 the calculation of the friction coefficient and the

[1]) Koganei, Tokyo 184, Japan.

recoil effect are performed with the Born approximation, phase shifts, and the Anderson model.

2. Theory

The Hamiltonian of the present system is decomposed as follows:

$$H = H_0 + H' = \hat{K} + H_c + H' , \tag{2}$$

where $\hat{K}$ is the kinetic energy of the moving atom, H_c describes the electronic state of the solid, and H' the interaction between the atom and the electrons of solids, respectively. $\hat{K}$ and H_c satisfy the Schrödinger equations,

$$\begin{aligned} \hat{K}\,|\alpha\rangle &= K_\alpha\,|\alpha\rangle , \\ H_c\,|n\rangle &= E_n\,|n\rangle , \end{aligned} \tag{3}$$

respectively. The energy loss rate per unit time can be written with the transition probability $W_{\beta n,\alpha 0}$ as follows:

$$-\frac{\mathrm{d}K}{\mathrm{d}t} = \sum_{n\beta} (K_\beta - K_\alpha)\, W_{\beta n,\alpha 0} , \tag{4}$$

where α, β and $0, n$ are the initial and the final states of the atom and the solids, respectively. $W_{\beta n,\alpha 0}$ is the transition probability (per unit time) which is written with the Heisenberg picture in the limit of the distorted wave Born approximation,

$$W_{\beta n,\alpha 0}(t) = \sum_{\bar{n}+\bar{m}} \int_{-\infty}^{t} \frac{1}{\hbar^2}\frac{1}{Q} [\langle \bar{m}\alpha|\, H'(t)\,|\bar{n}\beta\rangle\, \langle \bar{n}\beta|\, H'(t')\,|\bar{m}\alpha\rangle + \text{c.c}] \times$$

$$\times \exp(-\beta E_{\bar{m}})\, \mathrm{d}t' \qquad (\beta = 1/k_{\mathrm{B}}T) , \tag{5}$$

where $Q = \sum_m \exp(-\beta E_{\bar{m}})$ and T is the temperature of the solid. The distortion of the electronic state due to the moving atom is represented by $\bar{m}$, which satisfies

$$\bar{H}_c\,|m\rangle = (H_c + \langle\alpha|\, H'\,|\alpha\rangle\,|\bar{m}\rangle = E_{\bar{m}}\,|\bar{m}\rangle . \tag{6}$$

In the high-velocity case where the collision time t^c ($t^c \sim a/v$, a is the range of perturbation) is very small compared with $(E_n - E_0)/\hbar$, the golden-rule formula can be applied to the transition probability (5). On the other hand, one must treat (5) directly in the low-velocity case ($t^c > (E_n - E_0)/\hbar$). First we define $\dot{A}$ as the commutator of an operator A with K,

$$\dot{A} = \frac{1}{i\hbar}[A, \hat{K}] . \tag{7}$$

Then, the right-hand side (rhs) of (4) can be rewritten using (5) and (7) as

$$-\frac{\mathrm{d}K}{\mathrm{d}t} = \frac{1}{i\hbar} \int_{-\infty}^{t} \langle\!\langle [\dot{H}'(t), H'(t')] \rangle\!\rangle_{0\alpha}\, \mathrm{d}t' , \tag{8}$$

where

$$\langle\!\langle \ldots \rangle\!\rangle_{0\alpha} = \sum_{\bar{m}} \exp(-\beta E_{\bar{m}})\, \frac{1}{Q}\, \langle \bar{m}, \alpha|\, \ldots\, |\bar{m}\alpha\rangle .$$

Applying Kubo's identity

$$[A, \exp(-\beta \overline{H}_c)] = \exp(-\beta \overline{H}_c) \int_0^\beta \exp(\lambda \overline{H}_c) \times \times [A, \overline{H}_c] \exp(-\lambda \overline{H}_c) \, d\lambda \tag{9}$$

and the identity

$$\mathrm{Tr}\,[A, B]\, C = \mathrm{Tr}\, A[B, C]\,,$$

we rewrite the rhs of (8) as follows:

$$-\frac{dK}{dt} = \frac{-1}{i\hbar} \int_0^\beta \int_{-\infty}^t \langle\langle [H'(t' - i\hbar\lambda), \overline{H}_c]\, \dot{H}'(t) \rangle\rangle_{0\alpha} \, dt' \, d\lambda\,. \tag{10}$$

Now, we introduce an approximation to (10) in the low-velocity case where the time variation of the perturbation is small compared with $(E_n - E_0)/\hbar$, which gives

$$\frac{dH'}{dt} = \frac{1}{i\hbar} [H', H_0] \approx \frac{1}{i\hbar} [H', \overline{H}_c] + \dot{H}' \approx 0\,. \tag{11}$$

Then the energy loss rate of a slowly moving atom is described by

$$-\frac{dK}{dt} = \int_{-\infty}^t \int_0^\beta \langle\langle \dot{H}'(t' - i\hbar\lambda)\, \dot{H}'(t) \rangle\rangle_{0\alpha} \, d\lambda \, dt'$$

$$\equiv \beta \int_{-\infty}^t \left(\dot{H}'(t); \dot{H}'(t)\right)_{0\alpha} dt'\,. \tag{12}$$

Since $\dot{H}'(t)$ is represented, with (7), by

$$\dot{H}'(t) = -\left(\nabla_R H'(t)\right) \frac{i\hbar}{M} \nabla_R - \frac{i\hbar}{2M} \left(\nabla_R^2 H'(t)\right), \tag{13}$$

where M is the mass and $\boldsymbol{R}$ the position of moving atom, we obtain the energy loss rate as a function of the velocity of the atom as follows:

$$-\frac{dK}{dt} = \boldsymbol{v}\eta\boldsymbol{v} - \eta_0\,, \tag{14}$$

where the velocity of the atom v is defined by

$$\boldsymbol{v} = \left\langle \alpha \left| \frac{\hbar}{iM} \nabla_R \right| \alpha \right\rangle.$$

In (14) η is the friction tensor and η_0 represents the recoil effect both of which are defined by

$$\eta = \beta \int_0^\infty \left(\nabla_R H'(0); \nabla_R H'(t)\right)_{0\alpha} dt \tag{15}$$

and

$$\eta_0 = \left(\frac{\hbar}{2M}\right)^2 \beta \int_0^\infty \left(\nabla_R^2 H'(0); \nabla_R^2 H'(t)\right)_{0\alpha} dt\,, \tag{16}$$

respectively. The expression of the friction tensor (15) has been given by many authors previously. Because of the recoil term (16) it is emphasized that the stopping power which is defined as the energy loss rate of the moving atom in the homogeneous medium per unit path length, represented by

$$-\frac{\mathrm{d}K}{\mathrm{d}z} = \eta v - \frac{\eta_0}{v} \tag{17}$$

is not proportional to the velocity of the atom, but has the term inversely proportional to velocity. It is derived from (17) that the moving atom does not stop but has a final kinetic energy

$$K_\mathrm{f} = \frac{M\eta_0}{2\eta} \tag{18}$$

due to the recoil effect.

3. Calculation of the Electronic Stopping Power

In this section, we present the calculation of η and η_0 given by (15) and (16) for the electronic excitation of solids. First we consider only the single-electronic excitation because the plasmon excitation does not contribute to the energy loss of atoms moving more slowly than with Fermi velocity. We also assume that the wave function of the atom, $|\alpha\rangle$, is localized at $\boldsymbol{R} = \boldsymbol{0}$. Then, from (15), we have the expression of the friction coefficient

$$\eta = \pi\hbar \int \frac{\mathrm{d}^3\boldsymbol{k}}{(2\pi)^3} \frac{\mathrm{d}^3\boldsymbol{k}'}{(2\pi)^3} |\langle\psi_{\boldsymbol{k}}| \hat{F} |\psi_{\boldsymbol{k}'}\rangle|^2 \times$$

$$\times\, \delta\left(E_\mathrm{F} - \frac{(\hbar\boldsymbol{k})^2}{2m}\right) \delta\left(E_\mathrm{F} - \frac{(\hbar\boldsymbol{k}')^2}{2m}\right), \tag{19}$$

where m is the electron mass, E_F the Fermi energy, $\hat{F}$ the force operator

$$\hat{F} = -[\nabla_r, V] \qquad (V(r) = H'(r, R = 0)), \tag{20}$$

and $\psi_{\boldsymbol{k}}$ is the distorted wave of the electrons,

$$\psi_{\boldsymbol{k}} = \mathrm{e}^{i\boldsymbol{kr}} + \frac{1}{E - \dfrac{\hbar^2k^2}{2m} + i\delta} V\psi_{\boldsymbol{k}}. \tag{21}$$

Using (20) and (21), the rhs of (19) becomes

$$\eta = nmv_\mathrm{F}\sigma^{\mathrm{tr}}, \tag{22}$$

where n is the electron density of the solid, v_F the Fermi velocity $v_\mathrm{F} = (3\pi^2n)^{1/3}\,\hbar/m$, and σ^{tr} the transport cross-section of the Fermi electron defined by

$$\sigma^{\mathrm{tr}} = \int \mathrm{d}\Omega \left(1 - \frac{\boldsymbol{k}_\mathrm{F}\cdot\boldsymbol{k}'_\mathrm{F}}{k_\mathrm{F}^2}\right)\left(\frac{m}{2\pi\hbar^2}\right)^2 |\langle\boldsymbol{k}_\mathrm{F}|\, t\, |\boldsymbol{k}'_\mathrm{F}\rangle|^2. \tag{23}$$

t is the scattering t-matrix and Ω the solid angle of $\boldsymbol{k}'_\mathrm{F}$. The expression of the recoil

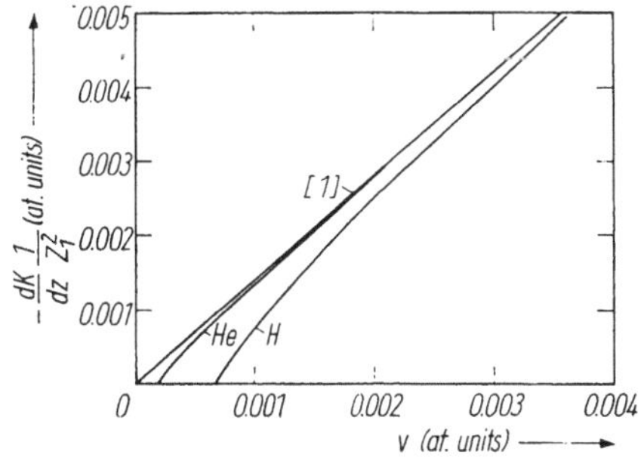

Fig. 1. Velocity dependence of the low-velocity stopping power for H and He. η and η_0 are calculated with (25) and (26), respectively, where the Fermi velocity is chosen as 1.0 (at. unit)

term η_0 is obtained with the similar procedure as follows:

$$\eta_0 = 3nv_F\sigma^{rec}\left(\frac{m}{M}\right)^2 E_F\,,$$

$$\sigma^{rec} = \int d\Omega\left(1 - \frac{\boldsymbol{k}_F\cdot\boldsymbol{k}_F'}{k_F^2}\right)^2\left(\frac{m}{2\pi\hbar^2}\right)^2 |\langle\boldsymbol{k}_F|\, t\, |\boldsymbol{k}_F'\rangle|^2\,. \tag{24}$$

When we evaluate (23) with the Born approximation for the t-matrix and use the Fermi-Thomas screened Coulomb potential for V, we have the famous Fermi-Teller formula [1] of the friction coefficient

$$\eta = \frac{2mZ_1^2e^2}{3\pi\hbar a_0}\left[\ln\,(1 + \pi a_0k_{FT}) - \frac{\pi k_{FT}a_0}{1 + \pi k_{FT}a_0}\right], \tag{25}$$

where a_0 is the Bohr radius and k_{FT} the Fermi-Thomas screening length. It should be noted that the dielectric response of (15) gives the same expression for the friction coefficient [4, 5, 7]. The recoil term is also calculated with the Born approximation as

$$\eta_0 = \left(\frac{m}{M}\right)^2\frac{4Z_1^2e^2}{\pi\hbar a_0}E_F\left[\frac{2 + \pi a_0k_{FT}}{1 + \pi a_0k_{FT}} - \frac{2}{\pi a_0k_{FT}}\,(\ln + \pi a_0k_{FT})\right]. \tag{26}$$

In the low-density limit ($n \to 0$), the final kinetic energy of the moving atom K_f is equal to the Fermi energy multiplied by the mass ratio

$$K_f = 2\left(\frac{m}{M}\right)E_F\,. \tag{27}$$

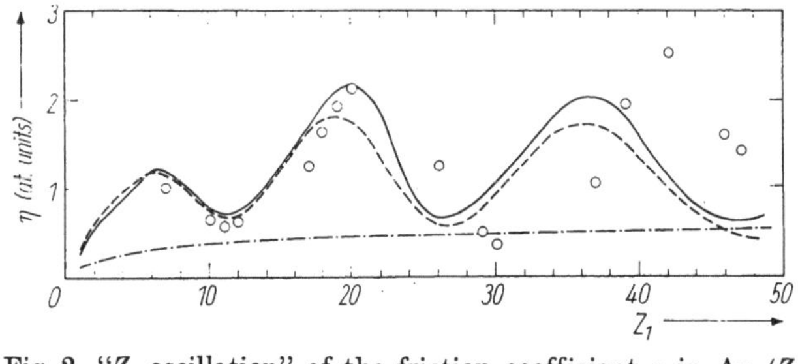

Fig. 2. "Z_1-oscillation" of the friction coefficient η in Au ($Z_2 = 79$) using the Molière potential $V(r) = Z_1Z_2e^2\sum_{j=1}^{3} A_j\exp(-a_jr/a_{TF})/r$. $A_j = (0.1, 0.35, 0.55)$, $a_j = (6.0, 0.3, 1.2)$, and a_{TF} the Thomas-Fermi radius. Phase shifts are calculated with spherical wave Born approximation (—·—), classical approximation (- - - -), and exactly (———). k_F is chosen as 1.0 (at. unit). Experiments of Bøttinger and Bason [13] are indicated by circles

In Fig. 1 the velocity dependence of the stopping power (17) with (25) and (26) is shown, H and He are chosen as the moving atoms.

To evaluate the stopping power for the heavier atoms, we cannot apply the Born approximation but should calculate the t-matrix in (23) and (24) exactly. Using the phase shift of δ's partial wave δ_l, (23) and (24) are rewritten by

$$\sigma^{\mathrm{tr}} = \frac{4\pi}{k_{\mathrm{F}}^2} \sum_{l=0} (l+1) \sin^2 (\delta_l - \delta_{l+1}) \tag{28}$$

and

$$\sigma^{\mathrm{rec}} = \frac{4\pi}{k_{\mathrm{F}}^2} \sum_{l=0} (l+1) \left[\frac{3l+4}{2l+3} \sin^2 \delta_l + \frac{3l+1}{2l+1} \sin^2 \delta_{l+2} - \right.$$
$$- 4 \sin \delta_l \sin \delta_{l+1} \cos (\delta_{l+1} - \delta_l) +$$
$$\left. + \frac{2l+4}{2l+3} \sin \delta_l \sin \delta_{l+2} \cos (\delta_{l+2} - \delta_l)\right], \tag{29}$$

respectively. The final kinetic energy K_{f} satisfies (27) when only the s-wave contribute to (28) and (29). Briggs and Pathak [6] successfuly explained the "Z_1-oscillation" of the stopping power [13] for the low-velocity atoms by calculating the phase shift δ_l. However, their calculation where the transport cross-section is calculated for the electrons moving with the velocity of the atom gives the stopping power in the intermediate velocity case. For the low-velocity atom, where the stopping power is proportional to the velocity, the transport cross-section should be that for the scattering of the Fermi electron by the atom as (28). In Fig. 2, the Z_1-dependence of the friction coefficient is shown. The atomic potential of the moving atom V is assumed as the Molière potential. Phase shifts are calculated [12] with the spherical wave Born approximation, classical approximation, and by solving the differential equations. We note that the classical mechanical calculation of the phase shifts also well explains the Z_1-oscillation of the stopping power.

Now, we calculate the electronic stopping power using the Anderson Hamiltonian

$$H = \sum_{k\sigma} E_k c_{k\sigma}^+ C_{k\sigma}^+ + \sum_{\sigma} \varepsilon_a a_\sigma^+ a_\sigma + \sum_{k\sigma} (V_{ka}(c_{k\sigma}^+ a_\sigma + \text{h.c.}) + \sum_{\sigma} U \hat{n}_{a\sigma} \hat{n}_{a-\sigma}, \tag{30}$$

where $c_{k\sigma}(c_{k\sigma}^+)$ and $a_\sigma(a_\sigma^+)$ are the annihilation (creation) operators of metal and atomic state $|k\rangle$ and $|a\rangle$, respectively, V_{ka} represents the coupling between those states and U the Coulomb integral. The theory of the friction coefficient of (15) using this model has been developed by various authors [8 to 10]. Bohnen et al. [9] presented the calculation of the friction coefficient taking account of U in (30) and Nourtier [10] extended the theory to the case where the atom passes through an inhomogeneous medium. Here we give the expression of η_0 with this model where the effect of inhomogeneity is ignored. The friction coefficient with this model is represented by

$$\eta = \sum_{\sigma} \frac{\hbar}{\pi} \alpha \sin^2 \delta ,$$
$$\alpha = \frac{2 \langle |\nabla_r V_{ka}|^2 \rangle}{\langle | V_{ka}|^2 \rangle}, \qquad \delta = \tan^{-1} \frac{\Delta}{\varepsilon_a - E_{\mathrm{F}} + i\delta}, \tag{31}$$
$$\Sigma = \Lambda - i\Delta = \sum_k \frac{|V_{ak}|^2}{E - E_k + i\delta}.$$

Similarly the recoil term is given by

$$\eta_0 = \sum_{\sigma} \frac{2\hbar}{\pi} \left(\frac{\hbar}{2M}\right)^2 \frac{\langle |\nabla_r^2 V_{ka}|^2 \rangle}{\langle | V_{ka}|^2 \rangle} \sin^2 \delta + \left(\frac{\langle |\nabla_r V_{ka}|^2 \rangle}{\langle | V_{ka}|^2 \rangle}\right)^2 \sin^2 \delta \cos (2\delta) . \tag{32}$$

4. Concluding Remarks

We have shown that the electronic stopping power for the slowly moving atom in solids contains a term inversely proportional to velocity (recoil term) in addition to the velocity-proportional friction term. This recoil term of the stopping power cannot be neglected for light atoms moving with very low energies. Because of the recoil term, the kinetic energy of atom remains of the constant value K_f (18) which is of the order of (27).

The contribution of the phonon excitation to the energy loss of the atom is more important than that of the electronic excitation in the very-low-energy case where the recoil effect discussed here is not negligible. In this energy region, the energy dissipation of a moving atom in a solid due to acoustic phonon excitation can be represented in the form of the friction force where the friction coefficient η^{ph} is written as [14]

$$\eta^{\text{ph}} = \frac{2\pi(\nabla_R F)^2}{3M_s\omega_D^3},$$

where F is the atom–atom interaction force, M_s the mass of solid atom, and ω_D the Debye frequency of the solid. We evaluate that η^{ph} is at least one order of magnitude larger than the electronic friction coefficient. The thermalization of the atomic motion due to phonons, which makes the kinetic energy of the atomto be $3k_BT/2$, is the dominant process to give the final energy of the atom. However, the recoil term η_0 given by (16) is nearly independent of the temperature of the solid and cannot be neglected to discuss the motion of slow atoms in the low-temperature solid.

The calculation of η and η_0 for atoms moving at surfaces are left for our next study.

Acknowledgements

The author would like to thank Prof. Y. H. Ohtsuki and Prof. N. Mishima for stimulating discussions.

References

[1] E. Fermi and E. Teller, Phys. Rev. 72, 399 (1947).
[2] O. B. Firsov, Soviet Phys. — J. exper. theor. Phys. **5,** 1192 (1957).
[3] J. Lindhard and M. Sharff, Phys. Rev. **124**, 128 (1961).
[4] J. Lindhard and A. Winter, Kong. Danske Vid. Selsk. mat.-fys. Med. **34**, No. 4, 1 (1964).
[5] B. A. Trubnikov and Yu. N. Yavlinskii, Soviet Phys. — J. exper. theor. Phys. **21**, 167 (1965).
[6] J. S. Briggs and A. P. Pathak, J. Phys. C **6**, L153 (1973).
[7] E. G. d'Agliano, P. Kumar, W. Schaich, and H. Suhl, Phys. Rev. B **11**, 2122 (1975).
[8] W. Schaich, Surface Sci. **49**, 211 (1975).
[9] K. P. Bohnen, K. Kiwi, and H. Suhl, Phys. Rev. Letters **34**, 1512 (1975).
[10] A. Nourtier, J. Physique **38**, 479 (1977).
[11] T. L. Ferrel, P. M. Echenique, and R. H. Ritchie, Solid State Commun. **32**, 419 (1979).
[12] N. F. Mott and H. S. W. Massey, The Theory of Atomic Collisions, Oxford University Press 1965 (p. 86).
[13] J. Bøttinger and F. Bason, Radiat. Eff. **2**, 105 (1970).
[14] S. Shindo, J. Phys. Soc. Japan **52**, 562 (1983).

(Received May 28, 1984)

phys. stat. sol. (b) **125**, 169 (1984)
Subject classification: 11; 22.1.2

High Energy Physics Institute, Academy of Sciences of the Kazakh SSR, Alma-Ata[1])

Simulation of Z_1-Oscillations in Low-Energy Heavy Ion Ranges in Solids

By
M. A. Chubisov and A. F. Akkerman

The Z_1-range oscillations of low-energy heavy ions in solids are investigated by computer simulation using an interatomic interaction based on the modified Green-Sellin-Zahor potential.

Die Z_1-Bereichs-Schwingungen von schweren Ionen niedriger Energie in Festkörpern werden mittels Computersimulation und einer interatomaren Wechselwirkung auf der Grundlage eines modifizierten Green-Sellin-Zahor-Potentials untersucht.

1. Introduction

The range oscillations of implanted keV ions with different atomic numbers which were revealed in recent experiments [1 to 4] have attracted much attention with the hope to deepen our understanding of ion interaction with solids. In the energy interval considered, elastic processes associated with some definite interatomic interaction potentials play the dominant role. The type of the potential represents the main initial information for various computer calculations describing low-energy ion penetration into a solid and for estimating the radiation effects caused by the ion interaction with atoms of solid.

The practical significance of such calculations is evident, and it is connected, first of all, with the problem of low-energy ion implantation in semiconductors. It also should be noted that the complete theoretical interpretation of Z_1-range oscillations is associated with the construction of some general approach capable to explain both the oscillations considered and the well-known Z_1-oscillations of the ion energy losses in inelastic processes [5, 6].

In this paper the suitability of the Green-Sellin-Zahor potential (V_{GSZ}) [7], modified in [8], for the calculation of Z_1-range oscillations of ions with $Z_1 = 20$ to 83 and $\varepsilon = 0.004$ to 0.08 (energy in Lindhard's system of units) in silicon has been examined.

Both the Monte Carlo simulation of low-energy ion penetration into a substance and numerical solutions of the transport equation carried out in the spirit of the binary collision approach [9] are incorrect, as many-body interactions of a slackening ion with a solid are neglected there. We have used the molecular dynamics method (MDM) [10, 11] accounting for the many-body effect automatically. The experiments [1, 2, 4] have been simulated with the block version of the GESID code [11, 12].

2. Calculational Model and Initial Information

The dynamical method describes the motion of individual atoms within the assembly of N atoms having their definite arrangement (amorphous or monocrystalline target) by virtue of Newton's equations of motion. It means that the potential enters directly into the expression for the force acting onto i-atom or ion from the side of its

[1]) 480082 Alma-Ata, USSR.

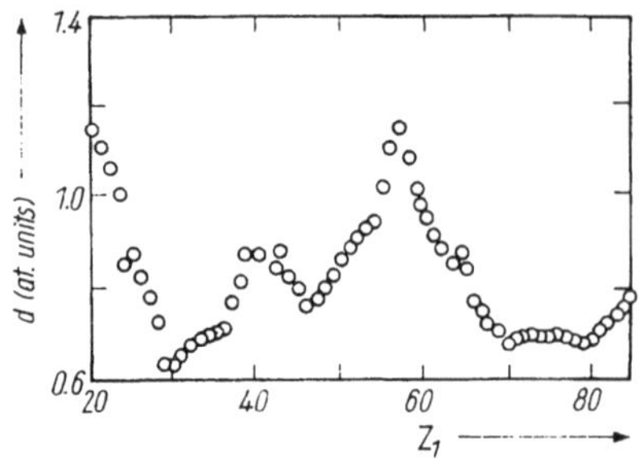

Fig. 1. Parameter d vs. atomic number Z [7]

j-neighbours,

$$\boldsymbol{F}_i = -\sum_{j \neq i} \frac{\partial}{\partial \boldsymbol{r}_i} V(|\boldsymbol{r}_i - \boldsymbol{r}_j|), \tag{1}$$

and in these calculations one can change the form of the potential easily.

In the GESID code the inelastic energy losses of ion are counted at the distance of the closest approach of the ion with the target atoms, although a trajectory is continuously constructed over the total time interval of their interaction by a finite number of iterations. Here inelastic energy losses are calculated according to the Oen-Robinson model [13]. The k parameter is defined by Lindhard et al.'s [14] formulae for the electronic stopping power.

Usually the description of projectile–target interactions for atoms is based on the screened Coulomb potential

$$V(r) = \frac{Z_1 Z_2 e^2}{r} \cdot \varphi\left(\frac{r}{a}\right), \tag{2}$$

with various screening functions φ.

Lindhard et al. [14] proposed a model for the stopping power with the ion–atom elastic interaction determined by a universal function of nuclear scattering in terms of the Thomas-Fermi-Firsov (TFF) potential. Often the results of the experiments are compared with the predictions of this model [14, 15]. However, the experiments on the ion ranges in solids [16] and on the angular distributions of the scattered ions passing through thin foils [17, 18] have shown that the most realistic potentials are to be shorter-ranged than V_{TFF} at mediate and large impact parameters. One type of such a potential has been deduced by Wilson, Haggmark, and Biersack (WHB) [19] from first principles. It has been calculated for a number of atom–atom combinations and then been averaged. The screening function of the averaged potential has the Moliere form,

$$\varphi\left(\frac{r}{a}\right) = \sum_{i=1}^{3} C_i \exp\left(-\frac{b_i r}{a}\right), \tag{3}$$

with the Firsov screening radius $a = a_{\mathrm{F}}$. The coefficients are $C_1 = 0.006905$, $C_2 = 0.166930$, $C_3 = 0.826165$, $b_1 = 0.131825$, $b_2 = 0.307856$, $b_3 = 0.916760$.

Another kind of potential has been proposed by Kalbitzer and Oetzmann (KÖ) [16] based on precision measurements of energy loss in elastic collisions utilizing the scheme: energy loss → cross-section → potential. In this case the form of the screening function is

$$\varphi\left(\frac{r}{a}\right) = \left(\frac{r}{a} + e\right)^{-\gamma \ln(1 + (r/a)/\gamma)}, \tag{4}$$

where the screening radius is chosen according to Lindhard, $a = a_{\mathrm{L}}$, and $\gamma = \frac{2}{3}$. The screening function (4) coincides with that of the quantum-mechanical potential,

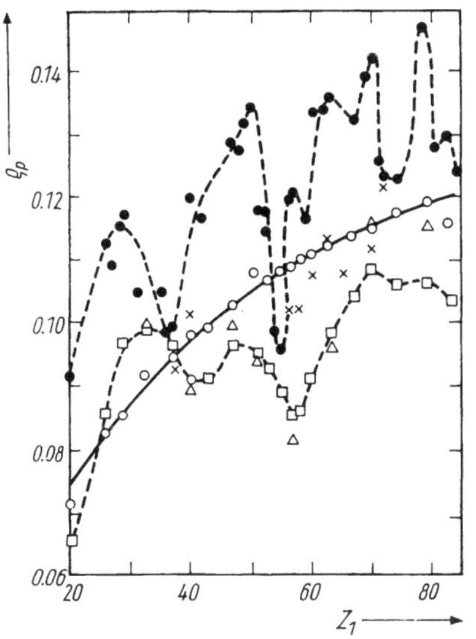

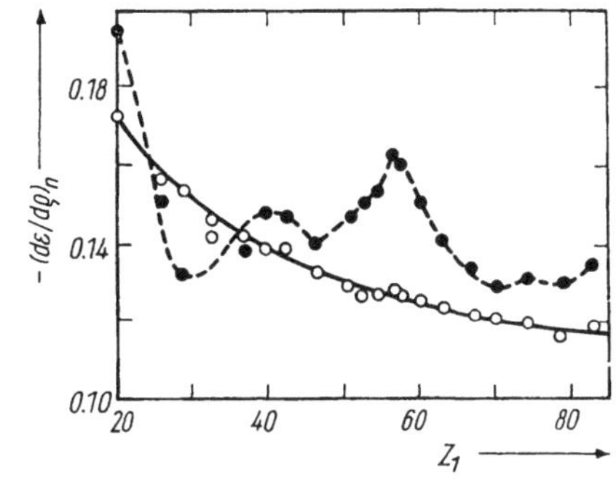

Fig. 2 Fig. 3

Fig. 2. Reduced projected range ϱ_p vs. atomic number Z_1 ($Z_1 \rightarrow$ Si, $\varepsilon = 0.015$). The experimental data: (1) ● [1], (2) × [25]. The calculated data: (3) ○ for the $V_{KÖ}$ potential; (4) □ for the V_{GSZ-K} potential with the parameter $d_{eff} = (d_1 + d_2)/2$; (5) △ for the V_{GSZ-K} potential with the parameter $d_{eff} = d_1 f + d_2(1 - f)$, where $f = Z_1/(Z_1 + Z_2)$

Fig. 3. Reduced elastic energy losses vs. atomic number Z_1($Z_1 \rightarrow$ Si, $\varepsilon = 0.015$). ○ for the $V_{KÖ}$ potential, ● for the V_{GSZ-K} potential

V_{WHB}, with the screening radius equal to a_L. Calculations with (3) and (4) have given satisfactory agreement with experimental data both for the ranges of ions in amorphous targets [16, 20, 21] and for the angular distributions of the scattered ions after transmission through thin foils [22].

The averaged potential V_{WHB} and the potential $V_{KÖ}$ produce ion scattering characteristics which monotonically depend on the atomic number of the projectile Z_1.

The experiments [1 to 4] have manifested large deviations (20 to 25%) from the average range value. It was suggested that the oscillations are caused by the shell structure of colliding atoms. The nonmonotonic dependence of the scattering characteristics for ions with reduced energy $\varepsilon < 0.1$ is revealed in their interactions with the stopping medium, when the elastic energy losses, $(-d\varepsilon/d\varrho)_n$, exceed the inelastic losses, $(-d\varepsilon/d\varrho)_e$, of the ions. The greater ε, the less is the oscillation amplitude [4].

One may study the Z_1-range oscillations directly by computer simulation in which the interatomic interaction potential represents the real distribution of the electron density.

The first attempt of a numerical simulation has been made in [23] applying the theoretical description of the elastic atom–atom scattering, as proposed in [24]. However, pronounced range oscillations have not been obtained. In our calculations Komarov's potential [8] has been used which is the Firsov-like modification of V_{GSZ} [7]. The screening function of the potential [8] is

$$\varphi(r) = \left[(Z_0 - 1)^{0.4}\, d_{eff}\left(\exp\left(\frac{1.13r}{d_{eff}}\right) - 1\right) + 1\right]^{-1}, \tag{5}$$

where $Z_0 = (Z_1^{1/2} + Z_2^{1/2})^2$, $d_{eff} = (d_1 + d_2)/2$.

The parameter $d(Z)$ is associated with the shell structure of an atom. Its magnitudes have been defined in [7] and are given in Fig. 1.

3. Results

With the above-mentioned potential, longitudinal spatial distributions of ions implanted into amorphous silicon characterized by Z_1 from 20 to 83 and reduced energies $\varepsilon = 0.004, 0.015, 0.08$ have been calculated. In Fig. 2 the experimental (1, 2) and the computed (3, 4) curves of the reduced projected ranges ϱ_p are shown. The range distribution parameters — the average projected range, R_p; the root-mean-squared deviation, ΔR_p; the skewness of the distribution S — are determined as follows [21]:

$$R_p = \frac{\int xN(x)\,dx}{\int N(x)\,dx}, \tag{6}$$

$$(\Delta R_p)^2 = \left[\frac{\int x^2N(x)\,dx}{\int N(x)\,dx}\right] - R_p^2, \tag{7}$$

$$S = \frac{\dfrac{\int x^3N(x)\,dx}{\int N(x)\,dx} - 3R_p(\Delta R_p)^2 - R_p^3}{(\Delta R_p)^3}, \tag{8}$$

where $N(x)$ is the computed concentration of implanted ions at depth x.

Earlier the GESID code was tested by comparing the computed with the experimental ranges [21] and with calculations of others [9, 21]. The parameters of the longitudinal distributions for antimony ions implanted into amorphous silicon are given in Table 1 as an example. Our data were obtained with the potential $V_{\text{KÖ}}$.

Table 1
Ranges of antimony ions in amorphous silicon

reference	range distribution parameters		E_0 (keV) 5.0	10.0	20.0	30.0
experiment [21]	R_p	(nm)	9.1	11.7	17.0	22.6
	ΔR_p	(nm)	4.7	5.4	6.7	8.6
	S		0.77	0.70	0.46	0.79
present calculation	R_p	(nm)	7.7	11.7	17.7	23.0
	ΔR_p	(nm)	2.4	3.8	5.7	7.5
	S		0.53	0.48	0.50	0.60
calculation by Winterbon's code [21]	R_p	(nm)	8.42	12.40	18.58	23.81
	ΔR_p	(nm)	2.54	3.74	5.60	7.17
	S		0.486	0.479	0.467	0.457
Burenkov et al. [9]	R_p	(nm)	9.2	13.3	19.4	24.4
	ΔR_p	(nm)	2.8	4.0	5.8	7.3
	S		0.49	0.48	0.48	0.47

The comparison of the experimental data with the results of three different calculations speaks in favour of the many-body approximation. Testing with other ion–atom combinations [20] has shown that the calculation technique used describes adequately the slowing down of ions in solids.

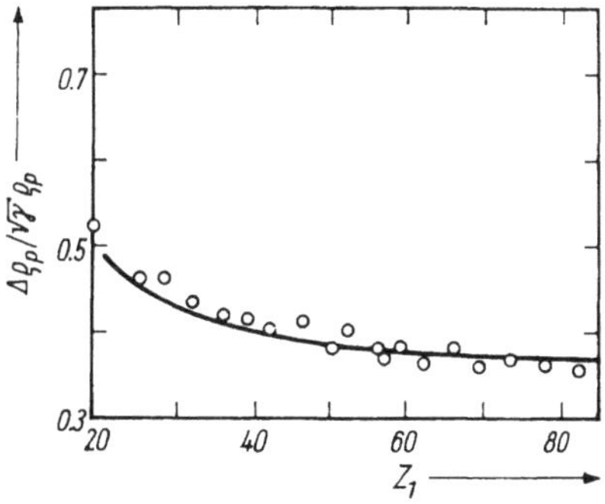

Fig. 4

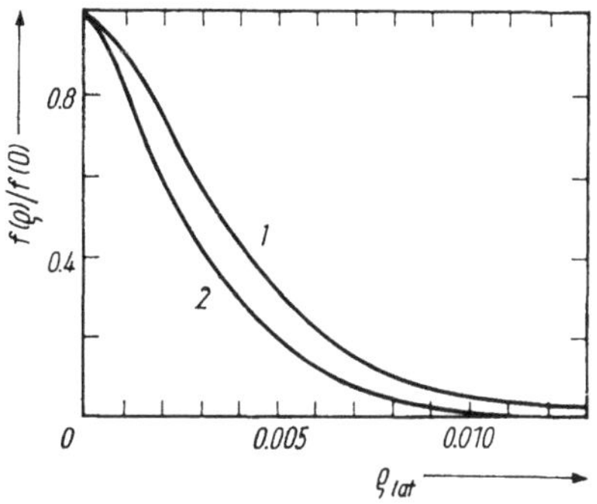

Fig. 5

Fig. 4. Relative reduced range straggling vs. atomic number Z_1 ($Z_1 \rightarrow$ Si, $\varepsilon = 0.015$). ○ calculation with the V_{GSZ-K} potential. The curve means Winterbon's data [26]

Fig. 5. Lateral range distributions of (1) La- and (2) Yb- ions in silicon for the V_{GSZ-K} potential. The reduced thickness of the amorphous silicon foil $\Delta\varrho$ is 0.0355 and the reduced ion energy ε is 0.015

As it follows from Fig. 2, the potential $V_{KÖ}$ results in a monotonous dependence of ϱ_p on Z_1, in contrary to V_{GSZ-K} resulting in the oscillatory dependence of ϱ_p on Z_1, the latter being correlated with the behaviour of the parameter $d(Z_1)$ [7] (see Fig. 1) characterizing the form of the screening function of V_{GSZ-K}. The dependence of the calculated elastic energy losses on the atomic number Z_1 correlates also with $\varrho_p(Z_1)$ (see Fig. 3). The experimental values of ϱ_p, depicted in Fig. 2, oscillate around ϱ_p for the Lenz-Jensen potential. Our computed ϱ_p, repeating the main features of these oscillations qualitatively, seems to be systematically smaller and, at least for $Z_1 \lessapprox 36$, to oscillate around ϱ_p for $V_{KÖ}$. Nevertheless, the extrapolated values of ϱ_p, obtained in [25], for $\varepsilon = 0.015$ reveal also oscillations. They occur also around the values computed with $V_{KÖ}$. The discrepancy can hardly be associated with a systematic computational error, as, e.g., computations of ϱ_p for V_{GSZ-K} for the ions of La and Ho with $\varepsilon = 0.08$ in Si give values overestimated by 7.6% and 11.1%, respectively, compared with the experiment [2].

The comparison of the computed straggling of the reduced relative longitudinal ranges, $\Delta\varrho_p/\sqrt{\gamma}\,\varrho_p$ ($\gamma = 4A_1A_2/(A_1 + A_2)^2$), using V_{GSZ-K} with the computed data [26] using V_{TFF} as well as in the experiment [4], shows that ϱ_p and $\Delta\varrho_p$ oscillate in the same manner. As a result, the monotonous dependence on Z_1 is observed for any potential (see Fig. 4).

The Z_1-oscillation of the lateral ion ranges for the potential V_{GSZ-K}, depicted in Fig. 5, has been obtained in computations. The transmission of La and Yb ions with the reduced energy $\varepsilon = 0.015$ through foils of amorphous silicon with the reduced thickness $\Delta\varrho = 0.0355$ has been calculated. One can see that the halfwidth of the lateral range distribution varies with Z_1. This fact agrees with the data [27] obtained in gaseous targets.

In Table 2 the computed energy dependence of the oscillation amplitude ratios for the closest maxima and minima for V_{GSZ-K} is presented, showing that oscillations decay when ε increases.

The computer simulation has demonstrated that the potential V_{GSZ} [7] may serve as the basis for the construction of the interatomic potential in a wide energy interval. The approximation for the calculation of d_{eff} proposed in [8], at least for $Z_1 > 36$,

Table 2

Energy dependence of the reduced relative amplitudes of range oscillations

reduced energy, ε	0.004	0.015	0.08
$\frac{\varrho_p^{La} - \varrho_p^{Ag}}{\varrho_p^{Ag}}$ (%)	−16.0	−12.8	− 8.2
$\frac{\varrho_p^{La} - \varrho_p^{Ho}}{\varrho_p^{Ho}}$ (%)	−23.7	−19.0	−11.7

needs further improvement. We can introduce a weighting factor f in the relation for d_{eff} ($d_{eff} = d_1 f + d_2(1 - f)$), where $f = Z_1/(Z_1 + Z_2)$, and d_1 and d_2. For such values of d_{eff} the new, calculated ranges are shown in Fig. 2 as points (5). At the same time an improvement is also needed in the procedure of cutting off the potential when comparing with precision measurements.

4. Discussion

As is pointed out, Z_1-range oscillations have been observed only for semiconductors: Si and Ge [1 to 4]. In [27] Z_1-oscillations are studied in the angular distribution of the ions passing through thin gaseous Xe targets. Altogether, this shows that the ocsillation effect takes place in substances where the spatial distribution of the atomic outer shell (valence) electrons is well localized. Seemingly, in this case one can use the screening technique quite surely and, hence, assume a definite potential, e.g., the Green-Sellin-Zahor modified potential. The latter confirms the possibility of performing computations of low-energy implanted ion distributions according to the scheme: potential → elastic cross-section → distribution characteristics. It should be noted that in the potential approach one may test the Z_1-oscillations of the electronic stopping power, too, as follows from [28].

On the other hand, the absence of Z_1-range oscillations for ions implanted into metals, as is shown in [4], means that the potential approach is not universal, because the a priori usage of V_{GSZ-K} in this case would produce range oscillations.

In fact, in slow atom collisions the impact parameters are large, and a weak overlapping of the atomic electron shells takes place during interaction. It is quite plausible that the elastic interaction of the screened nuclei becomes comparable with the scattering of the external electrons of the atom [27]. Just the latter contribution to the total cross-section of scattering is not realized in the case of metals (in contrast to semiconductors). An explanation quite close to that is given in [4]. If the abovementioned mechanism really takes place, it manifests itself most prominently in spatial distributions of the implanted ions inside aligned crystals. In this particular case further investigation of the V_{GSZ}-like potential application seems to be useful.

References

[1] F. Besenbacher, J. Bøttiger, T. Laursen, P. Loftager, and W. Möller, Nuclear Instrum. and Methods **170**, 183 (1980).

[2] D. J. O'Connor, B. W. Farmery, D. Chivers, and M. W. Thompson, Nuclear Instrum. and Methods **178**, 517 (1980).

[3] J. Berthold and S. Kalbitzer, Phys. Letters A **91**, 37 (1982).

[4] J. Berthold and S. Kalbitzer, Nuclear Instrum. and Methods **209/210**, 13 (1983).

[5] L. Eriksson, J. A. Davies, and P. Jespersgard, Phys. Rev. **161**, 219 (1967).

[6] F. H. EIZEN, Canad. J. Phys. **46**, 561 (1968).
[7] A. E. S. GREEN, D. L. SELLIN, and A. S. ZAHOR, Phys. Rev. **184**, 1 (1969).
[8] F. F. KOMAROV, Izv. vuzov, Ser. fiz. **3**, 70 (1973).
[9] A. F. BURENKOV, F. F. KOMAROV, M. A. KUMAKHOV, and M. M. TEMKIN, Tablitsy parametrov prostranstvennogo raspredeleniya ionno-implantirovannykh primesei, Izd. BGU, Minsk 1980.
[10] J. B. GIBSON, A. N. GOLAND, M. MILGRAM, and G. H. VINEYARD, Phys. Rev. **46**, 1229 (1960).
[11] M. A. CHUBISOV and A. F. AKKERMAN, phys. stat. sol. (a) **55**, K53 (1979).
[12] M. A. CHUBISOV, Preprint HEPI, Acad. Sci. Kazakh SSR, No. 47 – 77, Alma-Ata 1977.
[13] O. S. OEN and M. T. ROBINSON, Nuclear Instrum. and Methods **132**, 647 (1976).
[14] J. LINDHARD, M. SCHARFF, and H. E. SCHIØTT, Kong. Danske Vid. Selsk., mat.-fys. Medd. **33**, No. 14 (1963).
[15] J. LINDHARD, V. NIELSEN and M. SCHARFF, Kong. Danske Vid. Selsk., mat.-fys. Medd. **36**, No. 10 (1968).
[16] S. KALBITZER and H. OETZMANN, Proc. Conf. Ion Beam Modification of Materials, Vol. 1, Budapest, 4 to 8 Sept. 1978 (p. 3).
[17] S. SCHWABE and K. STOLLE, phys. stat. sol. (b) **47**, 111 (1971).
[18] P. KRYGEL, J. LIPPOLD, and S. SCHWABE, phys. stat. sol. (b) **53**, 255 (1972).
[19] W. D. WILSON, L. G. HAGGMARK, and J. P. BIERSACK, Phys. Rev. B **15**, 2458 (1977).
[20] A. F. AKKERMAN and M. A. CHUBISOV, Mater. Internat. Working Meeting Ion Implant. in Semicond. and Other Materials, Prague 1981 (p. 143).
[21] W. K. CHU, R. H. KASTL, and P. C. MURLEY, see [16] (p. 179).
[22] M. A. CHUBISOV and A. F. AKKERMAN, phys. stat. sol. (b) **102**, K55 (1980).
[23] K. GÄRTNER and K. HEHL, phys. stat. sol. (b) **108**, K109 (1981).
[24] K. GÄRTNER and K. HEHL, phys. stat. sol. (b) **94**, 231 (1979).
[25] M. KISIELEWICZ, see [20] (p. 147).
[26] K. B. WINTERBON, Ion Implantation Range Energy Deposition Distribution, Vol. 2, IFI/Plenum Data Co., New York 1975.
[27] H. KNUDSEN, F. BESENBACHER, J. HEINEMEIER, and P. HVELPLUND, Phys. Rev. A **13**, 2095 (1976).
[28] V. S. KESSELMAN, Zh. tekh. Fiz. **41**, 1708 (1971).

(Received February 28, 1984)

phys. stat. sol. (b) 125, 177 (1984)

Subject classification: 13.5 and 18.4

Sektion Physik der Karl-Marx-Universität Leipzig[1]) *(a) and Sektion Physik und Technik elektronischer Bauelemente der Technischen Hochschule Ilmenau (b)*

Two-Soliton Solution in a Modified Hubbard Chain[2])

By

U. LINDNER (a) and S. SCHERF (b)

In the case of constant electron density along the chain the two-soliton solution is given explicitly. For simplification the coupling between electronic system and lattice is neglected.

Für den Fall konstanter Elektronendichte entlang der Kette wird die Zwei-Solitonen-Lösung explizit angegeben. Zur Vereinfachung wird die Kopplung zwischen Elektronensystem und Gitter vernachlässigt.

1. Introduction

Experimental results in quasi 1-d antiferromagnets, e.g. TMMC, CPC, which suggest non-linear effects, have been interpreted next in terms of the antiferromagnetic Heisenberg model [1]. Because in these substances the distance between the paramagnetic ions along the chain, e.g. the Mn^{++} ions in TMMC, is the same as that of these ions in the pure metal (see, e.g. [1]), we have proposed a model with finite band width for describing these substances [2 to 4]. For this 1-d Hubbard model modified by coupling to the lattice in time-dependent, mean-field approximation in continuum limit we have shown that solitary excitations can exist if the ground state of the system has a magnetic order, and that these solitary excitations can be interpreted as domain walls.

Starting from an antiferromagnetic ground state order we have calculated the excitation energy, and with the dilute gas model in the same way as Krumhansl and Schrieffer [5] the free energy, the static correlation function of the lattice motion, and the width of the central peak of the dynamic structure factor connected with the one-soliton excitations [6].

In this paper we take the first step in order to study multi-soliton effects. We investigate the two-soliton excitations, this means, now we consider effects connected with the interaction of two domain walls, e.g. with the penetration of both walls, too. It turns out that in the case of constant electron density n along the chain the two-soliton solution can be given explicitly, whereby for simplification the coupling between electronic system and lattice is neglected, since this assumption does not change the qualitative behaviour of the solitary excitation of the electronic system.

2. One-Soliton Solution

In time-dependent, mean-field approximation and in the continuum limit our model [2] is represented by the Lagrangian density [4]

[1]) Karl-Marx-Platz, DDR-7010 Leipzig, GDR.

[2]) Report at the Joint Institute for Nuclear Research Dubna, June 20, 1983, and Preprint KMU-FKT, 3. 6. 1983.

$$\mathcal{L} = i \sum_\sigma \left(\Phi_\sigma^* \frac{\partial \Phi_\sigma}{\partial t} - \frac{\partial \Phi_\sigma^*}{\partial t} \Phi_\sigma \right) + Ma^2 \left(\frac{\partial x}{\partial t} \right)^2 - \overline{\mathcal{H}}$$

$$= i \sum_\sigma \left(\Phi_\sigma^* \frac{\partial \Phi_\sigma}{\partial t} - \frac{\partial \Phi_\sigma^*}{\partial t} \Phi_\sigma \right) + \frac{M}{2} a^2 \left(\frac{\partial x}{\partial t} \right)^2 -$$

$$- \frac{M}{2} a^2 \omega_0^2 \left(\frac{\partial x}{\partial \xi} \right)^2 + [\hat{t} + Ia] \sum_\sigma \frac{\partial \Phi_{-\sigma}^*}{\partial \xi} \frac{\partial \Phi_\sigma}{\partial \xi} -$$

$$- 2 \left[\hat{t} + Ia \left(1 - \frac{\partial x}{\partial \xi} \right) \right] \sum_\sigma \Phi_{-\sigma}^* \Phi_\sigma - \frac{U}{2} (\sum_\sigma \Phi_\sigma^* \Phi_{-\sigma})^2 + \mu \sum_\sigma |\Phi_\sigma|^2 \tag{1}$$

with the condition $(\partial/\partial\xi) \sum_\sigma \Phi_\sigma^* \Phi_{-\sigma} = 0$ at the boundary of the system if we start from an antiferromagnetic ground state order. $\overline{\mathcal{H}}$ is the Hamilton density, $x(\xi, t)$ is the deviation of a lattice particle related to the lattice constant a from the equilibrium position, $\omega_0^2 = \alpha/M$, where α is the force constant and M denotes the mass of a lattice particle. The transition amplitude of an electron between Wannier states of neighbouring atoms in the chain is denoted by $\hat{t}$, the repulsive interaction between electrons of the same atom by U, the chemical potential by μ, and I describes the coupling between electronic system and lattice [2]. For convenience instead of the complex probability amplitudes $\Phi_\sigma(\xi, t)$ and $\Phi_{-\sigma}(\xi, t)$ to find an electron with spin σ or $-\sigma$ at the time t at the place $a\xi$ in the chain we use the fields

$$C_\pm = \frac{1}{\sqrt{2}} (\Phi_\sigma \pm \Phi_{-\sigma}) \,. \tag{2}$$

Then

$$\sum_\sigma \Phi_\sigma^* \Phi_{-\sigma} = |C_+|^2 - |C_-|^2 \tag{3a}$$

plays the role of the spin density and

$$\sum_\sigma |\Phi_\sigma|^2 = |C_+|^2 + |C_-|^2 = n(\xi, t) \tag{3b}$$

that of the electron density.

For the equations of motion one obtains from (1)

$$M \frac{\partial^2 x}{\partial t^2} = M\omega_0^2 \frac{\partial^2 x}{\partial \xi^2} - \frac{2I}{a} \frac{\partial}{\partial \xi} (|C_+|^2 - |C_-|^2) \,, \tag{4a}$$

$$i \frac{\partial C_+}{\partial t} - T_0 \frac{\partial^2 C_+}{\partial \xi^2} - (2T_0 - \mu)\, C_+ + 2Ia \frac{\partial x}{\partial \xi} C_+ - U(|C_+|^2 - |C_-|^2)\, C_+ = 0 \,, \tag{4b}$$

$$i \frac{\partial C_-}{\partial t} + T_0 \frac{\partial^2 C_-}{\partial \xi^2} - (-2T_0 - \mu) C_- - 2Ia \frac{\partial x}{\partial \xi} C_- + U(|C_+|^2 - |C_-|^2)\, C_- = 0 \,, \tag{4c}$$

where $T_0 = \hat{t} + Ia$.

These differential system possesses solitary solutions [2 to 4] of which we regard here those solutions with constant electron density $n(\xi, t) = n =$ const (cf. (3b)) only. For this case the one-soliton solution (cf. [6]) reads

$$x(\xi, t) = x_0 \tanh \{\varkappa(\xi - \xi_{01} - \omega t)\} \,, \tag{5a}$$

$$C_+(\xi, t) = C_+^0 \operatorname{sech} \{\gamma(\xi - \xi_{01} - u_1 t)\}\, e^{i\beta(\xi - \xi_{01}) - iv_+ t} \,, \tag{5b}$$

$$C_-(\xi, t) = C_-^0 \tanh \{\gamma(\xi - \xi_{01} - u_1 t)\}\, e^{i\delta(\xi - \xi_{01}) - iv_- t} \,. \tag{5c}$$

(If the quantity $\gamma^2 < 0$, then $C_+(\xi, t)$ is of the type "tanh" and $C_-(\xi, t)$ is of the type "sech".) The solution (5) is written down in such a way that the centre position ξ_{01} and the frequency $\omega = u_1$ are chosen as free parameters. The other parameters read as follows [6]:

$$x_0 = \frac{\hat{I}n}{Ia\varkappa}, \qquad \varkappa = \gamma = \left(\frac{(U - \hat{I})\, n}{T_0}\right)^{1/2}, \qquad C_+^0 = C_-^0 = \sqrt{n}\,,$$

$$\beta = -\delta = -\frac{u_1}{2T_0}, \qquad v_+ = 2T_0 - \hat{I}n - \mu - \frac{u_1^2}{4T_0},$$

$$v_- = -2T_0 + Un - \mu + \frac{u_1^2}{4T_0}$$

with $\hat{I} = (2I)^2/M(\omega_0^2 - u_1^2)$.

One can interpret the one-soliton solution (5) as solitary spin distortion (domain wall) of the antiferromagnetic ground state order ((5b) and (5c)) which is accompanied by a solitary lattice distortion (5a). The free parameters describe the initial position $a\xi_{01}$ of the wall and its velocity au_1.

For simplification in the following we put $I = 0$, this means, we drop the coupling between electronic system and lattice. One sees from (5) immediately that in this simplification the solitary spin distortion (5b) and (5c) does not change qualitatively whereas, as one expects, the solitary lattice distortion vanishes. Furthermore, in the case of $n =$ const, where distortions of the spin ordering do not influence the electron density, it exists according to (3b) no coupling between the differential equations (4b) and (4c), and we obtain

$$i\,\frac{\partial C}{\partial t} + M\,\frac{\partial^2 C}{\partial \xi^2} - \lambda C + G\,|C|^2\, C = 0 \tag{6}$$

for $C \mathrel{\hat{=}} \{C_+, C_-\}$ with $M_+ = -M_- = -T_0, G_+ = G_- = -2U, \lambda_+ = 2T_0 - \mu - Un$, and $\lambda_- = -2T_0 - \mu - Un = \lambda_+ - 4T_0$. By the transformation

$$C(\xi, t) = A(\xi, t)\,\mathrm{e}^{-i\lambda t} \tag{7}$$

with λ real removing the term linear in C in (6) we get the non-linear Schrödinger equation (NLS)

$$i\,\frac{\partial A}{\partial t} + M\,\frac{\partial^2 A}{\partial \xi^2} + G\,|A|^2\, A = 0\,, \tag{8}$$

where both cases, e.g. G/M positive (attractive case) for $C_+(\xi, t)$ and G/M negative (repulsive case) for $C_-(\xi, t)$, have to be considered.

Now, in the following section, we can derive the two-soliton solution for the differential equations (6) for $C_+(\xi, t)$ and $C_-(\xi, t)$ coupled by (3b) with $n = n(\xi, t) =$ = const, considering the NLS (8) in attractive and repulsive cases.

3. Two-Soliton Solution

3.1 Attractive case

Using the transformation

$$\bar{t} = -t\,, \qquad \bar{x} = \xi \tag{9a}$$

(8) goes over into

$$i\,\frac{\partial A}{\partial \bar{t}} + T_0\,\frac{\partial^2 A}{\partial \bar{x}^2} + 2U\,|A|^2\,A = 0 \tag{9b}$$

which is the special case ($\alpha = \gamma = 0$) of Hirota's equation [7],

$$i\,\frac{\partial \psi}{\partial \bar{t}} + i3\alpha\,|\psi|^2\,\frac{\partial \psi}{\partial \bar{x}} + \beta\,\frac{\partial^2 \psi}{\partial \bar{x}^2} + i\gamma\,\frac{\partial^3 \psi}{\partial \bar{x}^3} + \delta\,|\psi|^2\,\psi = 0\,, \tag{10}$$

where α, β, γ, $\delta \geqq 0$ satisfying the relation $\alpha/\gamma = \delta/\beta$. For $\beta = T_0$ and $\delta = 2U$ we have $A(\bar{x}, \bar{t}) \equiv \psi(\bar{x}, \bar{t})$.

The two-soliton solution of (10) is given by Hirota [7],

$$\psi^{(2)}(\bar{x}, \bar{t}) = \frac{G^{(2)}(\bar{x}, \bar{t})}{F^{(2)}(\bar{x}, \bar{t})} \tag{11}$$

with

$$\begin{aligned} G^{(2)}(\bar{x}, \bar{t}) &= A_1 \operatorname{sech} \xi_1\, e^{i\zeta_1} (\cos \Phi_1 + i \sin \Phi_1 \tanh \xi_2) + \\ &\quad + A_2 \operatorname{sech} \xi_2\, e^{i\zeta_2} (\cos \Phi_2 + i \sin \Phi_2 \tanh \xi_1)\,, \\ F^{(2)}(\bar{x}, \bar{t}) &= \cosh \bar{a} + \sinh \bar{a} \left(\tanh \xi_1 \tanh \xi_2 - \operatorname{sech} \xi_1 \operatorname{sech} \xi_2 \cos (\zeta_1 - \zeta_2)\right), \end{aligned}$$

and

$$\begin{aligned} \xi_j &= \operatorname{Re}\,(P_j\bar{x} - \Omega_j\bar{t}) + \xi_j^0 = a_j\bar{x} - 2a_jb_jT_0\bar{t} + \xi_j^0\,, \\ \zeta_j &= \operatorname{Im}\,(P_j\bar{x} - \Omega_j\bar{t}) + \zeta_j^0 = b_j\bar{x} - (b_j^2 - a_j^2)\,T_0\bar{t} + \zeta_j^0\,, \\ P_j &= a_j + ib_j\,, \qquad \Omega_j = -i\beta P_j^2 = iT_0(a_j^2 + 2ia_jb_j - b_j^2)\,, \\ A_j &= \left(\frac{\gamma}{2\alpha}\right)^{1/2} |P_j + P_j^*| \qquad (j = 1, 2)\,, \\ \Phi_1 &= \arg\left[\frac{(P_1 - P_2)}{(P_1 + P_2^*)}\right] = \arctan\left(\frac{2(b_1 - b_2)\,a_2}{(b_1 - b_2)^2 + a_1^2 - a_2^2}\right), \\ \Phi_2 &= \Phi_1(1 \to 2, 2 \to 1)\,, \\ e^{\bar{a}} &= \frac{|P_1 - P_2|}{|P_1 + P_2^*|}\,. \end{aligned}$$

Therefore, because of (7) it yields, e.g.,

$$C_+^{(2)}(\xi, t) = \psi^{(2)}(\bar{x}, \bar{t})\, e^{-i\lambda_+ t}\,. \tag{12}$$

The constants in $\psi^{(2)}(\bar{x}, \bar{t})$ have to be fixed according to our model (1) and the condition $n = n(\xi, t) = \text{const.}$ But before we can do that we have to regard the two-soliton solution of the NLS (8) for $G/M < 0$, e.g. for $C_-(\xi, t)$.

3.2 Repulsive case

If we apply the transformation

$$\tilde{t} = 2Ut\,, \qquad \tilde{x} = \xi \left(\frac{2U}{T_0}\right)^{1/2} \tag{13a}$$

we obtain from (8)

$$i\frac{\partial A}{\partial \tilde{t}}+\frac{\partial^2 A}{\partial \tilde{x}^2}-|A|^2 A=0\,, \tag{13b}$$

i.e. the reduced form of NLS, where the non-linear term is negative.

Boiti, Laddomada, and Pempinelli (BLP) [8] found for this equation

$$i\frac{\partial q}{\partial \tilde{t}}+\frac{\partial^2 q}{\partial \tilde{x}^2}-|q|^2 q=0\,,\qquad q(\tilde{x},\tilde{t})\equiv A(\tilde{x},\tilde{t})\,, \tag{14}$$

the Bäcklund transformation (BT). They also tried to give the two-soliton solution [9], but unfortunately their expression for the two-soliton solution does not fulfil the BT and the differential equation (14) as would be necessary. One of the authors has shown [10] that it is possible to find from the BT of BLP the two-soliton solution which satisfies the BT and (14). This two-soliton solution reads

$$q^{(2)}(\tilde{x},\tilde{t})=q^{(1)}(\tilde{x},\tilde{t})-\sqrt{2}\,\Psi(\tilde{x},\tilde{t})\,\mathrm{e}^{i\theta(\tilde{x},\tilde{t})-i\omega(\tilde{x},\tilde{t})} \tag{15}$$

with

$$\Psi(\tilde{x},\tilde{t})=-\sqrt{2}\,\sigma_1 m\frac{y^2-2my+1}{1-my+d\exp\{-2x_2\}}\,,$$

$$q^{(1)}(\tilde{x},\tilde{t})=2\sigma_1 y\,\mathrm{e}^{i\theta}\,,$$

$$y=\tanh x_1\,,\qquad x_1=\sqrt{2}\,\sigma_1\eta_1\,,\qquad x_2=\sqrt{2}\,\sigma_2\eta_2\,,\qquad m=\frac{\sigma_2}{\sigma_1}\,,\qquad m^2\leqq 1\,,$$

$$\eta_1=\tilde{x}-\tilde{x}_{01}+4\hat{\xi}_1 t\,,\qquad \eta_2=\tilde{x}-\tilde{x}_{02}+2(\hat{\xi}_1+\hat{\xi}_2)\,\tilde{t}\,,$$

$$(\hat{\xi}_1-\hat{\xi}_2)\operatorname{ctg}\tfrac{1}{2}\omega=\sqrt{2}\,\sigma_1(y-m)\,,\qquad (\hat{\xi}_1-\hat{\xi}_2)^2=2\sigma_1^2(1-m^2)\,,$$

and

$$\theta(\tilde{x},\tilde{t})=-2\hat{\xi}_1\tilde{x}-4(\sigma_1^2+\hat{\xi}_1^2)\,\tilde{t}+\theta_{01}\,,$$

where d is a positive constant.

With (15) and (14) we get because of (7), e.g.

$$C_-^{(2)}(\xi,t)=q^{(2)}(\tilde{x},\tilde{t})\,\mathrm{e}^{-i\lambda_- t}\,. \tag{16}$$

Now, in principle we have found the two-soliton solution of (4) for $I=0$ and constant electron density n, but we have still to fix the constants A_j, a_j, b_j, ξ_j^0, and ζ_j^0 with $j=1,2$ in the function $\psi^{(2)}(\bar{x},\bar{t})$ in (12) and σ_j, $\hat{x}_{0j}$, $\hat{\xi}_j$ with $j=1,2$, and θ_{01} in $q^{(2)}(\tilde{x},\tilde{t})$ in (16) according to the parameters of our physical model U, T_0, n.

3.3 *Fixing of parameters*

In order to get a reasonable fixing of the constants A_j, a_j, b_j, ξ_j^0, ζ_j^0, σ_j, $\tilde{x}_{0j}$, $\hat{\xi}_j$, with $j=1,2$, and θ_{01} without contradictions and such that the initial positions $a\xi_{01}$ and $a\xi_{02}$ of two solitons at the chain, their velocities au_1 and au_2, and the ratio m of their amplitudes with $0<m\leqq 1$ are free parameters we proceed in two steps. Firstly, we compare the one-soliton solutions which are obtained from (12) and (16) (the second soliton does not exist) with the one-soliton solutions (5b) and (5c).

From (11), (12), and (5b) we obtain

$$\begin{aligned}\psi^{(1)}(\bar{x}, \bar{t}) &= A_1 \operatorname{sech} \xi_1 \, e^{-i\zeta_1} \\ &= C_+^{(1)}(\xi, t) \, e^{i\lambda_+ t} \\ &= \sqrt{n} \operatorname{sech} \left\{ \left(\frac{Un}{T_0} \right)^{1/2} (\xi - \xi_{01} - u_1 t) \right\} e^{-i \frac{u_1}{2T_0} (\xi - \xi_{01}) - i(v_+ - \lambda_+) t} . \end{aligned} \tag{17}$$

If we assume that amplitudes, arguments, and phases are equal, then it follows

$$\begin{aligned} A_1 &= \sqrt{n} , \\ \xi_1 &= a_1 \xi + 2 a_1 b_1 T_0 t + \xi_1^0 = \left(\frac{Un}{T_0} \right)^{1/2} (\xi - \xi_{01} - u_1 t) , \end{aligned}$$

and (18a)

$$\zeta_1 = b_1 \xi + (b_1^2 - a_1^2) \, T_0 t + \zeta_1^0 = - \frac{u_1}{2T_0} (\xi - \xi_{01}) - (v_+ - \lambda_+) \, t ,$$

thus we get

$$a_1 = \left(\frac{Un}{T_0} \right)^{1/2} , \qquad b_1 = - \frac{u_1}{2T_0} , \qquad \xi_1^0 = - \left(\frac{Un}{T_0} \right)^{1/2} \xi_{01} , \qquad \zeta_1^0 = \frac{u_1}{2T_0} \xi_{01} . \tag{18b}$$

Analogously, we obtain from (15), (16), and (5c)

$$\begin{aligned} q^{(1)}(\tilde{x}, \tilde{t}) &= 2\sigma_1 \tanh x_1 \, e^{i\theta(\tilde{x}, \tilde{t})} \\ &= C_-^{(1)}(\xi, t) \, e^{i\lambda_- t} \\ &= \sqrt{n} \tanh \left\{ \left(\frac{Un}{T_0} \right)^{1/2} (\xi - \xi_{01} - u_1 t) \right\} e^{i \frac{u_1}{2T_0} (\xi - \xi_{01}) - i(v_- - \lambda_-) t} \end{aligned} \tag{19}$$

and with

$$\begin{aligned} 2\sigma_1 &= \sqrt{n} , \\ x_1 &= \sqrt{2} \, \sigma_1 (\tilde{x} - \tilde{x}_{01} + 4 \hat{\xi}_1 \tilde{t}) = \left(\frac{Un}{T_0} \right)^{1/2} (\xi - \xi_{01} - u_1 t) , \end{aligned} \tag{20a}$$

and

$$\theta(\tilde{x}, \tilde{t}) = \frac{u_1}{2T_0} (\xi - \xi_{01}) + (\lambda_- - v_-) \, t ,$$

we find

$$\tilde{x}_{01} = \left(\frac{2U}{T_0} \right)^{1/2} \xi_{01} , \qquad \hat{\xi}_1 = - \frac{u_1}{4 \sqrt{2UT_0}} , \qquad \theta_{01} = - \frac{u_1}{2T_0} \xi_{01} . \tag{20b}$$

In order to evaluate the remaining constants, secondly, we take into account that for $|\xi_2 - \xi_1| \gg 2$ and $|x_2 - x_1| \gg 2$, respectively, the solutions $C_+^{(2)}(\xi, t)$ and $C_-^{(2)}(\xi, t)$ split up in two single solitons. This limiting case, where the distance between the solitons is large compared with the width of a soliton, corresponds to the limiting case $t \to \infty$ if we choose the free parameters of the two-soliton solution in such a way that $2T_0(a_2 b_2 - a_1 b_1)$ and $2U \sqrt{2} \, \sigma_1 (2m(\hat{\xi}_1 + \hat{\xi}_2) - 4\hat{\xi}_1)$ are positive.

Then we obtain from (12) and (11)

$$|C_+^{(2)}(\xi, t)|^2 = |\psi^{(2)}(\bar{x}, \bar{t})|^2 \underset{\substack{t\to\infty \\ (\bar{t}\to-\infty)}}{\to} \frac{A_1^2}{\cosh^2(\xi_1 + \bar{a})} + \frac{A_2^2}{\cosh^2(\xi_2 - \bar{a})} \tag{21}$$

with

$$\bar{a} = \frac{1}{2} \ln \left(\frac{(a_1 - a_2)^2 + (b_1 - b_2)^2}{(a_1 + a_2)^2 + (b_1 - b_2)^2} \right),$$

and from (16) and (15)

$$|C_-^{(2)}(\xi, t)|^2 = |q^{(2)}(\tilde{x}, \tilde{t})|^2 \underset{\substack{t\to\infty \\ (\tilde{t}\to\infty)}}{\to} n[\tanh^2(x_1 - \delta_1^+) + m^2 \tanh^2(x_2 - \delta_2^+) - m^2] \tag{22}$$

with

$$\delta_1^+ = \mathrm{Ar}\tanh m = \frac{1}{2} \ln \frac{1+m}{1-m}$$

and

$$\delta_2^+ = \frac{1}{2} \ln \frac{d}{1+m}.$$

Since we investigate solutions where the electron density n (cf. (3)) is not changed by one-soliton excitations ($n(\xi, t) = n = \text{const}$) in the "large distance" limit (cf. (21), (22)) the expressions $|C_+^{(2)}(\xi, t)|^2$ and $|C_-^{(2)}(\xi, t)|^2$ have to satisfy the condition

$$|C_+^{(2)}(\xi, t)|^2 + |C_-^{(2)}(\xi, t)|^2 = n \qquad (|x_2 - x_1|, |\xi_2 - \xi_1| \gg 2). \tag{23}$$

To fulfil (23) one possibility is to demand

$$A_2^2 = nm^2, \qquad \xi_1 + \bar{a} = x_1 - \delta_1^+, \qquad \xi_2 - \bar{a} = x_2 - \delta_2^+,$$

this means

$$A_2 = m\sqrt{n}, \tag{24a}$$

since A_2 has to be real (cf. (11)),

$$\bar{a} = -\delta_1^+ \tag{24b}$$

because of (18a) and (20a), and owing to (11) and (15)

$$a_2 = m \left(\frac{Un}{T_0}\right)^{1/2} = ma_1, \quad b_2 = \left(\frac{2U}{T_0}\right)^{1/2} (\hat{\xi}_1 + \hat{\xi}_2), \quad \xi_2^0 = -m\sqrt{\frac{n}{2}}\, \tilde{x}_{02}, \tag{24c}$$

and

$$\bar{a} = \delta_2^+. \tag{24d}$$

The conditions (24b) and (24d) are reasonable, and justify our way of fixing of parameters of the solutions (11) and (16), then using (18b), (20), (24c), and (15) one gets $(b_1 - b_2)^2 = a_1^2(1 - m^2)$. With this one sees that $\bar{a}$ and δ_1^+ according to (21) and (22) fulfil the condition (24b),

$$\bar{a} = \frac{1}{2} \ln \left(\frac{a_1^2(1-m)^2 + a_1^2(1-m^2)}{a_1^2(1+m)^2 + a_1^2(1-m^2)} \right) = \frac{1}{2} \ln \frac{1-m}{1+m} = -\delta_1^+.$$

If we put for the positive constant d in (15)

$$d = 1 - m, \tag{24e}$$

then δ_2^+ according to (22) fulfils (24d).

For convenience we write the argument ξ_2 for which we have now $\xi_2 \equiv x_2$ and the phase ζ_2 of the second soliton in the same form which we have used for the one-soliton solution (5):

$$\xi_2 = a_2\xi + 2a_2b_2T_0t + \xi_2^0 = m\left(\frac{Un}{T_0}\right)^{1/2} (\xi - \xi_{02} - u_2t)\,, \tag{25a}$$

$$\zeta_2 = b_2\xi + (b_2^2 - a_2^2)\,T_0t + \zeta_2^0 = b_2(\xi - \xi_{02}) + (b_2^2 - a_2^2)\,T_0t\,,$$

in other words, we put

$$b_2 = \left(\frac{2U}{T_0}\right)^{1/2} (\hat{\xi}_1 + \hat{\xi}_2) = -\frac{u_2}{2T_0}\,, \qquad \zeta_2^0 = \frac{u_2}{2T_0}\,\xi_{02}\,,$$

$$\xi_2^0 = -m\sqrt{\frac{n}{2}}\,\tilde{x}_{02} = -m\left(\frac{Un}{T_0}\right)^{1/2} \xi_{02}\,. \tag{25b}$$

Now we are able to write down the two-soliton solution of (4) for $n = n(\xi, t) = \text{const}$ and $I = 0$ in dependence on the parameters of our model U, T_0, and n and on five free parameters, namely, the initial positions ξ_{01}, ξ_{02}, the signal frequencies u_1, u_2, and the ratio of amplitudes $m = \sigma_2/\sigma_1 = A_2/A_1$ of solitons whereby we have taken $m^2 \leqq 1$ (cf. 15)) and

$$2T_0(a_2b_2 - a_1b_1) = 2U\sqrt{2}\,\sigma_1(2m(\hat{\xi}_1 + \hat{\xi}_2) - 4\hat{\xi}_1) = \left(\frac{Un}{T_0}\right)^{1/2} (u_1 - mu_2) > 0$$

(cf. (21), (22)). It is given by the equations (11), (12), (9a), (15), (16), (13a), (18), (20), (24), and (25).

Because the calculation of $\cosh \bar{a}$, $\sinh \bar{a}$, $\cos \Phi_1$, $\sin \Phi_1$, $\cos \Phi_2$, and $\sin \Phi_2$ in (11) for $C_+^{(2)}(\xi, t)$ according to (12) is rather tedious we give these expressions in a simple form here: From (24b) and (22) we get

$$m = -\tanh \bar{a} \tag{26a}$$

and therefore

$$\cosh \bar{a} = \frac{1}{\sqrt{1 - m^2}}\,, \qquad \sinh \bar{a} = \frac{-m}{\sqrt{1 - m^2}}\,. \tag{26b}$$

Furthermore, from (11), (18b), (20), (24c), and (15) follows

$$\cos \Phi_1 = \sqrt{1 - m^2}\,, \qquad \sin \Phi_1 = -m\,, \qquad \Phi_2 = \frac{\pi}{2}\,. \tag{26c}$$

4. Discussion

For the probability density $p_{\uparrow,\downarrow}^{(2)}(\xi, t) = |\Phi_{\uparrow,\downarrow}^{(2)}(\xi, t)|^2$ to find an electron with spin $\uparrow$ or $\downarrow$ at the time t at the place $a\xi$ in the chain (a is the lattice parameter) involved in a two-soliton excitation, i.e. owing to the existence of two domain walls and its interaction, we obtain with (2),

$$\begin{aligned} p_{\uparrow,\downarrow}^{(2)}(\xi, t) &= \tfrac{1}{2}\,|C_+^{(2)}(\xi, t) \pm C_-^{(2)}(\xi, t)|^2 \\ &= \tfrac{1}{2}\,|C_+^{(2)}|^2 + \tfrac{1}{2}\,|C_-^{(2)}|^2 \pm \tfrac{1}{2}\,(C_+^{(2)}C_-^{(2)*} + C_+^{(2)*}C_-^{(2)})\,, \end{aligned} \tag{27}$$

whereby $C_+^{(2)}(\xi, t)$ is given by (12), (11), and (9a) and $C_-^{(2)}(\xi, t)$ by (16), (15), and (13a) with (18), (20), (24), (25), and (26).

In the "large distance" limit (cf. (21), (22)) it yields $p_\uparrow^{(2)} = p_\downarrow^{(2)} = n/2$. If the solitons are not well separated we get from (27) a rather complicated expression, which has to be discussed numerically. The difference $p_{\uparrow,\downarrow}^{(2)}(\xi, t) - p_{\uparrow,\downarrow}^{(1)}(\xi, t; \sqrt{n}, \xi_{01}, u_1) - p_{\uparrow,\downarrow}^{(1)}(\xi, t; \sqrt{n}\, m, \xi_{02}, u_2)$ describes the influence of the interaction of walls.

References

[1] H. J. Mikeska, J. Magnetism magnetic Mater. **13**, 35 (1979).
[2] U. Lindner and V. K. Fedyanin, phys. stat. sol. (b) **89**, 123 (1978).
[3] U. Lindner and V. K. Fedyanin, phys. stat. sol. (b) **95** K83 (1979).
[4] U. Lindner, phys. stat. sol. (b) **106**, K77 (1981).
[5] J. A. Krumhansl and J. K. Schrieffer, Phys. Rev. B **11**, 3535 (1975).
[6] U. Lindner, phys. stat. sol. (b) **109**, K41 (1982).
[7] R. Hirota, J. Math. Phys. (New York) **14**, 805 (1973).
[8] M. Boiti, C. Laddomada, and F. Pempinelli, Nuovo Cimento **62B**, 315 (1981).
[9] M. Boiti, C. Laddomada, and F. Pempinelli, Nuovo Cimento, **65B**, 248 (1981).
[10] U. Lindner, Nuovo Cimento Letters **38**, 381 (1983).

(Received January 10, 1984)

phys. stat. sol. (b) **125**, 187 (1984)

Subject classification: 13.3 and 19; 21

Institute of Physics, Czechoslovakian Academy of Sciences, Prague[1])

Angular Dependence of SSWR in Thin Metallic Films

Approximate Analytical Solution

By

M. JIRSA

The spin-wave resonance spectrum characteristics, namely the mode position, linewidth, and intensity, for a thin homogeneous conducting film with an uniaxial surface anisotropy and an arbitrarily oriented external magnetic field are determined analytically using the spin-wave approximation. Comparison with the numerical data computed in the "full" model shows good agreement for all spin-wave modes including the lowest-order ones.

Die Charakteristiken eines Spinwellenresonanzspektrums — Position, Linienbreite und Intensität — in einer dünnen homogenen leitenden Schicht mit uniaxialer Oberflächenanisotropie und beliebig orientiertem äußeren Magnetfeld werden analytisch mit Hilfe der Spinwellenapproximation abgeleitet. Der Vergleich mit den numerischen Daten, die für das ,,Gesamtmodell" berechnet werden, zeigt eine gute Übereinstimmung für alle Spinwellenmoden einschließlich die der niedrigsten Ordnungszahlen.

1. Introduction

The interpretation of the standing spin-wave resonance (SSWR) experiments on thin metallic films has been hitherto made in two ways: by a numerical calculation of the spectrum point by point (based on the "full" model [1, 2]) or with the help of the simple analytical formulae for position [3], linewidth [4, 5], and intensity [6] of the SSWR modes derived for an insulator film.

It is well known that for a metallic film much thinner than the skin-depth δ the shift of the resonance mode position due to conductivity is significant only for the lowest-order modes [7]. The exchange-conductivity effect broadens the resonance line for all modes [8 to 10], by which the mode intensity (peak height) is also influenced. The analytical expressions for the latter two quantities incorporating the effect of the film conductivity were derived only for the parallel and perpendicular configurations of the experiment [7 to 10].

The aim of this paper is to determine analytically the SSWR mode position, the linewidth, and the intensity for an arbitrarily oriented external magnetic field. The influence of the conductivity of the film is taken into account.

Comparison with the "full" theory illustrates the relative error caused by the approximation.

2. Theory

A homogeneously magnetized, infinite, conductive film of thickness $2d \ll \delta$ is assumed to possess uniaxial surface anisotropy with an easy or hard axis perpendicular to the film plane. The driving field is assumed to be symmetric or antisymmetric, respec-

[1]) Na Slovance 2, 18040 Prague 8, Czechoslovakia.

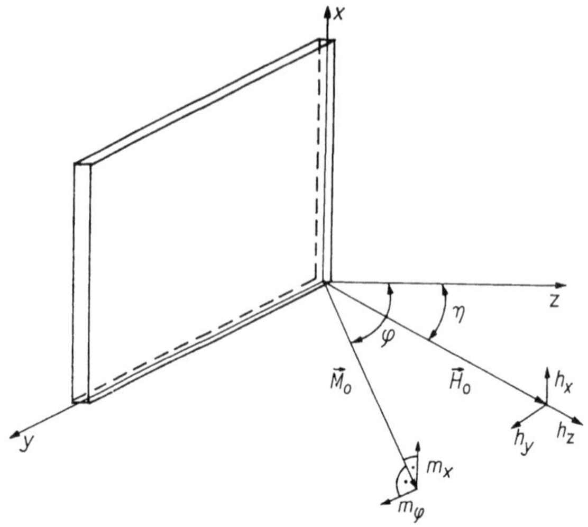

Fig. 1. Orientation of the film, the external magnetic field, and the magnetic polarization with respect to the coordinate system

tively, for modes symmetrical (S) and antisymmetrical (A) with respect to the central plane of the film. The orientation of the film with respect to the coordinate system, the external magnetic field, and the magnetization is shown in Fig. 1.

The Gilbert equation is used to describe the motion of the magnetic moment,

$$\frac{\mathrm{d}\boldsymbol{M}}{\mathrm{d}t} = -\gamma \boldsymbol{M} \times \left(\boldsymbol{H} + \frac{2A}{M_0^2} \nabla^2 \boldsymbol{M} - \frac{\alpha}{\gamma M_0} \frac{\partial \boldsymbol{M}}{\mathrm{d}t}\right), \tag{1}$$

where $\boldsymbol{H}$ is the internal magnetic field, $\boldsymbol{M}$ and $\boldsymbol{M}_0$ are the magnetic polarization vectors in the instantaneous and equilibrium positions, respectively, α is the Gilbert damping constant, γ the gyromagnetic ratio, and A the exchange stiffness constant.

Eliminating components of the electric field from the Maxwell equations we obtain

$$\nabla(\nabla \boldsymbol{H}) - \Delta \boldsymbol{H} = -\sigma \frac{\partial(\mu_0 \boldsymbol{H} + \boldsymbol{M})}{\partial t}, \tag{2}$$

where σ is the electric conductivity of the film and μ_0 the permeability of vacuum.

The microwave magnetic field and polarization are assumed to depend on time as $\boldsymbol{h}, \boldsymbol{m} \sim \mathrm{e}^{i\omega t}$.

Decomposition of (1) and (2) into components and linearization gives

$$-\frac{i\omega}{\gamma} m_\varphi = -\hat{P} m_x + M_0 h_x, \tag{3}$$

$$\frac{i\omega}{\gamma} m_x = -\hat{Q} m_\varphi + M_0 \cos\varphi h_y, \tag{4}$$

$$m_\varphi \cos\varphi = -\mu_0 \left(1 + \frac{i\delta^2}{2} \frac{\partial^2}{\partial z^2}\right) h_y, \tag{5}$$

$$m_x = -\mu_0 \left(1 + \frac{i\delta^2}{2} \frac{\partial^2}{\partial z^2}\right) h_x, \tag{6}$$

where $m_\varphi = \sqrt{m_x^2 + m_y^2}$. The operators $\hat{P}$ and $\hat{Q}$ are defined in accordance with Okochi and Nose [5] and others as

$$\hat{P} = H_0 \cos(\varphi - \eta) - \frac{M_0}{\mu_0} \cos^2 \varphi - \frac{2A}{M_0} \frac{\partial^2}{\partial z^2} + \frac{i\alpha\omega}{\gamma}, \tag{7}$$

$$\hat{Q} = \hat{P} + \frac{M_0}{\mu_0} \sin^2 \varphi, \tag{8}$$

where δ is the classical skin-depth, $\delta = \sqrt{2/\mu_0 \omega \delta}$.

The components of $\boldsymbol{h}$ and $\boldsymbol{m}$ are assumed to be periodic functions of z: h_j, $m_j \sim e^{ikz}$, $j = x, y, \varphi$; k is the complex wave vector. The condition for a nontrivial solution of (3) to (6) (secular equation) has the form (see [11])

$$\left[\left(\frac{\omega}{\gamma}\right)^2 - PQ\right]\left(1 - \frac{i\delta^2 k^2}{2}\right)^2 - \frac{M_0}{\mu_0}(P \cos^2 \varphi + Q)\left(1 - \frac{i\delta^2 k^2}{2}\right) - \left(\frac{M_0}{\mu_0} \cos \varphi\right)^2 = 0, \tag{9}$$

where P and Q are the characteristic values of the operators $\hat{P}$ and $\hat{Q}$.

In conducting films the solutions of (9) have a mixed spin-wave and electromagnetic character. In general two spin-wave-like roots $k_{1,2}$ exist, k_2 being always nearly purely imaginary. For the films with $2d \ll \delta$ the electro-magnetic-like roots k_3, k_4 are of the same order as δ^{-1} and, consequently, are much less in absolute value than $|k_1|$ and $|k_2|$ for all SSWR modes, except that for which $k_1 \approx \delta^{-1}$.

In the case of an insulator film ($\delta \to \infty$) only the term in the square brackets remains on the left of (9). Such equation has two spin-wave roots $k_{1,2}$, where k_2 is always purely imaginary.

The resonance condition (given by the real part of (9)) is then

$$\operatorname{Re}\left[\left(\frac{\omega}{\gamma}\right)^2 - PQ\right] = 0. \tag{10}$$

The position of the resonance mode is given by the shift $(2A/M_0)\, k_1^2$ of the resonance field from the position of the mode of homogeneous precession.

In metallic films the real part of the whole equation (9) must be solved. Nevertheless, if $|k_1| \gg \delta^{-1}$, the resonance condition is practically identical with (10) (the exchange-conductivity shift is negligible). The position of the SSWR modes in metallic films is then controlled practically only by the real part of the resonant spin-wave-like root k_1 as in the insulator film. This fact will be investigated in more detail in the following section.

3. Effective Surface Anisotropy

3.1 Insulator film

The value of k_1 depends on the effective spin-pinning on the film surfaces, which is given by the boundary conditions. In a homogeneous insulator film with uniaxial surface anisotropy the boundary conditions for the components of hf magnetic polarization have the form [3]

$$\begin{aligned} \frac{\partial m_\varphi}{\partial z} + \frac{K_s}{A} \cos 2\varphi \; m_\varphi &= 0, \\ \frac{\partial m_x}{\partial z} + \frac{K_s}{A} \cos^2 \varphi \; m_x &= 0, \end{aligned} \tag{11}$$

where K_s is the surface anisotropy constant.

These conditions can be rearranged using the effective surface anisotropy constant K_{eff} [11 to 13],

$$\begin{aligned} \operatorname{ctg} k_1 d &= \frac{k_1 A}{K_{eff}} \quad \text{for the S modes}, \\ -\operatorname{tg} k_1 d &= \frac{k_1 A}{K_{eff}} \quad \text{for the A modes}, \end{aligned} \tag{12}$$

for k_1 real (the volume modes) and

$$\begin{aligned} \operatorname{ctgh} |k_1 d| &= \frac{i k_1 A}{K_{eff}} \quad \text{for the S modes}, \\ \operatorname{th} |k_1 d| &= \frac{i k_1 A}{K_{eff}} \quad \text{for the A modes}, \end{aligned} \tag{12'}$$

for k_1 imaginary (the surface modes), where K_{eff} for a perfectly homogeneous insulator film [12] is

$$K_{eff} = K_s \frac{\Omega_1 \cos 2\varphi - \Omega_2 \cos^2 \varphi + \xi \cos 2\varphi \cos^2 \varphi \, (\Omega_1 - \Omega_2)}{\Omega_1 - \Omega_2 + \xi(\Omega_1 \cos^2 \varphi - \Omega_2 \cos 2\varphi)} \tag{13}$$

and the ellipticity factors $\Omega_{1,2}$ are

$$\Omega_{1,2} = \frac{\Omega}{\sin^2 \varphi \pm \sqrt{\sin^4 \varphi + \Omega^2}}. \tag{14}$$

Here Ω is the reduced frequency, $\Omega = 2\omega\mu_0/\gamma M_0$, and ξ is the pinning parameter, $\xi = K_s \operatorname{ctgh} |k_2 d|/A\,|k_2|$ for the S modes and $\xi = K_s \operatorname{th} |k_2 d|/A\,|k_2|$ for the A modes. K_{eff} is a function of the angle φ and of the ratio ω/M_0; it is mode-dependent through the value of k_2 as k_2 is related to k_1,

$$k_2^2 = k_1^2 - \frac{M_0^2}{2\mu_0 A} \sqrt{\sin^4 \varphi + \Omega^2}. \tag{15}$$

Because of the dependence of K_{eff} on the mode number, (12) to (15) must be solved simultaneously to obtain correct values of K_{eff} and k_1 for the given mode and given parameters.

The discussion of (12) and (12′) with respect to the value, angular and mode dependences of K_{eff} is reported in [12].

3.2 Metallic films

The constant of the effective surface anisotropy for metallic films was determined as follows.

The absorbed power P was calculated, using all four waves, separately for the S and A modes. From the dependence of dP/dH_0 on H_0 the mode positions H_{0n} were obtained. Then, K_{eff} was determined in two alternative ways.

First, the secular equation (9) was solved for the values of H_{0n} and the corresponding (complex) k_1 were evaluated. K_{eff} was then determined from (12) using the real part of k_1.

Secondly, the effective (real) wave vector k_{eff} was evaluated from the equation $2Ak_{eff}^2/M_0 = H_{0h} - H_{0n}$, where H_{0h} is the resonance field of the homogeneous mode (that with $k_1 = 0$) in an insulator film. The value of K_{eff} corresponding to k_{eff} was then determined from (12). It is compared with the other two values (the "insulator" one and that calculated from Re (k_1)) in Table 1.

Table 1

Comparison of K_{eff} calculated for an insulator film from (13) (I) and numerically for a metallic film from Re (k_1) (III) and k_{eff} (II)

	$\eta = 0°$			$\eta = 30°$			$\eta = 60°$			$\eta = 90°$		
n	I	II	III	I	II	III	I	II	III	I	II	III
0	3.750	3.778	0.128	0.695	0.705	0.065	−1.002	−1.005	+0.004	−1.306	−1.310	+0.066
1	3.750	3.752	4.307	0.675	0.675	1.023	−1.005	−1.005	−0.989	−1.306	−1.306	−1.000
2	3.750	3.676	3.758	0.626	0.486	0.545	−1.019	−0.859	−0.781	−1.307	−1.180	−1.090
3	3.750	3.741	3.768	0.538	0.518	0.538	−1.041	xx	xx	−1.309	−1.348	−1.224
4	3.750	3.736	3.748	0.402	0.289	0.309	−1.075	−1.049	−1.038	−1.311	−1.291	−1.280
5	3.750	3.748	3.755	0.197	0.181	0.187	−1.119	−1.092	−1.085	−1.314	−1.305	−1.298
6	3.750	3.746	3.751	−0.108	−0.035	−0.031	−1.176	−1.170	−1.166	−1.319	−1.315	−1.311
7	3.750	3.749	3.753	−0.559	−0.682	−0.679	−1.243	−1.239	−1.236	−1.324	−1.324	−1.321
8	3.750	3.749	3.751	−1.121	−1.112	−1.110	−1.311	−1.311	−1.309	−1.332	−1.333	−1.331

Frequency $\nu = 36.92$ GHz, film parameters (corresponding to the permalloy film) were $2d = 150$ nm, $M_0 = 1$ T, $g = 2.15$, $A = 1 \times 10^{-11}$ J/m, $K_s = 0.5$ mJ/m^2, $\alpha = 6 \times 10^{-3}$, and conductivity $\sigma = 7.69 \times 10^6$ S/m. The sign "xx" used in the table means that the corresponding mode was not detected. The number n denotes the sequence of SSWR modes starting from high resonance fields.

Whereas the real part of k_1 for a metallic film and k_1 for an insulator film (and corresponding values of K_{eff}) are in agreement only for the higher-order modes, k_{eff} negligibly differs from the "insulator" value even for the first two modes. The same is true for the corresponding values of K_{eff}.

The departure of the real part of k_1 from the value for an insulator film is evidently compensated by the electro-magnetic-like waves in such a way that the exchange-conductivity shift becomes small even for the mode $n = 0$.

A marked difference between the "insulator" values of K_{eff} and those obtained for k_{eff} exists only for modes strongly overlapped by the neighbouring ones; this shifts the mode position in the resonance spectrum of a metallic film.

Irrespective of this obstacle, the boundary conditions (12) or (12′) with K_{eff} given by (13) serve evidently as a very good approximation to the exact boundary conditions in metallic films.

4. Mode Linewidth and Intensity

The fact that the position of the SSWR modes in a metallic film is given by k_{eff} (which is practically equal to k_1 for an insulator film) suggests that k_{eff} will play a significant role also in the determination of the mode linewidth and the intensity.

The detailed analysis of the numerical calculation in the "full" model of SSWR in a metallic film shows that neither of the anti-Larmor waves contributes to the mode intensity and linewidth. The contribution of the Larmor electromagnetic-like wave is significant only for the lowest-order modes. The situation resembles the case of the mode positions. We express therefore the components of $\boldsymbol{h}$ and $\boldsymbol{m}$ in terms of trigonometric (or hyperbolic) functions of k_{eff}. As k_{eff} is practically equal to k_1 for an insulator film, k_{eff} will be in the following replaced by the "insulator" k_1. The mode dependence will be emphasized by the index $n: k_{1n}$,

$$m_{x,\varphi} = m_{nx,\varphi} \cos k_{1n}z \qquad \text{for the S modes}\,, \tag{16a}$$

$$m_{x,\varphi} = m_{nx,\varphi} \sin k_{1n}z \qquad \text{for the A modes}\,. \tag{16b}$$

The boundary conditions for the hf magnetic field, $h_x(d) = h_d$ and $h_y(d) = 0$, can be satisfied only by taking into account the contributions of all four waves. The contribution of the three nonresonant waves will be added in the form of a constant term (for the S modes) or a term linearly dependent on z (for the A modes),

$$\left.\begin{aligned} h_x &= h_d + h_{nx} (\cos k_{1n}z - \cos k_{1n}d) \\ h_y &= h_{ny}(\cos k_{1n}z - \cos k_{1n}d) \end{aligned}\right\} \qquad \text{for the S modes}\,, \tag{17a}$$

$$\left.\begin{aligned} h_x &= \frac{h_d z}{d} + h_{nx}(\sin k_{1n}z - \sin k_{1n}d) \\ h_y &= h_{ny}(\sin k_{1n}z - \sin k_{1n}d) \end{aligned}\right\} \qquad \text{for the A modes}\,, \tag{17b}$$

where h_d is the magnitude of the driving magnetic field on the film surface.

The above-mentioned approximation will be referred to as the spin-wave approximation. Its correctness will be verified by comparison with the results of the full theory (in which all four waves are correctly included).

Equating the absorption to the dissipation (both integrated over the film thickness) gives the values of the amplitudes m_{nx}. It is equivalent to the projection of (3) and (4) (with m_x, m_φ, h_x, and h_y substituted by (16) and (17), respectively) onto the characteristic functions $\cos k_{1n}z$ (for the S modes) or $\sin k_{1n}z$ (for the A modes). Taking into account (5) and (6) we obtain

$$m_{nx} = \frac{-M_0 h_{\rm d} \frac{J_n}{I_n} \left(Q + \frac{M_0 \cos^2 \varphi}{\mu_0 \left(1 - \frac{i\delta^2 k_{1n}^2}{2}\right)} \left(1 - \frac{J_n}{I_n} \cos y_{1n}\right)\right)}{\frac{\omega^2}{\gamma^2} - \left(P + \frac{M_0 \left(1 - \frac{J_n}{I_n} \cos y_{1n}\right)}{\mu_0 \left(1 - \frac{i\delta^2 k_{1n}^2}{2}\right)}\right)\left(Q + \frac{M_0 \cos^2 \varphi \left(1 - \frac{J_n}{I_n} \cos y_{1n}\right)}{\mu_0 \left(1 - \frac{i\delta^2 k_{1n}^2}{2}\right)}\right)}, \quad (18)$$

where

$$J_n = \frac{1}{2d} \int\limits_{-d}^{+d} \cos k_{1n} z \, dz = \frac{\sin y_{1n}}{y_{1n}}, \quad (19a)$$

$$I_n = \frac{1}{2d} \int\limits_{-d}^{+d} \cos^2 k_{1n} z \, dz = \frac{1}{2} \left(1 + \frac{\sin 2y_{1n}}{2y_{1n}}\right), \quad (19b)$$

and $y_{1n} = k_{1n} d$.

The linewidth of the resonance mode can be determined by equating the real and imaginary parts of the denominator in (18) for $H_0 = H_{0n} \pm \Delta H/2$. The inflection linewidth $\Delta H_{\rm inf}$ is then the $1/\sqrt{3}$-multiple of ΔH (the resonance line in thin films is Lorentzian),

$$\Delta H_{\rm inf} = \frac{2}{\sqrt{3}} \left[\frac{\alpha\omega}{\gamma} + \frac{2M_0}{\mu_0} \left(\frac{d}{\delta}\right)^2 \frac{1}{y_n^2} \left(1 - \frac{2F(\beta_n)}{y_n^2 + \beta_n^2 + \beta_n} \frac{Q_0 + P_0 \cos^2 \varphi}{Q_0 + P_0}\right)\right] K_0, \quad (20)$$

where P_0 and Q_0 are the real parts of P and Q, respectively, $\beta_n = K_{\rm eff} d/A$, $K_{\rm eff}$ corresponding to the n-th mode and $F(\beta_n) = \beta_n$ for the S modes and $F(\beta_n) = 1 + \beta_n$

Table 2

Dependence of ΔH_n (in 10^{-4} T) on the angle η of the external magnetic field with respect to the film normal

n	ΔH_n $\eta = 0°$		$\eta = 30°$		$\eta = 60°$		$\eta = 90°$	
	approx.	exact	approx.	exact	approx.	exact	approx.	exact
0	142.2	142.2	166.6	165.9	124.3	123.8	109.5	109.5
1	98.7	98.7	125.9	124.7	101.4	100.9	91.3	91.3
2	92.4	90.2	127.6	66.5	112.4	98.2	103.6	96.0
3	89.1	89.0	124.3	121.5	100.3	—	90.3	82.8
4	87.8	87.7	127.7	113.4	98.5	97.3	87.9	87.4
5	86.9	86.9	135.2	132.3	98.2	95.9	86.7	86.2
6	86.4	86.4	149.4	143.0	98.8	98.3	86.2	86.1
7	86.1	86.1	172.7	167.5	99.7	99.2	85.9	85.9
8	85.9	85.9	193.1	188.1	100.2	99.9	85.7	85.6

Table 3

The surface impedance P_n (in 10^{-3} W/m²) given for different angles η of the external magnetic field with respect to the film normal

	P_n							
n	$\eta = 0°$		$\eta = 30°$		$\eta = 60°$		$\eta = 90°$	
	approx.	exact	approx.	exact	approx.	exact	approx.	exact
0	217.7	217.6	276.3	276.8	443.0	442.4	508.3	505.3
1	139.3	139.3	157.6	158.5	229.9	228.2	260.0	256.6
2	30.90	30.05	2.83	0.83	19.17	15.75	37.57	33.97
3	20.55	20.51	4.42	4.28	0.01	—	0.39	0.44
4	5.87	5.86	0.10	0.06	1.22	1.42	2.11	2.38
5	4.78	4.78	0.37	0.32	0.01	0.03	0.05	0.10
6	1.64	1.64	0.00	0.01	0.28	0.40	0.41	0.56
7	1.54	1.54	0.01	0.00	0.01	0.03	0.01	0.05
8	0.60	0.60	0.04	0.08	0.11	0.20	0.13	0.24

for the A modes. K_0 is the geometrical factor entering through the non-collinearity of the vectors $\boldsymbol{H}$ and $\boldsymbol{M}$ at oblique angles η (see also [5]),

$$K_0 = \left[\cos(\varphi - \eta) - \frac{\sin(\varphi - \eta)\sin 2\varphi}{\sin^2\varphi + \sqrt{\sin^4\varphi + \Omega^2} - \dfrac{4A\mu_0 k_{1n}^2}{M_0^2}}\left(2 - \frac{\sin^2\varphi}{\sqrt{\sin^4\varphi + \Omega^2}}\right)\right]^{-1}. \tag{21}$$

The first term in (20) corresponds to the Gilbert damping, the additional term is the exchange-conductivity broadening.

Substituting (20) into (18) the amplitude m_{xn} at resonance can be written as

$$m_{nx} = -\frac{4}{\sqrt{3}}\frac{ih_\mathrm{d}}{\Delta H_\mathrm{inf}} M_0 K_0 \frac{Q_0}{Q_0 + P_0}\frac{J_n}{I_n}. \tag{19'}$$

The power absorbed per unit surface area of the film at resonance is

$$P_n = \frac{4h_\mathrm{d}^2}{\sqrt{3}}\frac{M_0 d}{\Delta H_\mathrm{inf}}\frac{1}{y_n^2}\frac{F^2(\beta_n)}{y_n^2 + \beta_n^2 + \beta_n}\frac{Q_0}{Q_0 + P_0} K_0 . \tag{22}$$

It is seen that the conductivity influences the mode intensities only through the line broadening.

Comparison of the approximate ΔH_inf and P_n with the results of the "full" theory is performed in Tables 2 and 3. A permalloy film 150 nm thick and the frequency 36.92 GHz are considered. The film parameters are $M_0 = 1$ T, $A = 1 \times 10^{-11}$ J/m, $2d = 150$ nm, $\alpha = 6 \times 10^{-3}$, $\varrho = 1.3 \times 10^{-7}$ Ωm, $K_s = 0.5$ mJ/m².

From Tables 2 and 3 it follows that the significant disagreement of both theories is connected only with the not well resolved modes (in the "full" theory). The lower the value of K_eff (the closer φ to the critical angle) is, the worse is the detectibility of the modes adjacent to the $n = 0$ and $n = 1$ modes. The disagreement is caused by overlapping of modes and can be removed by spectral analysis of the numerical or experimental spectra.

5. Discussion and Final Remarks

The inspection of (22) shows that at the angle φ_c^S (where $K_{eff} = 0$) the higher-order S modes vanish so that φ_c^S is the critical angle for the S modes. The critical angle for the A modes is evidently that, at which $K_{eff} = -A/d$. Consequently, the critical angles for the S and A modes are not necessarily identical. This serves as one possible explanation for the observation of the mode-dependent or non-existent critical angle [5, 13, 14]:

Any source of the antisymmetrical hf field in the film, e.g. asymmetrical spin-pinning, inhomogeneous magnetization, or effective internal field across the film thickness leads to the excitation of the antisymmetrical modes even in the symmetrical external driving field. Such a situation is met very frequently in experiment. Due to the different dependence of the intensity on K_{eff}, the A modes are then well excited at φ_c^S, where the S modes disappear, and vice versa. Moreover, the critical angle is in principle mode-dependent as K_{eff} is mode-dependent (see [12]).

The assumptions of the idealized model used in this paper can be hardly satisfied in experiment. Even the "full" theory does not take into account the noncollinearity of the spins near the film surfaces, which always exists for $K_s \neq 0$ at the oblique angles η. The noncollinearity of the surface spins contributes to the value of K_{eff} (for $0 < \eta < \pi/2$) and changes the value of the critical angle [15].

The above-mentioned effect can be incorporated into the spin-wave approximation by using the proper (more general) expression for K_{eff} (see [15, 16]).

The case of inhomogeneous saturation magnetization in the film interior (frequently met in evaporated metallic films) is more complicated and cannot be solved within the framework of the surface inhomogeneity model.

In summary, the spin-wave approximation gives a qualitative picture how the mode linewidth and the intensity depend on different film parameters and frequency and is in very good quantitative agreement with the "full" theory for all modes including the lowest-order ones. In application to experimental data, attention should be paid to the use of the proper expression for K_{eff} according to the real film properties.

Acknowledgement

The author would like to thank to Dr. V. Kambersky for helpful discussion.

References

[1] P. E. Wigen, C. F. Kooi, and M. R. Shanabarger, J. appl. Phys. **35**, 3302 (1964).
[2] C. Vittoria, R. C. Barker, and A. Yelon, J. appl. Phys. **42**, 1734 (1971).
G. C. Bailey and C. Vittoria, Phys. Rev. Letters **28**, 100 (1972).
[3] C. Kittel, Phys. Rev. **110**, 1295 (1958).
G. T. Rado and J. R. Weertmann, J. Phys. Chem. Solids **11**, 315 (1959).
R. F. Soohoo, Phys. Rev. **131**, 594 (1963).
[4] A. D. Berk, J. appl. Phys. **28**, 290 (1957).
[5] M. Okochi and H. Nose, J. Phys. Soc. Japan **27**, 312 (1969).
[6] C. Kittel, Phys. Rev. **110**, 1295 (1958).
M. Ondris and H. Gaertner, Z. angew. Phys. **24**, 200 (1968).
C. H. Wilts and O. G. Ramer, J. appl. Phys. **47**, 1154 (1976).
M. Jirsa and V. Kamberský, Czech. J. Phys. **B32**, 810 (1982).
[7] M. H. Seavey and P. E. Tannenwald, J. appl. Phys. **30**, 227S (1959).
[8] R. E. DeWames and T. Wolfram, Japan. J. appl. Phys. **13**, 68 (1974).
[9] M. Kaganov, Zh. eksper. teor. Fiz. **39**, 158 (1960).
T. Wolfram and R. E. DeWames, Phys. Rev. B **4**, 3125 (1971).

[10] M. Jirsa, phys. stat. sol. (b) **113**, 679 (1982).
M. Jirsa, Czech. J. Phys. **B34**, 236 (1984).
[11] A. Yelon, G. Sproken, T. Bui-Thieu, R. C. Barker, Y. J. Liu, and T. Kobayashi, Phys. Rev. B **10**, 1070 (1974).
[12] M. Jirsa, phys. stat. sol. (b) **124**, No. 2 (1984).
[13] F. Hoffmann, Phys. Rev. B **4**, 1604 (1971).
[14] L. J. Maksymowicz and D. Sendorek, J. Magnetism magnetic Mater. **37**, 177 (1983).
[15] T. Stakelon, J. appl. Phys. **49**, 1592 (1978).
[16] J. Spalek and A. Z. Maksymowicz, Solid State Commun. **15**, 559 (1974).

(Received May 2, 1984)

phys. stat. sol. (b) **125**, 197 (1984)

Subject classification: 14.1; 18.2; 21

L. V. Kirenskii Institute of Physics,
Academy of Sciences of the USSR, Siberian Branch, Krasnoyarsk[1])

Magnetoresistance of Ferromagnets with Domain Structure

By

Yu. V. Zakharov and Yu. I. Mankov

The dependence of the resistivity of a compensated ferromagnet with domain structure on a weak magnetic field is studied. It is shown that due to the change in periodicity of motion of the electrons crossing a domain wall, the conductivity near the wall has a non-monotonic dependence on the coordinate normal to the wall plane and is larger compared to the conductivity in a domain. The process of magnetization by domain wall motion is considered. The dependence of $\Delta\varrho/\varrho$ on relative magnetization is shown to have a negative minimum whose value is determined by the ratio of the cyclotron diameter to the domain width.

Исследована зависимость сопротивления компенсированного ферромагнетика с доменной структурой от слабого магнитного поля. Показано, что вследствие изменения периодичности движения электронов, пересекающих доменную границу, проводимость вблизи границы имеет немонотонную зависимость от координаты, нормальной к плоскости границы, и увеличивается по сравнению со значением проводимости в домене. Рассмотрен процесс намагничивания за счет смещения доменных границ и показано, что зависимость $\Delta\varrho/\varrho$ от относительной намагниченности имеет отрицательный минимум, величина которого определяется отношением циклотронного диаметра к ширине доменов.

1. Introduction

As is generally known [1 to 7] one can see the negative magnetoresistance in sufficiently pure metallic ferromagnets in weak magnetic fields at low temperatures. The effect is to decrease the electric resistivity of a sample under magnetization. Now we can consider that this phenomenon is mainly due to two reasons: (i) change in trajectories of electron motion caused by the magnetic induction near the domain wall [8 to 10] and (ii) the existence in neighbouring domains of the opposite Hall currents due to the possibility to compensate the Hall charges on the surface of the multidomain sample [11, 12]. Note that the latter will not occur in compensated ferromagnets when compensation becomes complete ($\tau\omega \gg 1$, where τ is the relaxation time, ω the cyclotron frequency due to the magnetic induction). The possibility to explain the negative magnetoresistance due to conduction electrons scattered by the domain walls was discussed in earlier experimental papers [1, 2]. However, as follows from the theoretical studies [13 to 17] and experiment [18], this mechanism gives the correct sign of the effect, but its magnitude is by two or three orders less than the experimentally found values [1 to 7]. In experiments one usually studies the dependence of the ratio $(\varrho^H - \varrho_0)/\varrho_0$ on a magnetic field, where ϱ_0, as a rule, is the resistivity of a demagnetized ferromagnet, ϱ^H is the resistivity of the sample in an external magnetic field H. At the same time, theoretical papers compare the resistivity of the demagnetized ferromagnet with the saturated one. The paper by Berger [11], in which the dependence of $\Delta\varrho/\varrho$ on magnetic field is obtained in the case $\tau\omega \ll 1$ ($\Delta\varrho = \varrho^H - \varrho_0$)

[1]) 660036 Krasnoyarsk, USSR.

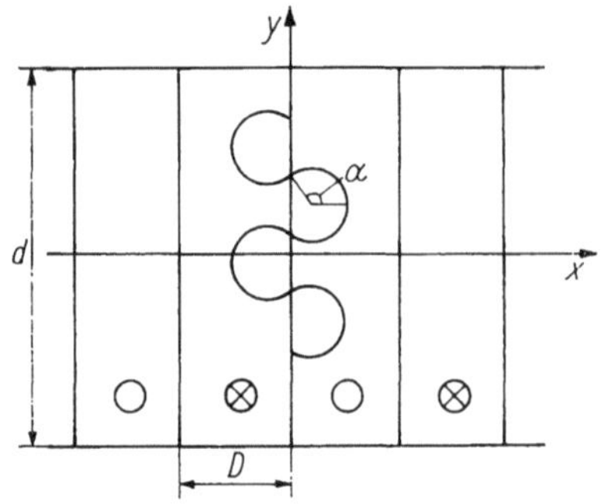

Fig. 1. Coordinate system. The easy axis is parallel to the z-axis; domain walls are parallel to the yz plane. The circles show the direction of the magnetic induction in the domains. The motion of the electrons crossing the domain wall is shown

is the only exception. The negative magnetoresistance effect achieves, however, the largest value in the opposite limit ($\tau\omega \gg 1$), when one should take into account peculiarities of orbital electron motion in the inhomogeneous magnetic field induced due to a domain structure. To obtain an expression for the resistivity of a ferromagnet in a weak external magnetic field in that limiting case may be valuable. We consider a compensated ferromagnet, thus the influence of the inhomogeneous Hall current in a multi-domain ferromagnet is neglected.

We consider a ferromagnetic layer (Fig. 1) with a plane-parallel domain structure. The layer thickness d is assumed to be much greater than the domain width D. The domains are separated by 180° Bloch walls of negligible thickness.

We describe the distribution of field and currents by the electrostatic equations

$$\text{rot}\,\boldsymbol{E} = \boldsymbol{0}\,, \qquad \text{div}\,\boldsymbol{j} = 0\,, \tag{1}$$

and the material equation

$$j_i(\boldsymbol{r}) = \sigma_{ij}(\boldsymbol{r})\, E_j(\boldsymbol{r})\,, \tag{2}$$

where $i, j = x, y, z$.

As is shown in [19] the inhomogeneity of the current in y-direction (see Fig. 1) is essential only in a layer of thickness $\approx D$ near the sample surfaces. Since we assume $d \gg D$, the electron contribution of this layer to the total current will be negligible. Thus, the sample is considered homogeneous in y-direction and, consequently, the physical values of the problem depend only on x. Then from (1) we have

$$j_x = \text{const}\,, \qquad E_y = \text{const}\,. \tag{3}$$

As an example of a compensated ferromagnet we consider a two-band model with equal concentrations n and effective masses m of electrons and holes. In this model $\sigma_{xy} = 0$. In a real metal this equality holds with an accuracy up to $(\tau\omega)^{-3}$ for $\tau\omega \gg 1$. Let us define the specific resistance of the ferromagnet ϱ_{kl} by the expression

$$\langle E_k \rangle = \varrho_{kl}\, \langle j_l \rangle\,, \tag{4}$$

where $k, l = x, y$, angular brackets denote averaging over sample size in the direction of the x-axis. From this definition as well as (2) and (3) when $\sigma_{xy} = 0$ one can get

$$\varrho_{xx} = \langle \sigma_{xx}^{-1}(x) \rangle\,, \qquad \varrho_{yy} = \langle \sigma_{yy}(x) \rangle^{-1}\,. \tag{5}$$

2. Conductivity near a Single Domain Wall; Coordinate Dependence

Let us now calculate the conductivity $\sigma_{ii}(x)$ and consider the domain wall in the plane $x = 0$. Conduction electrons in the sample may be divided into two groups: electrons moving under the effect of inhomogeneous magnetic induction along the trajectories crossing the wall (Fig. 1), and electrons in domains moving in a homogeneous field along closed trajectories. On the basis of earlier results [9] we find an expression for the conductivity at the point $x = x'$ in the form

$$\sigma_{ij}(x) = -\frac{e^2}{S_0}\int_0^\infty dt\,\frac{\partial f}{\partial \varepsilon}\,e^{-t/\tau}\left[\int_{\Gamma_1(x')} d\Gamma v_i^{(w)}(t)\,v_j^{(w)}(0) + \int_{\Gamma-\Gamma_1(x')} d\Gamma\,v_i(t)\,v_j(0)\right], \tag{6}$$

where e is the value of the electron charge, S_0 the area of the domain wall, f the equilibrium distribution function, ε the conduction electron energy, $d\Gamma = 2(2\pi\hbar)^{-3} \times \times\, d\boldsymbol{p}\,dy\,dz$, $\boldsymbol{p}$ is the electron quasimomentum, $\Gamma_1(x')$ the phase volume occupied by electrons the trajectories of which cross both planes: the plane $x = x'$ and the domain wall plane. Apparently, for $x' \geqq 2R$, where R is the cyclotron radius, the phase volume $\Gamma_1(x') = 0$, and in this case we obtain the expression for the conductivity in a uniform sample. Since the choice of the point x' is arbitrary, the prime will be omitted hereafter and thus (6) describes the dependence of conductivity on the x-coordinate. In (6) $v_i(t)$ designates the electron velocity in a domain in an inhomogeneous induction $B = 4\pi M_0$ (M_0 is the saturation magnetization). The Fermi surface of conduction electrons is supposed, for simplicity, cylindrical with the cylinder axis in z-direction, then

$$v_x(t) = v \sin(\omega t + \varphi)\,, \qquad v_y(t) = v\cos(\omega t + \varphi)\,. \tag{7}$$

Here φ is a phase characteristic and $v^2 = v_x^2 + v_y^2$.

For the components of the velocity of electron crossings the domain wall and exhibiting a new periodicity of motion, we use the expressions [9]

$$v_x^{(w)}(t) = \frac{2v\cos\alpha}{\alpha}\sum_{n=0}^{\infty}\frac{(-1)^n \sin[(2n+1)(\pi/2\alpha)(\omega t + \varphi)]}{1-(\pi/2\alpha)^2(2n+1)^2},$$
$$v_y^{(w)}(t) = \frac{v\sin\alpha}{\alpha}\left\{1 + 2\sum_{n=1}^{\infty}\frac{(-1)^n\cos[(n\pi/\alpha)(\omega t+\varphi)]}{1-(\pi/\alpha)^2 n^2}\right\}. \tag{8}$$

One sees that the components of the velocity of electrons bound to the wall are periodic functions with the period $4\alpha/\omega$. The parameter α is determined by the value of the momentum of the electron at the point $x = 0$ ($P_y = p\cos\alpha$) and is equal to half the arc the electron traverses in the uniform magnetic field of a domain between successive passages through the wall (see Fig. 1).

We substitute the components of electron velocity, determined in (7) and (8), into (6). Then for $\sigma_{xx}(x)$ we get

$$\sigma_{xx}(x) = \frac{\sigma_0}{1+\tau^2\omega^2}\left[1 - \frac{\varphi_1}{\pi} + \frac{1+\tau^2\omega^2}{2\pi}\int_{-\varphi_1}^{\varphi_1} d\varphi\, S(\alpha)\right]. \tag{9}$$

Here we carried out integrations with respect to time and momentum p with allowance for $\partial f/\partial\varepsilon = -\delta(\varepsilon - \varepsilon_F)$, where ε_F is the Fermi energy, as well as the usual averaging over the period of motion of an electron; $\sigma_0 = n_0 e^2\tau/m$, n_0 being the electron density in the sample. The first term in (9) is the result of the integration over the whole phase volume Γ and the rest of the terms describe the substitution in volume $\Gamma_1(x)$ of electrons moving along closed trajectories in the uniform induction for electrons crossing the domain wall.

To determine φ_1, note that the parameter α depends on φ,

$$\cos\varphi = \frac{x}{R} + \cos\alpha\,. \tag{10}$$

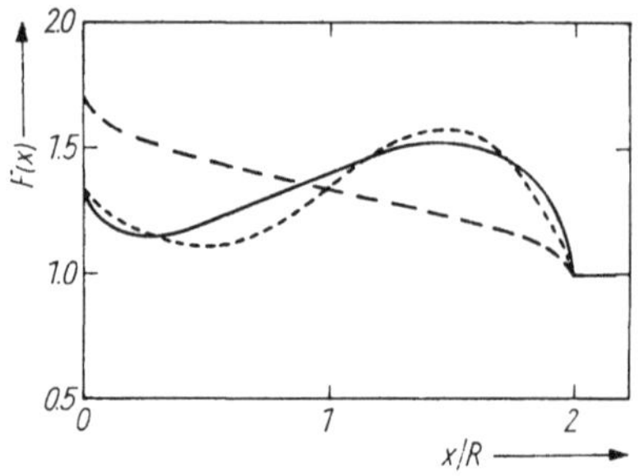

Fig. 2. Dependence of the conductivity σ_{xx} on the coordinate x normal to the wall plane ($F(x) \equiv \sigma_{xx}(x) \times \tau^2\omega^2/\sigma_0$). The continuous line shows the value F given by (15); the dashed one the value $\bar{F}$ by (17); the dotted line F by (18)

This follows directly from the expression for the momentum of the electron in a magnetic field, $p_y = P_y + (e/c)\,Bx$.

From (10) in the limiting case $\alpha = \pi$ we find the value of φ_1,

$$\varphi_1 = \arccos\left(\frac{x}{R} - 1\right). \tag{11}$$

The expression for $S(\alpha)$ follows from (6) and (8) and has the form

$$S(\alpha) = \frac{4\cos^2\alpha}{\alpha^2} \sum_{n=0}^{\infty} \frac{1}{[1 + s^2(\pi/2\alpha)^2\,(2n+1)^2]\,[1 - (\pi/2\alpha)^2\,(2n+1)^2]^2}, \tag{12}$$

where the notation $s = \tau\omega$ is adopted. Summing up by the Watson method, we find

$$S(\alpha) = \frac{\cos^2\alpha}{\alpha^2(1+s^2)}\left[\frac{2s^3\alpha\tanh(\alpha/s)}{1+s^2} + \frac{\alpha^2}{\cos^2\alpha} - \frac{1+3s^2}{1+s^2}\,\alpha\tan\alpha\right]. \tag{13}$$

We proceed to integration with respect to α from integration with respect to φ in (9), using (10) and (11). Substituting (13) into (9) and integrating with $\tau\omega \gg 1$ ($\tanh(\alpha/s) \approx \approx \alpha/s$) taken into account, we find

$$\sigma_{xx}(x) = \frac{\sigma_0}{s^2}\,[1 + \Delta\sigma(x)]\,, \tag{14}$$

where

$$\Delta\sigma(x) = \frac{1}{\pi}\left[\left(1 + \frac{2x^2}{R^2}\right)\arccos\left(\frac{x}{R} - 1\right) - \left(1 + \frac{3x}{R}\right)\sqrt{1 - \left(\frac{x}{R} - 1\right)^2}\,\right] - \frac{3}{\pi}\int_{\alpha_1}^{\pi} \frac{\cos\alpha\,\sin^2\alpha\,d\alpha}{\alpha\sqrt{1 - (x/R + \cos\alpha)^2}}. \tag{15}$$

Here $0 \leq x \leq 2R$, $D \geqq 4R$, $\alpha_1 = \arccos\big(1 - (x/R)\big)$ and the improper integral converges. Then integration in (15) is done numerically. The solid line in Fig. 2 shows the dependence of $F(x) = 1 + \Delta\sigma(x)$ on x for $0 \leq x \leq 2R$.

First of all we note that the conductivity near the wall increases as compared to its value in the domain for $x > 2R$. This is due to the fact that the number of electrons with frequency $\omega_1 = \pi\omega/2\alpha$ greater than ω (i.e. more localized in x-direction than the electrons in the domain) becomes less among the electrons crossing the wall than the number of electrons with frequency smaller than ω, i.e. less localized. Such a conclusion follows from the expression for the number of electrons crossing the wall at an angle of 0 to α,

$$n(\alpha) = \sin\alpha - \alpha\cos\alpha\,. \tag{16}$$

The non-monotonic behaviour of $\sigma(x)$ is connected with both the decrease of the total number of electrons that go across the wall as well as the plane x (the estimate shows that for $x \ll R$ $\sigma_{xx}(x) \sim n(x) \sim \arccos(x/R - 1)$, and the change of the ratio of the more or less localized electrons when x increases. It is obvious that the conductivity $\sigma_{xx}(x)$ for $x > 2R$ coincides with its expression in the uniform case, i.e. $\Delta\sigma(x) = 0$.

Note that one can derive an approximate expression for $\sigma_{xx}(x)$ by averaging the integrand over α in (9) with the state density $\partial n/\partial\alpha = \alpha \sin\alpha$. We have

$$\bar{\sigma}_{xx}(x) = \frac{\sigma_0}{s^2}\left[1 + \Delta\bar{\sigma}(x)\right] = \frac{\sigma_0}{s^2}\left[1 + \frac{2}{3\pi}\arccos\left(\frac{x}{R} - 1\right)\right]. \tag{17}$$

The obtained x-dependence of $\bar{\sigma}_{xx}(x)$ is due to the change in the number of electrons $n(x) \sim \arccos(x/R - 1)$ crossing both the wall and the plane x. The dashed line in Fig. 2 shows the plot of the function $\bar{F}(x) = 1 + \Delta\bar{\sigma}(x)$. The dotted line gives the plot of the expression

$$\tilde{F}(x) = 1 + \frac{1}{3\pi}\left(1 - \frac{2x}{R}\right)\arccos\left(\frac{x}{R} - 1\right) + \frac{x^2}{2R^2}\sin\frac{\pi x}{2R}, \tag{18}$$

approximating the behaviour of the exact function $F(x)$ determined by (14) and (15).

3. Conductivity of a Many-Domain Ferromagnet

On the basis of the earlier derived expressions for the conductivity of a ferromagnet with a single wall one can write expressions for the conductivity of a sample with a periodic domain structure. If the domain width $D > 4R$, $\sigma_{xx}(x)$ is determined by (14) in the region of width $4R$ near the wall and it remains constant outside $\left(\sigma_{xx}(x) = \sigma_0/s^2\right)$. If $D < 4R$, in the vicinity of the domain centre there exist both electrons which are related to a "left" wall and electrons related to the "right" one. Here the number of electrons that generally cross the wall is doubled as compared to the analogous value near a single wall. The expression for the conductivity in the region with overlapping orbits of the electrons going across the "left" and "right" walls has the form

$$\sigma_1(x) = \frac{\sigma_0}{s^2}\left[1 + \Delta\sigma(x) + \Delta\sigma(D - x)\right] \tag{19}$$

for $D - 2R \leqq x \leqq 2R$, $2R \leqq D \leqq 4R$.

In the regions without electron orbit overlapping, (14) (for $0 \leqq x \leqq D - 2R$) and the analogous expression with replacement $x \to D - x$ (for $2R \leqq x \leqq D$) remain valid.

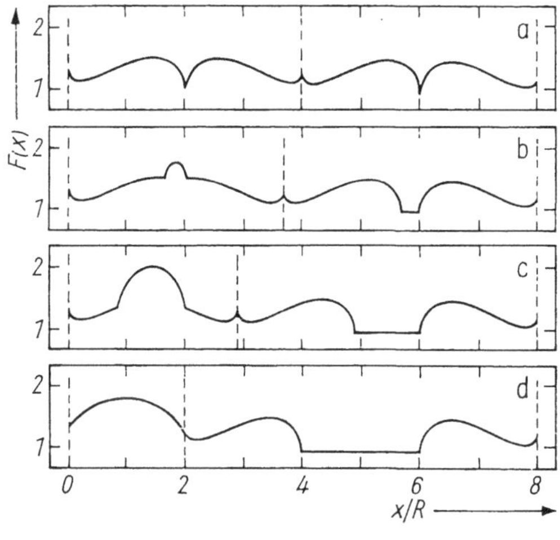

Fig. 3. Dependence $\sigma_{xx}(x)$ at different values of magnetization M/M_0: a) 0; b) 0.09; c) 0.29; d) 0.5 ($F(x) = \sigma_{xx}(x)\,\tau^2\omega^2/\sigma_0$). The dashed line stands for the domain walls. In the initial state $2R/D = 0.5$. The region of sharp increase of σ_{xx} corresponds to the overlapping of the trajectories of electrons crossing the left and right walls. Continuing the left domains symmetrically in both directions, we find $\sigma_{xx}(x)$ in the demagnetized state in dependence on different values of $2R/D$: a) 0.5; b) 0.55; c) 0.7; d) 1

The function $F(x) = (s^2/\sigma_0)\, \sigma_{xx}(x)$ originating from the expression for some definite values of the ratio D/R is depicted in Fig. 3. We should remark that $\sigma_{xx}(x)$ is always symmetric about domain wall and domain centre. To obtain the conductivity in the whole demagnetized sample it is necessary to extend symmetrically the left domains in Fig. 3 in both directions over the x-axis. The conductivity growth near the domain centre, for $D < 4R$, is due to the increasing number of electrons passing through the domain walls.

Thus, we have obtained (14), (15), and (19) describing the conductivity of the ferromagnet with domain structure in the region $D > 2R$. Substituting them into (5) and integrating with respect to x, one can get the value of resistivity of the multi-domain ferromagnet.

4. Magnetoresistance

In the demagnetized state, when $D \geqq 4R$, the resistance

$$\varrho_{xx}^{(w)} = a\Big(2 \int\limits_0^{2R} \sigma^{-1}(x)\, dx + D - 4R\Big)\,, \tag{20}$$

where $a \equiv s^2/\sigma_0 D$ and $\sigma \equiv \sigma_{xx}$. If $2R \leqq D \leqq 4R$, for $\varrho_{xx}^{(w)}$ we have

$$\varrho_{xx}^{(w)} = a\Big(2 \int\limits_0^{D-2R} \sigma^{-1}(x)\, dx + \int\limits_{D-2R}^{2R} \sigma_1^{-1}(x)\, dx\Big)\,. \tag{21}$$

In studying the magnetoresistance of a ferromagnet with domain structure, we restrict ourselves to the process of magnetization by the domain wall motion caused by a constant external field, $\boldsymbol{H}$, applied along the easy axis (z-axis). The period of the domain structure $2D$ is supposed to be independent of H, i.e. we consider only the fields $H \leqq 0.8\, H_s$ [20] in which the increase of the domain structure period may be neglected together with the effect of the field H on the cyclotron radius. Then the width of the increasing domain is $D_1 = D(1 + M/M_0)$, the width of the decreasing one $D_2 = D(1 - M/M_0)$, where M is the magnetization of the ferromagnet in the field H. The period of the domain structure $2D = D_1 + D_2$ is conserved.

The restrictions under which we have obtained expressions for the conductivity allow us to consider the magnetization process up to the state when $D_2 \geqq 2R$, i.e. the relative magnetization

$$\frac{M}{M_0} \leqq 1 - \frac{2R}{D}. \tag{22}$$

First we treat the case when there is no overlapping of the electron trajectories which cross the neighbouring walls. Such a situation will be conserved under magnetization until the decreasing domain width $D_2 \geqq 4R$, i.e.

$$\frac{M}{M_0} \leqq 1 - \frac{4R}{D}. \tag{23}$$

In cases when there are electron trajectories overlapping in the demagnetized state during the process of magnetization, the trajectory overlapping will be conserved in both domains till the width of the increasing domain $D_1 \leqq 4R$, i.e.

$$\frac{M}{M_0} \leqq -1 + \frac{4R}{D}. \tag{24}$$

The solid lines in Fig. 4 indicate limiting values of the relative magnetization versus $2R/D$, determined by inequalities (22) to (24) (lines 1 to 3, respectively).

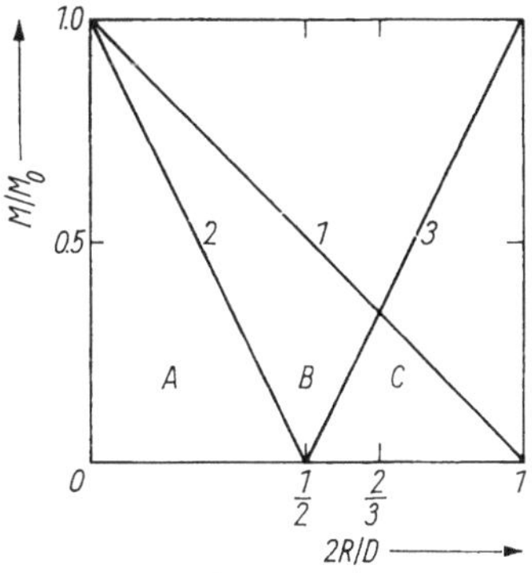

Fig. 4

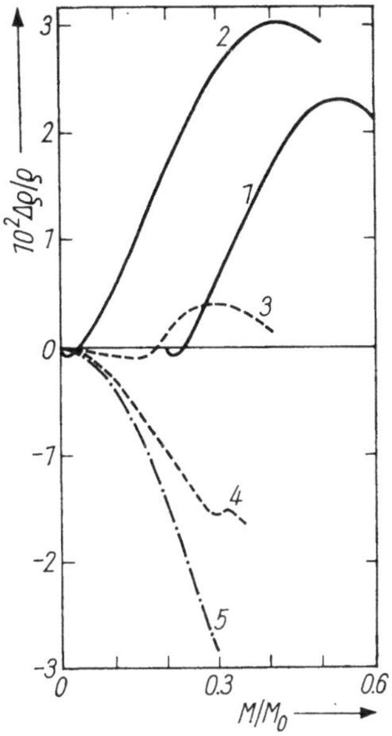

Fig. 5

Fig. 4. Phase diagram. Straight lines 1, 2, 3 are determined correspondingly by (22), (23), (24), respectively. Region A: the overlapping of electron trajectories going across the walls is absent in both domains; region B: the overlapping is only in the decreasing domain; region C: the overlapping takes place in both domains

Fig. 5. The dependence $\Delta\varrho/\varrho$ on M/M_0 for different values of the ratio $2R/D$: (1) $2R/D = 0.4$; (2) 0.5; (3) 0.6; (4) 0.65; (5) 0.7

In region A there is no overlapping of the trajectories in both domains during the magnetization process. Neglecting the H-dependence of $R(H)$, the resistivity of the ferromagnet under magnetization does not change and is determined by (20). (If we take into account a H-dependent $R(H)$, then $\Delta\varrho/\varrho \equiv (\varrho^H - \varrho^{(w)})/\varrho^{(w)} \sim HM/M_0^2$.)

After transition to region B, the overlapping occurs in the decreasing domain. This leads to a change in the resistivity ϱ^H due to the change in the conductivity $\sigma(x)$ (see Fig. 3),

$$\varrho_B^H = a\left[\frac{1}{2}\int\limits_{D_2-2R}^{2R} \sigma_{12}^{-1}(x)\,dx + \int\limits_0^{D_2-2R} \sigma^{-1}(x)\,dx + \int\limits_0^{2R} \sigma^{-1}(x)\,dx + \frac{D_1}{2} - 2R\right], \quad (25)$$

where $\sigma_{12}(x)$ is determined by (19) with substitution $D \to D_2$.

In region C the trajectory overlapping is conserved in both domains during the magnetization process. For the resistance we have

$$\varrho_C^H = a\left[\frac{1}{2}\int\limits_{D_2-2R}^{2R} \sigma_{12}^{-1}(x)\,dx + \int\limits_0^{D_2-2R} \sigma^{-1}(x)\,dx + \frac{1}{2}\int\limits_{D_1-2R}^{2R} \sigma_{11}^{-1}(x)\,dx + \int\limits_0^{D_1-2R} \sigma^{-1}(x)\,dx\right], \quad (26)$$

where $\sigma_{11}(x)$ is determined by (19) with substitution $D \to D_1$.

The magnetization process from the initial state characterized by $2R/D$ can be described in Fig. 4 by the movement along the ordinate from the point $2R/D$ to the straight line 1. It is obvious that the magnetization process may follow in three ways depending on the value of the ratio $2R/D$:

1. with magnetization increase there is a transition from region A to region B for $0 \leqq 2R/D \leqq \frac{1}{2}$ $(D \geqq 4R)$;
2. with magnetization increase there is a transition from region C to region B for $\frac{1}{2} \leqq 2R/D \leqq \frac{2}{3}$ $(3R \leqq D \leqq 4R)$;
3. the process of magnetization takes place in region C for $\frac{2}{3} \leqq 2R/D \leqq 1$ $(2R \leqq \leqq D \leqq 3R)$.

It is necessary to take the corresponding combination of expressions (20), (21), (25), and (26) in order to find the dependence $\Delta\varrho/\varrho = (\varrho^H - \varrho^{(w)}/\varrho^{(w)})$ on the magnetization at a given ratio $2R/D$. The results of computer calculations for some values of $2R/D$ are given in Fig. 5.

The different types of dependence $\Delta\varrho/\varrho$ on M/M_0 finally are due to the non-monotonic variation of $\sigma(x)$. In magnetizing the ferromagnet, its resistivity does not change (the part of curve 1 that coincides with the abscissa) till the trajectories start overlapping $(D_2 \geqq 4R)$. With further increase of field H the resistivity $\Delta\varrho/\varrho$ grows almost everywhere except in the small portion corresponding to the start of the overlapping of the trajectories in the decreasing domain D_2. Through the sharp steepness of growing $\Delta\sigma(x)$, the conductivity $\sigma_1(x)$ has a peak (see Fig. 3). This results in a more substantial decrease in the resistivity in domain D_2 compared to a resistivity increase linear in magnetization in domain D_1, at the expense of enlargening the region with $\sigma = \text{const}$ in domain D_1. In the same way the decrease of $\Delta\varrho/\varrho$ near the limiting $(D_2 \gtrapprox 2R)$ magnetization values M/M_0 for the given model at the end of the overlapping (see Fig. 2 and 3d) is explained as the sharp change of $\Delta\sigma(x)$.

The initial part of the plot where the magnetoresistance is negative broadens with the increase in the ratio R/D. This is due to the relatively large influence of the overlapping regions on $\Delta\varrho/\varrho$ on account of decreasing the domain structure period $2D$.

Simultaneously with the increase of R/D (for $2R/D > 0.5$) the increasing part of the plot of $\Delta\varrho/\varrho$ narrows. This is explained by the relative decrease of the contribution to the magnetization process of region B in Fig. 4 where the overlapping takes place only in the decreasing domain. Curve 5 in Fig. 5 which corresponds to the magnetization process going entirely through region C (Fig. 4) has no increase at all.

It is noteworthy that the curves $\Delta\varrho/\varrho$ obtained with the use of the approximating expression (18) are similar in character. However, the use of (17) for $\sigma(x)$ obtained after extra averaging over α does not give such a negative magnetoresistance.

5. Conclusion

We have considered the magnetization process up to the values of M/M_0 at which $D_2 = 2R$. With further increase in magnetization D_2 becomes less than $2R$ and a new type of electron trajectories crossing two domain walls appears. The influence of such a type of motion of electrons on the properties of the ferromagnet with domain structure will be considered in the following paper.

Here we may indicate the value of the maximum effect of magnetoresistance for a given ratio R/D, i.e. the ratio $\Delta\varrho^{\max}/\varrho = (\varrho_m^H - \varrho_0)/\varrho_0$ where $\varrho_m^H = s^2/\sigma_0$ is the resistivity of a ferromagnet in a completely magnetized state and $\varrho_0 = \varrho^{(w)}$ is determined by (20) or (21). The function $\Delta\varrho^{\max}/\varrho$ is shown in Fig. 6. The solid line depicts it when $\Delta\sigma(x)$ is calculated with the use of the exact expression (15), and the dashed line that with the use of (17) derived after averaging. One can see that in the compensated ferromagnet with a plane-parallel domain structure in the initial state the maximum magneto-

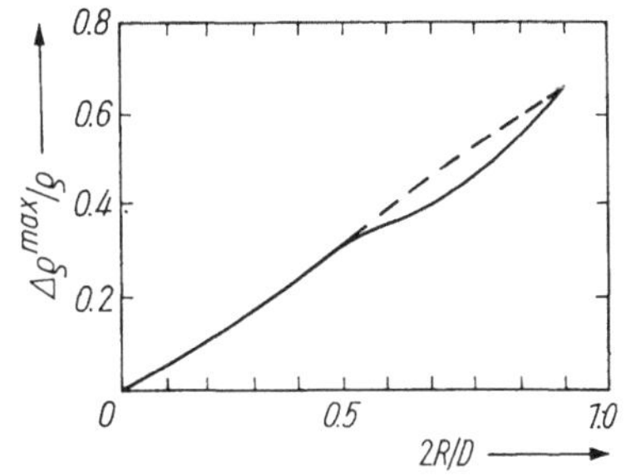

Fig. 6. The dependence of maximum value of magnetoresistance $\Delta\varrho^{max}/\varrho$ on the value of the ratio $2R/D$. The solid line corresponds to the calculations with the use of the exact expression (15); the dashed one to the averaged expression (17)

resistance effect related to the magnetization process is always positive. The dependence of $\Delta\varrho^{max}/\varrho$ on the ratio R/D is close to linear and is well described by (17).

The negative magnetoresistance for small values of M/M_0 and the positive magnetoresistance effect when magnetizing up to saturation indicate the presence of a minimum in the dependence of $\Delta\varrho/\varrho$ on M/M_0. Moreover, one should expect that when $2R > D_2$, the magnetoresistance $\Delta\varrho/\varrho$ will have a minimum along curves 4 and 5 in Fig. 5.

We thus have considered the simplest model of magnetizing a ferromagnet with a plane-parallel domain structure. The obtained results demonstrate that the dependence of $\Delta\varrho/\varrho$ on a magnetic field is essentially determined by the ratio $2R/D$.

Acknowledgements

The authors are grateful to V. A. Ignatchenko and V. E. Shapiro for stimulating and useful discussions.

References

[1] A. M. Sudovtsev and E. E. Semenenko, Zh. eksper. teor. Fiz. **35**, 305 (1958).
[2] E. E. Semenenko and A. M. Sudovtsev, Zh. eksper. teor. Fiz. **47**, 486 (1964); Soviet Phys. – J. exper. theor. Phys. **20**, 323 (1965).
[3] A. Isin and R. V. Coleman, Phys. Rev. **137**, A1609 (1965); **142**, 372 (1966).
[4] G. R. Taylor, A. Isin, and R. V. Coleman, Phys. Rev. **165**, 621 (1968).
[5] P. V. Shumate, R. V. Coleman, and R. C. Fivaz, Phys. Rev. B **1**, 394 (1970).
[6] R. V. Coleman, R. C. Morris, and D. J. Sellmyer, Phys. Rev. B **8**, 317 (1973).
[7] E. I. Kondorskii, O. S. Galkina, V. I. Ivanovskii, A. V. Cheremushkina, and U. T. Usarov, Zh. eksper. teor. Fiz. **65**, 1957 (1973).
[8] Yu. I. Mankov, Fiz. tverd. Tela **14**, 78 (1972).
[9] Yu. V. Zakharov, Yu. I. Mankov, and R. G. Khlebopros, Zh. eksper. teor. Fiz. **65**, 242 (1973).
[10] G. G. Cabrera and L. M. Falicov, phys. stat. sol. (b) **62**, 217 (1974).
[11] L. Berger, AIP Conf. Proc. **29**, 165 (1975).
[12] L. Berger, J. appl. Phys. **49**, 2156 (1978).
[13] E. A. Turov and A. N. Voloshinskii, Proc. X. Internat. Conf. Low-Temp. Phys., Moscow 1967, Vol. 4 (p. 105).
[14] R. G. Khlebopros, Yu. V. Zakharov, and Yu. I. Mankov, Fiz. tverd. Tela **16**, 115 (1974).
[15] Yu. V. Zakharov, Yu. I. Mankov, and R. G. Khlebopros, Preprint IFSO-20F, Krasnoyarsk 1974.
[16] G. G. Cabrera and L. M. Falicov, phys. stat. sol. (b) **61**, 539 (1974).
[17] G. G. Cabrera, J. Phys. F **7**, 827 (1977).
[18] V. P. Dyakina and N. V. Volkenshtein, Fiz. Metallov i Metallovedenie **26**, 628 (1968).
[19] L. Berger, J. Phys. Chem. Solids **35**, 947 (1974).
[20] V. A. Ignatchenko, I. F. Degtyarev, and Yu. V. Zakharov, Izv. Akad. Nauk SSSR, Ser. fiz. **25**, 1439 (1961).

(Received June 27, 1984)

phys. stat. sol. (b) **125**, 207 (1984)

Subject classification: 13.2; 12.1; 15; 21.1

Department of Physics, Harcourt Butler Technological Institute, Kanpur[1])

Effect of Cold Work on Hydrogenation of Nickel as Seen in the Thermoelectric Behaviour

By

A. C. GUPTA and A. N. NIGAM

Wires of nickel annealed at different temperatures are subjected to cathodic hydrogen diffusion with and without previous cold work. The d-band shape computed for differently annealed wires are different without cold work, but almost the same after cold work. Thus the cold work distorts the d-band in differently annealed samples to the same extent.

Nickeldrähte, die bei verschiedenen Temperaturen getempert werden, werden einer kathodischen Wasserstoffdiffusion mit und ohne vorheriges Kaltwalzen ausgesetzt. Die Form der d-Bande, die für unterschiedlich getemperte Drähte berechnet wird, ist ohne Kaltwalzen unterschiedlich, jedoch fast gleich nach Kaltwalzen. Somit deformiert das Kaltwalzen die d-Bande in unterschiedlich ausgeheilten Proben in gleichem Ausmaß.

1. Introduction

Thermo-e.m.f. is a property highly sensitive to the shape of the Fermi surface [1]. Cold work very slightly distorts the lattice. The Fermi surface shape should also get distorted. In principle the $n(E)$ versus E curve should also exhibit minor changes on cold working. Our object was to see if it is possible to observe such changes thermoelectrically. The principle involved is the following: In an earlier paper by Nigam and Garg [2] the thermoelectric behaviour of annealed nickel wire subjected to cathodic hydrogen diffusion was described and interpreted in terms of d-band filling by the electrons donated by the diffusing hydrogen. In Fig. 1a the $n(E)$ versus E curve is shown for the d-band of nickel (non-relativistic). The band is filled up to E_F. To the unfilled part PQ, diffusing hydrogen donates its electrons. When left to itself the metal loses hydrogen and the band PQ gets emptied. Changes in thermo-e.m.f. this period were recorded. Since thermo-e.m.f. is a property proportional to dn/dE [2], a plot of this quantity against energy E is shown in Fig. 1b for the part PQ of the band. The emptying process of the band begins from the higher energy side. Therefore, the mirror image of Fig. 1b is drawn in Fig. 1c and represents the form of the observed variation in thermo-e.mf. with time of hydrogen outgassing.

Conversely, if Fig. 1c is given from experiment, one can compute the part PQ of Fig. 1a by numerical integration of Fig. 1b. This is precisely what has been done in the present investigation. Annealed wires of nickel were subjected to hydrogen diffusion with and without could work. The computed PQ parts of the curve (Fig. 1a) were different for the two cases.

2. Experimental

Sample preparation, hydrogen diffusion procedure, and measurements of thermoelectric changes all have been described in [2]. We mention here only the cold work which was done by the method due to Wagler and Lorentz [3], i.e. winding and un-

[1]) Kanpur 208002, India.

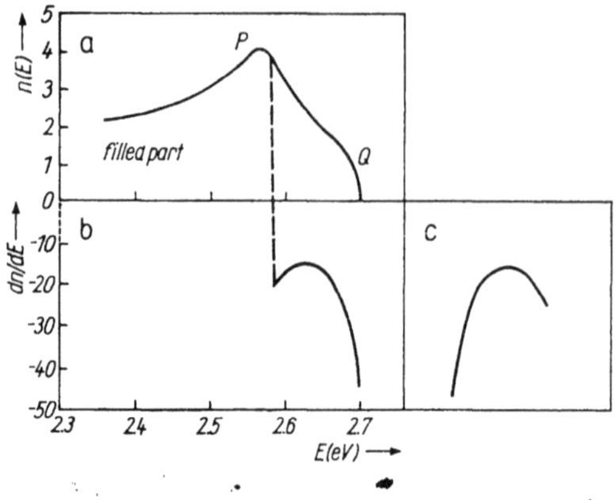

Fig. 1. a) Density of states curve for d-band of nickel, b) dn/dE curve for a) calculated by numerical differentiation, c) the mirror image of b)

winding the wire over a rod 0.7 cm diameter under a load of 1 kp. This method was found to be more effecive than the method of vertical or horizontal stretching of the nickel wire in a tensiometer. The procedure adopted is summarised below:

1. About 50 nickel wires (0.25 mm diameter and 1 m length) were sealed in glass/quartz capsules. Different capsules were annealed at different temperatures, 240, 290, 425, 600, and 750 °C. The wires were then electropolished.

2. The samples were divided into two groups. Wires of one group were subjected as such to electrolytic hydrogen diffusion in the presence of thiourea for 200 h, wires of the other group were first cold worked (25 to 30 cycles of winding–unwinding operations) and then hydrogenated.

3. After hydrogenation the changes in thermo-e.m.f. with time were measured for both kinds of wires and their curves used to compute the shape of the d-band (Fig. 1a).

3. Results and Discussions

Fig. 2 compares the experimentally determined variation in thermo-e.m.f. for hydrogenated nickel wires annealed at 240 °C. Curves for different samples (both with and without cold work) are drawn and show the range of reproducibility of our results. This range is 0.25 to 0.5 μV/K. Results for wires annealed at other temperatures are not given for the sake of brevity. However, these are described below in Table 1.

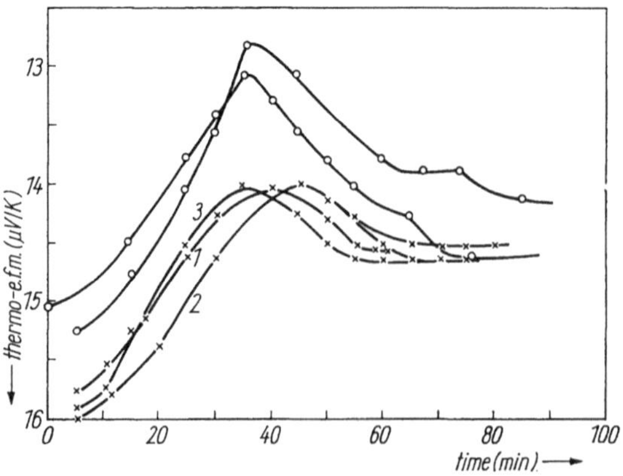

Fig. 2. Thermo-e.m.f. variation with time for annealed Ni samples. ○ hydrogenation without cold working, × hydrogenation with cold working; (1) 20, (2) 25, (3) 28 cycles. Annealing temperature 240 °C

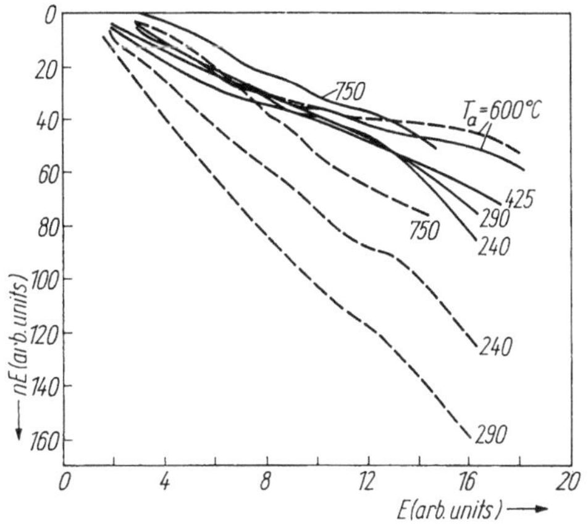

Fig. 3. Computed shapes $n(E)$ of PQ part of the d-band. Dashed curves are for samples without cold work, continuous curves are for cold-worked samples

Table 1
Thermo-e.m.f. data (in μV/K) for hydrogenated Ni wires

annealing temp. (°C)	H-diffusion without cold work			H-diffusion after cold work		
	initial value	peak value	final value	initial value	peak value	final value
240	15.2	13.2	14.8	15.8	14.3	15.0
290	17.3	16.3	17.5	16.3	15.4	16.3
425	15.3	15.0	15.8	16.5	15.3	16.4
600	16.0	13.5	15.5	16.8	14.5	15.8
750	16.1	13.8	15.8	17.8	15.0	15.8

Curves of the type shown in Fig. 2 were used to compute the PQ part of the curve shown in Fig. 1a. These computed curves are shown in Fig. 3. The d-band structures for wires annealed at different temperatures are different and well separated from each other, but the band shapes for cold-worked wires overlap each other, and seem to be grouped together. Thus irrespective of annealing temperature cold work has nearly the same effect on the d-band shapes. This can be reasoned as follows.

As mentioned in [2] wires annealed at different temperatures had different orientation factors. The limits for averaging of $n(E)$ in these wires are also different, and one gets slightly different d-band shapes for each case. On cold working (in which the wire length increases) the microcrystallites of the wire get oriented and distorted likely and we observe similar shapes of d-bands in all the cases.

What happens to the filled part of the band is not discernible from the present information.

References

[1] K. L. Chopra, Thin Film Phenomena, McGraw-Hill Publ. Co., New York 1969 (p. 419).
[2] A. N. Nigam and V. S. Garg, phys. stat. sol. (a) **12**, 589 (1972).
[3] O. Wagler and F. R. Lorenz, Z. tech. Phys. **11**, 242 (1930).

(Received January 17, 1984)

phys. stat. sol. (b) **125**, 211 (1984)

Subject classification: 13; 20; 21.4

Department of Physics, Madhav Science College, Vikram University, Ujjain[1])

Evidence of Open 4f Shells in Rare Earth Compounds

By

S. N. GUPTA and S. SHAH

A forbidden transition $^2P_{3/2}$–4f (unoccupied) is detected in the L-absorption spectra of compounds of rare earth metals Tb and Dy, indicating the presence of an open 4f shell in these compounds. The location of these 4f shells is found to shift towards the high-energy side when these metals form compounds.

Ein verbotener Übergang $^2P_{3/2}$–4f (unbesetzt) wird in den L-Absorptionsspektren von Verbindungen der Seltenerdmetalle Tb und Dy beobachtet, was auf eine offene 4f-Schale in diesen Verbindungen hinweist. Es wird gefunden, daß die Lage dieser 4f-Schalen sich nach der höherenergetischen Seite verschiebt, wenn diese Metalle Verbindungen bilden.

1. Introduction

The electronic structures and proportics of lanthanides provide an ever increasing research field due to their wide applications in modern technology and industry. Rare earth oxides are convenient as starting materials for most of synthetic work and this field has shown rapid increase in activity in recent years.

Lanthanides are characterized by progressive filling of 4f shells from La ($Z = 57$) to Lu ($Z = 71$). These 4f levels are highly localized and atom-like in character. Though they do not contribute to the conduction band, their influence on the conduction electrons has been found to be important and their presence cannot be ignored. In fact, the nature of 4f states and the conduction electrons are responsible for most of the unusual electrical, magnetic, and other transport properties of rare earths. Due to this importance, studies of 4f states are becoming increasingly interesting.

Several attempts have been made to know the exact position of these 4f levels using different experimental techniques, but their results are still unsatisfactory. Though several reports have appeared recently on the location of 4f occupied states [1 to 5], only few papers discuss the location of 4f unoccupied states.

Evidence of the presence of an open 4f shell is provided by X-ray absorption spectroscopy, XAS [6 to 9]. XAS studies of rare earths in compounds were carried out in our laboratory and an interesting observation is reported in this paper.

2. Results and Discussions

In XAS, we have recorded the L_3-absorption region of rare earth metals and some of their compounds. The L_3 absorption edge corresponds to an electron transition from $^2P_{3/2}$ to the first unoccupied level above the Fermi level E_F.

Typical microphotometer records of the L_3 absorption region for Tb and Dy compounds are shown in Fig. 1 and 2. Each record is normalized to unit absorption in the region of the L_3 absorption edge [11]. The intensity axis is shifted for the sake of clarity. Experimental and other details are described elsewhere [7 to 10].

[1]) Ujjain (M. P.), India.

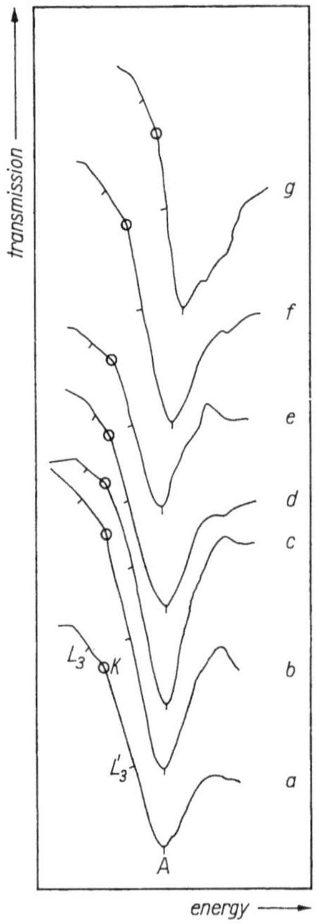

Fig. 1

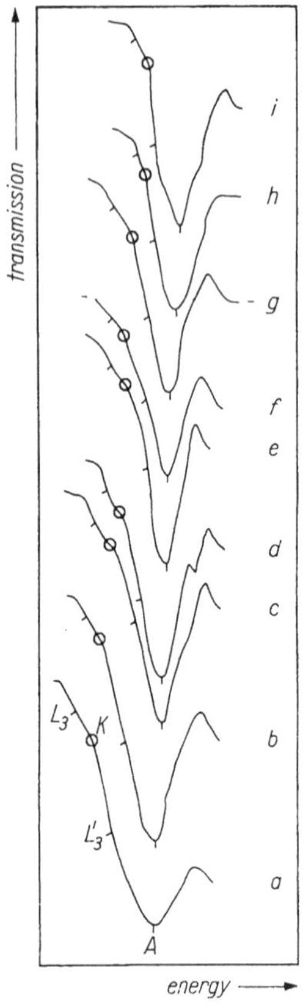

Fig. 2

Fig. 1. Microphotometer records of the L_3-absorption region of Tb compounds. (a) Metal, (b) carbonate, (c) sulphate, (d) acetate, (e) oxalate, (f) oxide, (g) chloride, (h) fluoride

Fig. 2. Microphotometer records of the L_3-absorption region of Dy compounds. (a) Metal, (b) acetate, (c) oxide, (d) oxalate, (e) nitrate, (f) sulphate, (g) chloride, (h) carbonate, (i) fluoride

A close examination of both the figures shows two abrupt rises in the L_3 absorption separated by the kink K. These are designated by L_3 and L_3'. On the high-energy side of the second rise L_3' there is an absorption resonance line A usually known as "white line" which is attributed to the high density of vacant states in the 5d 6s band [7, 8]. The white lines, therefore, reflect the condition of conduction/valance bands in these compounds and are sensitive to changes occurring due to compound formation [10, 11].

Earlier, in case of L-absorption spectra of heavy rare earth metals and their oxides [7, 8], we have denoted this first rise ending at K as the L_3-edge and attributed it to an electron transition $^2P_{3/2}$–4f (unoccupied) which has partially filled 4f states as the final state and is forbidden according to dipole selection rules. The second absorption rise L_3' ending at A was assigned to the electron transition $^2P_{3/2}$–5d 6s admixed states, which is of course allowed. The difference in transition probabilities is reflected in their relative intensities. The first rise L_3 is much weaker than the second rise L_3'.

It may be mentioned here that this is a more plausible interpretation in the light of recent renormalized atom calculations for rare earth elements which take account of

Table 1

Location of 4f unoccupied states, terbium compounds

compound		unoccupied 4f states K		Δf
		X.U.	(eV)	(eV)*)
1.	Tb metal	1647.05	7511.8	—
2.	$Tb_2(Co_3)_3 \cdot x\ H_2O$	1647.00	7512.1	0.3
3.	$Tb_2(SO_4)_3 \cdot x\ H_2O$	1646.88	7512.6	0.8
4.	$Tb(C_2H_3O_2) \cdot x\ H_2O$	1646.69	7513.5	1.7
5.	$Tb_2(C_2O_4) \cdot x\ H_2O$	1646.55	7514.1	2.3
6.	$Tb_2O_3 \cdot x\ H_2PO$	1646.05	7516.5	4.7
7.	$TbCl_3 \cdot x\ H_2O$	1646.02	7516.555	4.7
8.	$TbF_3 \cdot x\ H_2O$	1644.52	7523.4	11.6

*) $\Delta f = E\ (\text{kink K})_{comp} - E\ (\text{kink K})_{metal}$

screening and relaxation effects and are more appropriate to know the position of 4f states. X-ray photoelectron results of Baer and Bush [3] are in good agreement with this theory. The renormalized multielectron values for 4f unoccupied states are nearly 0.5 Ryd above E_F and fairly constant for rare earth metals [13]. Our experimental values are in satisfactory agreement with these theoretical calculations.

The kink K detected on the L_3 absorption curve of these rare earth compounds (oxide, sulphate, oxalate, nitrate, carbonate, acetate, chloride, and fluoride) of Tb and Dy can, therefore, be assigned to the forbidden transition $^2P_{3/2}$–4f (unoccupied). It, gives the location of open 4f shells in these compounds. The measured values of kink K are collected in Tables 1 and 2.

The detection of this weak forbidden transition may be attributed to the suitable dispersion and focussing conditions achieved in this work by employing a 400 mm curved-crystal vacuum spectrograph [10]. Similar features have been observed by Fischer and Baun [12] in the M_4- and M_5-absorption spectra of rare earth metals.

The electronic structures of Tb and Dy are Xe $4f^8\ 5d^1\ 6s^2$ and Xe $4f^9\ 5d^1\ 6s^2$. In solids the narrow 5d band hybridizes with 6s, while the 4f level remains localized and retains its atomic character lying below the 5d 6s band. A weak direct or indirect

Table 2

Location of 4f unoccupied states, dysprosium compounds

compound		unoccupied 4f states K		Δf
		X.U.	(eV)	(eV)*)
1.	Dy metal	1590.0	7781.3	—
2.	$Dy(C_2H_3O_2)_3 \cdot 4\ H_2O$	1589.6	7783.5	2.2
3.	Dy_2O_3	1589.1	7785.8	4.5
4.	$Dy_2(C_2O_4)_3 \cdot 10\ H_2O$	1589.0	7786.3	5.0
5.	$Dy(NO_3)_3 \cdot 5\ H_2O$	1588.9	7786.8	5.5
6.	$Dy_2(SO_4)_3 \cdot 8\ H_2O$	1588.8	7787.3	6.0
7.	$DyCl_3 \cdot x\ H_2O$	1588.5	7788.7	7.4
8.	DyF_3	1587.8	7792.2	10.9

*) $\Delta f = E\ (\text{kink K})_{comp} - E\ (\text{kink K})_{metal}$

interaction would have to exist between 5d 6s and 4f states to permit this P → f forbidden transition. A confirmation of this interaction from other experimental or theoretical studies would be valuable. One additional point speaking in favour of this interpretation is that we could easily detect the kink K at the L_2 edges also, whereas we failed to locate it at the L_1 edges.

Furthermore, it is seen from the tables that 4f unoccupied levels of the metals Tb and Dy shift towards the high-energy side when these metals enter chemical compounds. The shift depends on the chemical bond, ionic, covalent, or metallic. The shift is exceptionally large when a strong metal–metal bond appears.

The broad and diffuse shape of the transition makes the intensity and energy measurements difficult and uncertain. No quantitative conclusions can, therefore, be drawn from these measurements. This is very clear if we critically examine the tables. There is no definite order regarding the shift in these compounds. The shift varies from 0.3 to 11.6 eV in case of Tb and from 2.2 to 10.9 eV in case of Dy compounds. The order of the shift could not be explained by ionicity considerations. As band structure calculations for these compounds are not available in literature, no comparison of the results is possible, which could serve as a check for any future theoretical calculations.

Our results provide a direct experimental evidence of the presence of open 4f levels in compounds of Tb and Dy and show a positive shift (towards the high-energy side) of 4f levels in all these compounds. The L_3-absorption edge is also found to shift towards the high-energy side when these metals form compounds [10, 14]. Although we do not present a quantitative correlation between edge shifts and 4f levels, we can at least conclude that positive values of Δf correspond to positive shifts of the absorption edges and our results on the location of 4f levels will be useful for understanding the behaviour of these 4f states in rare earths.

Acknowledgements

Authors wish to express their sincere thanks to Dr. B. D. Padalia for valuable suggestions and one of us (S.S.) thanks UGC, New Delhi, for the award of a research fellowship.

References

[1] C. BONNELLE, R. C. KARNATAK, and C. K. JORGENSEN, Chem. Phys. Letters **14**, 145 (1972).
[2] V. A. FOMICHEV, S. A. GRIBOVAKII, and T. M. ZIMKINA, Proc. Internat. Symp. X-Ray Spectra and Electronic Structure of Matter, Vol. 2, Munich 1972 (p. 191).
[3] Y. BAER and G. BUSH, Internat. Conf. Electron Spectroscopy, Namur (Belgium), April 1974.
[4] B. D. PADALIA and D. J. FABIAN, J. Phys. **4**, L179 (1974).
[5] J. M. MARIOT, F. C. KARNATAK, and C. BONNELLE, J. Phys. Chem. Solids **35**, 657 (1974).
[6] B. D. PADALIA, S. N. GUPTA, V. P. VIJAYAVARGIYA, and B. C. TRIPATHI, J. Phys. F **4**, 938 (1974).
[7] V. P. VIJAYAVARGIYA, S. N. GUPTA, and B. D. PADALIA, phys. stat. sol. (b) **80**, 83 (1977).
[8] S. N. GUPTA, V. P. VIJAYAVARGIYA, B. D. PADALIA, B. C. TRIPATHI, and M. N. GHATIKAR, phys. stat. sol. (b) **82**, 603 (1977).
[9] V. P. VIJAYAVARGIYA, Ph. D. Thesis, University of Indore (India), 1977.
[10] S. SHAH, Ph. D. Thesis, University of Indore (India), 1983.
[11] G. H. VIA and J. H. SINFELT, J. chem. Phys. **70**, 11, 4849 (1979).
[12] D. W. FISCHER and W. L. BAUN, J. appl. Phys. **38**, 4830 (1967).
[13] J. F. HERBEST, D. N. LOWY, and R. E. WATSON, Phys. Rev. B **6**, 1931 (1972).
[14] B. K. AGRAWAL and V. BALKRISHNAN, J. Phys. C **12**, 1519 (1982).

(Received December 30, 1983)

phys. stat. sol. (b) **125**, 215 (1984)

Subject classification: 13.1 and 19; 16; 22.2.3

Service National des Champs Intenses, C.N.R.S., Grenoble[1])

Intraconduction Band Magneto-Optical Study of InSb under Hydrostatic Pressure

By

S. HUANT, L. DMOWSKI[2]), M. BAJ[3]), and L. C. BRUNEL

The cyclotron resonance and electron spin resonance of n-InSb under hydrostatic pressure up to 1 GPa are investigated using photoconductivity techniques. All the transitions observed are fitted using a modified Pidgeon-Brown model with a set of band parameters recently proposed by Goodwin and Seiler and a change of the band gap of $dE_g/dP = (0.140 \pm 0.004)$ meV/MPa, a value in good agreement with experimental results.

Nous avons étudié la résonance cyclotron et la résonance de spin électronique dans la bande de conduction de InSb soumis à des pressions hydrostatiques jusqu'à 1 GPa. Les transitions ont été observées par mesure de la photoconductivité. En utilisant un modèle de Pidgeon-Brown modifié, avec un ensemble de paramètres de bande récemment proposé par Goodwin et Seiler, nous obtenons un bon accord entre les résultats expérimentaux et les valeurs calculées en adoptant une variation de l'énergie de la bande interdite en fonction de la pression de $dE_g/dP = (0,140 \pm \pm 0,004)$ meV/MPa. Cette valeur est en bon accord avec les mesures expérimentales.

1. Introduction

Following the first observation of cyclotron resonance in InSb [1] a great number of magneto-optical experiments have been devoted to the study of this narrow-gap semiconductor. As a result, the band structure of InSb is now well understood from both an experimental and a theoretical point of view. Many intraconduction band, intravalence band, and interband optical transitions have been reported. The energy and the polarization of these transitions are, for most of them, well described by theoretical models.

Recently, Goodwin and Seiler [2], using a Pidgeon-Brown model [3] modified by Weiler et al. [4], were able to reproduce a huge amount of experimental data using a unique set of band parameters. They were able to fit intraband processes and two-photon interband transitions.

The data which Goodwin and Seiler fitted were obtained in experiments performed at atmospheric pressure. Much less work has been done concerning the properties of InSb under uniaxial or hydrostatic pressure. Kuchar and Kriechbaum [5] reported the electron spin resonance of n-InSb under uniaxial stress. The cyclotron resonance of n-InSb under hydrostatic pressure has been reported by Wasilewski et al. [6, 7].

Our purpose in the present investigation is to study the intraconduction band cyclotron resonance and electron spin resonance of InSb under hydrostatic pressure up to 1 GPa and to compare the experimental results with calculations using the set of band

---

[1]) 166X, 38042 Grenoble Cedex, France.

[2]) On leave of absence from the High Pressure Research Center Unipress, Warsaw, Poland.

[3]) On leave of absence from the Institute of Experimental Physics, The University of Warsaw, Poland.

parameters from Goodwin and Seiler [2]. It is particularly interesting to check if the pressure dependence of optical transitions can be explained by the known modification of the band gap with pressure [8].

2. Theory

The calculations of Landau levels in nearly degenerated bands are based upon the three-band Pidgeon and Brown model [3]. The 8×8 effective mass Hamiltonian treats exactly the interaction between conduction (Γ_6) and valence (Γ_7, Γ_8) bands, the interaction with other bands is introduced by second-order perturbation theory. Weiler [4] reduced the full Hamiltonian to two 4×4 matrices (one for each spin state) and presented a complete list of selection rules for intraconduction band processes. The cyclotron resonance ($\Delta n = +1$, $\Delta s = 0$) is allowed in Faraday configuration for σ_L polarized light for all the magnetic field orientations with respect to the crystallographic axis. Due to the inversion asymmetry of the crystal the electron spin resonance ($\Delta n = 0$, $\Delta s = -1$) is in particular allowed for the same configuration with a magnetic field parallel to the {111} crystallographic direction.

3. Experimental

Measurements were performed with an optically pumped far-infrared laser; we used laser lines between 7.6 ($\lambda = 163$ μm) and 32.7 meV ($\lambda = 37.9$ μm). The n-InSb samples were prepared from n-type material with a free carrier concentration of 1×10^{14} cm^{-3} and a mobility of 5.5×10^5 cm^2/Vs at 77 K. The samples were thin platelets (thickness = 0.1 mm) with an irradiated area of 1×3 mm^2. After etching in a bromine–methanol solution, four contacts to the sample were made by soldering thin copper wires with small indium dots. The sample was immersed in a liquid pressure cell with a sapphire optical window. The magnetic field provided by a Bitter magnet was parallel to the {111} crystallographic direction.

We use the photoconductivity technique which provides a very effective tool for the investigation of pure samples [9]. The current was kept at values low enough to work in the ohmic region of the current density electric field. With a free carrier concentration of 1×10^{14} cm^{-3} at the temperature and magnetic fields of observation the electrons are all in the lowest ($n = 0^+$) Landau level or in associated impurity states.

4. Results and Discussion

Typical photoconductivity spectra obtained at pressures up to 1 GPa are shown in Fig. 1. For a given pressure the peak at low magnetic field is the "impurity-shifted cyclotron resonance" (ICR) and corresponds to the intra-impurity level transition $(000)_+ \rightarrow (010)_+$ (notations are from [10]). The component at high magnetic field is the intraconduction band $0^+ \rightarrow 1^+$ cyclotron resonance (CR). As far as the fit between experimental data and theoretical calculations is concerned, we shall be interested only in the CR process. Davidson et al. [11] and Wasilewski et al. [6, 7] observed an important narrowing of the ICR and CR lines upon applying pressure. The narrowing observed in the present work is less marked.

Fig. 2 is a fan chart of the CR peaks observed at various pressures and for different incident radiations. The full lines represent theoretical results calculated using Weiler's model [4] with Goodwin and Seiler's parameters [2]. In order to take into account the change in energy with pressure we assume that the only parameter affected by pressure is the gap E_g. The best fit was obtained with a value of $dE_g/dP = (0.140 \pm 0.004)$ meV/MPa. It is reasonable to neglect the variations of the spin–orbit interaction Δ and of the momentum interband matrix element P [12 to 14]. The agreement between

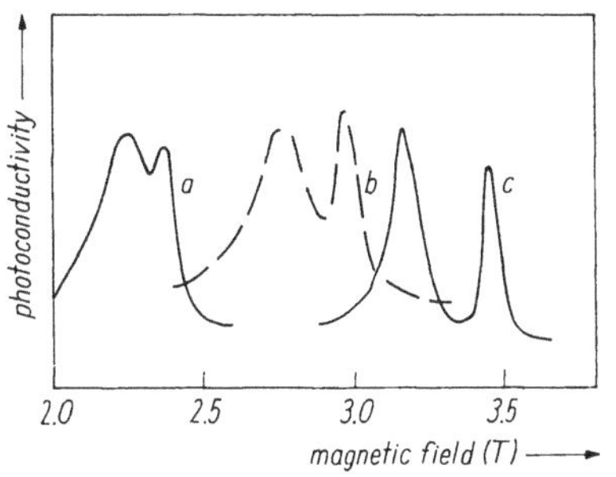

Fig. 1

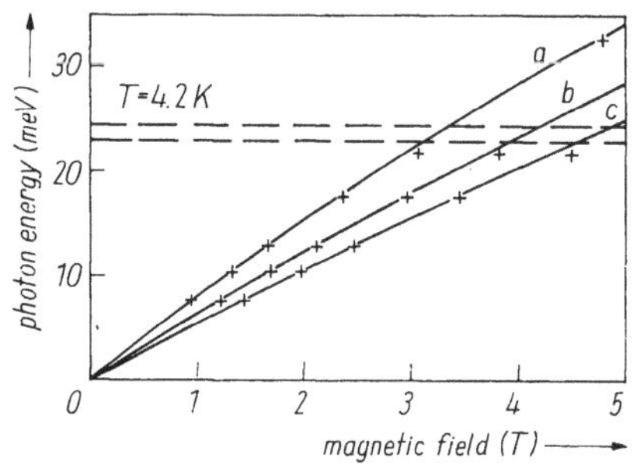

Fig. 2

Fig. 1. Cyclotron resonance photoconductivity spectra of n-InSb; $T = 4.2$ K; $\lambda = 70.51$ μm. Laser energy: 17.58 meV. (a) $P = 0$, (b) 545 MPa, (c) 1 GPa

Fig. 2. Fan chart of cyclotron resonance processes at different pressures. (a) $P = 0$, (b) 545 MPa, (c) 1 GPa. + experimental results, full lines model calculations, dashed lines optical phonon energies

experimental results and calculations is good in the pressure range we studied for all the transition energies. The slight deviation one observes for a photon energy of 21.7 meV is due to the increase of the effective mass caused by resonant polaron effect, i.e. the interaction between the free carrier motion and optical phonons (see [15] for optical phonon energies).

Fig. 3 and 4 display the experimental results concerning electron spin resonance (ESR) for two pressures. At 300 MPa each spectrum exhibits two sharp peaks with a half-width lower than 0.02 T. At 650 MPa, we observed three, also very sharp, com-

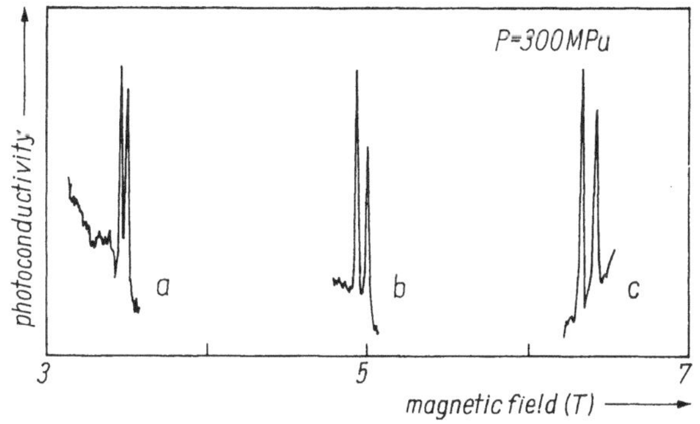

Fig. 3. Electron spin resonance photoconductivity spectra of n-InSb under a pressure of 300 MPa. Laser energy: (a) 7.60, (b) 10.43, (c) 12.85 meV

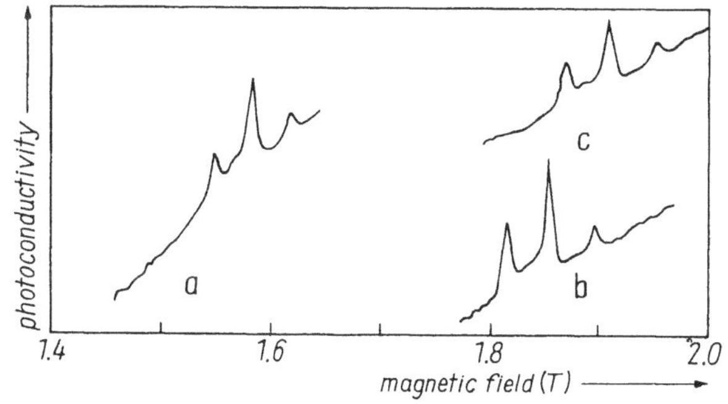

Fig. 4. Electron spin resonance photoconductivity spectra of n-InSb under a pressure of 650 MPa. Laser energy: (a) 3.05, (b) 3.49, (c) 3.66 meV

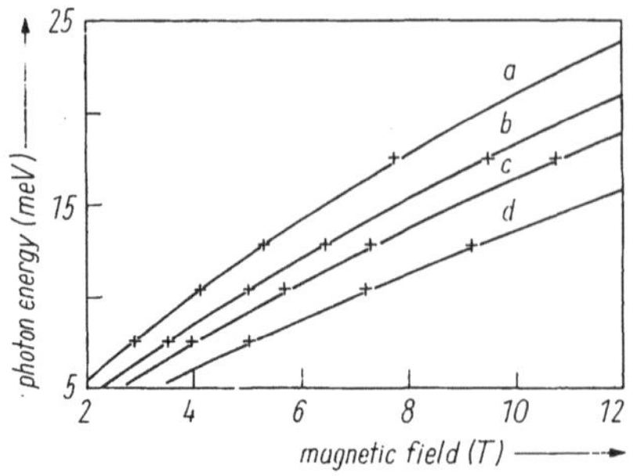

Fig. 5. Fan chart of electron spin resonance transitions under different pressures; $T = 4.2$ K. (a) $P = 0$, (b) 300 MPa, (c) 545 MPa, (d) 1 GPa. + experimental results, full lines model calculations

ponents (note the scale change between Fig. 3 and 4). The ESR in InSb has been studied at atmospheric pressure by McCombe [16] and under uniaxial stress by Kuchar and Kriechbaum [5]. These authors also observed two or three components according to experimental conditions. In the two-component spectra the peak at high magnetic field corresponds to $0^+ \to 0^-$ free carrier spin resonance transition, the other peak is associated with impurity states. In the three-component spectra the central peak is due to the $0^+ \to 0^-$ process, the other two peaks are impurity related. In Fig. 5 we compare the experimental results concerning the $0^+ \to 0^-$ ESR with calculations performed with the same parameters set used for cyclotron resonance. Here again a good fit to all data is very apparent.

The gap modification with pressure we adopted $dE_g/dp = (0.140 \pm 0.004)$ meV/MPa is in very good agreement with the values measured by laser emission which lead to changes between 0.140 and 0.142 meV/MPa [8]. We can conclude that the set of band parameters from Goodwin and Seiler [2] explains the intraconduction band CR and ESR processes of InSb under hydrostatic pressure up to 1 GPa. The accuracy we quote for $dE_g/dP(\pm$ 0.004 meV/MPa) is mainly limited by the sensibility of model calculations upon changes of that coefficient. We can estimate that in the pressure range 0 to 1 GPa, the variations of m_c^* and g_c^* can be calculated with an accuracy of about 3% using three-band model formulae (see for instance [4]).

Acknowledgement

Two of us, L. D. and M. B. have the pleasure to acknowledge the financial support they received from the "Université Scientifique et Médicale de Grenoble".

References

[1] G. DRESSELHAUS, A. F. KIP, C. KITTEL, and G. WAGONER, Phys. Rev. **98**, 556 (1955).
[2] M. W. GOODWIN and D. G. SEILER, Phys. Rev. B **27**, 3451 (1983).
[3] C. R. PIDGEON and R. N. BROWN, Phys. Rev. **146**, 575 (1966).
[4] M. H. WEILER, R. L. AGGARWAL, and B. LAX, Phys. Rev. B **17**, 3269 (1978).
M. H. WEILER, in: Semiconductors and Semimetals, Vol. 16, Ed. R. K. WILLARDSON and A. B. BEER, Academic Press, Inc., 1981 (p. 119).
[5] F. KUCHAR and M. KRIECHBAUM, in: Proc. Internat. Conf. Application of High Magnetic Fields in Semicond. Phys., Grenoble 1982, Springer Verlag, 1983 (p. 309).
[6] Z. WASILEWSKI, A. M. DAVIDSON, R. A. STRADLING, and S. POROWSKI, in: Proc. Internat. Conf. Application of High Magnetic Fields in Semicond. Phys., Grenoble 1982, Springer Verlag, 1983 (p. 233).
[7] Z. WASILEWSKI, A. M. DAVIDSON, P. KNOWLES, S. POROWSKI, and R. A. STRADLING, in: Proc. 4th Internat. Conf. Physics of Narrow Gap Semiconductors, Linz 1981, Springer Verlag, 1983 (p. 183).
[8] N. MENYUK, A. S. PINE, J. A. KAFALAS, and A. J. STRAUSS, J. appl. Phys. **45**, 3477 (1974).

[9] R. GRISAR, H. WACHERNIG, G. BAUER, J. WLASAK, J. KOWALSKI, and W. ZAWADZKI, Phys. Rev. B **18**, 4355 (1978).
[10] W. ZAWADZKI and J. WLASAK, in: Proc. 14th Internat. Conf. Phys. Semicond., Edinburgh 1978 (p. 413).
[11] A. M. DAVIDSON, P. KNOWLES, P. MAKADO, R. A. STRADLING, S. POROWSKI, and Z. WASILEWSKI, in: Proc. Internat. Seminar Physics in High Magnetic Fields, Oji 1980, Springer Verlag, 1981 (p. 84).
[12] D. E. ASPNES and M. CARDONA, Phys. Rev. B **17**, 726 (1978).
[13] G. MARTINEZ, in: Hdb. Semicond., Vol. 2, Ed. M. BALKANSKI, North-Holland Publ. Co., 1980 (p. 181).
[14] I. GORCZYCA, phys. stat. sol. (b) **112**, 97 (1982).
[15] S. S. MITRA and N. E. MASSA, in: Hdb. Semicond., Vol. 1, Ed. W. PAUL, North-Holland Publ. Co., 1982 (p. 81).
[16] B. D. MCCOMBE, in: Proc. Internat. Conf. Application of High Magnetic Fields in Semicond. Phys., Würzburg 1974, Lecture Notes (p. 146).

(Received March 8, 1984)

phys. stat. sol. (b) **125**, 221 (1984)

Subject classification: 13.1; 22.2.1; 22.2.3

Department of Theoretical Physics, University of Havana[1])

Fundamental Zero Energy Gap States in Superlattices

By

M. DE DIOS LEYVA, R. P. ALVAREZ, and J. L. GONDAR

The band structure of a model superlattice, in which the conduction band bottom in the first host semiconductor is located very close to or even below the valence band top in the other host semiconductor, is analyzed as a function of layer thickness and of the conduction band–valence band edge discontinuity. The analysis is carried out in the envelope-function approximation and in Kane's two-band framework. It is shown that with increasing magnitude of the discontinuity (or of layer thickness), the nature of the superlattice changes from a direct gap semiconductor to a direct zero gap semiconductor. Furthermore, it is also shown that in the direct zero energy gap states the valence and conduction bands may touch each other simultaneously over several circumferences in the Brillouin zone of the superlattice.

Зонная структура сверхрешетки, где край валентной зоны одного из полупроводников, составляющих сверхрешетку может быть расположен выше чем край зоны проводимости другого, рассматривается как функция характерных параметров сверхрешетки (толщина слоев и максимальное значение потенциальной энергии). Рассмотрение было сделано в приближениии эффективной массы и в рамках двухзонной модели Кейна. Показано, что при возрастании указанных выше параметров, характер сверхрешетки меняется от полупрводника с прямой запрещенной зоной до бесщелевого полупроводника. Кроме того в бесщелевых состояниях валентная зона и зона проводимости могут одновременно касаться по нескольким окружностям в зоне Бриллюэна сверхрешетки.

1. Introduction

In [1] it was pointed out that those superlattices, in which the conduction band edge in one host semiconductor lies below the valence band top in the other host semiconductor, form a new, special type of superlattices potentially important in practical applications. To this type belongs the InAs–GaSb superlattice.

Physical consideration [2], using the electron confinement concept, suggests that the semiconductor–semimetal transition is a characteristic phenomenon in these superlattices. This phenomenon has been established theoretically as well as experimentally in the InAs–GaSb superlattice: it was found in [2], based on the LCAO (linear combination of atomic orbitals) method, and in [3] it was observed that such a superlattice is a semiconductor when the layers of the constituent materials are thin, and becomes a semimetal when the layer thickness is increased; in [4], using the tight-binding method, it was shown that with increasing magnitude of the conduction band–valence band edge discontinuity, the nature of the superlattice changes from a direct gap semiconductor to a semimetal. However, it is incorrect to assert categorically the semiconductor–semimetal transition to be a common fact present in the above-mentioned superlattice type. In this connection there is not a rigorous demonstration: that is why we shall present, in detail, the analysis of this problem, using a model superlattice belonging to the type considered in [1].

---

[1]) San Lázaro y L, Habana, Cuba.

The analysis is carried out in the envelope-function approximation and in the two-band Kane model. In [5] was reported that the results obtained by means of this approximation are in excellent agreement with LCAO for the InAs–GaSb superlattice.

2. Effective Hamiltonian. Equation of Motion

Our theoretical calculations will be carried out on the basis of a model superlattice consisting of alternating layers of A and B host semiconductors which have band structures given by the simplified two-band Kane model (Fig. 1). The superlattice period d is the sum of the thicknesses a and b of the alternating layers. Along the z axis the energy was taken equal to zero; ε_A and ε_B are the energy gaps in A and B, respectively. The magnitudes V_A and V_B determine the valence and conduction band edge displacements for both materials.

Following Bastard [5] we neglect any phenomenon if rapidly varying on the atomic scale, and also assume the interface potential to be strongly localized in the vicinity of the A–B interfaces. These conditions allow the dispersion relation to be derived in the envelope-function approximation with a potential energy given by $V(z + d) = V(z)$ and

$$V(z) = \begin{cases} V_A & \text{if} \quad -a < z < 0, \\ V_B & \text{if} \quad 0 < z < b. \end{cases} \tag{1}$$

In this approximation the essential point is to draw attention to the effective Hamiltonian which governs the slowly varying envelope functions $F_j(\boldsymbol{r})$ in the superlattice. This Hamiltonian is the sum of the effective kinetic energy operator $D_{ij}(\boldsymbol{k}, z)$ and the potential energy $V(z)\,\delta_{ij}$, where δ_{ij} is the Kronecker symbol and $\boldsymbol{k} = -i\nabla$.

To describe the electron dynamics in the framework of the simplified two-band Kane model we will use, in each host material, the operator D_{ij} reported in [6], which has the same form as the kinetic energy operator in the relativistic Dirac equation,

$$D = P\left\{\begin{bmatrix} 0 & \sigma_x \\ \sigma_x & 0 \end{bmatrix}\hat{k}_x + \begin{bmatrix} 0 & \sigma_y \\ \sigma_y & 0 \end{bmatrix}\hat{k}_y + \begin{bmatrix} 0 & \sigma_z \\ \sigma_z & 0 \end{bmatrix}\hat{k}_z\right\} + \frac{1}{2}\,\varepsilon_G(z)\begin{bmatrix} I & 0 \\ 0 & -I \end{bmatrix}. \tag{2}$$

In (2) $\sigma_{x,y,z}$ are the Pauli matrices [7]; I is the 2×2 unit matrix; $\varepsilon_G(z)$ is a function with period d equal to ε_A and ε_B in A and B layers, respectively. It has been assumed that the parameter (matrix element) P is identical in both A and B host materials.

Using (1) and (2) the equation of motion may be written as

$$[D(\boldsymbol{k}, z) + V(z)\,I_4]\,F(\boldsymbol{r}) = EF(\boldsymbol{r}), \tag{3}$$

where E is the energy, I_4 the 4×4 unit matrix, and $F(\boldsymbol{r})$ a column matrix made up of four envelope functions $F_j(\boldsymbol{r})$. From (3) it follows that $\boldsymbol{k}_\perp = (k_x, k_y)$ is a good

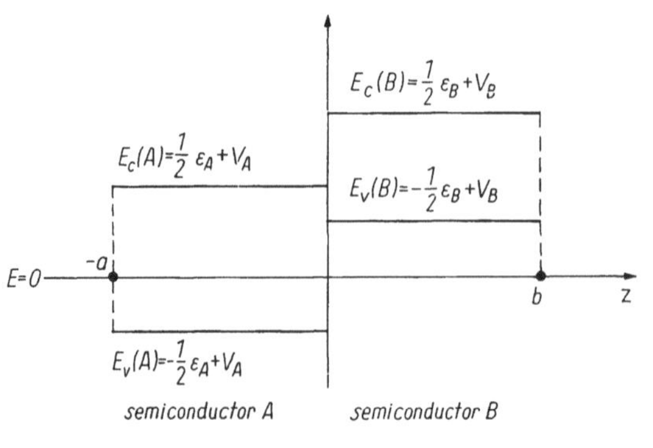

Fig. 1. The arrangement of the band edges of A and B host semiconductors with respect to a common energy reference. Along the z-axis the energy was taken equal to zero

quantum number, and then we can look for solutions of the form $F(\boldsymbol{r}) = \Phi(z) \times \times \exp i(xk_x + yk_y)$, where the function $\Phi(z)$ satisfies (3) replacing $\hat{k}_{x,y}$ by $k_{x,y}$.

Performing in (3) the unitary transformation $\varphi(z) = U\Phi(z)$ where $U = \frac{1}{\sqrt{2}}\begin{bmatrix} M & 0 \\ 0 & M^+ \end{bmatrix}$ and $M = I + i\sigma_z$, the function $\varphi(z)$ will be determined by the equation

$$H\varphi = \left\{P\begin{bmatrix} 0 & I \\ I & 0 \end{bmatrix}\hat{K} + P\begin{bmatrix} 0 & iI \\ -iI & 0 \end{bmatrix}\hat{k}_z + \frac{1}{2}\,\varepsilon_G(z)\begin{bmatrix} I & 0 \\ 0 & -I \end{bmatrix} + V(z)\,I_4\right\}\varphi = E\varphi\,, \tag{4}$$

where

$$\hat{K} = \begin{bmatrix} \sigma_x & 0 \\ 0 & \sigma_x \end{bmatrix} k_x + \begin{bmatrix} \sigma_y & 0 \\ 0 & \sigma_y \end{bmatrix} k_y\,.$$

The fact that operators H and $\hat{K}$ commute allows us to replace in (4) $\hat{K}$ by its eigenvalues $K = \pm\sqrt{k_x^2 + k_y^2}$, which lead the functions $\begin{bmatrix} \varphi_1 \\ \varphi_3 \end{bmatrix}$ and $\begin{bmatrix} \varphi_2 \\ \varphi_4 \end{bmatrix}$ to satisfy the same system of equations

$$\begin{bmatrix} \frac{1}{2}\,\varepsilon_G(z) + V(z) & PK + iP\hat{k}_z \\ PK - iP\hat{k}_z & -\frac{1}{2}\,\varepsilon_G(z) + V(z) \end{bmatrix}\begin{bmatrix} \psi_1 \\ \psi_2 \end{bmatrix} = E\begin{bmatrix} \psi_1 \\ \psi_2 \end{bmatrix}. \tag{5}$$

This relation incorporates finite $\boldsymbol{k}_\perp$ in a simple way.

Moreover it may be used to obtain the band structure for several superlattices. For example, if in PbTe–$Pb_{1-x}Sn_xTe$ superlattices [8] only the interaction between L_6 and $L_{6'}$ bands is taken into account, and the anisotropy of the bands in each host semiconductor is neglected, then the effective Hamiltonian is given by (2) [9]. It is not difficult to take into account the anisotropy in (5). If in GaAs–GaAlAs, InAs–GaSb, and InSb–CdTe superlattices only the interaction between Γ_6 and Γ_8 bands in each host material is taken into account, then the equation for the envelope functions of electrons and light holes coincides with (5) for $\boldsymbol{k}_\perp = \boldsymbol{0}$ [10].

3. Dispersion Relation

To obtain the dispersion relation we assume that the envelope functions in (5) are continuous functions satisfying the Bloch theorem. The first condition allows to consider the case where the interaction of the conduction band in one host semiconductor with the valence band of the other host semiconductor plays an important role [1]. The second condition allows us to write

$$\begin{bmatrix} \psi_1(z) \\ \psi_2(z) \end{bmatrix} = e^{ikz}\begin{bmatrix} u_1(z) \\ u_2(z) \end{bmatrix}, \tag{6}$$

where $\boldsymbol{k}$ is the wave vector along the superlattice axis, and $u_{1,2}(z)$ are periodical functions with period d.

The general solution of (5) in each host material $j =$ A, B, is given by

$$\begin{bmatrix} u_1^{(j)}(z) \\ u_2^{(j)}(z) \end{bmatrix} e^{ikz} = a_j\begin{bmatrix} 1 \\ \gamma_j \end{bmatrix} e^{izk_j} + b_j\begin{bmatrix} 1 \\ \beta_j \end{bmatrix} e^{-izk_j}\,. \tag{7}$$

In (7) a_j, b_j are constants (four) and

$$\gamma_j = \frac{E - V_j - \frac{1}{2}\varepsilon_j}{PK + iPk_j}; \qquad \beta_j = \frac{E - V_j - \frac{1}{2}\varepsilon_j}{PK - iPk_j};$$

$$k_j = \frac{1}{P}\sqrt{(E - V_j)^2 - (\tfrac{1}{2}\varepsilon_{j,\mathrm{eff}})^2}; \quad \tfrac{1}{2}\varepsilon_{j,\mathrm{eff}} = \sqrt{(\tfrac{1}{2}\varepsilon_j)^2 + P^2K^2}. \tag{8}$$

The functions $u_{1,2}^{(j)}(z)$ satisfy the boundary conditions

$$u_{1,2}^{(\mathrm{A})}(0) = u_{1,2}^{(\mathrm{B})}(0); \qquad u_{1,2}^{(\mathrm{A})}(-a) = u_{1,2}^{(\mathrm{B})}(b). \tag{9}$$

Substituting (7) into (9) and solving the resulting determinantal equation for the coefficients a_j and b_j gives the following dispersion relation:

$$\cos(a + b)\,k = \cos(ak_\mathrm{A})\cos(bk_\mathrm{B}) - Q\sin(ak_\mathrm{A})\sin(bk_\mathrm{B}), \tag{10}$$

where

$$Q = \frac{1}{2k_\mathrm{A}k_\mathrm{B}}\left\{\frac{E - V_\mathrm{A} - \frac{1}{2}\varepsilon_\mathrm{A}}{E - V_\mathrm{B} - \frac{1}{2}\varepsilon_\mathrm{B}}(K^2 + k_\mathrm{B}^2) + \right.$$
$$\left. + \frac{E - V_\mathrm{B} - \frac{1}{2}\varepsilon_\mathrm{B}}{E - V_\mathrm{A} - \frac{1}{2}\varepsilon_\mathrm{A}}(K^2 + k_\mathrm{A}^2) - 2K^2\right\}. \tag{11}$$

If we take the energy zero at the bottom of the A conduction band, this dispersion relation coincides with the one reported by Bastard [5] for the InAs–GaSb superlattice when $K = 0$. Furthermore, it is quite different from the one reported in [11] by Mukherji and Nag.

4. Fundamental Zero Energy Gap States

Relation (10) is a transcendental equation which gives E as an even function of k and K. Generally, in solving it numerical calculations are required. It is possible, however, to make a detailed qualitative analysis providing a clear picture about the possibility of a semiconductor–semimetal transition to take place in the model superlattice considered. For simplicity we choose a particular set of parameter values for the superlattice. Next we will see that the obtained qualitative results are analogous for an arbitrary choice of these values.

4.1 Symmetrical case

In this case $V_\mathrm{B} = -V_\mathrm{A} = V_0 > 0$, $\varepsilon_\mathrm{A} = \varepsilon_\mathrm{B} = \varepsilon_\mathrm{g}$, and $a = b$. Then

$$k_\mathrm{A} = \frac{1}{P}\sqrt{(E + V_0)^2 - (\tfrac{1}{2}\varepsilon_{\mathrm{g,eff}})^2}, \qquad k_\mathrm{B} = \frac{1}{P}\sqrt{(E - V_0)^2 - (\tfrac{1}{2}\varepsilon_{\mathrm{g,eff}})^2},$$

$$Q = \frac{E^2 - V_0^2 - (\frac{1}{2}\varepsilon_{\mathrm{g,eff}})^2}{P^2k_\mathrm{A}k_\mathrm{B}}. \tag{12}$$

With (12), (10) is invariant under the change $E \to -E$, i.e., the energy spectrum has inversion symmetry with respect to the plane $E = 0$. This means that the lowest conduction band and the highest valence band may touch each other only at the plane $E = 0$. This fact determines the occurrence of a fundamental zero energy gap (direct).

Setting $E = 0$ in (10) and (12) we obtain

$$\cos(2ak) = 1 + \frac{\frac{1}{2}\,\varepsilon_{g,\mathrm{eff}}^2}{V_0^2 - \left(\frac{\varepsilon_{g,\mathrm{eff}}}{2}\right)^2} \sin^2(aq)\,, \tag{13}$$

where

$$q = \frac{1}{P}\sqrt{V_0^2 - \left(\frac{\varepsilon_{g,\mathrm{eff}}}{2}\right)^2}\,.$$

From (13) it follows that the fundamental zero energy gap states may occur only for $V_0 > \frac{1}{2}\,\varepsilon_{g,\mathrm{eff}}$, $k = 0$ (in the reduced scheme) and $aq = n\pi$; where $n = 1, 2, \ldots$ The latter condition gives

$$V_0(n, a, K) = \sqrt{\left(\frac{\varepsilon_g}{2}\right)^2 + P^2K^2 + P^2\left(\frac{n\pi}{a}\right)^2} \quad (n = 1, 2, \ldots)\,. \tag{14}$$

The values of V_0 and a satisfying (14) will be called henceforth critical values. Then, from (14) it follows that for fixed values of a and K there exists a numerable set of critical values of V_0. This means that in each plane perpendicular to the K-axis the valence and conduction bands approach each other as V_0 increases from zero, until they touch each other at $(E, k, K) = (0, 0, K)$ and then move away. This process will occur a numerable set of times. Using (10) it is not difficult to prove that in each of such planes the conduction band bottom and the valence band top occur at $k = 0$. In Fig. 2 the numerical results are shown for the distance $E_g(0)$ (in ε_g units) from the valence band edge to the conduction band edge of the superlattice as a function of $v = V_0/\frac{1}{2}\varepsilon_g$ when $K = 0$. In this calculations, as well as in the following, we take $a\varepsilon_g/2P = 1$.

On the other hand, for fixed n and a, $V_0(n, a, K)$ is a continuous and increasing function of K, which reaches its minimum value at $K = 0$. The minimum critical value of V_0 is, therefore, $V_c = V_0(1, a, 0) = \sqrt{(\varepsilon_g/2)^2 + P^2(\pi/a)^2}$.

The fundamental zero energy gap state corresponding to this critical value occurs at $k = K = 0$. From (14) it follows that any value $V_0 \geqq V_c$ is a critical value and, hence, the superlattice is a direct zero energy gap semiconductor for $V_0 \geqq V_c$ and a direct gap semiconductor for $V_0 < V_c$.

Identical results may be obtained if V_0 is fixed (greater than $\frac{1}{2}\,\varepsilon_g$) and a varying. In this case the superlattice is a direct zero energy gap semiconductor for $a \geqq a_c = \pi P/\sqrt{V_0^2 - (\frac{1}{2}\,\varepsilon_g)^2}$.

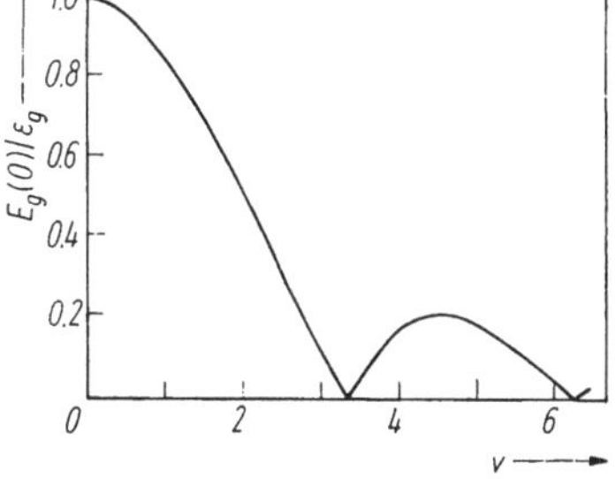

Fig. 2. Distance from the valence band edge to the conduction band edge of the superlattice as a function of $v = V_0/\frac{1}{2}\,\varepsilon_g$ when $K = 0$

4.2 General case

Now we shall consider briefly the case in which an arbitrary choice of the superlattice parameters is made. It is known that zero energy gap conditions, excluding those related to the periodicity of the effective mass [12], are $ak_A = n_1\pi$ and $bk_B = n_2\pi$, where $n_1, n_2 = 1, 2, \dots$ [13]. The corresponding zero energy gap states are direct gap states and, for a given K, it follows from (10) that these states occur at $k = 0$ if n_1 and n_2 have the same parity (both even or odd), and occur at $k = \pi/(a + b)$ if n_1 and n_2 have opposite parity.

Using (8) the zero energy gap conditions take the form

$$E = V_A \pm \sqrt{(\tfrac{1}{2}\,\varepsilon_{A,\mathrm{eff}})^2 + P^2\left(\frac{n_1\pi}{a}\right)^2}\,,$$

$$E = V_B \pm \sqrt{(\tfrac{1}{2}\,\varepsilon_{B,\mathrm{eff}})^2 + P^2\left(\frac{n_2\pi}{b}\right)^2}\,. \tag{15}$$

Furthermore, the fundamental zero energy gap states occur at $k = 0$ and when the valence band edge of one of the host semiconductors lies above the conduction band edge of the other host semiconductor. Additionally, k_A and k_B being real quantities, the energies of these states are localized between the mentioned edges. If we choose (without loss of generality) $V_A = 0$ and $V_B > 0$ the energies of these states are obtained taking the plus sign in the first relation of (15) and the minus sign in the second one. Eliminating E from these two relations we obtain the following relation for the critical values:

$$V_B(n_1, n_2, a, b, K) = \sqrt{(\tfrac{1}{2}\,\varepsilon_{A,\mathrm{eff}})^2 + P^2\left(\frac{n_1\pi}{a}\right)^2} + \sqrt{(\tfrac{1}{2}\,\varepsilon_{B,\mathrm{eff}})^2 + P^2\left(\frac{n_2\pi}{b}\right)^2}\,. \tag{16}$$

From here it is not difficult to arrive at analogous conclusions as in the symmetrical case. We have only to notice that fundamental zero energy gap states are obtained when $n_1 = n_2$.

5. Reconstruction of the Energy Spectrum

Now we have a geometrical picture about the changes experienced by the energy spectrum of the superlattices as a function of its parameters. We restrict our analysis to the symmetrical case when a is fixed and V_0 varies. A similar analysis may be done for fixed V_0 and variable a.

From (10) and (12) it follows that for $V_0 \to 0$ the band structure of the superlattice is given by the simplified two-band Kane model. If V_0 increases the valence and conduction bands approach each other and finally touch, for the first time, at the centre of the Brillouin zone of the superlattice, for $V_0 = V_c$. In Fig. 3 the numerical results for this process are shown for $K = 0$ and various values of $v = V_0/\frac{1}{2}\varepsilon_g$, with $\varepsilon = E/\frac{1}{2}\varepsilon_g$.

If V_0 increases further, but remaining lower than $V_2 = V_0(2, a, 0)$, the bands move away from each other in the centre of the Brillouin zone, but still touch each other at points $(k, K) = (0, \pm K_1)$ (K_1 depends on V_0 and is obtained from (14) for $n = 1$). These two contact points move away from the centre of the zone as V_0 increases. When $V_0 = V_2$ there will be three contact points. One of them is at $k = K = 0$ and the location of the other two is determined from (14) for $V_2 = V_0(1, a, K)$.

If V_0 increases even more, but remaining lower than $V_0(3, a, 0)$, there will be four contact points along the K-axis, that move away from the centre of the Brillouin zone

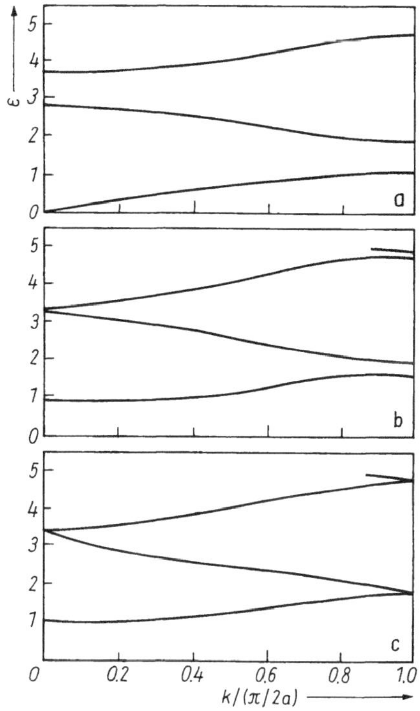

Fig. 3

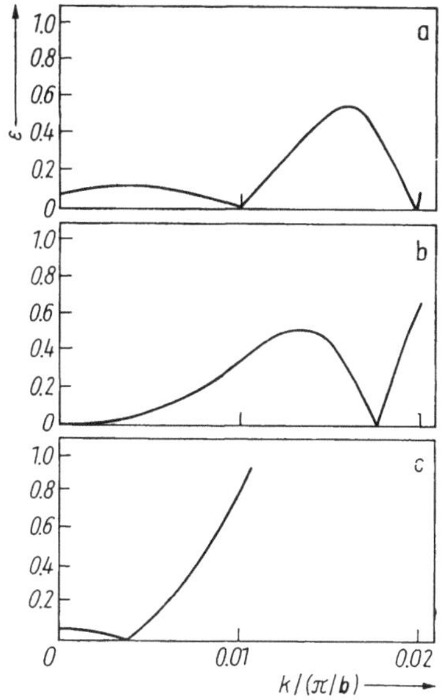

Fig. 4

Fig. 3. Electron energy $\varepsilon = E/\frac{1}{2}\varepsilon_g$ as a function of the wave number k when $K = 0$ and for various values of $v = V_0/\frac{1}{2}\varepsilon_g$. a) $v = V_c$, b) 0.5, c) 0

Fig. 4. Electron energy $\varepsilon = E/\frac{1}{2}\varepsilon_g$ as a function of K when $k = 0$ and for various values of $v = V_0/\frac{1}{2}\varepsilon_g$. a) $v = 7.0$, b) v_2, c) 3.5

as V_0 increases. The position of the most remote points may be obtained from (14) with $n = 1$, and that of the nearest points setting $n = 2$.

We can continue the analysis of the spectrum reconstruction as far as it is possible to increase V_0.

Fig. 4 shows the shape of the conduction band at $k = 0$, for three different values of $v = V_0/\frac{1}{2}\varepsilon_g$. The numerical calculation was made considering $\varepsilon = E/\frac{1}{2}\varepsilon_g$; $b = 10^{-2}a$, and $v_2 = V_2/\frac{1}{2}\varepsilon_g$. The fundamental zero energy gap states occur at the contact points of these curves with the K-axis.

5. Conclusions

The results obtained in this paper bring a clear picture of some of the properties which may have those superlattices where the valence band edge in one host semiconductor lies above the conduction band edge of the other host semiconductor.

Firstly, in the model superlattice considered, the semiconductor–semimetal transition, which is characteristic of the InAs–GaSb superlattice, does not really occur. Instead the direct gap semiconductor–direct zero energy gap semiconductor transition takes place. For the semiconductor–semimetal transition to be present in this model superlattice, the existence of a third valence band, such that its upper edge exceeds the lower edge of the conduction band when the superlattice parameters increase enough, is required. In the InAs–GaSb superlattice this third band is originated by the heavy-hole bands of the host materials.

Second, up to now induced zero gap energy states have been obtained for certain discrete values of some parameters: chemical composition in substitutional alloys, pressure in the case of semiconductors subject to this effect, etc. In superlattices, as we have shown, it is possible to obtain such states for a continuous range of values of the superlattice parameters. This is an interesting property which opens practical possibilities in the technological feasibility of zero energy gap semiconductors.

Finally, the band structure of superlattices, when the values of its characteristic parameters are greater than the critical values, shows unusual properties: valence and conduction bands may touch each other at several points along the K-axis or, equivalently, at more than one circumference in the (k_x, k_y) plane of the Brillouin zone. These properties, of course, suggest the need of carrying out a detailed theoretical study of different relevant physical quantities in this range of values of the superlattice parameters.

Acknowledgement

We wish to express our thanks to our colleages of the Department of Theoretical Physics, University of Havana, for helpful discussions.

References

[1] G. A. SAI-HALASZ, R. TSU, and L. ESAKI, Appl. Phys. Letters **30**, 651 (1977).
[2] G. A. SAI-HALASZ, L. ESAKI, and W. A. HARRISON, Phys. Rev. B **18**, 2812 (1978).
[3] L. L. CHANG, N. KAWAI, G. A. SAI-HALASZ, R. LUDEKE, and L. ESAKI, Appl. Phys. Letters **35**, 939 (1979).
[4] R. N. NUCHO and A. MADHUKAR, J. Vacuum Sci. Technol. **15**, 1530 (1978).
[5] G. BASTARD, Phys. Rev. B **24**, 5693 (1981).
[6] L. V. KELDYSH, Zh. eksper. teor. Fiz. **45**, 364 (1963).
[7] L. J. SCHIFF, Quantum Mechanics, 2. edition, McGraw-Hill Publ. Co., New York 1955 (p. 325).
[8] H. KINOSHITA and H. FUJIYASU, J. appl. Phys. **51**, 5845 (1980).
[9] G. MARTÍNEZ, Phys. Rev. B **8**, 4678 (1973).
[10] G. E. MARQUES and L. J. SHAM, Surface Sci. **113**, 131 (1982).
[11] D. MUKHERJI and B. R. NAG, Phys. Rev. B **12**, 4338 (1975).
[12] V. MILANOVIĆ and D. TJAPKIN, phys. stat. sol. (b) **110**, 687 (1982).
[13] S. T. LIN and J. SMIT, Amer. J. Phys. **48**, 193 (1980).

(Received March 27, 1984)

phys. stat. sol. (b) **125**, 229 (1984)

Subject classification: 13.1; 12.2; 22.2.2; 22.2.3

High Pressure Research Center, "Unipress", Polish Academy of Sciences, Warsaw[1])

Calculation of the Effect of Hydrostatic Pressure on the Band Structure of GaP and GaSb

By

Izabela Gorczyca

The band structure of GaP and GaSb is computed by the empirical pseudopotential method (EPM) including a nonlocal correction. The pressure dependence of the energy gaps and the wave functions are then calculated for these materials.

Die Bandstruktur von GaP and GaSb wird mittels empirischer Pseudopotentialmethode (EPM) einschließlich einer nichtlokalen Korrektur numerisch berechnet. Die Druckabhängigkeit der Energielücke und die Wellenfunktionen werden dann für diese Materialien berechnet.

1. Introduction

In the previous paper [1] the band structure of InSb in the Γ-, L-, and X-points of the Brillouin zone was calculated by the EPM with the nonlocal correction. The method of the band structure calculation was based on the approach of Chelikowsky et al. [2]. The pressure dependence calculations were performed by the scheme of Tsay and co-workers [3 to 5], but we have included the spin-orbit effect and a new set of pseudopotential parameters for InSb.

We present here the results of similar calculations for GaP and GaSb. As shown in the previous paper, disagreement of the calculations of Tsay and co-workers with the experimental data is mainly related to the pseudopotential parameters which are used by them. Namely, Tsay and co-workers use the Bergstresser-Cohen [6] pseudopotential parameters which describe the band structure of InSb and other considered materials only in the "first approximation". A small change of the pseudopotential form factors and correcting of EPM by adding the nonlocal parameter give much better agreement of the band structure with experimental data. Consequently, pressure coefficients calculated with such a new set of pseudopotential parameters are in better agreement with those given experimentally. Therefore, the first step in the pressure-dependence calculation is to choose the best possible set of form factors which will properly describe the band structure under normal pressure. The experimentally known energy gaps in Γ-, X-, and L-points of the Brillouin zone are taken as criterion.

The band structure of GaP and GaSb at normal pressure is described in Section 2. In Section 3 we present the results of calculation of the energy band structure under hydrostatic pressure.

2. Band Structure at Normal Pressure

Theoretical calculations of the band structure of GaP and GaSb have been described by many authors. One of the earliest calculations has been done by the EPM by Cohen and Bergstresser [6]. The authors determine first sets of pseudopotential form factors (PFF's) for the considered materials. Later Walter and Cohen [7] corrected the PFF's for GaP by fitting them to well-known energy gaps. The same was done for GaSb

[1]) ul. Sokolowska 29, 01-142 Warsaw, Poland.

Table 1

Pseudopotential parameters used in the present calculation and given by Chelikowsky et al. [2, 9] (in Ryd)

	GaP		GaSb	
	Chelikowsky et al. [2]	present work	Chelikowsky and Cohen [9]	present work
$V^S(3)$	−0.249	−0.223	−0.220	−0.212
$V^S(8)$	0.017	0.032	0.005	0.011
$V^S(11)$	0.083	0.081	0.045	0.048
$V^A(3)$	0.081	0.119	0.040	0.050
$V^A(4)$	0.055	0.064	0.030	0.039
$V^A(11)$	0.030	0.034	0.000	0.004
	—	—	—	0.0007
	—	—	0.0011	0.0014
m^*/m	1.075	1.012	—	1.06

by Cahn and Cohen [8]. Chelikowsky et al. [2] introduced new sets of PFF's for this type of materials. The PFF's has been fitted by comparing the calculated density of states with XPS and UPS data. Chelikowsky et al. showed that discrepancies between theoretical and experimental results are completely removed only after introducing nonlocal corrections to the pseudopotential. Firstly [2] they take the simplest approach to get the lowest-order corrections by allowing the electron mass to vary, i.e. $m \to m^*$. In the second paper Chelikowsky and Cohen [9] performed a similar calculation but with the full nonlocal pseudopotential. There were also other methods applied for band structure calculations of GaP and GaSb. For example Chadi [10] used a

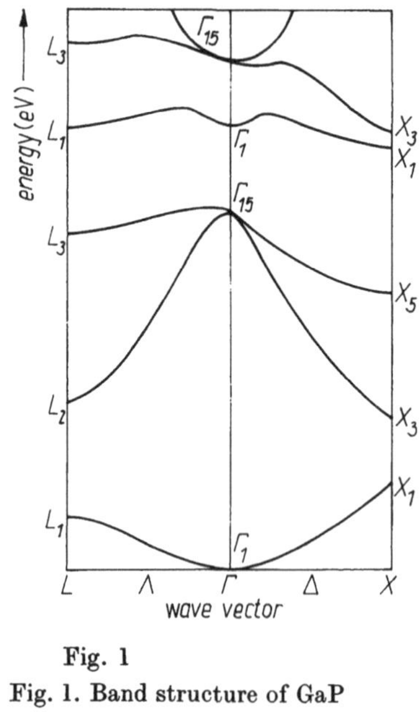

Fig. 1

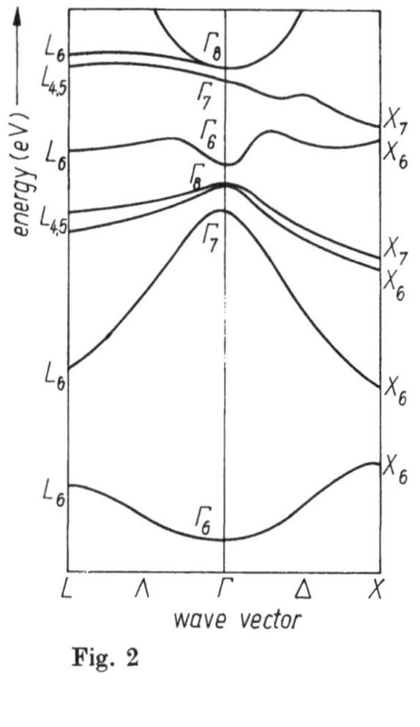

Fig. 2

Fig. 1. Band structure of GaP

Fig. 2. Band structure of GaSb

Table 2

Calculated energy gaps (in eV) of GaP in comparison with other theoretical and experimental results

energy gap	present work	Walter and Cohen [7]	Chelikowsky and Cohen [9]	Chadi [10]	experimental data
$E_0 = \Gamma_{15}^c - \Gamma_{15}^v$	5.17	5.23	5.24	5.14	5.19 [23]
$\Gamma_1^c - \Gamma_{15}^v$	2.82	2.79	2.88	2.88	2.77 [23], 2.92 [24]
$\Gamma_1^v - \Gamma_{15}^v$	−13.0	—	−13.0	−13.2	−11.8 ± 0.5 [25], −13.2 ± 0.4 [26]
$E_{\Gamma X} = X_1^c - \Gamma_{15}^X$	2.24	2.19	2.16	5.46	2.22 [27], 2.35 [28]
$\delta = X_3^c - X_1^c$	0.31*)	0.39	0.55	0.47	0.35 [29]
$\Gamma_{15}^v - X_1^v$	−10.3	—	−9.46	−9.69	− 9.7 ± 0.3 [25], − 9.6 ± 0.3 [26]
$E_{\Gamma L} = L_1^c - \Gamma_{15}^v$	2.62	—	2.79	3.27	—
$L_3^v - \Gamma_{15}^v$	−0.96	—	−1.10	−1.86	− 0.8 ± 0.2 [25], − 1.2 ± 0.3 [26], − 1.4 ± 0.3 [27]
$L_1^v - \Gamma_{15}^v$	−11.2	—	−10.6	−10.9	—

*) Corresponding value for InSb is 0.29.

parametrized tight binding method, Chen and Sher [11] a bond-orbital approximation. Their results, however, are satisfactory only in a limited energy range.

In the present work the EPM with the one-parameter nonlocal correction introduced by Chelikowsky is used. This method is relatively simple and yields sufficiently good results. A full extension of the nonlocal pseudopotential method requires much more computational time, while the results do not differ significantly from the one-parameter method. Differently from other authors [2, 6 to 9], we have fully exploited the crystal symmetry in our EPM calculation. Namely, the pseudowave functions were taken as symmetrized combinations of plane waves (SCPW's), partners of the irreducible representations of the crystal point group. The secular determinant calculated for a given symmetry point of BZ with the use of SCPW's is factorized into a few smaller determinants. Each of them belongs to the corresponding representation of the crystal group. In other methods which do not use a symmetrization procedure the original secular matrix is reduced to a smaller one by a Löwdin perturbation scheme and this requires the introduction of an additional fitting parameter [2, 6 to 9]. So, our method is more "exact" for the same number of plane waves taken into account, and it is obvious that the same pseudopotential parameters used in both methods give different results. In the first step we have used the values of PFF's given by Chelikowsky et al., but we changed slightly the PFF values so as to fit the well-known energy gaps in the Γ-, X-, and L-points of BZ. Moreover, application of the symmetrization procedure brings immediately the full classification of the energy levels with respect to the type of symmetry. Calculations performed by Chelikowsky et al. and by other authors [6 to 9] do not specify explicitly whether a given level calculated at X-point has X_3- or X_1-symmetry type. On the other hand, it is well

Table 3

Calculated energy gaps (in eV) of GaSb in comparison with other theoretical and experimental results

energy gap	present work	Zhang and Callaway [30]	Cahn and Cohen [8]	Chelikowsky and Cohen [9]	experimental data
$E_0 = \Gamma_5^c - \Gamma_8^v$	0.82	0.81	0.9	0.86	1.10 [31]
$\Gamma_8^v - \Gamma_7^v$	0.78	0.78	0.8	0.76	
$X_6^c - \Gamma_5^c$	0.26	0.32	0.48	0.86	0.315 ± 0.015 [31]
$L_6^c - \Gamma_6^c$	0.13	0.08	0.28	0.36	0.085 [31], 0.093 ± 0.003 [32]
$\delta = X_7^c - X_6^c$	0.26	0.38	—	0.07	0.37 [32]
$\Gamma_6^v - \Gamma_8^v$	−11.3	—	—	−12.0	-11.6 ± 0.3 [33]
$X_6^v - \Gamma_8^v$	−9.2	—	—	−9.33	-9.4 ± 0.2 [33]
$L_6^v - \Gamma_8^v$	−10.4	—	—	−10.17	—

proved by experiments that the s-like charge distribution of the lowest conduction band state of GaP and GaSb is concentrated mainly around the group-III atom. This indicates that the lowest conduction level has X_1-symmetry assuming that the origin of coordinates is at the Ga atom (positive antisymmetric pseudopotential form factors) [12 to 15]. Chelikowsky et al.'s parameters used in our method gave opposite ordering. In order to obtain the proper sequence of energy levels at the X-point it was necessary to change additionally the values of the antisymmetric PFF's. The resulting PFF's for GaP and GaSb are presented in Table 1.

In Fig. 1 and 2 the band structure of GaP and GaSb are presented. Spin-orbit interaction is not included for GaP.

A comparison of the energy gaps calculated in the present work with those measured experimentally and obtained theoretically by other authors is shown in Table 2 for GaP and in Table 3 for GaSb. It can be seen that all our values are in a good agreement with experimental data.

3. Pressure Dependence of the Band Structure

Following Tsay and co-workers [2 to 5], in the pressure dependence calculation we adopt the rigid-ion approximation [16 to 18]. In that model the ionic pseudopotential remains unchanged for small variations of the lattice constant. It follows that the symmetric and antisymmetric PFF's at pressure p are given by

$$V_p^{S,A}(\boldsymbol{G}_p) = \frac{V^{S,A}(\boldsymbol{G}_p)\ \Omega\varepsilon(\boldsymbol{G}_p)}{\Omega_p\varepsilon_p(\boldsymbol{G}_p)}, \tag{1}$$

where $\Omega = a/4$ is the primitive cell volume, $\varepsilon(\boldsymbol{G})$ the wave-vector-dependent dielectric function, and $\boldsymbol{G}_p$ a reciprocal lattice vector. Subscript p refers to quantities at pressure p. Using the model dielectric function $\varepsilon(\boldsymbol{q})$ introduced by Penn [19] and modified later by Brust [20], we have obtained finally

$$V_p^{S,A}(\boldsymbol{G}_p) = V^{S,A}(\boldsymbol{G}_p)\frac{a^3}{a_p^3}\frac{1 + 0.65\omega_p^2E_{gp}^2FQ_p^2(1 - f_x)}{1 + 0.65\omega_{pp}^2E_g^2F_pQ^2(1 - f_{xp})}. \tag{2}$$

Table 4
Input data for pressure calculation of band structure of GaP and GaSb

	K (10^{-2}/GPa)	E_g (eV)	dE_g/dp (10^{-2}eV/GPa)	ε_0	$d\varepsilon_0/dp$
GaP	1.13 [3]	5.16*)	3.8*)	9.1 [3]	—
GaSb	1.77 [3]	4.15*)	9.2*)	14.4 [3]	—

*) Values obtained self-consistently.

where

$$Q = 1 + \frac{E_F}{E_g} \frac{G^2}{k_F} \sqrt{F}\,, \qquad f_x(q) = \frac{0.5}{1 + \frac{4k_F^2}{\pi^2 G^2} + \frac{k_F^2}{G^2}}\,,$$

$$\omega_p^2 = \omega_g^2(\varepsilon_0 - 1)\,,$$

and

$$F = 1 - 0.25\,\frac{E_g}{E_F}\,.$$

In the above expressions ω_p, E_F, and k_F are the effective plasma frequency, Fermi energy, and Fermi wave vector, respectively, corresponding to the appropriate valence-electron density. E_g is the average Penn gap [19] identified as the gap at $\boldsymbol{k} = (2\pi/a)\,(\frac{1}{2}, \frac{1}{2}, 0)$ [20]. Subscript p refers to quantities at pressure p (for example ω_{pp} denotes ω_p at pressure p). The pressure dependence of E_g is determined self-consistently.

In the case of GaSb we include also the spin–orbit effect and its pressure dependence Following Bloom and Bergstresser [21] the spin–orbit interaction is expressed by the parameter

$$\lambda \approx \frac{12\pi N}{\Omega} E(\text{so}) \left[\int_0^\infty r^2 P_{nl}(r)\, dr\right]^2 a_B^{-5}\,, \tag{3}$$

Table 5
Pressure coefficients of important band gaps of GaP (in 10^{-2} eV/GPa)

	present work	Tsay et al. [3]	Camphausen et al. [22]	experimental data*)
dE_0/dp	11.7	8.6	9.2	10.5 ± 1.6
dE_X/dp	6.4	1.8	3.2	—
$dE_{\Gamma X}/dp$	−1.8	−1.7	−0.8	−1.1
dE_L/dp	4.8	3.5	4.6	5.8
$dE_{\Gamma L}/dp$	3.7	2.1	2.8	—
$d\delta/dp$	1.3	—	—	—

*) Taken after [3] and [22].

E_0 is taken as the transition $\Gamma_8^v \rightarrow \Gamma_6^c$, E_L is identified as the gap at $\boldsymbol{k} = (2\pi/a)$ (0.5, 0.5, 0.5), E_X as the gap at $\boldsymbol{k} = (2\pi/a)$ (1, 0, 0). The indirect gaps $E_{\Gamma L}$ and $E_{\Gamma X}$ are identified as the energies of the lowest conduction states at $\boldsymbol{k} = (2\pi/a)$ (0.5, 0.5, 0.5) and $(2\pi/a)$ (1, 0, 0), respectively, measured from the top of the valence band at the zone center, $\delta = X_7^c - X_6^c$.

Table 6

Pressure coefficients of important band gaps of GaSb (in 10^{-2} eV/GPa)

	present work	Tsay et al. [3]	Camphausen et al. [22]	experimental data*)
dE_0/dp	15.2	16.1	14.7	14.7
dE_X/dp	3.8	6.5	4.0	6
$dE_{\Gamma X}/dp$	−1.5	3.1	−0.6	—
dE_L/dp	7.0	10.0	6.6	7.5
$dE_{\Gamma L}/dp$	4.2	8.5	4.8	5
$d\delta/dp$	1.9	—	6	—

Conventions as in Table 5.
*) Taken after [3] and [22].

where P_{nl} are the atomic orbitals. It can be seen from the above expression that λ depends on pressure through the crystal volume Ω.

In Table 4 we present the values of the input data used for the pressure calculations. The results for the pressure coefficients of the important direct as well indirect band gaps are presented in Table 5 for GaP and Table 6 for GaSb. The agreement of calculated coefficients with experiments is good for all these energy gaps. For $dE_{\Gamma X}/dp$ we have obtained a small negative value in agreement with the results for InSb [1] and with the experimental data for similar materials. As compared with the results obtained by Tsay et al. [3], our "better" value of $dE_{\Gamma X}/dp$ is probably due to a more accurate determination of the band structure at normal pressure. Our results are compared also with the calculation of Camphausen et al. [22] by using the dielectric theory of chemical bonds. Analogously to InSb [1], the agreement is unexpectedly good, taking into account the quite different nature of the methods applied.

Additionally, we have got the pressure coefficient of the energy gap between X_1 and X_3 conduction levels (labelled often as δ). That gap and its pressure dependence is important when dealing with impurity problems (for example S, Se, Te donors in GaP). For comparison we give the value of the pressure coefficient of δ also for InSb, because it was not presented in the previous paper [1].

In the previous paper the pressure dependence of the wave functions of InSb was discussed. It was estimated as being very small. As an example, the momentum matrix element between valence and conduction bands in the Γ-point of BZ and its pressure coefficient were calculated. In the present paper we have calculated the momentum matrix element between the bottom of the conduction band and the top of the valence band at the X-point of GaP. This matrix element is especially interesting from the point of view of the impurity problems. The calculated value of the momentum matrix element is $P = 0.48$ at. units. Its pressure coefficient is almost the same as for InSb, i e. about 0.01 at. units/10 kbar. This result confirms the conclusions of the previous paper about a weak pressure dependence of the wave functions.

References

[1] I. Gorczyca, phys. stat. sol. (b) **112**, 97 (1982).
[2] J. Chelikowsky, D. J. Chadi, and M. L. Cohen, Phys. Rev. B **8**, 2786 (1973).
[3] Y. F. Tsay, S. S. Mitra, and B. Bendow, Phys. Rev. B **10**, 1476 (1974).
[4] Y. F. Tsay and B. Bendow, Phys. Rev. B **16**, 2663 (1977).
[5] B. Welber, M. Cardona, Y. F. Tsay, and B. Bendow, Phys. Rev. B **15**, 875 (1977).
[6] M. L. Cohen and T. K. Bergstresser, Phys. Rev. **141**, 789 (1966).

[7] J. P. WALTER and M. L. COHEN, Phys. Rev. **183**, 763 (1969).
[8] R. N. CAHN and M. L. COHEN, Phys. Rev. B **1**, 2569 (1970).
[9] J. R. CHELIKOWSKY and M. L. COHEN, Phys. Rev. B **14**, 556 (1976).
[10] D. J. CHADI, Phys. Rev. B **16**, 790 (1977).
[11] A. B. CHEN and A. SHER, Phys. Rev. B **17**, 4726 (1978).
[12] F. BASSANI and M. YOSHIMINE, Phys. Rev. **130**, 20 (1963).
[13] J. PHILLIPS, Phys. Rev. Letters **19**, 415 (1967).
[14] T. N. MORGAN, Phys. Rev. Letters **21**, 819 (1968).
[15] L. LAUDE, M. CARDONA, and R. POLLAK, Phys. Rev. B **1**, 1436 (1970).
[16] D. BRUST and L. LIU, Phys. Rev. **154**, 647 (1967).
[17] P. J. MELZ, J. Phys. Chem. Solids **28**, 1441 (1967).
[18] Y. W. TSANG and M. L. COHEN, Phys. Rev. B **3**, 1254 (1971).
[19] D. R. PENN, Phys. Rev. **128**, 2093 (1962).
[20] D. BRUST, Solid State Commun. **9**, 481 (1971).
[21] S. BLOOM and T. K. BERGSTRESSER, phys. stat. sol. **42**, 191 (1970).
[22] D. L. CAMPHAUSEN, G. A. N. CONNELL, and W. PAUL, Phys. Rev. Letters **26**, 184 (1971).
[23] C. VAREA DE ALVAREZ, J. P. WALTER, M. L. COHEN, J. STOKES, and Y. R. SHEN, Phys. Rev. B **6**, 1412 (1972).
[24] M. CARDONA, K. L. SHAKLEE, and F. H. POLLAK, Phys. Rev. **154**, 696 (1967).
[25] D. E. EASTMAN, W. D. GROBMAN, J. L. FREEOUT, and M. ERBUDAK, Phys. Rev. B **9**, 3473 (1974).
[26] R. POLLAK, L. LEY, S. KOWALCZYK, D. A. SHIRLEY, J. JOANNOPULOS, D. J. CHADI, and M. L. COHEN, Phys. Rev. Letters **29**, 1103 (1973)
[27] R. ZALLEN and W. PAUL, Phys. Rev. **134**, A1628 (1964).
[28] A. C. CARTER, P. J. DEAN, M. S. SKOLNIK, and R. A. STRADLING, J. Phys. C **10**, 5 (1977).
[29] A. A. KOPYLOV and A. N. PIKHIN, Soviet. Phys. – Semicond. **11**, 141 (1977).
[30] H. I. ZHANG and J. CALLAWAY, Phys. Rev. **181**, 1163 (1969).
[31] B. B. KOSICKI, Tech. Rep. No. HP-19, Harvard University, 1967.
[32] N. B. BRANDT, V. V. MOSHCHALKOV, A. S. RYLIK, and S. M. CHUDINOV, Proc. 15th Internat. Conf. Phys. Semicond., Kyoto 1980, in: J. Phys. Soc. Japan **49**, Suppl. A, 117 (1980).
[33] L. LEY, R. A. POLLAK, F. R. MCFEELY, S. P. KOWALCZYK, and D. A. SHIRLEY, Phys. Rev. B **9**, 600 (1974).

(Received January 16, 1984)

phys. stat. sol. (b) 125, 237 (1984)

Subject classification: 13.5.1

Institute of Physics, Academy of Sciences of the Azerbaidzhan SSR, Baku[1])

Coulomb Interaction and Excitons in a Superlattice

By

R. R. GUSEINOV

An exact expression is found for the potential of a point charge in a periodic layered system, consisting of alternating layers with two different values of the dielectric constant. The most interesting limiting cases for this potential are considered. Energy spectrum, bound energy, and the ground state radius of the exciton, which is formed by an electron and a hole, belonging to the same layer of the superlattice, are found.

Найдено точное выражение потенциала, создаваемого точечным зарядом в периодической слоистой системе, состоящей из чередующихся слоев с двумя различными значениями диэлектрической проницаемости. Рассмотрены наиболее интересные предельные случаи. Найдены спектр, энергия связи и радиус основного состояния экситона, образованного электроном и дыркой из одного и того же слоя сверхрешетки.

1. Introduction

At present the interest in superlattices especially in optical ones is increased due to the progress, which is achieved in the technology of their manufacturing. It is interesting from this point of view to solve the exciton problem in periodic structures with two different values of dielectric constants alternating in one direction, especially the case in which the dielectric constant of one layer considerably exceeds the dielectric constant of the other. It is clear from the physical consideration that in this case one can expect effects analogous to those predicted for a thin film which is surrounded by a medium the dielectric constant of which is much smaller than that of the film [1]. Namely, the effect of increasing Coulomb interaction of charges located in the layer with large value of dielectric constant and therefore the effect of considerably increasing binding energy of an exciton in comparison with that in the bulk material with the same value of dielectric constant. Besides, these effects must be the more expressed, the thicker the layers with small value of dielectric constant are.

The exact solution of the Poisson equation for the potential of a point charge in such a superlattice is found. Various approximations for the potential, especially those which are interesting for solving the exciton problem in the layer with large value of dielectric constant are obtained. In a certain region of parameter values the energy spectrum, binding energy, and radius of the ground state of an exciton formed of an electron and a hole belonging to the same layer of the superlattice, are found.

---

[1]) Prospekt Narimanova 33, 370143 Baku, USSR.

2. Coulomb Interaction in a Superlattice (Exact Solution)

We consider an ideal superlattice consisting of alternating layers with thickness d and dielectric constant ε and with thickness a and dielectric constant ε_1, respectively. We are interested in the potential at an arbitrary point of the superlattice, which is created by the charge e, located in some layer with dielectric constant ε.

Let us take the origin of coordinates at the boundary of the layers and the z-axis in the direction of the layer with dielectric constant ε_1 perpendicular to the boundary plane. For defineteness we consider the charge e located in the layer adjoining the origin of coordinates ($-d \leqq z' \leqq 0$). Then the Fourier component of the potential over $\boldsymbol{\varrho}$ ($\boldsymbol{\varrho}$ is the two-dimensional vector, which determines the position of point of observation relative to the charge e in the layer plane) satisfies the Poisson equation with the corresponding boundary conditions,

$$\varepsilon(z)\,\frac{\mathrm{d}^2\varphi(z,z',\boldsymbol{k})}{\mathrm{d}z^2} + \frac{\mathrm{d}\varepsilon(z)}{\mathrm{d}z}\,\frac{\mathrm{d}\varphi(z,z',\boldsymbol{k})}{\mathrm{d}z} - k^2\varepsilon(z)\,\varphi(z,z',\boldsymbol{k}) = -4\pi e\delta(z-z')\,. \quad (1)$$

Here $\varepsilon(z)$ is a periodic function of z with the period $a + d$. It takes two values ε and ε_1.

In order to find the solution of (1), let us at first solve the corresponding homogeneous equation

$$\varepsilon(z)\,\frac{\mathrm{d}^2\varphi(z,\boldsymbol{k})}{\mathrm{d}z^2} + \frac{\mathrm{d}\varepsilon(z)}{\mathrm{d}z}\,\frac{\mathrm{d}\varphi(z,\boldsymbol{k})}{\mathrm{d}z} - k^2\varepsilon(z)\,\varphi(z,\boldsymbol{k}) = 0\,. \quad (2)$$

This is a homogeneous equation of second order with periodic coefficients. It is well known from the theory of differential equations of second order that among all solutions of such an equation there must be two linearly independent ones, satisfying the following condition (Floquet theorem):

$$\varphi_{1,2}(z + a + d, \boldsymbol{k}) = \lambda_{1,2}\varphi_{1,2}(z, \boldsymbol{k})\,. \quad (3)$$

This property of solutions allows us to construct the solution of (2) over the superlattice, having found it in some "cell" (for example, in the "beginning cell", $-d \leqq z \leqq a$). One can easily find the solution in the "beginning cell",

$$\varphi_0(z,\boldsymbol{k}) = \begin{cases} (\alpha A_1 + \beta A_2)\,\mathrm{e}^{kz} + (\beta A_1 + \alpha A_2)\,\mathrm{e}^{-kz}; & -d \leqq z \leqq 0\,,\\ A_1\,\mathrm{e}^{kz} + A_2\,\mathrm{e}^{-kz}; & 0 \leqq z \leqq a\,, \end{cases} \quad (4)$$

where $\alpha = \frac{1}{2}\left(1 + (\varepsilon_1/\varepsilon)\right)$, $\beta = \frac{1}{2}\left(1 - (\varepsilon_1/\varepsilon)\right)$, A_1 and A_2 are arbitrary constants, and therefore, we have the solution in any "cell" of number n,

$$\varphi_n\big(z + n(a + d), \boldsymbol{k}\big) = \lambda^n\varphi_0(z, \boldsymbol{k})\,. \quad (5)$$

The constant λ can be found from the solvability condition of the system of equations, which arise when matching the solutions and their first derivatives at the boundaries of "cells",

$$\lambda^\pm = \mathrm{e}^{\pm k_0}\,, \qquad k_0 = \operatorname{Arch}\left[\operatorname{ch} k(a-d) + \frac{2\alpha^2}{2\alpha - 1}\operatorname{sh}(ka)\operatorname{sh}(kd)\right],$$
$$A_2^\pm = \gamma^\pm A_1^\pm\,, \quad \gamma^\pm = -\frac{\mathrm{e}^{ka} - \mathrm{e}^{\pm k_0}(\alpha\,\mathrm{e}^{-kd} + \beta\,\mathrm{e}^{kd})}{\mathrm{e}^{-ka} - \mathrm{e}^{\pm k_0}\,(\beta\,\mathrm{e}^{-kd} + \alpha\,\mathrm{e}^{kd})}\,. \quad (6)$$

Thus we have found two linearly independent solutions of (2) $\varphi_\pm$, satisfying Floquet theorem and each of them containing an arbitrary constant coefficient ($A_1^\pm$). The sum of these solutions is the general solution of the homogeneous equation (2).

The partial solution of the inhomogeneous equation (1) can be found by the Lagrange method. Then taking the general solution of the inhomogeneous equation as a sum of the general solution of homogeneous and partial solution of inhomogeneous one and demanding that it should vanish at $n \to \pm\infty$, we finally obtain

$$\varphi(z + n(a + d), z', \boldsymbol{k}) = \frac{2\pi e}{\varepsilon_1 k(\gamma^- - \gamma^+)} \times$$
$$\times \begin{cases} [(\alpha + \gamma^+\beta)\, e^{kz'} + (\beta + \gamma^+\alpha)\, e^{-kz'}]\, \varphi_-(z + n(a + d), \boldsymbol{k})\,; & z + n(a + d) \geqq z', \\ [(\alpha + \gamma^-\beta)\, e^{kz'} + (\beta + \gamma^-\alpha)\, e^{-kz'}]\, \varphi_+(z + n(a + d), \boldsymbol{k})\,; & z + n(a + d) \leqq z', \end{cases} \tag{7}$$

where

$$\varphi_\pm(z + n(a + d), \boldsymbol{k}) =$$
$$= e^{\pm nk_0} \begin{cases} (\alpha + \gamma^\pm\beta)\, e^{kz} + (\beta + \gamma^\pm\alpha)\, e^{-kz}; & -d \leqq z \leqq 0\,, \\ e^{kz} + \gamma^\pm\, e^{-kz}; & 0 \leqq z \leqq a\,. \end{cases} \tag{8}$$

Formulas (7) to (8) give us the solution of the problem. The Fourier component of the potential in the arbitrary point ($-d \leqq z \leqq a$) of the "cell" with number n which is created by the charge e, located in the layer with dielectric constant ε ($-d \leqq z' \leqq 0$) of the "beginning cell", is found.

We have to notice that, strictly speaking, we were solving equations (1) and (2) everywhere except at the boundary points, where the derivative of $\varepsilon(z)$ does not exist. But since we are looking for continuous solutions, our solutions can be used at these points, too.

3. Coulomb Interaction of Two Charges in One Layer of the Superlattice when $\varepsilon_1 \ll \varepsilon$ and $\varrho \gg d$

Bearing in mind the solution of the problem of an exciton formed of electron and hole belonging to the same layer, let us write down the explicit expression for the potential (7) at $n = 0$ and $z' \leqq z \leqq 0$ (when $-d \leqq z \leqq z'$ one must interchange z and z' in the expression, which will be written down). After cumbersome but simple transformations it is reduced to the following form:

$$\varphi(z, z', \boldsymbol{k}) = \frac{\pi e}{\varepsilon_1 k \operatorname{sh} k_0[\alpha \operatorname{sh} k(a + d) + \beta \operatorname{sh} k(a - d)]} \times$$
$$\times \{2(\alpha - \beta) \operatorname{ch} k_0 \operatorname{ch} k(z - z')\, [\alpha \operatorname{ch} k(a + d) - \beta \operatorname{ch} k(a - d)] -$$
$$- 2(\alpha - \beta) \operatorname{sh} k_0 \operatorname{sh} k(z - z')\, [\alpha \operatorname{sh} k(a + d) + \beta \operatorname{sh} k(a - d)] -$$
$$- 2 \operatorname{ch} k(z - z')\, (\alpha^2 + \beta^2 - 2\alpha\beta \operatorname{ch}^2 ka) + (\alpha - \beta)^2 \operatorname{ch} k(2d + z + z') -$$
$$- 2(\alpha - \beta) \operatorname{ch} k_0[\alpha \operatorname{ch} k(z + z' + d - a) - \beta \operatorname{ch} k(z + z' + d + a)] +$$
$$+ \operatorname{ch} k(z + z')\, [(\alpha^2 + \beta^2) \operatorname{ch} 2ka - 2\alpha\beta] -$$
$$- (\alpha - \beta) \operatorname{sh} k(z + z') \operatorname{sh} 2ka\}\,. \tag{9}$$

We consider the superlattice in which $\varepsilon_1 \ll \varepsilon$. Besides that, we are interested in the behaviour of the potential

$$\varphi(z, z', \boldsymbol{\varrho}) = \frac{1}{(2\pi)^2} \int e^{i\boldsymbol{k}\boldsymbol{\varrho}}\, \varphi(z, z', \boldsymbol{k})\, d^2k \tag{10}$$

when $\varrho \gg d$, since it is clear from the physical point of view, that the potential will be essentially different from the ordinary Coulomb potential in the medium with dielectric constant ε, especially at these distances. In this case the main contribution to the integral (10) is given by the values of k, for which $kd \ll 1$. Taking into account this fact and also the condition $\varepsilon_1 \ll \varepsilon$, one can essentially simplify (9),

$$\varphi(z, z', \boldsymbol{k}) \approx \frac{2\pi e}{\varepsilon_1 k} \frac{1}{\sqrt{1 + (kd/2\eta)^2 + (kd/\eta) \operatorname{cth} ka}}, \tag{11}$$

where $\eta = \varepsilon_1/\varepsilon$. When writing down (11), we have used the condition $a \gg \eta d$, too. The most interesting cases $a \approx d$ and $a > d$ satisfy this condition.

It is essential that the potential in our approximation does not depend on z and z'. This fact permits us to separate variables when solving the exciton problem and consider two-dimensional bound states in the layer plane.

Substituting (11) into (10) and integrating over the angle variable, we have

$$\varphi(\varrho) \approx \frac{e}{\varepsilon a} \int_0^\infty \frac{J_0((\varrho/a)\, x)\, dx}{\sqrt{\eta^2 + (d/2a)^2 x^2 + (\eta d/a)\, x \operatorname{cth} x}}, \tag{12}$$

where J_0 is the Bessel function.

Integral (12) has a singularity at the lower limit, when $\eta = 0$. This singularity can be easily isolated.

Let us consider two regions of values of parameter a separately.

3.1 $a \ll d/\eta$

In this case, taking into account that $\operatorname{cth} x$ behaves as $1/x$, when $x \to 0$, one can neglect η^2 under the square root and isolate the part which gives the main contribution to the integral (12),

$$\varphi(\varrho) \approx \frac{e}{\varepsilon a} \int_0^\infty \frac{J_0((\varrho/a)\, x)\, dx}{\sqrt{(d/2a)^2 x^2 + (\eta d/a)}} = \frac{2e}{\varepsilon d} I_0\left(\varrho \sqrt{\frac{\eta}{ad}}\right) K_0\left(\varrho \sqrt{\frac{\eta}{ad}}\right), \tag{13}$$

where I_0 is the Bessel function and K_0 the cylindrical function of imaginary argument.

We have used formula 6.552 from [2]. Using the well-known expressions for I_0 and K_0 at small values of the argument [2], we obtain

$$\varphi(\varrho) \approx -\frac{2e}{\varepsilon d}\left[\ln\left(\frac{\varrho}{2}\sqrt{\frac{\eta}{ad}}\right) + C\right] \quad \text{for} \quad d \ll \varrho \ll \sqrt{\frac{ad}{\eta}}, \tag{14}$$

where C is the Euler constant.

One can see from (14) that when $\eta \to 0$, $\varphi(\varrho)$ has a logarithmic singularity, which was told about above.

Asymptotic expressions of functions I_0 and K_0 at large values of the argument are known, too [2], therefore, in the opposite limiting case $\varrho \gg \sqrt{ad/\eta}$ we get

$$\varphi(\varrho) \approx \frac{e}{\sqrt{\varepsilon_1 \varepsilon}} \sqrt{\frac{a}{d}} \frac{1}{\varrho} \quad \text{for} \quad \varrho \gg \sqrt{\frac{ad}{\eta}}. \tag{15}$$

3.2 $a \gg d/\eta$

Such big values of the thickness of the layers with small dielectric constant ε_1 correspond to the case, when electrostatic potential in the layer with large dielectric constant in such a superlattice does not differ from the potential in the thin film of thickness d and dielectric constant ε, surrounded by the medium with dielectric constant $\varepsilon_1 \ll \varepsilon$ [1]. Indeed, the function $J_0((\varrho/a)\,x)$ in (12) strongly oscillates at $x \gg 1$ when $\varrho \gtrapprox a$ (and so $\varrho \gg (d/\eta)$) and, consequently, the large values of x give a small contribution to the integral (13). Therefore, taking into account the condition $a \ll d/\eta$, one can neglect the second and third terms under the square root, then

$$\varphi(\varrho) \approx \frac{e}{\varepsilon_1 \varrho} \quad \text{for} \quad \varrho \ll d/\eta\,. \tag{16}$$

But when ϱ satisfies the conditions $d \ll \varrho \ll d/\eta$, one can show that the main contribution to (12) is given by

$$\varphi(\varrho) \approx \frac{2e}{\varepsilon d} \int\limits_0^\infty \frac{J_0(x)\,dx}{x + 2(\eta\varrho/d)} \approx \frac{2e}{\varepsilon d}\left(\ln\frac{d}{\eta\varrho} - C\right). \tag{17}$$

Formulas (16) to (17) exactly coincide with those for the potential in a thin film surrounded by a medium with dielectric constant $\varepsilon_1 \ll \varepsilon$ [1]. Thus, the potential in the superlattice differs from that in a thin film for values a satisfying the double inequality $\eta d \ll a \ll d/\eta$, and besides, when $d \ll \varrho \ll \sqrt{ad/\eta}$ it has the form (14), which differs from (17) by a shift $-\ln\,[(1/2)\sqrt{\varepsilon d'/\varepsilon_1 a}]$, and when $\varrho \ll \sqrt{ad/\eta}$ it has the form (15), which differs from (16) by the factor $\sqrt{\varepsilon_1 a/\varepsilon d}$.

4. Exciton in the Superlattice ($\eta d \ll a \ll d/\eta$)

Knowing the form of the potential and the energy of interaction of charges, we can find the lower bound states formed of electron and hole from the same layer of superlattice. These states are the lowest ones unlike the bound states formed of electron and hole belonging to different layers.

As the dependence of the interaction energy on ϱ does not differ essentially from that in the thin film case, we can use the results of [1]. It must be noticed that as in [1] we consider two-dimensional bound states of an electron with a hole, taking the thickness of the layer d small enough, so that electron and hole are in their dimension-quantized ground states of motion in the direction perpendicular to the plane of the layer.

In the region of applicability of (14), using the units of distances $a_{ex} = (1/2)\sqrt{2a_2 d}$ ($a_2 = \varepsilon\hbar^2/2me^2$ is the radius of the two-dimensional exciton) and energies $E = 2e^2/\varepsilon d$, we transform the Schrödinger equation for the exciton to dimensionless form

$$(\nabla_{\boldsymbol{\varrho}} - \ln\varrho)\,\psi_n(\boldsymbol{\varrho}) = \gamma_n\psi_n(\boldsymbol{\varrho})\,, \tag{18}$$

where γ_n are the eigenvalues which do not depend on the parameters of our problem, and a_{ex} plays the role of the radius of the ground state of the exciton. It is necessary that the two inequalities

$$a \gg \eta a_2 \quad \text{and} \quad d \ll a_2 \tag{19}$$

are satisfied (and, certainly, $\eta d \ll a \ll d/\eta$, too) for the exciton radius to fall into the region of applicability of formula (14),

$$d \ll a_{\text{ex}} \ll \sqrt{\frac{ad}{\eta}} \,. \tag{20}$$

The excitonic spectrum in usual units

$$E_n^{\text{ex}} = -\frac{e^2}{\varepsilon d}\left[\ln\left(8\,\frac{\varepsilon a}{\varepsilon_1 a_2}\right) - 2C + 2\gamma_n\right] \tag{21}$$

differs from that in thin film [1] by the expression under the logarithm.

The binding energy of the exciton as in the case of a thin film strongly increases when the thickness d decreases due to the inverse dependence on d and the presence of the logarithm of a large quantity in (21) (in accordance with (19) $a \gg \eta a_2$).

In the region of extremely large values of ϱ, where (15) is applicable, the Schrödinger equation turns into the equation for the ordinary two-dimensional Coulomb exciton, the energy spectrum of which is well known. The radius of the ground state and the binding energy of such an exciton in our case are

$$a_{\text{ex}} = a_2 \sqrt{\frac{\varepsilon_1 d}{\varepsilon a}} \tag{22}$$

and

$$E_0^{\text{ex}} = E_2\,\frac{\varepsilon a}{\varepsilon_1 d}\,, \tag{23}$$

where $E_2 = 2me^4/\varepsilon^2\hbar^2$ is the binding energy of a two-dimensional exciton in a medium with dielectric constant ε. The applicability of (15) in this case is ensured by the condition

$$a_{\text{ex}} \gg \sqrt{\frac{ad}{\eta}} \tag{24}$$

or, taking into account (22),

$$a \ll a_2 \eta \,. \tag{25}$$

If one remembers that we consider the values a satisfying the double inequality $\eta d \ll a \ll d/\eta$, then one can see from (22) and (23) that the exciton radius in this case is much smaller and the binding energy is much larger those of the two-dimensional exciton in the medium with dielectric constant ε. However, (22) to (23) are unlikely to have a region of applicability, because for the feasibility of (25) already for $\eta \approx 0.1$, $a \lesseqq 0.01 a_2$, which is not realistic at all even for $a_2 \approx 10$ nm.

5. Conclusion

The conditions of applicability of the logarithmic law for the binding energy of the exciton

$$E_0^{\text{ex}} = -\frac{e^2}{\varepsilon d}\left[\ln\left(8\,\frac{\varepsilon a}{\varepsilon_1 a_2}\right) - 2C + 2\gamma_0\right]$$

put strong limitations to the region of permissible values of parameters η, d, a ($\eta \ll 1$, $d \ll a_2$, $a \gg \eta a_2$, $\eta d \ll a \ll d/\eta$) from the point of view of their realization. However, the obtained dependence of E_0^{ex} on η, d and a shows, apparently, a tendency in the behaviour of the binding energy in the real region of parameters.

The evaluation of binding energy using the numerical value $\gamma_0 = -0.528$ from [3], shows that although it is smaller than the exciton binding energy in the thin film with the same values of η and d, it is of the same order and approaches it when a rises. So, when $\eta = 0.001$, $d = 0.1a_2$, $a = 0.1a_2$ E_0^{ex} is $E_0^{ex} \approx 45E_2$, and when $a = a_2$, $E_0^{ex} \approx 70E_2$, while the binding energy in the thin film with the same values of the parameters η and d is $E_0^{ex} \approx 100E_2$.

The results of this paper could be used for the qualitative description of the situation in layered compounds in which the thickness of layers and the distances between them are big enough to consider them macroscopically. Besides, their dielectric constants must strongly differ. It is apparently premature to make any quantitative comparison with the excitonic spectrum of the layered compounds really investigated at present time.

In conclusion we must notice that excitons localized in one layer can in fact cross from layer to layer since there are nonvanishing matrix elements of Coulomb interaction, corresponding to the annihilation of the exciton in one layer with the simultaneous creation of it in the other. Such processes will lead to an excitonic band of "cross" motion in which exciton energy will depend on "cross" quasi-momentum. This problem will be considered in another paper.

Acknowledgement

The author wish to thank L. V. Keldysh for the attention to this work and helpful discussions of its results.

References

[1] L. V. Keldysh, Zh. eksper. teor. Fis., Pisma **29**, 716 (1979).
[2] I. S. Gradshtein and I. M. Ryjyk, Tabliki integralov, summ, ryadov i proisvedenii, Izd. Nauka, Moskva 1963.
[3] E. A. Andrjushin and A. P. Silin, Fiz. tverd. Tela **22**, 2676 (1980).

(Received April 9, 1984)

phys. stat. sol. (b) 125, 245 (1984)

Subject classification: 13.5 and 20; 7

Institute of Physics, Academy of Sciences of Vietnam, Hanoi[1])

Polariton Effect in Non-Linear Acousto-Optical Phenomena

By

NGUYEN VAN HIEU and HA VINH TAN

The role of polariton intermediate states in non-linear acousto-optical phenomena is studied. The matrix elements of the interaction processes acoustic phonon + photon → photon + acoustic phonon, photon → photon + 2 acoustic phonons are derived. The contribution of the effective photon–exciton transition is taken into account in any order. Then the exciton intermediate states are replaced by those of the corresponding polaritons. The damping is not neglected.

Изучается роль поляритонных промежуточных состояний в нелинейных акустооптических явлениях. Выводятся матричные элементы процессов взаимодействия акустический фонон + фотон → акустический фонон + фотон, фотон → 2 акустических фонона + фотон. Вклад эффективного экситон–фотонного перехода учитывается в любом порядке. Тогда промежуточные экситонные состояния становятся поляритонными. Затуханием не пренебрегается.

1. Introduction

The non-linear acousto-optical phenomena have been studied in many theoretical and experimental works. The quantum theory of the corresponding physical processes was developed by Guliaev and Shkerdin [1]. At light frequencies near the value of the band gap the contribution of the discrete-energy exciton states to the non-linear acousto-optical processes in the direct-band-gap semiconductors is very significant, as was shown in our recent work [2]. If we introduce the effective Hamiltonians of the exciton–photon transition and the phonon–exciton interaction, then the results concerning the discrete energy exciton states in [2] can be derived easily by means of the perturbation theory in the lowest order of these effective Hamiltonians. The pole terms of the considered physical quantities are due to the intermediate exciton states.

However, the stationary low-lying excited states in the semiconductors are not the excitons, but the polaritons. Therefore the physical intermediate states in the acousto-optical processes concerned must be those of the polaritons. The role of the polariton effect in many non-linear optical processes has been studied in [3 to 5]. It might be important also in the non-linear acousto-optical phenomena, and the investigation in our previous work [2] should be improved to take into account the polariton effect. This is done in our present work. We shall apply the method based on the renormalization procedure and proposed by one of the authors [6]. The Green function approach of [6] and the diagonalization procedure using the Bogolyubov transformation are equivalent if the damping is neglected, as was shown in [6, 7], but the former is more convenient if we want to study also the influence of the damping.

---

[1]) Nghia Do, Tu Liem, Hanoi, Vietnam.

The role of the polaritons in the first-order resonant Brillouin scattering process was investigated in many experimental works [8 to 14]. It has been used by the authors of these works as a powerful tool for determining the dispersion of the polaritons. In the present work we assume that the polariton dispersion is known, and we study the change of the scattering amplitude for the non-linear acousto-optical process due to the polariton effect.

Following the method of [6] we introduce the field operator $\hat{A}(\boldsymbol{r}, t)$ of the quantized electromagnetic field consisting of photons with a definite polarization vector and the field operator $\hat{B}(\boldsymbol{r}, t)$ for the excitons. The exciton-photon transition is described by the effective Hamiltonian

$$H^{\gamma\text{-Ex}} = g \int \hat{A}(\boldsymbol{r}, t)\, \frac{\partial \hat{B}(\boldsymbol{r}, t)}{\partial t}\, \mathrm{d}^3 r\,. \tag{1}$$

For the fundamental acoustic phonon–electron and acoustic phonon–hole interactione we use the Hamiltonian given in [1, 2]. Then we can derive the following effective acoustic phonon–exciton interaction Hamiltonian:

$$H^{\text{ph-Ex}} = \int \gamma(\boldsymbol{r} - \boldsymbol{r}')\, [\hat{B}(\boldsymbol{r}, t)]^2\, \hat{D}(\boldsymbol{r}, t)\, \mathrm{d}^3 r\, \mathrm{d}^3 r'\,, \tag{2}$$

where $\hat{D}(\boldsymbol{r}, t)$ is the longitudinal acoustic phonon field operator and $\gamma(\boldsymbol{r})$ the form-factor of the exciton–exciton–phonon vertex. The Fourier transform $\tilde{\gamma}(\boldsymbol{q})$ of this form-factor is expressed in terms of the exciton envelope wave functions and the matrix elements of the electron–phonon and hole–phonon interaction Hamiltonians.

2. Polariton Effect without Damping

The second-order non-linear acousto-optical phenomena are characterized by the scattering amplitudes of the processes: (i) acoustic phonon + photon → acoustic phonon + photon; (ii) photon → 2 acoustic phonons + photon, which are related by the well-known crossing symmetry relations: if we start from the amplitude of the first process (i) and change the sign of the momentum and the energy of the initial phonon, then we obtain that of the second one (ii). Therefore, we shall consider in detail only the first process (i). Denote by $\boldsymbol{k}$, $\boldsymbol{q}$ and ω, Ω the momenta and energies of the initial photon and phonon, by $\boldsymbol{k}'$, $\boldsymbol{q}'$, ω', Ω' the corresponding quantities in the final state. In the lowest order of perturbation theory with the effective interaction Hamiltonian

$$H^{\text{int}} = H^{\gamma\text{-Ex}} + H^{\text{ph-Ex}} \tag{3}$$

the scattering process (i) is represented by two Feynman diagrams in Fig. 1. The corresponding scattering amplitude has the form

$$\begin{aligned} &M_0(\boldsymbol{k}, \boldsymbol{q}, \omega, \Omega; \boldsymbol{k}', \boldsymbol{q}', \omega', \Omega') = \\ &= \frac{g^2}{4\sqrt{\omega\omega'\Omega\Omega'}}\, \frac{\omega}{\omega^2 - E^2(k)}\, \frac{\omega'}{\omega'^2 - E^2(k')}\, \tilde{\gamma}(\boldsymbol{q})\, \tilde{\gamma}(-\boldsymbol{q}') \times \\ &\times \left\{ \frac{1}{[\omega + \Omega(q)]^2 - E^2(k + q)} + \frac{1}{[\omega - \Omega'(q')]^2 - E^2(k - q')} \right\}. \end{aligned} \tag{4}$$

Each term on the right-hand side of (4) contains three exciton propagators which have poles at the photon energies in resonance with the corresponding exciton energies or at the photon energy equal to the sum or the difference of the energies of the exciton and the phonon. The scattering amplitude is divergent at these poles.

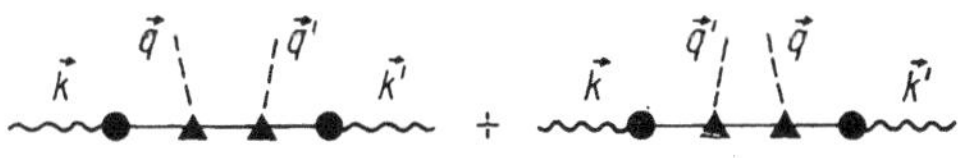

Fig. 1. Feynman diagrams for process (i) in lowest order of the photon–exciton transition Hamiltonian. Bare photon: wavy line; bare exciton: solid line; acoustic phonon: dashed line

Note that at higher orders in the exciton–photon transition Hamiltonian the Feynman diagrams contain new internal exciton lines whose two ends represent two exciton–photon transitions. Corresponding to each line of this type together with its two ends we have a factor of the form

$$\frac{g^2\omega^2}{\omega^2 - E^2(k)}$$

in the matrix element. At resonance these new factors may become very large, and we must take into account all higher-order approximations of the exciton–photon transition Hamiltonian $H^{\gamma\text{-Ex}}$. This can be done by means of the renormalization procedure proposed in [6]. Then each Feynman diagram in Fig. 1 is replaced by a corresponding series of diagrams of all orders in Fig. 2, and the scattering amplitude becomes

$$\begin{aligned} M(\boldsymbol{k}, \boldsymbol{q}, \omega, \Omega; \boldsymbol{k}', \boldsymbol{q}', \omega', \Omega') &= \\ = \frac{g^2}{4\sqrt{\omega\omega'\Omega\Omega'}} & \frac{\omega}{\{[\omega^2 - E^2(k)]^2 + g^2E^2(k)\}^{1/2}} \times \\ \times & \frac{\omega'}{\{[\omega'^2 - E^2(k')]^2 + g^2E^2(k)\}^{1/2}} \tilde{\gamma}(\boldsymbol{q})\,\tilde{\gamma}(-\boldsymbol{q}') \times \\ \times & \sum_{j=1}^{2} \left\{ \frac{|Z_j^{\text{Ex}}(q+k)|^2}{[\omega + \Omega(q)]^2 - \omega_j^{\text{pol}2}(q+k)} + \frac{|Z_j^{\text{Ex}}(k-q')|^2}{[\omega - \Omega'(q')]^2 - \omega_j^{\text{pol}2}(k-q')} \right\}, \end{aligned} \tag{5}$$

where the index $j = 1, 2$ labels two polariton branches, $\omega_j^{\text{pol}}(k)$ denotes the energy of tho polariton with momentum k in the branch j, $Z_j^{\text{Ex}}(k)$ is the corresponding wave function renormalization constant,

$$|Z_j^{\text{Ex}}(k)|^2 = \frac{g^2\omega_j^{\text{pol}2}(k)}{[\omega_j^{\text{pol}2}(k) - E^2(k)]^2 + g^2E^2(k)}\,, \tag{6}$$

$$\begin{aligned} \omega_{1,2}^{\text{pol}} = \tfrac{1}{2}\,[\omega^2(k) + E^2(k) + g^2] \pm \\ \pm \tfrac{1}{2}\,\{[\omega^2(k) + E^2(k) + g^2]^2 - 4\omega^2(k)\,E^2(k)\}^{1/2}\,. \end{aligned} \tag{7}$$

Note that instead of the bare photons in the initial and final states for the case when the exciton–photon transition is considered only in the lowest-order approximation, after the renormalization procedure we obtain the dressed initial and final photons, i.e. the photon-like polaritons in the appropriate branches. Therefore, we have

$$\omega' = \omega_{j'}^{\text{pol}}(k)\,, \qquad \omega = \omega_j^{\text{pol}}(k)\,; \qquad j, j' = 1, 2\,.$$

There exists a significant difference between the scattering amplitudes (4) and (5) instead of two common exciton propagators in the former,

$$\frac{1}{\omega^2 - E^2(k)}\,, \qquad \frac{1}{\omega'^2 - E^2(k')}$$

which are represented in the Feynman diagram by two exciton lines connected with

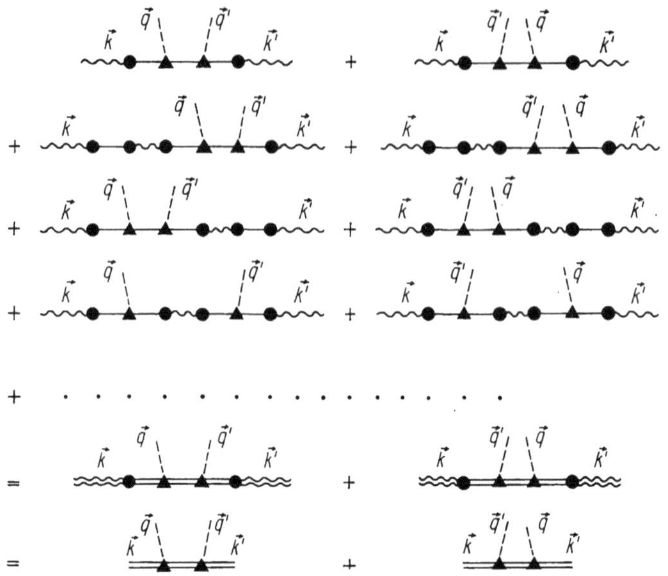

Fig. 2. Feynman diagrams for process (i) in any order of the photon–exciton transition Hamiltonian. Dressed photon: double wavy line; dressed exciton: double solid line

two external photon lines through the exciton–photon transition, we have two nonsingular functions,

$$\frac{1}{\{[\omega^2 - E^2(k)]^2 + g^2E^2(k)\}^{1/2}}, \qquad \frac{1}{\{[\omega'^2 - E^2(k')]^2 + g^2E^2(k')\}^{1/2}},$$

in the latter. Therefore, the scattering amplitude (5) in the polariton theory is less singular than (4) of this amplitude in the lowest-order approximation. Moreover, instead of the exciton propagators of the form

$$\frac{1}{[\omega \pm \Omega(q)]^2 - E^2(k \pm q)}$$

in the lowest-order scattering amplitude (4) after the summation over all higher-order matrix elements of the exciton–photon transition perturbation theory, we obtain the polariton propagators of the form

$$\sum_j \frac{|Z_j^{\mathrm{Ex}}(k \pm q)|^2}{[\omega \pm \Omega(q)]^2 - \omega_j^{\mathrm{pol}2}(k \pm q)}$$

in the scattering amplitude (5) of the polariton theory.

3. Polariton Effect with Damping

Now we generalize the results of the preceding section to the case of the exciton with damping. We assume the exciton propagator in the form

$$\frac{1}{\omega^2 - E^2(k) - i\omega\Gamma},$$

where Γ is the half-width of the exciton level. Instead of (4) we have then the folowing expression for the scattering amplitude in the lowest order of the perturbation theory for the exciton–photon transition:

$$M_0(\boldsymbol{k}, \boldsymbol{q}, \omega, \Omega; \boldsymbol{k}', \boldsymbol{q}', \omega', \Omega') =$$
$$= \frac{g^2}{4\sqrt{\omega\omega'\Omega\Omega'}} \frac{\omega}{\omega^2 - E^2(k) - i\omega\Gamma} \frac{\omega'}{\omega'^2 - E^2(k') - i\omega'\Gamma} \tilde{\gamma}(\boldsymbol{q}) \tilde{\gamma}(-\boldsymbol{q}) \times$$
$$\times \left\{ \frac{1}{[\omega + \Omega(q)]^2 - E^2(k+q) - i\omega\Gamma} + \right.$$
$$\left. + \frac{1}{[\omega - \Omega'(q')]^2 - E^2(k-q') - i\omega\Gamma} \right\}. \tag{8}$$

The behaviour of this amplitude at resonance has been analyzed in [2].

The polariton theory for the exciton with damping was developed by one of the authors [6]. It was shown that in the presence of the exciton damping each polariton branch is spread into a band. In the general case when the widths of the bands are large enough the theory is rather complicated. Therefore, we shall study the limiting case of very narrow polariton bands, when each band is considered as a line. Then in any non-singular function contained in the matrix elements we can let $\Gamma \to 0$ and we obtain the following formula for the scattering amplitude:

$$M(\boldsymbol{k}, \boldsymbol{q}, \omega, \Omega; \boldsymbol{k}', \boldsymbol{q}', \omega', \Omega') =$$
$$= \frac{g^2}{4\sqrt{\omega\omega'\Omega\Omega'}} \frac{\omega}{\{[\omega^2 - E^2(k)]^2 + g^2E^2(k)\}^{1/2}} \times$$
$$\times \frac{\omega'}{\{[\omega'^2 - E^2(k')]^2 + g^2E^2(k')\}^{1/2}} \tilde{\gamma}(\boldsymbol{q}) \tilde{\gamma}(-\boldsymbol{q}) \times$$
$$\times \sum_{j=1}^{2} \left\{ \frac{|Z_j^{\mathrm{Ex}}(k+q)|^2}{[\omega + \Omega(q)]^2 - \omega_j^{\mathrm{pol}2}(k+q) - i\omega\Gamma_j(k+q)} + \right.$$
$$\left. + \frac{|Z_j^{\mathrm{Ex}}(k-q')|^2}{[\omega - \Omega'(q')]^2 - \omega_j^{\mathrm{pol}2}(k-q') - i\omega\Gamma_j(k-q')} \right\}, \tag{9}$$

which has the same form as (4) does, but the propagator of the polariton in each branch now contains the corresponding damping constant $\Gamma_j(k)$, $j = 1, 2$. The expressions of the half-widths of the polaritons Γ_j in terms of that of the exciton Γ have been derived in [15]:

$$\Gamma_1(k) = \frac{\omega_1^{\mathrm{pol}2}(k) - \omega^2(k)}{\omega_2^{\mathrm{pol}2}(k) - \omega_1^{\mathrm{pol}2}(k)} \Gamma,$$
$$\Gamma_2(k) = \frac{\omega^2(k) - \omega_2^{\mathrm{pol}2}(k)}{\omega_2^{\mathrm{pol}2}(k) - \omega_1^{\mathrm{pol}2}(k)} \Gamma. \tag{10}$$

The behaviour of the scattering amplitude in the polariton theory at resonance is significantly different with that of this amplitude in the lowest order of the photon–exciton transition. In particular, the maximum value of the former is proportional to $1/\Gamma$, while the latter has a maximum value proportional to g^2/Γ^3. We have

$$\frac{\max|M|}{\max|M_0|} \approx \frac{\Gamma^2}{g^2}. \tag{11}$$

In the particular case of GaAs we have [2, 6]:

$$g^2 \approx 3 \times 10^{-4}\,(\mathrm{eV})^2\,, \qquad \Gamma^2 \approx 10^{-9}\,(\mathrm{eV})^2\,,$$

$$\frac{\Gamma^2}{g^2} \approx 3 \times 10^{-6}\,.$$

For many other direct-band-gap semiconductors:

$$g^2 \approx 10^{-4}\,(\mathrm{eV})^2\,, \qquad \Gamma^2 \leqq 10^{-6}\,(\mathrm{eV})^2\,,$$

$$\frac{\Gamma^2}{g^2} \leqq 10^{-2}\,.$$

We conclude that the polariton effect diminishes the maximum value of the photon–acoustic phonon scattering amplitude by several orders.

References

[1] YU. V. GULIAEV and G. N. SHKERDIN, Zh. eksper. teor. Fiz. 77, 1396 (1977).
[2] HOANG NGOC CAM, NGUYEN VAN HIEU, and HA VINH TAN, phys. stat. sol. (b) **116**, 25 (1983).
[3] F. HENNEBERGER and J. VOIGT, phys. stat. sol. (b) **76**, 313 (1976).
[4] F. BECHSTEDT and F. HENNEBERGER, phys. stat. sol. (b) **81**, 211 (1977).
[5] V. MAY and J. RÖSELER, phys. stat. sol. (b) **102**, 533 (1980).
[6] NGUYEN VAN HIEU, Ann. Phys. (U.S.A.) **127**, 179 (1980).
[7] HA VINH TAN and NGUYEN TOAN THANG, Czech. J. Phys., in the press.
[8] R. G. ULBRICH and C. WEISBUCH, Phys. Rev. Letters **38**, 865 (1977).
[9] G. WINTERLING and E. KOTELES, Solid State Commun. **23**, 95 (1977).
[10] C. HERMANN and P. Y. YU, Solid State Commun. **28**, 313 (1978).
[11] R. G. ULBRICH and C. WEISBUCH, in: Festkörperprobleme, Ed. J. TREUSCH, Vol. 18, Vieweg, Braunschweig 1978 (p. 217).
[12] B. SERMAGE and G. FISHMAN, Phys. Rev. Letters **43**, 1043 (1979).
[13] E. KOTELES and G. WINTERLING, Phys. Rev. Letters **44**, 948 (1980).
[14] E. KOTELESS, in: Excitons, Ed. E. I. RASHBA and M. D. STURGE, North-Holland Publ. Co., Amsterdam 1982.
[15] HOANG NGOC CAM, NGUYEN VAN HIEU, and NGUYEN AI VIET, to be published.

(Received May 2, 1984)

phys. stat. sol. (b) **125**, 251 (1984)

Subject classification: 11 and 20; 13; 22.1.3; 22.5.2

Institute of Nuclear Physics, Tomsk Polytechnical Institute[1])

Anomalous Radiation of Fast Channeled Electrons in Optically Transparent Crystals

By

D. E. Popov, E. I. Rozum, S. R. Uglov, and S. A. Vorobiev

The spectral and angular characteristics of radiation of channeled electrons in the mean energy region for optically transparent crystals are calculated. An anomalous radiation is shown to be generated in the ultraviolet, visible, and infrared ranges of the spectrum. Peculiarities of the experimental observations and investigation of the anomalous radiation are discussed.

Рассчитаны спектрально-угловые характеристики излучения каналирующих электронов средней энергии в оптически прозрачных кристаллах. Показано, что аномальное излучение генерируется в ультрафиолетовой, видимой и инфрактрасной областях спектра. Обсуждаются особенности экспериментального обнаружения и исследования аномального излучения.

1. Introduction

In the last few years the radiation of the channeled relativistic particles is investigated in many works (for example, see [1]). The experiments were made with observation of hard γ-radiation [2], quasi-linear X-ray radiation [3, 4], and optical radiation in the visible region [5, 6]. As was shown in [7, 8], at channeling radiation of charged particles the complex and anomalous Doppler effect is important. This effect determines the frequency spectrum and the angular distribution of the channeling radiation. For all this the frequency dispersion of the medium is of great importance, i.e., the dependence of the refraction coefficient of the medium upon the radiation wavelength, $n(\lambda)$. As was proposed by Frank [9] at oscillatory motion in the refractive medium with $n(\lambda) > 1$, the anomalous Doppler effect should be important. In this case the anomalous radiation is generated by the channeled particle due to the transition from the lower energy levels to the upper ones and the radiation wavelength decreases with the increase of the observation angle (in contrast to the normal Doppler effect).

This effect was not discovered up to the present time, because there was no real experimental situation. But now the channeling phenomenon of charged particles in a crystal gives the necessary unique situation. The system of channeled particles in a crystal medium represents a quasi-oscillator moving with relativistic velocity. In this case in the medium with $n(\lambda) > 1$ the anomalous Doppler effect should be revealed, because the channeled particle represents a source with own radiation frequency which moves through the medium with superlight velocity. Under real conditions the case $n(\lambda) > 1$ will be realized for visible radiation. Therefore, the investigation of the radiation properties in optically transparent crystals is of great interest for the observation of the anomalous Doppler effect.

---

[1]) P.O. Box 25, 634050 Tomsk, USSR.

2. Calculation of Transverse Energy Spectrum of Channeled Electrons

The realistic potential of atomic planes [10] does not allow to make an analytical consideration of the characteristics of channeled electrons. We used a model potential of the form

$$U(x) = -\frac{U_0}{\cosh^2 \alpha x}, \tag{1}$$

where U_0 is the depth and α the potential parameter, which characterizes the effective width of the potential well. In this case the value U_0 is set equal to the depth of the real potential well $U(x) = \sum_i V(x - id_p)$, where $V(x)$ is the potential of a single atomic plane [10], d_p the interplanar distance. The parameter α was chosen from the condition of conservation of the phase volume, i.e., from the conservation of a number of levels in the potential well. Table 1 represents the values U_0 and α for (110) atomic planes of some crystals. For potential (1) the eigenvalues of the transverse energy, defined by the Schrödinger equation for the transverse motion of channeled electrons, are equal to

$$\mathcal{E}_n = -\frac{(c\hbar\alpha)^2}{2E} (s - n)^2; \qquad n = 0, 1, 2, \dots, \tag{2}$$

where E is the total energy of the electron, c the light velocity, $\hbar$ the Planck constant, s a parameter characterizing the number of levels in the potential well,

$$N \leqq s = \frac{1}{2}\left(-1 + \sqrt{1 + \frac{8EU_0}{(c\hbar\alpha)^2}}\right). \tag{3}$$

Table 1 represents the number and position of the calculated subbarrier levels of the transverse energy for electrons with energy $E = 5.7$ MeV for (110) atomic planes of some crystals.

Table 1

Parameters of the effective potential (1) for (110) atomic planes of some crystals and values of sub-barrier transverse energy of 5.7 MeV electrons

target	z	U_0 (eV)	α $(\text{nm})^{-1}$	N	$\mathcal{E}$ (eV)			
					$n = 0$	$n = 1$	$n = 2$	$n = 3$
diamond	6	24.8	0.550	2	− 13.13	− 0.170	—	—
LiF	6	21.0	0.516	2	− 11.11	− 0.091	—	—
KCl	18	26.2	0.345	3	− 17.71	− 4.810	− 0.03	—
TlCl	49	132	0.569	3	− 98.93	−43.84	−10.86	—
TlBr	58	144	0.588	4	−108.51	−48.73	−12.56	−0.012

3. Calculation of Spectral and Angular Characteristics of Anomalous Channeling Radiation

It is known that at transition from the i-th to the j-th level of transverse energy the channeled electron radiates photons with the wavelength

$$\lambda_{ij} = \frac{2\pi c\gamma}{\omega_{ij}^0} \left| n(\lambda_{ij})\, \beta \cos \varphi - 1 \right|, \tag{4}$$

where $\omega_{ij}^0 = [\gamma(\mathscr{E}_i - \mathscr{E}_j)]/\hbar$ is the radiation frequency in the system with the electron at rest (the eigenfrequency of radiation), φ the observation angle, n the refraction parameter of the crystal medium, λ_{ij} the wavelength of channeling radiation outside the medium. The modulus $|n(\lambda_{ij})\,\beta\cos\varphi - 1|$ is determined by the frequency dispersion of the crystal medium $n = n(\lambda)$ in the direction of radiation expansion φ. At $n\beta\cos\varphi < 1$ the normal Doppler effect takes place and the normal radiation is emitted at the electron transition from a high-lying level to a lower one. At $n\beta\cos\varphi > 1$ the anomalous Doppler effect for the channeled particles must be realized and at the transition from a low-lying level to a higher one an electron must emit anomalous channeling radiation. At observation angles $\varphi < \varphi_{\check{C}}$ ($\varphi_{\check{C}}$ is the Čerenkov angle) for relativistic particles the condition of the anomalous Doppler effect reduces to the condition $n > 1$.

In the present calculations the frequency dispersion of the medium for a diamond crystal in ultraviolet and visible regions was set in the form

$$\frac{1}{n^2 - 1} = 0.2141 - \frac{2.678 \times 10^3}{\lambda^2}, \tag{5}$$

where λ should be taken in nm units. For KCl crystal we have interpolated the well-known dependence $n(\lambda)$ from the visible region to the short-wave one up to edge of the eigenabsorption band $\lambda = 163.2$ nm by the following expression:

$$n(\lambda) \approx 1.5 + \frac{4.8 \times 10^8}{\lambda^4}, \tag{6}$$

where λ should be taken in nm, too.

3.1 Angular dependence of the radiation wavelength

Fig. 1 shows the angular dependence of the wavelength of normal and anomalous radiation for electrons with energy $E = 5.7$ MeV at transitions $(0 \to 1)$ and $(1 \to 2)$ in case of (110) planar channeling in diamond (curves 1 to 3) and KCl crystals (curves 1′ to 3′). Here curves 1, 1′ show the angular dependence of anomalous radiation, 2, 2′ and 3, 3′ that of normal radiation in cases when $n > 1$ and $n = 1$, respectively. Curves 4, 4′ are the angular dependences of Čerenkov radiation. As is seen from Fig. 1, the anomalous radiation is emitted in the ultraviolet and visible regions. At $n > 1$ the normal radiation is emitted also in the ultraviolet and visible regions for observation angles $\varphi > \varphi_{\check{C}}$. At present only the case $n = 1$, $\varphi = 0$ has been investigated [2 to 4], i.e., only one point in the angular distribution in Fig. 1. The radiation regions $n > 1$, $\varphi > 0$ are practically not investigated.

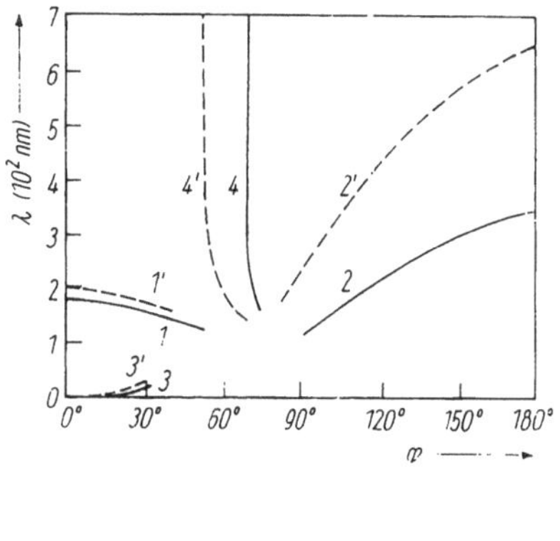

Fig. 1. Angular dependence of wavelength of anomalous (1, 1′), normal at $n > 1$ (2, 2′) and $n = 1$ (3, 3′), and Čerenkov (4, 4′) radiation. The energy of the electron is $E = 5.7$ MeV, transitions $(0 \leftrightarrow 1)$ in diamond and $(1 \leftrightarrow 2)$ in KCl crystal. ——— (110) diamond, – – – (110) KCl

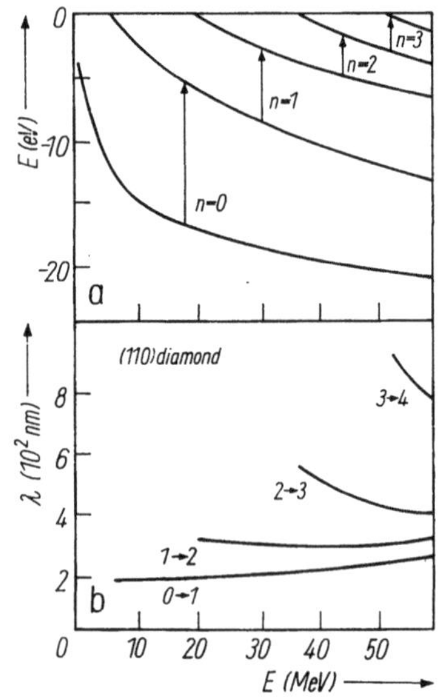

Fig. 2. Energetic dependence of a) transverse energy values and b) the wavelength of anomalous radiation in case $\varphi = 0$

3.2 Energetic dependence of the radiation wavelength

Fig. 2a shows the dependence of position and number of the transverse energy levels on the electron energy at (110) planar channeling in diamond crystal. The enhancement of the incident electron energy resulting in the increase of the number of sub-barrier levels of transverse energy and as a consequence leading to an increase of the number of spectral lines of anomalous radiation. Fig. 2b shows the energetic dependence of the wavelength of dipole ($0 \to 1$) spectral lines of anomalous radiation of channeled electrons in diamond at $\varphi = 0$. With the increase of electron energy anomalous radiation is generated at the transitions between the deep levels in the ultraviolet region, at the transitions between the middle levels — in the visible region, and at the transitions between the high-lying levels — in the infrared region of the radiation spectrum. For instance, in the ultrarelativistic limit for electrons with the energy $E \approx 1000$ MeV at (110) planar channeling in diamond the number of sub-barrier levels is $N \approx 20$, and the anomalous radiation will be generated in the visible region. At the transitions between the high-lying levels in this case the radiation will be generated in the infrared spectral region.

3.3 Above-barrier anomalous radiation

As is seen from Fig. 1 and 2, for electrons with energy 10 MeV the anomalous radiation is generated in the vacuum ultraviolet region at sub-barrier transitions under planar channeling of electrons, because $\mathcal{E}_i - \mathcal{E}_f \approx 10$ MeV. For the wavelength of the anomalous radiation to be in the visible region, e.g., for the direction $\varphi = 0$, the channeled electron with energy $E \leqq 10$ MeV must make a transition with a change of the transverse energy $\mathcal{E}_i - \mathcal{E}_f \approx 1$ to 2 eV in KCl and NaCl crystals and $\mathcal{E}_i - \mathcal{E}_f \approx 2$ to 4 eV in diamond. Such transitions of electrons with $E \leqq 10$ MeV are possible only in the above-barrier region of transverse energy, where the spectrum of the transverse energy eigenvalues is known to be quasi-continuous.

In the experiments [5, 6] the orientation and angular dependences of the optical radiation yield were measured under axial and planar channeling of electrons with energies $E = 2.0$ and 4.9 MeV in NaCl and KCl crystals, respectively. The light collection in this case was made in the angular region $|\varphi| \leqq \varphi_{\check{C}}$. Because the sub-barrier anomalous radiation propagating at $|\varphi| < \varphi_{\check{C}}$, as was mentioned above, is generated in the ultraviolet and the sub-barrier normal radiation in the visible region has emission angles $|\varphi| > \varphi_{\check{C}}$, evidently the electron transitions in the above-barrier region of transverse energy can generate the optical radiation which was observed in experiments [5, 6]. The intensity of the competitive kinds of optical radiation, such as Čerenkov and transition radiation, from the first estimations need not depend on the mutual orientation of crystal and incident electron beam. According to our estimations, Čerenkov radiation which is the most powerful one from those mentioned above, has an intensity comparable to the intensity of the anomalous above-barrier radiation.

4. Role of Refraction and Reflection of Channeling Radiation on the Boundary "Crystal–Vacuum"

Ultraviolet and visible radiation will undergo refraction and reflection while going out of the crystal medium to vacuum. This should result in the change of the really observed characteristics of the radiation outside the crystal in comparison with those inside the crystal.

4.1 Radiation refraction

If the crystal target is normally oriented to the direction of electron beam incidence, as is shown in Fig. 3a, the emission angle φ_{out} outside the crystal for photons with the propagation angle $0 < |\varphi_{in}| < \pi/2$ inside the crystal will be equal to

$$\varphi_{out} = \arcsin\left[n(\lambda) \sin \varphi_{in}\right], \tag{7}$$

and for "interior" angles $\pi/2 < |\varphi_{in}| < \pi$ the "exterior" angle is

$$\varphi_{out} = \pi - \arcsin\left[n(\lambda) \sin \varphi_{in}\right]. \tag{8}$$

Fig. 4 (curves 1, 2) shows the dependence of the wavelength of the anomalous ($1 \to 2$) spectral line on the emission angle inside and outside the (110) KCl crystal for electrons with energy $E = 5.7$ MeV. It is seen that the short-wave part of radiation undergoes the biggest influence of the refraction effect. For instance, the anomalous radiation with $\lambda \leqq 176.0$ nm undergoes total interior reflection ($\varphi_{out} \geqq 90°$), whereas the angle of propagation is $\varphi_{in} \geqq 30°$ inside the crystal. For the normal radiation outgoing into the back semi-sphere, with wavelength $\lambda \leqq 520.0$ nm, the "exterior" angle is $\varphi_{out} \leqq 90°$ (the total interior reflection), whereas $\varphi_{in} \geqq 135°$. Thus, for both normal and anomalous radiation the refraction effect leads to the phenomenon that the shortest-wave part is not observed outside the crystal, and in this case the true angular dependence of channeling radiation cannot be observed.

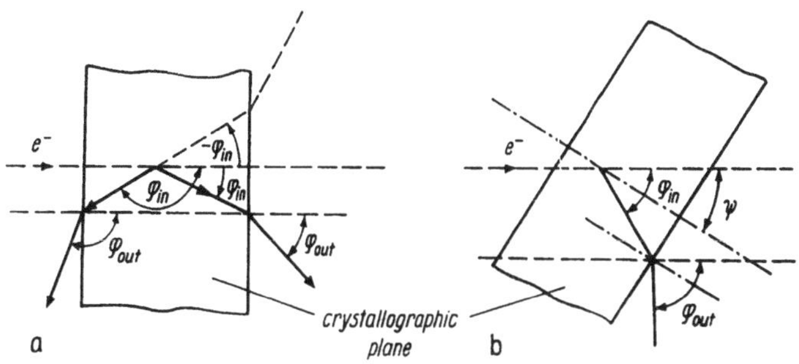

Fig. 3. Scheme of radiation refraction at the boundary "crystal–vacuum" for a) normal and b) inclined geometry. φ_{in} and φ_{out} are the angles of radiation emission inside and outside the crystal medium, respectively. ψ is the tilt angle of the crystal

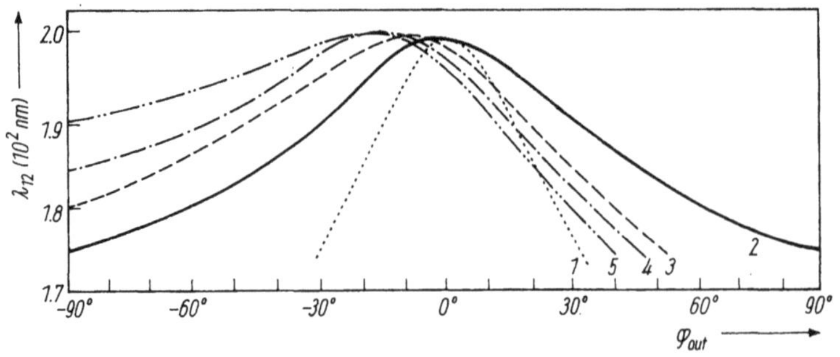

Fig. 4. Angular dependence of the wavelength of anomalous radiation. Curve 1 inside the crystal, curves 2 to 5 outside the crystal at values $\psi = 0°$, $10°$, $15°$, and $20°$, respectively

It is a most simple and realistic method to decrease the influence of the refraction effect and to approximate the angular distribution observed outside the crystal to the true distribution of channelling radiation. For this purpose it is necessary to tilt the crystal target in the plane which is perpendicular to the plane of the radiation emission at a definite angle. In this case the scheme of the beam refraction is shown in Fig. 3b. Then, the exterior angles of radiation propagation in contrast to (7) and (8) will be

$$\varphi_{out} = \arcsin\left[n(\lambda)\sin\left(\varphi_{in} - \psi\right)\right] + \psi\,, \tag{9}$$

$$\varphi_{out} = \pi - \arcsin\left[n(\lambda)\sin\left(\varphi_{in} - \psi\right)\right] + \psi\,, \tag{10}$$

where ψ is the tilt angle of crystal target. Fig. 4 shows the dependences of the wavelength of anomalous radiation ($1 \to 2$) for a (110) KCl crystal and 5.7 MeV electrons at some values of ψ. As is seen, the crystal tilting allows to reduce the influence of refraction on the short-wave part of radiation and to approach the values of the "exterior" angles of radiation propagation to their "interior" values. In particular, the refraction is evidently absent, i.e., $\varphi_{out} = \varphi_{in}$, for photon emission with the "interior" angles $\varphi_{in} = \psi$.

4.2 Radiation reflection

The reflection effect at the boundary "crystal–vacuum" results in a decrease of the radiation intensity outside the crystal target. As is generally known, at planar channeling the polarization vector of channeling radiation is perpendicular to the crystallographic plane. Therefore, the transmission coefficient in case when the plane of the

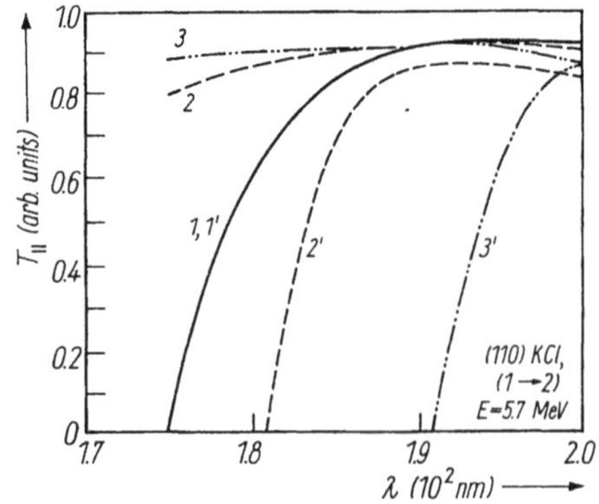

Fig. 5. Frequency dependence of the transmission coefficient of anomalous radiation ($1 \to 2$) for (110) KCl crystal and 5.7 MeV electron at the tilt angles of the crystal $\psi = 0°$ (1, 1′), $10°$ (2, 2′), $20°$ (3, 3′). Curves 1 to 3 are for $\varphi_{in} > 0$, curves 1′ to 3′ for $\varphi_{in} < 0$

radiation emission coincides with the crystallographic plane of electron channeling, will be

$$T_{\parallel} = 1 - \frac{\sin^2 (\varphi_1 - \varphi_2)}{\sin^2 (\varphi_1 + \varphi_2)} \tag{11}$$

and in the case when the plane of the radiation emission is perpendicular to the crystallographic plane,

$$T_{\perp} = 1 - \frac{\mathrm{tg}^2 (\varphi_1 - \varphi_2)}{\mathrm{tg}^2 (\varphi_1 + \varphi_2)}, \tag{12}$$

where $\varphi_1 = \varphi_{\text{in}} - \psi$ and $\varphi_2 = \varphi_{\text{out}} - \psi$ are the angles of radiation emission and refraction, respectively. Fig. 5 shows the dependence of the transmission coefficient $T_{\parallel}$ on the wavelength of channeling radiation for the $(1 \to 2)$ spectral line in case of a (110) KCl crystal and electrons with 5.7 MeV energy. The transmission coefficient $T_{\perp}$ has the same dependence and in the short-wave region this coefficient is slightly higher than $T_{\parallel}$. It is seen that the tilting of the crystal target leads to an increase of transmission of the short-wave radiation and the radiation intensity outside the crystal increases.

5. Concluding Remarks

The effect of the frequency dispersion $n(\lambda)$ for an optically transparent crystal leads to a new phenomenon, i.e., at channeling of mean-energy electrons besides an well-known X-ray radiation, anomalous ultraviolet and visible radiation should be emitted. Effects of reflection and refraction of the anomalous and normal radiation at the boundary "crystal–vacuum" essentially decrease the intensity of short-wave radiation and increase the real (outside the crystal) angle of observation. To decrease the influence of refraction and reflection effects it is reasonable to tilt the crystal target in the plane perpendicular to the plane of radiation emission. In this case the measured intensity of channeling radiation must be normalized by multiplying by a factor $\xi = \cos \psi$, where ψ is the tilt angle of crystal, because with crystal tilting their effective thickness is increased. The narrow spectral lines of channeling radiation occur as is known [3, 4] at sub-barrier transitions of the channeled particle. The radiation spectrum at above-barrier transitions is quasi-continuous. Therefore, to reduce the contribution of the continuous anomalous radiation and also the Čerenkov radiation, the thickness of the crystal target should not be greater than the dechanneling length of the sub-barrier states. In particular, for a two-level sub-barrier system the crystal thickness must be smaller than the dechanneling length of the ground state of the channeled particle.

In the case of rather thick crystals it is also the principal opportunity to increase the intensity of anomalous channeling radiation. For example, for the two-level system of the sub-barrier levels by using a well collimated electron beam, one may reach an inverted population of the ground state ($n = 0$) of the channeled electron. After a time τ_s^a the electron will make a spontaneous transition to the state $n = 1$ and emit the anomalous radiation quantum. From the state $n = 1$ to the state $n = 0$ the channeled electron will make a transition after a time τ_s^n with the emission of the normal radiation quantum. To decrease the time of the electron being in the state $n = 1$ it is necessary to stimulate the $(1 \to 0)$ transition, e.g., by the influence of an external electromagnetic field (laser radiation). In this case the direction of the external electromagnetic wave incidence on the crystal must be $\varphi_{\text{out}} > \varphi_{\check{C}}$ and connected with the radiation frequency according to (10). Then, the time of the electron being in the state $n = 1$ will essentially be reduced and the electron will get the possibility to emit a few anomalous quanta during its transition through the crystal. In this case the thickness of the crystal target must be a few lengths of the spontaneous anomalous radiation $L_s^a \approx c\tau_s^a$. Then the intensity of the anomalous radiation is increasing proportionally to the spectral density of the external electromagnetic field. For instance, we stimulate the $(1 \to 0)$ transition from 5.7 MeV electrons under (110) planar channeling in diamond, directing the external electromagnetic wave of $\lambda = 350.0$ nm at the angle $\varphi_{out} = 180°$. This procedure will give an increase of the anomalous radiation intensity in the vacuum ultraviolet region (see Fig. 1). In the case of a (110) KCl crystal by directing the laser ray of the visible light at the angle $\varphi_{out} \leqq 180°$ we get an enhancement of the anomalous radiation intensity in the vacuum ultraviolet.

In conclusion one should note that the observation and investigation of the anomalous radiation of channeled particles is promising for the study of phenomena connected with the motion of superlight radiators which have a set of eigenfrequencies, in comparison with Čerenkov radiation. The detection of the anomalous radiation and, therefore, the anomalous Doppler effect, has a basic physical importance and, probably, would find significant applications, e.g., in such scientific branches as astrophysics and plasma physics.

References

[1] R. Wedell, phys. stat. sol. (b) **99**, 11 (1980).
[2] B. N. Kalinin, V. V. Kaplin, A. P. Potylitsin, and S. A. Vorobiev, Phys. Letters A **70**, 447 (1979).
[3] M. I. Alguard, R. L. Swent, R. H. Pantell, B. L. Berman, S. D. Bloom, and S. Datz, Phys. Rev. Letters **42**, 1148 (1979).
[4] J. U. Anderson, K. R. Eriksen, and E. L. Laegsgaarol, Physica Scripta **24**, 588 (1981).
[5] A. A. Vorobiev, V. V. Kaplin, and S. A. Vorobiev, Nuclear Instrum. and Methods **127**, 265 (1975).
[6] Yu. N. Adishev, V. V. Kaplin, D. E. Popov, A. P. Potylitsin, E. I. Rozum, and S. A. Vorobiev, Phys. Letters A **81**, 40 (1981).
[7] V. G. Baryshevskii and I. Ya. Dubovskaya, Dokl. Akad. Nauk SSSR **231**, 1335 (1976).
[8] V. V. Beloshitskii, Phys. Letters A **64**, 95 (1977).
[9] I. M. Frank, Izv. Akad. Nauk SSSR, Ser. fiz. **6**, 3 (1942).
[10] B. R. Appleton, C. Erginsoy, and W. M. Gibson, Phys. Rev. **161**, 330 (1967).

(Received January 12, 1984)

phys. stat. sol. (b) **125**, 259 (1984)

Subject classification: 13

Laboratoire de Physique des Solides,
Equipe du Laboratoire de Cristallographie, Chimie et Physique des Solides, associé au C.N.R.S. n° 251, Universite de Caen[1])

Quantum Corrections to the Screening Length and Anisotropy in Magnetic Field

By

A. FORTINI

The self-consistent linearized Poisson equation leading to the screened potential of an ionized centre embedded in a free carrier plasma, is rewritten in a closed form taking into account the quantum character of the plasma. Complete resolution in the case of Boltzmann statistics clearly shows how quantum corrections cause a departure of the screening length from its usual semiclassical value. Quantum corrections are also responsible for the anisotropy of the screening in a static magnetic field. Calculations are worked out for the case of a strong quantizing field.

Die selbstkonsistente linearisierte Poissongleichung, die zu dem abgeschirmten Potential eines ionisierten Zentrums, das in einem Plasma freier Träger eingebettet ist, führt, wird in einer geschlossenen Form umformuliert unter Berücksichtigung des Quantencharakters des Plasmas. Eine vollständige Lösung für den Fall der Boltzmannstatistik zeigt deutlich, wie die Quantenkorrekturen eine Abweichung der Abschirmlänge von ihrem normalen halbklassischen Wert verursachen. Die Quantenkorrekturen sind auch verantwortlich für die Anisotropie der Abschirmung im statischen Magnetfeld. Berechnungen werden für den Fall eines starken quantisierenden Feldes durchgeführt.

1. Introduction

There has been much interest, for the last decades, in the theory of screening of ionized centres by free carriers in solid state physics. This is easily understood, on account, for instance, of the large number of problems which are concerned with screening in semiconductor physics. In many instances, it is quite sufficient to retain the usual value of the screening length, as given by the semiclassical self-consistent-field theory. In some cases, however, the semiclassical limit is unable to yield a correct evaluation of the screening. A typical example is exhibited by the anisotropic effects which occur in magnetic fields [1, 2]. As early pointed out by Horing [3], in the semiclassical model the magnetic field cannot have any influence on the static screening law, no matter how strong the field may be. Magnetic field effects only occur in conjunction with quantum effects arising from the quantum character of the free carrier plasma. Such quantum effects which may be of importance even in the absence of the field, result from the noncommutation of the free carrier Hamiltonian with the potential of the centre. They were calculated by Horing in the frame of the dielectric-function formalism [3]. This problem has been recently reviewed by Buot to within the distribution-function approach [4], and various results were derived therein in the form of expansions with respect to $\hbar^2$.

It is the purpose of this brief paper to show that quantum corrections can be included very simply, in a closed form, in the linearized contribution of free carriers to the electrostatic potential. The key of the derivation is the use of a suitable mathematical expansion giving, to first order, the exponential of two noncommuting

[1]) 14032 Caen Cédex, France.

operators, and the subsequent summation of matrix elements (Section 2). Two typical applications are next developed, assuming Boltzmann statistics. In the first one free electrons are taken in the effective mass approximation (Section 3). In the second one, the same electron gas is placed in a uniform magnetic field when the extreme quantum limit is reached (Section 4).

2. Quantum Expression of the Free Carrier Density Fluctuation

We start with a system of N free carriers of charge $-e$, of mean density $n_0 = N/\mathcal{V}$, in a perfect crystal of volume $\mathcal{V} = L_x L_y L_z$, whose motion is described by the unperturbed individual Hamiltonian H_0 with eigenvalues ϵ_α. For notational convenience, we will call $V(\boldsymbol{r})$ the individual electrostatic energy which is simply proportional (by a factor $-e$) to the electrostatic potential around an ionized centre which will be assumed to be placed at the origin. Let $+e$ be the charge of the centre, and denote by $V(\boldsymbol{q})$ the Fourier transform of $V(\boldsymbol{r})$,

$$V(\boldsymbol{q}) = \mathcal{V}^{-1} \int V(\boldsymbol{r}) \exp(-i\boldsymbol{q}\boldsymbol{r})\, d^3\boldsymbol{r}\,. \tag{1}$$

The Poisson equation is Fourier transformed into

$$V(\boldsymbol{q}) + \frac{e^2}{\mathcal{V}\varepsilon_l q^2}\,\delta n(\boldsymbol{q}) = -\frac{e^2}{\mathcal{V}\varepsilon_l q^2}, \tag{2}$$

where ε_l is the lattice dielectric constant and $\delta n(\boldsymbol{q})$ the q-th Fourier component of the free charge density fluctuation which is derived from the density matrix fluctuation $\delta\varrho$ by

$$\delta n(\boldsymbol{q}) = \mathrm{Tr}\,[\delta\varrho \exp(-i\boldsymbol{q}\boldsymbol{r})] \tag{3}$$

with

$$\delta\varrho = f\left[\frac{H_0 + V - \mu}{k_B T}\right] - f\left[\frac{H_0 - \mu}{k_B T}\right]. \tag{4}$$

f denotes the Fermi function and μ the Fermi energy which is defined in the unperturbed crystal as

$$\mathrm{Tr}\, f\left[\frac{H_0 - \mu}{k_B T}\right] = N\,. \tag{5}$$

$k_B T$ has its usual significance.

From a strictly classical point of view, if the term linear in $V/k_B T$ is only retained in $\delta\varrho$, (2) leads to the well-known Yukawa potential,

$$V(\boldsymbol{q}) = -\frac{e^2}{\mathcal{V}\varepsilon_l(q^2 + \lambda^{-2})},$$

where λ stands for the Thomas-Fermi or the Debye-Hückel length according to whether the free carrier plasma is degenerate or not. Instead, in a quantum-mechanical treatment, corrections arise because of the noncommutation rule $[H_0, V] \neq 0$, which leads to additional contributions, even to lowest order, in the expansion of $\delta\varrho$. We aim at compacting all of these lowest-order contributions into a simple expression.

First of all, we notice that, to first order in $V/k_B T$, the Fermi function of $H_0 + V$ can be expanded as follows:

$$f\left[\frac{H_0 + V - \mu}{k_B T}\right] - f\left[\frac{H_0 + V}{k_B T}\right] = -\left(\exp\frac{H_0 - \mu}{k_B T} + 1\right)^{-1} \times$$
$$\times \left(\exp\frac{H_0 + V - \mu}{k_B T} - \exp\frac{H_0 - \mu}{k_B T}\right)\left(\exp\frac{H_0 - \mu}{k_B T} + 1\right)^{-1} + \ldots. \tag{6}$$

We will next use the following expansion for the exponential of the sum of two non-commuting operators A, B [5]:

$$\exp(A + B) = \left[I + \sum_{n=0}^{\infty} \frac{(\text{ad } A)^n B}{(n+1)!}\right] \exp A$$

with $(\text{ad } A) B = [A, B]$ and $(\text{ad } A)^0 B = B$. This expansion is valid to first order in B. When applied to the Fermi function in (6), the trace appearing in (3) is given the following form:

$$\text{Tr}\,[\delta\varrho \exp(-i\boldsymbol{q}\boldsymbol{r})] = \sum_{\alpha} \langle\alpha|\,\delta\varrho \exp(-i\boldsymbol{q}\boldsymbol{r})\,|\alpha\rangle$$

$$= \sum_{\alpha,\beta} \langle\alpha|\,\delta\varrho\,|\beta\rangle\,\langle\beta|\exp(-i\boldsymbol{q}\boldsymbol{r})\,|\alpha\rangle$$

$$= -\sum_{\alpha,\beta}\left(\exp\frac{\epsilon_\alpha - \mu}{k_BT} + 1\right)^{-1} \langle\alpha| \sum_{n=0}^{\infty} \frac{(\text{ad } H_0)^n V}{(n+1)!\,(k_BT)^{n+1}}\,|\beta\rangle \exp\frac{\epsilon_\beta - \mu}{kT} \times$$

$$\times \left(\exp\frac{\epsilon_\beta - \mu}{k_BT} + 1\right)^{-1} \langle\beta|\exp(-i\boldsymbol{q}\boldsymbol{r})\,|\alpha\rangle\,.$$

It is shown in Appendix A that

$$\langle\alpha|\,(\text{ad } H_0)^n\,V\,|\beta\rangle = \epsilon_{\alpha\beta}^n V_{\beta'}^{\alpha} \qquad (\epsilon_{\alpha\beta} = \epsilon_\alpha - \epsilon_\beta)$$

whence

$$\langle\alpha| \sum \frac{(\text{ad } H_0)^n V}{(n+1)!\,(k_BT)^{n+1}}\,|\beta\rangle = \sum_{n=0}^{\infty} \frac{\epsilon_{\alpha\beta}^n}{(n+1)!} \frac{V_\beta^\alpha}{(k_BT)^{n+1}}$$

$$= [\exp(\epsilon_{\alpha\beta}/k_BT) - 1]\; V_\beta^\alpha/\epsilon_{\alpha\beta}\,,$$

and the subsequent final expression of the trace

$$\text{Tr}\,[\delta\varrho \exp(-i\boldsymbol{q}\boldsymbol{r})] = -\sum_{\alpha,\beta} \frac{e^{\frac{\epsilon_{\alpha\beta}}{k_B}} - 1}{\left(e^{\frac{\epsilon_\alpha - \mu}{k_BT}} + 1\right)\left(e^{-\frac{\epsilon_\beta - \mu}{k_BT}} + 1\right)} \frac{V_\beta^\alpha}{\epsilon_{\alpha\beta}} \langle\beta|\exp(-i\boldsymbol{q}\boldsymbol{r})\,|\alpha\rangle\,. \tag{7a}$$

In the Boltzmann limit ($|\mu/k_BT| > 1$), this result can be further simplified to

$$\text{Tr}\,[\delta\varrho \exp(-i\boldsymbol{q}\boldsymbol{r})] = -\sum_{\alpha\beta} e^{-\frac{\epsilon_\alpha - \mu}{k_BT}} \left(e^{\frac{\epsilon_{\alpha\beta}}{k_BT}} - 1\right) \frac{V_\beta^\alpha}{\epsilon_{\alpha\beta}} \langle\beta|\exp(-i\boldsymbol{q}\boldsymbol{r})\,|\alpha\rangle\,. \tag{7b}$$

The foregoing linearized expressions of the trace will now permit a convenient calculation of the free carrier contribution to the potential in (2).

3. Quasi-Free Electrons in the Effective Mass Approximation

The basic eigenstates are represented by $\boldsymbol{k}$-plane waves

$$\langle\boldsymbol{r}\mid\boldsymbol{k}\rangle = \mathcal{V}^{-1/2} \exp(-i\boldsymbol{q}\boldsymbol{r})$$

with eigenvalues $\epsilon_k = \hbar^2k^2/2m^*$. In that case

$$\langle\boldsymbol{k}'|\exp(-i\boldsymbol{q}\boldsymbol{r})\,|\boldsymbol{k}\rangle = \mathcal{V}^{-1}\delta_{\boldsymbol{k}',\boldsymbol{k}-\boldsymbol{q}}\,.$$

Analytic calculation of the trace, as given by (7), requires some approximation. In the strongly degenerate case, for example, Sommerfeld-type expansions can be used. In the following, we will restrict ourselves to Boltzmann statistics. Noting that

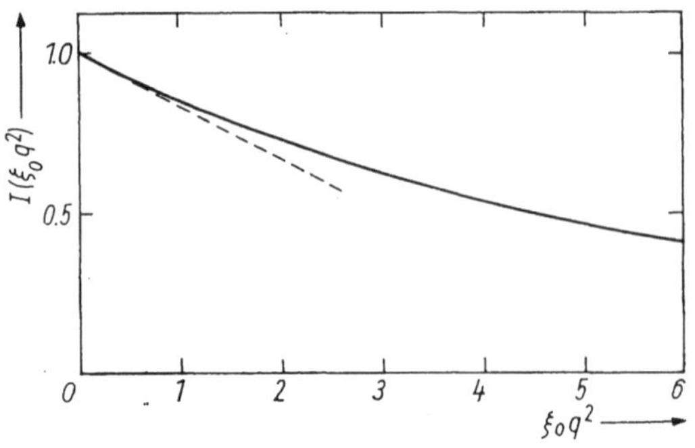

Fig. 1. Plot of the function $I(\xi_0 q^2)$ giving the quantum corrections to the squared inverse Debye length in the Fourier space. The slope of the dashed tangent is 1/6

$V^k_{k-q} = V(\boldsymbol{q})$, we obtain

$$\mathrm{Tr}\,[\delta\varrho \exp(-i\boldsymbol{q}\boldsymbol{r})] = -V(\boldsymbol{q}) \sum_k \mathrm{e}^{-\frac{\epsilon_k - \mu}{k_B T}} \left(\mathrm{e}^{\frac{\epsilon_{kk-q}}{k_B T}} - 1\right) \epsilon^{-1}_{kk-q} \,. \tag{8}$$

The summation over $\boldsymbol{k}$-states is performed in Appendix B, and leads to

$$\mathrm{Tr}\,[\delta\varrho \exp(-i\boldsymbol{q}\boldsymbol{r})] = \frac{\mathcal{V} V(\boldsymbol{q})\, I(\xi_0 q^2)}{(4\pi\xi_0)^{3/2}\, k_B T} \exp \frac{\mu}{k_B T}$$

with $\xi_0 = \hbar^2/2m^* k_B T$, and

$$I(\xi_0 q^2) = \sum_{n=0}^{\infty} \frac{(-\xi_0 q^2)^n}{2^n\,(n+1)!!} \,.$$

This function is represented in Fig. 1. On replacing the Fermi level from (5), which gives

$$\exp(\mu/k_B T) = (4\pi\xi_0)^{3/2}\, n_0 \,,$$

we finally obtain the solution of (2) in the form

$$V(\boldsymbol{q}) = -\frac{e^2}{\mathcal{V}\varepsilon_l[q^2 + \lambda_D^{-2} I(\xi_0 q^2)]}$$

with $\lambda_D^{-2} = n_0 e^2/\varepsilon_l k_B T$. It is observed that quantum corrections simply occur in the function $I(\xi_0 q^2)$ multiplying the squared inverse Debye length. For deriving approximate expressions, it is convenient to introduce the parameters

$$\alpha^2 = 2\xi_0 \lambda_D^{-2} = \frac{\hbar^2 n_0 e^2}{m^* \varepsilon_l (k_B T)^2} = \left(\frac{\hbar\omega_p}{k_B T}\right)^2$$

as in [3, 4] (ω_p denotes the electron plasma frequency). As long as $\alpha^2 \ll 1$, $I(\xi_0 q^2)$ can be truncated at second order which yields

$$V(\boldsymbol{q}) = -\frac{e^2}{\mathcal{V}\varepsilon_l[(1 - \alpha^2/12)\, q^2 + \lambda_D^{-2}]} \,. \tag{9}$$

Calculation of the original is now elementary and gives

$$V(\boldsymbol{r}) = -\frac{e^2 \exp(-r/\gamma^{1/2}\lambda_D)}{4\pi \mathcal{V}\varepsilon_l \gamma r} \tag{10}$$

with the notation $\gamma = 1 - \alpha^2/12$. The Debye length is approximately reduced to $\gamma^{1/2}\lambda_D$, i.e. in proportion of $\alpha^2/24$.

4. Anisotropic Screening by Quasi-Free Electrons in Quantizing Magnetic Field

Assuming the field to be applied along the z-axis, the motion of electrons is quantized in the xy plane according to the well-known Landau states specified with the quantum number N, k_y, k_z, (spin is ignored). Because of this quantization of basic states, anisotropic effects are expected to occur in the screening.

Let us restrict ourselves to the quantum limit where the only $N = 0$ states are occupied according to Boltzmann statistics. The relevant basic wave functions are written as

$$\langle \boldsymbol{r} \mid 0k_yk_z\rangle = \left(\sqrt{\pi}\, lVL_yL_z\right)^{-1/2} \exp\left[-\frac{(x+l^2k_y)^2}{2l^2} + i(k_yy + k_zz)\right], \tag{11}$$

where $l = (\hbar/eB)^{1/2}$ is the cyclotron radius. The related eigenvalues are given by

$$\epsilon_{0k_z} = \frac{1}{2}\,\hbar\omega_c + \frac{\hbar^2k_z^2}{2m^*},$$

where $\omega_c = eB/m^*$ is the cyclotron frequency. The redundant index 0 in ϵ_{0k_z} will be omitted from now on.

Using (7b) of the trace, the electron density fluctuation $\delta n(\boldsymbol{q})$ is now written as

$$\delta n(\boldsymbol{q}) = e^{\frac{\mu}{k_BT}} \sum_{\substack{k_yk_z\\ k_y'k_z'}} e^{-\frac{\epsilon_{k_z}}{k_BT}}\left(e^{\frac{\epsilon_{k_zk_z'}}{k_BT}} - 1\right) V^{0k_yk_z}_{0k_y'k_y'}\epsilon^{-1}_{k_zk_z'}\,\langle 0k_y'k_z'|\, e^{-i\boldsymbol{qr}}\,|0k_yk_z\rangle\,. \tag{12}$$

Matrix elements in the right-hand member are easily calculated by making use of the components of the Landau states on the plane waves

$$\langle \boldsymbol{k}_1 \mid 0k_yk_z\rangle = \left(\frac{2\pi l}{L_x}\right)^{1/2} \exp\left(il^2k_{1x}k_y - \frac{1}{2}\, l^2k_{1x}^2\right)\delta_{k_{1z}k_y}\delta_{k_{1z}k_z} \tag{13}$$

whence

$$\begin{aligned}\langle 0k_y'k_z'|\, e^{-i\boldsymbol{qr}}\,|0k_yk_z\rangle &= \sum_{\boldsymbol{k}_1} \langle 0k_y'k_z' \mid \boldsymbol{k}_1 - \boldsymbol{q}\rangle\, \langle \boldsymbol{k}_1 \mid 0k_yk_z\rangle\\ &= \exp\left[il^2\left(q_xk_y - \frac{1}{2}\,q_xq_y\right) - \frac{l^2}{4}\,(q_x^2 + q_y^2)\right]\delta_{k_y-q_y,\,k_y'}\delta_{k_z-q_z,\,k_z'}\end{aligned} \tag{14}$$

after some calculation and simplification. Similarly

$$\begin{aligned}V^{0k_yk_z}_{0k_y-q_yk_z-q_z} &= \sum_{k_1'k_2'} \langle 0k_yk_z \mid \boldsymbol{k}_1\rangle\, \langle \boldsymbol{k}_1|\, V\,|\boldsymbol{k}_2\rangle\, \langle \boldsymbol{k}_2 \mid 0k_y - q_yk_z - q_z\rangle\\ &= \frac{2\sqrt{\pi}\, l}{L_x} \sum_{k_{1x}k_{2x}} V(k_{1x} - k_{2x}, q_y, q_z)\, \times\\ &\quad\times \exp\left[-il^2(k_{1x} - k_{2x})\,k_y - il^2k_{2x}q_y - \frac{l^2}{2}\,(k_{1x}^2 + k_{2x}^2)\right].\end{aligned} \tag{15}$$

Substituting (14) and (15) into (12) gives

$$\begin{aligned}\delta n(\boldsymbol{q}) = \left(2\sqrt{\pi}\, l/L_x\right) e^{\frac{\mu}{k_BT}} &\sum_{k_yk_zk_{1x}k_{2x}} \left[e^{-\frac{\epsilon_{k_z}}{k_BT}}\left(e^{\frac{\epsilon_{k_zk_z-q_z}}{k_BT}} - 1\right)\epsilon^{-1}_{k_zk_z-q_z}\right]\times\\ &\times V(k_{1x} - k_{2x}, q_y, q_z)\exp\Bigg[-il^2(k_{1x} - k_{2x} - q_x)\,k_y - il^2\times\\ &\times\left(k_{2x} + \frac{q_x}{2}\right)q_y - \frac{l^2}{2}\,(k_{1x}^2 + k_{2x}^2) - \frac{l^2}{4}\,(q_x^2 + q_y^2)\Bigg].\end{aligned}$$

By summation over k_y, the constant $L_xL_y/2\pi l^2$ is factorized out with the condition

$$k_{1x} - k_{2x} = q_x$$

and the remaining integration over k_{2x} is elementary which gives

$$\delta n(\boldsymbol{q}) = (L_zL_y/2\pi l^2)\, V(\boldsymbol{q})\, e^{\frac{\mu}{k_BT}}\, e^{-\frac{l^2q_r^2}{2}} \sum_{k_z} e^{-\frac{\epsilon_{k_z}}{k_BT}} \left(e^{\frac{\epsilon_{k_zk_z-q_z}}{k_BT}} - 1\right) \epsilon^{-1}_{k_zk_z-q_z} \tag{16}$$

($q_r^2 = q_x^2 + q_y^2$). The integral in the right-hand member again reduces to the $I(\xi_0 q_z^2)$ function (Appendix B) and, introducing the electron concentration instead of the chemical potential, according to the relation

$$\exp\left[\left(-\tfrac{1}{2}\,\hbar\omega_c + \mu\right)/k_BT\right] = 4\pi^2 n_0 l^2 \hbar (2\pi m^* k_BT)^{-1/2}\,,$$

valid in Boltzmann statistics, we finally obtain

$$\frac{e^2}{\mathcal{V}\varepsilon_1}\,\delta n(\boldsymbol{q}) = \lambda_D^{-2} \exp\left(-\frac{1}{2}\, l^2q_r^2\right) I(\xi_0 q_z^2)\, V(\boldsymbol{q})\,,$$

leading to

$$V(\boldsymbol{q}) = -\frac{e^2}{\mathcal{V}\varepsilon_1[q^2 + \lambda_D^{-2} \exp\left(-\frac{1}{2}\, l^2 q_r^2\; I(\xi_0 q_z^2)\right)]}\,. \tag{17}$$

As expected, quantum corrections are the same as before along the z-direction, whereas in the xy plane, they are now strongly dependent on the magnetic field. Again approximate formulae can be written out, as long as $l^2 \ll \lambda_D^2$ and $\alpha^2 = (\hbar\omega_p/k_BT)^2 \ll 1$,

$$V(\boldsymbol{q}) \approx -\frac{e^2}{\mathcal{V}\varepsilon_1\left[q_r^2\left(1 - \frac{l^2}{2\lambda_D^2}\right) + q_z^2\left(1 - \frac{\alpha^2}{12}\right) + \lambda_z^2\right]}\,. \tag{18}$$

These lowest-order corrections were early obtained by Horing [3]. Using the simplified notations $\gamma_r = 1 - l^2/2\lambda_D^2$ and $\gamma_z = 1 - \alpha^2/12$, we find for the original of (18), in cylindrical (r, z) coordinates

$$V(r, z) = -\frac{e^2 \exp\left[-\frac{1}{\lambda_D}\left(\frac{r^2}{\gamma_r} + \frac{z^2}{\gamma_z}\right)^{1/2}\right]}{4\pi\mathcal{V}\varepsilon_1 \gamma_r \gamma_z^{1/2}\left(\frac{r^2}{\gamma_r} + \frac{z^2}{\gamma_z}\right)^{1/2}} \tag{19}$$

which is to be compared with (10). The screening length is still reduced by $1 - \alpha^2/24$ along the field, whereas in the transverse direction, the field-dependent reduction factor amounts to $1 - l^2/4\lambda_D^2$. Anisotropy can therefore be appreciated, in that limit, by the factor

$$\left(\frac{\gamma_r}{\gamma_z}\right)^{1/2} = 1 + \frac{1}{24}\left(\frac{\hbar\omega_p}{k_BT}\right)^2 \left(1 - 6\,\frac{k_BT}{\hbar\omega_c}\right).$$

Notice the particular value of the field, $\hbar\omega_c = 6k_BT$ for which isotropy is recovered. Such a critical field was also reported by Buot in the case of Fermi statistics [4].

5. Conclusion

We have been able to include, in a quite compact form, quantum corrections in the linearized equation obeyed by the self-consistent electrostatic field around a point charge screened by a free carrier plasma [(2) and (7)]. For the sake of brevity, applications have been restricted to two typical problems in Boltzmann statistics showing, on the one hand, the magnitude of quantum corrections in the Debye screening length and, on the other hand, anisotropic effects occurring under a quantizing field. A lot of physical situations of interest, such as degenerate statistics, weak magnetic field, magnetoresistance behaviour, could be investigated by means of the same equations. In many instances the method could likely permit a simple evaluation of the magnitude of quantum corrections, at least in the long-wavelength limit where mathematical simplifications are allowed.

Appendix A

Matrix elements of $(\mathrm{ad}H)^n$

We have for any operator A

$$\langle\alpha|\,[H_0, A]\,|\beta\rangle = \epsilon_{\alpha\beta}A^\alpha_\beta \qquad (\epsilon_{\alpha\beta} = \epsilon_\alpha - \epsilon_\beta)\,.$$

Hence, assuming that

$$\langle\alpha|\,(\mathrm{ad}\,H_0)^{n-1}\,V\,|\beta\rangle = \epsilon^{n-1}_{\alpha\beta}V^\alpha_\beta\,,$$

we can readily infer that

$$\langle\alpha|\,(\mathrm{ad}\,H_0)^n\,V\,|\beta\rangle = \langle\alpha|\,[H_0, (\mathrm{ad}\,H_0)^{n-1}\,V]\,|\beta\rangle\,,$$
$$= \epsilon_{\alpha\beta}\langle\alpha|\,(\mathrm{ad}\,H_0)^{n-1}\,V\,|\beta\rangle = \epsilon^n_{\alpha\beta}V^\alpha_\beta\,.$$

The result is thus established by a recursive procedure.

Appendix B

Intraband integrations in the calculation of the trace

In the case of parabolic dispersion law

$$\frac{\epsilon_k}{k_\mathrm{B}T} = \frac{\hbar^2k^2}{2m^*k_\mathrm{B}T} = \xi_0k^2\,,$$

the integration over $\boldsymbol{k}$-states appearing in (8) can be written as

$$\sum_k \frac{\mathrm{e}^{-\xi_0k^2}\,[\mathrm{e}^{\xi_0(2\boldsymbol{qk}-q^2)} - 1]}{\xi_0k_\mathrm{B}T(2\boldsymbol{qk} - q^2)} = \sum_k \frac{\int_0^{\xi_0} \mathrm{e}^{-\xi_0k^2+\xi(2\boldsymbol{qk}-q^2)}\,\mathrm{d}\xi}{\xi_0k_\mathrm{B}T}$$
$$= \frac{\mathcal{V}}{8\pi^3k_\mathrm{B}T\xi_0}\int_0^{\xi_0}\mathrm{e}^{-\xi q^2}\,\mathrm{d}\xi\int_0^\infty 2\pi\,\mathrm{e}^{-\xi_0k^2}\,k^2\,\mathrm{d}k\int_0^\pi \mathrm{e}^{-2\xi qk\cos\theta}\sin\theta\,\mathrm{d}\theta\,.$$

The integrations over θ and k are straightforward. We are left with a final integration over ξ which can be written as

$$\sum_k = \frac{\mathcal{V}}{8\pi^{3/2}\xi_0^{1/2}}\,I(\xi_0q^2)\,,$$

where

$$I(\xi_0q^2) = \int_0^1 \mathrm{e}^{(x^2-x)\,\xi_0q^2}\,\mathrm{d}x = \sum_{n=0}^\infty \frac{(-\xi_0q^2)^n}{2^n(2n+1)!!}$$

with $(2n+1)!! = 1.3.5\ldots(2n+1)$.

In the case of Landau's states (one-dimensional parabolic band) with

$$\epsilon_{k_z} = \frac{\hbar\omega_c}{2k_B T} + \xi_0 k_z^2 ,$$

the summation appearing in (16) is reduced to the same I function of $\xi_0 q_z^2$,

$$\mathrm{e}^{-\frac{\hbar\omega_c}{2k_B T}} \sum_{k_z} \frac{\mathrm{e}^{-\xi_0 k_z^2} [\mathrm{e}^{\xi_0(2q_z k_z - q_z^2)} - 1]}{\xi_0 k_B T(2q_z k_z - q_z^2)} =$$

$$= \frac{L_z}{2\pi k_B T} \mathrm{e}^{-\frac{\hbar\omega_c}{k_B T}} \int_0^{\xi_0} \frac{\mathrm{d}\xi}{\xi_0} \int_{-\infty}^{+\infty} \mathrm{e}^{-\xi_0 k_z^2 + \xi(2q_z k_z - q_z^2)} \,\mathrm{d}k_z$$

$$= \frac{L_z}{2\pi k_B T} \mathrm{e}^{-\frac{\hbar\omega_c}{k_B T}} \left(\frac{\pi}{\xi_0}\right)^{1/2} \int_0^{\xi_0} \mathrm{e}^{\left(\frac{\xi^2}{\xi_0} - \xi\right) q_z^2} \frac{\mathrm{d}\xi}{\xi_0} = \frac{L_z}{2\pi k_B T} \left(\frac{\pi}{\xi_0}\right)^{1/2} \mathrm{e}^{-\frac{\hbar\omega_c}{k_B T}} I(\xi_0 q_z^2) .$$

References

[1] P. R. WALLACE, J. Phys. C 7, 1136 (1974); 8, 3608 (1975).
[2] P. BOGUSLAWSKI and J. MYCIELSKI, J. Phys. C 13, 1019 (1980).
[3] N. J. HORING, Ann. Phys. 54, 405 (1969).
[4] F. A. BUOT, Phys. Rev. B 14, 977 (1976).
[5] N. BOURBAKI, Groupes et Algebre de Lie, Chap. 2, Hermann, Paris 1972.

(Received March 16, 1984)

phys. stat. sol. (b) 125, 267 (1984)

Subject classification: 13.1 and 17; 22.3

Sektion Physik der Martin-Luther-Universität Halle/Wittenberg, Halle (Saale)[1])

Angular-Resolved X-Ray Photoelectron Spectra of Valence and Core Levels from GaS

By

T. CHASSÉ, U. BERG, and O. BRÜMMER

Angular-resolved X-ray photoelectron spectra (ARXPS) are measured for valence bands and Ga core levels from the layer compound GaS. The spectra are interpreted taking into account initial and final state effects. Structures of dominating p_z- and $p_{x,y}$-like character are identified in the valence region. The core-level spectra are qualitatively understood on the basis of scattering of photoelectrons within the solid.

Es werden winkelaufgelöste Röntgenphotoelektronenspektren (ARXPS) der Valenzbänder und von Ga-Rumpfniveaus der Schichtverbindung GaS gemessen. Die Spektren werden unter Beachtung von Anfangs- und Endzustandseffekten ausgewertet. Im Valenzbandbereich werden Strukturen mit dominierendem p_z- bzw. $p_{x,y}$-Charakter identifiziert. Die Rumpfniveauspektren können in Zusammenhang mit der Photoelektronenstreuung im Festkörper qualitativ verstanden werden.

1. Introduction

During the last ten years GaS has been the subject of several investigations. The steadily growing interest in this substance was stimulated by both, the high anisotropy and the remarkable photoelectric properties of the layered III–VI semiconductors.

GaS has β-GaSe structure [1]. The crystals are built up of sandwiches of four atomic layers in the sequence S–Ga–Ga–S. A considerable knowledge about the electronic structure of GaS has been obtained based on photoelectron [2 to 7], X-ray emission [7, 8], and optical spectra (see, e, g., [9], as well as from calculations of band structure or density of states [10 to 13].

Here angular-resolved X-ray photoelectron spectra (ARXPS) of valence bands and selected Ga core levels of GaS are presented. The spectra are interpreted on the basis of simple models. In this manner they may lead to a refined knowledge about the substance of interest. On the other hand, it is possible to conclude about the quality of the models used for interpretation of the ARXPS because such experimental and theoretical work has been done until now.

2. Theoretical Considerations

The intensity I of photoelectrons emitted into direction $\boldsymbol{P}$ ($P = \sqrt{2mE}$) with energy E may be described by

$$I(E, \boldsymbol{P}) \sim \sum_i |M_i(\boldsymbol{P})|^2 \, \delta(E - E_i - \hbar\omega) \,, \tag{1}$$

$$M_i(\boldsymbol{P}) = \int d^3r \, \varphi_f^*(\boldsymbol{r}, \boldsymbol{P}) \, H'(\boldsymbol{r}) \, \varphi_i(\boldsymbol{r}) \,, \tag{2}$$

[1]) Friedemann-Bach-Platz 6, DDR-4020 Halle (Saale), GDR.

where $\varphi_i(\boldsymbol{r})$ are the wave functions of the crystal, $H'(r)$ is the operator of the perturbation by photons with energy $\hbar\omega$, and $\varphi_f(\boldsymbol{r}, \boldsymbol{P})$ represents the wave function of the final state of photoemission. It is defined by [14]

$$\varphi_f^*(\boldsymbol{r}, \boldsymbol{P}) = e^{-i\boldsymbol{P}\boldsymbol{r}} + \int d^3r' \, e^{-i\boldsymbol{P}\boldsymbol{r}'} \, V(\boldsymbol{r}') \, G(\boldsymbol{r}', \boldsymbol{r}, E) \, . \tag{3}$$

The first term in (3) represents a plane wave in the direction to the spectrometer. The second term gives the corrections due to the interactions of the electron waves with the potential $V(\boldsymbol{r})$ of the crystal. $G(\boldsymbol{r}', \boldsymbol{r}, E)$ is the Green's function of the crystal with surface. The intensity is determined by properties of the initial states as well as the final states.

2.1 Valence band spectra

Within the usual three-step model of the photoemission some attempts have been made to achieve improved approximations for the optical excitation step [15]. But in most cases a description of the XPS intensity has been done on the basis of plane wave final states (PWFS) as well as the neglect of electron scattering and diffraction at the surface [16 to 18]. That means a neglect of the second term in (3).

If the initial states $\varphi_i(\boldsymbol{r})$ are represented by a LCAO approach,

$$\begin{aligned} \varphi_i(\boldsymbol{r}) &= \varphi_{\boldsymbol{k},\nu}(\boldsymbol{r}) = \frac{1}{\sqrt{N}} \sum_{\boldsymbol{R}_n} \sum_j e^{i\boldsymbol{k}\boldsymbol{R}_n} \, \chi_j(\boldsymbol{r} - \boldsymbol{R}_n) \, c_j(\boldsymbol{k}, \nu) \, , \\ \chi_j(\boldsymbol{r}) &= R_j(r) \, Y_{l_j m_j}(\hat{\boldsymbol{r}}) \, , \end{aligned} \tag{4}$$

then the matrix elements may be expressed by

$$|M_i(\boldsymbol{P})|^2 = |M_{\boldsymbol{k},\nu}(\boldsymbol{P})|^2 \sim |\sum_j c_j(\boldsymbol{k}, \nu) \, Y^*_{l_j m_j}(\hat{\boldsymbol{P}})|^2 \, . \tag{5}$$

Here it is assumed that the solid contains only one atomic component, and that only electron states of one l quantum number give contributions to the ν-th band. This well-known result describes the fact that the angular dependence of matrix elements is characterized by atomic orbitals. At very low temperatures only direct transitions (DT) take place. Then the angular dependence of the intensity is strongly influenced by the density of states in that plane in the first Brillouin zone (BZ) that gives contributions to the observed intensity.

At higher temperatures indirect transitions (NDT) become dominant because of the increasing electron–phonon interaction. Thus, the temperature dependence of photoemission intensity may be understood in terms of a Debye-Waller model, as has been demonstrated recently [18, 19],

$$I(E, \boldsymbol{P}, T) = I_{DT}(E, \boldsymbol{P}) \, W(T) + [1 - W(T)] \, I_{NDT}(E, \boldsymbol{P}) \, , \tag{6}$$

$W(T)$ is the Debye-Waller factor.

XP spectra measured at room temperature are generally dominated by I_{NDT} [18]. That is why all states within the first BZ may contribute to the angular-resolved spectrum, and (5) may be simplified with respect to the symmetry of the crystals.

For a band of p-electrons in a hexagonal or trigonal solid we obtain [23]

$$I_p(E, \boldsymbol{P}) \sim 2N_{p_z}(E) \, p_z^2(\hat{\boldsymbol{P}}) + \left(N_{p_x}(E) + N_{p_y}(E)\right) \left(p_x^2(\hat{\boldsymbol{P}}) + p_y^2(\boldsymbol{P})\right) , \tag{7}$$

where N_{p_η} ($\eta = x, y, z$) are the orbital projections of the density of states with respect to the p_η-orbitals. The contributions of these different orbital projections are weighted by the angular parts of the orbitals themselves. The valence band spectra may be considered as a superposition of contributions like (7) from different compounds and different l-values (s, p, d electrons, etc.) each of them weighted with the characteristic total photoelectric cross section. The angular dependence of valence band peaks is determined by the symmetry of the dominating atomic orbital. Nevertheless it must be kept in mind that interferences and all final state effects are neglected within this picture. However, from experimental investigations of angular distributions of core level intensities it can be concluded that final state effects as scattering of the photoelectrons cause significant intensity variations. Recent experiments have shown the appearance of such effects in valence band spectra, too [16, 20]. Thus, the consideration of photoelectron scattering is of importance for the interpretation of angular-resolved valence band spectra.

2.2 Core-level spectra

Core levels are strongly localized and possess spherically symmetric charge distributions. Thus, they are of great interest for studies of final state effects. The initial states for core level emission may be described by atomic functions

$$\varphi_i(\boldsymbol{r}) = \varphi_c(\boldsymbol{r}) = \Phi_{l_c}(r)\; Y_{l_c m_c}(\hat{\boldsymbol{r}}) \tag{8}$$

and a partial wave description is used for the final states

$$\varphi_f(\boldsymbol{r}, \boldsymbol{P}) = \sum_{lm} B_{lm}^{*}(\boldsymbol{P})\; \Phi_l(r)\; Y_{lm}(\hat{\boldsymbol{r}})\,. \tag{9}$$

Summation over all emitters (at $\boldsymbol{R}_0$) gives the intensity

$$I(\boldsymbol{P}) \sim \sum_{\boldsymbol{R}_0} \sum_{m_c} \Big|\sum_{lm} B_{lm}^{\boldsymbol{R}_0}(\boldsymbol{P})\; \langle \Phi_l Y_{lm}|\; H'\; |\Phi_{l_c} Y_{l_c m_c}\rangle\Big|^2\,. \tag{10}$$

The angular dependence of the core level intensity is determined by the coefficients $B_{lm}^{\boldsymbol{R}_0}$. Within the framework of the small scattering centre approximation [21] and considering only single scattering events the $B_{lm}^{\boldsymbol{R}_0}$ coefficient becomes for an emitter at $\boldsymbol{R}_0 = \boldsymbol{0}$ [22]

$$B_{lm}^{0}(\boldsymbol{P}) \sim Y_{lm}(\hat{\boldsymbol{P}}) + \sum_{\boldsymbol{R} \neq \boldsymbol{0}} f^{\boldsymbol{R}}(\boldsymbol{P}, \boldsymbol{R})\; e^{-i\boldsymbol{P}\boldsymbol{R}}\; \frac{e^{-i\boldsymbol{P}\boldsymbol{R}}}{R}\; Y_{lm}(\hat{\boldsymbol{R}})\,. \tag{11}$$

The first term in (11) represents the direct contribution of the emitter itself. The second term takes into account all the electron waves scattered at atomic sites $\boldsymbol{R}$ travelling to the spectrometer. $f^{\boldsymbol{R}}(\boldsymbol{P}, \boldsymbol{R})$ is the scattering amplitude of the atoms at position $\boldsymbol{R}$. It should be noted that the arguments of the Y_{lm} in these two terms differ from each other. But at electron energies of about 1000 eV the forward scattering dominates.

In Fig. 1 the scattering amplitude for the Ga(3d) electron energy in GaS is shown as an example. The amount of the scattering amplitude possesses a sharp maximum for scattering into forward direction and has small values far from the maximum. In a first approximation only those contributions are essential for which the direction of emission is nearly parallel to the line between the emitting atom and a neighboured scatterer. Under these conditions the difference between the arguments of Y_{lm} in the direct and scattering term of (11) is negligible. Then the anisotropy of intensity due to scattering effects is independent of the anisotropies due to the initial states in this approximation.

Thus, structures in the angular resolved core level spectra may be identified on the basis of crystal structure. On the other hand, the use of the model described in Section 2.1 is justified for a qualitative interpretation of valence band spectra provided the final state effects have been eliminated by suitable normalization of intensities.

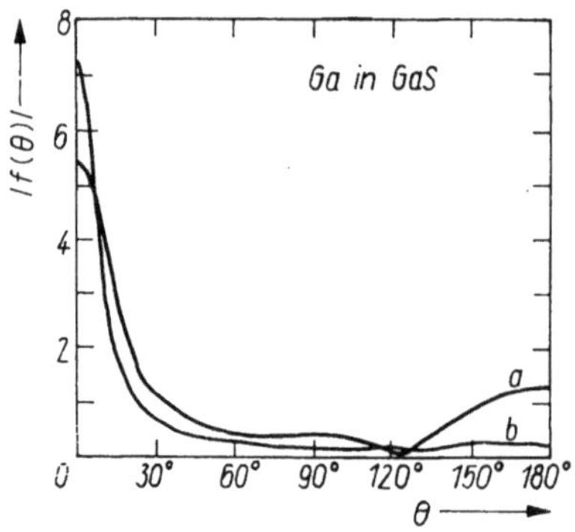

Fig. 1. Absolute value of the scattering amplitude for Ga core electrons in GaS (AlK$_\alpha$ excitation); (a) Ga($2p_{3/2}$), (b) Ga(3d) [22]

3. Experimental

The photoemission measurements were performed using a Hewlett Packard 5950 A photoelectron spectrometer modified to permit two axis rotations of the specimen in situ [17]. GaS samples were oriented by the help of Laue diffraction patterns. The crystals were cleaved under dry nitrogen and then inserted into the vacuum chamber (10^{-9} Torr). The surfaces were examined by core-level spectroscopy, yielding low levels of contaminants (<0.3 ML oxygen, <0.8 ML carbon).

4. Results and Discussion

4.1 Valence band spectra

Valence band spectra of GaS were measured in dependence on azimuthal angle φ and polar angle θ. These angles were chosen in such a manner that $\theta = 0°$ corresponds to [0001] and $\theta = 90°$, $\varphi = 30°$ corresponds to [1000].

In Fig. 2 a sequence of spectra measured at different azimuthal angles is shown. These angular-resolved spectra possess three main peaks, labelled A, B, and C. Although band structure calculations give a somewhat more differentiated picture, these peaks may be roughly characterized as p-like, Ga s-like, and S s-like structures of the valence bands. The strong rise of intensity below C is due to the tails of the Ga (3d) core peak. This typical shape of the GaS valence band spectra is well-known from angular-integrated photoemission measurements [2 to 4].

The variation of the azimuthal angle φ does not cause any significant changes of the spectra. Shape as well as relative intensities of the shown peaks and shoulders remain unchanged.

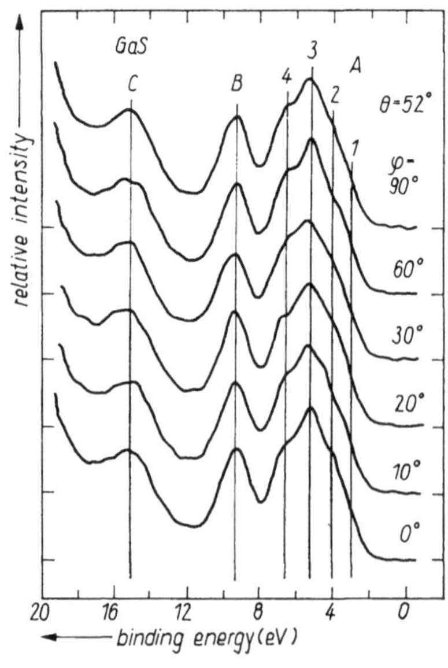

Fig. 2 Angular-resolved valence band spectra of GaS for different azimuthal angles

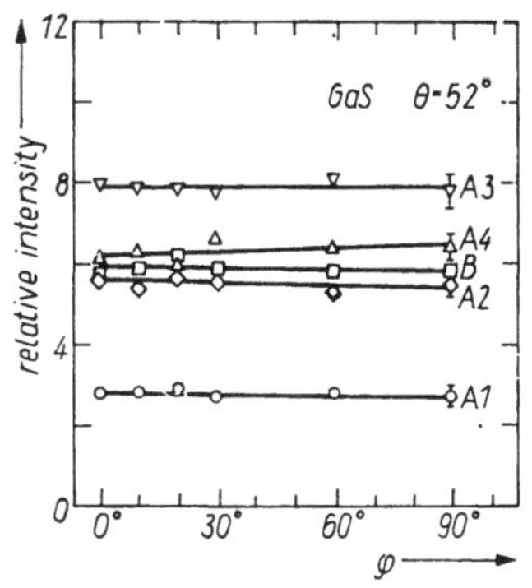

Fig. 3

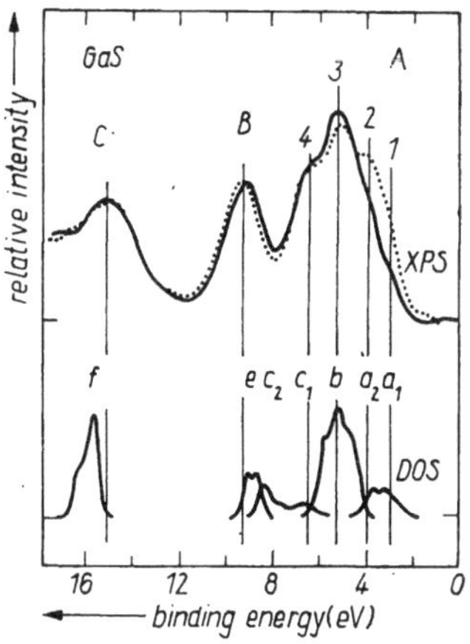

Fig. 4

Fig. 3. Intensity of significant valence band features in dependence on azimuthal angle (see Fig. 2)
Fig. 4. Angular-resolved valence band spectra of GaS for two different polar angles (—— $\theta = 75°$ and ..... 35°) and a theoretical density of states (Antonangeli et al. [4])

This is also demonstrated in Fig. 3. There the intensities of the valence band features are normalized with respect to the s-like peak C.

Thus, p_x- and p_y-like states cannot be distinguished experimentally and are degenerated in the GaS structure. This was to be expected from (7) because of the hexagonal symmetry of the crystals.

It should be noted that photoelectron scattering effects can be neglected approximately at the emission angles of interest for the valence band studies. Otherwise, the relative intensity of the Ga s-like peak B would not be constant in Fig. 3.

On the other hand, strong changes of relative intensities are observed after variation of the polar angle θ. In Fig. 4 a direct comparison of spectra taken at $\theta = 35°$ and $\theta = 75°$ ($\varphi = 0°$) is shown. These spectra were corrected with respect to inelastic losses as well as angular dependence of the instrumental resolution by the help of an inverse core loss function and an additional use of a numerical deconvolution filter [23].

A comparison of relative intensities shows that the structures A1 and A2 at the valence band edge possess dominating p_z-symmetry. This is in accordance with the results measured on GaSe [24]. But at A3 the p-states are nearly degenerated with only a slight dominance of $p_{x,y}$-like states. The other peaks and shoulders do not show significant anisotropies. Thus, these structures of the GaS valence bands seem to be due to bands with s- or p-character. In Fig. 4 the tight-binding (TB) DOS of Antonangeli et al. [4] is also shown. This DOS is resolved into the contributions of the different groups of bands (a to f). The maxima of DOS and ARXP spectra were set to equal binding energies to achieve a unique energy scale.

A comparison of experimental results and theoretical calculations allows a detailed insight into the structure of the GaS valence bands. After [4, 12] the a-bands are built up of S p_z-states as well as of hybridized s- and p_z-states of Ga. This is in good agreement with the anisotropy observed experimentally. On the other hand, the b-bands of [4, 12] are mainly based on S $p_{x,y}$-states. But as already discussed above, the changes of intensity at A3 are small and thus only slight dominance of $p_{x,y}$-like states is indicated by the photoemission measurements.

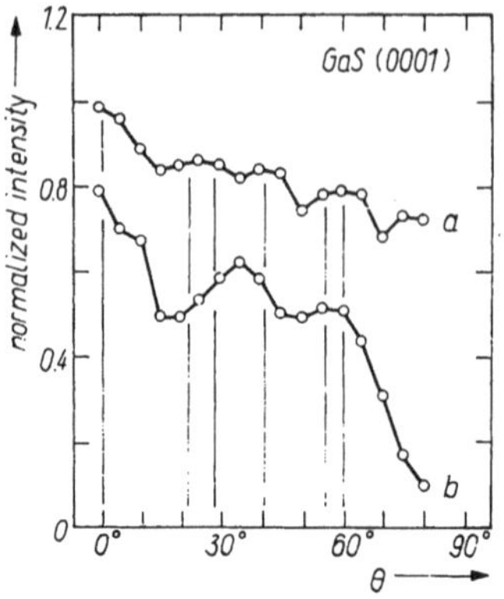

Fig. 5. Intensity of Ga core levels in dependence on polar angle; (a) Ga(3d), (b) Ga($2p_{3/2}$) (spectrum b has been shifted downwards by 0.2 units on the intensity scale)

The e- and c-bands are generated by bonding and antibonding combinations of Ga s-orbitals. This will be in accordance with the ARXPS results, if only the intensities are considered. But more recent calculations of Robertson [13] as well as polarized XES of Brümmer et al. [7] indicate significant S $p_{x,y}$ admixture to these bands. The contribution of these p-like states to the intensity of peak B might be covered because of the higher photoionization cross section of the Ga(4s) electrons.

On the other hand, a detailed analysis of the position of peak B shows a significant shift in dependence on polar angle. That means the difference between the actual binding energies of A3 and B will decrease about 0.5 eV if the angle of emission is raised from $\theta = 35°$ to $\theta = 75°$. This peak shift is an indication for p admixture to the s-like states within that binding energy region, as it was predicted theoretically by Robertson.

Although being not discussed in detail there exists generally a good agreement between the results of ARXPS and polarized XES [7, 8] as well as photoemission with synchrotron radiation [6]. Thus, we may conclude that the plane wave final state approximation allows the qualitative interpretation of ARXPS of the GaS valence bands.

4.2 Core levels

In Fig. 5 angular-resolved photoemission spectra of Ga(3d) and Ga($2p_{3/2}$) core levels are presented. The azimuthal angle φ corresponds to [1000]. The measured intensities were divided by the intensity response function of the spectrometer, and both curves were normalized to the normal emission intensity [23].

Table 1

Polar angles θ that are essential to forward scattering for the spectra of Fig. 5, related distance d between emitter and scatterer, and atomic nature of the scatterer

θ	d (nm)	scatterer
0°	0.24	Ga, (S)
22.1°	0.95	S
28.2°	0.76	S
40.5°	0.55	S
55.3°	0.44	Ga
59.7°	0.83	S

In Table 1 those emission angles θ are summarized, which are important in the concept of dominant forward scattering. As discussed in Section 2.2 at these polar angles structures in the spectra are expected. Additionally, the distances between emitting and scattering atoms are given in Table 1.

The Ga(3d) spectrum shows a considerable agreement between experimental peak positions and polar angles in Table 1. But all the structures of this angular-resolved Ga(3d) spectrum are only relatively weak.

The spectrum obtained from Ga($2p_{3/2}$) possesses one strong and broad peak at about $\theta = 35°$ instead of two peaks at 25° to 30° and 40°. The main reason for this striking difference between spectra of Ga(3d) and Ga($2p_{3/2}$) electrons is the energy dependence of the scattering amplitude. As presented in Fig. 1, the maximum of the scattering amplitude for Ga($2p_{3/2}$) photoelectrons with about 360 eV is much weaker and broader in the forward direction than that of photoelectrons emitted from Ga(3d) with about 1460 eV. Thus, the peak at 35° in the spectrum of Ga($2p_{3/2}$) electrons might be due to a superposition of broad neighboured structures corresponding to the distances of 0.76 nm ($\theta = 28.2°$) and 0.55 nm ($\theta = 40.5°$), respectively. Scattering at the S atom with a distance of 0.95 nm from the emitting Ga atom is of much less importance for the Ga($2p_{3/2}$) photoelectrons than for the Ga(3d) electrons. This effect can be understood on the basis of the energy dependence of the mean free path. At 1460 eV the mean free path is about two times greater than that at 360 eV electron energy. The strong decrease of the Ga($2p_{3/2}$) intensity at very high polar angles θ is due to the increasing influence of the contributions of surface contaminations.

Thus, the scattering concept is suitable for a qualitative interpretation of these core spectra for GaS. In the general case, careful calculations are needed to achieve a more quantitative interpretation of angular-resolved core level spectra.

5. Conclusions

The angular dependence of structures in the ARXP spectra is influenced by both initial and final states of the photoemission process.

The anisotropy of the GaS valence band spectra is due to the symmetry properties of crystal wave functions in the directions of investigation. The ARXPS results obtained are consistent, with polarized K-XES as well as further PES investigations.

Angular-resolved spectra of Ga core levels can be qualitatively interpreted on the basis of the scattering concept. Changes of the spectra in dependence on the kinetic energy of photoelectrons reflect changes in scattering amplitude and mean free path.

Acknowledgements

The authors would like to thank A. Chassé for making available some results prior to publication. Thanks are also due to J. Lärz for technical assistance.

References

[1] A. Kuhn, A. Chevy, and R. Chevalier, Acta cryst. **B32**, 983 (1976).
[2] J. M. Thomas, I. Adams, R. H. Williams, and M. Barber, J. Chem. Soc. Faraday Trans. II **68**, 755 (1972).
[3] F. R. Shepherd and P. M. Williams, Phys. Rev. B **12**, 5705 (1975).
[4] F. Antonangeli, M. Piacentini, A. Balzarotti, V. Grasso, R. Girlanda, and E. Doni, Nuovo Cimento **51B**, 181 (1979).
A. Balzarotti, R. Girlanda, V. Grasso, E. Doni, F. Antonangeli, and M. Piacentini, Canad. J. Phys. **56**, 700 (1978).

[5] J. F. Mc GILP and A. W. PARKE, phys. stat. sol. (b) **94**, 685 (1979).
[6] J. E. ROWE, G. MARGARITONDO, H. KASPER, and A. BALDERESCHI, Solid State Commun. **20**, 921 (1976).
G. MARGARITONDO, J. E. ROWE, and S. B. CHRISTMAN, Phys. Rev. B **15**, 3844 (1977).
G. MARGARITONDO, F. LEVY, N. G. STOFFEL, and A. D. KATNANI, Phys. Rev. B **21**, 5768 (1980).
[7] O. BRÜMMER, G. DRÄGER, U. BERG, W. CZOLBE, and T. CHASSÉ, Microchim. Acta, Suppl. **9**, 205 (1981).
[8] J. DRAHOKUPIL and A. SIMUNEK, in: Conf. Inner-Shell and X-Ray Physics of Atoms and Solids, Ed., D. J. FABIAN, H. KLEINPOPPEN, and L. M. WATSON, Plenum Press, New York/ London 1980 (p. 597).
[9] M. PIACENTINI, C. G. OLSON, A. BALZAROTTI, R. GRILANDA, V. GRASSO, and E. DONI, Nuovo Cimento **54B**, 248 (1979), and references therein.
[10] H. KAMINURA and K. NAKAO, J. Phys. Soc. Japan **24**, 1313 (1968).
[11] F. BASSANI and G. P. PARRAVICINI, Nuovo Cimento **50B**, 95 (1967).
[12] E. DONI, R. GIRLANDA, V. GRASSO, A. BALZAROTTI, and M. PIACENTINI, Nuovo Cimento **51B**, 154 (1979).
[13] J. ROBERTSON, J. Phys. C **12**, 4777 (1979); private communication.
[14] P. RENNERT and A. CHASSÉ, Proc. 10th Annual Internat. Symp. Electronic Structure of Metals and Alloys, Gaußig (GDR) 1980, TU Dresden (p. 87).
[15] S. M. GOLDBERG, C. S. FADLEY, and S. KONO, J. Electron. Spectroscopy and related Phenomena **21**, 285 (1981), and references therein.
[16] R. J. BAIRD, C. S. FADLEY, and L. F. WAGNER, Phys. Rev. B **15**, 666 (1977).
[17] U. BERG, T. CHASSÉ, and O. BRÜMMER, phys. stat. sol. (b) **108**, 507 (1981), and references therein.
[18] Z. HUSSAIN, C. S. FADLEY, S. KONO, and L. F. WAGNER, Phys. Rev. B **22**, 3751 (1980), and references therein.
[19] N. J. SHEVCHIK, J. Phys. C **10**, L 555 (1977).
[20] T. CHASSÉ and U. BERG, to be published.
[21] P. RENNERT, phys. stat. sol. (b) **105**, 213 (1981).
[22] P. RENNERT and A. CHASSÉ, to be published.
[23] T. CHASSÉ, Dissertation, MLU Halle, 1983.
[24] L. LEY, R. H. WILLIMAS, and P. C. KEMENY, Nuovo Cimento **39B**, 715 (1977).

(Received June 4, 1984)

phys. stat. sol. (b) **125**, 275 (1984)

Subject classification: 13.1

Zentralinstitut für Elektronenphysik der Akademie der Wissenschaften der DDR, Berlin[1])

Kink Lattice Solution in a Quasi-One-Dimensional Peierls-Fröhlich System with a Nearly One-Third-Filled Band

II. Thermodynamic Considerations

By

H. J. FISCHBECK

The thermodynamic meaning of the kink lattice solution derived in the first part of this work is analyzed. To this end the grand potential of this solution is calculated at zero temperature. Its minimum is self-consistent only if a special relation between the external parameters is realized. In general this solution describes the incommensurate phase only approximately. The phase transition into this approximate phase is for the most part discontinuous but in a certain region of the parameter plane continuous. This picture is expected to be essentially correct.

Die thermodynamische Bedeutung der im ersten Teil dieser Arbeit abgeleiteten Kink-Gitter-Lösung wird analysiert. Dazu wird das großkanonische Potential dieser Lösung bei $T = 0$ berechnet. Sein Minimum ist nur dann selbstkonsistent, wenn eine spezielle Relation zwischen den äußeren Parametern besteht. Im allgemeinen beschreibt diese Lösung die inkommensurable Phase nur approximativ. Der Phasenübergang in diese genäherte Phase ist vorwiegend unstetig, in einer gewissen Region der Parameterebene jedoch stetig. Es wird vermutet, daß dieses Bild im wesentlichen korrekt ist.

1. Introduction

In the first part [1] of this work (hereafter referred to as I) using the continuum approximation we have shown that the ground state of the quasi-1d Peierls-Fröhlich model with a nearly one-third-filled band is a lattice of amplitude-modulated $2\pi/3$ phase kinks. These periodically arranged kinks or solitons are local distortions of amplitude and phase of the trimerized Peierls lattice structure. They mediate the transition from one of the three degenerate Peierls vacua to another.

In the mean-field approximation and an additional low-energy approximation it was possible to calculate the static lattice distortion analytically and self-consistently and to solve the corresponding electronic problem exactly. It was found that the exceeding or missing electrons (compared to the one-third-filled band) are placed in a separate band, the so-called kink band which is formed within the Peierls gap. These findings confirm the results of other authors obtained by investigating single kinks numerically [2, 3] or a model kink analytically [4, 3] and generalize them to a finite density of kinks. In the frame of the same approximation as used in I and [4] Gammel and Krumhansl [5] recently considered a kink that has been given earlier by Brazovski [6]. It is easy to see that this kink is just the single-kink limit (elliptic modulus $k \to 1$) of our kink lattice.

---

[1]) Hausvogteiplatz 5–7, DDR-1086 Berlin, GDR.

[2]) Part I see phys. stat. sol. (b) **120**, 249 (1983).

In the present paper we mainly analyze the thermodynamic meaning of the kink lattice solution given in I. But first, several remarks are made as a supplement to I. In Section 2 the electronic energy spectrum is described in case of deficit of electrons. The parentage of kink band states is considered in Section 3. An explicit formula for the charge density revealing immediately the fractional kink charge $\pm 2\,|e|/3$ is given in Section 4.

By self-consistency we mean that the order parameter, the so-called gap function $\theta(x)$, satisfies the stationarity condition $\delta\Omega/\delta\theta^*(x) = 0$ of the grand potential Ω, which is called the gap equation. The two gap relations given in I are necessary for the gap function to fulfil the gap equation. For temperature $T = 0$ they are integrated in Section 5. Since the gap function $\theta(x)$ contains two adjustable parameters these are fixed by the gap relations. In order to study the phase transition from the commensurate (C) to the incommensurate (IC) phase, we calculate in Section 6 its grand potential at $T = 0$ and compare it with the grand potentials of two reference phases, the commensurate and an approximate "Peierls" (P) phase with a harmonic charge density wave (CDW).

We now are faced with the difficulty we already mentioned in I that the gap relations leave no freedom for minimizing the thermodynamic potential with respect to the kink density because they already fix the free parameters. This, however, is certainly necessary in order to underbid the thermodynamic potential of the P phase. For that reason we give up fulfilling both gap relations independently, but instead impose a suitable linear combination of them. The thermodynamic potential calculated in this way will then be minimized with respect to the kink density. Without doubt we therewith abandon the self-consistency of our solution. One can show, however, that self-consistency can be restored by manipulating one of the two external parameters of the model (electron–phonon coupling and wave number cut-off) so that both are brought into a special relation to each other which is given by a curve in the parameter plane. Therefore, the present solution (elliptic functions) is supposed to be the special case of a more general, yet unknow solution (hyperelliptic functions ?).

In Section 7 we discuss the order of the C–IC transition. It will be shown that there is a stripe in the parameter plane within which the phase transition is continuous in case of excess of electrons. Inside this stripe lies the curve of self-consistency at the transition. The value μ_0^+ of the chemical potential at which the continuous C–IC transition takes place is given. Outside this stripe and anywhere in case of a deficit of electrons the transition to our approximate, not fully self-consistent IC phase appears to be discontinuous permitting the conclusion that there is an attractive interaction between the kinks. The discontinuous transition occurs at a value $\mu_c^+ < \mu_0^+ (\mu_c^- > \mu_0^-)$ that is lower (higher) than the value $\mu_0^+ (\mu_0^-)$ at which a continuous transition would happen. In the concluding remarks it is guessed that these statements are relevant also for the C–IC transition to the fully self-consistent, yet unknown IC phase.

2. Antikinks and Their Energy Spectrum

The considerations of Part I of this paper concern the gap function $\theta(x)$ and the corresponding electronic energy spectrum E' which are determined by the coupled system of the gap equation (I 3.2) and the Schrödinger equation (I 2.16)

$$\begin{aligned} -i\chi_1' + \theta\chi_0 &= E'\chi_1\,, \\ i\chi_0' + \theta^*\chi_1 &= E'\chi_0\,, \end{aligned} \tag{2.1}$$

where (χ_1, χ_0) is a two-component wave function. The gap function $\theta(x)$ has the quasi-periodicity $\theta(x + 2\pi/3\varkappa) = e^{i2\pi/3}\,\theta(x)$ and is said to describe a lattice of kinks. Here $\varkappa$

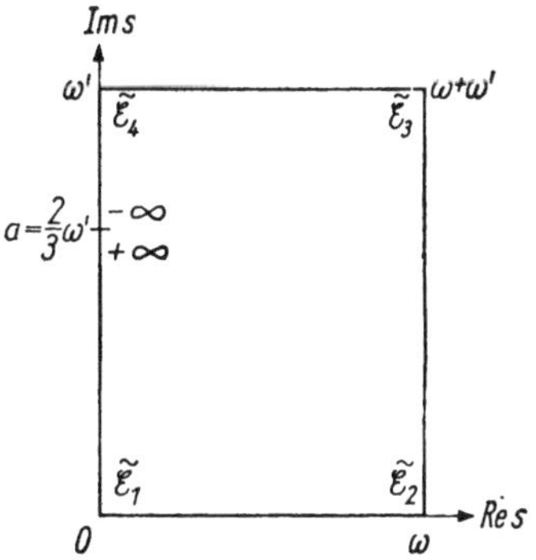

Fig. 1

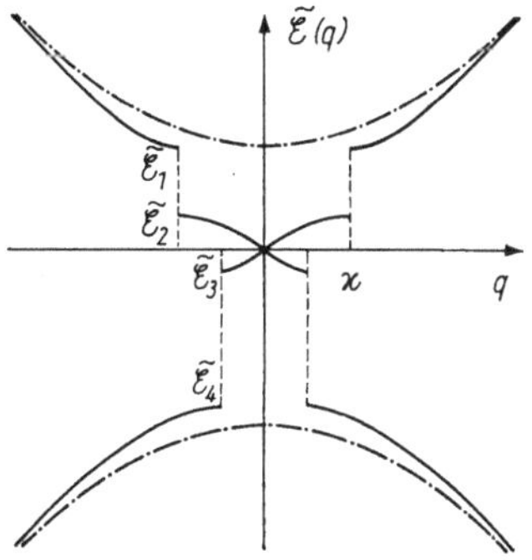

Fig. 2

Fig. 1. The fundamental rectangle of the Weierstrass function $\wp(s)$ with the corners $s = 0$, ω, $\omega + \omega'$, ω'. Within the rectangle the corresponding $\mathscr{E}$ values according to (2.5) are indicated

Fig. 2. Schematic plot of the IC energy spectrum in case of electron deficit as given by (2.6) and (2.7). The antikink band within the Peierls gap is empty at $T = 0$. The dash-dotted curves represent the C energy spectrum.

is the deviation of the critical wave vector from its commensurate value. Correspondingly $\tilde{\theta}(x) = \theta^*(x)$ with $\tilde{\theta}(x + 2\pi/3\varkappa) = \mathrm{e}^{-i2\pi/3}\,\tilde{\theta}(x)$ represents a periodic array of antikinks. By setting $\tilde{\chi}_0 = \chi_1$, $\tilde{\chi}_1 = -\chi_0$ we transform (2.1) into the antikink Schrödinger equation

$$
\begin{aligned}
-i\tilde{\chi}_1' + \tilde{\theta}\tilde{\chi}_0 &= -E'\tilde{\chi}_1 = \tilde{E}'\tilde{\chi}_1\,, \\
i\tilde{\chi}_0' + \tilde{\theta}^*\tilde{\chi}_1 &= -E'\tilde{\chi}_0 = \tilde{E}'\tilde{\chi}_0\,,
\end{aligned}
\tag{2.2}
$$

which tells us that the spectrum $\tilde{E}' = -E'$ is equal to the sign-reversed spectrum of the kinks. In the Appendix and Section 5 of I we tacitly assumed the parameters B and Q to be positive which was necessary for the kink spectrum. Since $\tilde{\theta}(x) = \theta^*(x) = \theta(x, -B, -Q)$ the formulas (I 5.4) and (I 5.5) remain valid for antikinks but B and Q are now negative,

$$
\begin{aligned}
&\frac{\mathrm{d}q}{\mathrm{d}\mathscr{E}} = \pm\,\frac{4\mathscr{E}^2 + 2\overline{\vartheta^2} - 9Q^2}{2c(\mathscr{E})}\,, \\
&c^2(\mathscr{E}) = 4\left(\mathscr{E} + \frac{3}{2}\,Q\right)\left[\left(\mathscr{E} - \frac{3}{2}\,Q\right)^2\left(\mathscr{E} + \frac{3}{2}\,Q\right) - B\right]; \qquad B, Q < 0\,.
\end{aligned}
\tag{2.3}
$$

As in (I 3.15) we have $\tilde{E}' = \tilde{\mathscr{E}} - Q/2$ ($\tilde{\mathscr{E}} = -\mathscr{E}, Q < 0$) but the tilde above $\mathscr{E}$ has been dropped in (2.3) because $\mathscr{E}$ is considered as an independent variable. The band edges are now

$$
\tilde{\mathscr{E}}_1 = -\mathscr{E}_4 = -\tfrac{3}{2}\,Q\,; \qquad \tilde{\mathscr{E}}_2 = -\mathscr{E}_3\,; \qquad \tilde{\mathscr{E}}_3 = -\mathscr{E}_2\,; \qquad \tilde{\mathscr{E}}_4 = -\mathscr{E}_1 \tag{2.4}
$$

and formulas (I5.9) read

$$
\tilde{\mathscr{E}}_1 - \tilde{\mathscr{E}}_4 = -\frac{B}{r_3}; \quad \tilde{\mathscr{E}}_1 - \tilde{\mathscr{E}}_3 = -\frac{B}{r_2}; \quad \tilde{\mathscr{E}}_1 - \tilde{\mathscr{E}}_2 = -\frac{B}{r_1}; \qquad B < 0\,.
$$

Because of $r_i = e_i + 3Q^2$ these values are interpolated by

$$
\tilde{\mathscr{E}}_1 - \mathscr{E} = -\frac{3}{2}\,Q - \mathscr{E} = \frac{-B}{\wp(s) + 3Q^2}\,, \tag{2.5}
$$

where $\wp(s)$ is Weierstrass' doubly-periodic $\wp$ function with the values $\wp(\omega) = e_1$, $\wp(\omega + \omega') = e_2$, $\wp(\omega') = e_3$ at the corners of its fundamental rectangle shown in Fig. 1. ω and ω' are the real and the imaginary half-periods of $\wp(s)$. This formula restores (I5.10) but the parameter s now runs in the reverse direction as compared to the kink case (Fig. 1). This means that (I5.11) has to be replaced by

$$\frac{d\mathcal{E}}{c(\mathcal{E})} = \pm i\, ds \tag{2.6}$$

with the upper sign referring to the antikink band and the lower one to the other bands. Integrating (2.3) by using (2.5) and (2.6) and choosing $q(\tilde{\mathcal{E}}_1) = \pi/3\omega$ we get

$$q(\mathcal{E}) = \frac{\pi}{3\omega} - i\left[\zeta(s) - \frac{\eta}{\omega}s + \frac{1}{2}\frac{\wp'(s)}{\wp(s) + 3Q^2}\right], \tag{2.7}$$

where $\zeta(s)$ is Weierstrass' ζ function and $\eta = \zeta(\omega)$. Together with (2.5) this is a parametric representation of the ascending branch of the antikink spectrum, which is schematically shown in Fig. 2.

3. The Parentage of the Gap States

At first sight the question for the parentage of the gap states seems trivially to be answered by a look at Fig. 2. One sees immediately that the upper (lower) band of the C phase (dash-dotted) contains $L\varkappa/\pi$ ($L\varkappa/2\pi$) states more than the corresponding bands of the IC phase (full). These states are in the antikink band, i.e. 2/3 of them split off from the upper and 1/3 come from the lower band of the C phase. In case of a kink lattice the opposite holds true as a look at Fig. 3 of I shows. But this is because we have chosen $q(\tilde{\mathcal{E}}_1) = \varkappa = \pi/3\omega$ or $q(\mathcal{E}_4) = -\varkappa$. This choice should be justified. To this end we calculate independently of this choice the differences of the numbers of states in the upper and lower bands by integrating over the densities of states. The commensurate limit of the incommensurate state is obtained for $B = 4Q^3$ in which case the quartic in (2.3) degenerates to

$$c_c^2(\mathcal{E}) = 4(\mathcal{E} + \tfrac{3}{2}Q)\,(\mathcal{E} - \tfrac{5}{2}Q)\,(\mathcal{E} + \tfrac{1}{2}Q)^2\,.$$

This means that the antikink band shrinks to a single level $\mathcal{E}_K = -Q/2$ ($Q > 0$ for kinks, $Q < 0$ for antikinks) within the energy gap between $-3Q/2$ and $5Q/2$. Therefore the commensurate band structure is described by

$$\begin{aligned} q_c(\mathcal{E}) &= \pm\sqrt{\left(\mathcal{E} + \frac{3}{2}Q\right)\left(\mathcal{E} - \frac{5}{2}Q\right)} = \mathcal{E}\sqrt{\left(1 + \frac{3Q}{2\mathcal{E}}\right)\left(1 - \frac{5Q}{2\mathcal{E}}\right)} \\ &\approx \mathcal{E} - \frac{1}{2}Q \quad \text{for} \quad \mathcal{E} \to \pm\infty \end{aligned} \tag{3.1}$$

which is π times the integrated density of states of the C phase per spin and length unit. The integrated density of states of the antikink phase is given by (2.7). It is now easy to evaluate

$$\int\limits_{\tilde{\mathcal{E}}_1}^{\infty} \left(\frac{dq_c}{d\mathcal{E}} - \frac{dq}{d\mathcal{E}}\right) d\mathcal{E} = [q_c(\mathcal{E}) - q(\mathcal{E})]_{\tilde{\mathcal{E}}_1 = -3Q/2}^{\infty}\,.$$

For large $\mathcal{E}$ from (3.1) and (2.5) we have $q_c(\mathcal{E}) \approx B/(\wp(s) + 3Q^2) - 2Q$. The limit $\mathcal{E} \to \infty$ is obtained by expanding $\wp(s)$ and the expression (2.7) about $s = a$. The values of the Weierstrass functions at $a = \frac{2}{3}\omega'$ are given in (IA3). We repeat and complement them here in a form being valid also for $B < 0$, $Q < 0$:

$$\begin{aligned} &\wp(a) = -3Q^2 \,; \qquad \wp'(a) = -2i\,|B| \,; \qquad \wp''(a) = 12BQ \,; \\ &\wp'''(a) = 72iQ^2\,|B| \,, \qquad \zeta(a) = \tfrac{2}{3}\eta' - i\,|Q| \,, \end{aligned} \tag{3.2}$$

where $\eta' = \zeta(\omega')$. With the help of Legendre's relation $\eta\omega' - \eta'\omega = i\pi/2$ [7] we straightforwardly obtain

$$\int_{\tilde{\mathcal{E}}_1}^{\infty} \left(\frac{dq_c}{d\mathcal{E}} - \frac{dq}{d\mathcal{E}}\right) d\mathcal{E} = \frac{2i}{3}\,(\eta'\omega - \eta\omega') = \frac{\pi}{3\omega} = \varkappa \,.$$

This justifies the integration constant $q(\tilde{\mathcal{E}}_1) = \varkappa = \pi/3\omega$ chosen in (2.7). In the same way one finds

$$\int_{-\infty}^{5Q/2} \frac{dq_c}{d\mathcal{E}}\, d\mathcal{E} - \int_{-\infty}^{\tilde{\mathcal{E}}_4} \frac{dq}{d\mathcal{E}}\, d\mathcal{E} = q_c(\mathcal{E})\big|_{-\infty}^{5Q/2} - q(\mathcal{E})\big|_{-\infty}^{\tilde{\mathcal{E}}_4} = \frac{\pi}{6\omega}$$

as it must be.

4. The Charge Density

The density of electrons in the IC phase is given by

$$n(x) = 2 \cdot 2 \sum_q f(E_q - \mu)\, \chi_q^+(x)\, \chi_q(x) = 4 \sum_q f(E_q - \mu)\, f_q(x) \,, \tag{4.1}$$

where $f(E) = (e^{\beta E} + 1)^{-1}$ is the Fermi function and $f_q(x) \equiv |\chi_{1q}|^2 + |\chi_{0q}|^2 = \alpha_q(2\vartheta^2 + 4\mathcal{E}_q^2 - 9Q^2)$ (see (I3.3), (I3.7)). The two factors 2 in front of (4.1) refer to spin and time reversal degeneracy, i.e. the sum runs over the ascending branch of the spectrum only. With regard to (I4.7) we have at $T = 0$

$$n(x) = \frac{4L}{2\pi} \int_{-q_s}^{\pm\varkappa/2} \frac{2\vartheta^2 + 4\mathcal{E}_q^2 - 9Q^2}{L(2\bar{\vartheta}^2 + 4\mathcal{E}_q^2 - 9Q^2)}\, dq \,,$$

where $\varkappa/2 = q(\mathcal{E}_2)$ and $-\varkappa/2 = q(\tilde{\mathcal{E}}_4)$ refer to the excess and deficit of electrons, respectively (see Fig. 2 and Fig. 3 of I). At $q = -q_s$ ($q_s \gg \varkappa$) the energy spectrum is cut off, which otherwise would be unbounded below. $\bar{\vartheta}^2$ is the spatial mean value of the squared amplitude of the gap function $\theta(x) = \vartheta(x)\, e^{i\varphi(x)}$ (see (I3.24) and (I4.8)). In the C phase the density of the electrons is constant and because of $|\chi_{0q}|^2 + |\chi_{1q}|^2 = 1/L$ given by $n_c = (4L/2\pi) \int_{-q_s}^{0} dq/L = 2q_s/\pi$. Therefore the excess (deficit) number density is

$$n(x) - n_c = \frac{2}{\pi}\left[\pm\frac{\varkappa}{2} + (\vartheta^2 - \bar{\vartheta}^2) \int_{-q_s}^{\pm\varkappa/2} \frac{dq}{\bar{\vartheta}^2 + 2\mathcal{E}_q^2 - 9Q^2/2}\right].$$

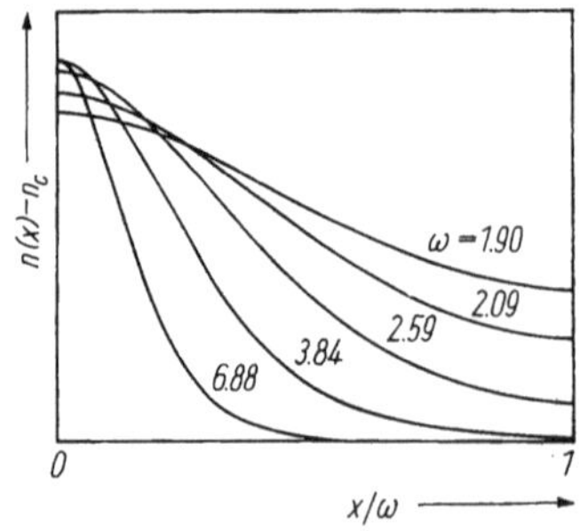

Fig. 3. The charge density vs. x/ω for several values of the kink density $1/2\omega = 3\varkappa/2\pi$

This can easily be integrated by means of (I 5.4) and (I 5.11) or (2.3) and (2.6). The result is

$$n(x) - n_c = \pm \frac{1}{3\omega} \pm \frac{2}{\pi} \left(\vartheta^2(x) - \bar{\vartheta}^2\right) i(\omega' - s_0) . \tag{4.2}$$

Here s_0 is the s value associated with the cut off $-q_s$ according to (I5.12) and (2.7) by

$$-q_s = \mp \frac{\pi}{3\omega} \pm i \left[\zeta(s_0) - \frac{\eta}{\omega} s_0 + \frac{1}{2} \frac{\wp'(s_0)}{\wp(s_0) + 3Q^2}\right]. \tag{4.3}$$

The upper (lower) sign refers to the excess (deficit) case. This convention holds throughout in the remainder of the paper. For $q_s \to \infty$ s_0 tends to $a = \frac{2}{3}\omega'$ so that in this limit

$$n(x) - n_c = \pm \frac{1}{3\omega} \pm \frac{2i\omega'}{3\pi} \left(\vartheta^2(x) - \bar{\vartheta}^2\right) . \tag{4.4}$$

The integral over the (real) period 2ω of $\vartheta^2(x) = \wp(x + \omega') + 3Q^2$ of course gives

$$\int_0^{2\omega} \left(n(x) - n_c\right) \mathrm{d}x = \pm \tfrac{2}{3} .$$

The number density (4.4) is shown in Fig. 3 for several kink distances 2ω over the normalized abscissa x/ω. As the kink density $1/2\omega$ varies from 0.145 to 0.528 a periodic array of exponentially localized charges $\frac{2}{3}e$ turns into a nearly harmonic CDW.

For later use we expand (4.3) as a Laurent series in powers of $\alpha = \pm (s_0 - a)$ because $|\alpha|$ is small for large q_s,

$$-q_s = -\frac{Q}{2} + \frac{i}{2\alpha} + i\alpha\left(\frac{3}{2} Q^2 - \frac{\eta}{\omega}\right) + \frac{B}{2}\alpha^2 - \frac{i\alpha^3}{10}(8BQ + 9Q^4) + \dots \tag{4.5}$$

This series can be inverted to give

$$\pm \frac{1}{s_0 - a} = \\ = i(2q_s - Q)\left[1 + \frac{2\bar{\vartheta}^2 - 3Q^2}{(2q_s - Q)^2} - \frac{B}{(2q_s - Q)^3} - \frac{(2\bar{\vartheta}^2 - 3Q^2)^2 - \frac{1}{5}(8BQ + 9Q^4)}{(2q_s - Q)^4} + \dots\right], \tag{4.6}$$

where (I 4.8) has been applied.

5. The Gap Relations

In I we have shown that the four parameters A, B, Q, b of our gap function $\theta(x) = \vartheta(x)\, e^{i\varphi(x)}$ are fixed by four relations (I 3.20), (I 3.21) which have to be fulfilled in order to satisfy the stationarity condition $\delta\Omega/\delta\theta^*(x) = 0$ of the grand-canonical potential Ω commonly called the gap equation. Two of these relations connect the parameters with each other: $A = 3Q^2$, $b = 6BQ$, so that A and b can be eliminated. The remaining parameters B, Q are determined by the other two relations (I 3.20),

$$3A_2 \pm 1 = 0\,, \qquad A_1 + A_2 + 3Q = 0\,. \tag{5.1}$$

Our aim is now to integrate these so-called gap relations for $T = 0$. Let us take the case of excess of electrons ($B > 0$, $Q > 0$, kink band occupied). The coefficients A_0, A_1A_0, A_2A_0 are given in (I 3.16). Using (I 4.7), the density of states (I 5.4), and (I 5.11) we get

$$A_0 = \frac{L}{2\pi}\int\limits_{-q_s}^{\varkappa/2} \alpha_q\, dq = \frac{1}{4\pi}\left[\int\limits_{-\mathcal{E}_s}^{\mathcal{E}_4} \frac{d\mathcal{E}}{c(\mathcal{E})} - \int\limits_{\mathcal{E}_3}^{\mathcal{E}_2} \frac{d\mathcal{E}}{c(\mathcal{E})}\right] = \frac{i}{4\pi}\left[\int\limits_{s_0}^{0} ds + \int\limits_{\omega}^{\omega+\omega'} ds\right] = $$
$$= \frac{i}{4\pi}\,(\omega' - s_0)\,, \tag{5.2}$$

where s_0 is defined by (4.3). In order to integrate

$$A_1A_0 = 2\,\frac{L}{2\pi}\int\limits_{-q_s}^{\varkappa/2} \alpha_q \mathcal{E}_q\, dq = \frac{1}{2\pi}\left[\int\limits_{-\mathcal{E}_s}^{\mathcal{E}_4} \frac{\mathcal{E}\, d\mathcal{E}}{c(\mathcal{E})} - \int\limits_{\mathcal{E}_3}^{\mathcal{E}_2} \frac{\mathcal{E}\, d\mathcal{E}}{c(\mathcal{E})}\right] \tag{5.3}$$

we additionally make use of (I 5.10) and obtain with the help of (IA1) and (3.2)

$$(A_1 + 3Q)\, A_0 = \frac{iB}{2\pi}\left[\int\limits_{s_0}^{0} \frac{ds}{\wp(s) + 3Q^2} + \int\limits_{\omega}^{\omega+\omega'} \frac{ds}{\wp(s) + 3Q^2}\right] =$$
$$= \frac{1}{4\pi}\left[\ln\frac{\sigma(a - s_0)}{\sigma(a + s_0)} - 2(\omega' - s_0)\,\zeta(a) + 2\eta' a\right]. \tag{5.4}$$

Here $\sigma(z)$ and $\zeta(z)$ are Weierstrass' σ and ζ function and $\eta' = \zeta(\omega')$. The expression for A_2A_0 reads at $T = 0$ simply

$$A_2A_0 = \frac{1}{2\pi}\left(\frac{\pi}{\lambda} - q_s - \frac{\varkappa}{2}\right) \equiv \frac{1}{2\pi}\left(P - \frac{\varkappa}{2}\right); \qquad P = \frac{\pi}{\lambda} - q_s\,, \tag{5.5}$$

where λ and q_s are the external parameters of our model. It is useful, however, to use $P = \pi/\lambda - q_s$ and q_s instead. We now multiply the gap relations (5.1) by A_0 and obtain by inserting (5.2), (5.4), and (5.5) their explicit form

$$\begin{aligned} &3\left(P \mp \frac{\varkappa}{2}\right) + \frac{i}{2}\,(\omega' - s_0) = 0\,; \\ &\frac{1}{2}\left[\ln\frac{\sigma(\pm a \mp s_0)}{\sigma(a + s_0)} - 2(\omega' - s_0)\,\zeta(a) + 2\eta' a\right] + P \mp \frac{\varkappa}{2} = 0\,. \end{aligned} \tag{5.6}$$

These relations already include the deficit case ($B < 0$, $Q < 0$, antikink band empty). The cut off parameter q_s enters these relations through (4.3) defining s_0.

6. The Thermodynamic Potential at $T = 0$

At zero temperature the grand potential (I2.20) of the system reads

$$\Omega = \frac{2L}{\pi} \int_{-q_s}^{\pm\varkappa/2} (E_q - \mu)\, \mathrm{d}q + \frac{2L}{\lambda} \overline{|\Delta|^2} = \frac{2L}{\pi} \left[\int_{-q_s}^{\pm\varkappa/2} (E'_q - \mu)\, \mathrm{d}q + \left(P \mp \frac{\varkappa}{2}\right) \overline{|\Delta|^2} \right], \tag{6.1}$$

where $E_q = E'_q - \overline{|\Delta|^2}$ and $\overline{|\Delta|^2}$ is the spatial mean value of the squared amplitude of the lattice distortion $\Delta(x)$ (see (I2.16) and (I2.19)). We first consider the grand potentials of two reference phases, namely the C phase and an approximate "Peierls" phase (P) with a harmonic CDW. The C phase has $\varkappa = 0$ and a constant gap "function" $\theta_0 = \vartheta_0\, \mathrm{e}^{i\varphi_0}$ so that (see (I2.19))

$$|\Delta|^2 = \overline{|\Delta|^2} = \vartheta_0^2 + 2\vartheta_0^3 \cos 3\varphi_0 = \vartheta_0^2 - 2\vartheta_0^3 \quad \text{for} \quad \varphi_0 = \pm\pi/3, \pi\,.$$

We put $\chi_0 = c_0 e^{iqx}$, $\chi_1 = c_1\, \mathrm{e}^{iqx}$ in (2.1) and get the energy spectrum

$$E'_q = \pm \sqrt{q_s^2 + \vartheta_0^2} \tag{6.2}$$

of the C phase. In (6.1) it leads to the grand potential

$$\tilde{\Omega}_c = -\frac{1}{2}\left[q_s \sqrt{q_s^2 + \vartheta_0^2} + \vartheta_0^2 \ln \frac{q_s + \sqrt{q_s^2 + \vartheta_0^2}}{\vartheta_0} \right] - \mu q_s + P(\vartheta_0^2 - 2\vartheta_0^3)\,. \tag{6.3}$$

Here and in the following $\tilde{\Omega}$ denotes $\tilde{\Omega} = (\pi/2L)\,\Omega$. Its minimum is determined by

$$\frac{\partial \tilde{\Omega}_c}{\partial \vartheta_0^2} = -\frac{1}{2} \ln \frac{q_s + \sqrt{q_s^2 + \vartheta_0^2}}{\vartheta_0} + P(1 - 3\vartheta_0) = 0\,. \tag{6.4}$$

The P phase has a harmonic CDW $\theta(x) = \vartheta_1\, \mathrm{e}^{i\varkappa x}$ which does not profit from the commensurability energy because

$$\overline{|\Delta|^2} = L^{-1} \int_{-L/2}^{L/2} (\vartheta_1^2 + 2\vartheta_1^3 \cos 3\varkappa x)\, \mathrm{d}x = \vartheta_1^2\,.$$

The energy spectrum follows from (2.1) by setting $\chi_0 = c_0\, \mathrm{e}^{i(q-\varkappa)x}$, $\chi_1 = c_1\, \mathrm{e}^{iqx}$,

$$E'_q = \frac{\varkappa}{2} \pm \sqrt{\left(q - \frac{\varkappa}{2}\right)^2 + \vartheta_1^2}\,. \tag{6.5}$$

With it (6.1) yields

$$\tilde{\Omega}_p = -\frac{1}{2}\left[\left(q_s + \frac{\varkappa}{2}\right) \sqrt{\left(q_s + \frac{\varkappa}{2}\right)^2 + \vartheta_1^2} + \vartheta_1^2 \ln \frac{q_s + \frac{\varkappa}{2} + \sqrt{\left(q_s + \frac{\varkappa}{2}\right)^2 + \vartheta_1^2}}{\vartheta_1} \right] - $$
$$- \left(\mu - \frac{\varkappa}{2}\right)\left(q_s + \frac{\varkappa}{2}\right) + \left(P - \frac{\varkappa}{2}\right) \vartheta_1^2\,. \tag{6.6}$$

In this case we have two minimum conditions

$$\frac{\partial\tilde{\Omega}_{\mathrm{p}}}{\partial\vartheta_1^2} = -\frac{1}{2}\ln\frac{q_{\mathrm{s}} + \frac{\varkappa}{2} + \sqrt{\left(q_{\mathrm{s}} + \frac{\varkappa}{2}\right)^2 + \vartheta_1^2}}{\vartheta_1} + P - \frac{\varkappa}{2} = 0\,, \tag{6.7}$$

$$\frac{\partial\tilde{\Omega}_{\mathrm{p}}}{\partial\varkappa/2} = -\sqrt{\left(q_{\mathrm{s}} + \frac{\varkappa}{2}\right)^2 + \vartheta_1^2} + q_{\mathrm{s}} + \varkappa - \mu - \vartheta_1^2 = 0\,. \tag{6.8}$$

In these relations (6.5) to (6.8) $\varkappa$ can be positive or negative.

Now we consider the grand potential of the incommensurate (anti) kink lattice phase and outline its calculation for deficit of electrons ($B < 0, Q < 0$, antikink band empty), in which case E_q' should be replaced by $\tilde{E}_q' = \tilde{\mathscr{E}}_q - Q/2$. Introducing the density of states (2.3) (the tilde over $\mathscr{E}$ is then dropped) and using (2.5) and (2.6) we obtain

$$\tilde{\Omega}_{\mathrm{IC}} = -i\int\limits_{s_0}^{\omega'}\left(\frac{B}{\wp(s) + 3Q^2} - \frac{3Q}{2}\right)\left[\frac{2B}{\wp(s) + 3Q^2}\left(\frac{B}{\wp(s) + 3Q^2} - 3Q\right) + \bar{\vartheta}^2\right]\mathrm{d}s -$$

$$-\left(\mu + \frac{Q}{2}\right)\left(q_{\mathrm{s}} - \frac{\varkappa}{2}\right) + \left(P + \frac{\varkappa}{2}\right)\overline{|\Delta|^2} =$$

$$= \frac{1}{2}\left(\frac{1}{4}\frac{\partial^2}{\partial a^2} - \bar{\vartheta}^2\right)\int\limits_{s_0}^{\omega'}\frac{\wp'(a)\,\mathrm{d}s}{\wp(s) - \wp(a)} + i\,\frac{3}{2}\,Q\bar{\vartheta}^2(\omega' - s_0) -$$

$$-\left(\mu + \frac{Q}{2}\right)\left(q_{\mathrm{s}} - \frac{\varkappa}{2}\right) + \left(P + \frac{\varkappa}{2}\right)\overline{|\Delta|^2}\,.$$

The second line is easily verified with the help of (3.2). The integration is now reduced to (IA1). The result is written in a form that comprises also the excess case ($B > 0$, $Q > 0$, kink band occupied) where the calculation runs similarly. The upper (lower) sign refers to the excess (deficit) case,

$$\tilde{\Omega}_{\mathrm{IC}} = \frac{1}{8}\left(\wp(s_0 - a) - \wp(s_0 + a)\right) -$$

$$-\,\bar{\vartheta}^2\left[\frac{1}{2}\ln\frac{\sigma(a + s_0)}{\sigma(\pm a \mp s_0)} - \eta' a + (\omega' - s_0)\,\zeta(a)\right] \mp$$

$$\mp\,\frac{i}{2}\,(\omega' - s_0)\,(3Q\bar{\vartheta}^2 - B) - \left(\mu + \frac{Q}{2}\right)\left(q_{\mathrm{s}} \pm \frac{\varkappa}{2}\right) + \left(P \mp \frac{\varkappa}{2}\right)\overline{|\Delta|^2}\,. \tag{6.9}$$

Finally we evaluate $\overline{|\Delta|^2}$. We first remark that the second-order differential equation (I3.10) defining the gap function $\theta(x)$ for $A = 3Q^2$ is equivalent to the system of two first-order differential equations

$$\theta' = \pm i\theta^{*2} + 2iQ\theta\,; \qquad \theta^{*\prime} = \mp i\theta^2 - 2iQ\theta^*\,. \tag{6.10}$$

By a little algebra we see that $|\theta|^2 + \theta^3 + \theta^{*3} = \left(1 - 4Q \pm 2\,\mathrm{Im}\,(\theta'/\theta)\right)|\theta|^2$. With $\theta = \vartheta\,\mathrm{e}^{i\varphi}$ and (I3.11) we get as spatial mean value

$$\overline{|\Delta|^2} = \frac{1}{L}\int\limits_{-L/2}^{L/2}(|\theta|^2 + \theta^3 + \theta^{*3})\,\mathrm{d}x = (1 - 6\,|Q|)\,\bar{\vartheta}^2 + 2\,|B|\,. \tag{6.11}$$

The thermodynamic potential (6.9) depends on the two adjustable parameters B and Q. It is, however, more convenient to use the roots $r_1 \geqq r_2 > r_3$ of the cubic (I3.22), which are related to each other by (I5.8), i.e. $\sqrt{r_1 r_2} - \sqrt{r_1 r_3} - \sqrt{r_2 r_3} = 0$. We choose as independent parameters

$$y = \sqrt{r_1} \quad \text{and} \quad x = \sqrt{\frac{r_3}{r_1}}\,; \qquad 0 < x \leqq 1/2\,, \tag{6.12}$$

so that $r_2 = y^2 x^2/(1-x)^2$, and obtain from (I5.7)

$$B = y^3 \bar{B} = \pm y^3 \frac{x^2}{1-x}; \qquad Q = y\bar{Q} = \pm \frac{y}{3}\left(1 + \frac{x^2}{1-x}\right). \tag{6.13}$$

The actual advantage of the parameters (6.12) is that the elliptic moduli

$$\begin{aligned} k^2 &= \frac{e_2 - e_3}{e_1 - e_3} = \frac{r_2 - r_3}{r_1 - r_3} = \frac{x^2(2x - x^2)}{(1-x^2)\,(1-x)^2}\,; \\ k'^2 &= 1 - k^2 = \frac{1 - 2x}{(1-x^2)\,(1-x)^2} \end{aligned} \tag{6.14}$$

depend on x only. Therefore, the real and imaginary half-periods of the $\wp$ function [7] are

$$\omega = \frac{1}{y}\,\bar{\omega} = \frac{1}{y}\,\frac{K(k)}{\sqrt{1-x^2}} \quad \text{and} \quad \omega' = \frac{1}{y}\,\bar{\omega}' = \frac{1}{y}\,\frac{iK(k')}{\sqrt{1-x^2}}\,, \tag{6.15}$$

where $\bar{\omega}$ and $\bar{\omega}'$ also depend on x only. $K(k)$ is the complete elliptic integral of the first kind. Now we are going to discuss the minimum condition of Ω_{IC} for fixed $\varkappa = \pi/3\omega = \pi y/3\bar{\omega}$. Let α be one of the parameters x, y. In order to obtain $\partial\Omega_{\text{IC}}/\partial\alpha$ from (6.1) we need $\partial|\Delta|^2/\partial\alpha$ and $\partial E'_q/\partial\alpha$. By means of (6.10) the derivative of the integrand of (6.11) is written as

$$\begin{aligned} &2\,\text{Re}\left[(\theta^* + 3\theta^2)\,\frac{\partial\theta}{\partial\alpha}\right] = 2\,\text{Re}\left[(\theta^* \mp 6Q\theta^* \pm i3\theta^{*\prime})\,\frac{\partial\theta}{\partial\alpha}\right] = \\ &= (1 - 6\,|Q|)\,\frac{\partial\vartheta^2}{\partial\alpha} \pm 3\varphi'\,\frac{\partial\vartheta^2}{\partial\alpha} \mp 3\vartheta^{2\prime}\,\frac{\partial\varphi}{\partial\alpha}. \end{aligned}$$

Integrating the last term by parts and applying (I3.11) we are led to

$$\begin{aligned} \frac{\partial\overline{|\Delta|^2}}{\partial\alpha} &= \frac{1}{L}\int\limits_{-L/2}^{L/2}\left[(1 - 6\,|Q|)\,\frac{\partial\vartheta^2}{\partial\alpha} \pm 3\,\frac{\partial}{\partial\alpha}\,(\varphi'\vartheta^2)\right]\mathrm{d}x = \\ &= (1 - 9\,|Q|)\,\frac{\partial\bar{\vartheta}^2}{\partial\alpha} + 3\,\frac{\partial\,|B|}{\partial\alpha} - 3\bar{\vartheta}^2\,\frac{\partial\,|Q|}{\partial\alpha}. \end{aligned} \tag{6.16}$$

With normalized eigenfunctions χ_{1q}, χ_{0q} (2.1) gives

$$\frac{\partial E'_q}{\partial\alpha} = \int\limits_{-L/2}^{L/2}\left(\chi_{1q}^*\,\frac{\partial\theta}{\partial\alpha}\,\chi_{0q} + \chi_{0q}^*\,\frac{\partial\theta^*}{\partial\alpha}\,\chi_{1q}\right)\mathrm{d}x = 2\int\limits_{-L/2}^{L/2}\text{Re}\left(g_q\,\frac{\partial\theta^*}{\partial\alpha}\right)\mathrm{d}x\,,$$

where $g_q(x) \equiv \chi_{0q}^*(x)\,\chi_{1q}(x)$ is given by (I3.14) in terms of ϑ. Inserting this and $\partial\theta^*/\partial\alpha = (\partial\vartheta/\partial\alpha - i\vartheta\,\partial\varphi/\partial\alpha)\,e^{-i\varphi}$ and integrating one of the arising terms by parts we find

$$\frac{\partial E_q'}{\partial\alpha} = 2\alpha_q L\left(\mathscr{E}_q\frac{\partial\bar{\vartheta}^2}{\partial\alpha} + \frac{1}{2}\frac{\partial B}{\partial\alpha} - \frac{1}{2}\bar{\vartheta}^2\frac{\partial Q}{\partial\alpha}\right), \tag{6.17}$$

where (I3.11) has been used again. If now (6.16) and (6.17) are combined to form the derivative of (6.1) at fixed $\varkappa$ and thereby the coefficients A_0, A_1, A_2 as given by (5.2), (5.3), (5.5) are used, the following expression is obtained:

$$\frac{\partial\tilde{\Omega}_{\mathrm{IC}}}{\partial\alpha} = 2\pi A_0\left[(A_1 + A_2 \mp 9QA_2)\frac{\partial\bar{\vartheta}^2}{\partial\alpha} + (1 \pm 3A_2)\left(\frac{\partial B}{\partial\alpha} - \bar{\vartheta}^2\frac{\partial Q}{\partial\alpha}\right)\right]. \tag{6.18}$$

As expected this expression vanishes if the two gap relations (5.1) are fulfilled. Unfortunately, these two relations fix both parameters x and y and with them also $\varkappa = \pi y/3\bar{\omega}$. Thus, contrary to the P phase there is no room for an additional minimization of the grand potential (6.9) with respect to $\varkappa$ which is certainly necessary for lowering the IC potential below the minimum of the P phase potential. Therefore, at an arbitrary value of $\varkappa$, we cannot satisfy both gap relations but only some linear combination of them. We write this linear combination as

$$A_1 + A_2 + 3Q + (\varrho - 2\,|Q|)\,(3A_2 \pm 1) = 0 \tag{6.19}$$

and choose ϱ so as to get the minimum of Ω_{IC} at a given value of $\varkappa = \pi y/3\bar{\omega}$. Therefore, we have $y = y(x) = 3\bar{\omega}\varkappa/\pi$ with $\bar{\omega} = \bar{\omega}(x)$ being given by (6.15) and (6.14). Let

$$\frac{\mathrm{d}}{\mathrm{d}x} = \frac{\partial}{\partial x} + \frac{\partial}{\partial y}\frac{\mathrm{d}y}{\mathrm{d}x} = \frac{\partial}{\partial x} + \frac{3\varkappa}{\pi}\frac{\mathrm{d}\bar{\omega}}{\mathrm{d}x}\frac{\partial}{\partial y}, \tag{6.20}$$

then from (6.18) we get

$$\frac{\mathrm{d}\tilde{\Omega}_{\mathrm{IC}}}{\mathrm{d}x} = 2\pi A_0\left[(A_1 + A_2 \mp 9QA_2)\frac{\mathrm{d}\bar{\vartheta}^2}{\mathrm{d}x} + (1 \pm 3A_2)\left(\frac{\mathrm{d}B}{\mathrm{d}x} - \bar{\vartheta}^2\frac{\mathrm{d}Q}{\mathrm{d}x}\right)\right]. \tag{6.21}$$

We now insert (6.19) and determine ϱ by demanding that this expression vanishes. We find

$$\varrho = \frac{\mathrm{d}(|B| - \bar{\vartheta}^2\,|Q|)/\mathrm{d}x}{\mathrm{d}\bar{\vartheta}^2/\mathrm{d}x}. \tag{6.22}$$

In the explicit form (5.6) the combined gap relation (6.19) reads

$$\frac{1}{2}\ln\frac{\sigma(\pm a \mp s_0)}{\sigma(a + s_0)} + \frac{2}{3}\eta' s_0 + \frac{i}{2}(\omega' - s_0)\,\varrho + \left(P \mp \frac{\varkappa}{2}\right)(1 + 3\varrho - 6\,|Q|) = 0. \tag{6.23}$$

The x value solving this equation and determining also $y = 3\bar{\omega}\varkappa/\pi$ gives the minimum of (6.9) at constant $\varkappa$. A subsequent minimization of Ω_{IC} with respect to $\varkappa$ leads to the true minimum of (6.9). Numerical computations have shown that it is always lower than the minimum value of the P phase potential (6.6) which is determined by (6.7) and (6.8). But certainly, when using the combined gap relation (6.23) only, we abandon the self-consistency of our solution, which is thereby no more exact. We guess that the present solution is only the special case of a more general, still unknown solution allowing both the satisfaction of the gap equation and an additional minimization with respect to $\varkappa$. In the frame of the present solution we have shown by a numerical procedure that self-consistency can be restored by a final adjustment of the external parameter $P = \pi/\lambda - q_s$. This can be understood as a preparation of the system. In the grand-canonical situation the length L of the system can be arbitrarily prescribed. Therefore, L can be chosen so that at given μ the special conditions are met under which self-consistency is retained because the effective electron-phonon coupling λ strongly depends on the length scale of the system (see (I2.21)). Exactly at the C–IC transition this adjustment can be made analytically, as is done in the next section.

7. The C–IC Transition

Our first task is to show that the grand potential (6.9) of the IC phase and its minimum condition at fixed $\varkappa$ (6.23) merge into the grand potential (6.3) of the C phase and its minimum condition (6.4) in the single-kink limit $x \to 1/2$ (i.e. $k \to 1$, $\bar{\omega} \to \infty$, $\varkappa \to 0$, y finite). It is generally possible and useful to extract from all x- and y-dependent quantities a power of y so that the remainder, which will be denoted by the same letter with an upper bar, depends on x only. This has already been done in (6.13) and (6.15), and this can be done with the Weierstrass functions by means of their homogeneity relations [7]. Setting $s = \bar{s}/y$ they give

$$\begin{aligned} \wp'(s, \omega\omega') &= y^3 \wp'(\bar{s}, \bar{\omega}\bar{\omega}') \,; & \wp(s, \omega\omega') &= y^2 \wp(\bar{s}, \bar{\omega}\bar{\omega}') \,; \\ \zeta(s, \omega\omega') &= y \zeta(\bar{s}, \bar{\omega}\bar{\omega}') \,; & \sigma(s, \omega\omega') &= \frac{1}{y}\, \sigma(\bar{s}, \bar{\omega}\bar{\omega}') \end{aligned} \tag{7.1}$$

and $\eta = y\bar{\eta}$, $\eta' = y\bar{\eta}'$. From (6.15) we have $a = \frac{2}{3}\,\omega' = \bar{a}/y$. The spatial mean value $\overline{\vartheta^2} = \overline{\wp(x + \omega') + 3Q^2} = r_1 + (r_1 - r_3)\, E/K$ [3]) is written as

$$\overline{\vartheta^2} = y^2 \overline{t^2} \,; \qquad \overline{t^2} = 1 - \frac{(1 - x^2)\, E}{K}, \tag{7.2}$$

where $E = E(k)$ and $K = K(k)$ are the complete elliptic integrals. This relation as well as (6.20), (6.13), and (6.15) are used to evaluate (6.22). With some effort we find

$$\varrho = y\bar{\varrho} \,;$$

$$\bar{\varrho} = \frac{2}{3}\, \frac{(1 + x^3)\, \dfrac{E^2}{K^2} + (1 - 2x)\, \dfrac{2x^3 + x^2 - 2}{(1 - x)^3}\, \dfrac{E}{K} + (1 - 2x)^2\, (1 + x)\, \dfrac{x^2 - 3x + 3}{3(1 - x)^4}}{2\, \dfrac{E}{K} - (1 - x^2)\, \dfrac{E^2}{K^2} - \dfrac{x^2(1 - 2x)}{3(1 - x)^3}}. \tag{7.3}$$

In the limit $x \to 1/2$ ($k \to 1$) we have

$$\begin{aligned} &\bar{B} = \bar{Q} = \frac{1}{2}, \qquad \frac{\bar{\eta}}{\bar{\omega}} = -\bar{e}_1 = -\frac{1}{4}, \qquad \bar{\omega}' = \frac{i\pi}{\sqrt{3}}, \\ &\bar{\eta}' = -\frac{i\pi}{4\sqrt{3}}, \qquad \overline{t^2} = 1\,, \qquad \bar{\varrho} = 0\,. \end{aligned} \tag{7.4}$$

---

[3]) see (4.16) in [8].

The relation (4.3) determining $s_0 = \bar{s}_0/y$ becomes in this limit[4])

$$-q_s = \pm\, iy\sqrt{3}\,\frac{\sinh\sqrt{3}\,\bar{s}_0}{1 + 2\cosh\sqrt{3}\,\bar{s}_0}\,.$$

It is solved by

$$\begin{aligned}\cos\left(-i\sqrt{3}\,\bar{s}_0\right) &= \frac{-2q_s^2 \pm 3y\sqrt{y^2 + q_s^2}}{3y^2 + 4q_s^2}\,;\\ \sin\left(-i\sqrt{3}\,\bar{s}_0\right) &= \sqrt{3}\,q_s\,\frac{2\sqrt{y^2 + q_s^2} \pm y}{3y^2 + 4q_s^2}\,,\end{aligned} \tag{7.5}$$

where the upper (lower) sign refers to the excess (deficit) case as usual. Using this result it is not difficult to derive the $x \to 1/2$ limit[4]) of the combined gap relation (6.23). One finds

$$-\frac{1}{2}\ln\frac{q_s + \sqrt{q_s^2 + y^2}}{y} + P(1 - 3y) = 0\,. \tag{7.6}$$

This is just the gap relation (6.4) of the C phase, i.e. we have $y = \vartheta_0$ for $x \to 1/2$. Using (7.1), (7.4), and (7.5) once more it is easy to realize[2]) that also the IC potential (6.9) together with (6.11) changes over to the C potential (6.3) for $x \to 1/2$, $\varkappa \to 0$, $y \to \vartheta_0$.

Also in this limit one can analytically write down the condition for P and q_s under which both gap relations (5.6) are fulfilled in the minimum of Ω_{IC}. One of the gap relations (5.6) should be satisfied in addition to the combined gap relation whose limit is (7.6). We take the first one for $x = 1/2$, i.e. $-i\sqrt{3}\,\bar{s}_0 = \pi - 6\sqrt{3}\,yP$, combine it with (7.5), and get

$$6\sqrt{3}\,yP = \arcsin\frac{\sqrt{3}\,q_s\left(2\sqrt{y^2 + q_s^2} \pm y\right)}{3y^2 + 4q_s^2}\,, \tag{7.7}$$

where the principal value of arcsin is to be taken. The explicitly solved form $q_s = y\sinh 2P(1 - 3y)$ of (7.6) and (7.7) can now be read as a parametric representation of the curve $P = P(q_s)$ at which for $x = 1/2$ self-consistency is preserved.

Now we consider the difference between the grand potentials (6.9) and (6.3) of the IC and C phase. Since an exact explicit solution of (4.3) for s_0 is possible only for $x = 1/2$ (i.e. (7.5)), we employ in the following the approximate solution (4.6) relying on an expansion in powers of $s_0 - a$. In the expansions[5])

$$\begin{aligned}\wp(s_0 - a) - \wp(s_0 + a) &= (s_0 - a)^{-2} + 3Q^2 - 2i\,|B|\,(s_0 - a) +\\ &\quad + \tfrac{1}{5}\,(27Q^4 - 36BQ)\,(s_0 - a)^2 + \dots\\ \ln\frac{\sigma(\mp s_0 \pm a)}{\sigma(s_0 + a)} &= \ln\left(\pm i(s_0 - a)\,(2\,|B|)^{1/3}\right) -\\ &\quad - \frac{8}{9}\,\eta'\omega' - \left(\frac{4}{3}\,\eta' + i\,|Q|\right)(s_0 - a) - \frac{3}{2}\,Q^2(s_0 - a)^2 + \dots,\end{aligned}$$

---

[4]) See [7], p. 339.

[5]) Because of $2a = 2\omega' - a$ the values of Weierstrass' functions at $2a$ can be obtained from the ones at a by means of the period relations [7]. Were mark at this place that $\sigma(a) = i\,e^{2\eta'\omega'/9}/(2|B|)^{1/3}$ [9].

we apply (4.6) and insert the results in (6.9) and (6.23),

$$\tilde{\Omega}_{\mathrm{IC}} = -\frac{1}{2}(q_s^2 + \bar{\vartheta}^2) + Q^2 + \left(\frac{i\omega'}{6} \mp \frac{1}{4q_s}\right)(|B| - |Q|\,\bar{\vartheta}^2) - \mu q_s \mp \frac{\varkappa}{2}\left(\mu + \frac{Q}{2}\right) +$$

$$+ \frac{\bar{\vartheta}^2}{8q_s^2}(2Q^2 - \bar{\vartheta}^2) + \frac{\bar{\vartheta}^2}{2}\ln\frac{(2\,|B|)^{1/3}}{2q_s} + \left(P \mp \frac{\varkappa}{2}\right)[(1 - 6\,|Q|)\,\bar{\vartheta}^2 + 2\,|B|] + \dots \tag{7.8}$$

$$\frac{1}{2}\ln\frac{(2\,|B|)^{1/3}}{2q_s} + \varrho\left(\frac{i\omega'}{6} \mp \frac{1}{4q_s}\right) + \frac{1}{8q_s^2}(4Q^2 - 2\bar{\vartheta}^2 - \varrho\,|Q|) +$$

$$+ \left(P \mp \frac{\varkappa}{2}\right)(1 + 3\varrho - 6\,|Q|) = 0\,. \tag{7.9}$$

Correspondigly we expand (6.3) and (6.4) up to second order in powers of ϑ_0/q_s and subtract the results from (7.8) and (7.9), respectively. We get

$$\tilde{\Omega}_{\mathrm{IC}} - \tilde{\Omega}_{\mathrm{c}} = Q^2 - \frac{\bar{\vartheta}^2}{2} + \frac{\vartheta_0^2}{4} + \left(\frac{i\omega'}{6} \mp \frac{1}{4q_s}\right)(|B| - |Q|\,\bar{\vartheta}^2) \mp$$

$$\mp \frac{\varkappa}{2}\left(\mu + \frac{Q}{2} + 2\,|B| - \bar{\vartheta}^2\,(1 - 6\,|Q|)\right) +$$

$$+ \frac{\bar{\vartheta}^2}{8q_s^2}(2Q^2 - \bar{\vartheta}^2 + \vartheta_0^2) - \frac{\vartheta_0^4}{16q_s^2} +$$

$$+ P\big(2\,|B| - \vartheta_0^3 + 3\bar{\vartheta}^2(\vartheta_0^2 - 2\,|Q|)\big) + \frac{\bar{\vartheta}^2}{2}\ln\frac{(2\,|B|)^{1/3}}{\vartheta_0}, \tag{7.10}$$

$$\frac{1}{2}\ln\frac{(2\,|B|)^{1/3}}{\vartheta_0} + \varrho\left(\frac{i\omega'}{6} \mp \frac{1}{4q_s}\right) + \frac{1}{8q_s^2}(4Q^2 - 2\bar{\vartheta}^2 - \varrho\,|Q| + \vartheta_0^2) \mp$$

$$\mp \frac{\varkappa}{2}(1 + 3\varrho - 6\,|Q|) + 3P(\varrho + \vartheta_0 - 2\,|Q|) = 0\,. \tag{7.11}$$

In (7.10) the quadratically expanded version

$$\ln\frac{2q_s}{\vartheta_0} + \frac{\vartheta_0^2}{4q_s^2} - 2P(1 - 3\vartheta_0) = 0 \tag{7.12}$$

of the C phase gap relation (6.4) has been used in order to eliminate $\ln 2q_s$. In view of (7.4) it is again obvious that for $x = 1/2$, $\varkappa = 0$ (7.11) is solved by $y = \vartheta_0$ and (7.10) becomes zero. When (7.11) with $y = y(x) = 3\bar{\omega}\varkappa/\pi$ is solved for x and this solution is then inserted in (7.10), the potential difference arises as a function of $\varkappa$, which vanishes for $\varkappa \to 0$. We have done this numerically. For $P = 3$, $q_s = 1.5$ the result is shown in Fig. 4 for several values of μ.

In order to clarify the order of the C–IC transition we at last consider the region $0.498 \lessapprox x \leqq 0.5$ where $1 - 2x \ll -\frac{1}{2}\ln(1 - 2x)$ so that we can put $x = 1/2$ everywhere except in $K(k) \approx \ln 4/k' \approx \frac{1}{2}\big(\ln 3 - \ln(1 - 2x)\big)$ which occurs in $\bar{\vartheta}^2 = y^2\bar{t}^2$ and $\varrho = y\bar{\varrho}$. Because of $\varkappa = \pi y/3\bar{\omega} \approx \pi y/(2\sqrt{3}\,K)$ we can write $1/K \approx 2\sqrt{3}\,\varkappa/(\pi y)$ and obtain from (7.2) and (7.3)

$$\bar{t}^2 \approx 1 - \frac{c\varkappa}{y}\,; \qquad \bar{\varrho} \approx \frac{c\varkappa}{2y - c\varkappa}\,; \qquad c = \frac{3\sqrt{3}}{2\pi}\,. \tag{7.13}$$

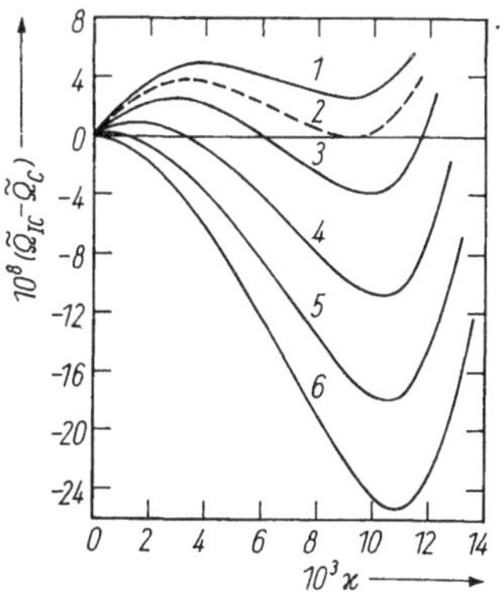

Fig. 4. The difference of the thermodynamic potentials in the IC and C phase at $P = 3$, $q_s = 1.5$ as a function of $\varkappa = \pi/3\omega$ for several values of μ. (1) $\mu = 0.002649$, (2) $\mu = \mu_c^+ = 0.002660$, (3) $\mu = 0.002676$, (4) 0.002704, (5) 0.002731, (6) $\mu = \mu_0^+ = 0.002759$

When inserting this in (7.11) and (7.10) we obtain in view of (6.13)

$$\frac{1}{2} \ln \frac{y}{\vartheta_0} + \frac{1}{8q_s^2} (\vartheta_0^2 - y^2) + 3P(\vartheta_0 - y) -$$

$$- \varrho\left(\frac{\pi}{6\sqrt{3}} \pm \frac{y}{4q_s} + \frac{y^2}{16q_s^2} \pm \frac{3}{2} \varkappa y - 3P_y\right) \mp \frac{\varkappa}{2}\left(1 - 3y \mp \frac{cy}{2q_s^2}\right) = 0\,, \tag{7.14}$$

$$\tilde{\Omega}_{IC} - \tilde{\Omega}_c \approx \frac{1}{4} (\vartheta_0^2 - y^2) - \frac{1}{16q_s^2} (y^2 - \vartheta_0^2)^2 + P\big(y^3 - \vartheta_0^3 - 3y^2(y - \vartheta_0)\big) + \frac{y^2}{2} \ln \frac{y}{\vartheta_0} +$$

$$+ \frac{\varkappa}{2}\Bigg[\mp\mu - \frac{y}{2} \mp y^2 \pm 2y^3 +$$

$$+ cy\left(1 - \ln \frac{y}{\vartheta_0} + 6P(y - \vartheta_0) \mp \frac{y}{4q_s} + \frac{3y^2}{8q_s^2} - \frac{\vartheta_0^2}{4q_s^2}\right)\Bigg] \pm$$

$$\pm \frac{\varkappa}{4}\left(2cy(1 - 3y) \mp \frac{c^2y^2}{2q_s^2}\right). \tag{7.15}$$

The next step is the expansion of the potential difference in powers of $\varkappa$ up to second order. To this end we put $y = \vartheta_0 + y_1$ in (7.14) and observe that terms linear in y_1 drop out of the first line of (7.15). This means that (7.14) must be solved for y_1 in first order only,

$$y_1 = \frac{\varkappa}{4} \frac{1 \pm 4\vartheta_0(1 - 3\vartheta_0) - c\vartheta_0(12P \mp 1/q_s + 7\vartheta_0/4q_s^2)}{1 - 6P\vartheta_0 - \vartheta_0^2/2q_s^2}. \tag{7.16}$$

It can be shown that the denominator of this relation is positive in the region of solvability of the C phase gap relation (7.12). The latter is defined as follows. The left-hand side of (7.12) has a minimum as a function of ϑ_0 which must be lower than zero for (7.12) to be solvable. The condition that the minimum equals zero defines a curve in the P, q_s plane which bounds the solvability region. It is shown in Fig. 5.

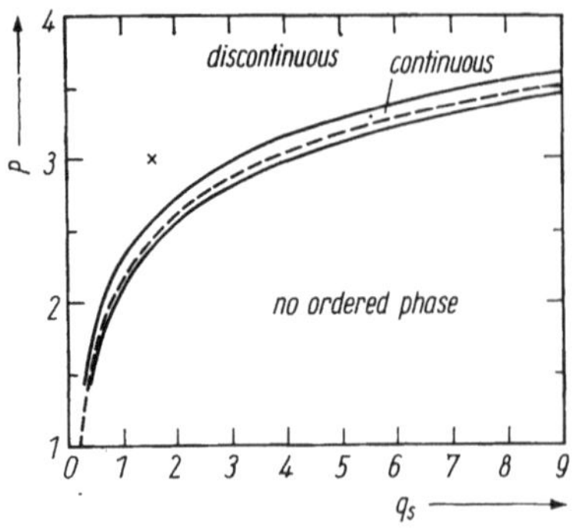

Fig. 5

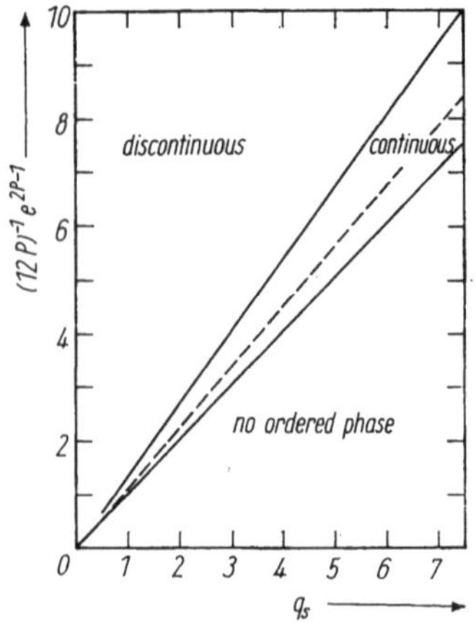

Fig. 6

Fig. 5. The stability region of ordered phases in the P, q_s plane. On the stripe between the two full curves the C–IC transition is continuous for excess of electrons. Along the dashed curve self-consistency is maintained at the transition. For electron deficit the phase transition is discontinuous everywhere in the stability region. The cross marks the parameter values for which the potential difference of Fig. 4 is calculated

Fig. 6. The same as in Fig. 5 with the ordinate $(12P)^{-1}\,e^{2P-1}$ instead of P

The quadratic expansion of (7.15) with respect to y_1 and the substitution of (7.16) yield

$$\tilde{\Omega}_{\mathrm{IC}} - \tilde{\Omega}_{\mathrm{c}} \approx \mp \frac{\varkappa}{2}(\mu - \mu_0^\pm) - \frac{\varkappa^2}{32}\left\{\frac{[1 \pm 4\vartheta_0(1-3\vartheta_0) - 12Pc\vartheta_0 \pm c\vartheta_0/q_s - 7c\vartheta_0^2/4q_s^2]^2}{1 - 6P\vartheta_0 - \vartheta_0^2/2q_s^2} \mp 16c\vartheta_0(1-3\vartheta_0) + \left(2c\frac{\vartheta_0}{q_s}\right)^2\right\}, \tag{7.17}$$

where

$$\mu_0^\pm = \pm\vartheta_0\left(c - \frac{1}{2}\right) - \vartheta_0^2 + 2\vartheta_0^3 - \frac{c\vartheta_0^2}{4q_s} \pm \frac{c\vartheta_0^3}{8q_s^2}; \qquad c = \frac{3\sqrt{3}}{2\pi}. \tag{7.18}$$

The C–IC transition into our approximate IC phase appears to be discontinuous if the coefficient of $\varkappa^2$ in (7.17) is negative. This is nearly always the case. In the deficit case (lower sign) this is obvious. In the excess case, however, there is a stripe in the P, q_s plane where the $\varkappa^2$ term is positive so that the C–IC transition becomes continuous. Then the transition takes place exactly at $\mu = \mu_0^+$ by formation of single kinks forming a lattice of repelling kinks for $\mu > \mu_0^+$. μ_0^+ is the energy required for adding one electron to the C phase. Since two electrons occupy three kinks the formation energy of a single kink is $\frac{2}{3}\mu_0^+$. Apart from higher order corrections in (7.18) the leading term $\frac{2}{3}\mu_0^+ = \frac{2}{3}\vartheta_0(c - \frac{1}{2})$ has already been found by Brazovski [6] and Gammel and Krumhansl [5]. The boundaries of the stripe within which the C–IC transition is continuous are defined by the condition that the curly bracket in (7.17) be zero. This condition can be read as a quadratic equation for P. Equating one of the two roots with the expression for P following from (7.12) we get an equation containing only just ϑ_0 and q_s. Solving for ϑ_0 and reexpressing P by this solution according to (7.12) we obtain the two curves $P = P(q_s)$ which bound the region of a continuous C–IC transition. It turns out that the lower curve within $2^0/_{00}$ coincides with the solvability boundary of the C phase gap relation so that they cannot be distinguished graphically in Fig. 5. In Fig. 5 we also show the curve $P = P(q_s)$ (dashed) along which the self-consistency of our solution in the limit $x \to 1/2$ is retained ((7.7) combined with (7.6)). It runs completely within the stripe where the phase transition is continuous. Therefore this curve (single-kink limit) is really relevant because single kinks appear at the continuous phase transition. In the deficit case, however, the corresponding curve (lower sign of (7.7)) is not relevant, because the C–IC transition is then always discontinuous and the single-kink limit is not reached. It is quite useful to draw the curves of Fig. 5 with the ordinate $(12P)^{-1} e^{2P-1}$ instead of P because they then nearly turn to straight lines. This is done in Fig. 6. It is easy to show that the curve bounding the stability region of the C phase (solvability of (7.12), lowest curve of Fig. 5) is approximately given by $q_s = (12P)^{-1} e^{2P-1}$ for $3Pq_s \gg 1$.

The cross in Fig. 5 marks the parameter values to which the potential difference shown in Fig. 4 belongs. It is placed in the discontinuous region as is obvious from Fig. 4, too. The lowest curve of Fig. 4 refers to $\mu = \mu_0^+ = 0.002759 = 0.317\vartheta_0$ whereas the phase transition (dashed curve) takes place at a value $\mu_c^+ = 0.306\vartheta_0$ which is 3.5% lower than μ_0^+. At the transition a kink lattice is formed with $\varkappa_c^+ = 0.00942$ and a lattice constant $2\omega_c^+ = 2\pi/3\varkappa_c^+ = 222.3$. This figure should be multiplied by the length unit $a/\sqrt{3}$ introduced in I to give 128 lattice units a of the reference lattice in its normal state. Compared to the Fermi wave number $\pi/3a$ of the one-third-filled band (in our length unit it is $q_F = \pi/3\sqrt{3}$) this kink lattice carries an excess of electrons of $\varkappa_c^+/2q_F = 0.0078$. Unfortunately, general formulas for $\mu_c^\pm$ and $\varkappa_c^\pm$ cannot be derived because the numerically determined minimum of (7.10) appears in a region where the approximation (7.13), (7.14), (7.15) is no longer valid. Especially it makes no sense to pursue the $\varkappa$ expansion (7.17), which is based on (7.14) and (7.15), to higher orders This does not invalidate the conclusion drawn from it.

8. Conclusion

At first a remark should be made about the asymmetry between excess and deficit of electrons as it is revealed by (7.17) and (7.18). The main source of this asymmetry is the last term of (6.1), i.e. $\mp\varkappa\,|\Delta|^2/2$ which on its part comes from the energetic shift $E_q = E_q' - |\Delta|^2$ inherent in (2.1). Using (6.11), (6.13), (7.2) and the approximations (7.13) together with $y \approx \vartheta_0$, $x \approx 1/2$ one easily gets $\mp\varkappa[\vartheta_0^2 - 2\vartheta_0^3 - c\varkappa\vartheta_0(1 - 3\vartheta_0)]/2$ and recognizes the second and third term of (7.18) as well as the second term in the curly bracket of (7.17) which is responsible for the fact that only in the excess case a region of a continuous phase transition exists. Another origin of asymmetry is that the occupied valence bands are different in both cases: the valence band in the deficit case is minus the conduction band in the excess case. This gives rise to the asymmetric fourth term of (7.18).

We have seen that the kink lattice solution derived in I is fully self-consistent only under special circumstances which can be realized by tuning the external parameters P, q_s. In the grand-canonical situation this can be done by suitably choosing the length L of the system which then must be held fixed. In general our solution can only be understood as an approximate description of the IC phase. Therefore, the chemical potential $\mu_c^+ \leqq \mu_0^+$ $(\mu_c^- > \mu_0^-)$ at which the transition into this approximate phase takes place is an upper (lower) bound to the true transition point on the μ scale. Nevertheless we believe that the conclusions about the order of the phase transition are essentially correct for the following reasons:

(i) The linear term of (7.17) is completely independent of the IC gap relations and, therefore, insensitive to self-consistency. It comes out of the square bracket of (7.15) if there y is replaced by its commensurate limit ϑ_0. It is therefore supposed to be correct. This means that if the C–IC transition were continuous it would occur at $\mu_0^{\pm}$.

(ii) We found, however, that the transition into the approximate phase happens discontinuously at a lower (higher) value μ_c^+ (μ_c^-) except for excess in a stripe of the P, q_s plane parallel to the stability boundary of the C phase. Thus, the transition into the exact phase must take place at an even lower (higher) μ value and will be all the more of first order.

(iii) Since the self-consistency curve of our solution lies within the stripe of a second-order phase transition (Fig. 5), we conclude that such a stripe also exists for the transition into the exact phase. Its lower edge will presumably exactly coincide with the stability boundary. Its upper edge will of course more or less deviate from Fig. 5 or 6.

Thus one can hope that the picture of the C–IC transition in a quasi-1d Peierls-Fröhlich system with a nearly one-third-filled band offered by our not fully self-consistent solution is nevertheless essentially correct.

Acknowledgements

It is a pleasure to thank my colleagues Prof. W. Brauer, Dr. H. W. Streitwolf, Dr. H. Puff, and Dr. J. Mertsching for many fruitful discussions.

References

[1] H. J. FISCHBECK, phys. stat. sol. (b) **120**, 249 (1983).
[2] W. P. SU and J. R. SCHRIEFFER, Phys. Rev. Letters **46**, 738 (1981).
[3] W. P. SU, Phys. Rev. B **27**, 370 (1983).
[4] J. MERTSCHING, phys. stat. sol. (b) **113**, 609 (1982).
[5] J. T. GAMMEL and J. A. KRUMHANSL, Phys. Rev. B **27**, 7659 (1983).
[6] S. A. BRAZOVSKII, Zh. eksp. teor. Fiz. **78**, 677 (1980).
[7] A. ERDELYI, W. MAGNUS, F. OBERHETTINGER, and F. TRICOMI, Higher Transcendental Functions, Bateman Manuscript Project, Vol. 2, McGraw-Hill Publ. Co., New York 1958.
[8] J. MERTSCHING and H. J. FISCHBECK, phys. stat. sol. (b) **103**, 783 (1981).
[9] M. ABRAMOWITZ and I. A. STEGUN, Handbook of Mathematical Functions, Dover Publ. Inc., New York 1968 (p. 634).

(Received June 13, 1984)

phys. stat. sol. (b) **125**, 293 (1984)

Subject classification: 13.5.1 and 20.3; 22.5.1

Fachbereich Physik, Universität-Gesamthochschule Paderborn[1])

Localized and Free Biexcitons in Silver Bromide

By

TH. KLEINEFELD[2]) and W. VON DER OSTEN

Recombination luminescence in AgBr at low temperature is studied for high power quasi-cw and pulsed excitation. The samples are nominally pure but vary systematically in iodine impurity content. Using time-resolving techniques several new luminescence transitions are detected and found to be correlated with iodine concentration. The results can be consistently explained assuming that, besides free excitons and free biexcitons, a biexciton bound to isoelectronic iodine is created and decays. The model is supported by energy positions and shape of the observed lines as well as by the dependence of luminescence intensity on excitation power and iodine concentration. To completely account for the observed spectra, the model also takes into account intervalley exciton scattering by- X-phonons known to be an effective exciton relaxation process in AgBr.

Die Rekombinationsstrahlung von AgBr bei tiefen Temperaturen und hohen Anregungsdichten wird für quasi-Gleichlicht- und Pulsanregung untersucht. Die nominell reinen Proben variieren systematisch in ihrem Jod-Störstellengehalt. Mit Hilfe zeitauflösender Spektroskopie werden zahlreiche neue Emissionsübergänge gefunden, die mit dem Jodgehalt korreliert sind. Die Ergebnisse lassen sich konsistent unter Annahme eines an Jod gebundenen Biexzitons deuten, das neben freien Exzitonen und freien Biexzitonen erzeugt wird und zerfällt. Diese Vorstellung wird gestützt durch Linienlagen und -form der beobachteten Emissionsübergänge sowie die Abhängigkeit der Emissionsintensität von Anregungsleistung und Jodkonzentration. Zur vollständigen Erklärung der Spektren muß die für AgBr bekannte Zwischental-Streuung der Exzitonen mit X-Phononen berücksichtigt werden.

1. Introduction

High density excitation effects in semiconducting materials have recently attracted considerable attention. At low temperatures and sufficiently high carrier concentrations in principle two different phenomena are observed: condensation of excitons into droplets of dense metallic electron–hole liquid (EHL) and formation of biexcitons or exciton molecules. Information on these excited states mostly originates from recombination luminescence under high intensity excitation which was extensively studied, both experimentally and theoretically, in various direct and indirect gap materials (for references see e.g. [1 to 3]).

Regarding indirect gap semiconductors, silicon and germanium are examples where both EHL and biexciton were found with the biexciton state being less stable than the EHL (see e.g. [3 to 5]). In AgBr that also exhibits an indirect edge the situation turned out to be more complex. Time-resolved luminescence spectroscopy at high power levels in this material revealed emission lines at 2.669 eV and in the energy range 2.62 to 2.50 eV. While from analyzing its lineshape and dependence of emission intensity on excitation power the former line could undoubtedly be attributed to biexciton decay [6 to 8], the interpretation of the lower energy luminescence in terms of EHL

[1]) D-4790 Paderborn, FRG.

[2]) New address: Fachbereich Physik–Technologie, Universität-Gesamthochschule Duisburg FRG.

renders more difficult [9, 10]. This difficulty arises from the fact that this band is found to be composed by contributions of different origin and that it exhibits a complex structure if studied with sufficiently high spectral resolution [10, 11]. Also the binding energy of 55 to 60 meV derived from analyzing the luminescence in terms of EHL considerably deviates from theoretical predictions (30 meV) even if the strong electron–LO coupling for AgBr is taken into account [9, 10, 12].

In this paper we describe in detail our results on recombination luminescence obtained in nominally pure AgBr crystals under high power quasi-cw and pulsed excitation. Some preliminary results were already published previously [13]. In particular, we consider the effect of ubiquitous iodine impurities in these crystals onto luminescence that we observe. The correlation between the intensity of some luminescence transitions in the highly excited spectra with iodine content leads us to propose that a biexciton bound to iodine impurity to a large part may be responsible for the recombination luminescence observed. Following Sections 2 and 3, in which experimental details and results are described, in Section 4 free and bound biexciton recombination is discussed with regard to energy positions, lineshapes and exciton kinetics. From the analysis binding energies and radius for both the free and bound biexcitons are derived. Finally in Section 5 a few remarks are made concerning multi-exciton complexes found in other materials.

2. Experimental

The crystals used in our experiments originated from the following sources: the Bulgarian Academy of Sciences in Sofia (Bulgaria), the University of North Carolina in Chapel Hill (USA), and from our own crystal growth facility. They were selected out of a number of nominally pure samples that usually contain small amounts of iodine. The iodine content was estimated from the integrated intensity of the well-known bound exciton luminescence band attributed to iodine impurity [14, 15] and using an intentionally doped sample with 1000 ppm as calibration point. The range of concentrations covered was $c_\mathrm{I} = 0.1$ to 500 ppm. All measurements were performed at 1.9 K with the samples immersed in pumped liquid helium.

Two types of experimental techniques were used allowing for measurements in different time regimes and differing in excitation power densities. One method employed uv excitation from a high power argon ion-laser (excitation wavelength $\lambda_\mathrm{L} \approx 360$ nm ($E_\mathrm{L} \approx 3.4$ eV)). The beam was chopped with a frequency of 22 kHz, which compared to the short decay time of the relevant states under investigation corresponds to quasi cw-excitation. By focussing the laser onto the sample, power densities up to 150 kW/cm^2 were obtained. A digital lock-in technique described in detail previously [10] was applied to separate luminescence components with different time behaviour. The time resolution in this case was about 100 ns.

The excitation source in the other technique applied was a narrow band dye-laser pumped by a pulsed frequency-tripled Nd–YAG laser. The power densities available for excitation in this case reached 150 kW/cm^2 up to several MW/cm^2. Special photon counting electronics were used for efficient sampling of data and simultaneously provided a time-resolution of about 2.5 ns [16]. The experiment was computer controlled. Time-resolved spectra with 0.1 nm (≈ 0.5 meV) maximum spectral resolution were obtained within several hours of measuring time.

The volume power density P_V generated by an exciting laser beam of wavelength λ_L and power density P_S at the sample surface is given by $P_\mathrm{V} = P_\mathrm{S}\alpha(\lambda_\mathrm{L}) \exp\{-\alpha(\lambda_\mathrm{L}) \cdot x\}$, where x is the coordinate inside the crystal parallel to the laser beam and α is the absorption coefficient. According to the strong dependence of $\alpha(\lambda_\mathrm{L})$ in AgBr, P_V may be varied by changing λ_L and keeping the number of incident photons almost constant.

For small penetration depths $P_V(\lambda_L) \sim \alpha(\lambda_L)$. To make the results obtained at different excitation wavelengths comparable, the surface power densities given in the following are corrected for a wavelength of 360 nm where $\alpha \approx 500\ cm^{-1}$ in AgBr. The maximum surface power density of 2 MW/cm² reached in the experiments corresponds to 5×10^{19} photons/cm² s.

3. Experimental Results

Typical luminescence spectra obtained for quasi-cw excitation with the uv lines of the argon-ion laser (excitation photon energy $E_L \approx 3.4$ eV) are represented in Fig. 1. The spectra are obtained by employing the digital lock-in technique (Section 2) and are shown time-resolved corresponding to decay times $\tau \gtrless 100$ ns. The long-lived component is due to the well-known recombination of the iodine bound exciton (BX, $\tau = 18\ \mu s$) consisting of a narrow zero-phonon line at 2.641 eV (I_{ZP}) and multiphonon structure [14, 15]. The fast component of the spectrum, besides the free exciton (FX) and several weakly bound exciton transitions near FX, contains a strong emission band with a characteristic double peak structure at around 2.59 eV (lines BXX_1) plus another weak and narrow line at 2.636 eV (line BXX_0). The lines BXX_1 and BXX_0 obviously occur at high excitation powers only, with a slightly sublinear increase of intensities up to 150 kW/cm².

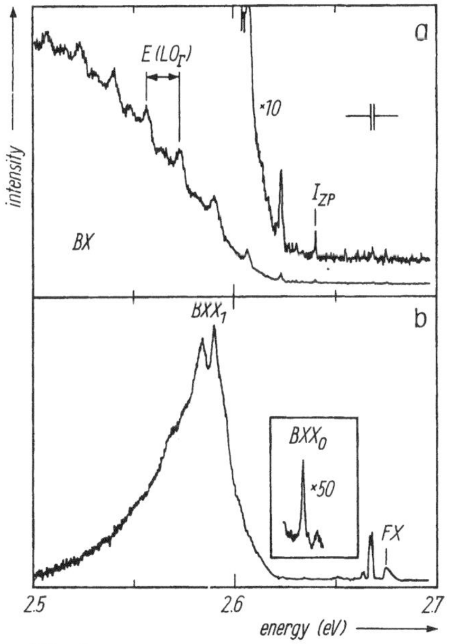

Fig. 1

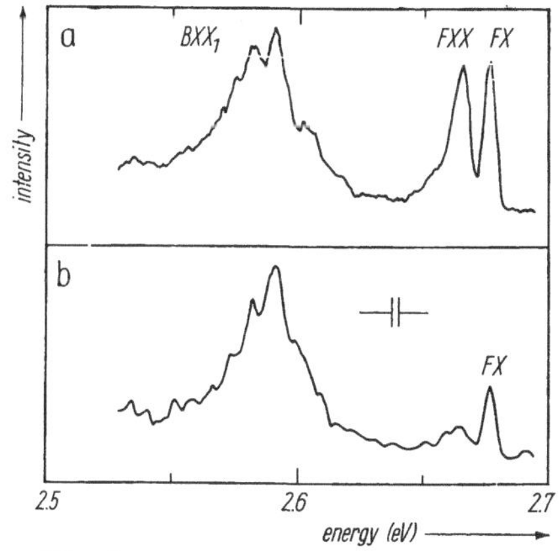

Fig. 2

Fig. 1. Time-resolved luminescence spectrum for quasi-cw excitation ($E_L = 3.4$ eV, $P = 150$ kW/cm²). Components with decay times $t \gtrless 100$ ns shown separately in a) and b). Inset represents BXX_0 line enlarged in intensity scale. Estimated iodine content of sample $c_I = 0.1$ ppm. For assignments see text. Narrow transitions near FX due to weakly bound excitons not discussed in this paper. a) $\tau > 100$ ns, b) $\tau < 100$ ns

Fig. 2. Luminescence spectra obtained for pulsed excitation at $E_L = 2.82$ eV. Power densities are a) $P = 1.8$ and b) 0.36 MW/cm². Estimated iodine content $c_I = 1.0$ ppm. For assignments see text

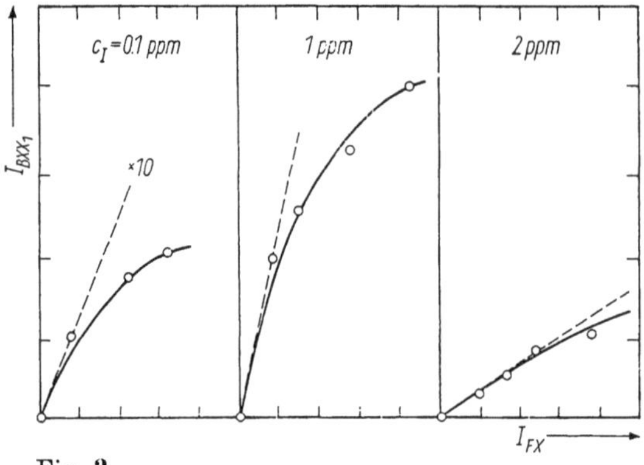

Fig. 3

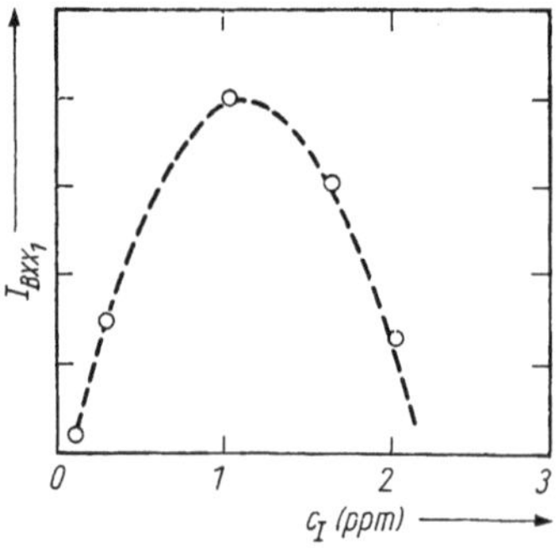

Fig. 4

Fig. 3. Dependence of BXX_1 band intensity on FX band intensity for pulsed excitation in samples of different iodine content. Integrated intensities in arbitrary units but in scale relative to each other for the three samples. Range of excitation powers $P = 0$ to $1.8\ MW/cm^2$

Fig. 4. Integrated intensity of BXX_1 band in arbitrary units vs. iodine concentration for quasi-cw excitation at constant excitation power of $100\ kW/cm^2$

The BXX_1 double peak is generally superimposed by another fast decaying luminescence band of unknown origin that is not necessarily due to a high excitation effect. This band ("Q-line") was reported previously by us and others [10, 11] and consists of a narrow zero-phonon line at 2.595 eV and pronounced LO(Γ) phonon structure. In Fig. 1 (lower spectrum) this band is noticeable as a weak shoulder only. As found earlier, it predominates the fast spectrum at low excitation levels down to $250\ W/cm^2$. Due to its different dependence on excitation intensity being sublinear, at high excitation powers it does not contribute much to the total intensity of the BXX_1 band.

The BXX_0 line is weak but clearly observed in the time-unresolved total spectrum. However, due to its small intensity and narrow halfwidth (about 1.5 meV), its intensity in the fast spectrum shows some variations that partly originate from the subtraction of spectra necessary to separate the different time components. To some extent this line is sample dependent.

Fig. 2 shows luminescence observed in AgBr under pulsed laser excitation for two different power densities in the MW/cm^2 regime. The excitation photon energy chosen was $E_L = 2.82$ eV which in view of the small absorption coefficient of AgBr at this energy ($\alpha \approx 20\ cm^{-1}$) guarantees volume excitation of the sample. According to the pulsed excitation (pulse width 10 ns), the iodine bound exciton luminescence is suppressed. Due to its small intensity and narrow halfwidth the BXX_0 line could not be detected in the pulsed-excited spectra. Like in the cw-excited spectrum of Fig. 1, both the free exciton (FX) and the strong BXX_1 recombination luminescence are observed. Although structure is indicated in the BXX_1 band maximum, the signal-to-noise ratio in this series of measurements, that covered a large range of photon energies and different excitation powers, is insufficient to definitely identify it with the double peak detected for cw-excitation. Measurements in a limited spectral range with improved signal-to-noise ratio and high spectral resolution, however, clearly prove the occurrence of the double peak also in the pulsed-excited spectra (see e.g. Fig. 7). Compared to Fig. 1, pulsed laser excitation leads to the occurrence of another strong band (FXX) at the low-energy side of the FX luminescence. This band was attributed before to the decay of the free biexciton (exciton molecule) mainly on the basis of its lineshape and its superlinear (although not quadratic) dependence of its intensity on excitation [6 to 8].

We tried to measure the decay times of all observed luminescence bands and found for the FX and FXX bands that in all samples they were below the time resolution (2.5 ns) of our multichannel photon counting system. The decay of the BXX_1 band is well described by an exponential function with a lifetime of $\tau_{BXX_1} = 7.9$ ns at 1.9 K.

Most important in interpreting the spectra observed under high intensity excitation is their correlation with iodine content c_I present in all samples. In Fig. 3, the dependence of the BXX_1 integral intensity in crystals with different c_I is represented as a function of the FX integrated intensity which is assumed to be proportional to excitation power. As shown for pulsed excitation, the BXX_1 band tends to saturate at higher excitation powers, and a similar result is obtained for quasi-cw excitation. For both modes of excitation, at constant excitation power, the BXX_1 intensity is found to exhibit a characteristic dependence on c_I. It is illustrated in Fig. 4 for 100 kW/cm^2 quasi-cw excitation, with the maximum of I_{BXX_1} occurring slightly shifted to higher c_I for increased values of excitation power. Qualitatively, a similar behaviour is observed for the BXX_0 line (in the cw-excited spectra) suggesting the BXX_1 and BXX_0 line to be of the same origin.

4. Discussion

4.1 Free biexciton recombination

In a free biexciton (FXX) two free excitons (FX) are bound essentially by Coulomb and exchange interaction [17].[3] A likely process for FXX radiative recombination is one in which one exciton remains in a real indirect FX state while the other recombines via a virtual indirect FX state. This process gives rise to a luminescence line that with regard to the FX emission is shifted to lower energy by about the FXX binding energy E^b_{FXX}. According to Cho [18], the FXX recombination lineshape in case of an indirect gap material is given by

$$I(E) = \int\int \mathrm{d}\boldsymbol{k}\, \mathrm{d}\boldsymbol{K} \exp\left(\frac{-\varepsilon_{FXX}}{k_B T}\right) |\tilde{M}|^2 \times$$
$$\times\, \delta\, \{E_{FXX}(\boldsymbol{K}) - \varepsilon_{FX}(\boldsymbol{k}) - E_{FX,0} - \hbar\omega_{\boldsymbol{K}-\boldsymbol{k}} - E\}\,. \tag{1}$$

In this expression, integration has to be carried out over all FX and FXX states with respective wave vectors $\boldsymbol{k}$ and $\boldsymbol{K}$. ε_{FX} and ε_{FXX} are the kinetic energies of the exciton

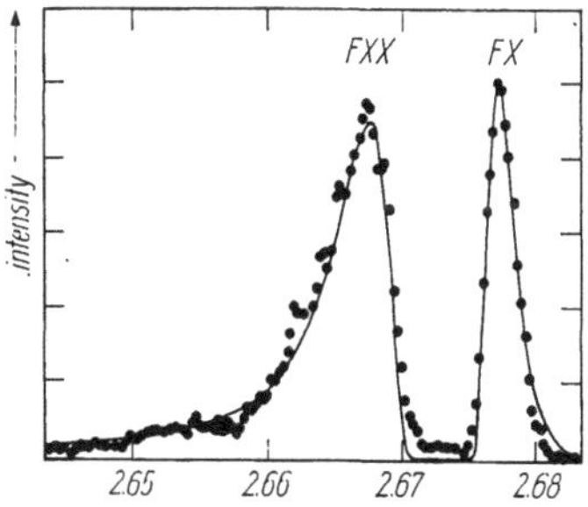

Fig. 5. FX and FXX recombination luminescence observed for pulsed excitation (E_L = 2.8 eV, P = 1.8 MW/cm^2). Points: experimental data. Full line: fits according to (1) and (2). FXX binding energy indicated with regard to the FX

[3]) In case of indirect FX, in principle two FXX states with different resulting wave vector are to be expected. In AgBr, where two indirect FX with $\boldsymbol{k} = \boldsymbol{k}_L$ may be combined, these are FXX states with $\boldsymbol{K} = 0$ and $\boldsymbol{K} = \boldsymbol{K}_X$ corresponding to the center and X-point of the Brillouin zone, respectively.

and biexciton which in case of AgBr may be written as $\varepsilon_{FX} = \hbar^2(\boldsymbol{k} - \boldsymbol{k}_L)^2/2M^*$ and $\varepsilon_{FXX} = \hbar^2(\boldsymbol{K} - 2\boldsymbol{k}_L)^2/4M^*$, respectively ($M^*$ effective single exciton mass). $E_{FX,0}$, $\hbar\omega_{\boldsymbol{K}-\boldsymbol{k}}$, and E are the energies of the indirect exciton gap, the momentum conserving (TO(L)-) phonon, and the photon. The exponential describes the population of FXX states with T representing the temperature of the FXX gas. According to [18], the matrix element is of the form

$$\tilde{M} = M_0 \left[c\left(\tfrac{1}{2}\boldsymbol{K} - \boldsymbol{k}\right) + c\left(\boldsymbol{k} - \tfrac{1}{2}\boldsymbol{K}\right)\right], \tag{2}$$

the overlap function $c(\boldsymbol{k}') = c_0/(\boldsymbol{k}'^2 + \varkappa_0^2)^2$, where $\varkappa_0 = 1/a_0$ is the reciprocal average distance a_0 of the two holes in the FXX (FXX "radius"). M_0 is a constant. Numerical integration of (1) allows to fit the FXX lineshape to the experimental data as shown in Fig. 5. From the fit the radius and the binding energy (corresponding to the transition $\boldsymbol{K} = 2\boldsymbol{k}_L \rightarrow \boldsymbol{k} = \boldsymbol{k}_L$) with regard to the FX state are obtained for the FXX. The derived values listed in Table 1 are in reasonable agreement with previous results

Table 1

Free (FXX) and bound (BXX_1) biexciton parameters in AgBr determined from fitting (1) and (6) to the experimental data (see Fig. 5, 7 and text)

	FXX	BXX_1
binding energy E^b (meV)	6.9 ± 0.2	83 ± 1
radius a_0 (nm)	0.28 ± 1	2.5 ± 2
temperature T (K)	20 ± 3	—

[6, 7]. The temperature of the FXX system resulting from the fit is $T_{FXX} = (20 \pm \pm 3)$ K.

Compared to earlier measurements, due to different experimental conditions (lower temperature, lower excitation photon energies) the FX and FXX luminescence lines are well separated in our spectra. This allows to individually fit the FX line, that formally may be described by a Maxwellian distribution [19, 20]. The temperature obtained in this way from the fit is $T_{FX} = (12 \pm 1)$ K, which is different from the FXX temperature in contrast to previous analyses [6, 7], where $T_{FXX} \approx T_{FX}$ was generally assumed.

As known from resonance Raman scattering with the indirect FX as intermediate state [20], the occurrence of the Maxwellian lineshape for the FX recombination does not necessarily imply thermal equilibrium between the exciton system and the lattice. As was shown in these experiments, at low temperature the lineshape is rather determined by various exciton relaxation processes involving acoustic and optical phonons. Due to the lack of any phonon absorption processes at low temperatures (lattice temperature = temperature of coolant = 1.9 K), thermal equilibrium cannot be established within the FX lifetime being of the order of nanoseconds in AgBr [21]. Also, at the low excitation densities in these experiments, there are no exciton–exciton collisions to establish thermal equilibrium among the excitons.

The difference in T_{FX} and T_{FXX} following from the analysis of the present high excitation experiments indicates that thermal equilibrium does also not exist between the FX and FXX systems. This suggests that even at the high excitation densities in our experiment exciton collisions are not sufficiently effective for thermalization. Consequently a temperature of the FX and the FXX gas cannot be defined, and T may rather be regarded as distribution parameter for the kinetic energies of the corresponding particles.

4.2 Bound biexciton recombination

Based on energy positions, lineshapes and dependence on iodine content the luminescence lines BXX_0 and BXX_1 that appear at high excitation powers (Fig. 1 and 2) can be consistently interpreted as due to recombination of a bound biexciton (BXX). The BXX is created by successive binding of two FX to an impurity trap that, on the basis of our results, is suggested to be isoelectronic iodine. Due to localization in real space the BXX state of energy E_{BXX} is spread out in $\boldsymbol{k}$-space.

For the decay of the BXX we consider two different processes. In the first one a bound single exciton (BX) is left behind, while the other exciton recombines via a virtual indirect FX state plus emission of a momentum-conserving L-point phonon. From the energy difference between final and initial states for this process the energy $h\nu_0$ of the emitted photon is given by

$$h\nu_0 = E_{BXX} - E_{BX} - \hbar\omega_{TO} , \tag{3}$$

where E_{BX} and $\hbar\omega_{TO}$ are the energies of the single exciton bound to iodine and the TO(L) phonon, respectively. Since only discrete energy amounts are involved in (3), this process results in a narrow recombination line. We attribute the BXX_0 line (at 2.636 eV) to this process.

The other decay channel leads to the double peak BXX_1. In this case we assume that both excitons become unbound so that one exciton remains in a real (indirect) FX state while the other decays via a virtual exciton state. This process is schematically illustrated in Fig. 6. As indicated there, the decay via the virtual exciton state may proceed in two different ways. One possibility is by means of emission of a momentum-conserving TO(L) phonon. The energy of the emitted photon in this case will be

$$h\nu_1 = E_{BXX} - E_{FX,0} - \varepsilon_{FX} - \hbar\omega_{TO} , \tag{4}$$

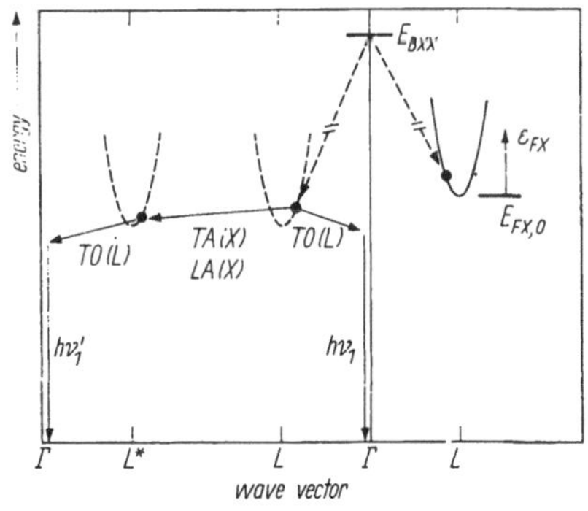

Fig. 6

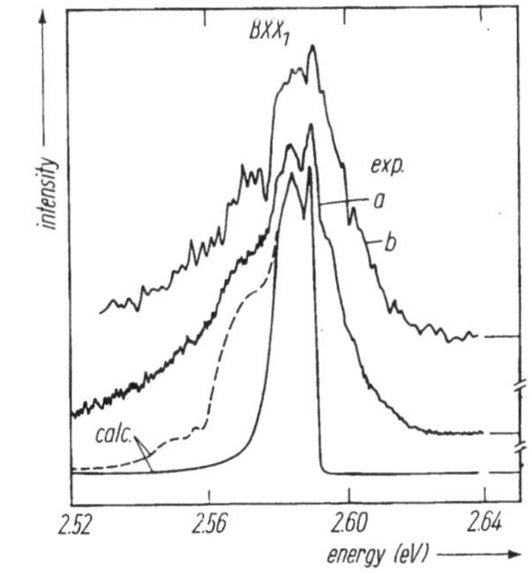

Fig. 7

Fig. 6. Schematic representation of the decay of the iodine BXX (dashed arrows) leaving behind two FX. Energy shown vs. wave vector. Full and dashed bands correspond to real and virtual states. Full arrows represent emitted photons and phonons

Fig. 7. BXX_1 recombination luminescence. Experimental curves shown are (a) for quasi-cw excitation ($E_L = 3.4$ eV, $P = 150$ kW/cm²) and (b) for pulsed excitation ($E_L = 2.8$ eV, $P = 1.8$ MW/cm²). The calculated spectrum (full line) represents recombination involving TA(X) and LA(X) scattering. Dashed line corresponds to calculation including LO(Γ) emission. Base lines of spectra shifted with regard to each other.

where $E_{\mathrm{FX,O}}$ is the FX gap energy (see Fig. 6). On the other hand, the exciton may be scattered by an intervalley X-phonon into a virtual FX state at a different point in the Brillouin zone (L → L*). As known from resonant Raman measurements [20], this process is highly efficient for exciton relaxation and involves TA(X) or LA(X) phonons. Energy conservation in this situation is analogous to (4), however, with the resulting photon energy $h\nu_1'$ being reduced with regard to $h\nu_1$ by the energy $\hbar\omega_{\mathrm{X}}$ of the X-phonon involved. Due to the variable energy term $\varepsilon_{\mathrm{FX}}$ in (4), both decay processes are expected to give rise to slightly broadened peaks. On the basis of their energetic distance from each other (equal to about $1\hbar\omega_{\mathrm{TA(X)}}$) and their widths, the BXX_1 lines (at 2.59 eV) are attributed to these decay channels.

To further support this model we have analyzed the lineshape of the BXX recombination in more detail. Since the BXX decay is similar to that of the FXX, we modified (1) neglecting the dependence on biexciton wave vector and taking into account the localization of the biexciton state by using zero kinetic energy. The energy dependence of the recombination intensity then may be written as

$$I(E) \sim \int\int \mathrm{d}\boldsymbol{k}\, |\tilde{M}|^2\, \delta\, \{E_{\mathrm{BXX}} - E_{\mathrm{FX,O}} - \varepsilon_{\mathrm{FX}} - \hbar\omega_{\mathrm{TO}} - E\}\,, \tag{5}$$

where we assume the same overlap function $c(\boldsymbol{k}')$ as for the FXX decay. Carrying out the integration one obtains

$$I(E) = I_0\, \frac{(E_{\mathrm{BXX,1}} - E)^{1/2}}{\left(E_{\mathrm{BXX,1}} + \dfrac{\hbar^2\varkappa_0^2}{2M^*} - E\right)^4}\,, \tag{6}$$

where I_0 is a constant, $E_{\mathrm{BXX,1}}$ is the energy of the onset of the luminescence band (at 2.593 eV), and $\varkappa_0^{-1}$ is the BXX radius.

Fig. 7 illustrates the fit according to our model in comparison to the experimental lineshape observed for both cw and pulsed excitation. Applying (6) the high energy peak of the BXX_1 band is well reproduced. For the fit a BXX binding energy of 83 meV (with regard to the FX) and a BXX radius of 2.5 nm are used (see Table 1). The lineshape of the low energy BXX_1 peak can be computed correspondingly taking into account intervalley scattering. In calculating the contribution of these processes to the luminescence band, we used the ratio of scattering probabilities for TA(X) and LA(X) scattering $c_{\mathrm{TA}}/c_{\mathrm{LA}} = 1.68$ as determined from resonant Raman scattering [20]. Also higher order intervalley processes 2TA(X) and TA(X) + LA(X) were included, for which the scattering probabilities are known to be large. As seen from Fig. 7, by inclusion of these processes the low-energy peak is either well reproduced both with respect to peak position and width, the full calculated line representing the linear superposition of all contributions. As suggested by the structure of the experimental lineshape at lower energies, interaction with LO(Γ) phonons also plays a role in the recombination process. The dashed line is an improved fit where we took a linear coupling to LO(Γ) phonons into account using a coupling constant S^n with $S = 0.3$ (n number of phonons involved).

A strong argument supporting our model in which the observed luminescence is attributed to the iodine BXX decay is the energy difference between the BXX_0 line position and the onset of the BXX_1 band. The experimental value from our spectra is $h\nu_0 - h\nu_1 = 43.5$ meV. On the other hand, this quantity can be determined independently as difference between the exciton gap energy ($E_{\mathrm{FX,O}} = 2.6845$ eV) and the zero-phonon transition energy ($E_{\mathrm{BX}} = 2.6415$ eV) of the iodine bound single exciton as obvious from subtracting (3) and (4) ($\varepsilon_{\mathrm{FX}} = 0$ in (4)). The value obtained in this way agrees within 1 meV.

Comparison of the BXX binding energy of 83 meV with that of the iodine bound single ($E^b_{BX} = 43$ meV [14]) shows that in the iodine BXX the two excitons are bound with about equal strength. The weak (integrated) intensity of the BXX_0 line relative to the BXX_1 band indicates that the BXX decay into two FX is more probable than the process in which a single BX is left behind.

4.3 Exciton kinetics

Starting from the ideas developed above, some of the characteristic features of the observed free and bound exciton and biexciton transitions can be described by means of a kinetic model in which the interaction of the various states is represented in form of rate equations. In particular, the model is able to explain the absence of the iodine-bound single exciton (BX) recombination in the pulsed excited spectra (Fig. 2), the saturating behaviour of the iodine BX and BXX (Fig. 3), and the dependence of the BXX_1 intensity on iodine content (Fig. 4). The relevant generation and decay processes for the various excitons and biexcitons are summarized schematically in Fig. 8. Correspondingly for the time derivations of the various intensities n_i the following relations exist:

$$\frac{dn_{FX}}{dt} = A_0 I_L - \frac{n_{FX}}{\tau_{FX}} - B_0 n_{FX}(c_I - n_{BX}) - \lambda_0 n_{FX}^2 - D_0 n_{FX} n_{BX} + \frac{n_{FXX}}{\tau_{FXX}} + \frac{n_{BXX}}{\tau_{BXX}}, \tag{7}$$

$$\frac{dn_{BX}}{dt} = B_0 n_{FX}(c_I - n_{BX}) - \frac{n_{BX}}{\tau_{BX}}, \tag{8}$$

$$\frac{dn_{FXX}}{dt} = \frac{1}{2} C_0 n_{FX}^2 - \frac{n_{FXX}}{\tau_{FXX}}, \tag{9}$$

$$\frac{dn_{BXX}}{dt} = D_0 n_{FX} n_{BX}(c_I - n_{BXX}) - \frac{n_{BXX}}{\tau_{BXX}}. \tag{10}$$

τ_i are the lifetimes of the different exciton states, I_L represents the laser excitation intensity creating free excitons, and $A_0 \dots E_0$ are the conversion coefficients for creation of the various particles as evident from Fig. 8. In deriving (8) and (10), we assumed $E_0 = 0$ according to the small probability of the BXX recombination process in which an iodine BX is left behind (see Section 4.2).

The stationary solutions ($dn_i/dt = 0$) of (8) and (10) are

$$n_{BX} = \frac{B_0 c_I n_{FX}}{\dfrac{1}{\tau_{FX}} + B_0 n_{BX}} \tag{11}$$

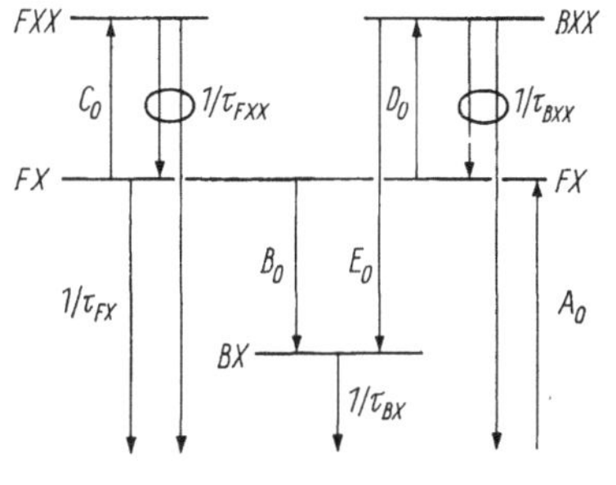

Fig. 8. Schematic representation of generation and recombination mechanisms of free and bound single excitons (FX, BX) and free and bound biexcitons (FXX, BXX). τ_i are the lifetimes of the various states, $A_0 \dots E_0$ are conversion coefficients for creation or annihilation

and

$$n_{\mathrm{BXX}} = \frac{D_0 c_{\mathrm{I}} n_{\mathrm{BX}} n_{\mathrm{FX}}}{\frac{1}{\tau_{\mathrm{BXX}}} + D_0 n_{\mathrm{BX}} n_{\mathrm{FX}}} . \tag{12}$$

They qualitatively reflect the saturation behaviour of the BXX_1 luminescence intensity (Fig. 3) as well as of the iodine BX intensity observed previously [22] at sufficiently high FX densities. In particular, (12) is consistent with our observation that for quasi-cw excitation (small n_{FX}) saturation of the BXX_1 band occurs only for samples with small iodine concentration ($c_{\mathrm{I}} = 0.1$ ppm). In contrast, for pulsed excitation (high n_{FX}) saturation is observed also for samples with higher c_{I}.

The characteristic dependence of the BXX intensity on iodine content qualitatively follows from reducing (7) and (10) to the relevant terms. Making the reasonable assumption that

1. the number of FX (n_{F}) is predominantly determined by generation via photons and trapping at iodine impurities, and

2. the number of BXX created under our excitation conditions is relatively small ($n_{\mathrm{BXX}} \ll c_{\mathrm{I}}$), their stationary solutions by means of (8) may be written

$$n_{\mathrm{FX}} = \tau_{\mathrm{FX}} A_0 I_{\mathrm{L}} - \tau_{\mathrm{FX}} \frac{n_{\mathrm{BX}}}{\tau_{\mathrm{BX}}} \tag{13}$$

and

$$n_{\mathrm{BXX}} = \tau_{\mathrm{BXX}} D_0 n_{\mathrm{FX}} n_{\mathrm{BX}} c_{\mathrm{I}} . \tag{14}$$

Replacing n_{FX} in (14) by (13) and assuming $n_{\mathrm{BX}} \sim c_{\mathrm{I}}$ for small n_{BX} according to (8), the density of BXX is approximately obtained as

$$n_{\mathrm{BXX}} \sim A_0 I_{\mathrm{L}} c_{\mathrm{I}}^2 - \frac{c_{\mathrm{I}}^3}{\tau_{\mathrm{BX}}} . \tag{15}$$

The dashed line in Fig. 4 is a fit of (15) to the experimental data at $I_{\mathrm{L}} =$ const, the good agreement qualitatively supporting our model. Also (15) reproduces the shift of the maximum of the curve at increased I_{L} observed experimentally.

Finally, also the lack of iodine BX luminescence for pulsed excitation may be understood from the kinetic model developed above. The quasi-cw spectra ($P < 150$ kW/cm^2) already show saturation of the number of iodine BX, so that further increase of excitation power in the pulsed spectra does not significantly increase n_{BX}. However, at these high excitation densities a sufficient number of FX is created to convert existing BX to BXX states. Consequently, for pulsed laser excitation n_{BX} decreases due to increasing n_{BXX}. This results in suppression of the iodine BX luminescence band and the observed predominance of the BXX luminescence (Fig. 2).

5. Concluding Remarks

Although we cannot exclude that luminescence due to EHL recombination contributes in the spectral range we have studied, the observed emission largely seems to be of different origin. The model of an iodine BXX developed above is in good accordance with all our experimental observations, but more specific measurements will have to be performed to further substantiate it. While multiexciton complexes bound at various donors and acceptors in Ge and Si are frequently reported in the literature (see e.g. [23]), the most closely related case to the present example is the biexciton bound to nitrogen isoelectronic trap in GaP [24]. In this material, like in AgBr, additional recombination lines appear at high excitation intensities. They are ascribed to recombination of one of the two excitons comprising the BXX, leaving a single exciton bound to nitrogen in the ground state. Multi-particle transitions including the analogue to the BXX_1 line in AgBr were sought but the evidence for these processes was found to be not conclusive. To understand the relative strengths of the various decay processes in AgBr and GaP, in particular the weak intensity of the BXX_0 compared to the BXX_1 line in AgBr, the proposed model requires further development. In principle, however, we believe that isoelectronic iodine plays a similar role for AgBr as nitrogen in GaP. In particular, it will be difficult to further purify the crystals and reduce the content of unintentionally present iodine in AgBr, which is required for further investigations of the EHL in this material.

Acknowledgements

The authors like to thank Dr. H. Stolz for many helpful discussions. The work was partly supported by the Minister for Wissenschaft und Forschung des Landes NRW.

References

[1] T. M. Rice, Solid State Phys. **32**, 1 (1977).
[2] J. C. Hensel, T. G. Phillips, and G. A. Thomas, Solid State Phys. **32**, 88 (1977).
[3] G. A. Thomas and V. B. Timofeev, in: Handbook of Semiconductors, Vol. 2, Ed. M. Balkanski, North-Holland Publ. Co., Amsterdam 1980 (p. 45).
[4] W. Schmid, A. Forchel, G. Moersch, and M. H. Pilkuhn, J. Phys. Soc. Japan **49**, 465 (1980).
[5] G. A. Thomas and T. M. Rice, Solid State Commun. **23**, 359 (1977).
[6] T. Baba and T. Masumi, Nuovo Cimento **39B**, 609 (1977).
[7] I. Pelant, A. Mysyrowicz, and C. Benoit a la Guillaume, Nuovo Cimento **39B**, 655 (1977); Phys. Rev. Letters **37**, 1708 (1976).
[8] I. Pelant, J. Hala, L. Parma, and K. Vacek, Solid State Commun. **36**, 729 (1980).
[9] D. Hulin, A. Mysyrowicz, M. Combescot, I. Pelant, and C. Benoit a la Guillaume, Phys. Rev. Letters **39**, 1169 (1977).
[10] Th. Kleinefeld, H. Stolz, and W. von der Osten, Solid State Commun. **31**, 59 (1979).
[11] T. Baba and T. Masumi, Proc. 14th Internat. Conf. Phys. Semicond., Ed. B. L. H. Wilson, Edinburgh 1978 (p. 833).
[12] G. Beni and T. M. Rice, Phys. Rev. Letters **37**, 874 (1976); Phys. Rev. B **18**, 768 (1978).
[13] Th. Kleinefeld, H. Stolz, and W. von der Osten, Physica (Utrecht) **117/118B**, 306 (1983).
[14] H. Kanzaki and S. Sakuragi, J. Phys. Soc. Japan **27**, 109 (1969).
[15] W. Czaja and A. Baldereschi, J. Phys. C **12**, 405 (1979).
[16] Th. Kleinefeld and H. Ziegler, J. Phys. E **15**, 888 (1982).
[17] D. B. Tran Thoai, Z. Phys. **B 26**, 115 (1976).
[18] K. Cho, Optics Commun. **8**, 412 (1973).
[19] W. von der Osten and J. Weber, Solid State Commun. **14**, 1133 (1974).
[20] J. C. Windscheif and W. von der Osten, J. Phys. C **13**, 6299 (1980).
[21] H. Stolz, private communication.
[22] J. Weber, phys. stat. sol. (b) **78**, 699 (1976).
[23] M. L. W. Thewalt, Proc. 14th Internat. Conf. Phys. Semicond., Edinburgh 1978 (p. 605).
[24] J. L. Merz, R. A. Faulkner, and P. J. Dean, Phys. Rev. **188**, 1228 (1969).

(Received February 28, 1984)

phys. stat. sol. (b) **125**, 305 (1984)

Subject classification: 20; 13; 21

Department of Physical Sciences, University of Turku[1])

The First Born Approximation in Estimates of the Threshold Anomalies in the X-Ray Spectra of Simple Metals

By

J. A. LEIRO

The XPS core line asymmetries and the XAS threshold singularities of Zn, Cd, In, Mg, and Na metals are considered. The singularity indices and threshold exponents, calculated by using the "$Z + 1$" approximation are compared with experimental values. Furthermore, the results obtained by a partial-wave method are compared with those estimated by the first Born approximation. It is found that the Born approximation values are in agreement with those within the partial-wave method.

Die XPS (Röntgenphotonenemissions)-Linienasymmetrien und die XAS-(Röntgenabsorptions)-Kantensingularitäten von Zn,- Cd-, In-, Mg- und Na-Metallen werden untersucht. Die Singularitätsindizes und die Kantenexponenten, die mit Hilfe der „$Z + 1$"-Approximation berechnet werden, werden mit experimentell gewonnenen Werten verglichen. Die Resultate der Schrödingergleichung werden mit der Born-Approximation verglichen und befriedigende Übereinstimmung wird gefunden.

1. Introduction

We consider X-ray absorption (XAS) and photoemission (XPS) by metals. On creation of an electronic hole state in the core level of the metal the conduction electrons screen the hole and relax toward it. The many-body effects caused by the deep localized hole on the shape of X-ray spectra have been investigated by several authors in recent years [1 to 4]. According to the model suggested by Mahan [5], Anderson [6], Nozières and De Dominicis [7] (MND) the readjustment of the conduction electrons to the suddenly created hole potential will manifest itself as a power law shape in the soft X-ray spectra of simple metals in the immediate vicinity of the Fermi level. In the case of the X-ray photoelectron (XPS) core lines Doniach and Šunjić (DS) [8] have given the asymmetric lineshape function which is essentially a convolution of the ND power law with the Lorentzian describing the lifetime effects of the core hole. The asymmetry of the XPS lineshape is determined by the singularity index [8]

$$\alpha = 2 \sum_{l=0}^{\infty} (2l + 1) \left(\frac{\delta_l}{\pi}\right)^2 , \tag{1}$$

where the scattering phase shifts δ_l of the core hole potential obey the Friedel sum rule

$$2 \sum_{l=0}^{\infty} (2l + 1) \frac{\delta_l}{\pi} = 1 \tag{2}$$

at the Fermi level.

[1]) Itäinen Pitkäkatu 1, SF-20520 Turku, Finland.

The X-ray absorption and emission edges are influenced by the power law which is dependent on the threshold exponent [7]

$$\alpha_l = \frac{2\delta_l}{\pi} - \alpha\,. \tag{3}$$

Because the DS lineshape function is the XPS power law convoluted with the Lorentzian it is easy to show that the following lineshape formula

$$\mu(E) \sim \frac{\cos\left[\dfrac{\pi(1-\alpha_l)}{2} - \alpha_l \arctan\left(\dfrac{E - E_0}{\gamma}\right)\right]}{[(E - E_0)^2 + \gamma^2]^{\alpha_l/2}} \tag{4}$$

can be deduced for the X-ray absorption (XAS) edge including the lifetime broadening, if the step function partial density of states is assumed. The energy E_0 is the threshold position and 2γ is the lifetime width.

Recently, it has been emphasized that the non-linear screening contribution to the XPS singularity index (1) is small [9]. The main purpose of this paper is to examine the Born approximation and show that we do indeed have a reasonable description of (1) and (3) in simple metals at usual metallic densities by calculating the phase shifts both using the partial-wave method, and using the first Born approximation. We calculate numerically the singularity indices and threshold exponents of Zn, Cd, In, Mg, and Na metals using the "$Z + 1$" approximation and the linear screening theory. A model potential is employed, by considering the electron gas screening separately with the Vashishta-Singwi dielectric function [10] and with the random phase approximation (RPA) dielectric function. Also a comparison is made between the exact partial-wave method and the first Born approximation. It is useful to estimate the sensitivity of the calculated exponents to the approximations made. Furthermore, these results are discussed in the light of experiments.

It is well known that the valence structure of the excited atom with the valence Z is almost identical with that of the $Z + 1$ atom in the ground state. If we remove an inner core electron, then this process is almost indistinguishable from a unit increase in nuclear change, the extra electron being now on the valence shell. We can therefore construct the core hole potentials by using the "$Z + 1$" approximation [11] in which the X-ray excited atom is replaced by the unexcited heavier atom including one additional nuclear charge. Now we turn our attention to the definition of the core hole potential. The static screening can be considered by redefining the potential. Because the Coulomb potential is too strong and the Thomas-Fermi screening is a rough approximation, a new definition for the core hole potential $V(q)$ and for the dielectric function is needed [12]. We consider the model potential approach. The interaction is defined in our case as the difference of screened model potentials [13, 14] of the $Z + 1$ and the Z ion. It should be emphasized that we have used the linear screening theory in order to treat the response of the conduction electrons to the sudden creation of the core hole by taking advantage of the weak model potentials. The linear screening response should be useful for a weak external potential and should be better for model potentials than for a small point charge potential because of the cancellation theorem, which eliminates the contribution from core states [15]. Instead of calculating the scattering phase shifts directly from the core hole potential we could also calculate the differences between the phase shifts with and without the core hole using the screened model potential of $Z + 1$ and Z valence ions separately. The results would be the same for our scattering potential provided the first Born approximation is used (the actual wave functions are replaced by plane waves). This indicates that the phase shifts do not differ very much in the cases, in which the model potentials and the linear screening are used, although the first Born approximation is only a qualitative method. In the nearly free electron like metals one would expect that the contribution from the phase shifts of the model potential in the absence of the core hole (the Z valence ion) is quite small [16]. Hence, the main contribution to the singularity indices and threshold exponents is most likely due to the $Z + 1$ valence ion potential (the core hole is present).

2. Results and Discussion

We consider two different screening functions. One proposed by Lindhard [17] (RPA) and the other proposed by Vashishta and Singwi [10] (V-S). The local field factor introduced by the latter authors is of the form

$$G(q) = A(1 - e^{-B(q/k_F)^2}), \tag{5}$$

which describes the exchange and correlation effects, A and B being two r_s-dependent parameters and k_F is the Fermi momentum. It should be noted that we have the RPA dielectric function if $A = 0$. The static dielectric function has the following property [18]:

$$\varepsilon(q, 0) = 1 + \frac{Q_0(q, 0)}{1 - G(q)\, Q_0(q, 0)} = \frac{1 + (1 - G(q))\, Q_0(q, 0)}{1 - G(q)\, Q_0(q, 0)} = \Lambda_q \varepsilon_e(q, 0)\,, \tag{6}$$

where

$$Q_0(q, 0) = \frac{k_s^2}{2q^2}\left(1 + \left(\frac{k_F}{q} - \frac{q}{4k_F}\right) \ln \left|\frac{2k_F + q}{2k_F - q}\right|\right).$$

and k_s is the adjustable screening constant. In the small q limit, Λ_q is proportional to the proper vertex function multiplied with the quasiparticle renormalization factor [19] and

$$\varepsilon_e(q, 0) = 1 + (1 - G(q))\, Q_0(q, 0)\,. \tag{7}$$

Table 1

Comparison of scattering phase shifts calculated by using the partial-wave method and the first Born approximation. Also shown are the results corresponding to inclusion of the static Vashishta-Singwi dielectric function and the RPA dielectric function, respectively. We can see that $\delta_1^B < \delta_1$. This means that the Born approximation underestimates the phase shift for the p-wave

	the partial-wave method						the first Born approximation					
	V-S			RPA			V-S			RPA		
	δ_0	δ_1	δ_2	δ_0	δ_1	δ_2	δ_0^B	δ_1^B	δ_2^B	δ_0^B	δ_1^B	δ_2^B
Zn	0.38	0.29	0.06	0.37	0.28	0.07	0.45	0.26	0.07	0.42	0.25	0.08
Cd	0.26	0.34	0.06	0.26	0.32	0.07	0.19	0.33	0.08	0.20	0.31	0.09
In	0.48	0.28	0.05	0.45	0.27	0.06	0.56	0.25	0.05	0.51	0.24	0.07
Mg	0.72	0.24	0.03	0.65	0.23	0.05	0.76	0.22	0.03	0.67	0.21	0.05
Na	0.98	0.19	0.01	0.89	0.20	0.02	1.03	0.18	0.00	0.87	0.19	0.03

The screened core hole potential can be now written in the momentum space,

$$\tilde{V}(q) = \frac{V(q)}{\varepsilon_e(q, 0)}, \tag{8}$$

where $V(q)$ is the bare hole potential involving the contribution from the excited atom and the unexcited atom. The core radii and parameters, which are needed for the potential, are obtained from [14] and the core parameters are extrapolated to the Fermi level taking into account the work functions.

In order to get the phase shifts δ_l at the Fermi level we must solve the Schrödinger equation

$$\frac{d^2}{dr^2} u(r) + \left[k_F^2 - \frac{l(l+1)}{r^2} - \frac{2m}{\hbar^2} \tilde{V}(r)\right] u(r) = 0, \tag{9}$$

where m is the mass of an electron and

$$\tilde{V}(r) = \left(\frac{1}{2\pi}\right)^3 \int e^{-\boldsymbol{r}\cdot\boldsymbol{q}} \tilde{V}(q) \, d^3q .$$

The obtained phase shifts can be seen in Table 1. Two sets of calculations are performed using the RPA dielectric function and the Vashishta-Singwi screening factor. This indicates that the s- and p-phase shifts reproduced by the latter dielectric function are usually slightly larger than those obtained using the former one. This is due to the fact that the local field factor (5) reduces the screening of the core hole potential.

Table 2

The threshold exponents and singularity indices obtained by the partial-wave method with the choice of the Vashishta-Singwi (V-S) dielectric function and the RPA dielectric function, respectively. The ratio of the singularity indices estimated by the different dielectric functions is also shown

	V-S				RPA					
	α	α_0	α_1	α_2	α	α_0	α_1	α_2	$\alpha_{RPA}/\alpha_{V-S}$	r_s
Zn	0.08	0.16	0.10	−0.05	0.08	0.16	0.10	−0.04	1.00	2.30
Cd	0.09	0.08	0.13	−0.05	0.08	0.09	0.12	−0.04	0.89	2.59
In	0.10	0.21	0.08	−0.06	0.09	0.20	0.08	−0.05	0.90	2.41
Mg	0.14	0.32	0.02	−0.12	0.12	0.30	0.03	−0.09	0.86	2.65
Na	0.21	0.41	−0.09	−0.21	0.18	0.38	−0.06	−0.17	0.86	3.93

experimental values

	α	α_0	α_1	α_2
Zn				
Cd	0.07 ± 0.02[a])			
In	0.11 ± 0.02[a])			
Mg	0.13 ± 0.015[b])	0.23 ± 0.02[b])	0.08 ± 0.02[c])	
Na	0.19 ± 0.015[a])	0.37 ± 0.03[b])		

[a]) Dow [22].
[b]) Citrin et al. [23].
[c]) Heinonen and Leiro [24].

In order to compare the partial-wave method and the first Born approximation for calculating the phase shifts, we have included the results of the latter approximation also in Table 1, which shows that the agreement is quite good between the two different methods. This can be partly explained by the crucial role of the Friedel sum rule which is important in the screening mechanism [20]. As is well known, the Born approximation is expected to be reliable when the true wave function in the scattering region does not differ significantly from the incident plane wave. Because of the centrifugal barrier the p- and d-scattering phase shifts may be approximated by the first Born approximation. Although the s-partial wave usually deviates considerably from corresponding plane wave, the phase shifts can be nevertheless estimated in some cases because the Born phase shifts fulfil the requirements of the Friedel sum rule [21].

Tables 2 and 3 show α_0, α_1, and α_2 together with α obtained with different methods and dielectric functions. It can be seen that the ratio $\alpha_{RPA}/\alpha_{V-S}$ between the singularity indices estimated using the partial-wave method with the RPA dielectric function and with that of Vashishta-Singwi is quite close to one (within 14%). This is due to the fact that the change in screening is partly compensated by the modification of the scattering wave function caused by the Friedel sum rule. If we compare the singularity indices calculated using the RPA screening, the Born approximation values do not deviate more than 12% from the corresponding partial-wave result. This shows that the p- and d-phase shifts have been accurately evaluated for our core hole potential. We have seen that the discrepancies between the results of the first Born approximation and the partial-wave method are of the same order of magnitude as those caused by the choice of the dielectric function.

The experimental exponents are included in Table 2. Unfortunately, there are no data for Zn available. For the other metals the agreement is quite good between theory and experiment as far as the XPS singularity index α is concerned. There is also a value for α_0 deduced from the $NaL_{2,3}$ absorption spectrum, which is nearly the same as that of the corresponding calculated exponent. However, only in the case of Mg there exists an experimental α_1 obtained from the K-absorption measurement. The data indicate that the $L_{2,3}$ and K edges are peaked. There is a reasonable agreement between theory and experiment although the "$Z + 1$" approximation predicts slightly too peaked $L_{2,3}$ absorption edge and less peaked K-edge compared with the experimental values of magnesium. It should be mentioned that the most part of the other

Table 3

The Born approximation results for the exponents calculated with the Vashishta-Singwi (V-S) dielectric function and with the RPA dielectric function. We can see also the ratio of the singularity indices between the Born approximation value and the corresponding partial-wave value evaluated with the same dielectric function

	V-S (Born)					RPA (Born)				
	α^B	α_0^B	α_1^B	α_2^B	α^B/α	α^B	α_0^B	α_1^B	α_2^B	α^B/α
Zn	0.09	0.20	0.08	−0.04	1.13	0.08	0.19	0.08	−0.03	1.00
Cd	0.08	0.04	0.13	−0.03	0.89	0.07	0.05	0.12	−0.02	0.88
In	0.10	0.25	0.06	−0.07	1.00	0.09	0.23	0.06	−0.05	1.00
Mg	0.14	0.34	0.00	−0.13	1.00	0.12	0.31	0.02	−0.09	1.00
Na	0.23	0.42	−0.12	−0.23	1.10	0.17	0.38	−0.05	−0.15	0.94

theoretical estimates concerning Mg and Na metals are collected by Citrin et al. [23] and therefore we do not consider them further.

We have also performed calculations for Na and Mg with a potential possessing a slightly different shape. In that case the Born phase shifts were nearly the same as those in Table 1 although the phase shifts obtained by the partial-wave method deviated considerably from the present values. This indicates that the choice of the core hole potential is very important especially in the case of partial-wave method.

Finally we make some remarks. As mentioned the derived threshold exponents and singularity indices are summarized in Tables 2 and 3. The Vashishta-Singwi and the RPA dielectric functions have been used. The partial-wave method and the first Born approximation gave very similar results in the case of Zn, Cd, In, Mg and Na metals. Also the results showed a fairly good agreement with the available experimental data. We may conclude that the "$Z + 1$" approximation was capable to interpret the XPS core line asymmetries in a quite satisfactory manner. It is apparent that the first Born approximation is sensitive to the choice of dielectric function at lower metallic densities, whereas it is likely that in the case of the partial-wave method the knowledge of the core hole potential is more important.

References

[1] P. MINNHAGEN, Phys. Letters A **56**, 327 (1976).
[2] K. SHUNG and D. C. LANGRETH, Phys. Rev. B **23**, 1480 (1981).
[3] Y. OHMURA and K. ISHIKAWA, J. Phys. Soc. Japan **49**, 1829 (1980).
[4] J. W. WILKINS, AIP Conf. Proc. **94**, 687 (1982).
[5] G. D. MAHAN, Phys. Rev. **163**, 612 (1967).
[6] P. W. ANDERSON, Phys. Rev. Letters **18**, 1049 (1967).
[7] P. NOZIÈRES and C. T. DE DOMINICIS, Phys. Rev. **178**, 1097 (1969).
[8] S. DONIACH and M. ŠUNJIĆ, J. Phys. C **3**, 285 (1970).
[9] C.-O. ALMBLADH and L. HEDIN, Handbook on Synchrotron Radiation, Ed. E. E. KOCH, Vol. 1b, North-Holland Publ. Co., 1983 (p. 607).
[10] P. VASHISHTA and K. S. SINGWI, Phys. Rev. B **6**, 875 (1972).
[11] J. D. DOW, D. R. FRANCESCHETTI, D. R. GIBBONS, and S. E. SCHNATTERLY, J. Phys. F **5**, L221 (1975).
[12] P. MINNHAGEN, J. Phys. F **7**, 2441 (1977).
[13] P. ASCARELLI, Solid State Commun. **21**, 205 (1977).
[14] G. D. MAHAN, phys. stat. sol. (b) **63**, 453 (1974).
[15] M. L. COHEN and V. HEINE, Solid State Phys. **24**, 235 (1970).
[16] S. M. GIRVIN and J. J. HOPFIELD, Phys. Rev. Letters **34**, 1320 (1975).
[17] J. LINDHARD, Kong. Danske Vid. Selsk., mat.-fys. Medd. **28**, No. 8 (1954).
[18] C. A. KUKKONEN and J. W. WILKINS, Phys. Rev. B **19**, 6075 (1979).
[19] L. HEDIN and B. I. LUNDQVIST, J. Phys. C **4**, 2064 (1971).
[20] R. SCHÖPKE, W. JOHN, and E. MROSAN, J. Phys. F **9**, 1205 (1979).
[21] P. LONGE, Phys. Rev. B **8**, 2572 (1973).
[22] J. D. DOW, Phil. Mag. B **35**, 837 (1977).
[23] P. H. CITRIN, G. K. WERTHEIM, and M. SCHLÜTER, Phys. Rev. B **20**, 3067 (1979).
[24] M. H. HEINONEN and J. A. LEIRO, Phil. Mag. B **46**, 669 (1982).

(Received May 15, 1984)

phys. stat. sol. (b) 125, 311 (1984)

Subject classification: 13.1 and 13.4; 22.9

Max-Planck-Institut für Festkörperforschung, Stuttgart[1])

A Crystal Orbital (CO) Model for Low-Dimensional Segregated Donor and Acceptor Stacks

By

M. C. BÖHM

A crystal orbital (CO) model for low-dimensional segregated donor (D) and acceptor (A) systems is developed. The approach can be traced back to two one-dimensional (1D) tight-binding calculations for "reference stacks" in the D and A manifolds that are coupled to the electrostatic fields created by the neighbouring 1D backbones. The Hartree-Fock (HF) operators in the self-consistent-field (SCF) approximation are defined by using the grand canonical (GC) averaging procedure. This ensemble averaging allows for a straightforward determination of the mean-field operators for fractional occupation schemes within the two 1D stacks.

Es wird ein Kristallorbital(CO)-Verfahren entwickelt, das die Behandlung von Donor(D)–Akzeptor(A)-Systemen ermöglicht, die in separaten Stapeln kristallisieren. Die Methode läßt sich auf zwei eindimensionale (1D)"Tight-Binding"-Ansätze für Referenzstapel innerhalb des Donor- und Akzeptorteilsystems zurückführen, die jeweils an die elektrostatischen Felder gekoppelt sind, die von den übrigen 1D-Fragmenten gebildet werden. Die Hartree-Fock(HF)-Operatoren in der SCF-Näherung werden mit Hilfe einer Mittlungsprozedur definiert, die auf der großkanonischen (GC) Gesamtheit aufbaut. Dieser Ensemble-Formalismus erlaubt eine einfache Berechnung der HF-Operatoren für fraktale Besetzungsschemata in den beiden 1D-Verbänden.

1. Introduction

The physical and chemical properties of low-dimensional solids composed by donor (D) and acceptor (A) moieties as the molecular building blocks have found considerable interest in recent years due to their semiconducting, conducting, or even superconducting properties [1 to 6]. On the basis of the mutual arrangement of the donor and acceptor fragments, it is possible to discriminate two different stacking patterns in the class of these charge transfer (CT) systems: mixed stacks with an alternating arrangement of the D and A units as well as segregated stacks, where a face-to-face crystallization is found within the D and A manifolds leading to one-dimensional (1D) backbones with columnar structures:

Mixed stacks: $\dots D^{+Q}A^{-Q}D^{+Q}A^{-Q}D^{+Q}A^{-Q}\dots$

Segregated stacks: $\dots D^{+Q}D^{+Q}D^{+Q}\dots$ and

$\dots A^{-Q}A^{-Q}A^{-Q}\dots,$

Q is the electronic charge that has been transferred from the donor sites to the acceptor units. Both topological arrangements of DA pairs are encountered in organic solids and in transition metal species.

A suitable theoretical tool to analyze the electronic structure of these low-dimensional systems is the crystal orbital (CO) formalism based on the well-known tight-binding approximation [7]. The state of art of this approach in combination with

[1]) Heisenbergstr. 1, D-7000 Stuttgart 80, FRG.

semiempirical and ab initio Hamiltonians has been reviewed by several authors for 1D (or 2D) solids with small or medium-sized unit cell dimensions [8 to 10]. The prohibitive computational expense of self-consistent-field (SCF) Hartree-Fock (HF) CO calculations, however, forms a severe restriction for numerical applications to solid state ensembles with extended unit cells. The first tight-binding studies on mixed donor–acceptor models, for instance, have been reported very recently [11, 12].

The theoretical and computational difficulties, that must be expected in tight-binding calculations on solids with segregated stacks, prevented until now CO studies on donor–acceptor systems with stoichiometries, where D and A moieties are both found in the crystallographic unit cells. In the previously published contributions on these systems it had been assumed implicitly, that interchain (DA, DD′, AA′) interactions can be neglected. The corresponding band structure calculations were thus restricted either to an electronically (isolated) donor or acceptor column. Such a model is strictly one-dimensional. Ground state configurations are derived in both regimes (D and A) that correspond to non-paramagnetic (i.e. closed shell systems in the language of molecular quantum mechanics) filling schemes. The dispersion curves $\varepsilon(k)$ associated to the two strands are either completely filled or empty.

The band structures of typical organic donor and acceptor strands crystallizing as segregated DA stacks have been analyzed (under the premises of the above-mentioned restrictions) by 1D tight-binding schemes in different degrees of sophistication [13 to 16]. The electronic structures of transition metal backbones with organic π ligands have been investigated by semiempirical SCF CO variants based on the ZDO (zero differential overlap) approximation [17 to 20]. A very recent tight-binding study is available for the I_3^- stack, an important acceptor group in the organometallic area [21]. To evade the numerical difficulties encountered in "real (DA) solids" various theoretical studies on CT systems have been published in the past that are based on the Hubbard Hamiltonian [22 to 25]. The simple model operators are often restricted to one or two orbitals and electron(s) per unit cell and can be solved analytically. They allow one to study the magnitude of charge transfer, stabilities or filling schemes as a function of a few sophisticated, highly formal parameters. The structural, electronic, and physical properties of the low-dimensional material, however, are not considered in such analytic descriptions.

It is the purpose of the present contribution to develop a crystal orbital model which allows for tight-binding studies of segregated donor and acceptor systems with extended unit cell dimensions and complicated stoichiometries. The key step of the approach is an electrostatic field approximation [26, 27] leading to a decoupling of the 3D solid state problem into two one-dimensional subproblems associated to the D and A regimes surrounded by the remaining A and D layers. The electronic structures of the reference chains are modified by the electrostatic potentials created due to the adjacent D and A columns. The tight-binding formalism with its k-dependence (k is the convenient wave vector in solid state theory describing the translational symmetry) is thus drastically simplified. Within the D and A regimes the grand canonical (GC) averaging procedure [28, 29] is employed. This approximation leads to a straightforward formulation of the SCF CO equations for systems with partially filled shells.

2. Physical Background and Basis Equations

The basic principles and the validity of the suggested model are schematically displayed in Fig. 1. In the following we will discuss only the most important aspects of the physical background (see [30, 31]). On the left of Fig. 1 the highest filled $\varepsilon(k)$ curve of the donor and the lowest empty energy band of the acceptor backbone are shown

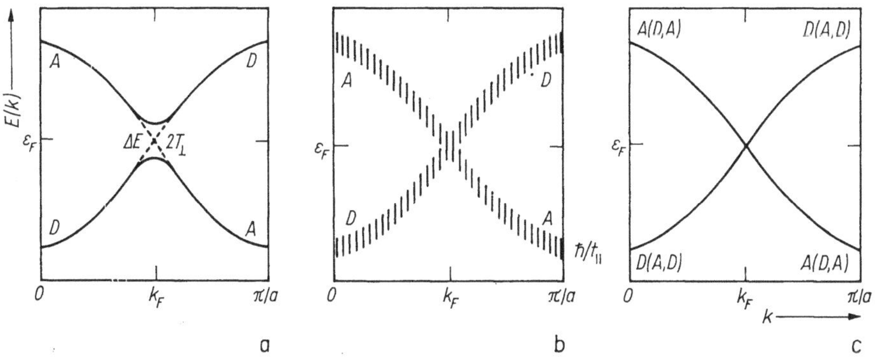

Fig. 1. The highest filled dispersion curve of a donor stack D and the lowest acceptor band A for different coupling conditions between the D and A strands. a) The $\varepsilon(k)$ curves symbolize the coherent 3D limit. ΔE is the covalency gap which is given by $2T_\perp$. b) The $\varepsilon(k)$ diagram represents the diffusive limit (see text). On c) HF dispersions for A and D, respectively, are shown that form the physical background of the present approach. The problem is divided into two one-dimensional tight-binding schemes for the 1D donor (D) and acceptor (A) chains that are coupled to the electrostatic fields of the remaining D and A backbones. This is symbolized by the labels (A, D) and (D, A), respectively. ε_F is the Fermi level in the segregated DA material and k_F the associated Fermi wave vector

for strong interchain interactions. The wave vector for such a system is characterized by three cartesian components k_x, k_y, and k_z, respectively, $\boldsymbol{k} = (k_x, k_y, k_z)$. The k-dependent interchain coupling parameters open in any case a covalency gap ΔE which is determined by the strength of the interaction. In the limit of comparable intrachain ($T_\parallel$) and interchain ($T_\perp$) "hopping integrals" coherent transfer processes of the charge carriers are encountered that must be described by the three-dimensional k-vector (coherent 3D solid). $T_\parallel$ and $T_\perp$ are tight-binding matrix elements of a crystal Hamiltonian which is not explicitly specified. $T_\parallel$ and $T_\perp$ numbers can be defined for both (D and A) backbones. In the following we assume for reasons of simplicity that averaged values can be employed in the stacking direction and in the direction of the perpendicular plane (k_x, k_y). The propagation for an injected carrier, which is localized on the donor stack at the time $t = 0$, is defined in (1) where φ_D is the wave function of the donor and φ_A is associated to the acceptor chain,

$$\varphi(t) = \varphi_D \cos\left(T_\perp \frac{t}{\hbar}\right) + \varphi_A \sin\left(T_\perp \frac{t}{\hbar}\right). \tag{1}$$

Equation (1) is not used in the subsequent approach to define time-independent (stationary) electronic (band) energies, etc. but has been given in order to get some insight into the dynamics of charge transfer processes.

The time constant for intrachain scattering events of the injected carriers is abbreviated by $t_\parallel$. The probability to find the carrier at $t = t_\parallel$ on the acceptor strand is then given in (2),

$$P_A = \sin^2\left(T_\perp \frac{t_\parallel}{\hbar}\right). \tag{2}$$

The initially prepared phase relation between the donor and acceptor is only conserved if inequality (3) is fulfilled.

Coherent 3D solids: $$(T_\perp t_\parallel/\hbar) \geqq 2\pi\,. \tag{3}$$

This condition characterizes the group of coherent 3D solids. Experimental investigations on low-dimensional organic and organometallic materials crystallizing in segregated lattices, however, have shown that the situation is much simpler in most of these compounds [1 to 6]. The smaller interchain transfer elements $T_\perp$ (in comparison to $T_{||}$) lead to (4) which prevents coherent motions of the charge carriers between adjacent 1D strands (DA, DD′, AA′).

$$\text{Incoherent 1D solids:} \qquad (T_\perp t_{||}/\hbar) \ll 1\,. \tag{4}$$

Only transport processes into the direction of the longitudinal axis (k_z) are associated to quantum numbers defined in terms of the k-vector. The transfer between the D and A backbones corresponds to a diffusive hopping motion described by the diffusion coefficient $D_{\rm DA}$ given in (5); $r_{\rm DA}$ is the separation between the two chains and $t_\perp$ is the hopping rate for the charge carriers,

$$D_{\rm DA} = \frac{r_{\rm DA}^2}{t_\perp}, \tag{5}$$

$$t_\perp = \frac{(2\pi/\hbar)^{-1}}{|T_\perp|^2\, t_{||}}. \tag{6}$$

The covalency gap displayed on Fig. 1a is smeared out as a direct consequence of the physical restrictions expressed in (4) to (6). The conditions that determine the diffusive carrier processes between D and A are symbolized in Fig. 1b. We want to assume that the partial charge transfer Q between the two types of the 1D stacks is determined by $D_{\rm DA}$. Furthermore it is supposed that the principle of quasi-stationarity (for the interchain transfer) is fulfilled once the SCF HF CO minimum has been reached in the allowed variational space; charge fluctuations (i.e. time-dependent Q variations) are neglected for a SCF convergent solution. The subsequent analysis corresponds therefore to a strictly time-independent approach.

The assumption of a quasi-stationary DA state leads to further simplifications in the theoretical approach. This is displayed on Fig. 1c. Here we have plotted the HF dispersions (highest filled donor and lowest empty acceptor bands) associated to two one-dimensional problems. The crossing of both $\varepsilon(k)$ curves in k-space is here allowed. The indices (A, D) at the HF bands have been adopted to symbolize the influence of the electrostatic potentials of the adjacent donor and acceptor strands in the solid; these interactions are taken into account in the 1D CO calculations on the "reference systems" D_R and A_R, respectively. The surrounding chains act as external perturbations perceived by the two reference backbones. The theoretical problem is therefore reduced to the determination of the band structures and population schemes within the D and A manifolds (which determine of course the Fermi energy ε_F), the magnitude of the charge transfer Q, as well as the mutual interaction energies between the reference strands D_R or A_R, respectively, and the neighbouring 1D strands.

The physical principles associated to the $\varepsilon(k)$ plot on Fig. 1c are schematized in Fig. 2. It is necessary to define two one-dimensional reference stacks D_R and A_R that lead to one-dimensional (k_z) crystal orbital eigenvalue problems. The band structures of D_R and A_R must be determined in separate SCF HF CO calculations that are mutually coupled via (k-independent) interaction elements H_{D_RA} and $H_{D_RD'}$ (reference system: donor strand) as well as H_{A_RD} and $H_{A_RA'}$ (reference chain: acceptor strand). The matrix elements H_{D_RA} symbolize the coupling between the (selected reference) donor and the acceptor moieties in the solid, while $H_{D_RD'}$ stands for the interaction between the reference donor and the "external" D backbones. Similar labels are used to define the coupling schemes with respect to the "reference" acceptor. In Fig. 2

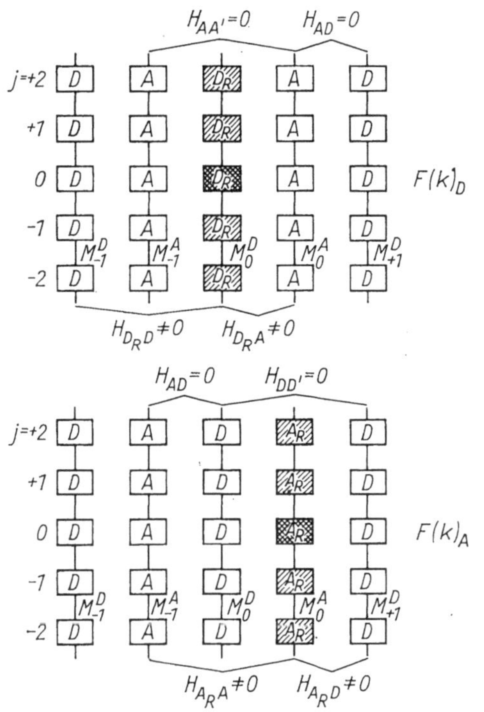

Fig. 2. Schematic representation of the basic principles employed in the CO model for segregated donor and acceptor solids. The theoretical approach is reduced to two subproblems for one dimensional donor and acceptor backbones that are coupled to the k-independent electrostatic fields of the remaining (D and A) strands in the low-dimensional material. For both tight-binding approaches it is necessary to define reference stacks D_R and A_R whose band structure is calculated in the direction of the stacking axis (k_z). The neighbouring layers in the donor and acceptor regime act as external electrostatic perturbations. These strands are labeled by M_i^R (R = D or A, $i =$ $= 0, \pm 1, \ldots \pm\infty$). The numbering scheme of the unit cells in the solid is determined by the index j ($j = 0, \pm 1, \ldots \pm\infty$). The "allowed" coupling elements in both one-dimensional subproblems are shown at the bottom of each symbolical representation, the interaction parameters that are set equal to zero in the present model are displayed at the top of the two sketches.

also those interaction elements are displayed that are set equal to zero in the present CO model for segregated stacks. The reduction to two one-dimensional reference chains (D_R and A_R) requires to neglect the mutual interactions between the "external" chains that create the electrostatic fields for the R (R = D, A) backbones,

$$H_{AA'} = H_{AD} = H_{DD'} = 0 \,. \tag{7}$$

A necessary condition for the approach symbolized in Fig. 2 is a well-defined k-vector (k_z) in the direction of the longitudinal axis. This, however, is only fulfilled in the case of D and A sublattices that are mutually commensurate. The simplest stacking pattern is obviously encountered in solids with identical translational symmetries into the stacking direction. The quantum number k_z is undefined in low-dimensional systems, where the Bragg lattice of D and A are incommensurate.

It is clear that tight-binding procedures beyond the conveniently adopted "pure state picture" have to be employed for the definition of crystal Hamiltonians that are suitable operators for the physical problems symbolized in Fig. 1c and 2. In the following we want to show that the grand canonical (GC) averaging procedure is a proper theoretical tool for this purpose.

In the field of statistical mechanics the GC ensemble is employed to describe thermodynamic equilibria. This classical definition is not adopted in this context where we will use the GC formalism as a rather formal computational scheme in the SCF HF (CO) framework allowing for the formation of average values over certain manifolds of one-electron states [28, 29]. Such procedures are widely used in MO (molecular orbital) studies on finite systems [32, 33]. Ensemble operators in combination with the crystal orbital formalism have been adopted to define HF operators that give an improved description of the empty Fermi sea [34].

To derive the most important properties of the GC ensemble, we will use the formalism of second quantization. For this purpose electron field operators $\psi(r)$ and $\psi^+(r)$ are expressed in terms of an orthonormal basis $\chi_i(r)$ and annihilation and creation operators c_i, c_i^+ that fulfil the convenient anticommutation rules (10) to (12),

$$\psi(r) = \sum_i \chi_i(r)\, c_i \,, \tag{8}$$

$$\psi^+(r) = \sum_i \chi_i^*(r)\, c_i^+ \,; \tag{9}$$

$$[c_i, c_j]_+ = c_i c_j + c_j c_i = 0 \,, \tag{10}$$

$$[c_i^+, c_j^+]_+ = c_i^+ c_j^+ + c_j^+ c_i^+ = 0 \,, \tag{11}$$

$$[c_i, c_j^+]_+ = c_i c_j^+ + c_j^+ c_i = \delta_{ij} \,. \tag{12}$$

In the case of a canonical one-particle basis diagonalizing the HF operator, it is possible to define a set of occupation numbers n_i (13) that are idempotent (14),

$$n_i = c_i^+ c_i \,, \tag{13}$$

$$n_i^2 = n_i \,. \qquad (14$$

The statistics of an N-electron system described by the n_i numbers is given by the grand canonical partition function Z,

$$Z = \mathrm{Tr}\, \mathrm{e}^{-(H_{\mathrm{HF}} - \mu N_i)/\theta} \,. \tag{15}$$

H_{HF} is the Hamiltonian in the mean-field description, μ the chemical potential, and θ the absolute temperature (in energy units). The number operator N_i in (15) reads

$$N_i = \sum_i n_i \,. \tag{16}$$

The density operator ϱ_{GC} in the GC ensemble is expressed in (17) and the GC averaging for any operator O is shown in (18),

$$\varrho_{\mathrm{GC}} = Z^{-1}\, \mathrm{e}^{-(H_{\mathrm{HF}} - \mu N_i)/\theta} \,, \tag{17}$$

$$\langle O \rangle_{\mathrm{GC}} = \mathrm{Tr}\, O \varrho_{\mathrm{GC}} \,. \tag{18}$$

The GC density operator in the mean-field approach can be factorized into the product (19) with parameters ω_i defined in (20),

$$\varrho_{\mathrm{GC}} = \prod_i \left[\frac{1 + \omega_i n_i}{2 + \omega_i}\right], \tag{19}$$

$$\omega_i = \mathrm{e}^{-(\varepsilon_i - \mu)/\theta} - 1 \,. \tag{20}$$

ε_i in the first equation is the i-th one-particle energy in the diagonal representation of the HF operator. The average occupation of the i-th one-particle state is given by the Fermi-like distribution law (21) and ϱ_{GC} thus can be written without reference to thermodynamic parameters (in (22)),

$$\langle n_i \rangle_{\mathrm{GC}} = \frac{1 + \omega_i}{2 + \omega_i} = (1 + \mathrm{e}^{(\varepsilon_i - \mu)/\theta)})^{-1} \,, \tag{21}$$

$$\varrho_{\mathrm{GC}} = \prod_i \left[1 - \langle n_i \rangle_{\mathrm{GC}} - (2\langle n_i \rangle_{\mathrm{GC}} - 1)\, n_i\right]. \tag{22}$$

Relations (23) to (26) follow directly from the definition of the GC density operator ϱ_{GC}:

$$\langle n_i n_j \rangle_{GC} = \langle n_i \rangle_{GC} \langle n_j \rangle_{GC} , \tag{23}$$

$$\langle c_i^+ c_j \rangle_{GC} = \langle n_i \rangle_{GC} \delta_{ij} , \tag{24}$$

$$\langle c_i^+ c_j^+ c_k c_l \rangle_{GC} = \langle n_i \rangle_{GC} \langle n_j \rangle_{GC} [\delta_{il}\delta_{jk} - \delta_{ik}\delta_{jl}] , \tag{25}$$

$$\langle N_i^2 \rangle_{GC} = \langle N_i \rangle_{GC}^2 + \sum_i \langle n_i \rangle_{GC} (1 - \langle n_i \rangle_{GC}) . \tag{26}$$

The last expression shows impressively that the GC ensemble corresponds to a pure state only for integer occupation numbers $\langle n_i \rangle_{GC}$; the width of the ensemble in the case $\langle n_i \rangle_{GC} \neq$ (integer) is determined by

$$\langle \langle N_i \rangle_{GC} - \langle N_i \rangle_{GC}^2 \rangle^{1/2} = [\sum_i \langle n_i \rangle_{GC} (1 - \langle n_i \rangle_{GC})]^{1/2} . \tag{27}$$

The k-dependent pseudoeigenvalue problem in the tight-binding formalism is shown in (28) and (29); the latter expression is based on the ZDO approximation [35] which will be the method of choice in solid state systems with larger unit cell dimensions (e.g., segregated DA stacks in organic or organometallic materials). $F(k)_R$ (R = D or A) is the one-dimensional (k_z) Fock operator associated to the donor or acceptor strand in the electrostatic field of the neighbouring backbones. $F(k)_R$ is decomposed into a k-dependent operator for the (isolated) reference stacks D_R or A_R and into a perturbational part due to the external electrostatic field (see below). $\varepsilon(k)_R$ symbolizes the one-electron energies that form the HF dispersions within the donor manifold ($\varepsilon(k)_D$) and the acceptor regime ($\varepsilon(k)_A$), respectively. The $c(k)_R$ parameters are the variational coefficients expressed in terms of atomic orbital (AO) basis functions. $S(k)_R$ in (28) is the convenient overlap (metric) matrix.

Full overlap basis: $$F(k)_R\, c(k)_R = S(k)_R\, \varepsilon(k)_R\, c(k)_R ; \tag{28}$$

ZDO basis: $$F(k)_R\, c(k)_R = \varepsilon(k)_R\, c(k)_R , \tag{29}$$

$$F(k)_R = F(k)_R^{1D} + F_R^{EF} , \tag{30}$$

$$R = D \text{ or } A .$$

The perturbational operator F_R^{EF} couples the external field(s) to the reference stacks D_R and A_R. As a result of the approximations shown in (4) to (6) F_R^{EF} is k-independent. The HF operators $F(k)_D^{1D}$ and $F(k)_A^{1D}$ can be expressed in terms of Fourier sums ((31)) where the index j has been used to label the unit cells in the 1D arrangement(s). $j = 0$ is the cell in the origin, $j = \pm 1, \pm 2, \ldots \pm\infty$ stands for the first, second, etc. nearest neighbours,

$$F(k)_R^{1D} = \sum_{j=-\infty}^{+\infty} e^{(ijk)}\, F(j)_R^{1D} . \tag{31}$$

The HF operator in the Bloch basis is composed by a core Hamiltonian $H(k)_R^{1D}$ of the one-electron type and the two-electron part $G(k)_{R,GC}^{1D}$ in the mean-field description. The latter contribution is written as a product of electron–electron (AO) basis integrals and elements of the charge-density-bond-order matrix which depend on the occupation pattern in the GC ensemble.

$$F(k)_{R,GC}^{1D} = H(k)_R^{1D} + G(k)_{R,GC}^{1D} , \tag{32}$$

$$G_{\mu_0\nu_j,GC}^{1D} = \sum_{j',j''=-\infty}^{+\infty} \sum_{\lambda,\sigma} P(j'' - j')_{\lambda\sigma,GC} \times$$
$$\times [(\mu_0\nu_j \mid \lambda_{j'}\sigma_{j''}) - (1/2)\,(\mu_0\sigma_{j''} \mid \lambda_{j'}\nu_j)] . \tag{33}$$

The two-electron part in (33) has been defined without any approximations (i.e. ab initio tight-binding ansatz in the full overlap basis). The matrix elements in (33) are dramatically simplified in CO variants based on the ZDO scheme. The charge-density bond-order matrix $P(j''-j')_{\lambda\sigma,\mathrm{GC}}$ of the GC ensemble is determined by numerical k-space integration within the first Brillouin zone (34); the parameter a in this relation stands for the unit cell dimension of the 1D material,

$$P(j''-j')_{\lambda\sigma,\mathrm{GC}} = (a/2\pi\,\langle N_i\rangle_{\mathrm{GC}}) \int_{-\pi/a}^{+\pi/a} \sum_i \langle n_i\rangle_{\mathrm{GC}}\, c^*(k)_{i\lambda}\, c(k)_{i\sigma} \times \times\, \mathrm{e}^{[ik(j''-j')]}\, \theta\big(\varepsilon_{\mathrm{F}} - \varepsilon(k)_{i,\mathrm{R}}\big)\, \mathrm{d}k\,. \tag{34}$$

The Heaviside step function θ is defined in (35) and (36),

$$\theta\big(\varepsilon_{\mathrm{F}} - \varepsilon(k)_{i,\mathrm{R}}\big) = 1\,; \qquad \varepsilon(k)_{i,\mathrm{R}} \leqq \varepsilon_{\mathrm{F}}\,, \tag{35}$$

$$\theta\big(\varepsilon_{\mathrm{F}} - \varepsilon(k)_{i,\mathrm{R}}\big) = 0\,; \qquad \varepsilon(k)_{i,\mathrm{R}} > \varepsilon_{\mathrm{F}}\,. \tag{36}$$

ε_{F} is the Fermi energy of the donor–acceptor system; this parameter is determined by the allowed curve crossing shown in Fig. 1. (34) displays immediately the simple determination of the charge-density bond-order matrices in the GC averaging procedure. It is only necessary to calculate in one of the two domains (D or A) the occupation pattern of the incompletely filled HF band. The statistical population in the 1D counterpart is then automatically given by the total number of electrons. In ionic DA pairs with narrow energy bands and a charge separation $Q = 1$ we observe $\langle n_i\rangle_{\mathrm{GC}} = 1$ for the "half-filled" bands in both backbones. All microstates in the first Brillouin zone are occupied by a single electron of one spin component (α), while the β states are unfilled. On the other hand, the GC ensemble allows also for the formation of averages where $\langle n_i\rangle_{\mathrm{GC}}$ is not restricted to primitive integers (0, 1, or 2). The k-dependent microstates must be filled by electrons of one spin-component ("ferromagnetic state") up to the Fermi level ε_{F} in "narrow" band materials; doubly (α and β spin electrons) occupied states are encountered in solids with "broad" dispersion curves. Theoretical criteria that define "narrow" and "broad" $\varepsilon(k)$ curves with their characteristic populations have been derived by several authors [36, 37 and references therein].

To summarize: the GC averaging procedure allows in a simple and straightforward way the definition of HF operators in the crystal orbital formalism for 1D systems with partially filled energy levels (i.e. nonintegral occupation numbers). The Fermi energy ε_{F} of the DA pair can be simply determined via the density of states distributions of the highest filled donor band and the lowest empty acceptor function. It is of course necessary to locate ε_{F} after each iterative step in the HF algorithm.

Next we discuss the determination of the perturbational operators $F_{\mathrm{D}}^{\mathrm{EF}}$ and $F_{\mathrm{A}}^{\mathrm{EF}}$ associated to the external fields that are coupled to the unperturbed HF Hamiltonians $F(k)_{\mathrm{D}}^{1\mathrm{D}}$ and $F(k)_{\mathrm{A}}^{1\mathrm{D}}$. The employed model is a self-consistent electrostatic-field (SCEF) approach that has been described in some detail in previous contributions [26, 27]. The degree of sophistication of this formalism lies between simple point charge approximations on one hand [38] and elaborate mutual self-consistent-field schemes, on the other [39]. In the present SCEF approximation electrostatic interactions between the one-dimensional backbones are considered (interchain elements). The self-consistent determination of the external fields leads furthermore to intrachain polarizations in the reference stacks. These two contributions are usually the dominating elements in electronic structure calculations on weakly coupled subsystems [40]; kinetic energy effects (hopping integrals), exchange and London interactions are neglected in the SCEF model.

The basic principles of the SCEF ansatz have been clarified in Fig. 2. The relations given in (7) lead immediately to the simple Hamiltonian $H_{\mathrm{R}}^{\mathrm{EF}}$ (R = D or A) for the external fields spanned by the D and A strands,

$$H_{\mathrm{R}}^{\mathrm{EF}} = \sum_{\mu_0=1}^{n} \sum_{M_i=-\infty}^{+\infty}{}^{\mathrm{D}} \sum_{j'=-\infty}^{-\infty}{}^{\mathrm{D}} \left[\sum_{B}{}^{\mathrm{D}} \frac{Z_{B,M_i,j'}}{r_{\mu_0 B}} - \sum_{B}{}^{\mathrm{D}} \sum_{\nu}{}^{\mathrm{D}} P_{\nu_0\nu_0,\mathrm{GC}} \times \times \int \frac{\nu(2)\,\nu(2)}{r_{\mu_0 B}}\right] + + \sum_{\mu_0=1}^{n} \sum_{M_i=-\infty}^{+\infty}{}^{\mathrm{A}} \sum_{j'=-\infty}^{+\infty}{}^{\mathrm{A}} \left[\sum_{B}{}^{\mathrm{A}} \frac{Z_{B,M_i,j'}}{r_{\mu_0 B}} - \sum_{B}{}^{\mathrm{A}} \sum_{\nu}{}^{\mathrm{A}} P_{\nu_0\nu_0,\mathrm{GC}} \times \times \int \frac{\nu(2)\,\nu(2)}{r_{\mu_0 B}}\right]; \tag{37}$$

R = D or A .

The expressions within the first parenthesis are associated to the 1D stacks in the donor regime (the reference stack $\mathrm{D_R}$ must be excluded in the summation for R = D), the latter terms to the acceptors ($\mathrm{A_R}$ must be neglected for R = A in $H_{\mathrm{R}}^{\mathrm{EF}}$). The first elements in the two partial sums are the core contributions ($Z_{B,M_i,j'}$: core charge of the B-th atomic site in the M_i-th neighbouring strand and the unit cell j' in this regime). The second integrals are symbols for the two-electron part of $H_{\mathrm{R}}^{\mathrm{EF}}$. μ_0 is the μ-th AO in the origin of the reference polymer R = D or A and $P_{\nu_0\nu_0,\mathrm{GC}}$ is the ν-th diagonal element of the charge-density bond-order matrix with $(j'' - j') = 0$. The summations in (37) can be simplified by adopting the classical central field approximation [41], which leads to the elimination of the elaborate two-electron integrals in the lattice summation. The $q_{B,M_i,j'}$ parameters in (38) are the atomic net charges at the B-th site in the building block j' belonging to the M_i-th chain.

$$H_{\mathrm{R}}^{\mathrm{EF}} = \Big(\sum_{\mu_0=1}^{n} \sum_{M_i=-\infty}^{+\infty}{}^{\mathrm{D}} \sum_{j'=-\infty}^{+\infty}{}^{\mathrm{D}} \sum_{B}{}^{\mathrm{D}} + \sum_{\mu_0=1}^{n} \sum_{M_i=-\infty}^{+\infty}{}^{\mathrm{A}} \sum_{j'=-\infty}^{+\infty}{}^{\mathrm{A}} \sum_{B}{}^{\mathrm{A}}\Big) \frac{q_{B,M_i,j'}}{R_{\mu_0 B}}, \tag{38}$$

R = D or A .

If the ZDO approximation is strictly applied (for extensions see [42]) $H_{\mathrm{R}}^{\mathrm{EF}}$ only gives raise to diagonal elements in the perturbational external field operator $F_{\mathrm{R}}^{\mathrm{EF}}$ which is defined in (39). $V_{\mu_0 B}^{B'B}$ stands for the electron-core interaction between the μ-th AO at the site B' in the origin ($j = 0$) of the reference stack and the B-th center with the cell labels j' and M_i, respectively. The restrictions in the two summations within the D and A manifolds are identical with the exclusions mentioned with reference to (37),

$$F_{\mu\mu,R}^{\mathrm{EF}} = \Big(\sum_{M_i=-\infty}^{+\infty}{}^{\mathrm{D}} \sum_{j'=-\infty}^{+\infty}{}^{\mathrm{D}} \sum_{B}{}^{\mathrm{D}} + \sum_{M_i=-\infty}^{+\infty}{}^{\mathrm{A}} \sum_{j'=-\infty}^{+\infty}{}^{\mathrm{A}} \sum_{B}{}^{\mathrm{A}}\Big)\, q_{B,M_i,j'} V_{\mu_0 B}^{B'B}. \tag{39}$$

The lattice sums in the last equation can be either calculated by real space monopole expansions (the validity of this scheme is discussed in [27]), by means of the Ewald technique in reciprocal space [43] or by well-defined multipole expansions with fast convergence properties [44].

The different computational steps of the present CO model for segregated DA stacks are displayed in the scheme and are summarized below: (i) Diagonalization of the unperturbed Fock operators $F(k)_{\mathrm{D}}^{1\mathrm{D}}$ and $F(k)_{\mathrm{A}}^{1\mathrm{D}}$ with non-paramagnetic ground states or of the GC ensemble operators $F(k)_{\mathrm{D,GC}}^{1\mathrm{D}}$ and $F(k)_{\mathrm{A,GC}}^{1\mathrm{D}}$ if experimental informations are available for a first guess of $\langle n_i \rangle_{GC}$ or Q; (ii) Calculation of the Fermi energy ε_F; (iii) Determination of the net charges $q_{B,M_i,j'}$ in the D and A stacks (as well as of the charge transfer Q); (iv) Construction of the external electrostatic fields F_D^{EF} and F_A^{EF} by means of the data spectrum derived in (iii); (v) Diagonalization of the complete HF operators (i.e. reference stack + external field) in the GC occupation scheme; (vi) Reiteration of the steps (ii) to (v) until a given convergence criterion is fulfilled.

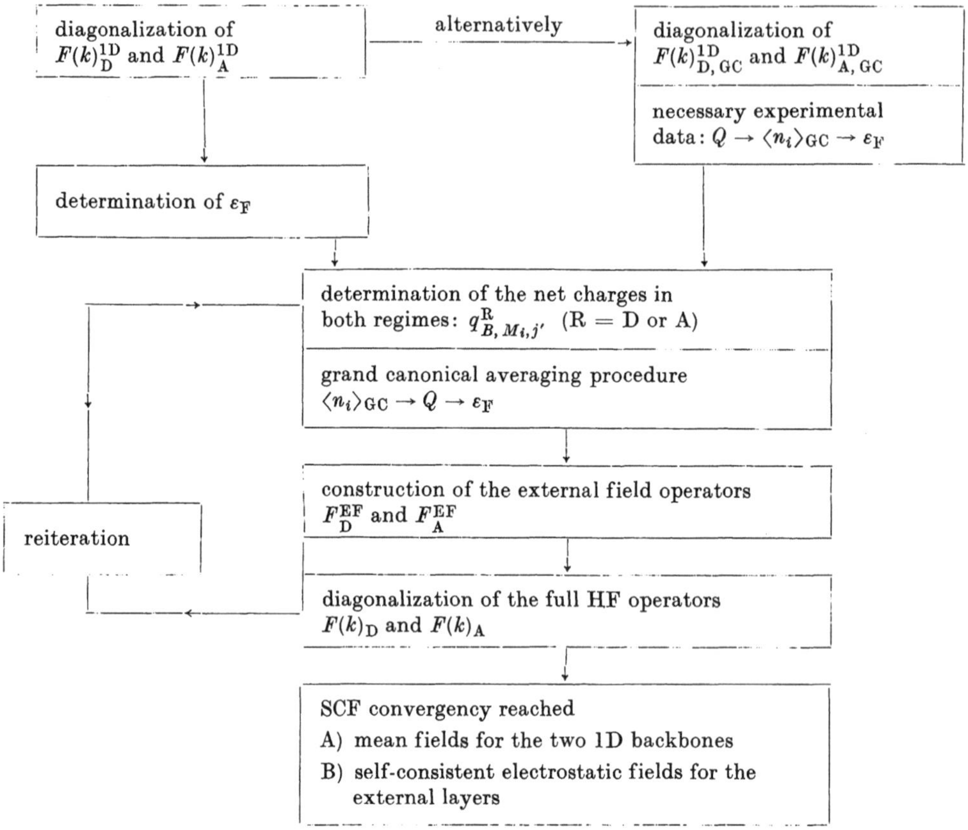

Acknowledgements

This work has been supported by the Stiftung Volkswagenwerk. The author is grateful to Dr. W. Borrmann for critically reading the manuscript.

References

[1] H. M. McConnell, B. M. Hoffman, and R. M. Metzger, Proc. Nat. Acad. Sci. U.S.A. **53**, 46 (1965).
Z. G. Soos, Ann. Rev. Phys. Chem. **25**, 121 (1974).
Z. G. Soos and D. J. Klein, Molecular Association, Ed. R. Foster, Academic Press, (Inc.), London/New York 1975.

[2] S. Barišić, A. Bjelis, and R. J. Cooper (Ed.), Quasi-One-Dimensional Conductors II, Vol. 96 of Lecture Notes in Physics, Springer Verlag, Berlin/Göttingen/Heidelberg/New York 1979.

[3] H. MÖHWALD, Phys. Bl. **35**, 617 (1979).
R. FOSTER, J. phys. Chem. **84**, 2135 (1980).
[4] H. J. KELLER (Ed.), Low-Dimensional Cooperative Phenomena, Plenum Press, New York 1975.
H. J. KELLER (Ed.), Chemistry and Physics of One-Dimensional Metals, Plenum Press, New York 1978.
J. T. DEVRESSE, R. P. EVRARD, and V. E. VAN DOREN (Ed.), Highly Conducting One-Dimensional Solids, Plenum Press, New York 1979.
J. S. MILLER (Ed.), Extended Linear Chain Compounds, Vol. 1 and 2, Plenum Press, New York 1982.
[5] W. A. HATFIELD (Ed.), Molecular Metals, Plenum Press, New York 1979.
L. ALCÁCER (Ed.), The Physics and Chemistry of Low Dimensional Solids, D. Reidel Publ. Co., Dordrecht 1980.
[6] J. A. IBERS, L. J. PACE, J. MARTINSEN, and B. M. HOFFMAN, Struct. Bonding **50**, 1 (1982).
B. M. HOFFMAN and J. A. IBERS, Accounts chem. Res. **16**, 15 (1983).
[7] T. E. PEACOCK and R. MCWEENY, Proc. Roy. Soc. London **74**, 385 (1959).
G. DEL RE, J. LADIK, and G. BICZÓ, Phys. Rev. **155**, 997 (1967).
J. M. ANDRÉ, L. GOUVERNEUR, and G. LEROY, Internat. J. Quantum Chem. **1**, 451 (1967).
[8] M. KERTÉSZ, Adv. Quantum Chem. **15**, 161 (1982).
[9] J. LADIK and S. SUHAI, Molecular Interactions, Ed. W. J. ORVILLE-THOMAS and H. RATAJACK, Wiley, New York 1980.
J. LADIK and S. SUHAI, Theoretical Chemistry, Vol. 4, Ed. C. THOMSON, Royal Society of Chemistry, London 1981.
[10] D. W. BULLETT, Solid State Physics, Vol. 35, Ed. H. EHRENREICH, F. SEITZ, and D. TURNBULL, Academic Press, Inc., London/New York 1980.
[11] M. C. BÖHM and H. VOGLER, Solid State Commun. **46**, 201 (1983).
M. C. BÖHM and H. VOGLER, Phys. Letters A **96**, 41 (1983).
[12] M. C. BÖHM, and H. VOGLER Phys. Rev. B **28**, 3342 (1983).
[13] A. KARPFEN, J. LADIK, G. STOLLHOFF, and P. FULDE, Chem. Phys. **8**, 215 (1975).
J. LADIK, Internat. J. Quantum Chem. **9**, 1563 (1975).
[14] R. D. SINGH and J. LADIK, Phys. Letters A **65**, 264 (1977).
S. SUHAI and J. LADIK, Phys. Letters A **77**, 25 (1980).
[15] M.-H. WHANGBO, W. M. WALSH, JR., R. C. HADDON, and F. WUDL, Solid State Commun. **43**, 637 (1982).
M.-H. WHANGBO, J. M. WILLIAMS, M. A. BENO, and J. R. DORFMAN, J. Amer. Chem. Soc. **105**, 645 (1983).
[16] J. F. KWAK, Phys. Rev. B **26**, 4789 (1982).
C. MINOT and S. G. LOUIE, Phys. Rev. B **26**, 4793 (1982).
[17] M. C. BÖHM, J. Phys. C **16**, 1631 (1983).
[18] M. C. BÖHM, Solid State Commun. **45**, 117 (1983).
M. C. BÖHM, Internat. J. Quantum Chem. **25**, 817 (1984).
[19] M. C. BÖHM, Phys. Letters A **94**, 371 (1983).
M. C. BÖHM, J. Phys. C **17**, 2091 (1984).
[20] M. C. BÖHM, Phys. Rev. B **28**, 6914 (1983).
[21] M. KERTÉSZ and F. VONDERVISZT, J. Amer. Chem. Soc. **104**, 5889 (1982).
[22] S. MAZUMDAR and Z. G. SOOS, Synthetic Metals **1**, 77 (1979).
S. R. BONDESON and Z. G. SOOS, J. chem. Phys. **71**, 3807 (1979).
[23] D. J. KLEIN and Z. G. SOOS, Mol. Phys. **20**, 1013 (1971).
S. MAZUMDAR and Z. G. SOOS, Phys. Rev. B **23**, 2810 (1981).
[24] Z. G. SOOS and S. MAZUMDAR, Phys. Rev. B **18**, 1991 (1978).
Z. G. SOOS, S. R. BONDESON, and S. MAZUMDAR, Chem. Phys. Letters **65**, 331 (1979).
[25] P. PETELENZ, Chem. Phys. Letters **47**, 603 (1977).
P. PETELENZ and T. ZYCZKOWSKA, Chem. Phys. Letters **67**, 149 (1979).
P. PETELENZ and B. PETELENZ, phys. stat. sol. (b) **81**, 49 (1977).
P. PETELENZ, phys. stat. sol. (b) **83**, 169 (1977).
[26] M. C. BÖHM, Chem. Phys. Letters **89**, 126 (1982).

[27] M. C. Böhm, phys. stat. sol. (b) **121**, 255 (1984).
M. C. Böhm, Physica B **124**, 327 (1984).
[28] S. F. Abdulnur, J. Linderberg, Y. Öhrn, and P. W. Thulstrup, Phys. Rev. A **6**, 889 (1972).
[29] P. Jørgensen, J. chem. Phys. **57**, 4884 (1972).
[30] M. Weger, J. Physique, C6-1456 (1978).
[31] J. Friedel and D. Jérome, Contemp. Phys. **23**, 583 (1982).
D. Jérome and H. J. Schulz, Adv. Phys. **31**, 299 (1983).
[32] M. J. S. Dewar, J. A. Hashmall, and C. G. Vernier, J. Amer. Chem. Soc. **90**, 1953 (1968).
P. Čarsky and R. Zahradnik, Topics Current Chem. **43**, 1 (1973).
[33] J. C. Slater, Internat. J. Quantum Chem. **S3**, 727 (1970).
[34] M. C. Böhm, Solid State Commun. **46**, 709 (1983).
[35] J. A. Pople and D. L. Beveridge, Approximate Molecular Orbital Theory, McGraw-Hill, Publ. Co., New York 1970.
[36] J. Hubbard, Proc. Roy. Soc. London A **277**, 237 (1964).
[37] M.-H. Whangbo, Accounts chem. Res. **16**, 95 (1983).
[38] E. Clementi, J. chem. Phys. **54**, 2492 (1971).
D. R. Armstrong, J. Jamieson, and P. G. Perkins, Theor. Chim. Acta **50**, 193 (1978).
[39] P. Otto and J. Ladik, Chem. Phys. **8**, 192 (1975).
P. Otto and J. Ladik, Chem. Phys. **19**, 209 (1977).
F. Beleznay, S. Suhai, and J. Ladik, Internat. J. Quantum Chem. **20**, 683 (1981).
[40] J. N. Murrell, M. Randić, and O. R. Williams, Proc. Roy. Soc. London A **284**, 566 (1965).
[41] E. U. Condon and G. H. Shortley, The Theory of Atomic Spectra, Cambridge University Press, Cambridge 1970.
[42] R. D. Brown, M. F. O'Dwyer, and K. R. Roby, Theor. Chim. Acta **11**, 1 (1968).
R. D. Brown and P. G. Burton, Theor. Chim. Acta **18**, 309 (1970).
[43] P. P. Ewald, Ann. Phys. (Leipzig) **64**, 253 (1921).
[44] M. P. Tosi, Solid State Physics, Vol. **16**, Ed. F. Seitz and D. Turnbull, Academic Press Inc., London/New York 1964.

(Received July 4, 1984)

phys. stat. sol. (b) 125, 323 (1984)

Subject classification: 6 and 13.5.1

Department of Physics, Neimenggu University, Huhehaote[1])

Properties of the Exciton–Phonon System in a Magnetic Field

By

SHI-WEI GU and YU-KUI ZHOU

The properties of excitons for a polar crystal in a weak magnetic field are discussed by using a perturbational method. When the interaction between phonons with different wave vectors in the recoil effect is neglected and a slow motion of the exciton is assumed, the electron–hole effective potential, the self-trapping energy of an exciton, and the renormalized masses are derived.

Применяя метод возмущиний, обсуждаем свойства экситона для полярного кристалла в слабом магнитном поле. При пренебрежении взаимодействия между фононами с неодинаковыми волновыми векторами в эффекте отдачи и предположении медленного движущего экситона, выводили электронно-дырочныи эффективныи потенциал, самозахватная энергия экситона и массы перенормировки.

1. Introduction

The exciton–phonon system has been a subject of great theoretical interest since the pioneering work of Haken [1] in 1956. The exciton–phonon system has been treated by many very elegant and ingenious methods [1 to 5]. Moreover, the properties of the exciton–phonon system in a magnetic field have been discussed by some authors. The properties of an exciton have been discussed by the adiabatic method for a high magnetic field [6], by the perturbational method for an arbitrary magnetic field [7]. But, in these works [6, 7], the effect of the phonon field on the exciton was neglected. Behnke et al. [8] have discussed the effect of the phonon field on the exciton by the variational method for weak magnetic field in 1976. Recently, Pang [9] has discussed this problem by a method with which Haga [10] studied the problems of polarons and Gu and Liang [11, 12] studied those of excitons. The numerical results of the ground-state energies of an exciton–phonon system in a magnetic field for GaAs, ZnO, and CdS have been evaluated variationally and compared with those of Behnke. It is found that, when $\alpha < 2$, satisfactory results can be obtained by the use of even the zero-order approximation effective potential.

In previous works [6 to 9], the exciton translation was not included and the renormalized masses were not derived. In this paper, we will discuss the properties of the exciton–phonon system in low magnetic field by a perturbational method, which is similar to the perturbation method [9, 11, 12]. But such improvements as the effects of the exciton translation and the magnetic field on the displacement amplitude f_k are being made here. When the interaction between phonons with different wave vectors in the recoil effect of the exciton is neglected, the effects of the lattice vibrations and the magnetic field on the exciton are discussed. We find the self-trapping energy and the effective potential of an exciton in magnetic field, and renormalized masses.

[1]) Huhehaote, Neimenggu, The Pèople's Republic of China.

2. The Effective Hamiltonian in a Magnetic Field

We start with the isotropic exciton–phonon system in a nondegenerate two-band model. The system consists of an electron and a hole interacting with each other through a dynamically screened Coulomb potential, and with the longitudinal-optical (LO) phonons due to the Fröhlich interaction and with a magnetic field $\boldsymbol{B}$. The Hamiltonian is written as [8]

$$H = \frac{1}{2m_1}\left[\boldsymbol{p}_1 + \frac{e}{c}\,\boldsymbol{A}(\boldsymbol{r}_1)\right]^2 + \frac{1}{2m_2}\left[\boldsymbol{p}_2 - \frac{e}{c}\,\boldsymbol{A}(\boldsymbol{r}_2)\right]^2 + \frac{e^2}{\varepsilon_\infty\,|\boldsymbol{r}_1 - \boldsymbol{r}_2|} + \\ + \sum_{\boldsymbol{k}} \hbar\omega a_{\boldsymbol{k}}^+ a_{\boldsymbol{k}} + \sum_{\boldsymbol{k}} \left[V_k a_{\boldsymbol{k}}\big(\exp(i\boldsymbol{k}\cdot\boldsymbol{r}_2) - \exp(i\boldsymbol{k}\cdot\boldsymbol{r}_1)\big) + \text{h.c.}\right], \tag{1}$$

where

$$V_k = i\left(\frac{2\pi e^2\hbar\omega}{V}\right)^{1/2}\left(\frac{1}{\varepsilon_\infty} - \frac{1}{\varepsilon_0}\right)^{1/2}\frac{1}{k} \tag{2}$$

describes the coupling of an electron to an optical phonon with energy $\hbar\omega$. The vector potential is $\boldsymbol{A}(\boldsymbol{r}_i) = \frac{1}{2}\boldsymbol{B}\times\boldsymbol{r}_i$, the index $i = 1(2)$ denotes the electron (hole). $a_{\boldsymbol{k}}^+(a_{\boldsymbol{k}})$ is the creation (annihilation) operator of the phonon with the wave vector $\boldsymbol{k}$.

It is useful to rewrite the Hamiltonian in the centre-of-mass system ($\boldsymbol{P}$, $\boldsymbol{p}$ and $\boldsymbol{R}$, $\boldsymbol{r}$ are the centre of mass (c.m.) and relative momenta and coordinates, respectively)

$$H = \frac{1}{2M}\left[\boldsymbol{P} + \frac{e}{c}\,\boldsymbol{A}(\boldsymbol{r})\right]^2 + \frac{1}{2\mu}\left[\boldsymbol{p} + \frac{e}{c}\,\boldsymbol{A}(\boldsymbol{R}) + (s_2^2 - s_1^2)\,\frac{e}{c}\,\boldsymbol{A}(\boldsymbol{r})\right]^2 + \\ + \frac{g}{2\mu}\left(\frac{e}{c}\,\boldsymbol{A}(\boldsymbol{r})\right)^2 - \frac{e^2}{\varepsilon_\infty r} + \sum_{\boldsymbol{k}} \hbar\omega a_{\boldsymbol{k}}^+ a_{\boldsymbol{k}} + \\ + \sum_{\boldsymbol{k}} \left[V_{\boldsymbol{k}} a_{\boldsymbol{k}} \exp(i\boldsymbol{k}\cdot\boldsymbol{R})\,\varrho_{\boldsymbol{k}}(\boldsymbol{r}) + \text{h.c.}\right], \tag{3}$$

with

$$\varrho_{\boldsymbol{k}}(\boldsymbol{r}) = \exp(-is_1\boldsymbol{k}\cdot\boldsymbol{r}) - \exp(is_2\boldsymbol{k}\cdot\boldsymbol{r}),$$

$$g = (s_1^3 + s_2^3) - s_1 s_2 - (s_1^2 - s_2^2)^2,$$

and

$$s_i = \frac{m_i}{M}, \qquad M = m_1 + m_2, \qquad \mu = s_1 s_2 M.$$

It is easily seen that the total momentum operator of the system

$$\boldsymbol{P} - \frac{e}{2c}\,\boldsymbol{B}\times\boldsymbol{r} + \sum_{\boldsymbol{k}} \hbar\boldsymbol{k} a_{\boldsymbol{k}}^+ a_{\boldsymbol{k}}$$

commutes with the Hamiltonian H and hence is a constant of the motion.

Carrying out the unitary transformation (in analogy to the free field case) with

$$U_1 = \exp\left[\frac{i}{\hbar}\left(\boldsymbol{K} + \frac{e}{c}\,\boldsymbol{A}(\boldsymbol{r}) - \sum_{\boldsymbol{k}} \hbar\boldsymbol{k} a_{\boldsymbol{k}}^+ a_{\boldsymbol{k}}\right)\cdot\boldsymbol{R}\right], \tag{4}$$

so that the total momentum, $\boldsymbol{K}$, becomes a "c" number, the Hamiltonian H is transformed to H_1,

$$H_1 = U_1^{-1} H U_1 = H_1^0 + H_1^1 + H_1^2, \tag{5a}$$

$$H_1^0 = \frac{\boldsymbol{Q}^2}{2M} + \sum_{\boldsymbol{k}} \hbar\Omega a_{\boldsymbol{k}}^+ a_{\boldsymbol{k}} + \sum [V_{\boldsymbol{k}} a_{\boldsymbol{k}} \varrho_{\boldsymbol{k}}(\boldsymbol{r}) + \text{h.c.}] , \tag{5b}$$

$$H_1^1 = \frac{1}{2\mu}\left[\boldsymbol{p} + (s_2^2 - s_1^2)\frac{e}{c}\boldsymbol{A}(\boldsymbol{r})\right]^2 + \frac{g}{2\mu}\left(\frac{e}{c}\boldsymbol{A}(\boldsymbol{r})\right)^2 - \frac{e^2}{\varepsilon_\infty r}, \tag{5c}$$

$$H_1^2 = \frac{\hbar^2}{2M}\sum_{\boldsymbol{k}\boldsymbol{k}'} \boldsymbol{k}\cdot\boldsymbol{k}' a_{\boldsymbol{k}}^+ a_{\boldsymbol{k}'}^+ a_{\boldsymbol{k}} a_{\boldsymbol{k}'} , \tag{5d}$$

with

$$\hbar\Omega = \hbar\omega + \frac{\hbar^2\boldsymbol{k}^2}{2M} - \frac{\hbar}{M}\boldsymbol{k}\cdot\boldsymbol{Q} , \tag{6}$$

$$\boldsymbol{Q} = \boldsymbol{K} + \frac{2e}{c}\boldsymbol{A}(\boldsymbol{r}) . \tag{7}$$

The relative motion of the exciton is coupled to the interacting phonon system through the $\boldsymbol{r}$- and $\boldsymbol{k}$-dependent coupling $V_{\boldsymbol{k}}\varrho_{\boldsymbol{k}}(\boldsymbol{r})$. The important new term is the expression $-\sum_{\boldsymbol{k}} (\hbar/M)\,\boldsymbol{k}\cdot\boldsymbol{Q} a_{\boldsymbol{k}}^+ a_{\boldsymbol{k}}$, describing the coupling of the exciton–phonon system to the magnetic field. H_1^2 describes the interaction between phonons with different wave vectors in the recoil effect of the exciton.

The coupling term in H_1^0 is linear in the creation (annihilation) operator for phonons. Therefore, H_1^0 does not conserve the total number of phonons. Thus the eigenstates of H_1^0 cannot be states with a fixed number of phonons. A reasonable approximation for the eigenstates of the interacting phonon system seems to be a coherent state, which has the form $U_2|0\rangle$. The structure of the unitary operator U_2 is well-known from the literature on coherent states (see e.g., [13]). We have

$$U_2 = \exp\sum_{\boldsymbol{k}} (f_{\boldsymbol{k}} a_{\boldsymbol{k}}^+ - f_{\boldsymbol{k}}^* a_{\boldsymbol{k}}) , \qquad f_{\boldsymbol{k}} = -\frac{1}{\hbar\Omega} V_{\boldsymbol{k}}^* \varrho_{\boldsymbol{k}}^*(\boldsymbol{r}) . \tag{8}$$

The displacement amplitudes $f_{\boldsymbol{k}}$ depend on both the relative coordinate r and the magnetic field $\boldsymbol{B}$. The transformed Hamiltonian

$$H_2 = U_2^{-1} H_1 U_2 = H_2^0 + H_2^1 + H_2^2 , \tag{8a}$$

$$H_2^0 = U_2^{-1} H_1^0 U_2 = \frac{\boldsymbol{Q}^2}{2M} + \sum_{\boldsymbol{k}} \hbar\Omega a_{\boldsymbol{k}}^+ a_{\boldsymbol{k}} - \sum_{\boldsymbol{k}} \hbar\Omega\, |f_{\boldsymbol{k}}|^2 , \tag{8b}$$

$$H_2^1 = U_2^{-1} H_1^1 U_2 = H^0 + H^1 + H^2 , \tag{8c}$$

$$H^0 = \frac{1}{2\mu}\left[\boldsymbol{p} + (s_2^2 - s_1^2)\frac{e}{c}\boldsymbol{A}(\boldsymbol{r})\right]^2 + \frac{g}{2\mu}\left(\frac{e}{c}\boldsymbol{A}(\boldsymbol{r})\right)^2 - \frac{e^2}{\varepsilon_\infty r} + \frac{\hbar^2}{2\mu}\sum_{\boldsymbol{k}} |\nabla f_{\boldsymbol{k}}|^2 , \tag{8d}$$

$$H^1 = -\frac{i\hbar}{\mu}\sum_{\boldsymbol{k}} (a_{\boldsymbol{k}}^+ \nabla f_{\boldsymbol{k}} - a_{\boldsymbol{k}} \nabla f_{\boldsymbol{k}}^*)\cdot\left[\boldsymbol{p} + (s_2^2 - s_1^2)\frac{e}{c}\boldsymbol{A}(\boldsymbol{r})\right] - \frac{\hbar^2}{2\mu}\sum_{\boldsymbol{k}} (a_{\boldsymbol{k}}^+ \nabla^2 f_{\boldsymbol{k}} - a_{\boldsymbol{k}} \nabla^2 f_{\boldsymbol{k}}^*) , \tag{8e}$$

$$H^2 = -\frac{\hbar^2}{2\mu}\sum_{kk'}(a_k^+a_{k'}^+\nabla f_k\cdot\nabla f_{k'} + \text{h.c.}) - \frac{\hbar^2}{2\mu}\sum_{kk'}(a_k^+a_{k'}\nabla f_k\cdot\nabla f_{k'}^*)\,, \tag{8f}$$

$$H_2^2 = U_2^{-1}H_1^2U_2 = \frac{\hbar^2}{2M}\sum_{kk'}\boldsymbol{k}\cdot\boldsymbol{k}'(a_k^+ + f_k^*)\,(a_{k'}^+ + f_{k'}^*)\,(a_k + f_k)\,(a_{k'} + f_{k'})\,. \tag{8g}$$

In the following, we will neglect the recoil energy H_2^2 and H^2 which is of the same order of magnitude as H_2^2 [11, 12]. For not too large field B and the exciton in slow motion, or $Q \ll u$, hence $u^2 + \boldsymbol{k}^2 \gg \boldsymbol{k}\cdot\boldsymbol{Q}$ [10], we may expand $(\hbar\Omega)^{-1}$ to the second-order of $\boldsymbol{k}\cdot\boldsymbol{Q}$,

$$\frac{1}{\hbar\Omega} = \frac{2M}{\hbar^2}\left(\frac{1}{u^2+\boldsymbol{k}^2} + \frac{2\boldsymbol{k}\cdot\boldsymbol{Q}}{\hbar(u^2+\boldsymbol{k}^2)} + \frac{4(\boldsymbol{k}\cdot\boldsymbol{Q})^2}{\hbar^2(u^2+\boldsymbol{k}^2)^3}\right), \tag{9}$$

where

$$u = \left(\frac{2M\omega}{\hbar}\right)^{1/2},$$

represents the wave number of an exciton with a translation energy which equals the phonon energy $\hbar\omega$.

As seen later, the diagonal element of H^1 corresponding to the zero-phonon state $|0\rangle$ of H_2^0 vanishes. We may regard H^1 as a perturbing term for the unperturbed Hamiltonian $H_2^0 + H^0$ [11, 12].

We are now going to calculate the perturbation energy due to the perturbing term H^1. Operating H^1 to $|0\rangle$, we have

$$H^1\,|0\rangle = -\frac{i\hbar}{\mu}\sum_k \nabla f_k\cdot\left[\boldsymbol{p} + (s_2^2 - s_1^2)\,\frac{e}{c}\,\boldsymbol{A}(\boldsymbol{r})\right]a_k^+\,|0\rangle - \frac{\hbar^2}{2\mu}\sum_k \nabla^2 f_k a_k^+\,|0\rangle\,. \tag{10}$$

The diagonal element of H^1 with respect to $|0\rangle$, i.e. $\langle 0|\,H^1\,|0\rangle$ vanishes, as is easily seen from (10), and hence the first-order perturbational correction due to H^1 vanishes.

By means of general perturbation theory we have the second-order perturbational correction

$$\Delta E = -\sum_n \frac{|(H^1)_{0n}|^2}{E_n - E_0}, \tag{11}$$

where the summation is taken over all n except $n = 0$. We will calculate the effective Hamiltonian to the second-order perturbational correction. From (8b), (8d), and (11), we have the effective Hamiltonian

$$H_{\text{eff}} = \frac{Q^2}{2M} + \frac{1}{2\mu}\left[\boldsymbol{p} + (s_2^2 - s_1^2)\,\frac{e}{c}\,\boldsymbol{A}(\boldsymbol{r})\right]^2 + \frac{g}{2\mu}\left(\frac{e}{c}\,\boldsymbol{A}(\boldsymbol{r})\right)^2 - $$
$$- \frac{e^2}{\varepsilon_\infty r} + \frac{\hbar^2}{2\mu}\sum_k |\nabla f_k|^2 - \sum_k \hbar\Omega\,|f_k|^2 + \Delta E\,,$$

the final three terms in H_{eff} can be calculated by replacing the summation by the integral,

$$\sum_k (\ldots) = \frac{V}{(2\pi)^3}\int_0^\infty\int_0^\pi\int_0^{2\pi} \boldsymbol{k}^2\sin\theta\,\mathrm{d}\boldsymbol{k}\,\mathrm{d}\theta\,\mathrm{d}\varphi\,(\ldots)\,,$$

and using (9) (see Appendix). We have the representation for H_{eff},

$$H_{\text{eff}} = E_s + \frac{K^2}{2M^*} + \frac{p^2}{2\mu^*} + V_{\text{eff}}\,, \tag{12}$$

where

$$E_s = -\alpha\hbar\omega\left(2 - \frac{1}{2}\,S + \frac{3}{8}\,S'\right) + \frac{1}{24}\,(4 + 3S)\,\frac{\alpha}{\mu}\left(\frac{e\boldsymbol{B}}{cu}\right)^2, \tag{13}$$

$$\frac{M}{M^*} = 1 - \frac{1}{3}\,\alpha + \frac{1}{4}\,\alpha S - \frac{5}{16}\,\alpha S' - 8(2 - 3\sin^2\theta_1)\,\frac{\alpha}{(ur)^3} + \\ + \alpha\exp(-ur)\,f_1(ur)\,, \tag{14}$$

$$\frac{\mu}{\mu^*} = 1 - \frac{1}{6}\,\alpha S + 8(2 - 3\sin^2\theta_2)\,\frac{\alpha}{(ur)^3} + \alpha\exp(-ur)\,f_2(ur)\,, \tag{15}$$

$$\begin{aligned} V_{\text{eff}} = {} & -\frac{e^2}{\varepsilon_0 r} + \frac{7}{4}\,\alpha\hbar\omega\exp(-ur) - \left(\frac{1}{\varepsilon_\infty} - \frac{1}{\varepsilon_0}\right)\frac{e^2}{r}\exp(-ur) - \\ & - \frac{1}{4}\,\alpha\hbar\omega ur\exp(-ur) + \frac{2e}{cM}\,\boldsymbol{A}(r)\cdot\boldsymbol{K} + \frac{s_2^2 - s_1^2}{\mu}\,\frac{e}{c}\,\boldsymbol{A}(r)\cdot\boldsymbol{p} + \\ & + \frac{2e\alpha}{cM}\,\boldsymbol{A}(\boldsymbol{r})\cdot\boldsymbol{K}\left[-\frac{1}{3} + \frac{1}{4}\,S - \frac{5}{16}\,S' - 8(2 - 3\sin^2\theta_1)\,\frac{1}{(ur)^3} + \right. \\ & \left. + \exp(-ur)\,f_1(ur)\right] - \\ & - \frac{2\alpha}{M}\left(\frac{e}{c}\,\boldsymbol{A}(r)\right)^2\left[8(2 - 3\sin^2\theta_1)\,\frac{1}{(ur)^3} - \exp(-ur)\,f_1(ur)\right] + \\ & + \frac{s_2^2 - s_1^2}{\mu}\,\frac{e\alpha}{c}\,\boldsymbol{A}(r)\cdot\boldsymbol{p}\left[-\frac{1}{6}\,S + 8(2 - 3\sin^2\theta_2)\,\frac{1}{(ur)^3} - \right. \\ & \left. - \exp(-ur)\,f_2(ur)\right] + \\ & + \frac{(s_1^2 - s_2^2)^2\,\alpha}{2\mu}\left(\frac{e}{c}\,\boldsymbol{A}(\boldsymbol{r})\right)^2\left[8(2 - 3\sin^2\theta_2)\,\frac{1}{(u\boldsymbol{r})^3} - \exp(-ur)\,f_2(ur)\right] + \\ & + \frac{\alpha}{\mu}\left(\frac{eB}{cu}\right)^2\left[4(2 - 3\sin^2\theta_3) + \exp(-ur)\,f_3(ur)\right] + \\ & + \left(\frac{e}{c}\,\boldsymbol{A}(\boldsymbol{r})\right)^2\left[\frac{1}{2\mu}\,(s_1^3 + s_2^3 - s_1 s_2) + \right. \\ & \left. + \frac{2}{M}\left(1 - \frac{1}{3}\,\alpha + \frac{1}{4}\,\alpha S - \frac{5}{16}\,\alpha S'\right) - \frac{(s_2^2 - s_1^2)^2}{12\mu}\,\alpha S\right], \end{aligned} \tag{16}$$

and

$$\begin{aligned} f_1(u\boldsymbol{r}) = {} & 8(2 - 3\sin^2\theta_1)\left(\frac{1}{(ur)^3} + \frac{1}{(ur)^2} + \frac{1}{2ur}\right) + \left(\frac{33}{8} - 4\sin^2\theta_1\right) + \\ & + \left(\frac{17}{8} - \sin^2\theta_1\right)ur - \left(\frac{3}{2} - \frac{11}{8}\sin^2\theta_1\right)u^2r^2 + \frac{1}{8}\,(1 - \sin^2\theta_1)\,u^3r^3\,, \end{aligned} \tag{17}$$

$$f_2(u\boldsymbol{r}) = -8(2 - 3\sin^2\theta_2)\left(\frac{1}{(ur)^3} + \frac{1}{(ur)^2} + \frac{1}{2ur}\right) - (3 - 4\sin^2\theta_2) - - (1 - \sin^2\theta_2)\, ur\,, \tag{18}$$

$$f_3(u\boldsymbol{r}) = (8 - 12\sin^2\theta_3)\left(\frac{1}{(ur)^3} + \frac{1}{(ur)^2} + \frac{1}{2ur}\right) + \left(\frac{7}{6} - \frac{1}{8}S - 2\sin^2\theta_3 - -\frac{1}{8}S\sin^2\theta_3\right) + \left(\frac{1}{6} - \frac{1}{8}S - \frac{1}{2}\sin^2\theta_3 - \frac{1}{8}S\sin^2\theta_3\right) ur - -\left(\frac{1}{24}S + \frac{1}{12}\sin^2\theta_3 + \frac{1}{48}S\sin^2\theta_3\right)u^2r^2 + \frac{1}{48}S\sin^2\theta_3\, u^3r^3\,, \tag{19}$$

$$S = \frac{s_1^2 + s_2^2}{s_1 s_2}\,, \qquad S' = \frac{s_1^4 + s_2^4}{s_1^2 s_2^2}\,,$$

where $\alpha = (Me^2/\hbar^2 u)\,(1/\varepsilon_\infty - 1/\varepsilon_0)$, to some extent it describes the exciton–phonon interaction, θ_1, θ_2, and θ_3 are the angles made by $\boldsymbol{K} + (2e/c)\,\boldsymbol{A}(\boldsymbol{r})$, $\boldsymbol{p} + (s_2^2 - s_1^2) \times \times (e/c)\,\boldsymbol{A}(\boldsymbol{r})$, and $\boldsymbol{B}$, respectively, with the relative coordinates.

E_s given by (13) describes the self-energy of an exciton interacting with the magnetic field $\boldsymbol{B}$. M^* and μ^* are the renormalized mass of M and μ, respectively, brought about by the exciton–LO-phonon interaction. V_{eff} describes the effective potential of the exciton. This is a representation which is far more complicated than that of Coulomb. In the following we will discuss these results in some detail.

3. Discussion

1. Self-trapping energy E_{tr} ($= -E_s$). From (13) we have

$$E_{\text{tr}} = \frac{3}{8}\left(\sigma + \frac{1}{\sigma} - \frac{2}{3}\right)^2 \alpha\hbar\omega + \frac{13}{12}\alpha\hbar\omega - \left(\frac{1}{6} + \frac{1}{8}S\right)\frac{\alpha}{\mu}\left(\frac{e\boldsymbol{B}}{cu}\right)^2, \tag{20}$$

where $\sigma = m_1/m_2 = s_1/s_2$ is the electron–hole mass ratio.

The self-trapping energy depends not only on both the interaction of the exciton–phonon and the electron–hole mass ratio but also on the magnetic field $\boldsymbol{B}$. For $B = 0$, it is obvious for the arbitrary electron–hole mass ratio that the exciton is always self-trapped ($E_{\text{tr}} > 0$). But for $B \neq 0$, it is possible for large electron–hole mass ratio that the exciton is not self-trapped ($E_{\text{tr}} < 0$). The $\boldsymbol{B}$-dependent term may be neglected for very weak fields and our result (20) is reduced to Pang's [9].

2. Renormalized masses M^* and μ^*. M^* and μ^* are given by (14) and (15), respectively.

The renormalized masses are related to both the interaction of the exciton–phonon and the electron–hole mass ratio, to the direction factors $\sin^2\theta_1$ and $\sin^2\theta_2$. For the large-radius exciton $ur \gg 1$, the terms including the factor $\exp(-ur)$ may be neglected. We have

$$\frac{M}{M^*} = 1 - \frac{1}{3}\alpha + \frac{1}{4}\alpha S - \frac{15}{16}\alpha S' - 8(2 - 3\sin^2\theta_1)\frac{\alpha}{(ur)^3}\,, \tag{21}$$

$$\frac{\mu}{\mu^*} = 1 - \frac{1}{6}\alpha S + 8(2 - 3\sin^2\theta_2)\frac{\alpha}{(ur)^3}\,. \tag{22}$$

If we calculate the perturbational correction to fourth order, the $\boldsymbol{B}$-dependent terms of M^* and μ^* can be obtained.

3. Effective potential V_{eff}. V_{eff} is the representation given by (16). The first three terms are a statically screened Coulomb potential, Born-Mayer potential (9), and Yukawa potential (11, 14), respectively. The $ru \exp(-ur)$-dependent term is a correction for the screened term in V_{eff}. The fifth term and the sixth term are the magnetic field-exciton translation-exciton interior motion and the magnetic field-exciton interior motion coupling energy, respectively. The next two terms are the magnetic field-exciton translation-exciton interior motion-phonon field coupling energy. The three-body coupling energy of the magnetic field-exciton interior motion-phonon field (9) is denoted by the ninth to eleventh terms. It depends on the electron–hole mass ratio and the direction factors $\sin^2\theta_2$ and $\sin^2\theta_3$, at the same time, it is effected by the phonon field. The final term is the diamagnetic shift of an exciton in a magnetic field. It expresses that the motion in direction perpendicular to the magnetic field is decreased, resulting from the effect of the phonon field on the exciton in the magnetic field.

Appendix

1. Let us first calculate $(\hbar^2/2\mu)\sum_{\boldsymbol{k}}|\nabla f_{\boldsymbol{k}}|^2$ by inserting $f_{\boldsymbol{k}}$ into it. We obtain

$$\frac{\hbar^2}{2\mu}\sum_{\boldsymbol{k}}|\nabla f_{\boldsymbol{k}}|^2 =$$

$$=\frac{\hbar^2}{2\mu}\left(\frac{2e\hbar}{Mc}\right)^2\sum_{\boldsymbol{k}}\left[\frac{\boldsymbol{A}(\boldsymbol{k})}{\hbar^2\Omega^2}\right]^2|V_{\boldsymbol{k}}\varrho_{\boldsymbol{k}}(\boldsymbol{r})|^2+\frac{\hbar^2}{2\mu}\sum_{\boldsymbol{k}}\left(\frac{1}{\hbar\Omega}\right)^2|V_{\boldsymbol{k}}\nabla\varrho_{\boldsymbol{k}}(\boldsymbol{r})|^2\,, \tag{A1}$$

where $\boldsymbol{A}(\boldsymbol{k}) = (\boldsymbol{B}\times\boldsymbol{k})/2$.

Substituting (9) into (A1) and replacing the summation by integration, (A1) may be written as

$$\frac{2}{\pi^2}\left(\frac{e\boldsymbol{B}}{uc}\right)^2\frac{\alpha u^5}{\mu}\int_0^\infty \mathrm{d}\boldsymbol{k}\,\frac{\boldsymbol{k}^2}{(u^2+\boldsymbol{k}^2)^4}\int_0^{2\pi}\mathrm{d}\varphi\int_0^\pi \sin\theta\,\mathrm{d}\theta\,(1-\cos^2\boldsymbol{B}\cdot\boldsymbol{k})\times$$

$$\times(1-\cos\boldsymbol{k}\cdot\boldsymbol{r})+\alpha\hbar\omega\,\frac{2u}{\pi}\left[S\int_0^\infty\frac{\boldsymbol{k}^2\,\mathrm{d}\boldsymbol{k}}{(u^2+\boldsymbol{k}^2)^2}+\frac{4u}{\pi\boldsymbol{r}}\int_0^\infty\frac{\boldsymbol{k}\sin\boldsymbol{k}\boldsymbol{r}}{(u^2+\boldsymbol{k}^2)^2}\,\mathrm{d}\boldsymbol{k}\right]+$$

$$+\alpha\hbar\omega\,\frac{6u}{\pi^2\hbar}\int_0^\infty\frac{\boldsymbol{k}^2\,\mathrm{d}\boldsymbol{k}}{(u^2+\boldsymbol{k}^2)^4}\int_0^{2\pi}\mathrm{d}\varphi\int_0^\pi\sin\theta\,\mathrm{d}\theta\,(\boldsymbol{k}\cdot\boldsymbol{Q})^2\,(S+2\cos\boldsymbol{k}\cdot\boldsymbol{r})\,. \tag{A2}$$

Integrating (A2), we have

$$\frac{\hbar^2}{2\mu}\sum_{\boldsymbol{k}}|\nabla f_{\boldsymbol{k}}|^2=\frac{\alpha}{6\mu}\left(\frac{e\boldsymbol{B}}{cu}\right)^2-\frac{4\alpha}{\mu}\left(\frac{e\boldsymbol{B}}{cu}\right)^2\frac{1}{(ur)^3}\,(2-3\sin^2\theta_3)+$$

$$+\frac{4\alpha}{\mu}\left(\frac{e\boldsymbol{B}}{cu}\right)^2\exp(-ur)\left[(2-3\sin^2\theta_3)\left(\frac{1}{(ur)^3}+\frac{1}{(ur)^2}+\frac{1}{2ur}\right)+\right.$$

$$\left.+\frac{1}{24}\,(7-12\sin^2\theta_3)+\frac{1}{24}\,(1-3\sin^2\theta_3)\,ur-\frac{1}{48}\sin^2\theta_3\,u^2r^2\right)+$$

$$+\alpha\hbar\omega\left[\frac{1}{2}\,S+\exp(-ur)\right]+$$

$$+\frac{Q^2}{2M}\,\alpha\left[\frac{1}{4}\,S+\frac{1}{2}\exp(-ur)\left(1+ur-(1-\sin^2\theta_1)\,u^2r^2\right)\right]. \tag{A3}$$

By similar treatment for $-\sum_{\boldsymbol{k}} \hbar\Omega\, |f_{\boldsymbol{k}}|^2$, we may obtain

$$-\sum_{\boldsymbol{k}} \hbar\Omega\, |f_{\boldsymbol{k}}|^2 =$$
$$= -2\alpha\hbar\omega \left[1 - \frac{1}{ur}\left(1 - \exp\left(-ur\right)\right)\right] - \frac{Q^2}{2M}\alpha\left[\frac{1}{3} + 8(2 - 3\sin^2\theta_1)\frac{1}{(ur)^3}\right] +$$
$$+ \frac{Q^2}{2M}\alpha \exp\left(-ur\right)\left[8(2 - 3\sin^2\theta_1) \times\right.$$
$$\left.\times\left(\frac{1}{(ur)^3} + \frac{1}{(ur)^2} + \frac{1}{2ur}\right) + (3 - 4\sin^2\theta_1) + (1 - \sin^2\theta_1)\, ur\right], \tag{A4}$$

where θ_1 and θ_3 are the angles that $\boldsymbol{r}$ makes with $\boldsymbol{Q}$ and $\boldsymbol{B}$.

2. To calculate the second-order perturbational correction ΔE, we substitute the following expression for $\langle n_{\boldsymbol{k}}|\, H^1\, |0\rangle$ ($|n_{\boldsymbol{k}}\rangle = (a_{\boldsymbol{k}}^+)^n\, |0\rangle$),

$$\langle n_{\boldsymbol{k}}|\, H^1\, |0\rangle =$$
$$= -\frac{i\hbar}{\mu}\sum_{\boldsymbol{k}} \nabla f_{\boldsymbol{k}}\cdot\left[\boldsymbol{p} + (s_2^2 - s_1^2)\frac{e}{c}\boldsymbol{A}(\boldsymbol{r})\right] - \frac{\hbar^2}{2\mu}\sum_{\boldsymbol{k}} \nabla^2 f_{\boldsymbol{k}} \quad \text{for} \quad n = 1\,,$$
$$\langle n_{\boldsymbol{k}}|\, H^1\, |0\rangle = 0 \qquad \text{for} \quad n \neq 1\,, \tag{A5}$$

and (9) into ΔE, and obtain

$$\Delta E = -\left(\frac{\hbar}{\mu}\right)^2 \sum_{\boldsymbol{k}} \frac{1}{\hbar\Omega}\left|\nabla f_{\boldsymbol{k}}\cdot\left[\boldsymbol{p} + (s_2^2 - s_1^2)\frac{e}{c}\boldsymbol{A}(\boldsymbol{r})\right]\right|^2 - \left(\frac{\hbar^2}{2\mu}\right)^2 \sum_{\boldsymbol{k}} \frac{1}{\hbar\Omega}\, |\nabla^2 f_{\boldsymbol{k}}|^2\,. \tag{A6}$$

Here we have neglected higher-order perturbational terms than second-order ones.

Inserting the representation for $f_{\boldsymbol{k}}$ into ΔE, replacing summation by integration, and integrating it, we obtain

$$\Delta E = -\frac{1}{2\mu}\left[\boldsymbol{p} + (s_2^2 - s_1^2)\frac{e}{c}\boldsymbol{A}(\boldsymbol{r})\right]^2\left[\frac{1}{6}S - 8(2 - 3\sin^2\theta_2)\frac{1}{(ur)^3}\right] -$$
$$- \frac{1}{2\mu}\left[\boldsymbol{p} + (s_2^2 - s_1^2)\frac{e}{c}\boldsymbol{A}(\boldsymbol{r})\right]^2 \exp\left(-ur\right)\alpha\left[8(2 - 3\sin^2\theta_2)\times\right.$$
$$\left.\times\left(\frac{1}{(ur)^3} + \frac{1}{(ur)^2} + \frac{1}{2ur}\right) + (3 - 4\sin^2\theta_2) + (1 - \sin^2\theta_2)\, ur\right] -$$
$$- \alpha\hbar\omega\left[\frac{3}{8}S' - \frac{1}{4}\exp\left(-ur\right)(3 - ur)\right] - \frac{Q^2}{2M}\frac{5}{16}\alpha S' + \frac{Q^2}{2M}\times$$
$$\times \exp\left(-ur\right)\frac{\alpha}{8}\left[5(1 + ur) - (8 - 7\sin^2\theta_1)\, u^2r^2 + (1 - \sin^2\theta_1)\, u^3r^3\right] +$$
$$+ \frac{\alpha S}{8\mu}\left(\frac{e\boldsymbol{B}}{cu}\right)^2 - \frac{\alpha S}{48\mu}\left(\frac{e\boldsymbol{B}}{cu}\right)^2 \times$$
$$\times \exp\left(-ur\right)\left[6(1 + \sin^2\theta_3)(1 + ur) + (2 + \sin^2\theta_3)\, u^2r^2 - \sin^2\theta_3\, u^3r^3\right]. \tag{A7}$$

References

[1] H. HAKEN, Z. Phys. **146**, 527 (1956); in: Polarons and Excitons, Ed. C. G. KUPER and G. D. WHITFIELD, Oliver & Boyd, Edinburgh 1963 (p. 302) and references therein.
[2] K. HATTORI, phys. stat. sol. (b) **76**, 281 (1976).
[3] J. SAK, Phys. Rev. B **6**, 2226 (1972).
[4] J. POLLMANN and H. BÜTTNER, Phys. Rev. B **16**, 4480 (1977).
[5] H. BARENTZEN, phys. stat. sol. (b) **71**, 245 (1975).
[6] H. HASEGAWA and R. E. HOWARD, J. Phys. Chem. Solids **21**, 179 (1961).
[7] W. EKARDT, Solid State Commun. **16**, 233 (1975).
[8] G. BEHNKE, H. BÜTTNER, and J. POLLMANN, Solid State Commun. **20**, 873 (1976).
[9] Q. J. PANG, Chinese J. Semicond. **4**, 215 (1983).
[10] E. HAGA, Progr. theor. Phys. (Kyoto) **11**, 449 (1954).
[11] S. W. GU, Acta phys. Sinica **28**, 751 (1979).
[12] S. W. GU and X. X. LIANG, Scientia Sinica **A3**, 234 (1982).
[13] K. E. CAHILL and R. J. GLAUBER, Phys. Rev. **177**, 1857 (1969).
[14] A. B. TULUB, Zh. eksper. teor. Fiz. **34**, 1641 (1958).

(Received March 16, 1984)

phys. stat. sol. (b) **125**, 333 (1984)

Subject classification: 13.4 and 19; 20.3; 22.6.1

Department of Physics, University of British Columbia, Vancouver[1])

ESR Study of Cu^{2+} Diffused into Sodium β''-Alumina

By

L. Abello, S. J. Schwerdtfeger and C. F. Schwerdtfeger

ESR spectra due to Cu^{2+} ions diffused into sodium β''-alumina are studied. The symmetry of the spectra place the Cu^{2+} ions at the mid-oxygen sites of the conduction planes. The complex spectra are a superposition from several different configurations of ions around this site which change as a function of the amount of annealing time. The angular variations of the spectra can be divided into two main types; an anisotropic site symmetry with spin-Hamiltonian parameters $g_{||} \approx 2.355$, $g_{\perp} \approx 2.087$, $A_{||} \approx 12.0$ mT, $A_{\perp} \approx 2.0$ to 4.0 mT, and a more isotropic site symmetry with $g_{||} \approx 2.003$, $g_{\perp} \approx 2.240$, $A_{||} \approx 9.9$ mT, $A_{\perp} \approx 8.8$ mT. These spectra are only visible at liquid-helium temperatures. In addition an anisotropic, random-type spectrum is observed to replace the Cu^{2+} spectra at the temperature of annealed crystals is increased. This is attributed to thermal motion of the Cu^{2+} ions due to their weak binding to the site.

EPR-Spektren von in Natrium-β''-Aluminiumoxid diffundierten Cu^{2+}-Ionen werden untersucht. Die Symmetrie der Spektren zeigt, daß die Cu^{2+}-Ionen die Sauerstoffmittelplätze der Leitungsebenen besetzen. Die komplexen Spektren stellen eine Überlagerung von mehreren verschiedenen Konfigurationen der Ionen um diese Plätze dar, die sich in Abhängigkeit vom diffundierten Cu^{2+}-Anteil ändern. Die Winkeländerungen der Spektren lassen sich in zwei Haupttypen einteilen, eine anisotrope Platzsymmetrie mit Spin-Hamiltonparameter $g_{||} \approx 2{,}355$, $g_{\perp} \approx 2{,}087$, $A_{||} \approx 12{,}0$ mT, $A_{\perp} \approx 2{,}0$ bis 4,0 mT, und eine anisotrope Platzsymmetrie mit $g_{||} \approx 2{,}003$, $g_{\perp} \approx 2{,}240$, $A_{||} \approx 9{,}9$ mT, $A_{\perp} \approx 8{,}8$ mT. Diese Spektren sind nur bei der Temperatur des flüssigen Heliums beobachtbar. Zusätzlich wird ein anisotropes statistisches Spektrum beobachtet, das die Cu^{2+} Spektren ersetzt, wenn die Temperatur der Temperaturkristalle erhöht wird. Dies wird der Thermalbewegung der Cu^{2+} Ionen zugeschrieben, wegen ihrer schwachen Platzbindung.

1. Introduction

The sodium β and β''-aluminas with their exceptionally high ionic conductivities (10^{-2} to 10^{-1} $(\Omega \text{cm})^{-1}$ at 25 °C) [1] are very promising candidates for ambient-temperature, high-energy-density batteries such as the proposed M'-doped β''-alumina/undoped β''-alumina/M''-doped β''-alumina solid state cell. Such a solid state cell would have anode and cathode portions differing only by incorporation of a foreign metal ion in the β''-alumina [2]. Although the single-crystal conductivities of these two structures can vary over a rather wide range depending upon crystal preparation techniques, single crystals of sodium β''-alumina usually have an ionic conductivity at 25 °C which exceeds the conductivity of the β structure by a factor of about four.

Electron spin resonance (ESR) provides a powerful tool to study lattice sites occupied by dopant atoms introduced either during growth of the single crystals or by diffusion. Although several dopants have been investigated in sodium β-alumina (e.g., Cr^{3+} [3], Mn^{2+} [4 to 6], Cu^{2+} [7, 8]), the only report of an ESR study in sodium β''-alumina is that of Mn^{2+} by the present authors [9].

The structural difference between sodium β-alumina and sodium β''-alumina is that every second spinel block is inverted in the sodium β''-alumina. This difference brings

[1]) 6224 Agriculture Road, Vancouver, B.C., Canada V6T 2A6.

about a different atomic arrangement in the sodium–oxygen layers of the two structures and it affects their stability and transport properties. The structures are compared in Fig. 1 of [9] and for a complete picture of the difference in the repetition of the spinel blocks the reader is referred to Fig. 1 of [10].

2. Experimental

Cu^{2+} was introduced into the Na β''-alumina single crystals, which typically measured about 4 × 4 mm along the edges and 0.5 mm in thickness, by heating them to 627 °C in a molten mixture of 98% CuCl– 2% $CuCl_2$ by weight and maintaining this temperature for about 2 min. The crystals were submerged in the molten mixture at the bottom of a crucible which was partitioned by a tightly-fitting disk so that the portion of the crucible above the disk could be filled with setter sand to prevent excessive sodium loss from the crystal. The crucible and its contents were maintained in an argon atmosphere. The same arrangement was used for annealing but, in this case, the crystals were submerged in CuCl only and were maintained at 627 °C for two hours. Attempts to anneal the crystals only in setter sand destroyed the Cu^{2+} signals.

ESR spectra were obtained with a standard X-band spectrometer. The cavity frequency was measured with a Hewlett Packard 5255A counter and the field, with an NMR probe outside the dewar system. The field was calibrated with a LiF:Li crystal. The sample crystal was mounted on a rotatable, fused quartz rod and the cavity was immersed in liquid helium which was maintained at ~ 1.6 K.

Fig. 1 displays a conduction plane with its adjacent oxygen planes (the A plane above and the B or C plane below) of a partial unit cell of sodium β''-alumina. Also shown are column-oxygen atoms (denoted by cO) in the same conduction plane but for each of two adjacent unit cells. The sites situated halfway between adjacent column oxygen atoms, the mid-oxygen sites, are denoted by mO. There are thus three equivalent mO sites rotated 120° with respect to one another about the *c* axis of the crystal.

Laue X-ray patterns were used to orient the crystals and ESR spectra were recorded as a function of angle for three different planes. The planes are: (i) containing the *c* axis and perpendicular to the cO–mO–cO direction for one of the three equivalent sites; (ii) perpendicular to the *c* axis; (iii) containing the *c* axis and two column oxygens, i.e. along a cO–mO–cO direction, which is also parallel to one of the crystal symmetry axes.

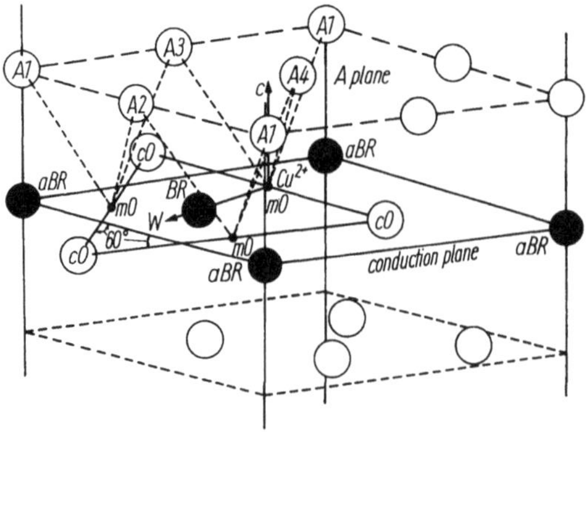

Fig. 1. Conduction plane with its adjacent oxygen planes of a partial unit cell of sodium β''-alumina. Also shown are column–oxygen atoms (denoted by cO) in the same conduction plane but for each of two adjacent unit cells. The sites situated halfway between adjacent column–oxygen atoms, the mid–oxygen sites, are denoted by mO. The bonds between the Cu^{2+} ion and the three nearest neighbour oxygen atoms in the B or C plane are not indicated. o a BR and BR mean anti-Beevers-Ross and Beevers-Ross sites, respectively

3. Results and Discussion

The analysis of the ESR spectra obtained as a function of angle in these planes indicates that the Cu^{2+} occupies the mO site as shown in Fig. 2a and has an axially symmetric g tensor. That $g_x \approx g_y$ is evident from spectra obtained when the crystal was rotated about the c axis such that the magnetic field sampled plane 2. Bleaney et al. [11] obtained similar angular data for Cu^{2+} in a regular octahedral site in $(Cu, Mg)_3 \cdot La_2(NO_3)_{12} \cdot 24\, D_2O$. In the present case the bonds which a Cu^{2+} ion forms with the three oxygen nearest neighbours above and below the conduction plane constitute an irregular octahedron because the distance between nearest-neighbour oxygen atoms (0.28 nm) is less than the O–Cu^{2+} distance (0.294 nm). Moreover, there are in addition the two column–oxygen atoms that are 0.28 nm from the Cu^{2+} ion in the conduction plane. Thus the three mO sites are irregular, dodecahedral sites.

Owing to the fact that within experimental accuracy $g_x \approx g_y$ the results are described with the axially symmetric spin Hamiltonian

$$\mathcal{H} = \beta g_{\parallel} H_z S_z + \beta g_{\perp}(H_x S_x + H_y S_y) + A_{\parallel} I_z S_z + A_{\perp}(I_x S_x + I_y S_y)\,.$$

The spectra differ markedly with the amount of Cu^{2+} diffused through annealing. Similar diffusion and annealing times on Na β''-alumina crystals from different batches yielded similar ESR spectra. The main effect of heating the pure crystals is to create Na vacancies in the conduction plane by diffusing the Na out of the crystals. The differences in the ESR parameters are attributed to sites having one or both of the nearest-neighbour Na sites vacant (i.e. the corner or anti-Beevers-Ross and the Beevers-Ross sites (see, e.g., p. 1829 in [12]) labelled aBR and BR in Fig. 1). This leads to two main types of ESR sites, a site with more anisotropic A values when only one Na is missing and a site with more isotropic A values when both Na neighbours are missing.

Fig. 3 compares two spectra taken (a) after 2 min of diffusion of Cu^{2+} ions into the sample and (b) after 2 h of subsequent annealing at 627 °C. The spectra are very complex, as is immediately evident, and a part of the more isotropic spectrum which

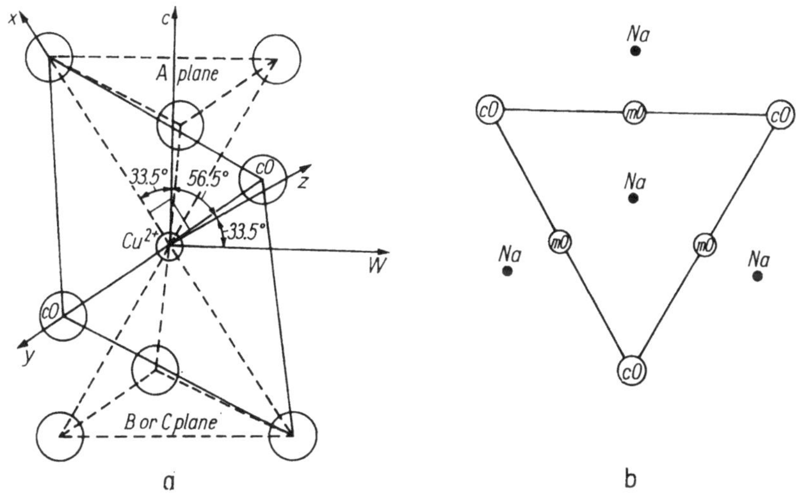

Fig. 2. a) Irregular dodecahedral site comprising three oxygen atoms above and below the mO site and two column-oxygen atoms in the conduction plane perpendicular to the c axis. The shaded area represents the $g_{\perp}$ plane. b) View along the c axis of the conduction plane displayed in Fig. 1

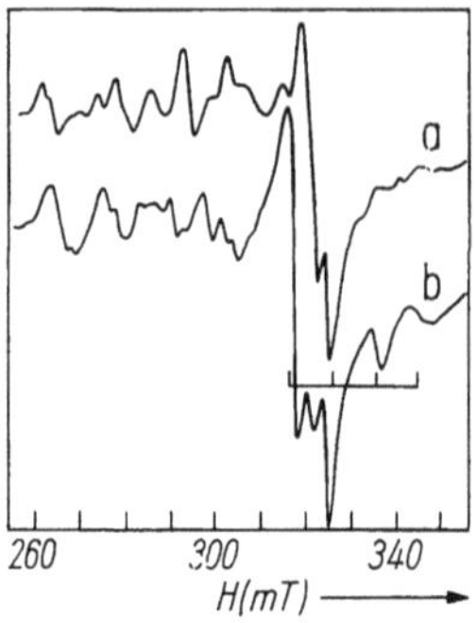

Fig. 3. Spectra from an (a) unannealed crystal, and from an (b) annealed crystal. Both spectra were taken at 0° ($H \parallel z$) where the spectra are most separated. The shorter scale at the right of curve (b) indicates the more isotropic set of lines which "grows" in during anneailing. The gain in (a) is higher than that in (b) thereby offsetting the greater intensity of the lines in (b). Also, the strong line at ≈ 292.5 mT in (a) is due to a superposition of lines which, due to the change of g and A values through annealing, are more spread out in (b)

"grows" in intensity with annealing is indicated. The set of more isotropic lines has a "reversed" g value, that is, its $g_{\parallel}$ value is smaller than its $g_{\perp}$ value. A "reversed" g value is usually attributed to a compression of charges along the z axis of an octahedral-type site with tetragonal distortion (see, e.g., p. 182 in [13]). This could easily occur in the neighbourhood of an mO site when both Na ions are missing. Spectra from two different mO site configurations which have "reversed" g values with respect to one another would then exhibit the same angular dependence but opposite field extrema. When one neighbouring Na site is vacant the mO site becomes not only more anisotropic but also tilted in one direction or the other depending on which of the two neighbouring Na is missing. Observed angular shifts of spectra caused by these tilts will be discussed below.

The resultant spectrum is thus a superposition, not only of these spectra already mentioned, but also of the corresponding spectra from the two equivalent sites which are rotated by $\pm 120°$ with respect to the site described (see Fig. 1 and 2a). There is, in addition, a shift in the spin parameters as the sample is annealed which is attributed to additional diffusion of copper ions into the conduction planes and/or diffusion of the sodium ions out of the conduction planes thus respectively stretching or contracting the bonds between the spinel blocks. In contrast to the spin-Hamiltonian parameters already given for the annealed sample, the spin-Hamiltonian parameters for the more anisotropic sets of lines, the only ones clearly visible in the spectra from the unannealed sample, are the following: $g_{\parallel} \sim 2.310$, $g_{\perp} \sim 2.053$, $A_{\parallel} \sim 14.5$ mT, $A_{\perp} \sim$

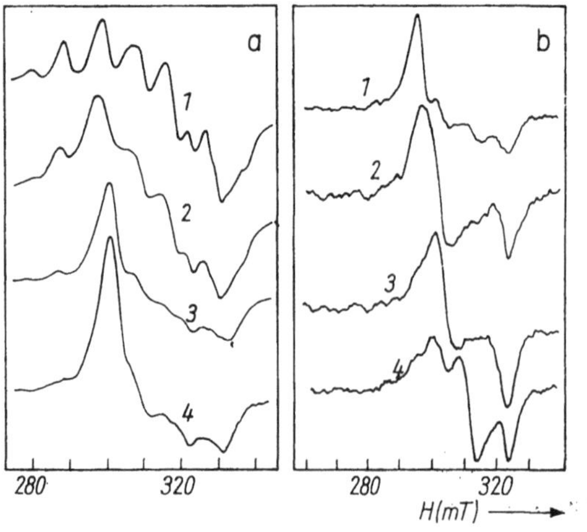

Fig. 4. a) Spectra from an annealed crystal with $H \perp c$ at (1) $T = 1.6$ K, (2) 4, (3) 5, and (4) 77.3 K. b) Spectra from an annealed crystal at 77.3 K as the crystal is rotated from $H \perp c$ at (1) 0° (top spectrum) to (2) 30°, (3) 60°, and (4) 90°, where $H \parallel c$

~ 1.7 mT. The ease of diffusion of the Cu^{2+} ions would seem to suggest that the ions are not bound very tightly to the mO sites. This is supported by the temperature dependence of the spectra.

As the temperature of the annealed crystal is increased the Cu^{2+} line spectra first broaden (see Fig. 4a). Upon further increase in temperature the spectra broaden to the extent that individual lines are no longer discernible but are replaced by an anisotropic, random-type spectrum. A somewhat similar effect is observed in many compounds containing Cu^{2+} ions (see, for example, [11]) and is attributed to a dynamic Jahn-Teller effect. These random-type spectra, however, are not usually as anisotropic as the present spectra shown in Fig. 4b. The anisotropic, random character of the spectrum is most likely due to thermal motional averaging caused by the fact that the Cu^{2+} ions are not tightly bound to the mO sites and move more easily in the conduction plane than perpendicular to it.

As the temperature of the unannealed crystal is increased, the Cu^{2+} line spectra decrease much more slowly than in the case of the annealed crystal. Moreover, no anisotropic, random-type spectrum is detected at any temperature in the case of the

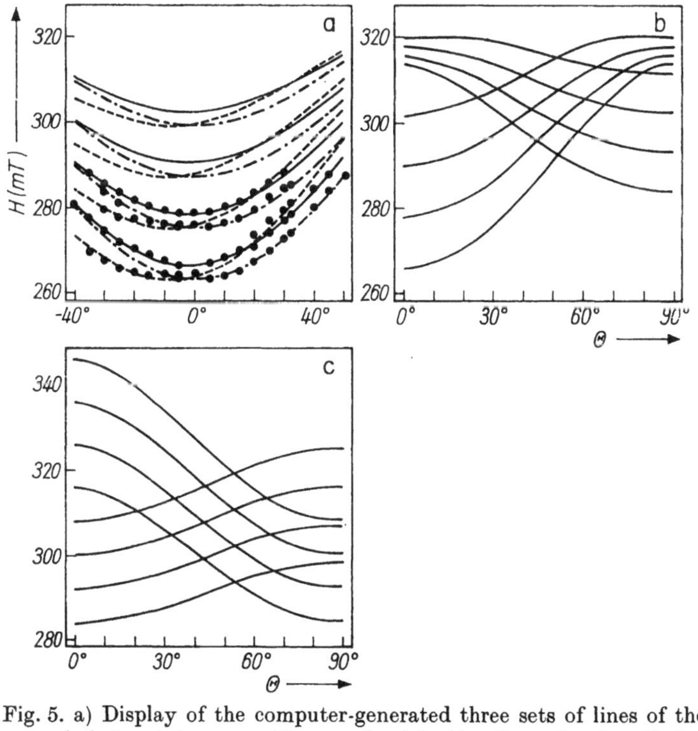

Fig. 5. a) Display of the computer-generated three sets of lines of the more anisotropic spectra around their minimum positions and originating from the described sites in plane 1. Note how the set of lines shifted by +7° (- - - -), crosses the set shifted by −2.5° (—·—), and then crosses the unshifted set of lines (———). The dots represent the best eye-estimate of the line positions in the original data around the maximum positions of the g values; overlapping at intermediate fields makes it impossible to distinguish individual lines. b) Angular variation of the unshifted line set of the more anisotropic spectra from the described site in plane 1 (yielding the lines at minimum field), and from the two equivalent sites which are rotated ±120° with respect to the described site. The g and A values used in Fig. 5a, 5b, and 5c were obtained from the computer fit. c) Angular variation of the spectra from the described and equivalent sites of the more isotropic type with "reversed" g values

unannealed crystal. This could be due to the much lower concentration of impurities in its conduction planes.

Annealing the crystal also decreases its spin–lattice relaxation constant. ESR measurements at Q-band yielded spectra with individual Cu^{2+} lines from an annealed crystal but only broad lines from an unannealed one, although the individual Cu^{2+} lines were obtained from both these crystals at X-band. Experiments at very high power levels at X-band indicated that saturation occurred much more readily in the case of the unannealed crystal than in the case of the annealed one. The decrease of the spin–lattice relaxation constant due to annealing is attributed to the increased impurity concentration in the conduction planes. The high impurity concentration in the annealed crystal is also accompanied by a darker colour.

As already discussed and demonstrated in the figures, the ESR spectrum is a complex superposition of several characteristic four-line, Cu^{2+} spectra. It is difficult, if not impossible, to identify all four lines of any given set except possibly at extremum positions of the g values. In order to obtain the best possible g and A parameters and, in fact, in hindsight to see the need for including the slightly tilted configurations caused by aBR or BR vacancies, the spectra were fit by computer simulation in an iterative way. A Lorentzian line shape and identical line width (1.5 mT) were chosen for each line. Owing to the underlying, residual, random-type spectrum and most likely some other contributions, which were not taken into account, the fit is not perfect. However, the general features are reasonably well reproduced. Due to overlap, the positions of the individual lines on the original spectra can only be approximated. The fit of the simulated spectra gives a much better determination of the individual line positions. This is seen in Fig. 5a which displays the three sets of lines of the more anisotropic spectra around their minimum positions. Two of these sets are at tilt angles of $+7°$ and $-2.5°$ where $0°$ is taken as the angle at which the more isotropic spectrum reaches its maximum and an unshifted line set of the more anisotropic spectra reaches its minimum. Fig. 5b shows the angular variation in plane 1 of the hyperfine spectra for the more anisotropic site and Fig. 5c shows the angular variation for the more isotropic site.

Shown for comparison in Fig. 6a, b, and c are the simulation and experimental data for three different angles in plane 1. These are for an annealed sample and again it must be emphasized that the g and A values do vary as a function of annealing time and hence these values should be taken as typical values allowing for at least a 10%

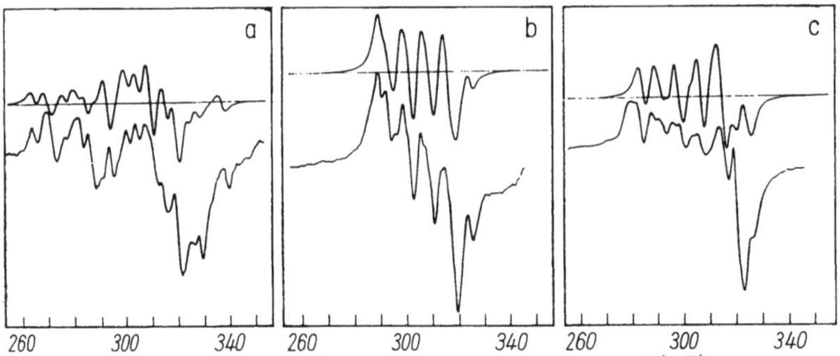

Fig. 6. Comparison of computer-generated results (upper curves) and ESR spectra with a) H in plane 1 and making an angle of 24° with z and an angle of 9.5° with the conduction plane; b) with H along the c axis; c) with H in plane 1 and perpendicular to the z axis (i.e. in the $g_\perp$ plane)

variation. The approach taken was to use as few parameters as possible to obtain a meaningful and reasonable fit of the computer-generated spectra. The rather close correspondence between the data and the simulation in Fig. 6a and b is typical of the close correspondence achieved over the whole range of angles, except for the somewhat poorer correspondence when H is within a few degrees from being parallel to the $g_\perp$ plane, the worst fit being given in Fig. 6c. This more noticeable discrepancy seems to be due to an extra set of lines at minimum field. An unexplained distortion affecting only some of the more isotropic-type of sites would cause a doubling of the more isotropic lines which, at this orientation are at their minimum field but there is no apparent reason why such a doubling would not appear at other orientations. Since there is no firm explanation for what appear to be extra lines within this narrow range of angles, it was felt that the addition of more parameters to the computer programme to simulate this localized discrepancy was not justified.

One extra line which is independent of angle but appears in most of the spectra at $g = 2.035$ is due to an impurity in the crystal and was included in the simulation for esthetic purposes.

The computer-generated spectra are thus a superposition of (i) the three anisotropic sites in the plane of rotation and their respective two equivalent sites at $\pm 120°$ (Fig. 5b); (ii) the isotropic site as well as its two equivalent sites (Fig. 5c); (iii) the impurity line. Thus each computer-generated spectrum is the sum of 49 Lorentzian lines.

From the analysis, one would expect that rotating the crystal such that the magnetic field samples plane 3 would yield very symmetric results. A highly symmetric angular variation of the ESR spectrum is indeed obtained when the crystal is rotated in plane 3.

Acknowledgements

The authors wish to thank Dr. J. B. Bates of the solid state division of Oak Ridge National Laboratories, Oak Ridge, Tennessee, USA, for providing the crystals used in this study. Donation of the setter sand by Johnson Controls, Inc., Milwaukee, Wisconsin, USA, is also gratefully acknowledged. This work was supported by a grant from the National Science and Engineering Research Council of Canada.

References

[1] I. Wynn-Jones and L. J. Miles, Proc. Brit. Ceram. Soc. **19**, 161 (1971).
[2] J. R. Akridge, Ph. D. Thesis, University of California, Santa Barbara 1978.
[3] H. Sixl and R. Hundhausen, Z. Phys. B **38**, 299 (1980).
[4] R. C. Barklie and K. O'Donnell, J. Phys. C **10**, 4127 (1977).
[5] R. C. Barklie, K. O'Donnell, and A. Murtagh, J. Phys. C **10**, 4815 (1977).
[6] J. Antoine, D. Vivien, J. Livage, J. Thery, and R. Collongues, Mater. Res. Bull. **10**, 865 (1975).
[7] D. Gourier, J. Antoine, D. Vivien, J. Thery, J. Livage, and R. Collongues, phys. stat. sol. (a) **41**, 423 (1977).
[8] D. Gourier, D. Vivien, J. Thery, J. Livage, and R. Collongues, phys. stat. sol. (a) **45**, 599 (1978).
[9] L. Abello, S. J. Schwerdtfeger, and C. F. Schwerdtfeger, Solid State Commun. **44**, 497 (1982).
[10] J. P. Boilot, J. Kahn, J. Thery, R. Collongues, J. Antoine, D. Vivien, C. Chevrette, and D. Gourier, Electrochim. Acta (London) **22**, 741 (1977).
[11] B. Bleaney, F. R. S., K. D. Bowers, and R. S. Trenam, Proc. Roy. Soc. A **228**, 157 (1955).
[12] C. R. Peters, M. Bettman, J. W. Moore, and M. D. Glick, Acta Cryst. B **27**, 1826 (1971).
[13] B. R. McGarvey, Transition Metal Chem. **3**, 89 (1966).

(Received March 13, 1984)

Note added in proof see on page **340**.

Note added in proof:

A recent paper on divalent cation exchange in sodium β''-alumina by G. C. Farrington and B. Dunn (Solid State Ionics **7**, 267 (1982)) reported observation of complete ionic exchange of Na^{1+} by many divalent cations such as Ba^{2+}, Sr^{2+}, Ca^{2+}, Cd^{2+}, etc. but an anomalous behaviour for Cu^{2+}. Using gravimetric and radiometric techniques they found only a 37% "exchange" in 24 h of diffusion for Cu^{2+}. Our paper now explains this result since the Cu^{2+} does not exchange for the Na^{1+} but resides in the mO sites. In addition some Na^{1+} diffuses out of the crystal leaving the observed vacancies and a net 37% gravimetric result.

phys. stat. sol. (b) **125**, 341 (1984)

Subject classification: 13.4; 20.1; 22

Institute of Theoretical Physics, Warsaw University[1])

Shallow Impurity States in Strong Magnetic Field

By

T. R. GAWRON and J. MYCIELSKI

Assuming that the shallow-impurity wave function is composed of the states of a single Landau subband, an effective one-dimensional wave equation is derived and discussed in detail, allowing for the band anisotropy. The optical transitions from the bound impurity states to both, bound and free-carrier states are considered. For strong effective-mass anisotropy, e.g. for acceptors in semimagnetic semiconductors, the energies and wave functions are explicitly given for all impurity states.

Supposant que la fonction d'onde d'impûreté peu profond est composé des états d'une seule subbande de Landau, une unidimentionelle équation d'onde effective est derivée et discutée en detail dans le cas général d'une bande anisotrope. Les transitions optiques entre les états liés et les autres états liés ou états delocalisés sont considérées. Dans le cas d'une forte anisotropie de la masse effective, e.g. pour les accepteurs dans les semiconducteurs semimagnétiques, les energies et les functions d'onde sont presenté pour tous les états d'impûreté.

1. Introduction

Shallow impurity states in semiconductors in the presence of a strong external magnetic field have been studied for many years. A useful tool in finding the impurity eigenstates is the adiabatic approximation consisting in the separation of the longitudinal and transverse motion of the carriers (see, e.g., [1, 2]). If the magnetic field is strong enough, the wave function of the impurity state is composed of the wave functions of a single Landau subband ("quantum limit"). Several authors have constructed effective one-dimensional wave equations for the impurity states in the quantum limit, and discussed those originating from the ground Landau subband (see, e.g., [3 to 7]). It was also observed that the quantum-limit problem is very much simplified for the transverse effective mass being much lower than the longitudinal one, e.g. for $Bi_{1-x}Sb_x$ [4] or for acceptors in semimagnetic semiconductors [6]. The impurity levels in such materials were discussed in [4 to 6, 8] (mainly those originating from the ground Landau subband), and the selection rules for optical transitions between them in [5].

The purpose of the present paper is to discuss in more detail the quantum-limit impurity wave equations and their solutions, originating from both ground and higher Landau subbands. In particular, we present explicit formulas for the energies and the wave functions of all states in the case of a strong effective-mass anisotropy. In Section 2 we deal with the wave equations and their eigenstates, Section 3 is devoted to optical transitions between impurity states (photoionization included), and in Section 4 we consider the case of a strong effective-mass anisotropy.

---

[1]) Ul. Hoża 69, 00-681 Warsaw, Poland.

2. Quantum-Limit Approximation

In a strong external magnetic field the Landau level spacing may be larger than the binding energy of a Coulomb impurity. In this case ("quantum limit") we can construct the wave function of an impurity state from the orthonormal wave functions corresponding to a single Landau subband (i.e. with a given N),

$$\Psi_{iNmk_z}(\boldsymbol{r}) = L_z^{-1/2} \exp(ik_z z)\, \chi_{Nm}(\boldsymbol{r})\, \Phi_i(\boldsymbol{r})\,. \tag{1}$$

Here the spinor Φ_i is the Bloch function corresponding to the band extremum (the mean value of $|\Phi_i|^2$ is unity), and $N \geqq 0$ is the quantum number determining the energy of the Landau level E_{iN}. The magnetic induction $\boldsymbol{B}$ is in z-direction. In the symmetric gauge the vector potential is $\boldsymbol{A} = \frac{1}{2} B(-y, x, 0)$ and

$$\chi_{Nm}(\boldsymbol{r}) = \beta \binom{n + |m|}{n}^{1/2} (\pi |m|\,!)^{-1/2} \times$$
$$\times \exp(im\varphi)\, (\beta\varrho)^{|m|} \exp(-\beta^2\varrho^2/2)\, M[-n, |m| + 1, (\beta\varrho)^2]\,, \tag{2}$$

where $N \geqq m > -\infty$, $n = N - m + (m - |m|)/2$, $M(a, b, x)$ is the confluent hypergeometric series (for special functions we use the notation of Abramowitz and Stegun [9]), $\beta^2 = eB/2\hbar c$ $\big((2^{1/2}\beta)^{-1}$ is the magnetic length$\big)$, $-e$ is the electron charge, ϱ and φ are the cylindric coordinates, $k_z = 2\pi K_z/L_z$, L_z is the dimension of the periodicity box in z-direction, and K_z is an integer. χ_{Nm} are orthonormal. The energy of the carrier in state (1) is $E_{iNk_z} = E_{iN} + \hbar^2 k_z^2/2m_{iN}$, where m_{iN} is the carrier (longitudinal) effective mass of the Landau subband i, N.

If in the vicinity of E_{iN} there is another Landau level $E_{i'N'}$, our approximation is justified only for $\langle \Phi_{i'} \mid \Phi_i \rangle = 0$ (in particular, for $\Phi_{i'} = u\downarrow$ and $\Phi_i = u\uparrow$).

The electron states of a donor (or the hole states of an acceptor) are described by the equation

$$\left[H - \frac{e^2}{\varkappa(\varrho^2 + z^2)^{1/2}}\right] \Psi = E\Psi\,, \tag{3}$$

where H is the crystal Hamiltonian of a free carrier in the magnetic field, and $\varkappa$ is the crystal dielectric constant.

In the quantum-limit approximation, we take the impurity wave function in the form

$$\Psi_{iN}(\boldsymbol{r}) = \sum_{m,\, k_z} a_m(k_z)\, \Psi_{iNmk_z}(\boldsymbol{r})\,. \tag{4}$$

Inserting this into (3), multiplying by $\Psi^*_{iNm'k'_z}$, integrating over $\boldsymbol{r}$, multiplying by $\exp(ik'_z z)$, and summing over k'_z, we show that a given impurity state is composed of Ψ_{iNmk_z} with a given m. The normalized impurity wave function has the form

$$\Psi_{iNms}(\boldsymbol{r}) = T_{iNms}(z)\, \chi_{Nm}(\boldsymbol{r})\, \Phi_i(\boldsymbol{r})\,, \tag{5}$$

where the normalized envelope function $T_{iNms}(z)$ and the impurity state energy E_{iNms} fulfil the following "effective" one-dimensional wave equation:

$$\left[-\frac{1}{2}\frac{\mathrm{d}^2}{\mathrm{d}\xi^2} - \delta_{iN} V_{Nm}(|\xi|) - \varepsilon_{iNms}\right] T_{iNms}(\xi) = 0\,, \tag{6}$$

where $\xi = \beta z$, $\delta_{iN} = (2^{1/2}\beta a_{iN})^{-1}$, $a_{iN} = \hbar^2\varkappa/e^2 m_{iN}$ (the effective Bohr radius for the longitudinal effective mass), $\varepsilon_{iNms} = 2(E_{iNms} - E_{iN})/\hbar\omega_{iN}$, $\omega_{iN} = eB/cm_{iN}$, and

$$V_{Nm}(|\xi|) = \frac{2^{1/2}(|m| + n)!}{(|m|!)^2\, n!} \int\limits_0^\infty \frac{t^{|m|}}{(t + \xi^2)^{1/2}} \exp(-t)\, [M(-n, |m| + 1, t)]^2\, dt\,, \tag{7}$$

where n is defined as in (2). It is worth noting that δ_{iN} is the ratio of the magnetic length to the Bohr radius.

Performing the integration over t, we obtain from (7)

$$V_{Nm}(|\xi|) = \frac{2^{1/2} n!}{(|m| + n)!} \sum_{k,l=0}^{n} (-1)^{k+l} \binom{|m| + n}{n - k} \binom{|m| + n}{n - l} \frac{1}{k!\, l!} \times$$
$$\times \left[\Gamma\left(p + \frac{1}{2}\right) M\left(\frac{1}{2}, -p + \frac{1}{2}, \xi^2\right) + \right.$$
$$\left. + \frac{p!}{\pi^{1/2}} \Gamma\left(-p - \frac{1}{2}\right) |\xi|^{2p+1} M\left(p + 1, p + \frac{3}{2}, \xi^2\right)\right], \tag{8}$$

where $p = |m| + k + l$.

In particular,

$$V_{00}(|\xi|) = (2\pi)^{1/2} \exp(\xi^2)\, [1 - \operatorname{erf}(|\xi|)]\,, \tag{9}$$

where erf (x) is the error function.

Equation (6) with (8) may be rather easily solved numerically.

It follows from (7) that $V_{Nm}(|\xi|) > 0$, $V_{Nm}(|\xi|)$ decreases with increasing $|\xi|$, and for $|\xi| \gg 1$ there is $V_{Nm}(|\xi|) \approx 2^{1/2}/|\xi|$. As $V_{Nm}(|\xi|)$ is symmetric, a bound eigenstate of (6) is symmetric or antisymmetric.

In the following T_{iNms} will denote the envelope functions of the bound (localized) states. Strictly speaking, the higher of those states are resonant states, as they overlap in energy with the continuum states of the Landau subbands [7]. The free-carrier solutions of (6) can be written approximately as

$$T^{(c)}_{iNmk_z}(z) = L_z^{-1/2} \left[\exp(ik_z z) - \sum_{s=0}^{\infty} \langle T_{iNms}(z) \mid \exp(ik_z z)\rangle\, T_{iNms}(z)\right]. \tag{10}$$

They are orthogonal to all T_{iNms} (with the same i, N, and m). With an accuracy of the order of $1/\beta L_z$, the functions $T^{(c)}_{iNmk_z}$ form an orthonormal set (for given i, N, and m), and the energy corresponding to T_{iNmk_z} is E_{iNk_z}.

For small $|\xi|$ there is

$$V_{N0}(|\xi|) \approx V_{N0}(0) - 2^{3/2}\, |\xi|\,, \tag{11}$$

$$V_{Nm}(|\xi|) \approx V_{Nm}(0) - A_{Nm}\xi^2 \qquad (m \neq 0)\,, \tag{12}$$

where

$$V_{Nm}(0) = \frac{2^{1/2}\, n!}{(|m| + n)!} \sum_{k,l=0}^{n} (-1)^{k+l} \binom{|m| + n}{n - k} \binom{|m| + n}{n - l} \frac{\Gamma(p + \frac{1}{2})}{k!\, l!}\,, \tag{13}$$

$$A_{Nm} = \frac{2^{1/2} n!}{(|m| + n)!} \sum_{k,l=0}^{n} (-1)^{k+l} \binom{|m| + n}{n - k} \binom{|m| + n}{n - l} \frac{\Gamma(p + \frac{1}{2})}{(2p - 1)\, k!\, l!} \qquad (m \neq 0)\,. \tag{14}$$

In particular,

$$V_{00}(0) = (2\pi)^{1/2}\,, \qquad V_{0m}(0) = (2\pi)^{1/2} \prod_{t=1}^{|m|} \left(1 - \frac{1}{2t}\right) \qquad (m < 0)\,, \tag{15}$$

$$A_{0m} = \frac{(2\pi)^{1/2}}{(2|m| - 1)} \prod_{t=1}^{|m|} \left(1 - \frac{1}{2t}\right) \qquad (m < 0)\,. \tag{16}$$

It follows from (15) and (16) that $V_{0m}(0)$ and A_{0m} decrease with increasing $|m|$ and tend to 0 for $|m| \to \infty$.

3. Optical Transitions

Let us now consider the transitions from bound impurity states to other bound states or to free-carrier states (photoionization), induced by the electric field of the radiation

$$\text{Re}\,[\boldsymbol{E} \exp(-i\omega t)]\,, \tag{17}$$

where $\boldsymbol{E}$ is a complex vector and ω the radiation frequency. We use in the following the notation

$$E_{\pm} = 2^{-1/2}\,(E_x \pm iE_y)\,. \tag{18}$$

The perturbation of the microscopic Hamiltonian of the carrier by the radiation field has the form

$$H_r = H_1 \cos \omega t + H_2 \sin \omega t\,, \tag{19}$$

where H_1 and H_2 are Hermitian and time-independent operators. For transitions with photon absorption, the transition matrix element is the matrix element of the operator [10]

$$H_+ = H_1 + iH_2\,. \tag{20}$$

In the following we consider separately intra- ($i' = i$) and interband ($i' \neq i$) transitions. The upper sign in (21), (22), (24), (25), (27), (28), (32), and (33) corresponds to electrons, and the lower to holes.

3.1 Intraband transitions

Using the gauge in which the radiation field (17) is expressed by a scalar potential, we have

$$H_r = \pm\, e\boldsymbol{r} \cdot \text{Re}\,[\boldsymbol{E} \exp(-i\omega t)] \tag{21}$$

and

$$H_+ = \pm\, e(E_+\varrho_- + E_-\varrho_+ + E_z z)\,, \tag{22}$$

where we have denoted

$$\varrho_{\pm} = 2^{-1/2}\,(x \pm iy)\,. \tag{23}$$

For transitions between bound states we obtain, using (2), (5), and (22),

$$\begin{aligned} \langle \Psi_{iN'm's'}|\, H_+\, |\Psi_{iNms}\rangle &= \pm\, eE_+ \frac{\text{sgn}\,(m - \frac{1}{2})}{2^{1/2}\beta}\, \delta_{m'm-1} \times \\ &\times \langle T_{iN'm-1s'} \mid T_{iNms}\rangle\, [N^{1/2}\delta_{N'N-1} - (N - m + 1)^{1/2}\, \delta_{N'N}] \pm \\ &\pm\, eE_- \frac{\text{sgn}\,(m + \frac{1}{2})}{2^{1/2}\beta}\, \delta_{m'm+1} \langle T_{iN'm+1s'} \mid T_{iNms}\rangle\, [(N + 1)^{1/2}\, \delta_{N'N+1} - \\ &- (N - m)^{1/2}\, \delta_{N'N}] \pm eE_z\, \langle T_{iNms'}|\, z\, |T_{iNms}\rangle\, \delta_{m'm}\delta_{N'N}\,. \end{aligned} \tag{24}$$

For transitions from bound to free carrier states we have, using (2), (5), (10), and (22),

$$\langle \Psi^{(c)}_{iN'm'k_z}| H_+ |\Psi_{iNms}\rangle = \pm eE_+ \frac{\operatorname{sgn}(m - \frac{1}{2})}{(2L_z)^{1/2}\beta} \delta_{m'm-1} \times$$

$$\times [\langle \exp(ik_z z) \mid T_{iNms}\rangle - \sum_{s'=0}^{\infty} \langle \exp(ik_z z) \mid T_{iN'm-1s'}\rangle \langle T_{iN'm-1s'} \mid T_{iNms}\rangle] \times$$

$$\times [N^{1/2}\delta_{N'N-1} - (N - m + 1)^{1/2} \delta_{N'N}] \pm eE_- \frac{\operatorname{sgn}(m + \frac{1}{2})}{(2L_z)^{1/2}\beta} \delta_{m'm+1} \times$$

$$\times [\langle \exp(ik_z z) \mid T_{iNms}\rangle - \sum_{s'=0}^{\infty} \langle \exp(ik_z z) \mid T_{iN'm+1s'}\rangle \langle T_{iN'm+1s'} \mid T_{iNms}\rangle] \times$$

$$\times [(N+1)^{1/2} \delta_{N'N+1} - (N-m)^{1/2} \delta_{N'N}] \pm eE_z L_z^{-1/2} [\langle \exp(ik_z z)| z \,| T_{iNms}\rangle -$$

$$- \sum_{s'=0}^{\infty} \langle \exp(ik_z z) \mid T_{iNms'}\rangle \langle T_{iNms'}| z \,| T_{iNms}\rangle] \delta_{m'm}\delta_{N'N} . \tag{25}$$

In the first and second sums over s' in (25) only terms with s' and s corresponding to the same parity of T do not vanish. In the third sum only terms with s' and s corresponding to opposite parities of T contribute.

From (24) and (25) we obtain the following selection rules for intraband transitions:

$$\begin{aligned} &E_+: \quad m' = m - 1\,, \qquad N' = N \quad \text{or} \quad N - 1\,, \\ &E_-: \quad m' = m + 1\,, \qquad N' = N \quad \text{or} \quad N + 1\,, \\ &E_z: \quad m' = m\,, \qquad N' = N\,. \end{aligned} \tag{26}$$

Moreover, for transitions between bound states there is a selection rule for s: the parity of T should be conserved for $E_\pm$, and reversed for E_z.

3.2 Interband transitions

Using the gauge in which the radiation field (17) is expressed by the vector potential $(c/\omega)\,\mathrm{Im}\,[\boldsymbol{E} \exp(-i\omega t)]$, we have

$$H_r = \mp i \frac{e\hbar}{m_0\omega} \mathrm{Im}\,[\boldsymbol{E} \exp(-i\omega t)]\,(\nabla_x - i\beta^2 y\,, \quad \nabla_y + i\beta^2 x\,, \quad \nabla_z)\,, \tag{27}$$

where m_0 is the free-electron mass, and

$$H_+ = \mp \frac{e\hbar}{m_0\omega} [E_+(\nabla_- + \beta^2\varrho_-) + E_-(\nabla_+ - \beta^2\varrho_+) + E_z\nabla_z]\,, \tag{28}$$

where we have denoted

$$\nabla_\pm = 2^{-1/2} (\nabla_x \pm i\nabla_y)\,. \tag{29}$$

We denote also

$$P^\pm_{i'i} = \frac{\hbar}{m_0} \langle \Phi_{i'}| \nabla_\pm |\Phi_i\rangle\,, \tag{30}$$

$$P^z_{i'i} = \frac{\hbar}{m_0} \langle \Phi_{i'}| \nabla_z |\Phi_i\rangle\,. \tag{31}$$

Integration is here over the unit volume.

For transitions between bound states we obtain, using (5) and (28),

$$\langle \Psi_{i'N'm's'} | H_+ | \Psi_{iNms} \rangle = \mp \frac{e}{\omega} (E_+ P^-_{i'i} + E_- P^+_{i'i} + E_z P^z_{i'i}) \times$$
$$\times \langle T_{i'Nms'} | T_{iNms} \rangle \, \delta_{m'm} \delta_{N'N} \,. \tag{32}$$

For transitions from bound to free carrier states we have, using (5), (10), and (28),

$$\langle \Psi^{(c)}_{i'N'm'k_z} | H_+ | \Psi_{iNms} \rangle = \mp \frac{e}{\omega L_z^{1/2}} (E_+ P^-_{i'i} + E_- P^+_{i'i} + E_z P^z_{i'i}) \times$$
$$\times [\langle \exp(ik_z z) | T_{iNms} \rangle - \sum_{s'=0}^{\infty} \langle \exp(ik_z z) | T_{i'Nms'} \rangle \langle T_{i'Nms'} | T_{iNms} \rangle] \, \delta_{m'm} \, \delta_{N'N} \,. \tag{33}$$

In the sum over s' in (33) only terms with s' and s corresponding to the same parity of T do not vanish.

From (32) and (33) we obtain simple selection rules for interband transitions,

$$m' = m \,, \qquad N' = N \,, \tag{34}$$

for all polarizations of radiation. Moreover, for transitions between bound states there is a selection rule for s: the parity of T should be conserved.

4. The Case of Strong Effective-Mass Anisotropy

For an isotropic effective mass, the quantum-limit regime — Landau level spacing larger than the impurity binding energy — means $\delta_{iN} \ll 1$. If, however, the effective mass of the carrier in the direction perpendicular to the magnetic field is much lower than that in parallel direction, it is possible — for not too strong magnetic field — to have still $\delta_{iN} \gg 1$ being already in the quantum limit. This follows from the fact that the Landau level spacing is determined by the transverse effective mass, not by the longitudinal one involved in δ_{iN}.

The quantum-limit regime with $\delta_{iN} \gg 1$ was considered in $Bi_{1-x}Sb_x$ — a material with strong effective-mass anisotropy [4]. The acceptor ground state in this regime was observed also in the narrow-gap semimagnetic semiconductor $Hg_{1-x}Mn_xTe$ [6]. In such zincblende-structure materials the exchange interaction of the band-electron spin with the spins of $3d^5$-electrons of the manganese ions, ordered by the external magnetic field, produces a very strong anisotropy of the effective mass in the valence band [11]. This, in turn, blows up the Landau level spacing [12, 13]. The exchange-induced decrease of the transverse mass yields a decrease of the acceptor binding energy in the presence of an external magnetic field- the so-called "boil-off effect" [6].

It follows from (6) and (11) to (14) that if $\delta_{iN} \gg 1$, the envelope function T_{iNms} does not vanish only for $|\xi| \ll 1$, at least for the ground state of (6) and not too high N and $|m|$. One can then replace V_{Nm} in (6) by its expansions (11) (for $m = 0$) or (12) (for $m \neq 0$).

4.1 $m = 0$

Substituting (11) into (6) we obtain an equation of the Airy type. For both $\xi > 0$ and $\xi < 0$ it has a solution vanishing for $|\xi| \to \infty$,

$$\mathrm{Ai}\,\{2^{5/6} \delta_{iN}^{1/3} [|\xi| - 2^{-3/2} V_{N0}(0) - 2^{-3/2} \delta_{iN}^{-1} \varepsilon_{iN0s}]\} \,, \tag{35}$$

where $\mathrm{Ai}(x)$ is the Airy function.

In order to have $T_{iN0s}(\xi)$ continuous with the first derivative at $\xi = 0$, it must be

$$\varepsilon_{iN0s} = -\delta_{iN} V_{N0}(0) - (2\delta_{iN})^{2/3} x_s , \tag{36}$$

$$T_{iN0s}(\xi) = (\operatorname{sgn} \xi)^s \operatorname{Ai} (2^{5/6}\delta_{iN}^{1/3} |\xi| + x_s) C_{iNs} , \tag{37}$$

where $s = 0, 1, 2, \ldots$, C_{iNs} are normalization factors, and x_s are zeros of the Airy function (odd s) or of its derivative (even s): $x_s \approx -1.02, -2.34, -3.25, -4.09, -4.82, -5.52, \ldots$ for $s = 0, 1, 2, 3, 4, 5, \ldots$.

T_{iN0s} is symmetric for even s and antisymmetric for odd s. As one can see from (36), the ground state is the symmetric state $s = 0$. In particular, using (15) we obtain

$$\varepsilon_{i00s} = -(2\pi)^{1/2} \delta_{i0}(1 + 2^{1/6}\pi^{-1/2}\delta_{i0}^{-1/3}x_s) \tag{38}$$

which can be also written as[2])

$$E_{i00s} - E_{i0} = -\frac{(2\pi)^{1/2}}{\delta_{i0}} (1 + 2^{1/6}\pi^{-1/2}\delta_{i0}^{-1/3}x_s) E_\mathrm{I} , \tag{39}$$

where $E_\mathrm{I} = e^4 m_{i0}/2\hbar^2\varkappa^2$ is the shallow impurity binding energy (for $\boldsymbol{B} = \boldsymbol{0}$) calculated for the longitudinal effective mass. As $\delta_{i0} \gg 1$, the binding energy $E_{i0} - E_{i000} < E_\mathrm{I}$. However, it follows from the quantum-limit condition and from (39) that $E_{i0} - E_{i000}$ is higher than the impurity binding energy calculated for the transverse effective mass. It is also interesting to observe that $E_{i0} - E_{i000}$ increases with magnetic field and depends only weakly on m_{i0}, as E_I/δ_{i0} is mass-independent.

4.2 $m \neq 0$

Substituting (12) into (6) we obtain an equation of the harmonic-oscillator type. Its eigenvalues and eigenfunctions are

$$\varepsilon_{iNms} = -\delta_{iN} V_{Nm}(0) + (2\delta_{iN}A_{Nm})^{1/2} (s + \tfrac{1}{2}) , \tag{40}$$

$$T_{iNms}(\xi) = \exp [-\tfrac{1}{2} (2\delta_{iN}A_{Nm})^{1/2} \xi^2] H_s[(2\delta_{iN}A_{Nm})^{1/4} \xi] C_{iNms} , \tag{41}$$

where $s = 0, 1, 2, \ldots$, $H_s(x)$ is the Hermite polynomial, and C_{iNms} is a normalization factor. T_{iNms} is symmetric for even s and antisymmetric for odd s. It follows from (15), (38), and (40) that $E_{i0} - E_{i0m0}$ is at least about twice lower than the binding energy $E_{i0} - E_{i000}$.

Acknowledgement

This work was supported by the Institute of Physics, Polish Academy of Sciences.

References

[1] Y. Yafet, R. W. Keyes, and E. N. Adams, J. Phys. Chem. Solids 1, 137 (1956).
[2] A. G. Zhilich and B. S. Monozon, Fiz. tverd. Tela 8, 3559 (1966).
[3] M. von Ortenberg, J. Phys. Chem. Solids 34, 397 (1973).
[4] S. D. Beneslavskii, N. B. Brandt, E. M. Golyamina, S. M. Chudinov, and G. D. Yakovlev, Zh. eksper. teor. Fiz., Pisma 19, 256 (1974).
[5] S. D. Beneslavskii and E. Entralgo, Zh. eksper. teor. Fiz. 68, 2271 (1975).

[2]) In the corresponding formula for $s = 0$ in [6] the coefficient $2^{1/6}\pi^{-1/2}x_0 \approx -0.65$ was erroneously replaced by -0.73.

[6] A. MYCIELSKI and J. MYCIELSKI, J. Phys. Soc. Japan **49**, Suppl. A 807 (1980).
[7] O. P. GUPTA, B. JOOS, and P. R. WALLACE, Phys. Rev. B **25**, 1101 (1982).
[8] O. AKIMOTO and H. HASEGAWA, J. Phys. Soc. Japan **22**, 181 (1967).
[9] M. ABRAMOWITZ and I. A. STEGUN (Ed.), Hdb. Mathematical Functions, Dover Publ., Inc., New York 1965.
[10] J. MYCIELSKI, G. BASTARD, and C. RIGAUX, Phys. Rev. B **16**, 1675 (1977).
[11] J. A. GAJ, J. GINTER, and R. R. GAŁĄZKA, phys. stat. sol. (b) **89**, 655 (1978).
[12] G. BASTARD, C. RIGAUX, Y. GULDNER, J. MYCIELSKI, and A. MYCIELSKI, J. Physique **39**, 87 (1978).
[13] M. JACZYŃSKI, J. KOSSUT, and R. R. GAŁĄZKA, phys. stat. sol. (b) **88**, 73 (1978).

(Received March 29, 1984)

phys. stat. sol. (b) **125**, 349 (1984)

Subject classification: 13.4 and 20.1; 22.8

Istituto di Chimica Fisica, Università di Parma[1])

The Electronic Spectrum of the CrO_4^{3-} Ion in the $Ca_2SiO_4 \cdot CaCl_2$ Host Lattice

By

R. Borromei

The polarised absorption spectrum of single crystals of $Ca_2SiO_4 \cdot CaCl_2$ containing chromium as CrO_4^{3-} or CrO_4^{4-} is measured at 4.2 K in the range 8000 to 46000 cm^{-1}. By a careful analysis based on a $T_d \rightarrow D_{2d}$ site symmetry perturbation and on the band polarisation behaviour, it is concluded that the observed spectrum is due to CrO_4^{3-}, the ground state is $^2E_{z^2}$ and the system of bands in the region 10000 to 20000 cm^{-1} corresponds to the ligand field transition. Moreover all the observed bands except that at 37730 cm^{-1} correspond to a $^2T_2$ excited state.

Le spectrum électronique polarisé de CrO_4^{3-} ou CrO_4^{4-} dans les crystals de $Ca_2SiO_4 \cdot CaCl_2$ a été mesuré à la température de 4,2 K dans la région 8000 à 46000 cm^{-1}. Avec une analyse précise sur le comportement déduit de la polarisation en introduisant la perturbation de site $T_d \rightarrow D_{2d}$ on peut conclure que le spectre observé est dû au CrO_4^{3-}, que l'état fondamental est $^2E_{z^2}$ et que toutes les bandes dans la région 10000 à 20000 cm^{-1} sont dues à une transition de "ligand field". Outre cela toutes les bandes, excepté celle à 37730 cm^{-1}, sont dues aux états excités $^2T_2$.

1. Introduction

It is already several years that we are interested in the electronic spectra of tetroxo-ions of transition metals [1 to 8].

In this research program we have tried to prepare single crystals of $Ca_2SiO_4 \cdot CaCl_2$ "doped" with CrO_4^{4-} since this ion could substitute the SiO_4^{4-} ions and in this way could be "stable" because of the large interaction with the host lattice. It is worth to point out that there is no spectrum of the CrO_4^{4-} ion reported in literature and very likely it is not yet known. Unfortunately in the present work we were unsuccessful. In fact from a comparison of the obtained absorption spectrum with those already known for CrO_4^{3-} in other host lattices and from a careful analysis of the expected behaviour of the CrO_4^{4-} and CrO_4^{3-} ions we show that in $Ca_2SiO_4 \cdot CaCl_2$ chromium is present as CrO_4^{3-}. However, we are able to give some interesting results. Particularly we give a complete interpretation of the spectrum by taking into account the site perturbation according to a scheme already used for other host lattices.

2. Experimental

Single crystals of $Ca_2SiO_4 \cdot CaCl_2$ doped with CrO_4^{3-} have been grown from a saturated solution of a flux of $CaCl_2$ by slow cooling. The best crystals, up to 15 mm in length, were grown from a starting mixture of composition 15 mol% Ca_2SiO_4, 85 mol% $CaCl_2 \cdot 2\,H_2O$, and 0.03 to 0.1 mol% Cr_2O_3.

The monoclinic crystals of green colour were separated easily from $CaCl_2$ by leaching with hot water. Their orientation was checked by rotating crystal X-ray diffractometry after optic-goniometry measurements had been made.

---

[1]) Via M. D'Azeglio 85, 43100 Parma, Italy.

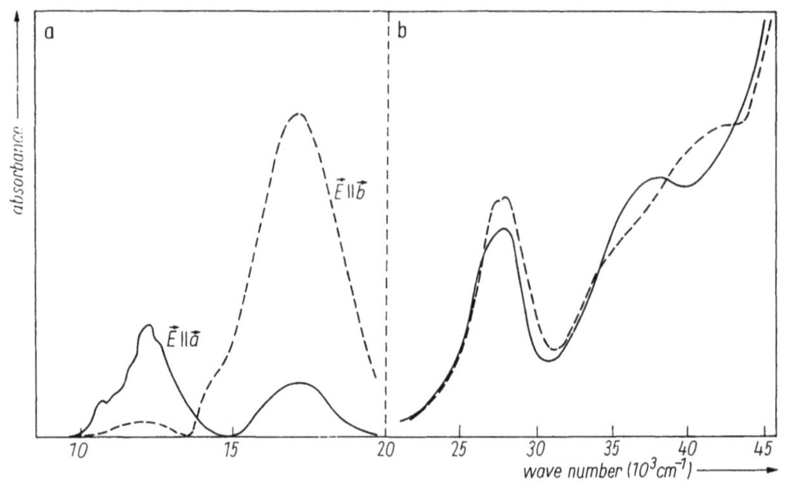

Fig. 1. Absorption spectra of CrO_4^{3-} ion in $Ca_2SiO_4 \cdot CaCl_2$ at 4.2 K with light incident normal to the face (001) and with $\boldsymbol{E} \parallel \boldsymbol{b}$ (– – –) and $\boldsymbol{E} \parallel \boldsymbol{a}$ (———); a) the ligand field band; b) the charge transfer bands

The spectra were measured at 4.2 K with polarized light incident orthogonal to the face (001) and with the electric vector $\boldsymbol{E}$ parallel to a and b crystallographic axes.

3. Description of the Spectrum

The spectrum obtained from our measurements is reported in Fig. 1.[2]) The description can be summarized as follows:

(a) In the range 10000 to 20000 cm^{-1} there is a system of two bands with maxima at 12200 cm^{-1} and 17290 cm^{-1} with an overall intensity weaker than that of the bands at higher energy. In particular the band at 12200 cm^{-1} shows the highest intensity when $\boldsymbol{E} \parallel \boldsymbol{a}$ and almost disappears when $\boldsymbol{E} \parallel \boldsymbol{b}$. Conversely the band at 17250 cm^{-1} is more intense when $\boldsymbol{E} \parallel \boldsymbol{b}$ and becomes rather weak when $\boldsymbol{E} \parallel \boldsymbol{a}$.

(b) In the region 20000 to 31000 cm^{-1} we have an intense broad band sharply separated from the others. The overall intensity is higher when $\boldsymbol{E} \parallel \boldsymbol{b}$ with respect to that when $\boldsymbol{E} \parallel \boldsymbol{a}$.

(c) Above 31000 cm^{-1} there is strong absorption and the spectrum cannot be resolved sharply into specific bands. However, at 37750 cm^{-1} with $\boldsymbol{E} \parallel \boldsymbol{a}$ and at 42550 cm^{-1} with $\boldsymbol{E} \parallel \boldsymbol{b}$ there are one maximum and one shoulder respectively which probably correspond to two bands.

4. Discussion

For the interpretation of the spectrum we have examined the hypotheses of the presence of the chromium either as CrO_4^{4-} or CrO_4^{3-} in SiO_4^{4-} crystal sites. In particular we have examined which of these species is more consistent with the main features of the spectrum when the site perturbation of the host lattice is properly taken into account. Of course in the case of CrO_4^{3-} we must assume also some charge compensation. Other hypotheses as Cr^{2+} or Cr^{3+} in Ca^{2+} sites have been ruled out by considering that the observed spectrum presents very intense bands immediately above 20000 cm^{-1} which certainly correspond to charge transfer (CT) transitions. Such bands are too

[2]) Parts a and b concern crystals with different chromium concentration.

low in energy to be ascribed to Cr^{2+} or Cr^{3+} ions as "CT bands". In fact in all cases reported in literature concerning such ions with oxygen and/or Cl^- ions in the first coordination sphere the CT bands appear above 32000 to 35000 cm^{-1}. Moreover both Cr^{2+} and Cr^{3+} should present several ligand field (LF) bands which cannot be recognized in the observed bands below 20000 cm^{-1}.

The compound $Ca_2SiO_4 \cdot CaCl_2$ crystallizes in the monoclinic system with four formula units per cell ($a = 0.9782 \pm 0.0012$, $b = 0.6738 \pm 0.0006$, $c = 1.0799 \pm \pm 0.0014$ nm, and $\beta = (106.01 \pm 0.09)°$). The space group is $P2_1/c$. A specific description of the crystal structure is reported in [9]. There are four equivalent SiO_4^{4-} tetrahedra per cell which in strict sense have no symmetry elements. However the Si–O bond lengths are rather close (0.1625, 0.1620, 0.1625, 0.1638 nm). In addition two of the O–Si–O bond angles are larger (116.6° and 115.8°) than the angle of a regular tetrahedron and concern two different oxygen pairs, whereas the other four bond angles are slightly smaller (105.6°, 107.1°, 108.4°, and 103.9°). Approximately each SiO_4^{4-} corresponds to a tetrahedron distorted by a compression along the z axis (i.e. $S_{4,z}$). It follows that the assumption of D_{2d} symmetry is a good approximation. The principal two-fold axis in D_{2d} symmetry makes angles of about 82°, 24°, and 114° with the crystallographic axes a, b, and c.

The ions CrO_4^{3-} and CrO_4^{4-} have the configurations d^1 and d^2. The possible transitions for these configurations in T_d symmetry have been widely discussed in previous papers [1 to 5, 8]. According to a molecular orbital description the following energy order can be assumed:

$$1t_2 < t_1 < e < 2t_2$$

where t_1 and, to a large extent, $1t_2$ are molecular orbitals localized on the oxygen atoms, whereas e and $2t_2$ are d-type molecular orbitals mainly localized on the metal. It follows that the d^1 and d^2 systems have the following properties:

4.1 d^1 systems

1. The ground state is ^2E with configuration $1t_2^6t_1^6e^12t_2^0$.
2. Only $^2T_1$ and $^2T_2$ excited states are accessible via transitions allowed by the electric dipole mechanisms.
3. Only one LF transition ($^2E \rightarrow {}^2T_2$) is possible corresponding to the excitation $e \rightarrow 2t_2$.
4. Four doublet excited states are possible for each of the CT transitions: $t_1 \rightarrow e$, $1t_2 \rightarrow e$. Two of them are $^2T_1$ and two are $^2T_2$. Moreover in connection with each of the excitations $t_1 \rightarrow 2t_2$ and $1t_2 \rightarrow 2t_2$ we have four excited states $^2T_1$ and four $^2T_2$ [1].
5. The first CT excited state corresponds to $^2T_2(t_1 \rightarrow e)$. Such state is rather lower in energy and sharply separated from the other CT excited states. This is because in this state the e^2 shell of the configuration $1t_2^6t_1^5e^22t_2^0$ is represented by a $^3A_2$ term which is particularly lower in energy than the other terms ^1E and $^1A_1$ [1, 8].

4.2 d^2 systems

1. The ground state is $^3A_2$ with configuration $1t_2^6t_1^6e^22t_2^0$.
2. Only transitions to $^3T_1$ states are allowed by the electric dipole mechanism.
3. Three LF excited states of the triplet are possible: $^3T_1(e^12t_2^1)$, $^3T_2(e^12t_2^1)$, and $^3T_1(e^02t_2^2)$. But in regular T_d symmetry and with negligible mixing of the $^3T_1(e^02t_2^2)$ state with other $^3T_1$ states only one of them corresponds to an "allowed" transition: $^3A_2(e^22t_2^0) \rightarrow {}^3T_1(e^12t_2^1)$.

Table 1

Comparison of the ranges of the bands of CrO_4^{3-} in various host matrices with those of the spectrum of $Ca_2SiO_4 \cdot CaCl_2$ doped with chromium (10^3 cm^{-1})

	Li_3PO_4	$Sr_5(PO_4)_3Cl$	Ca_2PO_4Cl	$Ca_2SiO_4 \cdot CaCl_2$
ligand field	10.0 to 16.0	10.0 to 12.0 12.0 to 17.0	10.0 to 14.0 14.0 to 21.0	10.0 to 14.5 14.0 to 20.5
1st charge transfer	22.0 to 32.0	23.0 to 31.0	24.0 to 32.0	22.0 to 31.0
2nd charge transfer	32.0 to 40.0	32.0 to 40.0	32.0 to 40.0	32.0 to 40.0

4. One $^3T_1$ and one $^3T_2$ CT excited states are possible in connection with each of the excitations $t_1 \to e$ and $1t_2 \to e$. With regard to the excitations $t_1 \to 2t_2$ and $1t_2 \to 2t_2$ a rather large number of triplet CT excited states is possible.

The spectrum of CrO_4^{3-} has been investigated in many host lattices [1, 4, 10 to 12]. As it has been pointed out in some previous papers [1, 4] the bands below 20000 cm^{-1} correspond to the components of the LF transition $^2E(e^1 2t_2^0) \to {}^2T_2(e^0 2t_2^1)$ which are split apart in different ways according to the site perturbation of the host matrix. Moreover the bands above 20000 cm^{-1} correspond to CT transitions which do not present any splitting because they are much less sensitive to the site perturbation. If we take into account these points and we make a comparison of the spectra of CrO_4^{3-} in the various matrices [1, 4, 10 to 12] with our spectrum we have a very good correspondence as it appears in Table 1. In particular both in $Ca_2SiO_4 \cdot CaCl_2$ and in the various host matrices of CrO_4^{3-} the first CT band is separated sharply from the other CT bands in agreement with what is expected from the previous considerations.

Unfortunately no spectrum of CrO_4^{4-} is known to make a similar comparison. However, by considering the isoelectronic ion MnO_4^{3-} as in [3, 5] one can easily deduce that the hypothesis of CrO_4^{4-} is very unlikely. In fact it is known that for the configurations d^0 and d^1 the first CT band of the chromate ions CrO_4^{n-} ($n = 2, 3$) is about 10000cm^{-1} higher in energy than the corresponding isoelectronic ions $MnO_4^{(n-1)-}$. Moreover the first CT band of MnO_4^{3-} appears at about 32000 cm^{-1}. Therefore, cer-

Table 2

Splitting of d^1 and d^2 terms with lowering of symmetry $T_d \to D_{2d}$

CrO_4^{3-} T_d	D_{2d}	CrO_4^{4-} T_d	D_{2d}
^2E	$^2A_1 = {}^2E_{z^2}$ $^2B_1 = {}^2E_{x^2-y^2}$	$^3A_2$	$^3B_1$
$^2T_2$	$^2B_2 = {}^2T_{2,xy}$ ^2E: $^2E_x = {}^2T_{2,yz}$; $^2E_y = {}^2T_{2,xz}$	$^3T_1$	$^3A_2 = {}^3T_{1z}$ ^3E: $^3E_x = {}^3T_{1x}$; $^3E_y = {}^3T_{1y}$
$^2T_1$	$^2A_2 = {}^2T_{1z}$ ^2E: $^2E_x = {}^2T_{1x}$; $^2E_y = {}^2T_{1y}$		

tainly the first CT band of CrO_4^{4-} has an energy higher than 32000 cm^{-1}. This is irreconcilable with the observed spectrum where the first CT band appears at 27000 cm^{-1}.

A confirm to the hypothesis of CrO_4^{3-} can be found by considering the splitting of the LF band system and the polarization behaviour of its components due to the site perturbation.

The splitting of the T_d states with the lowering of symmetry $T_d \to D_{2d}$ for CrO_4^{3-} and CrO_4^{4-} is summarized in Table 2. It is clear that in the case of CrO_4^{3-} we can have two different symmetries of the ground state, i.e. $^2A_1 = {}^2E_{z^2}$ or $^2B_1 = {}^2E_{x^2-y^2}$. The excited states $^2T_2$ are split into two components $^2B_2$ and ^2E and the $^2T_1$ excited states into $^2A_2$ and ^2E.

In the case of CrO_4^{4-} only the transition to the state $^3T_1(e^12t_2^1)$ is allowed. Such a state is split into the components $^3A_2$ and ^3E.

By applying the expression reported in [4, 3] which gives the oscillator strength relative to the unit cell as a function of the transition moment and of the direction of the electric vector of the incident light and taking into account the mutual dependence of the transition moments [1, 4, 5], we have obtained in the case of $Ca_2SiO_4 \cdot CaCl_2$ the expressions reported in Table 3. With m_1, m_2, and m_3 we denote the moments relative to the transitions $^2E_{z^2} \to {}^2T_{2z}$, $^2E_{x^2-y^2} \to {}^2T_{1z}$, and $^3A_2 \to {}^3T_{1z}$ when $\boldsymbol{E} \parallel z$. η_a, η_b, $\eta_{c'}$, are the direction cosines of the electric vector with respect to the crystallographic axes a and b and to a new axis c' orthogonal to a and b. Moreover α_z, β_z, γ_z are the direction cosines of the principal two-fold axis in D_{2d} symmetry (denoted by z) with respect to the axes a, b, and c': $\alpha_z = \cos 82°$, $\beta_z = \cos 24°$, $\gamma_z = \cos 112.47°$. In the same table we report also the expected intensity ratios both for the singular components and for all the ligand field band system when the light is incident orthogonally to the face (001), i.e. $\eta_{c'} = 0$, and the electric vector is parallel to the crystallographic axes a and b.

It is evident that the observed spectrum is consistent only with the case of CrO_4^{3-} with $^2E_{z^2}$ as ground state. In fact, as expected from Table 3 for the LF transitions we

Table 3
Oscillator strengths

transitions T_d	transitions D_{2d}	oscillator strength	intensity ratio $f_a : f_b$	intensity ratio overall intensity ratio
		CrO_4^{3-}		
$^2E_{z^2} \to {}^2T_2$	$^2A_1 \to {}^2B_2$	$f \sim m_1^2\omega^2$*)	0.032:0.835	0.277:0.876
	$^2A_1 \to {}^2E$	$f \sim \frac{1}{4} m_1^2(1-\omega^2)$	0.242:0.041	
$^2E_{z^2} \to {}^2T_1$	$^2A_1 \to {}^2A_2$	$f = 0$		
	$^2A_1 \to {}^2E$	$f \sim \frac{3}{4} m_2^2(1-\omega^2)$	0.726:0.124	
$^2E_{x^2-y^2} \to {}^2T_2$	$^2B_1 \to {}^2B_2$	$f = 0$		
	$^2B_1 \to {}^2E$	$f \sim \frac{3}{4} m_1^2(1-\omega^2)$	0.726:0.124	
$^2E_{x^2-y^2} \to {}^2T_1$	$^2B_1 \to {}^2A_2$	$f \sim m_2^2\omega^2$	0.032:0.835	0.277:0.876
	$^2B_1 \to {}^2E$	$f \sim \frac{1}{4} m_2^2(1-\omega^2)$	0.242:0.041	
		CrO_4^{4-}		
$^3A_2 \to {}^3T_1$	$^3B_1 \to {}^3A_2$	$f \sim m_3^2\omega^2$	0.032:0.835	1.000:1.000
	$^3B_1 \to {}^3E$	$f \sim m_3^2(1-\omega^2)$	0.968:0.165	

*) $\omega^2 = \alpha_z^2\eta_a^2 + \beta_z^2\eta_b^2 + \gamma_z^2\eta_{c'}^2 - 2\alpha_z\gamma_z\eta_a\eta_{c'}$.

observe two bands which behave in reversed manners: one is rathe rintense when $\boldsymbol{E} \parallel \boldsymbol{b}$ and very weak when $\boldsymbol{E} \parallel \boldsymbol{a}$ and the other is exactly the opposite. Moreover the overall intensity decreases to almost one third from $\boldsymbol{E} \parallel \boldsymbol{b}$ to $\boldsymbol{E} \parallel \boldsymbol{a}$.

With regard to the case of CrO_4^{3-} with $^2E_{x^2-y^2}$ as ground state there is a remarkable discrepancy. In fact from Table 3 we expect one single band with highest intensity when $\boldsymbol{E} \parallel \boldsymbol{a}$, whereas we observe two bands and the most intense of them appears when $\boldsymbol{E} \parallel \boldsymbol{b}$. The observed spectrum disagrees also with the expected behaviour of the case of CrO_4^{4-}. In fact according to Table 3 we should have two bands with intensities varying in complementary way, so that the overall intensity should be independent of the polarization of the light. Clearly this behaviour is not observed.

A further confirm to the hypothesis of CrO_4^{3-} is given by the intensity behaviour of the first CT band at 27770 cm^{-1}, which in CrO_4^{3-} must correspond to a $^2T_2$ excited state and in CrO_4^{4-} to a $^3T_1$. From Table 3 we have that in the case of CrO_4^{4-} the overall intensity must be independent of the polarization of the light. Conversely in the case of CrO_4^{3-} the intensity must vary with the polarization and particularly when the ground state is $^2E_{z^2}$ the intensity must decrease from $\boldsymbol{E} \parallel \boldsymbol{b}$ to $\boldsymbol{E} \parallel \boldsymbol{a}$ as it is really observed in the spectrum.

We have to point out that a ground state $^2E_{z^2}$ well agrees with the assumption that the CrO_4^{3-} ions undergo the same kind of distortion as the SiO_4^{4-} ions. In fact even from the simple LF theory we have that a compression of a tetrahedral coordination system along the z-axis (i.e. $S_{4,z}$) makes the e_{z^2} molecular orbital lower in energy than $e_{x^2-y^2}$. Moreover a ground state $^2E_{z^2}$ is in perfect agreement with previous results on the CrO_4^{3-} ion in Ca_2PO_4Cl [1, 10 to 12] where the distortion of the CrO_4^{3-} ions is rather similar to that in $Ca_2SiO_4 \cdot CaCl_2$ and a $^2E_{z^2}$ ground state has been proved by ESR measurements.

5. Conclusions

From all the above considerations these conclusions follow:

1. Chromium is present as CrO_4^{3-}. The other hypotheses, i.e. Cr^{2+}, Cr^{3+}, and CrO_4^{4-}, are inconsistent with the main features of the spectrum.

2. The CrO_4^{3-} ion undergoes a site perturbation which approximately corresponds to a lowering of symmetry $T_d \rightarrow D_{2d}$ and the ground state is $^2E_{z^2}$.

3. The bands of the spectrum below 20000 cm^{-1} correspond to the various components of the LF transition $^2E_{z^2} \rightarrow {}^2T_2$, i.e. $e_{z^2} \rightarrow 2t_2$. Particularly the $^2T_2$ excited state under the site perturbation is split into $^2B_2$ and ^2E. The band at 12200 cm^{-1} corresponds to the ^2E component and the band at 17250 cm^{-1} to $^2B_2$.

4. The bands above 20000 cm^{-1} arise from CT transitions. Such bands do not present any splitting due to the site perturbation. Particularly the bands at 27700 and at 42550 cm^{-1}, which have higher intensity when $\boldsymbol{E} \parallel \boldsymbol{b}$ with respect to $\boldsymbol{E} \parallel \boldsymbol{a}$, correspond to $^2T_2$ excited states. The band at 37700 cm^{-1} has an opposite behaviour and corresponds to a $^2T_1$ excited state.

Acknowledgement

I thank Prof. L. Oleari for helpful discussions and the CNR (Italian Research Council) for financial support.

References

[1] L. Di Sipio, L. Oleari, and P. Day, J. Chem. Soc. Faraday Trans. II 68, 1032 (1972).
[2] P. Day, L. Di Sipio, G. Ingletto, and L. Oleari, J. Chem. Soc. Dalton Trans. 2595 (1973).
[3] R. Borromei, L. Oleari, and P. Day, J. Chem. Soc. Faraday Trans. II 73, 135 (1977).

[4] R. BORROMEI, L. OLEARI, and P. DAY, J. Chem. Soc. Faraday Trans II **75**, 401 (1979).
[5] R. BORROMEI, L. OLEARI, and P. DAY, J. Chem. Soc. Faraday Trans. II **77**, 1563 (1981).
[6] R. BORROMEI and G. INGLETTO, Chem. Phys. Letters **81**, 62 (1981).
[7] R. BORROMEI, G. INGLETTO, and L. DI SIPIO, Chem. Phys. Letters **92**, 288 (1982).
[8] R. BORROMEI, L. OLEARI, and P. DAY, phys. stat. sol. (b) **124**, 707 (1984).
[9] V. R. CZAYA and G. BISSERT, Acta cryst. **B27**, 747 (1971).
[10] E. BANKS, M. GREENBLATT, and S. HOLT, J. chem. Phys. **49**, 1431 (1968).
[11] E. BANKS, M. GREENBLATT, and R. McGARVEY, J. chem. Phys. **47**, 3772 (1967).
[12] J. B. MILSTEIN, J. ACKERMAN, S. HOLT, and B. R. McGARVEY, Inorg. Chem. **11**, 1178 (1972).

(Received April 19, 1984)

phys. stat. sol. (b) 125, 357 (1984)

Subject classification: 13.4 and 20.3; 22.1.2

Laboratoire de Physique des Solides, associé au C.N.R.S., et Institute National des Sciences Appliques, Toulouse [1]) *(a) and Le Silicium Semiconducteur, Usine de Tours* [2]) *(b)*

The No-Phonon Luminescence at 0.793 eV from Si:Au

Uniaxial Stress and Zeeman Spectroscopy Experiments

By

D. THÉBAULT (a), J. BARRAU (a), G. ARMELLES (a), N. LAURET (a), and J. P. NOGUIER (b)

Uniaxial stress and Zeeman spectroscopy experiments on the no-phonon luminescence at 0.793 eV from Si:Au confirm the hypothesis of a transition from shallow "hydrogen-like" levels towards the deep donor level. The initial states are the split valley–orbit 1s: ($^2T_2$ + ^2E). The experimental results suggest a D_{2d} symmetry group or one of its subgroups.

Untersuchungen bei uniaxialer Spannung und Zeeman-Spektroskopie der Null-Phonon-Lumineszenz bei 0,793 eV von Si:Au bestätigen die Hypothese eines Übergangs von flachen „wasserstoffähnlichen" Niveaus zu einem Niveau eines tiefen Donators. Die Ausgangszustände sind die aufgespaltenen Tal–Bahn-1s-Zustände: ($^2T_2$ + ^2E). Die experimentellen Ergebnisse weisen auf eine D_{2d}-Symmetriegruppe oder auf eine ihrer Untergruppen hin.

1. Introduction

In a first paper (Mazzaschi et al. 1981) [1], we reported on a sharp line structure attributed to phonon-assisted radiative emission at a gold centre in silicon. The luminescence band was peaking at 0.78 eV and a zero-phonon line was identified at 0.793 eV.

In a second paper (Thébault et al. 1983) [2] we reported on the photoluminescence spectrum of the zero-phonon luminescence recorded at high resolution. Four lines were revealed and attributed to transitions from shallow "hydrogen-like" states towards the famous deep donor state at about 0.83 meV below the conduction band. From the experiment the final level is unsplit while the initial level contains four sublevels. We suggested that the initial states are 1s: $^2T_2$ or 1s: ($^2T_2$ + ^2E) split by the central-cell potential and the spin–orbit perturbation. The spin–orbit perturbation alone being not sufficient to induce the observed splitting we concluded that the defect symmetry is not tetrahedral but lower.

The present report relates the results of uniaxial stress and Zeeman spectroscopy of the 0.793 eV no-phonon emission. They confirm and contribute to precise the interpretation.

2. The No-Phonon Luminescence

In these experiments we have studied many different samples coming from different n-type ingots. A 80 nm gold layer was deposited by evaporation technique and the eutectic alloy with silicon was formed at 720 °C. Finally gold was diffused for 3 h at different temperatures (700 to 1200 °C).

[1]) Avenue de Rangueil, 31077 Toulouse-Cédex, France.

[2]) Rue M. et P. Curie, B.P. 0155, 37001 Tours-Cédex, France.

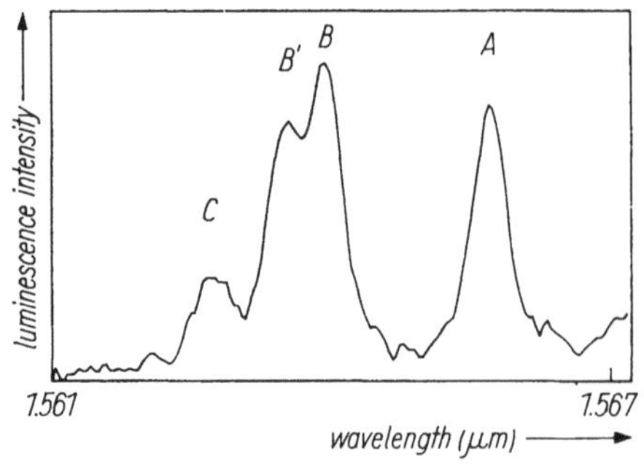

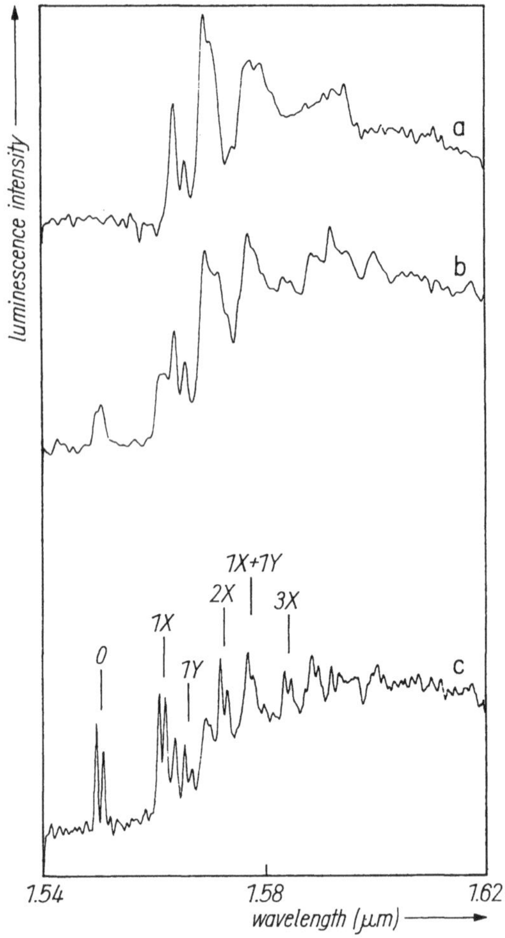

Fig. 1

Fig. 2

Fig. 1. The no-phonon photoluminescence spectrum associated with the 0.78 eV band from Si:Au at 4.2 K

Fig. 2. The photoluminescence spectra for three samples coming from the same n-type 100 Ωcm silicon ingot and diffused with gold during 3 h at (a) 1000, (b) 1050, (c) 1100 °C, respectively. As the temperature increases a new structure superimposes on the "low-temperature spectrum". This new structure, dominant at 1100 °C, is characterized by a pair or zero-phonon lines (0) and its first replica labelled 1X/1Y/2X/1X + 1Y/3X ...

We did not observe luminescence associated with gold for diffusion temperatures below 850 °C. Between 850 and 1000 °C we always observed the type of spectra previously reported, the intensity increasing with temperature. The quality of the associated no-phonon spectrum presented in Fig. 1 is better than in Thébault et al.. The B' line is now clearly resolved. This spectrum is the sum of nine spectra, each point is integrated for three seconds, the slit width is 300 μm providing a 0.12 meV resolution.

When the diffusion temperature increases above 1000 °C, a new structure appears in the spectrum. This is displayed in Fig. 2 for three samples coming from the same 100 Ω cm n-type ingot. This new structure appears as superimposed to the "low temperature spectrum". Above 1100 °C it becomes the dominant contribution. This new photoluminescence spectrum is characterized by a pair of zero-phonon lines (0) at 800 meV accompanied by a series of identified phonon relpica labelled 1X/ 1Y/ 2X/ 1X + 1Y/ 3X . . . and involving two types of phonons $X = 5.9$ meV and $Y = 8.3$ meV. We believe that this new spectrum is also related to gold. The possibility of a complex can be considered since contamination is always important in silicon after a high-temperature treatment. But for the moment we leave this complex for a further study.

For uniaxial stress and Zeeman spectroscopy experiments we worked on n-type [110], 20 Ω cm silicon wafers, doped with gold during 3 h at 1000 °C. These samples exhibit a pure "low-temperature" spectrum. Notice that our resolution during these experiments is not so good as for the spectrum shown in Fig. 1.

3. The Uniaxial Stress Spectroscopy

The zero-phonon luminescence spectra under [100] and [111] uniaxial stress are presented in Fig. 3 a and b.

We summarize in Fig. 4, 5, 6 the behaviour of the 0.793 eV no-phonon line under [100], [111], [110] uniaxial stress. The data points correspond to different samples. We have drawn in the insets of figures the theoretical splittings of 1s: ($^2T_2$ + ^2E) hydrogen-like states assuming coincidence of the $^2T_2$ and ^2E levels in the free crystal. The agreement between theory and experiment is naturally achieved by considering a zero splitting.

During the experiments under [100] uniaxial stress we observed the sudden quenching of the stress-induced hot luminescence when the splitting reached the intervalley

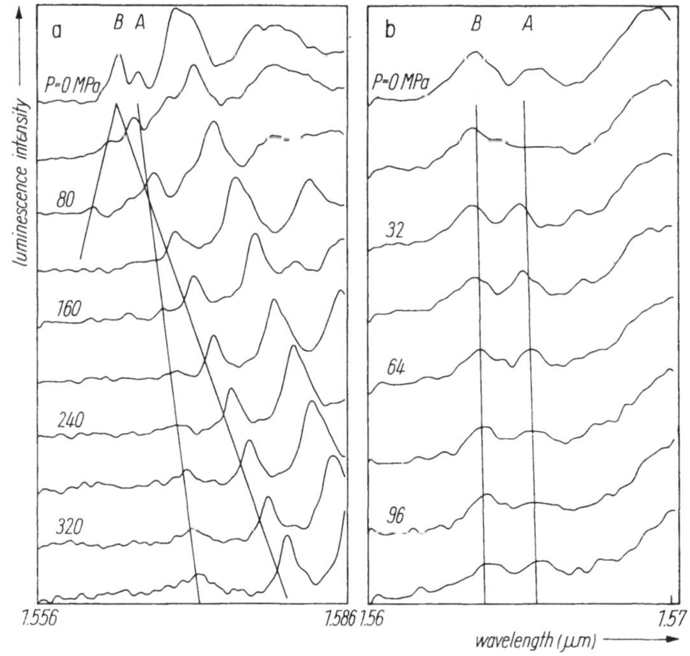

Fig. 3. The luminescence spectra under a) [100] and b) [111] uniaxial stress: The straight lines follow the zero-phonon splitting

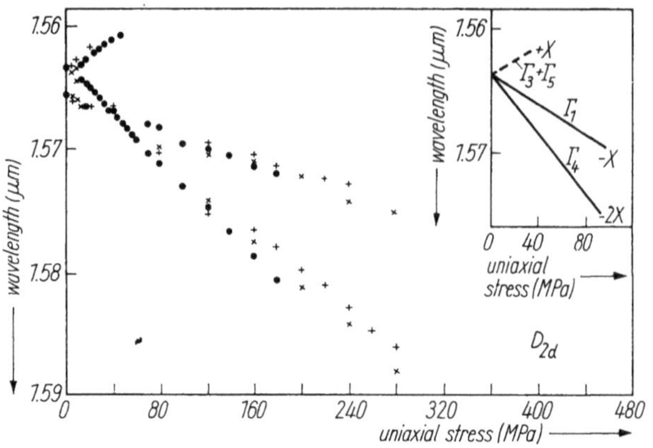

Fig. 4. Uniaxial stress data points for the no-phonon luminescence showing the splitting for stress (T) along [100] direction. The results, corresponding to three different samples, are reported (different types of points). The samples are immersed in liquid helium. The inset shows the theoretical splitting of the valley–orbit 1s: ($^2T_2 + {}^2E$) states for uniaxial stress along the [100] direction (the zero-stress splitting is neglected). The full lines correspond to cold luminescence and the dashed line to hot luminescence. $x = \frac{1}{3} T \Xi_\mu (s_{11} - s_{12})$, where s_{11} and s_{12} are the elastic compliance constants and Ξ_μ is the shear-deformation potential constant

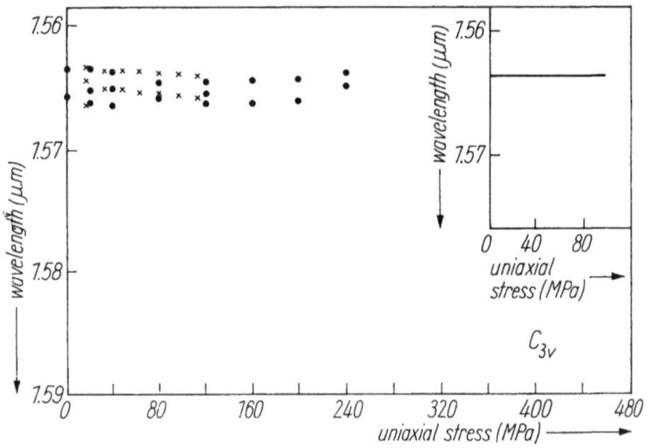

Fig. 5. Uniaxial stress data points for the no-phonon luminescence and the corresponding theoretical behaviour for the 1s: ($^2T_2 + {}^2E$) states when the stress is along the [111] direction. Same comments as for Fig. 4

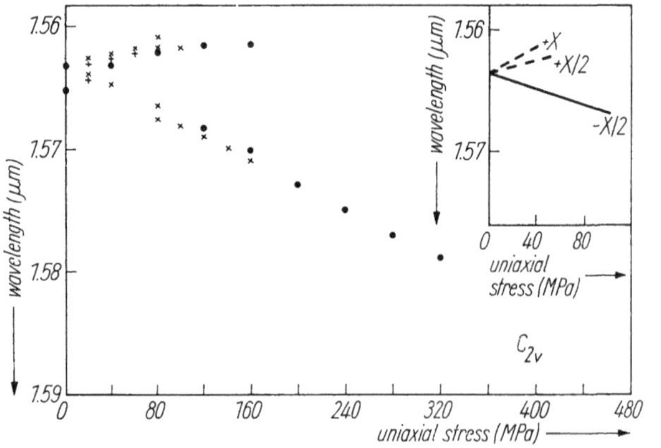

Fig. 6. Uniaxial stress data points for the no-phonon luminescence and the corresponding theoretcal splitting for the 1s: ($^2T_2 + {}^2E$) states when the stress is applied along the [110] direction. Same comments as for Fig. 4

TA phonon energy: this is a known signature of localized states composed of conduction-band wave functions. Further the comparison between experimental and theoretical splittings is very favourable: we deduce a shear-deformation potential constant of the 100 conduction-band minima $\Xi_\mu = (7.5 \pm 1)$ eV while Tekippe et al. [3]

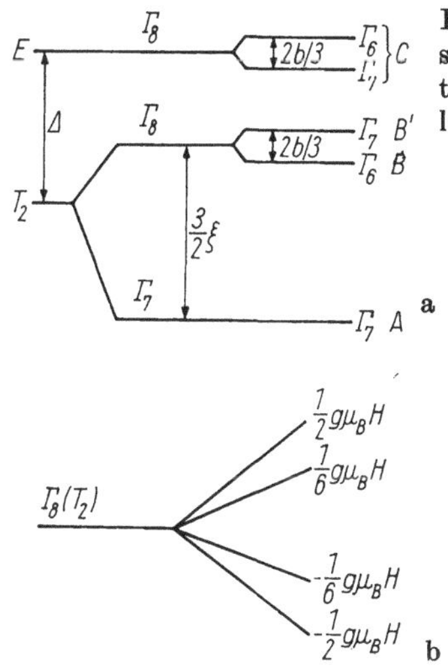

Fig. 7. a) The splitting scheme for the valley–orbit 1s: ($^2T_2$ + ^2E) states showing the effects of spin–orbit and tetragonal perturbtaions. b) The magnetic field splitting of the $\Gamma_8(T_2)$ sub-level (the B, B′ lines)

reported 8.72 eV for an average value deduced from a piezo-spectroscopic study of As, Sb, and P donors in silicon.

We observe no splitting under [111] uniaxial stress, which is the expected result. Moreover there is no shift of the spectrum. This proves that the final level of the transitions is pinned to the conduction band as the initial level. This result agrees with a previous observation by Wong et al. (1975) [4] concerning the deep donor level of the gold centre.

The experimental points for the [110] direction confirm the model but do not provide any new information. The main conclusion of the present stress experiment is that the initial electronic states of the luminecencc are really the valley–orbit 1s: ($^2T_2 + {}^2E$). As a result, we conclude that any interpretation of the zero-stress splitting has to take into account a central cell potential of symmetry lower than T_d in addition of the possible spin–orbit perturbation.

The spectra do not show ividence of splitting of the final level. However, such splitting cannot be definitively excluded at high stress where the insufficient quality of the recordings prevent us from a precise analysis. We can only state that if a splitting exists the stress splitting parameter relative to the final level is much smaller than this one relative to the initial level.

Polarization measurements are difficult because the line intensity is weak: the only very clear result is a *total* polarization perpendicular to the stress of the low-energy line induced by [100] stress, (initial state $\Gamma_4(z)$, Fig. 4). This information is very important for the model. Now two possible final states of the luminescence at zero-stress are consistent with the previous hypothesis of a defect symmetry lower than T_d:

(i) an electronic orbital doublet belonging to $\Gamma_5(D_{2d})$, the wave functions transforming like yz, zx for [001] centres and xy, yz for [010] or [100] centres; (ii) an electronic orbital singlet generated by an additional reduction of symmetry of $\Gamma_5(D_{2d})$.

Since we have not detected the effect of orientational degeneracy we suggest that the spin–orbit perturbation is stronger than the tetragonal field perturbation. The corresponding splitting scheme is shown in Fig. 7 for the excited states of the defect in the free crystal. It agrees with our luminescence spectrum, Fig. 1, taking

the tetragonal parameter $\quad b = 0.03$ meV ,

the spin–orbit parameter $\xi = 0.66$ meV ,

the valley–orbit parameter $\Delta = 0.8$ meV ,

and considering that the broader line C is really an unresolved doublet. The values of ξ and Δ are in agreement with the known tendencies reported by Grimmeiss et al. [5, 6] for group V and VI donors in silicon.

4. The Zeeman Spectroscopy

The luminescence being very weak we will consider only the behaviour of the two main lines A and (BB′).

In Fig. 8 the variation of the Zeeman spectrum is displayed versus the magnetic field up to 6 T for $\boldsymbol{H} \parallel [100]$. We observe the splitting of the (BB′) line into two components, one is shifted to lower energy, the other is practically unshifted. Line A splits into several components which cannot be separated.

The splittings in the [111] and [110] magnetic field directions are different: here we observe the splitting of (BB′) into three components while A broadens as the magnetic field intensity increases. Fig. 9 reports the spectra recorded at 4 T for the three important directions. The observed splitting patterns of the (BB′) line are summarized on Fig. 10 (dots).

The present Zeeman study is crude, due to the very weak luminescence intensity. However, it is possible to attempt to interpret the observed pattern in Fig. 10 from the conclusions reached in Section 3. The D_{2d} symmetry is consistent with the fact that the finest Zeeman spectrum occurs for $\boldsymbol{H} \parallel [111]$: this point is confirmed when

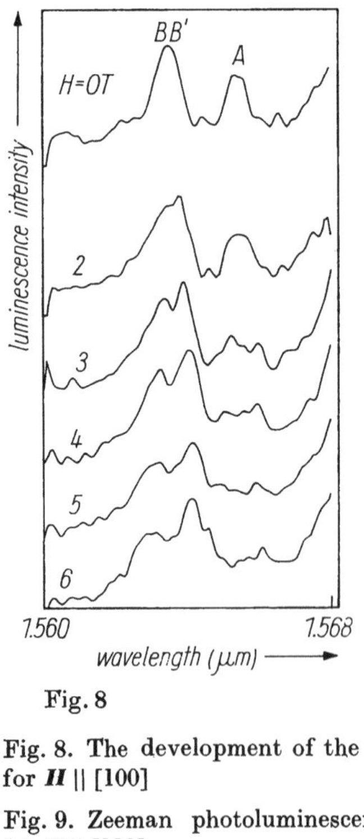

Fig. 8

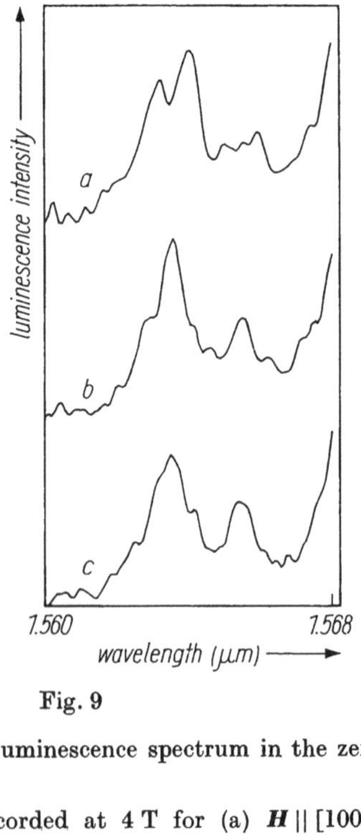

Fig. 9

Fig. 8. The development of the Zeeman photoluminescence spectrum in the zero-phonon region for $\boldsymbol{H} \parallel [100]$

Fig. 9. Zeeman photoluminescence spectra recorded at 4 T for (a) $\boldsymbol{H} \parallel [100]$, (b) $\boldsymbol{H} \parallel [111]$, (c) $\boldsymbol{H} \parallel [110]$

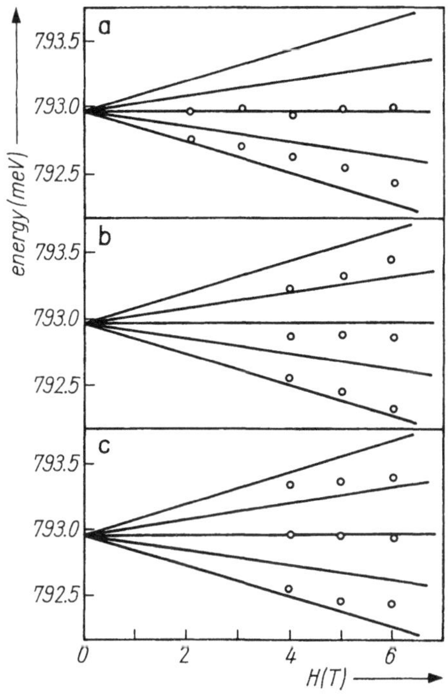

Fig. 10. Splitting patterns of the (BB′) line a) $\boldsymbol{H} \parallel [100]$, b) $\boldsymbol{H} \parallel [111]$, c) $\boldsymbol{H} \parallel [110]$. The dots are experimental and the full curves are theoretical

the magnetic field is rotated away from the [100] direction in the [1$\bar{1}$0] plane. Returning to the splitting scheme (Fig. 7), the effect of the magnetic field on the (BB′) line is shown in the insot: as B and B′ remain unresolved in the Zeeman analyse we neglect here the tetragonal field and describe the splitting from the first-order effect of the effective Hamiltonian $g\mu_B \boldsymbol{HS}$ on $\Gamma_8(T_2)$. (The magnetic splitting remains always small compared to the spin–orbit splitting $\Gamma_8(T_2)/\Gamma_7(T_2)$.) The required hypothesis for the ground state, which leads likely to the interpretation of the patterns in Fig. 10, lies on a spin doublet $S = 1/2$ which splits into $\pm (1/2)\, g\mu_B H$. On this basis, the theoretical splitting of the BB′ line has been represented by the full lines, Fig. 10 ($g = 2$). The experimental data points approximatively fall on the theoretical lines while it can be only verified that the broadened experimental band coming from BB′ spreads well over the five theoretical lines at all magnetic field and for each direction.

5. Conclusion

Uniaxial stress and Zeeman spectroscopies on the no-phonon luminescence at 0.793 eV from Si:Au confirm the preceding hypothesis of a transition from shallow hydrogen-like states towards the deep donor level. Further we arrive to the conclusion that the ground state wave functions belong to the $\Gamma_5(D_{2d})$ representation.

This result can be considered as an argument for a gold atom embedded in a vacancy-like defect. According to Watkins [7] the structure predicted for "Au^0" in silicon is a $5d^{10}$ shell with the remaining electron in a vacancy-like t_2 deep level ($5d^{10} + V^-$). Here V^0 (D_{2d} symmetry) becomes V^- (C_{2v} symmetry) by capture of an electron into a $\Gamma_2(C_{2v})$ ground state, a result which is in complete agreement with the conclusion above. However, more investigations must be carried out to reach definitive conclusions.

References

[1] J. MAZZASCHI, J. C. BRABANT, B. BROUSSEAU, J. BARRAU, M. BROUSSEAU, F. VOILLOT, and P. BACUVIER, Solid State Commun. **39**, 1091 (1981).
[2] D. THÉBAULT, J. BARRAU, M. BROUSSEAU, DO XUAN THANH, J. C. BRABANT, F. VOILLOT, and A. RIBAULT, Solid State Commun. **45**, 645 (1983).
[3] V. J. TEKIPPE, H. R. CHANDRASEKHAR, P. FISHER, and A. K. RAMDAS, Phys. Rev. B **6**, 2348 (1972).
[4] D. C. WONG and C. M. PENCHINA, Phys. Rev. B **12**, 5840 (1975).
[5] H. G. GRIMMEISS, E. JANZEN, and K. LANSSON, Phys. Rev. B **25**, 2627 (1982).
[6] E. JANZEN and H. G. GRIMMEISS, J. Phys. C **15**, 5791 (1982).
[7] G. D. WATKINS, Proc. 16th Internat. Conf. Phys. Semicond., Ed. M. AVEROUS, Montpellier 1982 (p. 9).

(Received March 2, 1984)

phys. stat. sol. (b) **125**, 365 (1984)

Subject classification: 20.1 and 20.3; 13.4; 22.8.2

Institute of Physics, Academy of Sciences of the Belorussian SSR, Minsk[1])

Two-Photon Luminescence Excitation Spectra of Eu^{3+} in Yttrium Aluminium Garnet

By

L. E. KHOLODENKOV and A. G. MAKHANEK

The results are reported of the first study of the $^5D_2$ manifold of Eu^{3+} in the noninversion and inversion centres of YAG using the method of two-photon luminescence excitation. Based on the number of Stark components and the two-photon polarization dependences it is shown that the set of Stark components of ^5D multiplet corresponds to tetragonal local symmetry. The components of $^5D_2$ manifold are identified and the crystal field parameters are found for the noninversion centre. The europium ion distribution over the centres is determined showing that at least 60% of the ions are in the inversion centre.

Впервые представлены результаты исследования методом двухфотонного возбуждения люминесценции $^5D_2$ уровня в инверсионном и неинверсионном центрах YAG. На основе числа наблюдаемых компонент и двухфотонных поляризационных зависимостей показано, что совокупность штарковских компонент ^5D мультиплета соответствует тетрагональной локальной симметрии. Проведена идентификация компонент $^5D_2$ уровня и найдены параметры кристаллического поля в случае неинверсионного центра. Проведена оценка распределения ионов европия по центрам, показавшая, что не менее 60% ионов находятся в инверсионном центре.

1. Introduction

The purpose of this paper is to study some peculiarities of two-photon luminescence excitation (TPLE) of Stark components in Eu^{3+}:YAG and to employ them for the solution of spectroscopic problems associated with the yttrium aluminum garnet. The expediency of choosing Eu^{3+} as a spectroscopic object was noted in [1] and will also be made clear from the following. In recent years the TR^{3+}:YAG system (where TR^{3+} is a rare-earth activator) has been the subject of much investigation as a promising laser material. Therefore, it is important to obtain information providing a more complete and adequate description of spectroscopic properties of such a system. The choice of europium as an activator with all peculiarities of its optical spectrum does not restrict the study of the main properties of TR^{3+} (at least those with effective radius close to europium in the oxygen environment) entering into the garnet lattice and the manner of their distribution over the lattice sites and can, in a number of cases, make some disputable questions much easier.

As follows from some works [2, 3], the question of a proper choice of the symmetry type to describe the splitting of manifolds belonging to different multiplets and occupying essentially different energy positions still remains to be solved. Thus, for instance, in describing the manifold Stark splitting of the Eu^{3+} ^7F multiplet in the garnet lattice, orthorhombic symmetry D_2 is used along with the tetragonal symmetry S_4. In this respect the question arises of the principal possibility to use one universal

[1]) Leninskii Prospekt 70, 220602 Minsk, USSR.

symmetry type and the corresponding set of crystal-field parameters to describe adequately the manifold Stark splitting of different multiplets for a given impurity-crystal pair.

The X-ray analysis data [4] indicate that the yttrium and aluminum cations in the garnet lattice occupy positions with the coordination number (c.n.) of the nearest oxygen environment 8 and 6, 4, respectively. The data of the same work on interionic distances show that the ions of the TR^{3+} impurity can replace, besides yttrium ions, aluminium ions in the position with c.n. 6 and 4. It follows from [5] that the local symmetry of the positions with c.n. 6 and 4 can be C_{3i} and S_4. The results of [6] show that erbium ions replace aluminium ions in octahedric positions. When aluminium ions are replaced by impurity ions the octahedron can be distorted, but not as a rule, causing a change in the initial local symmetry. Therefore, the investigation of a local centre by the TPLE method along with the OPLE method is of interest.

As applied to the material just outlined, the present paper gives the results of TPLE and OPLE spectra of possible centres and their angular dependences in linearly polarized light. Based on the selection rules for TPLE, analytical angular expressions, and experimental data, an analysis is performed of the optical centres formed when Eu^{3+} enters the yttrium aluminium garnet.

2. Experimental

The experiment was performed at 77 K with a Eu^{3+}:YAG sample (the crystal was obtained from a high-temperature melt at the Institute of Crystallography of the USSR Academy of Sciences) which contained 0.5 wt% Eu^{3+}. The TPLE was performed with a single laser beam [1]; the linearly polarized beam was rotated in the limits of 0° to 180 ° by means of two pairs of crystalline quartz wedges. All investigations were carried out at the $^7F_0 \rightarrow {}^5D_2$ transition with luminescence registration at the magnetic dipole $^5D_0 \rightarrow {}^7F_1$ transition.

According to [7], the optical centres in the garnet must be oriented in six and eight spatially nonequivalent directions. Due to this disordering of the centre orientation with respect to the crystallographic axes of the garnet cubic lattice, the experimentally boserved angular dependences of the TPLE spectrum are the result of the superposition of nonequivalent centres. Therefore, the use of a specified set of crystallographic directions of the sample with respect to the laboratory system of coordinates is sometimes not justified. In the present work, measurements were made on a sample with arbitrary orientation of axes. The TPLE spectrum components are within the frequency range 21190 to 21540 cm^{-1}. The correlation of the TPLE components with the centres involved was determined by the spectrum correspondence of the excitation to the luminescence one. First, the entire TPLE spectrum was recorded under the conditions $\Delta\lambda = 20$ nm, $\lambda = 592$ nm. Then, the entire luminescence spectrum was recorded within the range 590 to 630 nm with $\Delta\lambda = 1.5$ nm under selective TPLE of individual components which were identified by checking for the absence of either excitation or luminescence spectrum components. The procedure was facilitated by the fact that one of the centres involved was the inversion.

3. Theoretical Grounds

In interpreting the spectra of TR^{3+} excitation in crystals one should use those values which are less effected by unwanted factors. Among these factors are such as different types of energy migration between the centre in question and another centre or the lattice. In practice, all these energy migration processes are selective in energy relaxation rates for Stark components of a given manifold; they mask the immediate effect of the crystal field of specified symmetry on the manifold in question. For this reason, the distribution of the relative intensities in the observed spectrum often does not agree with the calculated distribution. The most reliable interpretation in this respect is that based on the study of the number of observed components and their polarization dependences, provided it is exactly known that all components belong to the same optical centre. Based on this approach, we shall briefly outline the necessary theoretical premises as applied to transitions from $^7F_0$ to other manifolds of Eu^{3+}.

It is known that the character of the splitting of a manifold of a selected multiplet and the intensity distributions between the Stark components in the electric dipole transition are determined by the set of even and odd parameters B_{tp}, respectively, of the crystal field series expansion. For the symmetry types S_4 and D_2 we are interested in, these sets are of the following form [8], p. 246:

$$\begin{aligned} S_4&: \begin{array}{l} B_{20}\ B_{40}\ B_{44}^{+}\ B_{44}^{-}\ \dots \\ B_{32}^{+}\ B_{32}^{-}\,, \end{array} \\ D_2&: \begin{array}{l} B_{20}\ B_{22}^{+}\ B_{40}\ B_{42}^{+}\ B_{44}^{+}\ \dots \\ B_{32}^{-}\ \dots\,. \end{array} \end{aligned} \tag{1}$$

Presented here are only those parameters which must be taken into account at the transitions for which $J = 0, 1 \rightarrow J' = 2$. As follows from (1), the set of parameters S_4 is in many ways similar to that of D_2. This similarity must show itself in the number of observed Stark components in a one-photon electric dipole transition as given in Table 1. In both cases the spectrum consists of three components, i.e. the spectra of the crystal field of symmetries S_4 and D_2 are similar in this respect. The difference in the number of observed components can show up itself only in the two-photon transition (TPT) spectrum. This is due to the fact that according to the selection rules for TPT, the number of allowed components exactly corresponds to that of split components in a field of specified symmetry [8], p. 288. It is equal to 4 and 5 for the symmetries S_4 and D_2, respectively, as given in Table 1. However, this approach has a disadvantage in studying similar types of low symmetry, for instance, C_2 and C_s, where there are five components in OPT and TPT spectra. This disadvantage can be overcome by using the polarization dependences of these spectra.

Table 1

OPLE and TPLE Stark spectra depending on the symmetries S_4 and D_2 for transition $J, M = 0 \rightarrow J' = 2$, $M' = 0, \pm 1, \pm 2$. Quantum states given in brief form as Dirac symbols

S_4	$\lvert 2^-\rangle$, (B″)	$\lvert 2^+\rangle$, (B′)	$\lvert 0\rangle$, (A)	$\lvert \pm 1\rangle$, (E)	
OPLE	+	+	−	+	
TPLE	+	+	+	+	
D_2	$\lvert 2^-\rangle$, (B_1)	$\lvert 0, 2^+\rangle$, (A′)	$\lvert 0, 2^+\rangle$, (A″)	$\lvert 1^+\rangle$, (B_2)	$\lvert 1^-\rangle$, (B_3)
OPLE	+	−	−	+	+
TPLE	+	+	+	+	+

In a number of cases, it is possible to uniquely identify the Stark components from the polarization dependences according to the quantum states determined by the given symmetry type and calculate, on this basis, the even parameters B_{tp} of the crystal field. On the other hand, it is possible to determine the symmetry type of a given optical centre by selecting analytical expressions corresponding to a definite symmetry for the observed polarization dependences. We shall make use of the analytical expressions given in [9] for the two-photon angular dependences of Eu^{3+} in tetragonal and rhombic symmetry fields.[2]) They are listed in Table 2 together with angular expressions for OPT.

Table 2

The OPLE and TPLE angular parts according to Table 1. The expressions are given for the case of one quantization axis

S_4	B″	B′	A	E	
OPLE	$\cos^2\theta$	$\cos^2\theta$	—	$\sin^2\theta$	
TPLE	$\frac{6}{5}\sin^4\theta\sin 2\varphi$	$\frac{6}{5}\sin^4\theta\cos^2 2\varphi$	$\frac{2}{5}(2\cos^2\theta-\sin^2\theta)$	$\frac{6}{5}\sin^2 2\theta$	
D_2	B_1	A′	A″	B_2	B_3
OPLE	$\cos^2\theta$	—	—	$\sin^2\theta\cos^2\varphi$	$\sin^2\theta\sin^2\varphi$
TPLE	$\frac{6}{5}\sin^4\theta\sin^2 2\varphi$	$(aP_1+bP_2)^2$*)	$(-bP_1+aP_2)^2$	$\frac{6}{5}\sin^2 2_2\sin^2\varphi$	$\frac{6}{5}\sin^2 2\theta\cos^2\varphi$

*) $a^2+b^2=1$, $P_1=(\frac{6}{5})^{1/2}\sin^2\theta\cos 2\varphi$, $P_2=(\frac{2}{5})^{1/2}(2\cos^2\theta-\sin^2\theta)$.

As follows from Table 2, with properly selected pairs of angles θ, φ it is possible to obtain OPT and TPT spectra which make possible a unique assignment of the spectrum to one of the symmetries given in Table 2 by the number of nonzero components. In practice, however, one has often to deal with a combination of centres oriented in different manners as in the case of the garnet lattice. Then, there remains the possibility of analysis by correlating the analytical expressions with the experimentally observed angular dependences. This will be shown in the following section.

4. Noninversion Centre

In the ranges 21050 to 21880 cm^{-1} and 10525 to 10940 cm^{-1}, OPLE and TPLE spectra, respectively, have been obtained. Both spectra are shown in Fig. 1. The OPLE spectrum consists of three components, which is in agreement with both symmetries involved. The TPLE spectrum consists of four components, which is indicative of symmetry S_4. For these spectra, the polarization dependences shown in Fig. 2, 3 have also been obtained. As might be expected, the minima in the spectra do not reach zero values. On the other hand, two-photon diagrams have clearly defined

[2]) In [9, 11] (based on [12, 13]) the relative TPT probabilities are calculated and tabulated taking into account angular expressions for the majority of TR^{3+} transitions in crystal fields of all symmetries. In [14], similar polarization characteristics are tabulated for four-photon parametric processes. The theoretical and experimental results obtained independently in [15 to 17] confirm convincingly these tabulated polarization expressions and the idea of the necessity of taking into account the scalar part of the TPT tensor for $\Delta J=0$ suggested by us in [8]. The details of calculations of the scalar part of the TPT tensor given in [14] have shown that there is an error in [16] lowering the TPT line intensities by a factor of four. This error was later pointed out in [17].

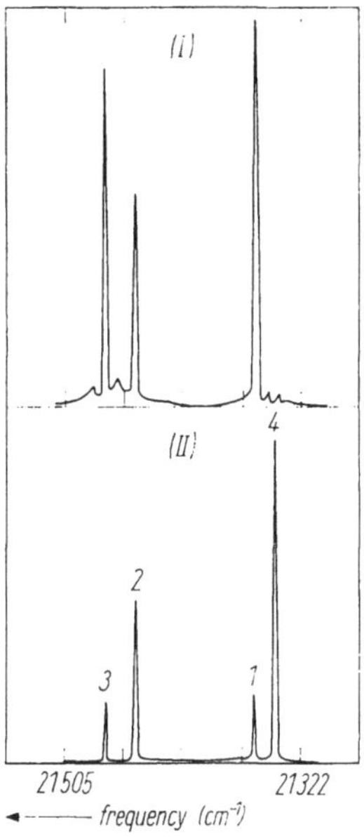

Fig. 1. The OPLE (I) and TPLE (II) spectra of the noninversion centre. The luminescence intensities normalized to exciting laser pulse energy (I/Q_L) (I) and to the squared one (I/Q_L^2) (II) are in arbitrary units along the ordinate axis

maxima and minima near some angles of laser beam polarization. The value of α between the two maxima of curves 3, 2, 1 is approximately 90°, while that for curve 4 between the three maxima is approximately 60°. This is indicative of some spatial ordering of the nonequivalent optical centres. It may primarily be due to the small degree of disorientation of the centres located along each of the directions [001], [010], [100], which makes further utilization of the analytical expressions of Table 2 more valid.

According to Table 2, component 4 in Fig. 1 can most easily be identified, it may be classified as the state A (symmetry S_4). Thus, the identification amounts now to the analysis of components 3, 2, 1. Whatever the orientation choice of the garnet lattice axes, to correlate polarization curves 3, 2, 1 with analytical expressions one simple and common feature of the latter may be used: the similar or opposite behaviour of their angular parts as a function of θ or φ. Opposite behaviour means that for some θ or φ for which the cosine modulus has a maximum, the sine modulus will have a minimum. As follows from Table 2, the states B′ and B″ are similar for symmetry S_4 at OPT and opposite to the state E as to their angular parts, whereas for TPT they behave oppositely. This condition is satisfied for curves 1, 2, and 3. As for symmetry D_2, B_2 and B_3 behave oppositely both at OPT and TPT and they could be assigned to the pair of curves 2 and 3. However, as follows from Fig. 3, the correspondence of extrema is not so good for them as for curves 2 and 1. It is impossible for curves 2, 1 to the up their one-to-one correspondence with the states B′, B″ on the basis of given experimental conditions. This can be done by using data obtained from the splitting in the field of rhombic symmetry where the states B′ and A get mixed with the formation of a pair of states A′,″ whose polarization dependences, as follows from Table 2, must have similar behaviour. In the manner described above in this symmetry one can identify some components, in particular, the component of the state B_1. If a change in symmetry is not followed by a change in position of components, the state B_1 position found will correspond to that of B″ in symmetry S_4. In this respect it is interesting to compare the given splitting of the $^5D_2$ manifold with that in the field of tetragonal and rhombic symmetries obtained from calculation for a given manifold employing the parameters used to describe the $^7F_2$ manifold splitting. We make use of the parameters A_{tp} defined in [10] which are of the following form: $A_{20} = 268$, $A_{22} = 51$, $A_{42} = 100$, $A_{40} = -190$, $A_{44} = 950$. Fig. 4 shows the splitting scheme in these fields. The field parameters A_{tp} obtained from scheme 4b are $A_{20} = 80$, $A_{40} = -383$, $A_{44} = \pm 1300$. A comparison between schemes 4b and 4c shows that the relative positions of the components remain the same with a high degree of accuracy,

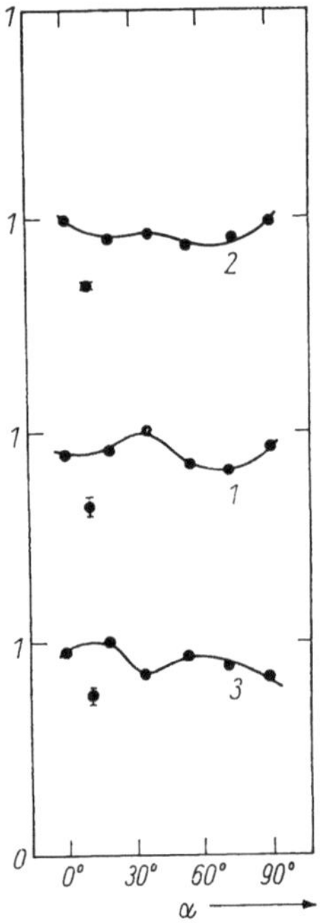

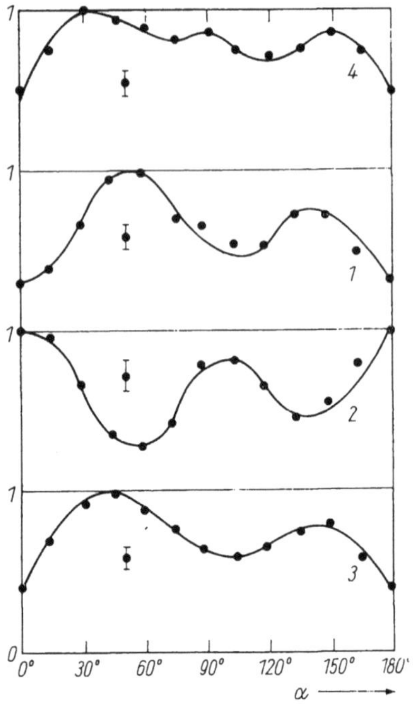

Fig. 2 Fig. 3

Fig. 2. Polarization diagrams (I/Q_L) of the OPLE spectrum of the noninversion centre

Fig. 3. Polarization diagrams (I/Q_L^2) of the TPLE spectrum of the noninversion centre. Here, as in Fig. 2, the relative changes normalized on unit of the component intensities are given. α is the angle of laser beam polarization rotation in the laboratory system

but the general splitting changes almost by a factor of two and A and B interchange their positions. Due to the fact that A_{22} and A_{42} are small as compared to other A_{tp}, the splitting in scheme 4c naturally changes to that of scheme 4d. This corresponds to a slight distortion of the Thomson cube (position with c.n. 8) to a dodecahedron, which is in agreement with the slight difference between the interionic distances d_{88} and d_{68}, d_8 and d_{80} [4]. Taking into account the interchange of the positions of states A and B″, it may be concluded from scheme 4d that B_1 corresponds to B″ in scheme 4b. However, this procedure would stricter if the components of the $^7F_2$ manifold were identified by their two-photon polarization dependence.

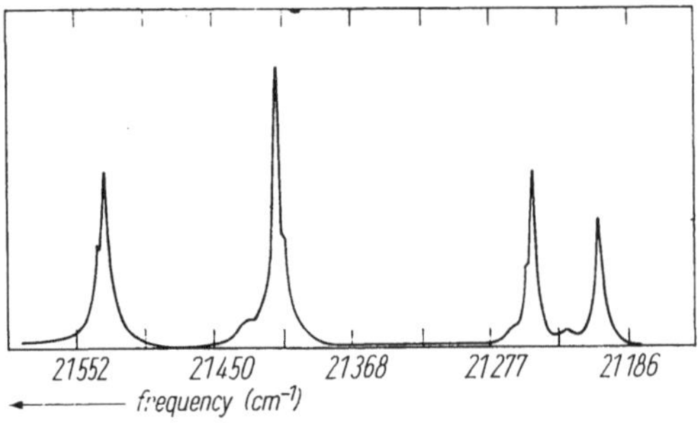

Fig. 4. The TPLE spectrum of the inversion centre. The luminescence intensities normalized to the squared exciting laser pulse energy (I/Q_L^2) are given in arbitrary units

5. Inversion Centre

The TPLE spectrum is given in Fig. 5. The inversion centre components are broadened as compared to those of the noninversion centre. The luminescence spectrum of the transition $^5D_0 \rightarrow {^7F_1}$ consists of three groups of inhomogeneously broadened lines with group centres at 604, 611, 618 nm. When each of the four observed components of the TPLE spectrum is excited the proportion of the intensities in the luminescence spectrum is almost preserved, being on the average 0.35:0.30:1.00 with increasing wavelength. Polarization measurements of the TPLE spectrum have indicated that the relative intensity changes for all components do not exceed 10 to 15% and the positions of extrema are not ordered in the polarization diagrams.

The number of observed components is indicative of the presence of a tetragonal symmetry and the grouping of components is characteristic of the nearest surrounding with c.n. not less than 6 [8, p. 284]. It follows that the europium ions enter the octahedral sites of the garnet lattice. In the presence of a distortion along the fourth-order axis of the octahedron the local symmetry will be D_{4h}. The centres in octahedral position are disposed in four directions [111] and lie within the manifold of planes {110} and {112} [4]. Proceeding from the above, there must be a high degree of disorientation of the centres, which is confirmed by the character of the polarization dependences observed.

It has also been found that there is energy migration between two given centres. Thus, when the components of the $^5D_2$ manifold of the noninversion centre are excited, selective energy transfer to the group of levels in the region of 611 nm of the inversion centre is observed, maximum transfer occurring under excitation of component 2. In this case the ratio between the brightest line intensity in the $^5D_0 \rightarrow {^7F_1}$ spectrum for the noninversion centre and that for the inversion centre is 2.5. Due to this, energy transfer occurs most likely in the luminescence stage between two optical centres. It is noteworthy that the inverse migration was not observed.

Making use of the ratio between the integrated TPLE intensities of both centres it is possible to evaluate the concentration distribution of europium ions over these centres. The investigation of the europium luminescence spectrum in YAG at 77 K has shown that all luminescence is practically concentrated at the transitions $^5D_0 \rightarrow {^7F_J}$ ($J = 1$ to 6), the most intense luminescence being observed between the components of the transition $^5D_0 \rightarrow {^7F_1}$ [2].

Let us find the expression for the integrated TPLE intensity under excitation of the component j of the $^5D_2$ manifold, the integrated luminescence spectrum of the $^5D_0 \rightarrow {^7F_1}$ transition being recorded. The balance equation for the concentration of particles on the $^5D_0$

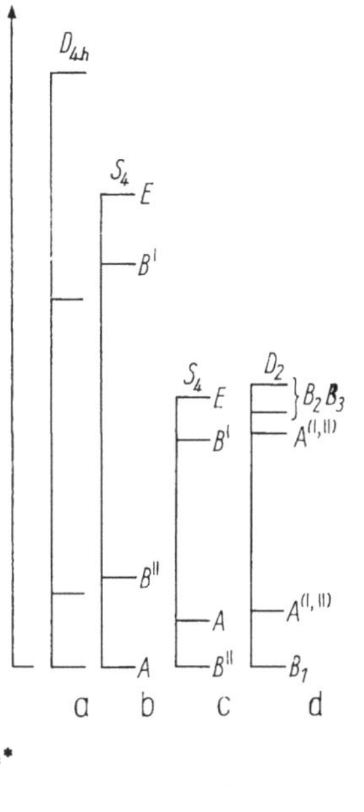

Fig. 5. $^5D_2$ manifold splitting; a) in inversion centre; b) in noninversion centre (experimental); c), d) calculated on the basis of the parameters from [10]. The ratio of the scale of a) to that of b), c), d) is 1:2

level is of the form

$$\dot{n}(t) = S_j n_j(t) - \sum_k (A_k + S_k)\, n(t)\,, \tag{2}$$

where S_j is the probability of nonradiative energy transfer from the component j to the $^5D_0$ level; A_k and S_k are the probabilities of radiative and nonradiative energy transfer, respectively, from the $^5D_0$ level to the components of the ^7F_J manifold; $n_j(t)$ is the concentration of particles on the component j. The relaxation times $\tau_j = 1/S_j$. for the manifolds of the ^5D multiplet are usually within the limits of 10^{-6} to 10^{-3} s and the time of two-photon absorption under the conditions of our experiment is limited by 3×10^{-8} s. Therefore, it may be assumed to a sufficient accuracy that

$$n_j(t) = n_j(t_0) \exp(-S_j t)\,,$$

where

$$n_j(t_0) = \sum_l N_l \delta_{jl}^{(2)}(\theta_l, \varphi_l)\, F_L(t_0)\,, \qquad F_L(t_0) = \int_0^{t_0} I_L^2(t')\, dt'\,,$$

$I_L(t')$ is the laser pulse intensity; $\delta_{jl}^{(2)}$ is the two-photon absorption efficiency; N_l is the particle concentration in the l-th position of a given centre.

Then, the solution of (2) will be

$$n(t) = n_j(t_0)\, F(\tau, \tau_j)\,,$$

where

$$F(\tau, \tau_j) = \frac{\tau}{\tau - \tau_j}\left[\exp\left(\frac{-t}{\tau}\right) - \exp\left(\frac{-t}{\tau_j}\right)\right], \qquad \tau = \left[\sum_k (A_k + S_k)\right]^{-1},$$

τ being the $^5D_0$ level lifetime.

The integrated TPLE intensity in terms of the number of quanta is

$$I(\text{TPLE}) = \text{const} \sum_{k',j} A_{k'} n_j(t_0)\, F(\tau, \tau_j)\,.$$

k' is the index of summation over the $^7F_1$ manifold components. Since τ_j is sufficiently small as compared to τ, $F(\tau, \tau_j)$ is weakly dependent on τ_j and this function may be removed from the sign of summation over j by taking a certain mean value $\tau_j = \tilde{\tau}$. Then, taking into account the equiprobable europium ion distribution over the spatially nonequivalent positions of the given centre we can write

$$\sum_l N_l \delta_l(\theta_l, \varphi_l) = N_0 \delta_{\text{int}}^{(2)}\,,$$

where N_0 is the particle concentration in the given centre; $\delta_{\text{int}}^{(2)}$ is the integrated two-photon absorption efficiency depending only upon the choice of the quantum transition [8, p. 288]. As follows from Table 2, this efficiency for the case of one quantization axis is independent of angular variables. Then,

$$I(\text{TPLE}) = \text{const} \left(\sum_{k'} A_{k'}\right) N_0 F_L(t_0)\, F(\tau, \tilde{\tau})\, \delta_{\text{int}}^{(2)}\,.$$

The radiative probabilities of electric dipole transitions being equal to zero in the inversion centre, there is a slight increase in τ (15 to 20% on the average) which must be taken into account in changing $F(\tau, \tilde{\tau})$. By summing the TPLE intensities over curves 3, 2, 1, 4 in Fig. 3 we can make sure that $\delta_{\text{int}}^{(2)}$ is anisotropic. Therefore, we use the integrated TPLE value corresponding to the maximum of $\delta_{\text{int}}^{(2)}$.

Taking into account all these facts and the presence of energy transfer from one centre to another under known conditions of excitation and recording with respect to the integrated TPLE intensities normalized to the squared laser pulse energy, we have found the ratio of europium ion concentration in the inversion centre to that of the noninversion centre to approximately 0.6.

6. Conclusion

The results obtained for the two centres considered in this paper speak in favour of the choice of the manifolds of the ^5D multiplet, the tetragonal type of symmetry rather than the rhombic one by which the manifold splitting of the low-lying ^7F multiplet is described. This is confirmed by the $^5D_1$ and $^7F_1$ manifold splitting in the noninversion centre. For $^5D_1$ the number of observed components is two, whereas for $^7F_1$ it is three [2]. The presence of three inhomogeneously broadened components of the $^7F_1$ manifold of the inversion centre indicates that the manifolds of the ^7F multiplet of this centre in a first approximation (neglecting inhomogeneous splitting) are described by rhombic symmetry possessing now an inversion centre (symmetry D_{2h}). All this points to the fact that the crystal field symmetry within the 4f shell is some function of the energy position of the split manifold. At the same time the values of the crystal field parameters change. Thus, when moving from the ^7F_J to the ^5D_J manifold the fourth-order parameters increase and the second-order parameters decrease.

The polarization dependences obtained for the noninversion centre in the presence of some disorientation and the number of observed TPLE components (except for the observed anisotropy of $\delta_{\rm int}^{(2)}$) are in agreement with the theory and can sometimes be a good means for the determination of the symmetry type and identification of the Stark splitting components.

The evaluation of the distribution of europium ions over the centres has shown that despite the different conditions of aluminium and europium ion incorporation, the latter effectively enter the octahedral positions of the garnet matrix during its growth from the high-temperature melt. This fact is worth noticing since an increase in TR^{3+} ion concentration in the inversion centre decreases the laser generation efficiency in the TR^{3+}:YAG system.

Acknowledgement

The authors wish to thank Dr. Kh. S. Bagdasarov for supplying the crystal sample used in these investigations.

References

[1] L. E. Kholodenkov and A. G. Makhanek, phys. stat. sol. (b) **112**, K 149 (1982).
[2] J. A. Koningstein, Phys. Rev. **136**, A 717 (1964).
[3] Yu. K. Voronko and A. A. Sobol, Trudy Fiz. Inst. Akad. Nauk SSSR **98**, 41 (1977).
[4] F. Euler and J. A. Bruce, Acta cryst. **19**, 971 (1965).
[5] S. C. Abrahams and S. Geller, Acta cryst. **11**, 437 (1958).
[6] M. N. Ashurov, Yu. K. Voronko, V. V. Osiko, and A. A. Sobol, in: Collection „Spektrosko, piya kristallov", Ed. P. P. Feofilov, Izd. Nauka i Tekhnika, Leningrad 1978.
[7] S. Geller and M. A. Gilleo, J. Phys. Chem. Solids **3**, 30 (1957).
[8] A. G. Makhanek and V. S. Korolkov, Analiticheskie metody v kvantomekhanicheskoi teorii vozmushchenii, Izd. Nauka i Tekhnika, Minsk 1982.
[9] L. A. Jugurian, Preprint No. 232, Inst. Fiz., Akad. Nauk BSSR, Minsk 1980.
[10] J. A. Koningstein, J. chem. Phys. **42**, 3195 (1965).
[11] L. A. Jugurian, Preprint No. 233, Inst. Fiz., Akad. Nauk BSSR, Minsk 1980.

[12] P. A. APANASEVICH, R. I. GINTOFT, V. S. KOROLKOV, A. G. MAKHANEK, and G. A. SKRIPKO, phys. stat. sol. (b) **58**, 745 (1973).
[13] A. G. MAKHANEK and G. A. SKRIPKO, phys. stat. sol. (a) **53**, 243 (1979).
[14] L. A. JUGURIAN and A. G. MAKHANEK, Preprint No. 301, Inst. Fiz., Akad. Nauk BSSR, Minsk 1983.
[15] M. DAGENAIS, M. DOWNER, R. NEUMANN, and N. BLOEMBERGEN, Phys. Rev. Letters **46**, 561 (1981).
[16] B. R. JUDD and O. R. POOLER, J. Phys. C **15**, 59 (1982).
[17] M. C. DOWNER and A. BIVAS, Phys. Rev. B **28**, 3677 (1983).

(Received March 15, 1984)

phys. stat. sol. (b) **125**, 375 (1984)

Subject classification: 4; 1.2; 22

Sektion Physik der Karl-Marx-Universität Leipzig[1])

Charge Ordering in the Anti-Th_3P_4 Structure

By

B. Lorenz

The existence of charge ordering in rare-earth compounds crystallizing in the anti-Th_3P_4 structure is investigated. On the basis of Madelung energy calculations it is shown that for instance in Eu_4As_3 the electrostatic interactions stabilize a charge ordered low temperature phase. The thermodynamics and the transition to the disordered phase are described by a lattice gas model. The transition is found to be of first order. The transition temperature estimated for Eu_4As_3 lies well above room temperature. The theoretical results are compatible with recent experiments, although the transition itself has to be confirmed by high temperature measurements.

Das Problem der Ladungsordnung wird in Verbindungen der seltenen Erden untersucht, die in der Anti-Th_3P_4-Struktur kristallisieren. Die Berechnung der Madelungenergie zeigt, daß beispielsweise in Eu_4As_3 eine ladungsgeordnete Tieftemperaturphase durch die elektrostatische Wechselwirkung stabilisiert wird. Die Thermodynamik und der Übergang zur ungeordneten Phase werden mit Hilfe eines Gittergasmodells beschrieben und ein Phasenübergang erster Art wird erhalten. Die für Eu_4As_3 berechnete Übergangstemperatur liegt oberhalb der Zimmertemperatur. Die theoretischen Ergebnisse sind mit neueren Experimenten verträglich, die Existenz eines Phasenübergangs muß jedoch noch durch Hochtemperaturmessungen bestätigt werden.

1. Introduction

In recent years the mixed valence (MV) compounds have attracted increasing experimental and theoretical attention. In particular, the inhomogeneous MV systems are of special interest because of cation ordering due to the electrostatic interaction and the existence of an order–disorder transition (Verwey transition) which is generally connected with a jump of the dc conductivity. These compounds are characterized by the presence of cations in different valence states on equivalent crystallographic sites and the absence of hybridization of localized and band states at the Fermi energy. Verwey transitions have been observed, for instance in Fe_3O_4 [1], Ti_4O_7 [2], and Eu_3S_4 [3]. While the transitions in Fe_3O_4 and Ti_4O_7 have been investigated extensively ([4, 5] and references therein), the transition in the Th_3P_4 structure of Eu_3S_4 originally suggested by Carter in 1972 [6] was confirmed experimentally only very recently [3]. Theoretical investigations show that in the materials quoted above the Coulomb interactions of cations with different valency play a major role [4, 6]. For instance, in Eu_3S_4 a unique ordered structure below the transition temperature ($T_V = 186$ K) is stabilized by the electrostatic interaction. The ordering is accompanied by a small rhombohedral lattice distortion [6, 7].

In view of the fact that Verwey transitions are observed in very different materials it is interesting to investigate other MV compounds. Recently, Mössbauer and magnetic susceptibility measurements in Eu_4As_3 have revealed the presence of divalent and trivalent Eu ions in the ratio of 3 : 1 [8] corresponding to the ionic formula $Eu_3^{2+}Eu^{3+}As_3$. Furthermore, a rhombohedral distortion similar to that in Eu_3S_4 below T_V is observed

[1]) Karl-Marx-Platz, DDR-7010 Leipzig, GDR.

in Eu_4As_3 [9]. On the basis of the experimental results and in comparison with the similar situation in Eu_3S_4 it was suggested that Eu_4As_3 will undergo a Verwey transition at sufficiently high temperatures [8]. However, the transition to the cubic (disordered) phase could not be detected below room temperature and high temperature studies have not yet been performed.

This paper is devoted to the problem of charge ordering in Eu_4As_3 or, more generally, in the anti-Th_3P_4 structure. In Section 2 we investigate the stability of charge order due to Coulomb interactions among divalent and trivalent cations by calculating the Madelung energies for different ionic configurations. The order–disorder transition is described by a lattice gas model in the mean-field approximation (Section 3). In Section 4 we discuss the results and estimate the transition temperature for Eu_4As_3.

2. Stability of Charge Order

Eu_4As_3 is one of the rare-earth compounds which crystallize in the cubic anti-Th_3P_4 (Gd_4Bi_3) structure [10] where the cations occupy the crystallographically equivalent P-(Gd-)sites. The b.c.c. lattice of Th_3P_4 is built up on a basis of two formula units per primitive cell [11]. The eight phosphorous sites closest to a thorium form the octaverticon shown in Fig. 1. In the following we will consider only the lattice of P-sites occupied by the excess electrons of the divalent cations. A given site is surrounded by three nearest, two next-nearest, and six third-nearest neighbour sites [12]. The whole lattice is divided into four sublattices consisting of lines parallel to the four directions [1, 1, 1], [$\bar{1}$, 1, 1], [1, $\bar{1}$, 1], and [1, 1, $\bar{1}$] marked in Fig. 1. The ground state configuration is determined by that distribution of the $\frac{3}{4}N$ electrons over the N lattice sites which minimizes the electrostatic energy.

Considering the Coulomb interactions up to sixth-nearest neighbours we obtain the most stable configuration given by an alternative arrangement of charges along the lines of the two sublattices (1 and 2) shown in Fig. 2. The remaining two sublattices (3 and 4) are fully occupied. This charge ordering minimizes the Coulomb energy among nearest and next-nearest neighbours and is stable with respect to a transfer of electrons from sublattices 3 and 4 to sublattices 1 or 2. This result is confirmed by

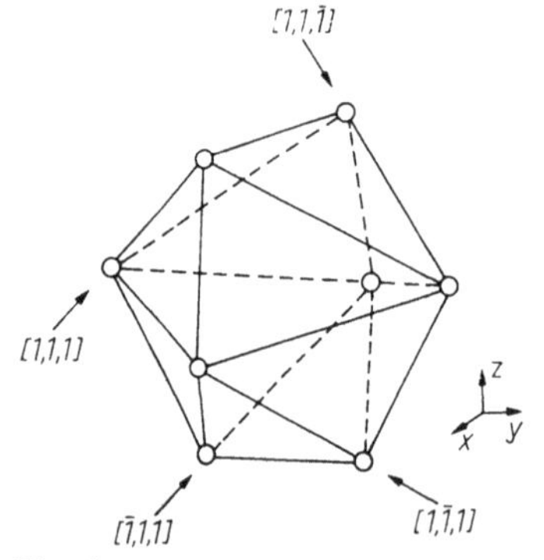

Fig. 1

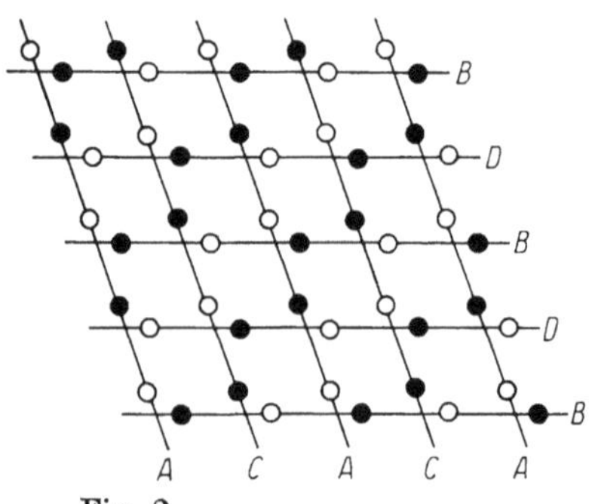

Fig. 2

Fig. 1. Basis of eight P ions closest to a Th ion forming an octaverticon in the anti-Th_3P_4 structure

Fig. 2. Electron ordering scheme for Eu_4As_3 as obtained from Madelung calculations. The two fully occupied sublattices are not shown. Four subsequent layers are denoted by A, B, C, and D. Lines A, C and lines B, D belong to the first and second sublattice, respectively. Full circle: occupied site (Eu^{2+} ion), empty circle: unoccupied site (Eu^{3+} ion)

Madelung energy calculations for various kinds of charge order. With $a_0 = 0.9176$ nm (edge length of the cubic unit cell) for Eu_4As_3 the energy of the ordered ground state (Fig. 2) with respect to the completely disordered state (Madelung energy) is obtained as

$$\frac{\Delta E}{N} = 0.42 \frac{\text{eV}}{\varepsilon} = 0.021 \text{ eV} . \tag{1}$$

In (1) we have assumed a static dielectric constant of $\varepsilon \approx 20$. Obviously, the ordered ground state is sixfold degenerate.

3. Verwey Transition

For the description of thermodynamic properties and the order–disorder transition we use the nearest neighbour lattice gas model given by

$$H = \frac{U}{2} \sum_{\langle i,j \rangle} n_i n_j . \tag{2}$$

n_i is the occupation number operator for an electron at the site i. The thermodynamic average $\langle n_i \rangle$ is the probability of finding an Eu^{2+} ion at the site i. The sum in (2) is over all nearest neighbour pairs of the lattice described in the previous section. The Coulomb integral U is considered as a renormalized parameter including the effects of long-range interaction, screening, and lattice polarization.

The thermodynamics of (2) is calculated in the mean field approximation where the two-particle operator is substituted by

$$n_i n_j \approx n_i \langle n_j \rangle + \langle n_i \rangle n_j - \langle n_i \rangle \langle n_j \rangle . \tag{3}$$

In the following we denote the occupied and empty sites by A and B, respectively. The charge order parameter is defined as the charge difference between A- and B-sites

$$m = n_\text{A} - n_\text{B} . \tag{4}$$

The total electron number is conserved by the condition

$$3 = 3n_\text{A} + n_\text{B} . \tag{5}$$

In (4) and (5) we have used the notation $n_{\text{A(B)}} = \langle n_i \rangle$, $i \in \text{A(B)}$. With (3) to (5) the Hamiltonian (2) is written as

$$H = \frac{U}{4} \Big(\sum_{i \in \text{A}} (9 - m) \, n_i + \sum_{j \in \text{B}} (9 + 3m) \, n_j \Big) - \frac{3}{32} \, UN(9 - m^2) . \tag{6}$$

From the mean field model (6) the charge densities n_A, n_B are obtained as

$$n_\text{A} = f\left(\frac{U}{4}(9 - m) - \mu\right), \qquad n_\text{B} = f\left(\frac{U}{4}(9 + 3m) - \mu\right), \tag{7}$$

where $f(E) = \big(1 + \exp(\beta E)\big)^{-1}$ is the Fermi function and $\beta = 1/k_\text{B}T$ denotes the inverse temperature. The chemical potential μ and the charge order parameter m are determined self-consistently by (4) and (5)

$$3f\left(\frac{U}{4}(9 - m) - \mu\right) + f\left(\frac{U}{4}(9 + 3m) - \mu\right) = 3 ,$$

$$f\left(\frac{U}{4}(9 - m) - \mu\right) - f\left(\frac{U}{4}(9 + 3m) - \mu\right) = m . \tag{8}$$

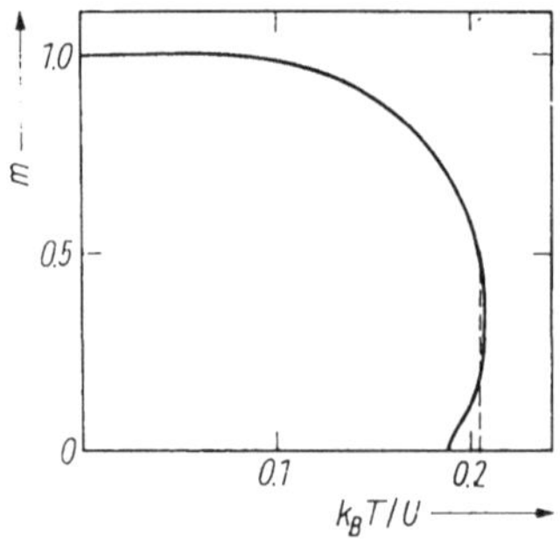

Fig. 3. Temperatur dependence of the charge order parameter. The first-order transition point is marked by the dashed line

Solving the coupled system of equations (8) we get

$$\mu = \frac{3}{4} U(3 + m) + \frac{1}{\beta} \ln \left\{\frac{3(1 - m)}{1 + 3m}\right\} \tag{9}$$

and the implicit expression for the order parameter

$$U\beta = -\frac{1}{m} \ln\left\{\frac{3(1 - m)^2}{(3 + m)(1 + 3m)}\right\}. \tag{10}$$

The order parameter as function of temperature is shown in Fig. 3. For $k_BT > 0.207U$ there is only the trivial solution $m = 0$. In the range $3/16U < k_BT < 0.207U$ two nontrivial solutions exist. Obviously, the transition to the disordered phase is a first-order transition and the free energies of the different phases have to be compared in order to determine the transition point. The free energy per site is given by

$$\frac{F}{N} = -\frac{1}{\beta N} \ln \operatorname{Tr} e^{-\beta(H - \mu\hat{N})} + \frac{3}{4}\mu ,$$

where $\hat{N}$ is the particle number operator. Performing the trace and inserting (9) we get

$$\frac{F}{N} = \frac{3}{32} U(3 + m)^2 - \frac{1}{4\beta} \ln \left\{\frac{1 + 3m}{4}\right\} - \frac{3}{4\beta} \ln \left\{\frac{1 + 3m}{3(1 - m)} + e^{\beta U m}\right\}. \tag{11}$$

The transition temperature is found to be

$$k_BT_c = 0.205U . \tag{12}$$

At T_c the order parameter jumps from $m(T_c) = 0.48$ to zero (dashed line in Fig. 3).

4. Discussion and Conclusions

For Eu_4As_3 the parameter U is estimated with the help of the Madelung energy (1). Due to electron ordering the ground state energy is lowered with respect to the energy of the disordered phase by $\Delta E/N = (3/32)\, U$, and comparing with (1) we get $U \approx 0.22$ eV. From (12) the transition temperature is found to be $T_c \approx 520$ K. This temperature might be reduced by a few per cent due to correlation effects as in the Verwey transition in Eu_3S_4 [7]. In the mean field theory of Section 3 correlations among the electrons have been neglected. However, the estimated T_c is of a reasonable magnitude and lies well above room temperature in accordance with experiments [8, 9]. High-temperature studies in the transition region are of considerable interest and may reveal the order–disorder transition. From our theoretical investigation we expect a first-order phase transition. Furthermore, neutron diffraction experiments in the ordered phase should give some information on the charge order. The results can be compared with the scheme proposed in Section 2 (Fig. 2).

Note that we have not considered explicitly the interaction of the electrons with the lattice. In Eu_3S_4 and probably in Fe_3O_4 small polarons are formed [4, 7] and the Coulomb interaction is renormalized. Whether small polaron formation takes place in Eu_4As_3 is not yet clear. We have roughly incorporated this effect in using a static dielectric constant.

In summary, we have shown that in Eu_4As_3 the electrostatic interaction can stabilize a charge-ordered low-temperature phase. The estimated transition temperature is compatible with experimental results, but the transition has yet to be confirmed by high temperature studies. It is also interesting to consider other materials with the anti-Th_3P_4 structure. For instance, Yb_4Bi_3 is postulated to have a similar ionic distribution as Eu_4As_3 [10] and should be investigated with respect to a distortion from the cubic structure and the existence of a Verwey transition.

References

[1] E. J. W. Verwey and P. W. Haayman, Physica (Utrecht) 8, 979 (1941).

[2] M. Marezio, D. B. McWhan, P. D. Dernier, and J. P. Remeika, Phys. Rev. Letters 28, 1390 (1972).

[3] R. Pott, G. Güntherodt, W. Wichelhaus, M. Ohl, and H. Bach, Phys. Rev. B 27, 359 (1983).

[4] N. F. Mott, Festkörperprobleme 19, 331 (1979); Phil. Mag. B 42, 327 (1980).

[5] C. Schlenker and M. Marezio, Phil. Mag. B 42, 453 (1980).

[6] F. L. Carter, J. Solid State Chem. 5, 300 (1972).

[7] D. Ihle and B. Lorenz, phys. stat. sol. (b) 116, 539 (1983).

[8] R. Nagarajan, E. V. Sampathkumaran, R. Vijayaraghavan, and Bhaktdarshan, phys. stat. sol. (a) 75, K149 (1983).

[9] F. Hullinger, Mater. Res. Bull. 14, 33 (1979).

[10] R. J. Gambino, J. less-common Metals 12, 344 (1967).

[11] K. Meisel, Z. anorg. allg. Chem. 240, 300 (1939).

[12] F. Holtzberg and S. Methfessel, J. appl. Phys 37, 1433 (1966).

(Received May 25, 1984)

phys. stat. sol. (b) **125**, 381 (1984)

Subject classification: 20.3; 10.2; 22.5.2

Institute for Low Temperature and Structure Research, Polish Academy of Sciences, Wrocław[1])

Low-Temperature Photoluminescence of the Aggregate Centres in Eu^{2+}-Doped KCl Crystals

By

E. MUGEŃSKI and R. CYWIŃSKI

The emission spectra and fluorescence lifetimes of different aggregation products of Eu^{2+} dipoles in KCl matrix are examined in the range 15 to 300 K. Two aggregation types are formed in dependence on the annealing temperature. The optical properties of the aggregates (i) and (ii) formed at RT and 125 °C are very similar to those characteristic of isolated dipoles. Above 125 °C both aggregates undergo decomposition. A third aggregation product is formed between 200 and 250 °C. Differently from (i) and (ii) the pertinent absorption and emission spectra do not show any vibronic structure. The effect of the above aggregation products on the fluorescence lifetime of isolated dipoles present in crystals is ascertained.

Die Emissionsspektren und Fluoreszenzlebensdauern von verschiedenen Aggregationsprodukten der Eu^{2+}-Dipole in der KCl-Matrix werden im Temperaturbereich von 15 bis 300 K untersucht. Abhängig von der Temperungstemperatur werden zwei verschiedene Aggregationstypen festgestellt. Die optischen Eigenschaften der Aggregate (i) und (ii), die bei Zimmertemperatur bzw. bei 125 °C entstehen, sind den Eigenschaften der isolierten Dipole sehr ähnlich. Beide Aggregate erleiden Zersetzung bei Temperaturen über 125 °C. Das dritte Aggregationsprodukt entsteht zwischen 200 und 250 °C. Es ist von den beiden oben erwähnten Produkten verschieden, da seine Absorptions- und Emissionsspektren keine vibronische Struktur zeigen. Ein Einfluß der Aggregationsprodukte auf die Fluoreszenzlebensdauer der in dem Kristall vorhandenen isolierten Dipole wird festgestellt.

1. Introduction

Photoluminescence of isolated Eu^{2+}-cation vacancy dipoles in KCl matrix has been examined by several authors [1 to 5]. There are also some incomplete data concerning the appearance and optical properties of the pertinent aggregate centres [5 to 7].

In general, two types of Eu^{2+} aggregate centres can be distinguished. The aggregates formed during room temperature storage are characterized, among others, by a distinct shift of the vibronic sideband towards lower energies [5] as compared with that found for the isolated Eu^{2+} dipoles. A quite different emission spectrum is observed for the association products formed during annealing between 200 and 250 °C. Their optical properties were examined so far at room temperature [6, 7]. These latter aggregate centres can be considered as the nuclei of some precipitates [6, 7] responsible, among others, for the appearance of diffused reflections in X-ray diagrams [8].

2. Experimental

Single Eu^{2+}-doped KCl crystals were grown in vacuum-sealed quartz ampoules by the Bridgman technique using prepurified material containing the dopant and treated in the melt by CCl_4 saturated argon [9].

[1]) P. O. Box 937, 50-950 Wrocław, Poland.

The measurements were performed using samples quenched after annealing at 500 °C and stored at room temperature or additionally annealed at temperatures between 100 and 250 °C.

The emission spectra were recorded in a conventional system. The excitation was performed with the 366 nm line of a high pressure mercury lamp HBO 200, passed through a monochromator. The emitted light was modulated by a chopper (400 Hz) and after passing a grid monochromator (SPM-2, Zeiss-Jena) recorded by the system consisting of a M10FQS29 photomultiplier, selective preamplifier (INTiBS), lock-in nanovoltmeter (type 232 UNIPAN), and a multichannel analyser NTA-1024 (EMG) used as digital oscilloscope. The fluorescence lifetime was measured in a system consisting of a nitrogen pulse laser (337 nm line) operated at repetition frequency of 10 Hz, analysing monochromator, photomultiplier, preamplifier (LM 733), boxcar integrator (BCI 280, AdW DDR), and multichannel analyser. Additionally, the systems of reference and triggering voltage as well as programmable calculator (type 666/B EMG) cooperating with multichannel analyser were used.

The measurements were performed in the temperature range 15 to 300 K using a microcooler Spectrim (CTI) and a temperature controller (Oxford Instruments).

3. Results and Discussion

The formation of the dopant aggregate centres in KCl:Eu^{2+} crystals is conditioned by two agents: dopant concentration and annealing temperature within the range up to about 250 °C. The effect of Eu^{2+} concentration on the emission spectra of the samples stored at room temperature during one year at least (long enough for approaching the equilibrium state) is shown in Fig. 1. For a freshly quenched crystal (16 ppm) the emission spectrum is characteristic of the isolated Eu^{2+}-cation vacancy dipoles (curve 1). Room temperature storage induces in this crystal a slight shift of the low-energy tail showing a hardly detectable structure. With increasing Eu^{2+} concentration this structure develops and dominates in the 128 ppm sample. The appearance and continuous growth of characteristic vibronic bands suggest that at room temperature only one type of the simplest aggregate centres (i) is formed (presumably dimers).

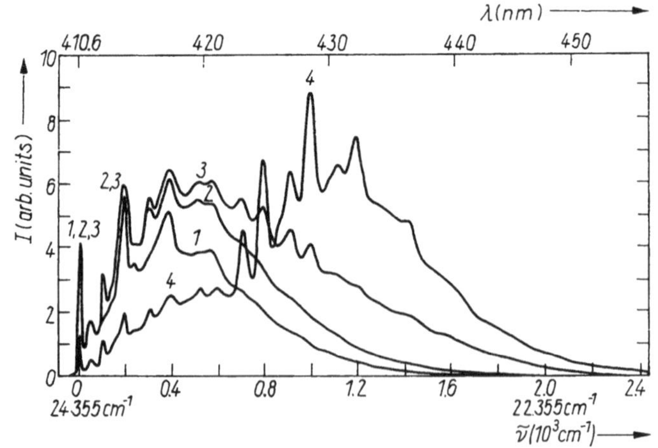

Fig. 1. Emission spectra of KCl:Eu^{2+} crystals (T = 15 K, λ_{exc} = 366 nm): (1) freshly quenched 16 ppm sample, (2, 3, 4) samples stored at RT, concentration 16, 53, 128 ppm, respectively

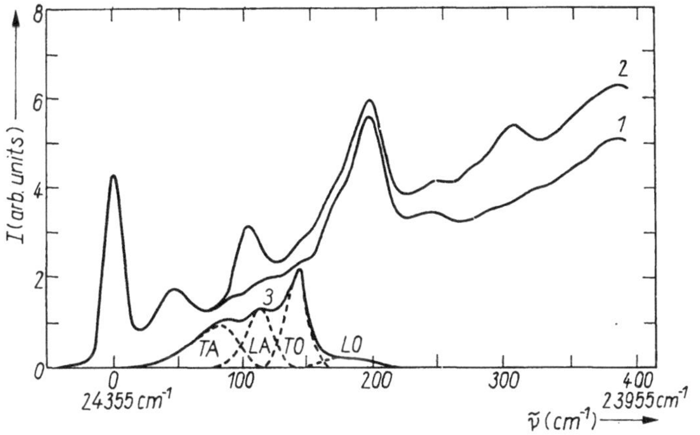

Fig. 2. One-phonon range of 16 ppm crystal ($T = 15$ K, $\lambda_{exc} = 366$ nm); (1) after quenching, (2) stored at RT, (3) calculated phonon spectrum of KCl crystal [11]

In the emission spectra of RT stored crystals, interesting changes appear also in the dipole-related one-phonon range (Fig. 2). The emission bands peaking at $\tilde{\nu}_2 \approx 50$ and $\tilde{\nu}_1 \approx 200\,cm^{-1}$ are according to [1, 3, 4] ascribed to resonance modes of A_1 symemetry. In agreement with [10, 11] the bands localized in the middle part of this range should be ascribed to the phonon structure of the lattice (curve 3 reflects the pertinent calculated spectrum [11] for an ideal KCl crystal). The changes appearing in this range in the room temperature stored crystal (curve 2) suggest that the aggregates (i) induce a distinct rise of the longitudinal acoustic modes (LA).

The effect of temperature on the formation of the aggregates is shown in Fig. 3. The annealing of the crystals at temperatures up to about 125 °C results in the formation of new aggregate centres (ii) not recorded so far. Their emission spectrum (curve 2) displays a vibronic structure similar to that found for the aggregates (i) but shifted towards lower energies by about 150 cm⁻¹. In the high doped crystal stored first at room temperature the concentration of these new centres reaches saturation at 125 °C

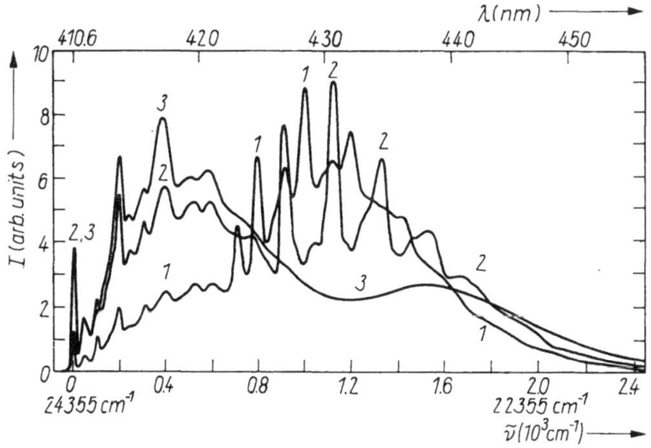

Fig. 3. Emission spectra of KCl:Eu^{2+} crystals ($T = 15$ K, $\lambda_{exc} = 366$ nm): (1) after long RT storage, (2) after annealing at 125 °C, (3) after annealing at 225 °C, concentration 128 ppm

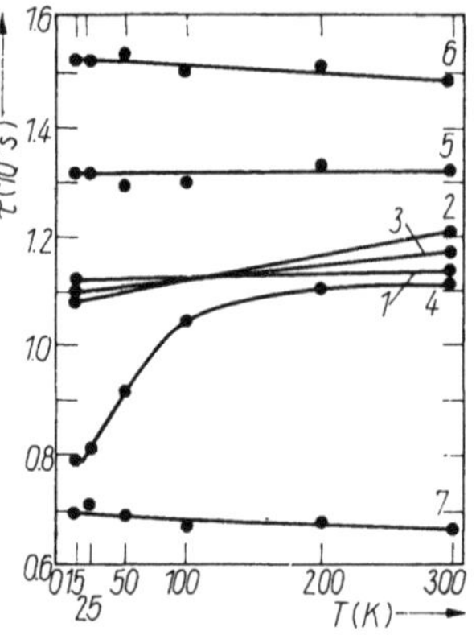

Fig. 4. Temperature dependence of the fluorescence lifetime for the emission related with isolated dipoles ($\lambda_{em} = 410$ nm): (1, 2, 3, 4) for samples containing isolated dipoles only or containing also aggregation products (i), (ii), and (iii), respectively, (5, 6) for the emission ($\lambda_{em} = 440$ nm) related with aggregates (i) and (ii), respectively, (7) for the emission ($\lambda_{em} = 460$ nm) related with aggregation product (iii)

after about 24 h. At the same time a fraction of preexisting aggregates (i) undergoes decomposition resulting in the rise of the dipole-related bands. In freshly quenched samples annealed at 125 °C the same state is reached after a few hundred hours.

The annealing of the crystals at temperatures above 125 °C results in practically complete decomposition of the aggregate centres (i) and (ii) and recovery of the simple isolated dipoles. Annealing of the crystals at temperatures between 200 and 250 °C induces, independently of the original state of the impurity ions, the formation of the third aggregation product (iii). As shown in Fig. 3 the emission spectrum (curve 3) is shifted still further towards lower energies and as different from the former ones does not exhibit any vibronic structure. This feature suggests a strong interaction of the excited electron with the lattice.

For all the aggregation forms the fluorescence lifetime has been measured at appropriate wavelengths in the temperature range 15 to 300 K (Fig. 4). In the quenched sample (curve 1) the dipole fluorescence lifetime (τ) is practically independent of temperature and, in agreement with [5], amounts to about 1.1 μs. The presence of the aggregate centres (i) and (ii) induces some small changes in the τ value characteristic of the dipole emission (curve 3, 4). The fluorescence lifetimes related with the aggregates (i) and (ii) are distinctly higher and at 15 K amount to 1.3 μs [5] and 1.5 μs, respectively (curves 5, 6). In contrast with the former values that obtained for the aggregation product (iii) amounts at 15 K to 0.7 μs.

The aggregate centres (i) and (ii) do not distinctly affect the major absorption bands of Eu^{2+} ions, compared with that obtained for the quenched sample. In the range of the low-energy tail of the absorption spectrum at 15 K, vibronic structures appear shifted with respect to each other in dependence on the impurity state [12]. These shifts show the same features as those found in the emission spectrum. Similarly to the isolated dipoles, in the vibronic spectrum of the above aggregates one can distinguish the bands related with the resonance modes and the phonon spectrum of the lattice. In Fig. 5 appropriate ranges are compared for both isolated dipoles (best in agreement with calculated phonon spectrum) and aggregates. It should be noticed that probably due to somewhat higher electron–lattice coupling no zero-phonon lines related with the aggregates were observed. Hence, in Fig. 5 the fraction of the spectra enclosed between the closest $\tilde{\nu}_1$ bands for (i) and (ii) aggregates are compared with that between 0–0 and $\tilde{\nu}_1$ for the isolated dipoles. The results obtained suggest that the aggregates (i) and (ii) are composed of weakly interacting dipoles (in the nearest neighbourhood of Eu^{2+} ions, but weakly disturbed), and that the number of the component dipoles is higher in the aggregate (ii). The rise of the fluorescence lifetime value in the

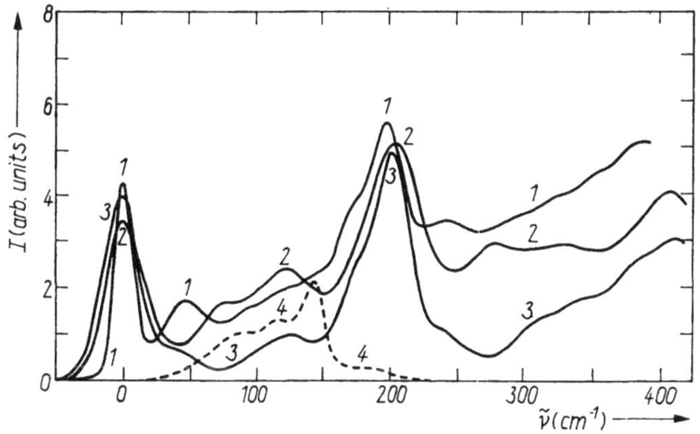

Fig. 5. Vibronic structure of the crystals containing aggregates (i) and (ii), curves (2) and (3) respectively, compared with (1) freshly quenched 16 ppm crystal, and (4) phonon spectrum of KCl according to [11]

sequence of isolated dipoles, aggregate (i), and aggregate (ii) suggests that accompanying changes of the magnetic perturbation, such as a spin–orbit interaction, or/and electrostatic interaction are responsible for the decreasing probability of nonradiative processes [13, 14].

The absorption spectrum related with the presence of aggregation product (iii) entirely differs from that obtained for aggregates (i) and (ii), and suggests a strong disturbance of the electron configuration. The fluorescence lifetime is here the lowest one (Fig. 4, curve 7) and points to a strong interaction between component Eu^{2+} ions responsible for the desactivation resembling the concentration quenching effect.

The presence of the latter aggregation form affects in a peculiar way the temperature dependence of the fluorescence lifetime of the fraction of isolated dipoles still present in the sample (Fig. 4, curve 4). Significant lowering of τ observed at lower temperatures suggests an effective excitation energy transfer. This process is possibly of a selective character, i.e. it is concerned with the dipoles at the lowest energy level. The temperature course of the curve beginning with 25 K can be reasonably well described by the equation

$$\tau^{-1} = (\tau_{300\,\mathrm{K}})^{-1} + s\left[1 - \exp\left(-\frac{E}{kT}\right)\right] \tag{1}$$

yielding the activation energy of 27.5 cm^{-1} and the frequency equal to $0.46 \times 10^6\ s^{-1}$.

Acknowledgement

The authors thank Prof. J. Z. Damm for a critical reading of the manuscript.

References

[1] M. Wagner and W. E. Bron, Phys. Rev. **139**, A223 (1965).
[2] W. E. Bron and M. Wagner, Phys. Rev. **139**, A233 (1965).
[3] W. E. Bron and M. Wagner, Phys. Rev. **145**, 689 (1966).
[4] K. Nasu, A. Tamura, K. Kojima, and T. Kojima, J. Phys. Soc. Japan **36**, 55 (1974).
[5] L. D. Merckle, R. C. Powell, and T. M. Wilson, J. Phys. C **11**, 3103 (1978).
[6] F. J. Lopez, H. Murietta S., J. Hernandez A., and J. Rubio O., Phys. Rev. B **22**, 6428 (1980).

[7] J. RUBIO O., H. MURIETTA S., and J. HERNANDEZ A., Phys. Rev. B **24**, 4847 (1981).
[8] J. STĘPIEŃ-DAMM, E. MUGEŃSKI, and K. ŁUKASZEWICZ, Acta crystall. A **34**, S245 (1978).
[9] R. VOSZKA, I. TARJAN, L. BERKES, and J. KRAJSOVSZKY, Kristall und Technik **1**, 423 (1966).
[10] A. M. KARO, J. chem. Phys. **33**, 7 (1960).
[11] A. M. KARO and J. R. HARDY, Phys. Rev. **129**, 2024 (1963).
[12] H. OPYRCHAŁ and K. NIERZEWSKI, unpublished results.
[13] G. W. ROBINSON and R. P. FROSCH, J. chem. Phys. **37**, 1962 (1962).
[14] G. W. ROBINSON and R. P. FROSCH, J. chem. Phys. **38**, 1187 (1963).

(Received April 9, 1984)

phys. stat. sol. (b) **125**, 387 (1984)

Subject classification: 20.3; 22.8

Solid State Department, Physical Laboratory, Utrecht State University[1])

Luminescence and Energy Transfer Characteristics of TbOCl

By

P. A. M. BERDOWSKI, J. VAN HERK, L. JANSEN, and G. BLASSE

Luminescence and energy transfer properties of the system $Gd_{1-x}Tb_xOCl$ ($0 < x \leqq 1$) are reported, with special stress on those of TbOCl. Concentration quenching occurs above $x = 0.15$ at 300 K. Transfer from intrinsic Tb^{3+} ions to extrinsic Tb^{3+} ions and to rare earth ion impurities are observed. The emission spectra of TbOCl are dominated by trap emission, which makes a quantitative analysis of the intrinsic Tb^{3+} luminescence properties in TbOCl cumbersome.

Es wird über die Lumineszenz und die Energieübertragungseigenschaften des Systems $Gd_{1-x}Tb_xOCl$ ($0 < x \leqq 1$) berichtet. Konzentrationstilgung erfolgt für $x > 0{,}15$ bei 300 K. Energieübertragung von Tb^{3+} intrinsic-Ionen auf fehlgeordnete Tb^{3+}-Ionen und auf Seltenerdfremdionen wird beobachtet. Die Emissionsspektren des TbOCl werden von Haftstellenemission bestimmt, was eine quantitative Analyse der Lumineszenzeigenschaften der Tb^{3+}-Ionen auf Eigenstellen in TbOCl schwer macht.

1. Introduction

The luminescence properties of Tb^{3+}-activated rare earth oxyhalides have been studied extensively. An early report is due to Blasse and Bril [1] and was followed by many others (see e.g. [2] and [3]). The crystal structure of GdOCl is isomorphous with that of PbFCl [4] and consists of alternating $(Gd_2O_2)_2$ and $(Cl)_2$ layers perpendicular to the c-axis. There is only one crystallographic site for Gd^{3+}. The site symmetry is C_{4v}.

In the course of our investigations on energy migration in two-dimensional systems, we investigated the luminescence properties of powder samples of $Gd_{1-x}Tb_xOCl$ ($0 < x \leqq 1$) using laser site-selection and time-resolved spectroscopy. The measurements of emission and excitation spectra of samples with low Tb^{3+} concentrations were in agreement with the results reported earlier [1 to 3]. Concentration quenching of the $^5D_3 \rightarrow {}^7F_J$ emission due to cross relaxation could be observed for samples containing Tb^{3+} concentrations $x > 0.01$. Concentration quenching of the $^5D_4 \rightarrow {}^7F_J$ emission due to energy migration to quenching centres becomes explicit above $x = 0.15$ at 300 K. Energy migration to Tb^{3+} traps and other rare earth ions in TbOCl is reported in this paper.

2. Experimental

All measurements described were performed on powdered samples. The compounds were prepared by dissolving the rare earth oxides (Highways International, 99.99%) in HCl, evaporating to dryness and firing in air at 500 °C for 6 h. The resulting products had a white colour. Samples were checked by X-ray powder diffraction. For the low-temperature measurements the samples were immersed into liquid helium in a Thor bath cryostat (S-100). The temperature of the samples could be varied between 4.2 K and room temperature. The emission spectra were obtained with a tunable dye laser (Molectron DL 200) pumped with a nitrogen laser (Molectron UV 14) as an

[1]) P.O. Box 80000, 3508 TA Utrecht, The Netherlands.

excitation source. The generated laser pulse had a peak power of 30 kW and a width of about 10 ns. The resolution of this equipment amounts to 1 cm^{-1}. The emission was detected using a Spex 1704 × 1 m monochromator in combination with a cooled photomultiplier (RCA C31034). To record the time-resolved spectra and the decay of the luminescence a PAR model 162/165 boxcar averager was used.

3. Results

The excitation spectrum of the Tb^{3+} emission in GdOCl consists of a broad band with a maximum at 245 nm and lines in the region 270 to 500 nm. Part of these peaks correspond to the Gd^{3+} absorption lines and indicate energy transfer from Gd^{3+} to Tb^{3+}. These peaks were not observed in the excitation spectra of the Tb^{3+} emission in TbOCl. The other lines in the region 270 nm to 500 nm correspond to transitions within the $4f^8$ configuration of Tb^{3+}.

In the emission spectra of samples with low Tb^{3+} concentration ($x < 0.1$) the lines corresponding to the transitions from $^5D_3$ levels as well as those from $^5D_4$ levels can be observed after excitation in the broad band around 245 nm (4f–5d transition). Above $x = 0.01$ the intensity of the emission from $^5D_3$ levels decreases rapidly in favour of that of the emission from $^5D_4$ levels. The concentration quenching of the $^5D_3 \rightarrow {}^7F_J$ emission is analogous to the results found for $La_{1-x}Tb_xOBr$ [5], and is due to cross relaxation.

Fig. 1 shows the emission spectrum of $Gd_{0.95}Tb_{0.05}OCl$ in the $^5D_4 \rightarrow {}^7D_J$ spectral region at 4.2 K. A dye laser was used as an excitation source and the excitation wavelength was 486.3 nm ($^7F_6 \rightarrow {}^5D_4$). The spectrum is in agreement with the results reported in [6]. The number of peaks as well as their positions in the emission spectrum of Tb^{3+} in TbOCl differ from those in the spectrum of GdOCl:Tb. It was also observed that this emission spectrum depends strongly on sample history and is hard to reproduce. In addition the emission spectra of TbOCl show a strong temperature dependence. This is illustrated in Fig. 2, which shows the $^5D_4 \rightarrow {}^7F_5$ emission spectra at various temperatures upon narrow line excitation into the $^5D_4$ level (exc. 486.3 nm) for the sample with the highest luminescence intensity. It is clear that the spectra differ from those of GdOCl:Tb at all temperatures. Only a few peaks in the emission spectrum of the latter can be recognized in the spectra of TbOCl. The small peak at 567.4 nm is probably due to a Sm^{3+} impurity as will be discussed below.

Fig. 3 presents the integrated emission intensities of $Gd_{1-x}Tb_xOCl$ as a function of x at 300 K. Concentration quenching can be observed for $x > 0.15$. The integrated emission intensity of TbOCl is only 25% of that of the sample with $x = 0.2$. The latter gives a bright yellow-green luminescence at room temperature. Concentration quenching suggests that the excitation energy migrates through the Tb^{3+} sublattice to quenching centres (killers) [7, 8]. The strong temperature dependence of the emission

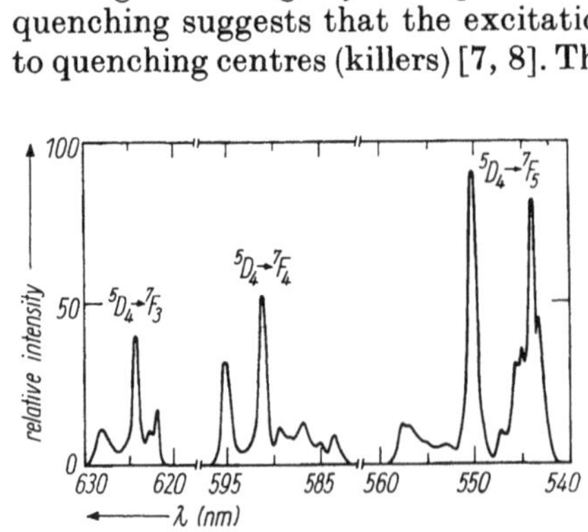

Fig. 1. Emission spectrum of $Gd_{0.95}Tb_{0.05}OCl$ recorded at 4.2 K (exc. 486.3 nm)

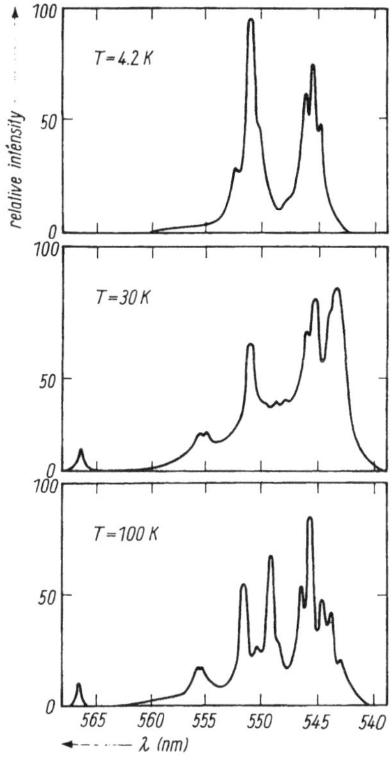

Fig. 2. Emission spectrum of TbOCl, recorded at various temperatures in the $^5D_4 \rightarrow {}^7F_5$ spectral region (exc. 486.3 nm)

spectra of TbOCl can also be explained by excitation energy migration. The migrating excitation energy can be trapped by Tb^{3+} ions which occupy sites different from the ideal crystallographic sites. The temperature dependence of the emission arises from thermally activated back-transfer from the traps to the intrinsic ions. Back-transfer depends strongly on the energy difference between the levels of the intrinsic ions and the traps. At higher temperatures the shallower traps will be emptied and the deeper traps will dominate the emission spectrum. Fig. 2 suggests that in the temperature region under study trap emission dominates.

In addition to migration to Tb^{3+} traps migration to rare earth impurities can occur. This explains the appearance of the characteristic Sm^{3+} line at 567.4 nm in the emission spectra recorded at $T > 25$ K. Also the other Sm^{3+} emission lines were observed under these conditions. These peaks were not observed in the spectra recorded at room temperature. At this temperature back-transfer to Tb^{3+} ions seems to have become effective.

In order to obtain information about the energy migration in TbOCl we tried to investigate the decay characteristics of the $^5D_4$ emission for GdOCl:Tb and TbOCl. For the former the decay curves after excitation in $^5D_4$ are exponential in the whole temperature region (4.2 to 300 K). The decay time at 4.2 K was 2.5 ms and remained constant up to 300 K. It was not possible to measure the decay characteristics of the intrinsic Tb^{3+} ions in our TbOCl samples because of the dominant role of the traps. The decay time of the most important trap at 4.2 K was $\approx$ 2 ms. However, this decay time decreased rapidly with increasing temperature due to back-transfer.

In order to investigate energy migration to traps with their own characteristic emission, we investigated $Tb_{0.99}Eu_{0.01}OCl$. Fig. 4 shows the $^5D_0 \rightarrow {}^7F_2$ Eu^{3+} emission and the $^5D_4 \rightarrow {}^7F_5$ Tb^{3+} emission spectra of this sample at various temperatures upon excitation into the $^5D_4$ level of the intrinsic Tb^{3+} ions. At 4.2 K emission from Tb^{3+} traps and from Eu^{3+} ions occurs. With increasing temperature the Tb^{3+} emission intensity decreases rapidly relative to that of the Eu^{3+} emission. The Tb^{3+} emission can hardly be observed above 8.5 K. In passing we note that the Tb^{3+} traps in this sample (Fig. 4) are different from those in the sample of Fig. 2, illustrating the variation in the Tb^{3+} traps from sample to sample.

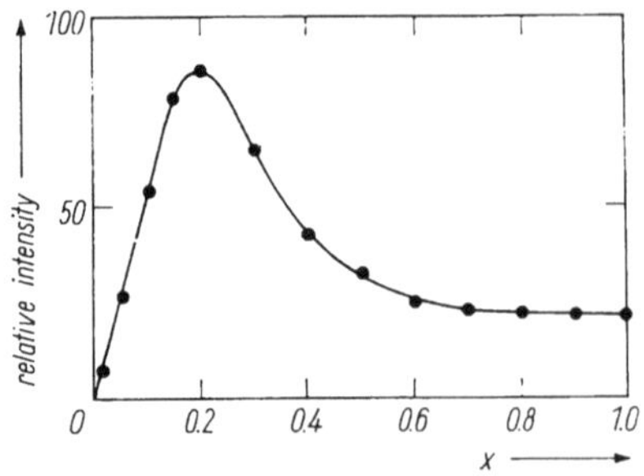

Fig. 3. The relative intensities of the Tb^{3+} emission of $Gd_{1-x}Tb_xOCl$ as a function of x at 300 K

To monitor any possible energy transfer from the Tb^{3+} ions to the Eu^{3+} ions we measured at 4.2 K emission spectra as a function of time in the range 0.1 to 1 ms after the excitation pulse. Fig. 5 shows the time dependence of the $^5D_0 \rightarrow {}^7F_2$ Eu^{3+} emission and the $^5D_4 \rightarrow {}^7F_5$ Tb^{3+} emission upon selective laser excitation in the $^5D_4$ level of the intrinsic Tb^{3+} ions. Immediately after the pulse we observe mainly intrinsic Tb^{3+} emission (characterized by the 550.9 nm peak; compare Fig. 1). With increasing time a build-up of the lines associated with the Eu^{3+} ions at the expense of those of the

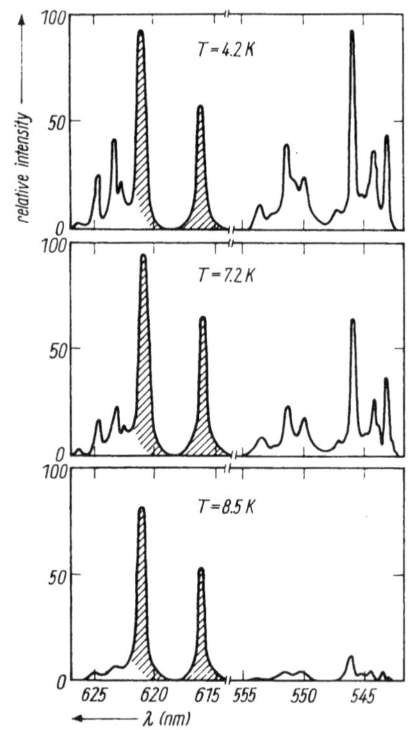

Fig. 4

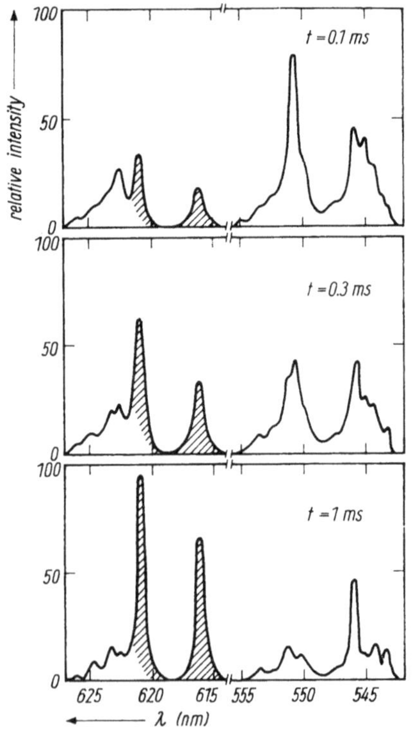

Fig. 5

Fig. 4. Emission spectra of $Tb_{0.99}Eu_{0.01}OCl$, recorded at various temperatures in the $^5D_0 \rightarrow {}^7F_2$ and $^5D_4 \rightarrow {}^7F_5$ spectral regions (exc. 486.3 nm). The Eu^{3+} peaks are hatched. The Tb^{3+} peaks are open

Fig. 5. The time dependence of the emission spectrum after excitation in the intrinsic Tb^{3+} ions in $Tb_{0.99}Eu_{0.01}OCl$ at 4.2 K. The Eu^{3+} peaks are hatched. The Tb^{3+} peaks are open. t indicates the time after the excitation pulse

Tb^{3+} ions can be observed. It seems reasonable to conclude that energy transfer from the Tb^{3+} ions to the Eu^{3+} ions is responsible for the transient behaviour of the emission in $Tb_{0.99}Eu_{0.01}OCl$ as has been proposed for the first time for $Tb_3Al_5O_{12}$ [9]. The Tb^{3+} emission at longer times after the laser pulse is dominated by Tb^{3+} trap emission. This explains the relatively strong Tb^{3+} emission at 1 ms after the pulse. This experiment shows that even 1% Eu^{3+} is not able to overcome the influence of the Tb^{3+} traps.

4. Discussion

From the spectra and from the time-resolved measurements it is obvious that energy transfer from intrinsic Tb^{3+} ions to extrinsic Tb^{3+} ions (Tb^{3+} traps) and other rare earth ions occurs in TbOCl. From Fig. 3 it follows that the quenching of the luminescence in $Gd_{1-x}Tb_xOCl$ powders occurs above $x = 0.15$ at 300 K. Apparently energy migration through the Tb^{3+} sublattice becomes of influence above this concentration. The quenching can be ascribed to the loss of migrating excitation energy at killer sites. Possible killer centres are Tb^{4+} ions. Also in $GdOF:Tb^{3+}$ the Tb^{4+} ion has been shown to act as a killer centre [10]. Since the ratio oxygen/halogen is hard to control, Tb^{4+} is a very likely defect, charge compensation taking place by an increase of the oxygen/halogen ratio.

As stated above, the emission spectra of the concentrated samples are dominated by traps. Because of the strong influence of the traps, it was not possible to make a quantitative analysis of the decay curves of the intrinsic emission in order to distinguish between several types of energy migration (see e.g. [11]). This has recently been performed for the concentrated europium oxychloride system [12].

The strong influence of the Tb^{3+} traps in the luminescence properties of TbOCl suggests that the trap concentration is high. A very high trapping rate can be excluded, since we are dealing with rare earth ions. The extrinsic Tb^{3+} ions originate from a deviation from the perfect crystal structure. There are many possibilities for imperfections, e.g. anion disorder or stacking faults (see also [13]).

In conclusion we have found that TbOCl shows pronounced deviations from the ideal crystal structure. This results in the presence of Tb^{3+} traps which, in turn, determine the luminescence properties of TbOCl.

Acknowledgements

The investigations were supported by the Netherlands Foundation for Chemical Research (SON) with financial aid from the Netherlands Organisation for Advancement of Pure Research (ZWO).

References

[1] G. Blasse and A. Bril, Philips Res. Repts. **22**, 481 (1967).
[2] J. Hölsä and M. Leskelä, phys. stat. sol. (b) **103**, 797 (1981).
[3] L. H. Brixner, J. F. Ackerman, and C. M. Foris, J. Lum. **26**, 1 (1981).
[4] R. W. G. Wyckoff, in: Crystal Structures, Vol. 1, Wiley & Sons, New York 1963.
[5] J. Hölsä, M. Leskelä, and L. Niniströ, Mater. Res. Bull. **14**, 1403 (1979).
[6] J. Hölsä and P. Porcher, J. chem. Phys. **76**, 2798 (1982).
[7] F. Kellendonk and G. Blasse, J. Phys. Chem. Solids **43**, 481 (1982).
[8] M. J. Joubert, B. Jacquier, R. Moncorgé, and G. Boulon, J. Physique **43**, 893 (1982).
[9] W. W. Holloway, Jr. and M. Kestigan, Phys. Letters (Netherlands) **21**, 364 (1966).
[10] R. Lammers and G. Blasse, to be published.
[11] D. L. Huber, Phys. Rev. **20**, 2307 (1979).
[12] P. A. M. Berdowski, J. van Herk, and G. Blasse, to be published.
[13] A. F. Halff and J. Schoonman, Z. phys. Chem., Neue Folge **109**, 161 (1978).

(Received May 14, 1984)

phys. stat. sol. (b) 125, 393 (1984)

Subject classification: 13; 20.3; 22

Sektion Mathematik der Karl-Marx-Universität Leipzig[1])

Instabilities and δ-Distributions in Solid State Physics

By

H. ENGLISCH

It is not admissible to use δ-distributions as potentials, e.g. as central cell corrections, as it has been done by Lucovsky and Kiefer for doped semiconductors considered in effective mass approximation (EMA). Rigorous theorems like Kato's one for the wave function and Kimball's one for the pair distribution function are due to the δ-terms contained not appropriate to test the quality of trial functions. The intensity formula for the radiative recombination contains the dipole matrix element having in EMA two instabilities, one arising from a δ-term. The other one is connected with the asymptotic decay of the trial functions which also makes the usefulness of the notion of the localization length doubtful.

Es ist unzulässig, δ-Distributionen als Potential zu verwenden, z. B. als Zentralzellenkorrektur wie das von Lucovsky und Kiefer für dotierte Halbleiter in Effektivmassen-Approximation (EMA) getan wurde. Strenge Theoreme wie das von Kato für Wellenfunktionen oder von Kimball für die Paarverteilungsfunktion sind aufgrund der enthaltenen δ-Terme nicht geeignet, die Qualität von Probefunktionen zu testen. Die Intensitätsformel für die strahlende Rekombination enthält das Dipolmatrixelement, das in EMA zwei Instabilitäten besitzt. Eine davon tritt in Folge eines δ-Terms auf. Die andere ist mit dem asymptotischen Abfall der Probefunktionen verknüpft, der auch die Nützlichkeit des Begriffes der Lokalisierungslänge zweifelhaft erscheinen läßt.

1. Introduction

The simple rule for the calculation of matrix elements

$$\langle \psi, \delta\psi \rangle = \psi^2(0) \tag{1}$$

seems to recommend δ-potentials for the description of short-range effects. These potentials have been successfully used in one-dimensional problems of solid state physics and in liquid models [1]. In more than one dimension the singularities of the δ-potential become so strong that they produce unphysical properties like nonselfadjoint Hamiltonians and infinitely negative energies.

Some known mathematics to the δ-potential is presented in Section 2, and in Section 3 it is shown how far the applications of δ-potentials to central cell corrections by Lucovsky [2] and Kiefer [3] are incorrect. Sections 4 and 5 deal with a stability problem connected with the δ-distribution occurring in the formulas for the pair distribution function and the dipole matrix element in EMA, respectively. The notion of the localization length used especially for disordered systems is analysed in Section 6.

Before we can consider the stability of a physical quantity with respect to small perturbations we have to clarify the meaning of "small", i.e. we have to determine the topology of the underlying state space. The usual L^2-norm of the Hilbert space is too crude: If we work with the N-particle Hamiltonian

$$H(V) = \sum_{i=1}^{N} - \Delta_i + V_i(\boldsymbol{r}_i) + \sum_{i<j} V_{ij}(\boldsymbol{r}_{ij}) ; \qquad \boldsymbol{r}_{ij} = \boldsymbol{r}_i - \boldsymbol{r}_j , \tag{2}$$

[1]) Karl-Marx-Platz 10, DDR-7010 Leipzig, GDR.

then a perturbation of the wave function such that $||\psi - \overline{\psi}||_2 < \varepsilon$ can produce an arbitrarily large change in the expectation value of $H(V)$. Therefore, we use the Sobolev-space norm ([4], § 9.6)

$$||\psi||_s = \langle \psi, (1 + \sum_{i=1}^{N} - \Delta_i)\, \overline{\psi}\rangle^{1/2} . \tag{3}$$

The topology (3) is also not too fine, which can be demonstrated as follows: If $H(V)$ has an isolated non-degenerate ground state ψ_1 with the energy $E(V)$, then $\langle\psi, H(V)\psi\rangle < E(V) + \varepsilon$ implies that $||\psi - \psi_1||_s$ is small ([5], Section 6). (Strictly speaking, the potentials V_i and V_{ij} have to be form-bounded with respect to $-\Delta$, cf. ([4], § 10.2). A sufficient condition for the form-boundedness of V is the existence of a partition $V = V_1 + V_2$, $V_1 \in L^{3/2}(R^3)$, $V_2 \in L^\infty(R^3)$. E.g. the Coulomb potential is form-bounded.) In Sections 4 to 6 some examples of quantities being unstable with respect to the norm (3) are presented. These quantities cannot be approximately calculated from trial functions (even not from good ones).

2. Mathematical Results for the δ-Distribution

In one dimension the δ-potential is form-bounded with respect to $-\Delta$, i.e. for every $\varepsilon > 0$ there is a constant c_ε such that

$$\langle\psi, \delta\psi\rangle \leqq \varepsilon\langle\psi, -\Delta\psi\rangle + c_\varepsilon\langle\psi, \psi\rangle \tag{4}$$

for every ψ with finite $||\psi||_s$. Thus $-\Delta + \lambda\delta$ can be defined as a semi-bounded self-adjoint operator having for $\lambda < 0$ exactly one bound state ([4], § 10.2).

For higher dimensions $-\Delta + \lambda\delta$, $\lambda < 0$, is unbounded from below and (4) holds no longer. Then $-\Delta + \lambda\delta$ cannot be defined as a selfadjoint operator (cf. below), and the arguments in [6] indicate that it has no bound states: If $\psi \in L^2(R^d)$ satisfies

$$(-\Delta + \lambda\delta)\, \psi = E\psi , \tag{5}$$

then the Fourier transformation yields

$$k^2\hat{\psi}(\boldsymbol{k}) + \lambda \int \hat{\psi}\,(\boldsymbol{k}')\, \mathrm{d}^d\boldsymbol{k}' = E\hat{\psi}(\boldsymbol{k}) \tag{6}$$

and

$$\frac{\lambda \int \hat{\psi}(\boldsymbol{k}')\, \mathrm{d}^d\boldsymbol{k}'}{E - k^2} = \hat{\psi}(\boldsymbol{k}) . \tag{7}$$

Integrating over $\boldsymbol{k}$ and dividing by $\int \hat{\psi}(\boldsymbol{k})\, \mathrm{d}^d\boldsymbol{k}$ we get

$$\lambda \int \frac{1}{E - k^2}\, \mathrm{d}^d\boldsymbol{k} = 1 . \tag{8}$$

But for $d \geqq 2$ the integral in (8) diverges, i.e. $-\Delta + \lambda\delta$ has no eigenvalues.

For dimensions $d \geqq 3$ it can be seen very easily that $-\Delta + \lambda\delta$, $\lambda < 0$, is unbounded from below: Take the hydrogen functions $\psi_\alpha(\boldsymbol{r}) = \exp(-\alpha r)/||\exp(-\alpha r)||_2$, $\alpha \to \infty$, as trial functions. The same trial functions show the boundlessness of $-\Delta + \lambda\delta$ for $d = 2$ and sufficiently negative λ. Working with refined trial functions we see that $-\Delta + \lambda\delta$ is unbounded from below also for $d = 2$ and arbitrary $\lambda < 0$: Take the Fourier transform of

$$\hat{\psi}(\boldsymbol{k}) = \begin{cases} k^{-2} \ln^{-1} k & \text{for} \quad k > 2 , \\ 0 & \text{for} \quad k \leqq 2 . \end{cases} \tag{9}$$

Then $\psi \in L^2(R^2)$,

$$\psi^2(0) = (\int \hat{\psi}(\boldsymbol{k})\, d^2\boldsymbol{k})^2 = (2\pi \int_2^\infty k^{-1} \ln^{-1} k\, dk)^2 = (2\pi \ln \ln k|_2^\infty)^2 = \infty\,, \tag{10}$$

but

$$\langle -\Delta\psi, \psi\rangle = \int k^2 \hat{\psi}^2(\boldsymbol{k})\, d^2\boldsymbol{k} = 2\pi \int_2^\infty k^{-1} \ln^{-2} k\, dk = \frac{2\pi}{\ln k|_2^\infty} < \infty\,. \tag{11}$$

Nevertheless for $d = 2$ and $d = 3$ there are local perturbations of $-\Delta$ ("point interactions"), which yield a selfadjoint operator bounded from below ([4], Th. 10.11). (For $d \geqq 4$ the Laplacian is selfadjoint on $C_0^\infty(R^d \setminus \{0\})$; thus such local perturbations cannot exist for three dimensions.) The admissible point interactions for $d = 2$ and $d = 3$ can be interpreted as δ-potentials with infinitesimal coupling constant λ [7]. Since all selfadjoint extensions of $-\Delta$ restricted to $C_0^\infty(R^d \setminus \{0\})$, $d = 2, 3$, are bounded from below we conclude that $-\Delta + \lambda\delta$, $\lambda < 0$, is not only unbounded from below, but cannot be defined as a selfadjoint operator. (The Hamiltonian $-\Delta + x$ is an example of an unbounded operator which is nevertheless selfadjoint ([4], Ex. 5 to Th. 10.38).)

The perturbation theory deceives one on the difficulties with a multidimensiona δ-distribution. For example, the first order of the perturbation theory applied to $H = H_0 + \lambda\delta$ predicts an energy shift

$$E - E_0 \approx \lambda\psi_1^2(0)\,, \tag{12}$$

where ψ_1 is the ground state of H_0 with the energy E_0. Since eigenfunctions of usual Hamiltonians are bounded ([4], Th. 13.38) the right-hand side of (12) is bounded. But (12) is without meaning, since perturbation theory cannot be applied to perturbations not being form-bounded ([4], § 12.2). (Nevertheless Fermi successfully used a δ-pseudopotential for describing the scattering of slow nucleons in the Born approximation — a variant of perturbation theory ([23], § 115).)

It should be noted that the criticism against δ-potentials in higher dimensions is not directed against the following version of a δ-potential smeared out over the sphere with radius R:

$$\langle\psi(\boldsymbol{r}), \delta(r - R)\, \psi(\boldsymbol{r})\rangle = \int_{r=R} \psi^2(\boldsymbol{r})\, d^d\boldsymbol{r}\,. \tag{13}$$

Due to ([4], Th. 9.39) $\delta(r - R)$ is form-bounded with respect to $-\Delta$, i.e. for arbitrary $d \geqq 1$ the operator $-\Delta + \lambda\, \delta(r - R)$ is selfadjoint and bounded from below.

3. Central Cell Corrections by δ-Potentials

Short-range potentials occur at neutral impurities in solids as well as within the central cell of charged impurities. Since the informations on the short-range potentials are usually rare and since δ-distributions are easily to handle, it is tempting to use them as model potentials. Lucovsky [2] used them and claimed that

$$\psi(\boldsymbol{r}) = \frac{\left(\dfrac{\alpha}{2\pi}\right)^{1/2} \exp(-\alpha r)}{r} \tag{14}$$

is the ground state of $-\Delta - \alpha^2\delta$. Many authors [8 to 11] referred this paper without pointing to the error. Lucovsky did not prove his assertion but probably it had been established since $\exp(-\alpha r)/4\pi r$ is known as being the free Green's function ([4],

Ex. 1 to Th. 9.29), i.e. the solution of

$$(-\Delta + \alpha^2)\,\psi = 2(2\pi\alpha)^{1/2}\,\delta\,. \tag{15}$$

But it is not the solution of

$$(-\Delta + \alpha^2)\,\psi = -\lambda\delta\psi(0) \tag{16}$$

for some appropriately chosen $\lambda < 0$. Remember that for the choice (14) $\psi(0)$ is infinite!

Kiefer [3] independently made calculations with the δ-potential, unfortunately also incorrect ones: She considered the Hamiltonian $H_\lambda = -\Delta + \lambda\delta - 1/r$ and used trial functions $\psi(\boldsymbol{r}) = \exp(-\alpha r)/||\exp(-\alpha r)||_2$. This yields

$$E_{\alpha,\lambda} = \langle \psi_\alpha, H_\lambda \psi_\alpha \rangle = -\alpha + \alpha^2 + \lambda\alpha^3\,. \tag{17}$$

In order to find the minimum she solved the equation

$$\frac{\mathrm{d}E_{\alpha,\lambda}}{\mathrm{d}\alpha} = -1 + 2\alpha + \lambda\alpha^2 = 0\,. \tag{18}$$

For $-1 < \lambda < 0$ (18) has two solutions belonging to the maximum and a local minimum of $E_{\alpha,\lambda}$. Kiefer thought the latter one as being the global minimum searched for. But for all $\lambda < 0$ $E_{\alpha,\lambda}$ is unbounded from below!

In order to circumvent the difficulties arising by using δ-distributions, short-range potentials can be modeled e.g. by a Yukawa potential, a muffin-tin potential [3], or by the smeared δ-potential (13). They have, however, the disadvantage of possessing two degrees of freedom. Usually there is only one energy known (the ground-state energy of the separated impurities) which can be used for the fit of the short-range potential. Therefore, the more or less arbitrary fixation of one parameter of the potential is indispensable. A possible choice for the range of a muffin-tin potential or of the smeared δ-potential would be a volume equal to the atomic volume of the impurity within the lattice or its surface, respectively.

4. Kato's Cusp Condition

The eigenfunctions of a Hamiltonian with Coulomb interaction

$$H = -\sum \Delta_i + \sum_{i<j} \frac{1}{r_{ij}} - \sum_i \sum_j \frac{Z_j}{|\boldsymbol{r}_i - \boldsymbol{R}_j|} \tag{19}$$

satisfy for coinciding arguments Kato's cusp condition,

$$\psi(\boldsymbol{r}_1, \boldsymbol{r}_2, \dots, \boldsymbol{r}_N)_{|\boldsymbol{r}_1=\boldsymbol{r}_2} = \frac{\mathrm{d}\displaystyle\int_{r=R} \frac{\psi(\boldsymbol{r}_1, \boldsymbol{r}_1 + \boldsymbol{r}, \boldsymbol{r}_3, \dots, \boldsymbol{r}_N)\,\mathrm{d}^3\boldsymbol{r}}{4\pi R^2}}{\mathrm{d}R_{|R=0}} \tag{20}$$

as well as a related condition for $\boldsymbol{r}_i = \boldsymbol{R}_j$ [12]. Hoffmann-Ostenhof and Seiler [13] (cf. also [14]) derived analogous statements for the coalescence of more than two particles. The integration over $(N-1)$ variables in (20) yields for the pair distribution function

$$g(R) = \binom{N}{2} \frac{\displaystyle\int_{r=R} \psi^2(\boldsymbol{r}_1, \boldsymbol{r}_1 + \boldsymbol{r}, \boldsymbol{r}_3, \dots, \boldsymbol{r}_N)\,\mathrm{d}^3\boldsymbol{r}_1\,\mathrm{d}^3\boldsymbol{r}\,\mathrm{d}^3\boldsymbol{r}_3 \dots \mathrm{d}^3\boldsymbol{r}_N}{4\pi R^2} \tag{21}$$

the condition

$$g(0) = g'(0)/2 . \tag{22}$$

A related equation for the one-particle density

$$n(\boldsymbol{r}) = N \int \psi^2(\boldsymbol{r}, \boldsymbol{r}_2, \dots, \boldsymbol{r}_N) \, \mathrm{d}^3\boldsymbol{r}_2 \dots \mathrm{d}^3\boldsymbol{r}_N \tag{23}$$

follows from the condition for $\boldsymbol{r}_i = \boldsymbol{R}_j$ mentioned above [15]. Kimball [16] derived (22) without referring to Kato's paper. Some authors [16 to 18] recommend Kato's formulas, their corrollaries or an inequality containing $n(0)$ [18] as tests for the accuracy of the wave functions. But for $d > 1$ the left-hand sides of (20) and (22) are unstable since the condition $\boldsymbol{r}_1 = \boldsymbol{r}_2$ corresponds to a δ-distribution which is not form bounded (cf. Section 2). Additionally the right-hand sides of (20) and (22) are highly unstable, for they correspond to a term with the derivative of the δ-distribution. Pack and Byers Brown [19] and Gilbert [20] mentioned that both the cusp condition and the correct value of $g(0)$ are of little help for the construction of good approximate wave functions. But they did not work out the reasons for instability. Gilbert conjectured that good trial functions have to possess a nearly correct behaviour of $g(R)$ over a wide range of R. This conjecture is correct for every $R > 0$: Since $g(R)$ can be expressed by a smeared δ-distribution (13), small differences of the wave functions in the Sobolev-space norm (3) imply small differences of $g(R)$ for every $R > 0$. Further, the instability of $g(R)$ is so weak that

$$\int \frac{g(R)}{R} \, \mathrm{d}R = \binom{N}{2} \langle \psi(\boldsymbol{r}_1, \dots, \boldsymbol{r}_N) \, , \, |\boldsymbol{r}_1 - \boldsymbol{r}_2|^{-1} \, \psi(\boldsymbol{r}_1, \dots, \boldsymbol{r}_N) \rangle \, , \tag{24}$$

the expectation value of the electron–electron interaction, is stable in the Sobolev-space norm. This stability extends to the whole potential energy, such that the virial theorem ([1], Th. 13.59) can be recommended as a test for trial functions, working only with stable quantities.

5. The Dipole Matrix Element and the Intensity of Radiative Recombination in EMA

The formula for the intensity of radiative recombination of excitons (or of electrons and holes localized at impurities) [21] contains the probability of radiative recombination, which is proportional to the square of the dipole matrix element. In EMA this term includes, according to Hopfield [22], the overlap integral

$$I_{\mathrm{ov}} = \int \psi(\boldsymbol{r}_{\mathrm{e}}, \boldsymbol{r}_{\mathrm{h}}) \, \delta(\boldsymbol{r}_{\mathrm{e}} - \boldsymbol{r}_{\mathrm{h}}) \, \mathrm{d}^3\boldsymbol{r}_{\mathrm{e}} \, \mathrm{d}^3\boldsymbol{r}_{\mathrm{h}} \, . \tag{25}$$

($\boldsymbol{r}_{\mathrm{e/h}}$ is the coordinate of the electron or the hole, respectively, ψ is the envelope function.) The intensity is usually only calculated with uncorrelated envelope functions $\psi(\boldsymbol{r}_{\mathrm{e}}, \boldsymbol{r}_{\mathrm{h}}) = \psi_{\mathrm{e}}(\boldsymbol{r}_{\mathrm{e}}) \, \psi_{\mathrm{h}}(\boldsymbol{r}_{\mathrm{h}})$ [21] such that (25) reduces to $I_{\mathrm{ov}} = \int \psi_{\mathrm{e}}(\boldsymbol{r}) \, \psi_{\mathrm{h}}(\boldsymbol{r}) \, \mathrm{d}^3\boldsymbol{r}$. The δ-distribution in (25) causes an instability of I_{ov} for the dimension $d = 3$ (cf. Section 2). This is not surprising, for the right-hand side of (25) looks like $g(0)$ as defined by (21), whose instability was discussed in Section 4.

The instability of I_{ov} is a result of EMA: The condition for the derivation of EMA is that the envelope function varies "very slowly". If we assume that the envelope function is constant over every basis cell, then the δ-term no longer produces any instability.

The overlap integral is not only unstable with respect to local perturbations of the envelope function, but is also unstable with respect to smooth global perturbations:

Functions in $L^2(R^6)$ which tend sufficiently slowly to zero do not lie in $L^1(R^6)$. This instability does not occur in $g(0)$, since there ψ enters with the second power instead of the first one as in (25). A global instability already appears after a first transformation of the exact formula for the dipole matrix element: The matrix element between the initial state ψ_i and the final state ψ_f reads ([23], (78.9))

$$m_d = \langle \psi_i, p\psi_f \rangle \,. \tag{26}$$

The Schwarz inequality implies that this term is stable with respect to small per-perturbations in the Sobolev-space norm,

$$\langle \psi_i, p\psi_f \rangle \leqq \langle \psi_i, \psi_i \rangle^{1/2} \langle p\psi_f, p\psi_f \rangle^{1/2} < \|\psi_f\|_s \,. \tag{27}$$

Due to the commutation relation, (27) can be transformed into

$$\langle \psi_i, p\psi_f \rangle = i(E_i - E_f) \langle \psi_i, r\psi_f \rangle \tag{28}$$

([23], (78.10)), where E_i and E_f are the energies of the states ψ_i and ψ_f, respectively. (The formula (28) only holds for the exact eigenfunctions of the Hamiltonian considered!) The multiplication operator r is not form-bounded with respect to $-\Delta$, i.e. small deviations of ψ_i and ψ_f, changing their asymptotic behaviour, can cause arbitrarily large changes of $\langle \psi_i, \psi_f \rangle$. Thus one should not try to estimate the dipole matrix element by the right-hand side of (28) where ψ_i and ψ_f have been approximated. If the particle is weakly bound, the small factor $(E_i - E_f)$ in (28) is compensated for by a large $\langle \psi_i, r\psi_f)$-term. To consider for weak binding only the second factor in (28) leads to the confusing impression that weak binding is connected with high recombination intensities [22, 24]. But in pseudopotential theory (cf. [32], §5) the pseudo-wave function does not describe well the momentum for small r. Therefore, in this case it seems to be better to work with the right-hand side of (28).

6. The Asymptotic Decay of Wave Functions

The localization length L for a wave function $\psi(\boldsymbol{r}_1, \ldots, \boldsymbol{r}_N) \in L^2(R^{dN})$ is defined by

$$L = \limsup_{r\to\infty} \frac{r}{\max\left(-\ln|\psi(\boldsymbol{r})|\,;\,0\right)}, \qquad \boldsymbol{r} = (\boldsymbol{r}_1, \ldots, \boldsymbol{r}_N) \tag{29}$$

and tells us that $\psi(r)$ decays in any direction at least as rapidly as $\exp\left(-r/(L+\varepsilon)\right)$ for arbitrarily small $\varepsilon > 0$. Another definition of the localization length reads

$$L' = \inf\left(l\colon \exp(rl)\,\psi(\boldsymbol{r}) \in L^2(R^{dN})\right)\,; \tag{30}$$

for sufficiently smooth ψ it holds $L = L'$. Obviously, $L \geqq L' \geqq 0$. Hamiltonians with decaying potential (such as the Coulomb potential) or with random potential have for arbitrary dimension d and particle number N ([4], § 13.11, [25]), or for $d = N = 1$ [26], exponentially localized eigenfunctions, where at least estimates for L (or L') are known. But these theorems are not appropriate for testing the quality of trial functions, since L (and L') are extremely unstable with respect to small perturbations (in the Sobolev-space norm (3)). The high instability can be demonstrated by two comparisons:

(i) L and L' are, in contrast to $\psi^2(0)$, also unstable for $N = d = 1$ (4).

(ii) They are, in contrast to $\mathrm{d}g(R)/\mathrm{d}R$, unstable for $d = 1$ with respect to perturbations being small in a norm defined by replacing in (3) $-\Delta_i$ by $\mathrm{d}^4/\mathrm{d}x_i^4$.

Rotenberg and Stein [28] have already remarked that L cannot be used for testing trial functions, although they found that variational calculations with trial functions showing the correct asymptotic behaviour sometimes lead to very good results. Another difficulty of the notion of the localization length reveals in a paper by Kimball [27], which has been discussed in [29]: The eigenfunctions of $-d^2/dx^2 + V^\omega$, $V^\omega \geqq 0$ random, are, for energies near the band edge $E = 0$, nearly constant over a large region, where V^ω is nearly zero, and they decay at least as rapidly as $\exp(-|x|/c)$ (the constant $c < \infty$ does not depend on the energy), but only asymptotically. Kimball calculated that the variance in the spatial distribution of the eigenfunctions tends to infinity for energies converging to the band edge.

The instability and the pathological behaviour at the band edges require an alternative definition of the localization length, which will correspond better to the intuitive ideas the physicists connect with this notation. The variance in the spatial distribution seems to be no way out, since it is also unstable in the $||...||_s$-norm. But the confidence intervals, known from statistics, represent a stable quantity measuring the extension of a wave function. The localization lengths L_a defined implicitly by

$$\int_{r \leqq L_a} \psi^2(\boldsymbol{r})\, d^{dN}\boldsymbol{r} = a \tag{31}$$

depend on an arbitrary parameter $\alpha \in (0, 1)$. Usually $a = 0.95$ or $a = 0.99$ are taken in statistics. Two disadvantages of definition (31) (and of the variance in the spatial distribution) should be mentioned:

(i) There are many rigorous statements about L (and L'), while one cannot expect related results for L_a.

(ii) If V^ω is a random (ergodic with respect to space translations) one-dimensional potential, L, but not L_a, is self-averaging, i.e. it almost surely does not depend on the realization of the potential.

Thus the value of L_a consists mainly of the possibility of comparing numerical calculations after averaging over many realizations.

In order to avoid confusion, we stress that the Lyapunov exponent J, which for random one-dimensional potentials is the inverse of the localization length L [26], is a stable quantity with respect to perturbations of the *potential* [30].

The definition of J is analogous to (29)

$$J = \limsup_{|x| \to \infty} \frac{\ln |\psi(x)|}{|x|}, \tag{32}$$

where $\psi(x)$ is an arbitrary solution (not an eigenfunction!) of

$$\left(-\frac{d^2}{dx^2} + V^\omega(x)\right)\psi(x) = 0\,. \tag{33}$$

For almost all initial conditions and realizations of the potential the limits in (32) coincide; this value is denoted by J [31].

Acknowledgements

I would like to thank Dipl.-Phys. R. Englisch for her participation in the initial stages of this work and Prof. K. Unger, Prof. W. Weller, and Prof. P. Ziesche for stimulating discussions.

References

[1] P. Erdös and R. C. Herndon, Adv. Phys. 31, 65 (1982).
[2] G. Lucovsky, Solid State Commun. 3, 299 (1965).
[3] E. Kiefer, Dissertation, Univ. Regensburg, 1977.

[4] M. REED and B. SIMON, Methods of Modern Mathematical Physics II/IV, Academic Press, New York/London 1975/1978.
[5] H. ENGLISCH and R. ENGLISCH, phys. stat. sol. (b) **123**, 711 (1984); **124**, 373 (1984).
[6] J. AVRON, A. GROSSMAN, and R. RODRIGUEZ, Ann. Phys. (USA) **103**, 47 (1977).
[7] S. ALBEVERIO and R. HOEGH-KROHN, J. Operator Theory **6**, 313 (1981).
[8] R. A. CHAPMAN and W. G. HUTCHINSON, Phys. Rev. **157**, 615 (1966).
[9] C. N. BERGLUND and H. J. BRAUN, Phys. Rev. **164**, 790 (1967).
[10] J. S. BLAKEMORE and C. E. SARVER, Phys. Rev. **173**, 767 (1968).
[11] H. B. BEBB, Phys. Rev. **185**, 1116 (1969).
[12] T. KATO, Commun. pure appl. Math. **10**, 151 (1957).
[13] M. HOFFMANN-OSTENHOF and R. SEILER, Phys. Rev. A **23**, 21 (1981).
[14] B. R. JOHNSON, Phys. Rev. A **24**, 2339 (1981).
[15] W. A. BINGEL, Z. Naturf. **18a**, 1249 (1963).
[16] J. C. KIMBALL, J. Phys. A **8**, 1513 (1975).
[17] N. H. MARCH, Self-Consistent Fields in Atoms, Pergamon Press, Oxford/New York/Toronto/Sydney/Paris/Braunschweig 1975 (p. 105).
[18] M. HOFFMANN-OSTENHOF, T. HOFFMANN-OSTENHOF, and W. THIRRING, J. Phys. B **11**, L571 (1978).
[19] R. T. PACK and W. BYERS BROWN, J. chem. Phys. **45**, 556 (1966).
[20] T. L. GILBERT, Rev. mod. Phys. **35**, 491 (1962).
[21] R. BINDEMANN and K. UNGER, phys. stat. sol. (b) **66**, 133 (1974).
[22] J. J. HOPFIELD, in: Internat. Conf. Semiconductors, Dunod, Paris 1964.
[23] A. S. DAWYDOW, Quantenmechanik, VEB Deutscher Verlag der Wissenschaften, Berlin 1967.
[24] E. I. RASHBA and G. E. GURGENISHVILI, Fiz. tverd. Tela **4**, 1029 (1962).
[25] R. CARMONA and B. SIMON, Commun. Math. Phys. **80**, 59 (1981).
[26] R. CARMONA, One-Dimensional Schrödinger Operators with Random Potentials, Preprint, Irvine 1983.
[27] J. C. KIMBALL, J. Phys. C **13**, 5701 (1980).
[28] M. ROTENBERG and J. STEIN, Phys. Rev. **182**, 1 (1969).
[29] H. ENGLISCH, J. Phys. C **17**, L17 (1984).
[30] YU. L. DALETSKII and M. G. KREIN, Stability of Solutions of Differential Equations in Banach Spaces, Izd. Nauka, Moscow 1970 (in Russian).
[31] L. A. PASTUR, Commun. Math. Phys. **75**, 179 (1980).
[32] R. SANDROCK, Festkörperprobleme **10**, 283 (1970).

(Received April 23, 1984)

phys. stat. sol. (b) **125**, 401 (1984)

Subject classification: 6 and 13.4; 10.2; 22.5.2

Institute of Chemistry, Academy of Sciences of the Moldavian SSR, Kishinev[1])

Multimode Pseudo Jahn-Teller Effect for Off-Centre Impurities in Crystals

By

G. I. Bersuker and V. Z. Polinger

Systems with off-centre impurity in the orbitally nondegenerate ground state are discussed. The off-centre position of the impurity is shown to result from the mixing of the ground state and one of the excited electronic states by odd crystalline vibrations, displacing the impurity from the lattice point (pseudo Jahn-Teller effect, PJTE). At high enough temperatures transitions of the impurity between the equivalent off-centre positions occur due to the coupling with the PJTE active phonons. The qualitative estimation of the transition probability obtained for this process is in agreement with the line broadening data for the Li^+:KCl system.

Рассмотрены системы с нецентральными примесями, находящяиеся в орбитально невырожденном основном состоянии. Нецентральность обусловлена смешиванием локализованных вблизи примеси основного и одного из возбужденных электронных состояний системы с нечетными кристаллическими колебаниями выводящими примесный ион из узла решетки-псевдоэффектом Яна-Теллера (ПЭЯТ). С ростом температуры примесный ион совершает переходы между различными эквивалентными нецентральными положениями, индуцированные переопределенными вибронной связью фононами активными в ПЭЯТ. Получено выражение для вероятности таких процессов, качественные оценки для которой находятся в хорошем согласии с данными по ширинам линий для Li^+:KCl.

1. Introduction

A number of substituted impurities are known to be of off-centre type [1]. Usually the off-centre position of the impurity is explained in terms of the difference between the ionic radii of the impurity and the substituted atom. In definitely established cases this difference is of the order of 0.02 to 0.03 nm or more. As can be seen from the table of ionic radii, quite a number of impurities may be expected to be off-centre ones. The case should be pointed out when the impurity can anharmonically vibrate with a big amplitude in its cavity, the average position being in a lattice site. Such a "pseudo off-centre" impurity position also can result in some interesting experimental features similar to those of the "real off-centre" ones.

The displacement of the impurity is due to the lack of coincidence of the host lattice site with the position of the minima on the potential energy surface of the impurity system. In other words, the substitution of an atom of the host crystal by an impurity ion results in new minima of the potential energy (or, as in the case of the pseudo off-centre impurity, in an essential deviation of the well shape from the parabolic one), their number and positions being determined by the symmetry of the nearset environment. This feature of the system under consideration can be related to the peculiarities of the electronic structure of the impurity centre. In particular, the interaction of impurity electrons with lattice vibrations leads to an instability of the ground electronic state in the high-symmetry nuclear configuration (pseudo Jahn-Teller effect

[1]) 277028, Kishinev USSR.

(PJTE)), resulting in a displacement of the impurity from the lattice site. The PJTE nuclear configuration instability occurs, if the ground electronic state of the system formed by the impurity and the host crystal is coupled strongly enough to the excited electronic states by the appropriate nuclear displacements (when the PJTE active nuclear displacements lead to severe changes of the electron distribution in two electronic states). This is true when the states mentioned above (not necessarily being the nearest in energy) involve bonding and antibonding orbitals. Moreover, the vibronic interaction is the only possible reason for the off-centre positions of the impurities, since the nonvibronic contribution to the curvature of the adiabatic potential in the high-symmetry nuclear configuration was shown to be always positive [2].

In the previous paper the off-centre impurities (mainly transition metal ions, in particular Cu^{2+}:SrO) with degenerate ground electronic states (E-term) have been investigated in the framework of the vibronic approach [3]. In the present paper (see also [4]) we consider the wide-spread case of off-centre impurities with a singlet electronic ground state. A typical example is the well-known case of the Li^+ ion in the KCl crystal. The adiabatic potential of the system is investigated and the expression for the transition probability of the impurity between the off-centre positions is obtained. The qualitative estimation of the transition probability based on these results is in good agreement with the experimental data on the temperature dependence of the band width for the Li^+:KCl system.

2. Adiabatic Potential

The displacements taking the impurity off the lattice site are of T_{1u} type and mix the nondegenerate A_{1g} ground state with the T_{1u} type excited ones. The Jahn-Teller interaction of the T_{1u} term with e_g and t_{2g} vibrations also have to be taken into account. This problem is a multimode one since the impurity in a crystal interacts with the whole range of lattice vibrations of appropriate (e_g and t_{2g}) symmetries. Consider therefore the multimode PJTE in the $(A_{1g} + T_{1u}) \otimes (t_{1u} + e_g + t_{2g})$ system. The Hamiltonian of the impurity–phonon system in the basis of the A_{1g} and T_{1u} states has the following form:

$$H = H_{\text{cryst}} + H_{\text{PJT}} + H_{\text{JT}} + H_{\text{el}}\,, \tag{1}$$

where

$$H_{\text{cryst}} = \frac{1}{2} \sum_{\Gamma = t_{1u}, e_g, t_{2g}} \{\sum_{n\gamma} P^2_{n\Gamma\gamma} + \sum_{n,m,\gamma} (\omega^2_\Gamma)_{nm}\, Q_{n\Gamma\gamma} Q_{m\Gamma\gamma}\} + H' \tag{2}$$

is the Hamiltonian of the lattice harmonic vibrations given in the symmetrized displacements $Q_{n\Gamma\gamma}$ of the atoms of all coordination spheres, n numbering both the different coordination spheres and the vibrational representations of the same Γ type, and γ are the rows of the Γ representation [5]. In (2) mass-weighted coordinates are used. The Hamiltonian H' corresponds to all the rest of vibrations of interest. The Hamiltonian

$$H_{\text{PJT}} = t\left[\begin{pmatrix}0&1&0&0\\1&0&0&0\\0&0&0&0\\0&0&0&0\end{pmatrix} Q_{1x} + \begin{pmatrix}0&0&1&0\\0&0&0&0\\1&0&0&0\\0&0&0&0\end{pmatrix} Q_{1y} + \begin{pmatrix}0&0&0&1\\0&0&0&0\\0&0&0&0\\1&0&0&0\end{pmatrix} Q_{1z}\right] \tag{3}$$

describes the pseudo Jahn-Teller (PJT) mixing of the ground and excited terms; t is the coupling constant for the PJT active t_{1u} vibrations. Q_{1x}, Q_{1y}, Q_{1z} are the t_{1u} type atomic displacements of the first coordination sphere. From here on the index of representation Γ will be omitted for simplicity. The Hamiltonian

$$H_{\mathrm{JT}} = e\left[\begin{pmatrix} 0 & 0 & 0 & 0 \\ 0 & -1 & 0 & 0 \\ 0 & 0 & -1 & 0 \\ 0 & 0 & 0 & 2 \end{pmatrix} Q_{1\theta} + \sqrt{3}\begin{pmatrix} 0 & 0 & 0 & 0 \\ 0 & 1 & 0 & 0 \\ 0 & 0 & -1 & 0 \\ 0 & 0 & 0 & 0 \end{pmatrix} Q_{1\varepsilon}\right] +$$
$$+ v\left[\begin{pmatrix} 0 & 0 & 0 & 0 \\ 0 & 0 & 0 & 0 \\ 0 & 0 & 0 & 1 \\ 0 & 0 & 1 & 0 \end{pmatrix} Q_{1\xi} + \begin{pmatrix} 0 & 0 & 0 & 0 \\ 0 & 0 & 0 & 1 \\ 0 & 0 & 0 & 0 \\ 0 & 1 & 0 & 0 \end{pmatrix} Q_{1\eta} + \begin{pmatrix} 0 & 0 & 0 & 0 \\ 0 & 0 & 1 & 0 \\ 0 & 1 & 0 & 0 \\ 0 & 0 & 0 & 0 \end{pmatrix} Q_{1\zeta}\right] \tag{4}$$

corresponds to the JT interaction of the T_{1u} term, with the symmetrized displacements $Q_{1\theta}$, $Q_{1\varepsilon}$ and $Q_{1\xi}$, $Q_{1\eta}$, $Q_{1\zeta}$ of the nearest atoms transforming according to the e_g and t_{2g} representations; e and v being the corresponding vibronic coupling constants. It is assumed that the impurity electrons are localized within the first coordination sphere. The Hamiltonian

$$H_{\mathrm{el}} = \Delta\begin{pmatrix} -3 & 0 & 0 & 0 \\ 0 & 1 & 0 & 0 \\ 0 & 0 & 1 & 0 \\ 0 & 0 & 0 & 1 \end{pmatrix} \tag{5}$$

describes the effect of crystal field where Δ is the energy gap between the ground and excited terms under consideration.

Let us investigate the potential energy surface of the system. It should be noted firstly that the interaction of the impurity ion with the host crystal cannot be considered as a small perturbation because it is this interaction that results in the large displacements of the impurity ion from the lattice site.

For this purpose one has to solve the secular equation for the potential energy matrix in (1) neglecting the Jahn-Teller interaction H_{JT}, i.e. the JTE is assumed to be much smaller than the PJTE. In this starting approximation ($e = v = 0$) the lowest sheet of the adiabatic potential possesses an equipotential continuum of minima in the subspace of the t_{1u} displacements. These minima are placed on the surface of a sphere,

$$\varepsilon = \tfrac{1}{2}\sum_{n,m,\gamma} (\omega^2_{\mathrm{T}_{1u}})_{nm}\, Q_{n\gamma}Q_{m\gamma} - \Delta - \sqrt{4\Delta^2 + t^2(Q^2_{1x} + Q^2_{1y} + Q^2_{1z})}\,. \tag{6}$$

In (6) the PJTE is assumed to be strong enough, $E_{\mathrm{PJT}} \gg \hbar\omega$, where ω is the maximum frequency of vibrations in the minima of the lowest sheet, thus the latter can be considered only, having the physical meaning of the potential energy of the system. In first order of perturbation theory, considering the interaction H_{JT} to be small enough, one gets the following expression for the potential energy:

$$\varepsilon = \tfrac{1}{2}\sum_{n,m}\ \sum_{\Gamma = t_{1u}, e_g, t_{2g}}\ \sum_{\gamma} (\omega^2_{\Gamma})_{nm}\, Q_{n\gamma}Q_{m\gamma} + V\,, \tag{7}$$

$$\begin{aligned} V = &- \Delta - \sqrt{4\Delta^2 + t^2\varrho_1^2} + a\varrho_1^2\,\{e[(3\cos^2\chi_1 - 1)\,Q_{1\theta} + \\ &+ \sqrt{3}\sin^2\chi_1(\cos^2\varphi_1 - \sin^2\varphi_1)\,Q_{1\varepsilon}] + v[\sin 2\chi_1 \sin\varphi_1 Q_{1\xi} + \\ &+ \sin 2\chi_1\cos\varphi_1 Q_{1\eta} + \sin^2\chi_1 \sin 2\varphi_1 Q_{1\zeta}]\}\,, \end{aligned} \tag{8}$$

where $a = t^2/2T(T + 2\Delta)$, $T = \sqrt{4\Delta^2 + t^2\varrho_1^{(0)2}}$, $\varrho_1^{(0)}$ is the equilibrium value of ϱ_1 for the potential energy ε from (6). In (8) spherical coordinates are used: $Q_{nx} = \varrho_n \cos \varphi_n \sin \chi_n$, $Q_{ny} = \varrho_n \sin \varphi_n \sin \chi_n$, $Q_{nz} = \varrho_n \cos \chi_n$.

It can be seen that e_g and t_{2g} vibrations reduce the symmetry to the cubic one, i.e. with account of JT coupling with e_g and t_{2g} vibrations minima placed on the second-, third-, and fourth-order symmetry axes appear. The eight-well potential of XY_8-type with minima placed on the third-order axes are of interest in the present paper and our consideration will now be restricted to potentials of this type (Fig. 1).

The minima placed on the third-order axes must be absolute ones for the model of the eight-minima potential to be available. This condition results in some restrictions on the vibronic coupling constants and frequencies of Hamiltonian (1) which can be obtained approximately by the determination of the extrema of (8) or, more exactly, by the method of Opik and Pryce [6]. One can carry out successively two rotations by φ_1 and χ_1 in the space of the t_{1u} displacements of the all coordination spheres with the help of the unitary shift operators S,

$$S = \exp \left\{ \chi_1 \sum_{n \neq 1} \frac{\partial}{\partial \chi_n} \right\} \exp \left\{ \varphi_1 \sum_{n \neq 1} \frac{\partial}{\partial \varphi_n} \right\}. \tag{9}$$

The Hamiltonian of the system, determined by (7) with the addition of the kinetic energy operator $(1/2) \sum_{\Gamma \gamma n} P^2_{n\Gamma\gamma}$, when effected by the operator S, takes the following form:

$$\begin{aligned}\tilde{H} = SHS^{-1} \approx & \frac{1}{2} \sum_{\Gamma} \sum_{\gamma \neq x, y} \{ \sum_{n} P^2_{n\Gamma\gamma} + \sum_{n, m} (\omega^2_{\Gamma})_{nm} Q_{n\Gamma\gamma} Q_{m\Gamma\gamma} \} + \\ & + \frac{1}{2} \sum_{n \neq 1} \sum_{\gamma = x, y} P^2_{n\gamma} + \frac{1}{2} \sum_{n, m \neq 1} \sum_{\gamma = x, y} (\omega^2_{T_{1u}})_{nm} Q_{n\gamma} Q_{m\gamma} + \\ & + \frac{1}{2Q^2_{1z}} \left[- i\hbar \frac{\partial}{\partial \chi_1} - \sum_{n \neq 1} (L_{ny} \cos \varphi_1 + L_{nx} \sin \varphi_1) - \frac{i\hbar}{2 tg \chi_1} \right]^2 + \\ & + \hbar^2 (8Q^2_{1z} \, tg^2 \chi_1)^{-1} + (2Q^2_{1z} \sin^2 \chi_1)^{-1} \left[-i\hbar \frac{\partial}{\partial \varphi_1} - \sum_{n \neq 1} L_{nz} \right]^2 + V \,. \end{aligned} \tag{10}$$

Here $L_{n\gamma}$ ($\gamma = x, y, z$) are the operators of the moments of the t_{1u} displacements of the n-th coordination sphere, $L_{nz} = Q_{nx} P_{ny} - Q_{ny} P_{nx} = -i\hbar (\partial / \partial \varphi_n)$, and so on. To obtain the coordinates of the extrema of interest, corresponding to the displacement of the impurity ion along the [111] direction the following set of equations has to be solved:

$$\left. \frac{\overline{\partial H}}{\partial P_{n\gamma}} \right|_{P_{n\gamma} = P^{(0)}_{n\gamma}} = 0 \,, \qquad \left. \frac{\overline{\partial H}}{\partial Q_{n\gamma}} \right|_{Q_{n\gamma} = Q^{(0)}_{n\gamma}} = 0 \,. \tag{11}$$

In (11) the quantum mechanical averaging was performed, $P^{(0)}_{n\gamma}$, $Q^{(0)}_{n\gamma}$ are the equilibrium values of the momenta $P_{n\gamma}$ and the conjugated displacements $Q_{n\Gamma}$. The values of $P^{(0)}_{nx}$, $P^{(0)}_{ny}$, $Q^{(0)}_{nz}$ determined by (11) differ from zero, while $P^{(0)}_{nz} = Q^{(0)}_{nx} = Q^{(0)}_{ny} = 0$, i.e. we get the coordinates of the minimum placed on the z-axis in the new coordinate system rotated by the operator S. The calculations are simplified essentially for a minimum on one of the coordinate axes. The z-axis can be oriented in the direction of one of the eight minima, and the positions of the other ones can be obtained by symmetry transformation.

Performing the substitution $Q_{n\gamma} = Q_{n\gamma}^{(0)} + q_{n\gamma}$; $P_{n\gamma}^{(0)} = P_{n\gamma}^{(0)} + p_{n\gamma}$ and reading the variables from the new equilibrium positions $Q_{n\gamma}^{(0)}$, $P_{n\gamma}^{(0)}$ in the senior order with regard to the values $Q_{n\gamma}^{(0)}(P_{n\gamma}^{(0)} \sim 1/Q_{n\gamma}^{(0)})$, we get

$$\begin{aligned}\tilde{H} \approx & -\frac{\hbar^2}{2\varrho_0^2}\left[\frac{\partial}{\partial\chi_1} + \frac{1}{2\,\mathrm{tg}\,\chi_1} - \frac{i}{\hbar}\sum_{n\neq 1}(Q_{nz}^{(0)}p_{nx}\cos\varphi_1 - Q_{nz}^{(0)}p_{ny}\sin\varphi_1)\right]^2 + \\ & + \frac{\hbar^2}{8\varrho_0^2\,\mathrm{tg}^2\,\chi_1} - \frac{\hbar^2}{2\varrho_0^2\sin^2\chi_1}\frac{\partial^2}{\partial\varphi_1^2} + 2a\varrho_0^2 v\,(\sin\chi_1\cos\chi_1\sin\varphi_1 Q_{1\xi}^{(0)} + \\ & + \sin\chi_1\cos\chi_1\cos\varphi_1 Q_{1\eta}^{(0)} + \sin^2\chi_1\cos\varphi_1\sin\varphi_1 Q_{1\zeta}^{(0)}) + E_0 + \\ & + \frac{1}{2}\sum_{n\neq 1}\sum_{\gamma=x,y} p_{n\gamma}^2 + \frac{1}{2}\sum_{n,m\neq 1}\sum_{\gamma=x,y}(\omega_{\mathrm{T_{1u}}}^2)_{nm}\,q_{n\gamma}q_{m\gamma} + \\ & + \frac{1}{2}\sum_{\Gamma}\sum_{\gamma\neq x,y}\{\sum_n p_{n\Gamma\gamma}^2 + \sum_{n,m}(\omega_\Gamma^2)_{nm}\,q_{n\Gamma\gamma}q_{m\Gamma\gamma}\}\,. \end{aligned} \tag{12}$$

Here, in the square brackets the strongest part ($\sim Q_{nz}^{(0)}$) of the φ_1 dependence is kept only and the coupling of the T_{1u} term with e_g vibrations is omitted for brevity. All the terms independent of the dynamical variables are included in the energy E_0. In (12) for convenience the notation $Q_{1z}^{(0)} = \varrho_0$ is introduced where ϱ_0 is the displacement of the impurity from the centre in the direction [111]. This displacement is caused by the interaction of the impurity ion with the electronic cloud, distributed according to a low symmetry within the first coordination sphere. As mentioned above the coordinate system is oriented in such a way that the impurity is displaced along the z-axis. In this case the equilibrium positions $Q_{nz}^{(0)}$ of atoms of all the other coordination spheres are changed due to the elastic interaction. Thus in the case under consideration a multielectron formation of polaron type appears similar to the case of the multimode Jahn-Teller effect [7].

3. The Transitions of the Impurity between the Off-Centre Positions

The motion of the system under consideration (polaron formation) which is formed by the impurity ion with surrounding electrons and lattice phonons of t_{1u}-type, can be regarded as small harmonic vibrations near equilibrium positions at the bottom of one of the eight minima and the motion of the whole system along the angular variables φ_1 and χ_1 between these minima. The above vibrations ($\omega > 10^{12}\ \mathrm{s}^{-1}$) can be considered as a fast subsystem compared with the transitions of the polaron formation between the minima (for example in the case of Li^+:KCl the transition frequency is of the order of $\approx 10^9\ \mathrm{s}^{-1}$). One can separate these subsystems adiabatically if the operator of non-adiabaticity, which mixes them, is small enough. This condition is assumed to be fulfilled. As will be shown below for an example, the approximation employed is valid in some cases.

The Hamiltonian of the fast subsystem can be obtained by eliminating the terms with $\partial/\partial\varphi_1$ and $\partial/\partial\chi_1$ from (12) and fixing the values φ_1 and χ_1 in the minimum (on the z-axis) of the adiabatic potential,

$$\begin{aligned} H_{\mathrm{vib}} = & \tfrac{1}{2}\sum_{n\neq 1}\sum_{\gamma=x,y} p_{n\gamma}^2 + \tfrac{1}{2}\sum_{n,m\neq 1}\sum_{\gamma=x,y}(\omega_{\mathrm{T_{1u}}}^2)_{nm}\,q_{n\gamma}q_{m\gamma} + \\ & + \tfrac{1}{2}\,\varrho_0^{-2}(\sum_{n\neq 1} Q_{nz}^{(0)}p_{nx})^2\cos^2\varphi_1 + \tfrac{1}{2}\,\varrho_0^{-2}(\sum_{n\neq 1} Q_{nz}^{(0)}p_{ny})^2\sin^2\varphi_1 - \\ & - \tfrac{1}{2}\,\varrho_0^{-2}(\sum_{n\neq 1} Q_{nz}^{(0)}p_{nx})(\sum_{n\neq 1} Q_{nz}^{(0)}p_{ny})\sin 2\varphi_1 + \tfrac{1}{2}\sum_n\sum_\Gamma\sum_{\gamma\neq x,y} p_{n\Gamma\gamma}^2 + \\ & + \tfrac{1}{2}\sum_{n,m}\sum_\Gamma\sum_{\gamma\neq x,y}(\omega_\Gamma^2)_{nm}\,q_{n\Gamma\gamma}q_{m\Gamma\gamma}\,. \end{aligned} \tag{13}$$

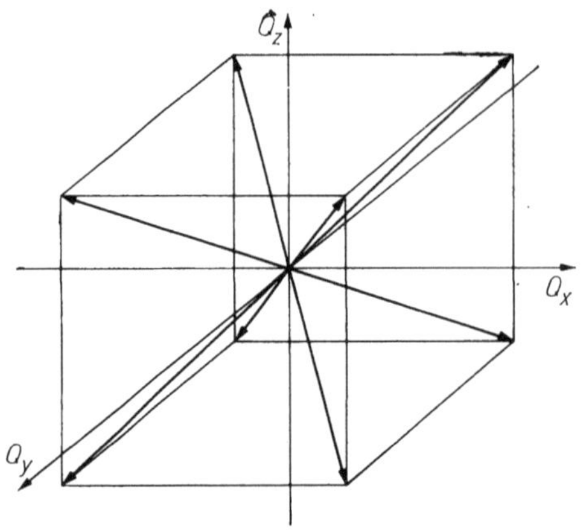

Fig. 1. The directions of displacements of the impurity ion corresponding to the XY_8 potential model

The continious energy spectrum of the Hamiltonian (13) can be investigated with the help of the two-time Green's functions [8]. The account of vibronic interaction is seen to change essentially the vibrations of x and y types in comparison with the initial ones (see Hamiltonian (2)), the frequencies of $q_{n\gamma}$ vibrations ($\gamma = z, \xi, \eta, \zeta$) being the same. In particular the Green's function $\langle\langle p_{nx} \mid p_{my} \rangle\rangle$ (it is convenient to take its scalar product with $Q^{(0)}_{nz}Q^{(0)}_{mz}$) is of the following form:

$$G(\omega) = \sum_{n,m\neq 1} Q^{(0)}_{nz}Q^{(0)}_{mz} \langle\langle p_{nx} \mid p_{my} \rangle\rangle = K^2\pi^{-1}(\overline{\omega^{-2}})^2 \, (\overline{\omega^{-2}} + g_0 - \omega^2\overline{\omega^{-2}}g_0)^2 \times$$

$$\times \, (\overline{\omega^{-2}} + g_0 + \omega^2\overline{\omega^{-2}}g_0)^{-1} \, [\gamma(\overline{\omega^{-2}} + g_0 - \omega^2\overline{\omega^{-2}}g_0) + 2\overline{\omega^{-2}}\omega^2 g_0]^{-1} \,, \quad (14)$$

where g_0 is Green's function of the t_{1u} vibrations of the initial (without vibronic interaction) crystal,

$$\gamma = 2 \, \frac{\omega^2\overline{\omega^{-2}}g_0}{g_0 + \overline{\omega^{-2}} + \omega^2\overline{\omega^{-2}}g_0} \,, \quad K \approx \left(t^2\overline{\omega^{-2}} - \frac{4\Delta^2}{t^2\overline{\omega^{-2}}}\right)^{1/2},$$

$$\overline{\omega^{-2}} = \text{Re}\, g_0(0) = -\frac{1}{\pi} \int \text{Im}\, g_0(z) \, \frac{dz}{z} \,. \quad (15)$$

Here Re $g_0(\omega)$ is the real part of $g_0(\omega)$, Im $g_0(\omega)$ is the initial density of the t_{1u} state of the crystal. The terms caused by the weak Jahn-Teller interaction of the T_{1u} term with t_{2g} vibrations are omitted from the above expression for K.

It follows from (14) that the vibronic coupling results in the radical redetermination of the vibrational spectrum for two components (of x and y types for the above choice of the operator S (9) in the case under consideration) of the t_{1u} vibrations. As will be found these redetermined vibrations are the most important ones in the migration of the off-centre impurity.

Averaging the Hamiltonian (12) over the ground state of the fast subsystem we get the Hamiltonian of the slow one, which corresponds to transitions of the polaron forma-

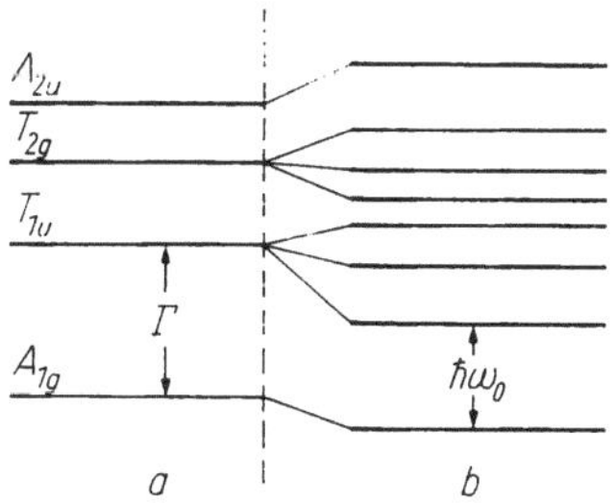

Fig. 2. The multiplet of the lowest energy vibronic levels of the slow subsystem (Hamiltonian (16). a) The case of negligible values of random strains, b) the case of strong random strains as compared with the tunneling gap Γ

tion between the eight minima of the potential surface,

$$H_{\text{rot}} = -\frac{\hbar^2}{2\varrho_0^2}\left[\frac{\partial}{\partial\chi_1}+\frac{1}{2\,\text{tg}\,\chi_1}+l_0\sin 2\varphi_1\right]^2+\frac{\hbar^2}{8\varrho_0^2\,\text{tg}^2\,\chi_1}-$$
$$-\frac{\hbar^2}{2\varrho_0^2\sin^2\chi_1}\frac{\partial^2}{\partial\varphi_1^2}-2va\varrho_0^2\,(\sin\chi_1\cos\chi_1\sin\varphi_1 Q_{1\xi}^{(0)}+$$
$$+\sin\chi_1\cos\chi_1\cos\varphi_1 Q_{1\eta}^{(0)}+\sin^2\chi_1\sin\varphi_1\cos\varphi_1 Q_{1\zeta}^{(0)})\,, \tag{16}$$

where $l_0 = \int_{-\infty}^{\infty} \text{Im}\, G(\omega)\, d\omega$, $\text{Im}\, G(\omega)$ is the imaginary part of $G(\omega)$ of [14].

The Hamiltonian (16) is equivalent up to the omitted small terms to the model Hamiltonian of the eight-well potential XY_8, which has been investigated beginning with [9]. All the differences in features are due to the unitary transformation S. But in contrast to the phenomenological approaches used earlier the values of parameters in (16) are determined by the electronic structure of the impurity centre and the phonon dispersion law of the host crsytal (for instance $\varrho_0^2 = K\overline{\omega^{-2}}$, where K and $\overline{\omega^{-2}}$ are determined by (15)).

Thus the energy spectrum and the wave functions obtained earlier in the frame of the XY_8 model are valid (taking into account the unitary transformation S) in the present approach, too. The eightfold degeneracy of the ground state is lifted due to tunneling and a multiplet of four terms is formed, two of them being threefold degenerated, cf. Fig. 2. It should be noted that the crystal structure is not perfect. The random strains caused by the dislocations, vacancies, impurities, and so on [10] result in the nonequivalency of the eight minima, thereby lifting the residual threefold degeneracy of the lowest states of the Hamiltonian (16). The external electric field, randomly oriented affects the energy spectrum of the system under consideration in the same way. Every one of the resulting singlet states $|i\rangle$ can be considered as localized in its i-th minimum. Thus the transitions of the system between these localized states are accompanied by the hopping of the impurity between the off-centre positions and by the corresponding distortion of the lattice.

The transition mentioned above between the discrete states of the slow subsystems is caused by a small nonadiabacity. The operator of nonadiabacity as follows from (10), in the senior order with regard to $Q_{nz}^{(0)}$ is of the following form (for brevity this term has been omitted in (12)):

$$W \approx \frac{i\hbar}{\varrho_0^2}\frac{\partial}{\partial\chi_1}\sum_{n\neq 1} Q_{nz}^{(0)}(p_{nx}\cos\varphi_1 - p_{ny}\sin\varphi_1)\,. \tag{17}$$

As one can see from (17), the transitions of interest are determined by the x and y components of the t_{1u} type lattice vibrations, which are strongly affected by the vibronic coupling. The system is locked by the random strains in the i-th minimum and thereby the starting state $|i\rangle$ is prepared. Considering the transition to the state $|j\rangle$ localized in the j-th minimum as radiationless and limiting the treatment by the one-phonon processes, one can get for the transition probability R_{ij} the following expression:

$$R_{ij} = \frac{\hbar^2 M_{ij}^2}{\pi \varrho_0^2} \frac{\operatorname{Im} g_0(\omega)}{2\omega^2[\operatorname{Re} g_0(\omega)^2 + \operatorname{Im} g_0(\omega)^2]} \left[\exp\left(\frac{\hbar\omega}{kT}\right) - 1\right]^{-1}_{\omega=\omega_0}, \tag{18}$$

where $M_{ij} = \langle i| \, \partial/\partial\chi_1 \, |j\rangle$, $\hbar\omega_0$ is the energy of the transition. Taking the Debye law for the crystalline t_{1u} type vibrations, taking $\hbar\omega_0 \approx 10^{12}\ s^{-1}$, $M_{ij} \approx 0.1$ (analogous to the vibronic reduction factor of the angular momentum operator) and taking into account that the magnitude of the impurity displacement is of the order of 0.01 nm and $\hbar\omega_0 > kT$, one gets $R_{ij} \approx 10^{-9} T$ (s K)$^{-1}$. The value of R_{ij} is of the same order of magnitude as the experimental data for Li^+ ions in alkali halides [11]. The vibronic approach used above offers the opportunity to separate adiabatically the system under consideration and thereby to single out the interaction causing the motion of the impurity ion. The value of R (in energy units) as compared with $\hbar\omega_0$ evidences that the operator of nonadiabacity is small enough (in terms of the perturbation theory), i.e. the separation of the system into the slow and fast ones is justified.

References

[1] M. F. Deigen and M. I. Glinchuk, Uspekhi fiz. Nauk **114**, 185 (1974). V. Naraynamurti and R. O. Pohl, Rev. mod. Phys. **42**, 201 (1970).
[2] I. B. Bersuker, Nouveau J. Chem. **4**, 139 (1980).
[3] G. I. Bersuker, V. Z. Polinger and, V. P. Khlopin, Fiz. tverd. Tela **24**, 2471 (1982).
[4] G. I. Bersuker and V. Z. Polinger, Internat. Symp. Synergetics and Cooperative Phenomena, Tallin 1982.
[5] V. Z. Polinger and G. I. Bersuker, phys. stat. sol. (b) **95**, 403 (1979).
[6] U. Opik and M. H. L. Pryce, Proc. Roy. Soc. **A 238**, 425 (1957).
[7] G. I. Bersuker and V. Z. Polinger, Solid State Commun. **38**, 795 (1981); Soviet Phys.—J. exper. theor. Phys. **53**, 930 (1981).
[8] D. N. Zubarev, Uspekh. fiz. Nauk **71**, 71 (1960).
[9] M. Gomez, S. P. Bowen, and J. A. Krumhansl, Phys. Rev. **53**, 1009 (1967).
[10] R. W. Timme, B. Dischler, and T. L. Estle, Phys. Rev. **B 1**, 1610 (1970).
[11] R. Osswald and H. C. Wolf, phys. stat. sol. (b) **50**, K93 (1972).

(Received March 12, 1984)

phys. stat. sol. (b) **125**, 409 (1984)

Subject classification: 2 and 14.4; 22.9

Departamento de Física, Facultad de C. Químicas, Universidad del País Vasco, San Sebastian[1])

Dielectric Relaxation at the Glass Transition as a Free Volume Process

II. A Continuous Distribution of Relaxation Times[2])

By

A. ALEGRÍA, J. M. BARANDIARÁN[3]), and J. COLMENERO

A previous work describing the α dielectric relaxation in glasses by means of a single, free volume-dependent, relaxation time, is generalized by allowing for a distribution of relaxation times. Such a distribution is assumed to arise from differences in free volume within the glass. A phenomenological parameter is introduced for describing the interaction among the different zones. By using a temperature independent, symmetric, distribution in the inverse of the free volume expansion coefficient the experimentally found, asymmetric and temperature- as well as thermal-story dependent distribution of dielectric relaxation times is reproduced. The physical meaning of the assumed distribution is discussed together with some implications on the dielectric response of a glass. Comparison with measurements performed on PVC shows good agreement.

Dans ce papier on fait une généralisation d'un travail précédent où on donnait une description de la relaxation diélectrique α dans des verres en utilisant un seul temps de relaxation qui dépendait du volume libre. Nous supposons maintenant une distribution de temps de relaxation qui apparaît comme consequence des différences de volume libre dans le verre. Les interactions entre les zones avec une différente quantité de volume libre sont décrites à l'aide d'un paramètre phénomenologique. On est capable de reproduire les distributions des temps de relaxation diélectrique telles qu'on les trouve expérimentalement, c'est à dire asymétriques et dépendentes de la température aussi bien que de l'histoire thermique, tout en utilisant une distribution en l'inverse du coefficient d'expansion du volume libre qui est, par contre, symétrique et indépendente de la température. On discute la signification physique d'une telle distribution aussi bien que ses implications pour la réponse diélectrique d'un verre. Des mesures faites dans le PVC sont en bon accord avec le modèle proposé.

1. Introduction

In a previous paper [1] reported as the first part of this work, we have shown that the mean features of dielectric relaxation around the glass transition can be accounted for by a free volume dependent relaxation time. However, a single relaxation time process has never been found in practice. The treatment of dielectric data is usually done in terms of empirical expressions like those of Cole-Cole, Cole-Davidson, or the most general of Havriliak-Negami [2] which relate the complex dielectric constant to frequency for a given temperature. It must be emphasized that both the mean relaxation time and the spread of the distributions are found to be temperature dependent in most cases [3]. The compensation laws arising in TSDC are clear examples of this behaviour [4].

---

[1]) B° de Alza, San Sebastián, Spain.

[2]) Part I see phys. stat. sol. (b) **120**, 349 (1983).

[3]) Present address: Departamento de Electricidad y Electrónica, Universidad del País Vasco, Lejona, Vizcaya, Spain.

More recently, Jonscher and others have reviewed the existing data and proposed a "universal" dielectric response law, which is fulfilled in all solids [5]. Although the Jonscher law seems to be directly related to the physical mechanism of the dielectric relaxation in solids [6] it is almost equivalent to the empirical Havriliak-Negami formula description [7].

In any case, an independent hypothesis is needed when the temperature is taken into account. Arrhenius or Vogel-Fulcher behaviour is assumed for the mean time of the distribution. In our previous paper [1], we succeeded in explaining the complex thermal behaviour of the relaxation time around the glass transition by using a configurational dependence for it. This is a usual procedure when describing structural relaxation in glasses which is responsible for the dielectric relaxation at the glass transition. The aim of the present work is to account for the distribution in dielectric relaxation times by assuming a related distribution in the parameters describing the configuration and structural relaxation of the glass.

2. Phenomenological Distributions of Structural Relaxation Times

2.1 *Survey of usual formulations*

The most common multiparameter models used in describing structural relaxation in glasses can be divided in two main groups:

(i) those developed by Kovacs and coworkers [8, 9] based on a distributed free volume description of the configuration of the glass,

(ii) the model of Moynihan and others [10, 11] based on the fictive temperature representation and the response function.

These two formulations, as well as other more recent ones [12, 13], have been shown [14] to be equivalent and can be reduced to the multiparameter theory of Kovacs et al., a short description of which will be presented hereafter.

Apart from the usual thermodynamic variables, the description of the actual state of a glass involves a discrete set $\{\xi_i\}$ of N "internal degrees of freedom" or "order parameters". (No objection is made against a continuous distribution of ξ, although the discretized form has to be used for calculations.) At a given temperature the glass has an occupied volume v_0 and a free volume: $v_f = v_{eq} + v_{ex}$, where v_{eq} is the value of v_f at equilibrium and v_{ex} a "frozen-in", excess free volume.

The out-of-equilibrium glass is characterized by the value of the fractional excess free volume: $\delta = v_{ex}/v_\infty$, where $v_\infty = v_0 + v_{eq}$ is the equilibrium volume at the temperature under consideration.

This excess free volume is composed of the contributions δ_i of each order parameter ξ_i, so that $\delta = \sum_i \delta_i$. The evolution of each δ_i is given by the kinetic equation

$$\frac{d\delta_i}{dt} = -\frac{\delta_i}{\tau_i} - \Delta\alpha_i q\,, \tag{1}$$

which can be deduced from fundamental thermodynamic considerations [9].

Here q is the heating ($q > 0$) or cooling ($q < 0$) rate, $\Delta\alpha_i$ the contribution of ξ_i to the difference between the expansion coefficients of the liquid and the glass (as δ, the total $\Delta\alpha$ is given by $\Delta\alpha = \sum \Delta\alpha_i$). Finally τ_i is the relaxation time associated to ξ_i and δ_i, and is arbitrarily assumed to depend on both temperature and configuration through

$$\tau_i = \tau_{ri} a_T(T)\, a_\delta(\delta) = \tau_{ri} \exp\left[-\theta(T - T_r)\right] \exp\left[-(1 - x)\, \theta \frac{\delta}{\Delta\alpha}\right], \tag{2}$$

where θ and x are adequate parameters.

The dependence of each τ_i on the whole δ establishes the coupling between the N differential equations (1). The equilibrium relaxation times at the reference temperature, τ_{ri}, are assumed to fill a double box distribution extending over four decades in time.

Three main criticisms can be made to this expression for τ_i: First of all, the linearized expression (2) is only valid in a narrow temperature range around T_r and cannot reproduce the equilibrium ($\delta = 0$) behaviour of the glass which is experimentally found to be of the Vogel-Fulcher type.

A more general expression for a single τ was used in [1] to account for the observed temperature dependence at equilibrium, i.e.

$$\tau = \tau_0 \exp\left(\frac{1}{f + \delta}\right) = \tau_0 \exp\left(\frac{\gamma v^*}{v_f}\right), \tag{3}$$

where $f = v_{eq}/v_\infty$ is the equilibrium fractional free volume which, as a first approximation, can be taken to vary linearly with temperature as

$$f = \alpha_f(T - T_0)\,, \tag{4}$$

and $\gamma v^* \approx v_\infty$.

On the other hand, the distribution of relaxation times in (2) relates only to the pre-exponential factor (τ_{ri} being the only distributed parameter) so that it shifts as a whole when changing the temperature, in contrast with the observed experimental behaviour [14, 15].

This feature is in common with the treatment of Moynihan et al., and will be reproduced by taking a distribution in τ_0 when using (3).

Finally the dependence of each relaxation time (2) on the total excess free volume seems rather arbitrary. In the following we shall consider these points in detail and propose a new expression for the distribution of relaxation times.

2.2 Distributions in the free volume

By assuming an expression like (3) for each of the relaxation times in the distribution, we have in the most general case

$$\tau_i = \tau_{0i} \exp\left[\frac{1}{f_i + \delta_i}\right] = \tau_{0i} \exp\left(\frac{\gamma v^*}{v f_i}\right) \tag{5}$$

in which both τ_0 and the fractional free volume are distributed.

Distributions of the pre-exponential factor τ_0 can arise from differences in the relaxing units, but there will be discrete distributions spreading over a reduced range (no distribution is expected in a monoatomic glass), so that this possibility will not be taken into account for simplicity.

Regarding the free volume distribution, only f_i has to be specified, for each δ_i vanishes at equilibrium and is determined out of equilibrium by the thermal history through the kinetic equations (1).

From (5) both α_f and T_0 can be taken as distributed parameters. The inverse of the expansion coefficient of the free volume, $1/\alpha_f$, can be viewed as an energy barrier for the movement of the atoms or molecules of the glass. For instance, if a Lennard-Jones interacting potential is assumed, the anharmonic term, responsible for thermal expansion, has a direct relationship with the depth of the potential well. A distribution in energy barriers has been extensively used in very recent descriptions of structural relaxation in glasses. In this sense Harris [16] considers model of N non-interacting particles which can take two different states separated by energy barriers statistically distributed in height. Chen [17] uses an energy surface representation in the configurational space of Goldstein [18], and Gibbs et al. [19] propose the idea of a spectrum of available processes, whose activation energies are widely distributed.

So, we shall consider only $1/\alpha_f$ as distributed and take τ_0 and T_0 as fixed parameters. This procedure results in a temperature-dependent distribution of relaxation times which, as will be seen latter, is in good agreement with experiment.

Let $\Phi(1/\alpha_f)$ be the distribution function of α_f. In discrete form, we can choose N different values $1/\alpha_f$ at regular intervals of magnitude h, and give them a weight factor

$$g_i = \Phi\left(\frac{1}{\alpha_{fi}}\right) h\,. \tag{6}$$

With (3) for τ_i, (1) represents now a set of N independent differential equations to be solved for each δ_i. Coupling between them can be introduced by considering some influence of the average free volume of the glass $\langle v_f\rangle$ in the individual τ_i. In this way, we define an "effective" free volume $\bar{v}_{fi}$ for the motion of the structural units,

$$\bar{v}_{fi} = v_{\infty i}(\bar{f}_i + \bar{\delta}_i) = p v_{fi} + (1 - p)\,\langle v_f\rangle\,, \tag{7}$$

where p is a phenomenological parameter taking into account the interactions between zones with different amounts of free volume.

If $p = 0$, $\bar{v}_{fi} = \langle v_f\rangle$ for all the values of i and the glass relaxes as a whole with a single τ. For $p = 1$ we have $\bar{v}_{fi} = v_{fi}$ and the glass can be viewed as composed by N independent zones, each of them relaxing with a single time τ_i. Values of p between 0 and 1 are expected to occur in practice.

As a particular case, at equilibrium we can write (7) as

$$\frac{\bar{v}_{fi}}{v_{\infty i}} = \frac{\bar{v}_{eqi}}{v_{\infty i}} = \bar{f}_i = p\alpha_{fi}(T - T_0) + (1 - p)\,\langle\alpha_f\rangle\,(T - T_0) =$$

$$= \bar{\alpha}_{fi}(T - T_0)\,, \tag{8}$$

where $\bar{\alpha}_{fi}$ is an "effective" expansion coefficient for the corresponding $\bar{f}_i$ and $\langle\alpha_f\rangle$ the average one in the sample. The relationship between them is

$$\bar{\alpha}_{fi} = \langle\alpha_{fi}\rangle + p(\alpha_{fi} - \langle\alpha\rangle)\,. \tag{9}$$

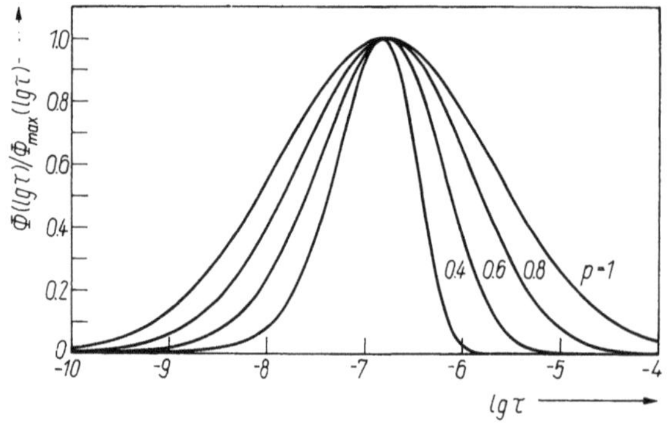

Fig. 1. Distribution of relaxation times from (12) showing the influence of parameter p. Values of other parameters are $T_0 = 300$ K, $T = 400$ K, $\alpha_f^* = 7\times10^{-4}$ K^{-1}, $\tau_0 = 10^{-13}$ s, $\sigma = 4$

As a summary, in this picture, the glass is composed of N different kinds of infinitesimal, but macroscopic, interacting regions. Each of them has a fractional excess free volume $\delta_i = (v_i - v_{\infty i})/v_{\infty i}$, and an equilibrium one $f_i = \alpha_{fi}(T - T_0)$. Relaxation of δ_i is driven by the equation

$$\frac{d\delta_i}{dt} = -\frac{\delta_i}{\tau_i} - \Delta\alpha_i q\,, \tag{10a}$$

where $\Delta\alpha_i$ is the "anelastic" part of the expansion coefficient of such a region [1]. The relaxation time

$$\tau_i = \tau_0 \exp\left(\frac{1}{f_i + \delta_i}\right) \tag{10b}$$

includes the interaction between different zones through the definition of the "effective" fractional free volume in (7). Finally, the "average" or total fractional excess and expansion coefficient of the free volume are

$$\delta = \langle\delta\rangle = \sum g_i\delta_i\,,$$
$$\alpha_f = \langle\alpha_f\rangle = \sum g_i\alpha_{fi}\,, \tag{10c}$$

where the normalized weight (g_i) gives the relative number of zones of each kind.

2.3 Some implications of the model

The development of the model needs an appropriate distribution function for $1/\alpha_f$ to be chosen. A simple one, like a Gaussian, can be tentatively used so if we write

$$\Phi\,\frac{1}{\alpha_f} = \frac{\sigma}{\sqrt{\pi}} \exp\left[-\sigma^2\left(\frac{\alpha_f - \alpha_f^*}{\alpha_f}\right)^2\right], \tag{11}$$

where $1/\alpha_f^*$ is the central value of the distribution. The corresponding distribution of relaxation times at equilibrium $(\delta_i = 0)$ is

$$\Phi \lg\frac{\tau}{\tau_0} = \frac{\sigma}{\sqrt{\pi}} \exp\left\{-\sigma^2\left[1 - \frac{\alpha_f^*}{\langle\alpha_f\rangle}\,\frac{p\lg(\tau/\tau_0)}{\lg(\tau/\tau_0) + (p-1)\lg(\tau/\tau_0)}\right]^2\right\}. \tag{12}$$

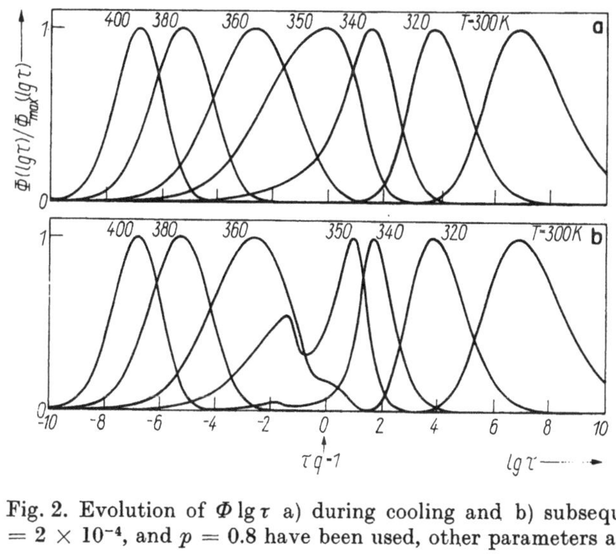

Fig. 2. Evolution of $\Phi \lg\tau$ a) during cooling and b) subsequent heating at $q = 1\ \mathrm{K\,s^{-1}}$. $\alpha_T = 2 \times 10^{-4}$, and $p = 0.8$ have been used, other parameters as in Fig. 1

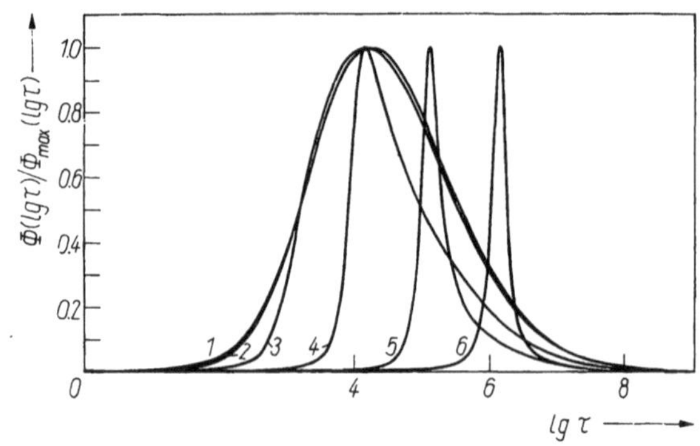

Fig. 3. Isothermal changes in Φ lg τ at 320 K after cooling from equilibrium at 1 K s^{-1} (1) as-quenched (2) after 10, (3) 10^2, (4) 10^3, (5) 10^4, (6) 10^5 s. Same parameters as above

As shown in Fig. 1, Φ lg τ is more or less asymmetric depending on the value of p. This will allow empirical, asymmetric distributions to be fitted by our model.

It must be noticed that the main value, α_f^*, and the average one, $\langle\alpha_f\rangle$, appearing in (12) are not the same, due to the fact that the distribution is symmetric with respect to $1/\alpha_f$, but not with respect to α_f. The same is true with the corresponding relaxation times τ^* and $\langle\tau\rangle$ whose difference depends also on the value of p.

Out-of-equilibrium behaviour can be obtained by numerical solution of the set of equations (10). For the calculation we have assumed that the "elastic" part of the free volume expansion coefficient [1], α_T, is the same, whichever the value of α_{fi}. Then, appropriate values of $\Delta\alpha_i$ are obtained as

$$\Delta\alpha_i = \alpha_{fi} - \alpha_T \,. \tag{13}$$

A discrete set of 40 relaxation times has been obtained from 40 equally spaced values of $1/\alpha_f$ in (11). The corresponding 40 differential equations (10a) have been solved using the procedure outlined in the previous paper [1], when dealing with a single relaxation time, in a HP-85 Desk Computer. Fig. 2 shows the evolution of the spectrum of relaxation times at constant rate cooling and subsequent heating. Starting from equilibrium at high temperature, the distribution spreads when cooling until the glass transition $\tau q \approx 1$ is reached, at this point the shape of Φ lg τ changes to a narrower one, but spreads again by further cooling in the glassy zone. When heating, the same behaviour is observed, but the effects are more pronounced, as can be seen in the very narrow curve at 340 K and the distortion around $\tau q = 1$.

Annealing at a constant temperature in the glassy state has also a dramatic effect on the distribution, as shown in Fig. 3. An interesting feature is the sharpening of the distribution without change in the main relaxation time at the first stages of annealing.

Although not represented, the main relaxation time in the distribution shows the same behaviour as a single relaxation time does both when cooling or heating and under isothermal conditions. This behaviour was shown in [1].

3. Dielectric Relaxation

As done in [1], we shall assume that, concerning the α relaxation, dielectric and structural behaviours are governed by the same distribution in relaxation times Φ lg τ.

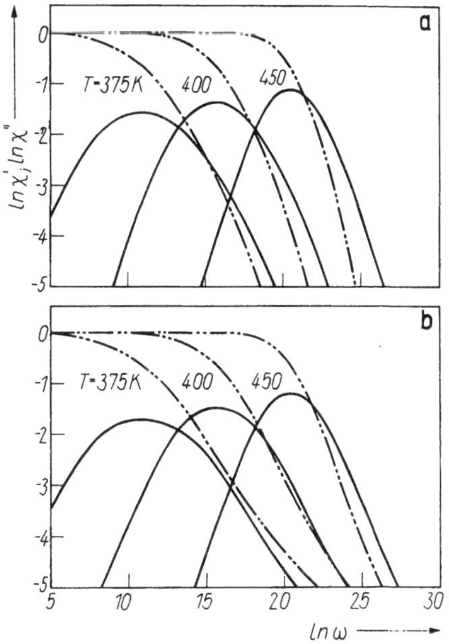

Fig. 4. Electric susceptibility (ln χ' (—··—), ln χ'' (———)) vs. frequency at different temperatures as predicted by the model for two distributions of $1/\alpha_f$. a) Gaussian expression (11) b) Lorentzian expression (18)

Dielectric relaxation is then viewed as a superposition of elementary processes, each of them obeying a Debye equation,

$$\frac{dP_i}{dt} = \frac{P_0 - P_i}{\tau_i}, \tag{14}$$

the total polarization is

$$P(t) = \sum g_i P_i(t) = \chi(t)\, \varepsilon_0 E \;, \tag{15}$$

where g_i are the normalized weight factors and E the applied field. P_0 is the equilibrium polarization, i.e. the asymptotic value of P when $t \to \infty$, and χ the electric susceptibility.

For sinusoidally varying fields, a frequency-dependent, complex, dielectric susceptibility can be defined as

$$\chi(\omega) = \chi' - i\chi'' = \frac{\varepsilon(\omega) - \varepsilon_\infty}{\varepsilon_s - \varepsilon_\infty} = \sum \frac{g_i}{1 + i\omega\tau_i}, \tag{16}$$

where ε_s and ε_∞ are the static ($\omega = 0$) and instantaneous ($\omega = \infty$) values of the dielectric constant. Expression (16) is the discretized form of the more general one [2]

$$\chi(\omega) = \int_{-\infty}^{+\infty} \frac{\Phi \lg \tau}{1 + i\omega\tau}\, d \lg \tau \;, \tag{17}$$

which corresponds to a continuous distribution of relaxation times. However, the discretized form has to be used for numerical calculation.

The equilibrium behaviour is usually studied by ac techniques, so that (10b) with $\bar{\delta}_i = 0$ can be used in (16) to obtain the model prediction in such a situation. Fig. 4a shows the curves calculated with a Gaussian distribution for $1/\alpha_f$. At high frequencies,

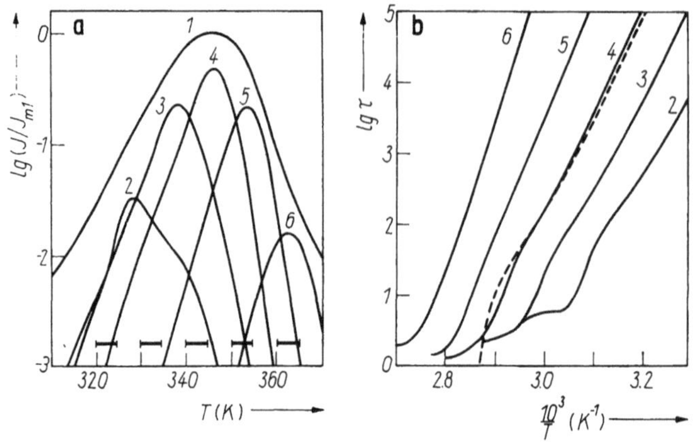

Fig. 5. a) Simulated TSDC spectra and b) corresponding Bucci times. Polarization interval is 380 to 300 K for peak 1 and is represented by thick lines in a) for other peaks. The dashed line in b) correspond to the evolution of a single relaxation time with the same parameters as the main time of the distribution. Cooling and heating rates are 0.3 and 0.2 K s^{-1}, respectively

$\chi'(\omega)$ varies as ω^{-2} although χ'' varies as ω^{-1} showing a typical Debye behaviour. However, it has been experimentally established that both $\chi'(\omega)$ and $\chi''(\omega)$ vary as ω^{-n} where $0 \leqq n \leqq 1$ [5] (Jonscher law). Such a behaviour can be reproduced if long tails are allowed in the distribution of relaxation times. The exact $\Phi(1/\alpha_f)$ which reproduces the experimental behaviour has to be obtained from a deconvolution process. A very simple distribution giving approximately the Jonscher law is

$$\Phi \frac{1}{\alpha_f} = \frac{2}{\pi} \frac{\sigma}{1 + \left[\sigma\left(\frac{\sigma_f^*}{\alpha_f} - 1\right)\right]^4}, \tag{18}$$

i.e., an high-order Lorentzian function. Fig. 4b shows the results obtained with such an expression. As can be observed, higher temperatures result in a enhancement of the value of n in the Jonscher law. This effect reflects the narrowing in the distribution of relaxation times, which has been experimentally observed in many circumstances [3, 20, 21] as the temperature increases.

The use of a particular distribution function for $1/\alpha_f$ has no influence on the general results obtained in the last paragraph, so that we shall use (18) instead of (11) through the discussion of dielectric relaxation.

The out-of-equilibrium dielectric behaviour is experimentally accessible through thermally stimulated depolarization currents (TSDC) which is a "time-domain" technique [1]. Expressions (14) and (15) have to be used with τ_i obtained by solving the set of equations (10). Apart from the global behaviour, the technique of "thermal sampling", allows the contribution of a narrow interval of relaxation times to be experimentally obtained [22]. This technique can be also numerically simulated and Fig. 5a shows some results. Fig. 5b represents the "Bucci times" [23], $\tau_B = -P(t)/(dP/dt)$, for the different contributions isolated in Fig. 5a, as a function of the inverse temperature. The behaviour of a single relaxation time with the parameters of τ^* is also shown for comparison. At low temperatures, the behaviour of each τ_B is very similar to that of a single relaxation time, but at high temperature

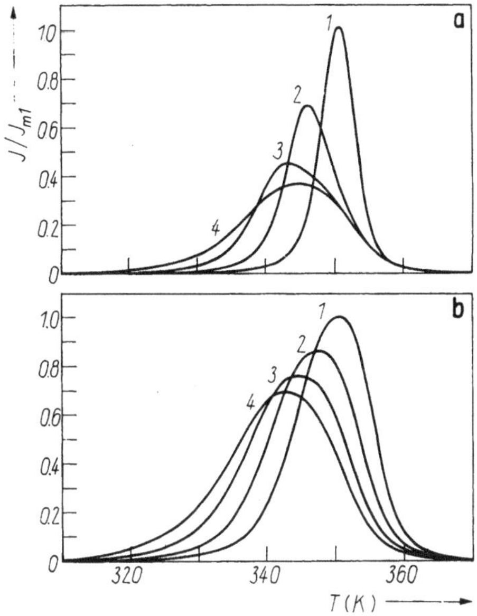

Fig. 6. a) Effect of annealing at 320 K on TSDC, annealing times are: (1) 10^5, (2) 10^4, (3) 10^2 s, (4) as-quenched. Cooling and heating as in Fig. 5; b) shows the influence of the cooling rate. (1) 1, (2) 10^{-1}, (3) 10^{-2} (4) 10^{-3} K s^{-1}. Heating rate is 10^{-1} K s^{-1}

they remain at a constant value, in contrast to the fast descent shown by τ^*. This is in accordance with the experimental behaviour [1] and shows that the distribution cannot be fully decomposed in elementary processes by "thermal sampling" [24]. Another feature, shown in Fig. 5b, is the convergence of the different Bucci times when extrapolating its low-temperature behaviour. This will result in a "compensation law" [4] even if the same thermal history is followed when performing the "thermal sampling".

The effect of annealing at a temperature of about 40 K below the glass transition is shown in Fig. 6a for the TSDC peak. Both the position and the shape of the peak are strongly influenced by the annealing. The general features are the same observed when dealing with a single relaxation time [1]. It must be noticed, however, that when a distribution of relaxation times is present, annealing is no longer equivalent to any cooling rate, see Fig. 2 and 3.

The effect of varying the cooling rate on the TSDC peak is shown in Fig. 6b.

Annealing effects can also be observed in low-frequency ac measurements [25]. They have been simulated and the result is similar to that obtained with a single relaxation time, although in this case the distribution smoothes the sudden step appearing in the losses peak.

Finally, numerical calculation has also shown that the effect of the heating rate on the average relaxation time $\langle\tau\rangle$ is the same as was obtained for a single relaxation time. However, the main relaxation time τ^* is usually lower than the average one. So, when the generalized Kissinger or Ozawa expressions [1] are applied to the maximum of the TSDC peak (relating to τ^*) the resulting values of α_f are slightly higher than the starting ones.

4. Comparison with Experiment

Dielectric measurements performed on PVC have been used as a test of the proposed model. Both the characteristics of the samples and the experimental procedures have been reported in our previous work [1] and in the references therein.

As stated there, the main relaxation time τ^*, defined as $\tau^* = \omega_m^{-1}$ where ω_m is the frequency of the maximum of the losses peak, can be described at equilibrium by (10b) with $\alpha_f^* = 7.37 \times 10^{-4}$ K^{-1}, $T_0 = 311$ K, and $\tau_0 = 8.3 \times 10^{-14}$ s. The parameters σ and p of the distribution were obtained by fitting the experimental points

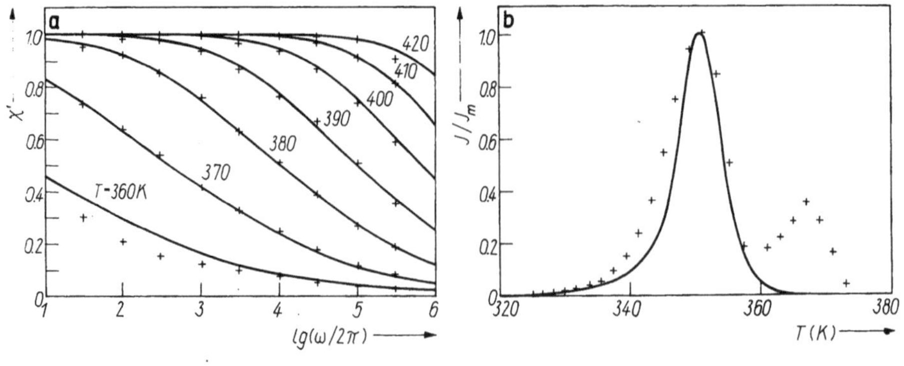

Fig. 7. Fitting of experimental a) ac and b) TSDC data from PVC (crosses) to the model (full line). Parameters are given in the text

for $\chi'(\omega)$ at 380 K using a distribution function like (18). We obtained $\sigma = 3.2$ and $p = 0.74$. Fitting between experimental points at other temperatures and the model prediction with such parameters, is then good enough, except for the lower-temperature curve (see Fig. 7a). The $\chi'(\omega)$ data have been preferred to the $\chi''(\omega)$ ones because some uncertainty arises in the latter when subtracting the conductivity contribution from the experimental curves [1].

The only parameter to be determined is α_T which can be obtained from TSDC measurements. From the low-temperature behaviour of the Bucci time isolated at the centre of the distribution, and using the expression

$$\alpha_T = \frac{E}{kT^2} \left| \ln \frac{\tau_B}{\tau_0} \right|^{-2}, \tag{19}$$

which can be deduced from (16) in [1] if $\tau q \gg 1$, we have obtained $\alpha_T = 1.5 \times 10^{-4}\ K^{-1}$. Then, the average value of $\Delta\alpha$ is $6.5 \times 10^{-4}\ K^{-1}$ in good agreement with direct measurements performed at our laboratory: $\Delta\alpha \approx 7 \times 10^{-4}$ [26].

Now, the dielectric behaviour can be numerically reproduced. Fig. 7b represents the calculated and experimental TSDC peaks. Some contribution of the ϱ peak, associated with the resistivity [27] is still present in the experimental curve. As can be seen, the temperature of the simulated peak coincides with the experimental one, but the latter is wider than the former. This discrepancy can be accounted for by the experimental conditions. The fast cooling can give rise to temperature gradients across the sample, resulting in a wider polarization window than the nominal one. The influence of the ϱ peak can also contribute both to the TSDC peak and to the low-temperature ac behaviour.

On the other hand, some amount of crystallinity is always present in PVC and can be responsible for some anomaly at T_g, as suggested by Schik et al. [28]. Finally, rough simplifications which are not inherent in the model, have been used. They are the very simple distribution function (18) and the consideration of a constant elastic contribution to the free volume expansion coefficient.

In any case, the general features of dielectric relaxation in glasses can be explained by this simple model. Moreover, the model itself can also be applied to the evolution of other physical magnitudes during structural relaxation around the glass transition. Work on this line is actually in progress.

References

[1] A. Alegría, J. M. Barandiarán, and J. Colmenero, phys. stat. sol. (b) **120**, 349 (1983).
[2] C. J. F. Böttcher and P. Bordewijk, Theory of Electric Polarization, Vol. 2, Elsevier Publ. Co., Amsterdam 1978.
[3] M. G. McCrum, B. E. Read, and G. Williams, Anelastic and Dielectric Effects in Polymeric Solids, Wiley, London 1967.
[4] C. Lacabanne, Ph. D. Thesis, Univ. Paul-Sabatier, Toulouse 1974.
[5] A. K. Jonscher, phys. stat. sol. **83**, 585 (1977); **84**, 150 (1977).
[6] L. A. Dissado, Physica Scripta **T1**, 110 (1982).
[7] S. Yagihava, R. Nozaki, S. Takeishi, and S. Mashimo, J. chem. Phys. **79**, 2419 (1983).
[8] A. J. Kovacs, J. M. Hutchinson, and J. J. Aklonis, Structure of Non-Crystalline Materials, Ed. P. H. Gaskell, Taylor and Francis, London 1977.
[9] A. J. Kovacs, J. J. Aklonis, J. M. Hutchinson, and A. R. Ramos, J. Polym. Sci. **17**, 1097 (1979).
[10] M. A. De Bolt, A. J. Easteal, P. B. Macedo, and C. T. Moynihan, J. Amer. Ceram. Soc. **59**, 16 (1976).
[11] C. T. Moynihan, P. B. Macedo, C. J. Montrose, P. K. Gopta, M. A. De Bolt, J. F. Dill, B. E. Dom, P. W. Drake, A. J. Easteal, P. B. Elterman, R. P. Moeller, H. Sasabe, and J. A. Wilder, Ann. New York Acad. Sci. **279**, 15 (1976).
[12] J. T. Bendler and K. L. Nagai, Polymer Preprints **22**, 287 (1981).
[13] K. L. Nagai, Comments Solid State Phys. **9**, 127 (1979).
[14] T. S. Chow and W. M. Prest, Jr., J. appl. Phys. **53**, 10 (1982).
[15] C. K. Chai and N. J. McCrum, Polymer **21**, 706 (1980).
[16] R. Harris, Solid State Commun. **46**, 711 (1983).
[17] H. S. Chen, Amorphous Metallic Alloys, Ed. F. E. Luborsky, Butterworths, London 1983.
[18] M. Goldstein, J. chem. Phys. **51**, 3728 (1969).
[19] M. R. J. Gibbs, J. E. Evetts, and J. A. Leake, J. Mater. Sci. **18**, 278 (1983).
[20] A. Alegría, J. M. Barandiarán, J. Colmenero, and J. J. del Val, Proc. 1st ICSD, Internat. Conf. on Conduction and Breakdown in Solid Dielectrics Toulouse 1983, IEEE Service Center, New York 1983 (p. 464).
[21] E. Scholsser, Polymer Bull. **8**, 461 (1982).
[22] R. A. Creswell and M. H. Perlman, J. appl. Phys. **41**, 2365 (1970).
[23] G. Bucci, R. Fieschi, and G. Guidi, Phys. Rev. **148**, 816 (1966).
[24] N. G. McCrum, J. Polymer Sci. **23**, 161 (1982).
[25] M. Abrowitz and D. F. Pochan and J. M. Pochan J. appl. Phys. **51**, 1539 (1980).
[26] A. Alegria et al. to be published.
[27] P. Müller, phys. stat. sol. (a) **67**, 11 (1981).
[28] C. Schik, H. Tanneberger, E. Donth, Macromol. Chem. Rapid Commun. **1**, 407 (1980).

(Received March 5, 1984)

phys. stat. sol. (b) **125**, 421 (1984)

Subject classification: 15; 6; 13; 22

Institute of Physics, Academy of Sciences of the Azerbaidzhan SSR, Baku[1])

Thermopower and Thermomagnetic Effects of Hot Carriers under Conditions of Mutual Electron–Phonon Prag

By

M. M. BABAEV, T. M. GASYMOV, and A. A. KATANOV

The thermopower and the thermomagnetic effects of hot carriers in non-degenerate semiconductors are studied taking into account both, a mutual drag and heating of the electrons and phonons. The cases when a thermal reservoir for the long-wavelength phonons is or is not present are considered. It is shown that the mutual drag leads to a strong increase of the thermopower $V = V_e + V_p$ and of the electron part of the Nernst-Ettingshausen (NE) voltage U_e. Moreover the drag produces a very strong dependence of these effects on an external heating electric field E. For instance if a thermal reservoir is present $V_e \sim E^{4/3}$ and $V_p \sim E^{10/3}$ under the conditions of electron and phonon strong mutual drag. Presence or absence of the thermal reservoir influences the phonon part of the NE voltage U_p essentially. For example in presence of the thermal reservoir $U_p = 0$, while in its absence U_p is non-zero and grows rapidly on raising E: $U_p \sim E^{26/9}$.

Die Thermospannung und die thermomagnetischen Effekte heißer Ladungsträger in nichtentarteten Halbleitern werden unter Berücksichtigung sowohl von gegenseitigem Elektronen- und Phononendrag als auch des Aufheizens von Elektronen und Phononen untersucht. Die beiden Fälle, wo ein thermisches Reservoir für langwellige Phononen vorhanden ist oder nicht, werden betrachtet. Es wird gezeigt, daß der gegenseitige Drageffekt zu einem starken Ansteigen der Thermospannung, $V = V_e + V_p$ und des Elektronenanteils der Nernst-Ettingshausen (NE)-Spannung U_e führt. Darüberhinaus ruft der Drageffekt eine sehr starke Abhängigkeit dieser Effekte von einem äußeren aufheizenden, elektrischen Feld E hervor. Zum Beispiel variiert $V_e \sim E^{4/3}$ und $V_p \sim E^{10/3}$ unter den Bedingungen eines starken gegenseitigen Elektronen- und Phononen-Drags, wenn ein thermisches Reservoir vorhanden ist. Die An- oder Abwesenheit des thermischen Reservoirs beeinflußt den Phononenanteil der NE-Spannung U_p wesentlich. Zum Beispiel ist mit thermischem Reservoir $U_p = 0$, während bei seiner Abwesenheit U_p ungleich null ist und mit wachsendem E rapide zunimmt, $U_p \sim E^{26/9}$.

1. Introduction

The distribution of the long-wavelength (LW) phonons in a semiconductor exposed to a strong nonuniform electric field deviates from the equilibrium one because of three reasons [1, 2]: (i) phonon heating, (ii) phonon temperature gradient which leads to a phonon–electron drag (thermal drag), (iii) phonon drift following that of electrons due to their mutual drag. The fact that the phonons are in a non-equilibrium state influences all the transport phenomena in semiconductors, especially the thermopower and the thermomagnetic effects of hot carriers due to the gradient of the heating electric field.

The influence of phonon heating and phonon–electron drag upon the thermopower and thermomagnetic effects of hot carriers was studied elsewhere [1, 3] under the assumption that the short-wavelength phonons produce a thermal reservoir for the

[1]) Ul. Narimanova 33, 370143 Baku-143, USSR.

long-wavelength ones. However, the mutual electron–phonon drag was not taken into account in [1, 3]. On the other hand, electrons and phonons are scattered mainly by each other at low lattice temperatures in the comparatively pure samples, therefore the mutual drag should necessarily be taken into account. As shown in this paper (see also [4]) the mutual drag influences essentially the thermopower and thermomagnetic effects.

The present paper deals with a theoretical study of the thermopower and the Nernst–Ettingshausen (NE) effect in nondegenerate n-type semiconductors with regard to all the above-mentioned reasons producing the non-equilibrium state of phonons. We consider both the cases of a thermal reservoir being present and being absent (PTR and ATR, respectively). The electron dispersion law is assumed to be parabolic, $\varepsilon = p^2/2m$.

2. Basic Equations

The basic equations of the problem under consideration are the coupled Boltzmann transport equations for electrons and phonons. We consider the case of quasi-elastic electron scattering by acoustic phonons, when the electron distribution function $f(\boldsymbol{p}, \boldsymbol{r})$ and that of the LW phonons $N(\boldsymbol{q}, \boldsymbol{r})$ may be presented in the form

$$f(\boldsymbol{p}, \boldsymbol{r}) = f_0(\varepsilon, \boldsymbol{r}) + \boldsymbol{f}_1(\varepsilon, \boldsymbol{r}) \frac{\boldsymbol{p}}{p}\,; \qquad |\boldsymbol{f}_1| \ll f_0\,, \tag{1}$$

$$N(\boldsymbol{q}, \boldsymbol{r}) = N_0(q, \boldsymbol{r}) + \boldsymbol{N}_1(q, \boldsymbol{r}) \frac{\boldsymbol{q}}{q}\,; \qquad |\boldsymbol{N}_1| \ll N_0\,. \tag{2}$$

We consider the case of high electron concentration when the isotropic part of the electron distribution function may be assumed to be a Fermi one with a temperature $T_e(\boldsymbol{r})$ [5],

$$f_0(\varepsilon, \boldsymbol{r}) = \left[1 + \exp\left(\frac{\varepsilon - \xi(T_e)}{T_e}\right)\right]^{-1}. \tag{3}$$

Here $\xi(T_e)$ is the chemical potential of hot electrons. The isotropic part of the LW phonon distribution function $N_0(q, \boldsymbol{r})$ was found in [6]. In the PTR case ($q_{\max} \approx 2\bar{p} \ll \ll T/s$, T is the lattice temperature, $\bar{p}$ the average electron momentum, and s the sound velocity in the crystal) N_0 has the form

$$N_0(q, \boldsymbol{r}) \approx \frac{T_p(\boldsymbol{r})}{sq}\,, \tag{4}$$

where T_p is the effective temperature of the LW phonons. In the ATR case ($2\bar{p} \gg T/s$) the major contribution to the expressions of the transport coefficients was provided by the phonons with a crystal momentum within the interval $q_{ep} < q < T_p/s$. Here q_{ep} was obtained from the condition $\beta_e(q_{ep}) = \beta_p(q_{ep})$, where β_e and β_p are the collision frequencies of the LW phonons with electrons and SW phonons, respectively. For LW phonons $N_0(q, \boldsymbol{r})$ has the form [6]

$$N_0(q, \boldsymbol{r}) \approx \frac{T_p(\boldsymbol{r})}{sq} + \left(\frac{q_{ep}}{q}\right)^{5-t} \frac{T_e(\boldsymbol{r})}{sq}\,. \tag{5}$$

Starting from the Boltzmann transport equations we obtain the following ones for $\boldsymbol{f}_1$ and $\boldsymbol{N}_1$:

$$\frac{\partial \boldsymbol{f}_1}{\partial t} + \frac{p}{m}\nabla f_0 - e\boldsymbol{E}\frac{p}{m}\frac{\partial f_0}{\partial \varepsilon} - \Omega[\boldsymbol{h}\boldsymbol{f}_1] + \nu(\varepsilon)\,\boldsymbol{f}_1 +$$

$$+ \frac{2\pi}{(2\pi\hbar)^3}\frac{m}{p^2}\frac{\partial f_0}{\partial \varepsilon}\int_0^{2p} \boldsymbol{N}_1(q)\, W(q)\, \hbar\omega_q q^2\, \mathrm{d}q = 0\,, \tag{6}$$

$$\frac{\partial \boldsymbol{N}_1}{\partial t} + s\nabla N_0 + \beta(q)\,\boldsymbol{N}_1 - \frac{4\pi m}{(2\pi\hbar)^3} W(q)\, N_0(q)\int_{q/2}^{\infty} \boldsymbol{f}_1\, \mathrm{d}p\,. \tag{7}$$

Here $\Omega = eH/mc$ is the electron cyclotron frequency, $\boldsymbol{h} = \boldsymbol{H}/H$, $\hbar\omega_q$ is the phonon energy, $\boldsymbol{E} = \boldsymbol{E}_\mathrm{h} + \boldsymbol{E}_\mathrm{T}$ the total electric field acting on the electron ($\boldsymbol{E}_\mathrm{h}$ is the external field and $\boldsymbol{E}_\mathrm{T}$ is the thermoelectric field), $W(q) = W_0 q^t$ the square matrix element of the electron–phonon interaction ($t = 1$ for deformation (d) and $t = -1$ for piezoelectric (p) electron–acoustic phonon interaction), $\beta(q) = \beta_\mathrm{e} + \beta_\mathrm{p} + \beta_\mathrm{b} + \beta_\mathrm{d}$ and $\nu(\varepsilon) = \nu_\mathrm{p} + \nu_\mathrm{i}$ are the total phonon and electron momentum scattering rates, the indices e, p, b, d, and i meaning the electrons, phonons, crystal boundaries, inperfections, and ionized impurities.

For the individual scattering mechanisms the frequency of the momentum-transfer electron collisions, $\nu(\varepsilon)$ may be written in the form

$$\nu(\varepsilon) = \nu_0 \theta_\mathrm{e}^{-r} \theta_\mathrm{p}^{l} \left(\frac{\varepsilon}{T_\mathrm{e}}\right)^{-r} = \nu(T_\mathrm{e}) \left(\frac{\varepsilon}{T_\mathrm{e}}\right)^{-r}. \tag{8}$$

For the scattering of electrons by impurity ions $r = 3/2$, $l = 0$. If the electrons are scattered by acoustic phonons, then $r = 3/2$ and $l = 4 + t$ in the ATR case and $r = -t/2$ and $l = 1$ in the PTR case. θ_e and θ_p are the dimensionless electron and phonon temperatures: $\theta_{\mathrm{e,p}} = T_{\mathrm{e,p}}/T$ in presence of a thermal reservoir, and $\theta_{\mathrm{e,p}} = T_{\mathrm{e,p}}/T^*$ in absence of a thermal reservoir, where $T^* = \hbar s/a$ with a being the lattice constant.

The phonon collision frequency takes the form

$$\beta(q) = \beta(T)\, \theta_\mathrm{e}^{-n} \left(\frac{sq}{T}\right)^{k}, \tag{9}$$

where $k = 0$ when the LW phonons are scattered by the crystal boundaries, $k = t$ when they are scattered by electrons, $k = 1$ when they are scattered by the SW phonons (in the PTR case), and $k = 5$ when the phonons are scattered by phonons in the ATR case [6]. For the scattering of LW phonons by non-degenerate electrons $n = 3/2$, while in the other cases $n = 0$.

Noting that $\boldsymbol{f}_1 = \boldsymbol{f}_{1\mathrm{h}} + \boldsymbol{f}_{1\mathrm{T}}$, and taking account of the relation $\boldsymbol{f}_1 = -p(\partial f_0/\partial\varepsilon)\, \boldsymbol{v}(\varepsilon)$, (6) and (7) lead to the integral equation for $\boldsymbol{v}_\mathrm{T}(\varepsilon)$

$$\boldsymbol{v}_\mathrm{T}(\varepsilon) - \frac{\Omega}{\nu(\varepsilon)}[\boldsymbol{h}\boldsymbol{v}_\mathrm{T}(\varepsilon)] + \frac{e\boldsymbol{E}_\mathrm{T}'}{m\nu(\varepsilon)} + \frac{\varepsilon - \xi(T_\mathrm{e})}{m\nu(\varepsilon)\, T_\mathrm{e}}\nabla T_\mathrm{e} + \frac{2\pi}{(2\pi\hbar)^3}\frac{m}{p^3\nu(\varepsilon)} \times$$

$$\times \int_0^{2p} \mathrm{d}q \frac{W(q)}{\beta(q)}\hbar\omega_q q^2 \left[s\nabla N_0 - \frac{4\pi m^2}{(2\pi\hbar)^3} W(q)\, N_0(q) \int_{\varepsilon(q/2)}^{\infty} \boldsymbol{v}_\mathrm{T}(\varepsilon)\left(-\frac{\partial f_0}{\partial \varepsilon}\right) \mathrm{d}\varepsilon\right] = 0\,, \tag{10}$$

where $\boldsymbol{E}'_{\mathrm{T}} = \boldsymbol{E}_{\mathrm{T}} + (1/e)\,\nabla\xi(T_{\mathrm{e}})$. Taking account of the fact that the major contribution to the integral $\int \boldsymbol{v}_{\mathrm{T}}(\varepsilon)\,\big(-\partial f_0/\partial\varepsilon)\big)\,\mathrm{d}\varepsilon$ is provided by the electrons with energies near T_{e}, this integral may be replaced approximately by $\boldsymbol{v}_{\mathrm{T}}(T_{\mathrm{e}})\,f_0\big(\varepsilon(q/2)\big)$. Just the term associated with this factor is responsible for the mutual drag, whereas the term proportional to ∇N_0 describes the thermal drag. The substitution of $\boldsymbol{v}_{\mathrm{T}}(\varepsilon)$ in the integrand by $\boldsymbol{v}_{\mathrm{T}}(T_{\mathrm{e}})$ transforms (10) to an algebraic one with respect to $\boldsymbol{v}_{\mathrm{T}}(\varepsilon)$. Solving this equation it is easy to calculate the electric current density of the electrons,

$$\boldsymbol{j}_{\mathrm{T}} = -\frac{e(2mT_{\mathrm{e}})}{3\pi^2\hbar^3}\int_0^\infty \boldsymbol{v}_{\mathrm{T}}(x)\left(-\frac{\partial f_0}{\partial x}\right)x^{3/2}\,\mathrm{d}x\,,\qquad x = \frac{\varepsilon}{T_{\mathrm{e}}}\,. \tag{11}$$

Let us direct $\nabla T_{\mathrm{e,p}}$ along the z-axis, and let both $\boldsymbol{E}_{\mathrm{h}}$ and $\boldsymbol{H}$ be directed along the y-axis. Furtheron consider the isothermal case $\nabla_x T_{\mathrm{e,p}} = 0$. Then using $j_{\mathrm{T}z} = j_{\mathrm{T}x} = 0$ we obtain the following expressions for the thermal field $E_{\mathrm{T}z}$ and the NE field $E_{\mathrm{T}x}$:

$$E_{\mathrm{T}z} + \frac{1}{e}\nabla_z\xi(T_{\mathrm{e}}) = \alpha_{\mathrm{e}}\nabla_z T_{\mathrm{e}} + \alpha_{\mathrm{p}}\nabla_z T_{\mathrm{p}}\,;\qquad \alpha_{\mathrm{e,p}} = -\frac{\sigma_{11}\beta_{11}^{(\mathrm{e,p})} + \sigma_{12}\beta_{12}^{(\mathrm{e,p})}}{\sigma_{11}^2 + \sigma_{12}^2}\,, \tag{12}$$

$$E_{\mathrm{T}x} = -H(Q_{\mathrm{e}}\nabla_z T_{\mathrm{e}} + Q_{\mathrm{p}}\nabla_z T_{\mathrm{p}})\,;\qquad Q_{\mathrm{e,p}} = \frac{1}{H}\,\frac{\sigma_{11}\beta_{12}^{(\mathrm{e,p})} - \sigma_{12}\beta_{11}^{(\mathrm{e,p})}}{\sigma_{11}^2 + \sigma_{12}^2}\,. \tag{13}$$

Here α_{e} and α_{p} are the electron (e) and phonon (p) parts of the thermopower, whereas Q_{e} and Q_{p} are the electron and phonon parts of the NE coefficient, respectively.

The transport coefficients σ_{1i} and β_{1i} are given by the following expressions:

$$\sigma_{1i} = \int_0^\infty \alpha(x)\left(\frac{\Omega}{\nu(x)}\right)^{i-1}[1 + b_i(x)]\,\mathrm{d}x\,,\qquad \alpha(x) = \frac{4n_0e^2}{3\sqrt{\pi}\,m}\,\frac{\nu(x)\,x^{3/2}e^{-x}}{\nu^2(x) + \Omega^2}\,, \tag{14}$$

$$\beta_{1i}^{(\mathrm{e})} = \frac{1}{e}\int_0^\infty \alpha(x)\left(\frac{\Omega}{\nu(x)}\right)^{i-1}\left[x - \frac{\xi}{T_{\mathrm{e}}} + \left(1 - \frac{\xi}{T_{\mathrm{e}}}\right)b_i(x)\right]\mathrm{d}x\,, \tag{15}$$

$$\beta_{1i}^{(\mathrm{p})} = \frac{1}{e}\int_0^\infty \alpha(x)\left(\frac{\Omega}{\nu(x)}\right)^{i-1}[\lambda(x) + \lambda(T_{\mathrm{e}})\,b_i(x)]\,\mathrm{d}x\,. \tag{16}$$

Here n_0 is the electron concentration,

$$b_1(x) = \gamma(x)\,\frac{\nu(x)\,\nu(T_{\mathrm{e}})\,(1 - \gamma_0) - \Omega^2}{\nu^2(T_{\mathrm{e}})\,(1 - \gamma_0)^2 + \Omega^2}\,,$$

$$b_2(x) = \gamma(x)\,\nu(x)\,\frac{\nu(x) + \nu(T_{\mathrm{e}})\,(1 - \gamma_0)}{\nu^2(T_{\mathrm{e}})\,(1 - \gamma_0)^2 + \Omega^2}\,, \tag{17}$$

$$\gamma(x) = \frac{\nu_{\mathrm{p}}(x)}{\nu(x)}\cdot\left\langle\frac{\beta_{\mathrm{e}}}{\beta}\right\rangle = \gamma_0 x^j\,,\quad \lambda(x) = \frac{ms^2}{T_{\mathrm{p}}}\,\nu_p(x)\cdot\left\langle\frac{1}{\beta}\left(\frac{T_{\mathrm{p}}}{N_0}\cdot\frac{\partial N_0}{\partial T_{\mathrm{p}}}\right)\right\rangle = \lambda_0 x^h\,, \tag{18}$$

$$\langle A\rangle = \frac{\int_0^{2\mathrm{p}} A(q)\,W(q)\,N_0(q)\,q^3\,\mathrm{d}q}{\int_0^{2\mathrm{p}} W(q)\,N_0(q)\,q^3\,\mathrm{d}q}\,. \tag{19}$$

The coefficient $\lambda(x)$ characterizes the efficiency of the thermal drag, whereas $\gamma(x)$ describes that of the mutual drag. If $\lambda(T_e) \gg 1$, then the phono ncontributions of the thermomagnetic effects are much higher than the electron ones. The mutual drag becomes essential if the electrons and phonons are scattered mainly by each other. In this case $\gamma(x) \to 1$. However, the other scattering mechanisms leading to a decrease of $\gamma(x)$ exist as well; hence in fact $\gamma(x) < 1$. In the PTR case we have $h = (t - k)/2$, $j = r + t - (k/2)$, and

$$\gamma_0 = \frac{\beta_e(T)}{\beta(T)} \frac{\nu_{p0}}{\nu_0} \theta_p^{1-l} \times \begin{cases} \dfrac{3+t}{3+2t-k} \left(\dfrac{8ms^2}{T}\right)^h \theta_e^{n-(3/2)+j} & \text{if } 3+2t-k>0\,, \\ \left(\dfrac{8ms^2}{T}\right)^{-1} \theta_e^{r-3} \ln \dfrac{8mT_e}{q_{ep}^2} & \text{if } 3+2t-k=0; \end{cases} \tag{20}$$

$$\lambda_0 = \frac{(3+t)\cdot 2^{-(3k/2)}}{3+t-k} \left(\frac{ms^2}{T}\right)^{1-(k/2)} \theta_e^{n+h} \frac{\nu_{p0}}{\beta(T)}\,. \tag{21}$$

In the ATR case the coefficient γ does not depend on the electron energy ($j = 0$) and $h = -3/2$, then

$$\gamma_0 = \frac{\beta_e(T^*)}{\beta(T^*)} \frac{\nu_{p0}}{\nu_0} \theta_e^{n-3/2} \theta_p^{1-l} \times \begin{cases} \dfrac{3+t}{3+2t-k} \theta_p^{3+2t-k} & \text{if } 3+2t-k>0\,, \\ (3+t) \ln \dfrac{T_p}{sq_{ep}} & \text{if } 3+2t-k=0\,, \\ \dfrac{3+t}{k-3-2t} \left(\dfrac{T^*}{sq_{ep}}\right)^{k-3-2t} & \text{if } 3+2t-k<0\,. \end{cases} \tag{22}$$

$$\lambda_0 = \frac{\nu_{p0}}{\beta(T^*)} \frac{ms^2}{T^*} \theta_e^{n-3/2} \times \begin{cases} \dfrac{3+t}{3+t-k} \theta_p^{3+t-k} & \text{if } 3+t-k>0\,, \\ \dfrac{3+t}{k-3-t} \left(\dfrac{T^*}{sq_{ep}}\right)^{k-3-t} & \text{if } 3+t-k<0\,. \end{cases} \tag{23}$$

In a special case when the condition of strong mutual drag ($\gamma \to 1$) is satisfied one obtains in presence and in absence of a thermal reservoir, respectively,

$$\lambda_0 \approx e^{-(\xi/T_e)} = \frac{(2mT)^{3/2}}{4\pi^{3/2}\hbar^3 n_0} \theta_e^{3/2} \quad \text{and} \quad \lambda_0 = \frac{T^{*3}}{24\pi^{3/2}\cdot \hbar^3 s^3 n_0} \theta_p^3\,.$$

3. Thermopower and NE Coefficient

Consider first the weak magnetic field case ($\Omega \ll \bar{\nu}$). Then taking account of (14) to (19) one obtains from (12) and (13),

$$\alpha_e = -\frac{1}{e}\left[1 + \frac{\gamma_0}{1-\gamma_0} \frac{\Gamma(\frac{5}{2}+j)}{\Gamma(\frac{5}{2}+r)}\right]^{-1} \times \left[\frac{5}{2} + r - \frac{\xi}{T_e} + \left(1 - \frac{\xi}{T_e}\right) \frac{\gamma_0}{1-\gamma_0} \frac{\Gamma(\frac{5}{2}+j)}{\Gamma(\frac{5}{2}+r)}\right], \tag{24}$$

$$\alpha_p = -\frac{1}{e}\left[1 + \frac{\gamma_0}{1-\gamma_0} \frac{\Gamma(\frac{5}{2}+j)}{\Gamma(\frac{5}{2}+r)}\right]^{-1} \times \left[\frac{\Gamma(\frac{5}{2}+r+h)}{\Gamma(\frac{5}{2}+r)} + \frac{\gamma_0}{1-\gamma_0} \frac{\Gamma(\frac{5}{2}+j)}{\Gamma(\frac{5}{2}+r)}\right]\lambda_0\,, \tag{25}$$

$$Q_e = \frac{1}{e}\frac{\mu_0}{c}\theta_e^r\theta_p^{-l}\frac{\Gamma(\frac{5}{2})}{\Gamma^2(\frac{5}{2}+r)}\left[1+\frac{\gamma_0}{1-\gamma_0}\frac{\Gamma(\frac{5}{2}+j)}{\Gamma(\frac{5}{2}+r)}\right]^{-2}\times$$

$$\times\left\{r\Gamma\left(\frac{5}{2}+2r\right)+\frac{\gamma_0}{1-\gamma_0}\left(\left(\frac{3}{2}+2r\right)\frac{\Gamma(\frac{5}{2}+2r)\Gamma(\frac{5}{2}+j)}{\Gamma(\frac{5}{2}+r)}-\right.\right.$$

$$\left.\left.-\left(\frac{3}{2}+r\right)\left[\Gamma\left(\frac{5}{2}+r+j\right)+\frac{1}{1-\gamma_0}\Gamma\left(\frac{5}{2}+j\right)\right]\right)\right\}, \tag{26}$$

$$Q_p = \frac{1}{e}\frac{\mu_0}{c}\theta_e^r\theta_p^{-l}\lambda_0\frac{\Gamma(\frac{5}{2})}{\Gamma^2(\frac{5}{2}+r)}\left[1+\frac{\gamma_0}{1-\gamma_0}\frac{\Gamma(\frac{5}{2}+j)}{\Gamma(\frac{5}{2}+r)}\right]^{-2}\times$$

$$\times\left\{\Gamma\left(\frac{5}{2}+2r+h\right)-\frac{\Gamma(\frac{5}{2}+2r)}{\Gamma(\frac{5}{2}+r)}\Gamma\left(\frac{5}{2}+r+h\right)+\right.$$

$$+\frac{\gamma_0}{1-\gamma_0}\left(\Gamma\left(\frac{5}{2}+r+j\right)\left[1-\frac{\Gamma(\frac{5}{2}+r+h)}{\Gamma(\frac{5}{2}+r)}\right]+\right.$$

$$\left.+\frac{\Gamma(\frac{5}{2}+j)}{\Gamma(\frac{5}{2}+r)}\left[\Gamma\left(\frac{5}{2}+2r+h\right)-\Gamma\left(\frac{5}{2}+2r\right)\right]\right)+$$

$$\left.+\frac{\gamma_0}{(1-\gamma_0)^2}\Gamma\left(\frac{5}{2}+j\right)\left[1-\frac{\Gamma(\frac{5}{2}+r+h)}{\Gamma(\frac{5}{2}+r)}\right]\right\}. \tag{27}$$

Here

$$\mu_0 = \frac{\Gamma(5/2+r)}{\Gamma(5/2)}\frac{e}{mv_0} \tag{28}$$

is the mobility of "cold" ($T_e = T$) electrons.

In particular, in the ATR case at $\gamma_0 \to 0$ we obtain from (24) to (27)

$$\alpha_e = -\frac{1}{e}\left(4-\frac{\xi}{T_e}\right), \qquad \alpha_p = -\frac{1}{e}\frac{\sqrt{\pi}}{8}\lambda_0, \tag{29}$$

$$Q_e = \frac{945\sqrt{\pi}}{128}\frac{1}{e}\frac{\mu_0}{c}\theta_e^{3/2}\theta_p^{-(4+t)}, \qquad Q_p = -\left(\frac{315\pi}{512}-1\right)\frac{1}{e}\frac{\mu_0}{c}\theta_e^{3/2}\theta_p^{-(4+t)}\lambda_0. \tag{30}$$

Thus contrary to the PTR case, in the ATR case Q_e and Q_p are positive and negative, respectively, irrespective of the electron and phonon scattering mechanisms.

Turning now to the conditions of a strong mutual electron–phonon drag we obtain from (24) to (27)

$$\alpha_e = -\frac{1}{e}\left(1-\frac{\xi}{T_e}\right), \quad \alpha_p = -\frac{1}{e}\lambda_0, \quad Q_e = \frac{1}{e}\frac{\mu_0}{c}\theta_r^e\theta_p^{-l}\frac{\Gamma(\frac{5}{2})}{\Gamma(\frac{5}{2}+j)},$$

$$Q_p = \frac{1}{e}\frac{\mu_0}{c}\theta_e^r\theta_p^{-l}\lambda_0\frac{\Gamma(\frac{5}{2})}{\Gamma(\frac{5}{2}+j)}\left[1-\frac{\Gamma(\frac{5}{2}+r+h)}{\Gamma(\frac{5}{2}+r)}\right].$$

In a strong magnetic field ($\Omega \gg \bar{\nu}$) the calculations lead to the following expressions:

$$\alpha_e = -\frac{1}{e}\left(\frac{5}{2}-\frac{\xi}{T_e}\right), \qquad \alpha_p = -\frac{1}{e}\frac{\Gamma(\frac{5}{2}+h)}{\Gamma(\frac{5}{2})}\lambda_0, \tag{31}$$

$$Q_{\mathrm{e}} = \frac{1}{e} \frac{c}{\mu_0 H^2} \theta_{\mathrm{e}}^{-r} \theta_{\mathrm{p}}^{l} \frac{\Gamma(\frac{5}{2}+r)}{\Gamma^2(\frac{5}{2})} \left[r\Gamma\left(\frac{5}{2}-r\right) - \frac{3}{2}\gamma_0 \Gamma\left(\frac{5}{2}+j-r\right) \right], \tag{32}$$

$$Q_{\mathrm{p}} = \frac{1}{e} \frac{c}{\mu_0 H^2} \theta_{\mathrm{e}}^{-r} \theta_{\mathrm{p}}^{l} \frac{\Gamma(\frac{5}{2}+r)}{\Gamma^3(\frac{5}{2})} \times$$

$$\times \left[\Gamma\left(\frac{5}{2}-r\right)\Gamma\left(\frac{5}{2}+h\right) - \Gamma\left(\frac{5}{2}\right)\Gamma\left(\frac{5}{2}+h-r\right) + \right.$$

$$\left. + \gamma_0 \Gamma\left(\frac{5}{2}+j-r\right)\left[\Gamma\left(\frac{5}{2}\right) - \Gamma\left(\frac{5}{2}+h\right)\right]\right]\lambda_0 . \tag{33}$$

From (25), (27), (31), and (33) it follows that if the LW phonons are scattered by electrons, then in the PTR case the phonon parts of the transverse and longitudinal NE effects (due to thermal drag) are equal to zero for any value of γ. In the ATR case $Q_{\mathrm{p}} \neq 0$ and $\Delta\alpha_{\mathrm{p}}(H) \neq 0$ for this mechanism of the phonon scattering. Furtheron the magnetic field leads to a decrease of the mutual drag influence on the thermopower of hot electrons.

4. Field Dependence of the Thermopower and of the NE Voltage

In the most interesting case when the electrons and phonons are strongly heated ($\theta_{\mathrm{e}0} \gg 1$, $\theta_{\mathrm{p}0} \gg 1$) at one end of the sample ($z = 0$), whereas at the other end ($z = L_z$) they are not heated at all ($\theta_{\mathrm{e}L_z} \approx \theta_{\mathrm{p}L_z} \approx 1$), one obtains for the total thermopower (V) and the NE voltage (U)

$$V_{\mathrm{e}} \approx \alpha_{\mathrm{e}}(T_{\mathrm{e}0})\, T_{\mathrm{e}0}\,, \qquad V_{\mathrm{p}} \approx \alpha_{\mathrm{p}}(T_{\mathrm{e}0}, T_{\mathrm{p}0})\, T_{\mathrm{p}0}\,,$$
$$U_{\mathrm{e}} \approx \frac{L_x}{L_z} H Q_{\mathrm{e}}(T_{\mathrm{e}0}, T_{\mathrm{p}0})\, T_{\mathrm{e}0}\,, \qquad U_{\mathrm{p}} \approx \frac{L_x}{L_z} H Q_{\mathrm{p}}(T_{\mathrm{e}0}, T_{\mathrm{p}0})\, T_{\mathrm{p}0}\,. \tag{34}$$

The field dependence of V and U in absence of mutual drag were considered elsewhere [3]. Thus we consider here just the strong mutual drag regime. Neglecting the weak (logarithmic) dependence upon $T_{\mathrm{e}0}$ one obtains from (24) to (34) $V_{\mathrm{e}} \sim \theta_{\mathrm{e}0}$ and $V_{\mathrm{p}} \sim \theta_{\mathrm{e}0}^{5/2}$ in the PTR case and $V_{\mathrm{p}} \approx \theta_{\mathrm{p}0}^{4}$ in the ATR case. In a weak magnetic field $U_{\mathrm{e}} \sim \theta_{\mathrm{e}0}^{-t/2}$ and $U_{\mathrm{p}} \sim \theta_{\mathrm{e}0}^{(3-t)/2}$ in the PTR case, whereas in the ATR case $U_{\mathrm{e}} \sim \theta_{\mathrm{e}0}^{5/2} \times \theta_{\mathrm{p}0}^{-(4+t)}$ and $U_{\mathrm{p}} \sim \theta_{\mathrm{e}0}^{3/2} \cdot \theta_{\mathrm{p}0}^{-t}$. In a strong magnetic field $U_{\mathrm{e}} \sim \theta_{\mathrm{e}0}^{2+t/2}$, $U_{\mathrm{p}} \sim \theta_{\mathrm{e}0}^{(7+t)/2}$ and $U_{\mathrm{e}} \sim \theta_{\mathrm{e}0}^{-1/2} \cdot \theta_{\mathrm{p}0}^{4+t}$, $U_{\mathrm{p}} \sim \theta_{\mathrm{e}0}^{-3/2} \cdot \theta_{\mathrm{p}0}^{8+t}$ for the PTR and ATR cases, respectively.

The field dependence of θ_{e} and θ_{p} may be found using the energy-balance equations. In the PTR case under the strong mutual drag conditions ($k = t$, $r = -t/2$) $\theta_{\mathrm{e}} = \theta_{\mathrm{p}}$ and the balance equation takes the form

$$\theta_{\mathrm{e}}^{3+t/2}(\theta_{\mathrm{e}} - 1) = \frac{\pi^2 \hbar^3}{8} \frac{n_0 e^2}{m^3 T^2 s} \frac{E^2}{v_0 \beta(T)} \frac{1}{1-\gamma_0} . \tag{35}$$

In the ATR case the electrons receiving energy from the electric field transfer it to the LW phonons and the latter pass it to the surrounding medium through the crystal boundaries. Then $\theta_{\mathrm{p}} < \theta_{\mathrm{e}}$ both in presence and in absence of the mutual drag; the energy-balance equations take the form

$$\theta_e^{t-1/2}\theta_p^l\left[1-\frac{2^{-(3+t)}}{\Gamma((5+t/2))}\left(\frac{\theta_p}{\theta_e}\right)^{((5+t)/2)}\left(\frac{T^*\theta_p}{ms^2}\right)^{(3+t)/2}\right]=$$

$$=\frac{3+t}{2^t\Gamma((5+t)/2)}\,\frac{\pi^2e^2\hbar^3E^2}{(2mT^*)^{1+t/2}\,m^2sW_{40}\nu_0}\,\frac{1}{1-\gamma_0}, \tag{36}$$

$$\theta_p=\left[\frac{2^t\Gamma((5+t)/2)}{3+t}\,\frac{W_0n_0}{\beta_b^{(\varepsilon)}}\left(\frac{ms^2}{T^*}\right)^2\left(\frac{m}{T^*}\right)^{t/2}\right]^{1/4}\theta_e^{(2+t)/8}, \tag{37}$$

where $\beta_b^{(\varepsilon)}$ is the LW phonon energy loss rate due to crystal boundaries.

If the electrons and phonons are scattered by each other, then $\gamma_0 \to 1$ and it follows from (35) and (36) that $\theta_e \to \infty$. Thus to calculate γ from (18) one should take into account as well some other non-basic mechanisms of electron scattering (by impurity ions) and phonon scattering (by phonons and crystal boundaries). We consider the following three cases:

(i) $\dfrac{\beta_p+\beta_b}{\beta_e} \ll \dfrac{\nu_i}{\nu_p}$;

(ii) $\beta_p \gg \beta_b$, $\quad \dfrac{\beta_p}{\beta_e} \gg \dfrac{\nu_i}{\nu_p}$;

(iii) $\beta_b \gg \beta_p$, $\quad \dfrac{\beta_b}{\beta_e} \gg \dfrac{\nu_i}{\nu_p}$.

The E-dependences of θ_e and θ_p for the PTR and ATR cases under the conditions of strong mutual drag are tabulated at $\theta_e \gg 1$ (see Table 1). Using these results one can easily obtain the E-dependences of V and U. At $t = 1$ the quantities V_e, V_p, and U_p (as well as U_e in a strong magnetic field) increase with θ_e; hence the mutual drag leads to these effects becoming more and more important. Besides that the mutual drag leads to a change of the very form of these quantities as the functions of E. For instance in the PTR case at a strong mutual drag one obtains $V_e \sim E^{4/3}$ and $V_p \sim E^{10/3}$ if inequality (i) is satisfied, while in the absence of mutual drag we have [3]: $V_e \sim E^{4/9}$ and $V_p \sim E^{10/9}$ at $t = 1$, and $V_e \sim E^{4/7}$ and $V_p \sim E^{10/7}$ at $t = -1$.

Table 1
Dependences of the electron and phonon temperatures upon electric field strength

	PTR case	ATR case	
		$t = 1$	$t = -1$
subcase (i)	$\theta_e = \theta_p \sim E^{4/3}$	$\theta_e \sim \left[1-\left(\frac{E}{E_k}\right)^2\right]^{-8/13}$, $\theta_p \sim \left[1-\left(\frac{E}{E_k}\right)^2\right]^{-3/13}$	$\theta_e \sim E^{-2}$, $\theta_p \sim E^{-1/4}$
subcase (ii)	$\theta_e = \theta_p \sim E^{2/3}$	$\theta_e \sim E^{16/27}$, $\theta_p \sim E^{2/9}$	$\theta_e \sim E^{16/9}$, $\theta_p \sim E^{2/9}$
subcase (iii)	$\theta_e = \theta_p \sim E^{4/11}$	$\theta_e \sim E^{2/3}$, $\theta_p \sim E^{1/4}$	$\theta_e \sim E^2$, $\theta_p \sim E^{1/4}$

In the ATR case using inequality 3) and putting $\gamma \to 1$, one obtains $V_p \sim E$. Then $V_e \sim E^{2/3}$ at $t = 1$ and $V_e \sim E^2$ at $t = -1$. In the subcase (i) under the ATR conditions V_e and V_p fall off with rising E if $t = -1$ and $\gamma \to 1$: $V_e \sim E^{-2}$ and $V_p \sim E^{-1}$.

The absence or presence of a thermal reservoir influences the NE voltage as well. For example in the PTR case under conditions (ii) we obtain $U_e \sim E^{1/3}$ if $\gamma \to 1$, $t = -1$, and $\Omega \ll \bar{\nu}$, while in the ATR case $U_e \sim E^{34/9}$. Besides, as noted above, in the PTR case $U_p = 0$, while in the ATR case U_p is non-zero increasing strongly with rising E, i.e. $U_p \sim E^{26/9}$.

References

[1] L. E. GUREVICH and T. M. GASYMOV, Fiz. tverd. Tela **9**, 3493 (1967).
[2] J. E. PARROT, Proc. Phys. Soc. B **70**, 590 (1957).
[3] M. M. BABAEV and T. M. GASYMOV, phys. stat. sol. (b) **84**, 473 (1977).
[4] T. M. GASYMOV, A. A. KATANOV, and M. M. BABAEV, phys. stat. sol. (b) **119**, 391 (1983).
[5] V. L. BONCH-BRUEVICH and S. G. KALASHNIKOV, Fizika poluprovodnikov, Izd. Nauka, Moskwa 1977.
[6] L. E. GUREVICH and T. M. GASYMOV, Fiz. tverd. Tela **9**, 106 (1967).

(Received May 25, 1984)

phys. stat. sol. (b) 125, K1 (1984)

Subject classification: 1 and 4; 22

Department of Physics, S.V. College, Aligarh[1)]

Phillips Ionicity and Structural Classification of Layered Transition Metal Dichalcogenides

By

V. P. GUPTA and O. P. SINGH

Introduction Transition metal dichalcogenides having layered structures are a very interesting group of materials. In these materials, the intralayer interaction is very strong in comparison to that of interlayer interaction and therefore these materials can be cleaved in very thin sheets. Keeping in view their technological and industrial applications these materials have been investigated for the determination of their various electronic, optical, and structural properties /1 to 5/. Gamble /6/ has classified these compounds on the basis of the Pauling ionicity and the ionic radii ratio while Madhukar /7/ gave a classification on the basis of Pauling ionicity and metal to chalcogen covalent radii ratio. As these materials are less ionic therefore the use of ionic radii is not justified. Pauling's definition of ionicity works well for molecular systems /8, 9/, in case of crystalline solids it does not work satisfactorily because of the fact that it is based on the molecular approach and bonding properties and not on band structure. Our view point is further supported by the fact that according to the Pauling definition the ionicity for Ti and Zr dichalcogenides comes out to be the same while they have different values of effective charges /10/. On the other hand, ionicity by Phillips /9/ is based on band structure and therefore a more appropriate parameter. It is therefore worthwhile to investigate Phillips' ionicity of these materials and to look into their classification on the basis of the Phillips definition of ionicity.

Theory and results Phillips' ionicity (f_i) has been defined in terms of ionic gap (E_c) and average gap (E_p) as follows /9/:

$$f_i = E_c^2/E_p^2 ,$$

1) Aligarh 202001, India.

where

$$E_p^2 = E_c^2 + E_h^2 ,$$

$$E_c^2 = b e^2 \left(\frac{Z_A}{r_A} - \frac{Z_B}{r_B} \right) \exp -K_s \frac{r_{AB}}{2} ,$$

$$E_h = 40.468/r_{AB}^{2.5} ,$$

where Z_A, Z_B and r_A, r_B are the valencies and covalent radii of atoms A and B, respectively. K_s is the Thomas-Fermi screening parameter, r_{AB} is the interatomic distance between atoms A and B, and b is an adjustable parameter. The value of b depends on the coordination number around the cation, i.e. for tetrahedral coordination (four bonds) the value of b is 1.5 while for octahedral coordination (six bonds) it is 2.25 (1.5 times the ratio of the coordination numbers of two structures i.e. 6/4). Based on detailed investigations on different systems, Lavin /11/ gave an empirical relation between b and the average coordination number $\bar{N}_c$ ($b = 0.89 \bar{N}_c^2$). The values of E_h and E_c depend on the bond length and number of bonds coming out of the cation and not on the angles between bonds. This viewpoint is further supported by the fact that the above theory has been successfully applied to distorted tetrahedral structures like chalcopyrites. In this presentation the above theory has been applied to determine Phillips' ionicity for transition metal dichalcogenides having layer structure.

The chemical formula of these materials is MX_2 compared to binary compounds, but as far as the nearest surrounding around the cation is concerned there is no difference between the octahedral MX_2 and a binary octahedral compound semiconductor. The structure in case of trigonal prismatics is slightly different compared to the octahedral one but the coordination number is still six.

Based on the above approach and using reported data /12, 13/ the values of E_h, E_c, and E_p hence the values of f_i have been evaluated. The calculated values have been presented in Table 1. The derived values of f_i (Phillips' ionicity) have been plotted against R_M/R_X (metal to chalcogen covalent radii ratio) (see Fig. 1). This plot clearly devides the octahedral and trigonal prismatic structures. The occurrence of different types of structures in different materials can be explained on the basis of sp hybridization as explained by Madhukar /7/.

Table 1

material	E_c (eV)	E_h (eV)	E_p (eV)	f_i
TiS_2 (O)	3.91	4.44	5.91	0.44
$TiSe_2$ (O)	3.65	3.93	5.36	0.46
$TiTe_2$ (O)	2.31	3.34	4.06	0.32
ZrS_2 (O)	4.89	3.85	6.23	0.61
$ZrSe_2$ (O)	4.26	3.50	5.18	0.54
$ZrTe_2$ (O)	2.70	3.03	3.96	0.41
HfS_2 (O)	4.85	3.25	6.20	0.61
$HfSe_2$ (O)	3.82	5.50	5.18	0.54
$HfTe_2$ (O)	2.72	3.03	4.07	0.44
VS_2 (O)	3.07	4.61	5.60	0.30
VSe_2 (O)	1.87	4.22	4.61	0.16
VTe_2 (O)	0.80	3.47	3.56	0.05
NbS_2 (T)	3.34	4.22	5.38	0.38
$NbSe_2$ (T)	2.38	3.74	4.44	0.28
$NbTe_2$ (O)	1.17	3.14	3.36	0.12
TaS_2 (O)	3.29	4.17	5.38	0.37
(T)	3.46	4.35	5.56	0.38
$TaSe_2$ (O)	2.40	3.74	4.45	0.29
(T)	2.49	3.89	4.62	0.28
$TaTe_2$ (O)	1.17	3.15	3.36	0.12
MoS_2 (T)	1.99	4.44	4.86	0.16
$MoSe_2$ (T)	1.06	4.13	4.27	0.06
$MoTe_2$ (T)	9.44×10^{-3}	3.28	3.28	8.21×10^{-6}
WS_2 (T)	1.04	4.42	5.10	0.94
WSe_2 (T)	1.04	4.05	4.18	0.06
WTe_2 (O)	0.03	3.24	3.24	1.07×10^{-4}

O stands for octahedral structure, T stands for trigonal prismatic structure.

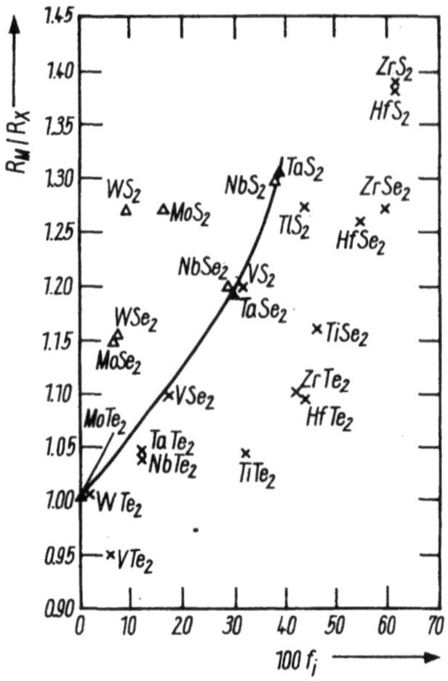

Fig. 1. Plot of Phillips' ionicity f_i versus R_M/R_X. Δ trigonal prismatic, × octahedral, ▲ both (trigonal prismatic and octahedral)

The plot clearly shows that the materials having high ionicity and large metal to chalcogen covalent radii ratio occur in octahedral structure while the materials having the same value of metal to chalcogen covalent radii ratio but less ionicity occur in trigonal prismatic structure. The materials having a high value of ionicity but low metal to chalcogen covalent radii ratio will occur in octahedral structure.

Conclusion The Phillips' and Van Vechten theory has been used to evaluate the Phillips ionicity for layer type transition metal dichalcogenides. The values of the Phillips ionicity so obtained clearly classify the two structural types found in these materials.

References

/1/ P.A. LEE (Ed.), Optical and Electronic Properties, D. Reidel Publ. Co., Boston 1916.

/2/ J.A. WILSON and A.D YOFFE, Avd. Phys. 18, 193 (1969).

/3/ V.P. GUPTA, J. Phys. Chem. Solids 41, 757 (1980).

/4/ V.P. GUPTA and V.K. SHRIVASTAVA, J. Phys. Chem. Solids 42, 1071 (1981).

/5/ V.P. GUPTA, A. GUPTA, and V.K. SHRIVASTAVA, phys. stat. sol. (b) 108, 323 (1981).

/6/ F.R. GAMBLE, J. Solid State Chem. 9, 358 (1974).

/7/ A. MADHUKAR, Solid State Commun. 16, 383 (1975).

/8/ L. PAULING, The Nature of Chemical Bond, Cornell Univ. Press, New York 1960.

/9/ J.C. PHILLIPS, Bond and Bands in Semiconductors, Academic Press, New York 1973 (Chap. 12).

/10/ S. UCHIDA and T.J. SHOJI, J. Phys. Soc. Japan 45, 153 (1978).

/11/ B.F. LAVIN, Phys. Rev. B 7, 2591 (1973).

/12/ A. ZUNGER and A.J. FREEMAN, Phys. Rev. B 16, 906 (1977).

/13/ G. LUCOVSKY, R.M. WHITE, J.A. BENDA, and J.P. REVELLI, Phys. Rev. B 7, 3859 (1973).

(Received April 2, 1984)

phys. stat. sol. (b) 125, K7 (1984)

Subject classification: 6 and 20.1; 22.8

Dipartimento di Fisica dell' Università and GNSM-CISM del C.N.R., Parma[1]

Comments on the Optic Vibrational Modes in Defect Chalcopyrites

By

A. PARISINI, P.P. LOTTICI, and C. RAZZETTI

Ternary and pseudoternary (i.e. mixed) compounds of the $A^{II}B_2^{III}X_4^{VI}$ type are receiving an increasing attention due to their fundamental interest and to the potential applications as, for example, non-linear optics for photoconductive devices /1, 2/.

Of the 27 possible AB_2X_4 compounds with A = Zn, Cd, Hg; B = Al, Ga, In; X = S, Se, Te, 23 crystallize in a structure in which all metal atoms are tetrahedrally coordinated to anions, 3 ($ZnAl_2S_4$, $CdIn_2S_4$, $HgIn_2S_4$) are spinels, and $ZnIn_2S_4$ crystallizes in a C_{3v}^5 layered structure /3/. In the two last cases the cation coordination is both tetra- and octahedral. Of the 23 compounds with tetrahedral coordination $CdIn_2Se_4$ and the tellurides may show a pseudocubic structure with c = a. The remaining compounds crystallize in a tetragonal $I\bar{4}$ structure with $c \approx 2a$, called defect chalcopyrite structure.

The optical phonon spectrum of defect chalcopyrites at k = 0 is given by

$$3A + 5B + 5E$$

modes, all Raman active. The E and B polar modes are also infrared active, with dipole moment in the xy plane and along the z(c) axis, respectively.

From a fundamental point of view of particular interest is the defective structure of the tetragonal compounds /1/. Their vibrational dynamics is strongly influenced by the vacancies: a very strong Raman active mode of A symmetry has been detected in all the spectra of these compounds and is currently described as a breathing motion of the anions against a vacancy /1, 2/. This picture has been confirmed by experiments and calculations on $InPS_4$ /4, 5/ where, in correspondence of two vacancies per unit cell, two breathing modes have been observed.

In addition, our previous Raman results on gallium defect chalcopyrites have evidenced a close correspondence between the spectra /6 to 10/. The

1) via d' Azeglio 85, 43100 Parma, Italy.

frequencies of the corresponding modes proved to be proportional: to account for this proportionality law a simple model, based on the vibrational decoupling of AX_4 and BX_4 tetrahedral units in the tetragonal AB_2X_4 lattice, has been proposed /9/.

Literature now reports, in addition to our Raman results, IR and Raman data on $HgGa_2S_4$ and $HgGa_2Se_4$ /11/ and IR data on $CdAl_2S_4$ and $ZnAl_2Se_4$ /12/. Extending to these compounds the correlation analysis of the spectra, one obtains a rather wide phenomenological basis from which to gain a general picture of the optical phonon spectra in defect chalcopyrites. The correlation procedure, as already mentioned, is greatly simplified by the similarity of the spectra. The comparison between the experimental results suggests the mode-mode correspondence shown in Table 1.

The frequencies of the Hg and Al compounds were obtained from powdered polycrystalline samples and these data may be affected by a larger uncertainty with respect to the previous ones, where the frequencies were determined rather accurately by Raman scattering on good quality single crystals and in controlled geometries.

The mode-mode correspondence obtained by inspection may be checked against the average proportionality factors as given by the tetrahedral model

Table 1

Frequencies in cm^{-1} of the optic vibrational modes as detected by IR and Raman spectroscopy in defect chalcopyrites. The mode-mode correspondence in different compounds is apparent

	mode frequencies (cm^{-1})									
$CdAl_2S_4$	94	-	184	206	261	295	-	412	425	453
$ZnGa_2S_4$	108	-	136	169	259	281	339	362	371	380
$CdGa_2S_4$	86	-	136	165	237	257	291	320	365	377
$HgGa_2S_4$	59		126	157	228	243	-	317	361	376
$ZnAl_2Se_4$	-	126	138	-	200	221	-	336	-	380
$ZnGa_2Se_4$	84	94	109	128	201	-	237	255	-	266
$CdGa_2Se_4$	68	76	105	124	178	196	222	241	262	264
$HgGa_2Se_4$	52	-	-	120	160	179	229	-	240	254

/9/. The frequencies of the polar E and B modes, in the bond stretching limit, are given by /9/

$$\nu^2_{E,B} = \frac{k}{m_X}\left(1 + \frac{4m_X}{3m_C}\right) , \qquad (1)$$

where k is a cation-anion force constant, m_X is the anion mass, and m_C is a cation mass.

By using $CdGa_2Se_4$ /7/ and $ZnGa_2S_4$ /8/ as standards for compounds containing Se and S as anion, respectively, and hypothizing similar force constants in isoanionic compounds /9/, we "project" the frequencies of corresponding modes by means of (1) with three different choices for the cation mass: m_A, m_B, and an average cation mass $\bar{m}_C = (m_A + 2m_B)/3$. As shown in /6, 7/, due to the relatively weak anisotropy of these compounds, for each E symmetry phonon the anisotropy-split B symmetry phonon has only a few cm^{-1} different frequency: we can then group the polar phonon spectrum of defect chalcopyrites in five couples (I to V) of E and B modes. The results of the projection are shown in Table 2.

As regards couple I, whose discussion has been reported elsewhere /7, 10/ the frequency data suggest a dependence of the squared frequency on $(m_A + m_X)^{-1/2}$. This behaviour is evidenced in cation A mixed compounds /13/.

Couple II and V are well reproduced by $m_C = m_B$, couple IV by $m_C = \bar{m}_C$, whereas couple III may be equally described by $m_C = m_A$ or $m_C = \bar{m}_C$.

In addition to this tentative assignments to particular cations, the mode-mode correspondence allows to label symmetry unassigned (for Al compounds) or uncertain modes (in Hg compounds). These spectra may then be described as follows.

$CdAl_2S_4$: eight polar modes have been detected in /12/. The 94 cm^{-1} mode is the lowest frequency E phonon. As in all other sulphides the corresponding B mode seems to be undetectable. The 184 and 284 cm^{-1} modes are E and B phonons of couple II, connected mainly to the motion of group B cations. The 412 cm^{-1} mode belongs to couple IV and the 458 cm^{-1} feature is the highest frequency Γ_{15}-like /7/ B symmetry phonon. As to the 425 cm^{-1} mode, we label it as the E symmetry phonon of the couple V, even if its anisotropy splitting is quite large in comparison with the other defect chalcopyrites.

Table 2

Projected frequencies, as expected from (1) with different cation masses, obtained from $CdGa_2Se_4$ for the selenides and from $ZnGa_2S_4$ for the sulphides. Couple notation, symmetry assignment, and experimental values are indicated. All frequencies are in cm^{-1}

couple	symmetry	standard exp	m_A	m_B	$\bar{m}_C$	exp	m_A	m_B	$\bar{m}_C$	exp	m_A	m_B	$\bar{m}_C$	exp
		$CdGa_2Se_4$	$ZnAl_2Se_4$				$HgGa_2Se_4$				$ZnGa_2Se_4$			
I	E	68	79	95	86	-	60	68	63	52	79	68	77	84
	B	76	88	106	97	126	67	76	70	-	88	76	86	94
II	E	105	122	147	134	138	93	105	97	-	122	105	118	109
	B	124	144	173	158	-	110	124	114	120	144	124	140	128
III	E	178	207	249	226	200	158	178	165	160	207	178	201	201
	B	196	228	274	249	221	174	196	181	179	228	196	221	-
IV	B	222	258	310	282	≈	197	222	205	-	258	222	250	237
	E	241	280	337	307	336	214	241	223	222	280	241	272	255
V	E	262	304	366	333	-	233	262	242	240	304	262	295	-
	B	264	306	369	336	380	234	264	244	254	306	264	298	266
		$ZnGa_2S_4$	$CdGa_2S_4$				$CdAl_2S_4$				$HgGa_2S_4$			
I	E	108	99	108	104	86	99	137	113	94	92	108	99	59
	B	-	-	-	-	-	-	-	-	-	-	-	-	-
II	E	136	124	136	131	136	124	172	142	184	116	136	125	126
	B	169	154	169	163	165	154	214	176	206	145	169	156	157
III	E	259	237	259	250	237	237	328	270	261	222	259	238	228
	B	281	257	281	271	257	257	356	293	295	241	281	259	243
IV	B	339	310	339	327	291	310	429	354	-	290	339	312	-
	E	362	331	362	349	320	331	458	378	412	310	362	333	317
V	E	371	339	371	357	365	339	470	387	425	318	371	341	361
	B	380	347	380	366	377	347	481	397	453	325	380	350	376

$ZnAl_2Se_4$: six polar modes have been reported in /12/. The 126 cm^{-1} feature, reported as the lowest frequency phonon, has a doubtful origin. It is unlikely the B mode of the couple I. The 138 cm^{-1} phonon could be the E mode of the couple II. Couple III is given by the 200 and 221 cm^{-1} modes; the B mode of couple IV escaped observation, whereas the 336 cm^{-1} phonon is of E symmetry and the 380 cm^{-1} feature is the sphalerite-like B mode /7/.

$HgGa_2S_4$ and $HgGa_2Se_4$: the symmetry assignment, very different from that reported in /11/, may be inferred from Table 2.

The general picture up to now obtained is rather consistent and satisfactory. For more quantitative models, better results from oriented single crystals would be needed. We are presently working by Raman scattering along these lines.

References

/1/ A. MILLER, A. MAC KINNON, and D. WEAIRE, Solid State Phys. 36, 119 (1981).

/2/ C. RAZZETTI, P.P. LOTTICI, and L. ZANOTTI, Mat. Chem. Phys., in the press.

/3/ E. PARTHE, Crystal Chemistry of Tetrahedral Structures, Gordon and Breach, New York 1964.

/4/ V.S. DORDYAI, I.I. NEBOLA, E.YU. PERESH, and V.YU. SLIVKA, Soviet Phys. - Solid State 22, 1347 (1982).

/5/ I.I. NEBOLA, L.M. SUSLIKOV, V.S. DORDYAI, YU.M. VYSOKANSKII, N.N. KHARKHALIS, and V.YU. SLIVKA, Soviet Phys. - Solid State 24, 2069 (1984).

/6/ R. BACEWICZ, P.P. LOTTICI, and C. RAZZETTI, J. Phys. C 12, 3603 (1979).

/7/ C. RAZZETTI, P.P. LOTTICI, and R. BACEWICZ, J. Phys. C 15, 5657 (1982).

/8/ P.P. LOTTICI and C. RAZZETTI, Nuovo Cimento D2, 2050 (1983).

/9/ P.P. LOTTICI and C. RAZZETTI, J. Phys. C 16, 3449 (1983).

/10/ P.P. LOTTICI and C. RAZZETTI, Solid State Commun. 46, 681 (1983).

/11/ H. HAEUSELER, J. Solid State Chem. 26, 367 (1978).

/12/ H. HAEUSELER and A. CANSIZ, Z. Naturf. 38b, 311 (1983).

/13/ C. RAZZETTI, P.P. LOTTICI, L. ZANOTTI, and M. CURTI, phys. stat. sol. (b) 118, 743 (1983).

(Received June 5, 1984)

phys. stat. sol. (b) 125, K13 (1984)

Subject classification: 6 and 20.1; 22.8.1

Laboratoire de Physique Cristalline, U.E.R. de Physique, Université de Nantes[1] (a)
et Laboratoire de Physique Cristalline de l' Académie Hongroise des Sciences, Budapest[2] (b)

Influence des impuretés cationiques (Cu, Cr, Ni, Er) sur les spectres Raman du niobate de lithium

Par

G. CORRADI[3] (a), E. RZEPKA (a) et K. POLGÁR (b)

Introduction De part ses propriétés ferroélectriques, piézoélectriques et de non-linéarité optique, le niobate de lithium est un cristal présentant un potentiel technologique important. Par ailleurs, dans certaines applications comme la réalisation de guides d' onde ou de mémoires holographiques, un dopage avec certains cations est indispensable. De ce fait, l' étude de l' influence de ces impuretés sur les propriétés vibrationnelles du cristal s' avère nécessaire.

Il y a quelques années déjà, des études portant sur la relaxation spin-réseau d' impuretés paramagnétiques, comme les ions des métaux de transition et de terres rares, ont été entreprises /1, 2/. Il est montré en particulier que dans la majorité des cas, en ce qui concerne le temps de relaxation spin-réseau T_1, la relaxation est due à la modulation des champs électriques internes (champ cristallin) par des vibrations de réseau. Or la dépendance de T_1 avec la température, que l' on observe au-dessus de 20 K, ne permet pas d' exclure à priori le rôle joué par des modes localisés ou quasi-localisés dus éventuellement aux impuretés.

Des études de diffusion Raman sur le $LiNbO_3$ dopé ont été seulement décrites dans le cas du dopage au Fe /3/. Ces mesures portent sur les variations d' intensité d' une raie à ≈ 628 cm^{-1}, dans le cas d' observation X(ZX)Y, en fonction de la concentration en Fe et par comparaison au cristal pur. Or, ces résultats ont été discutés en considérant cette raie comme permise dans le cas

1) 2, rue de la Houssinière, F-44072 Nantes, France.

2) BP 132, H-1502 Budapest, Hungary.

3) Adresse permanente: Laboratoire de Physique Cristalline de l' Académie Hongroise des Sciences, BP 132, H-1502 Budapest, Hungary.

d' observation indiqué, contrairement aux attributions des modes de vibration publiées par Barker et Loudon /4/, Claus et al. /5, 6/ et Schuller et al. /7/. Des études détaillées de la réflectivité infrarouge et de la diffusion Raman, incluant la dispersion angulaire de ces phénomènes /6, 7/, y compris pour les spectres de polaritons /5/, ont définitivement confirmé ces attributions. Une tentative de Kaminov et Johnston /8/ de traiter une faible raie à 621 cm^{-1} comme un mode de type E (LO), a été réfutée /5, 7/ et le caractère symétrique (mode de type A_1 (TO)) d' une raie intense à 630 cm^{-1}, ayant néanmoins tendance à apparaître en dépolarisation dans les spectres Raman de symétrie différente, a été clairement mis en évidence.

Pour ces différentes raisons, l' étude de l' influence des impuretés cationiques sur la structure vibrationnelle de $LiNbO_3$, par diffusion Raman, nous a paru intéressante.

Détails expérimentaux Les échantillons ont été préparés au Laboratoire de Physique Cristalline de l' Académie Hongroise des Sciences, Budapest, en prenant comme produit de départ du Li_2CO_3 Merck Suprapur et du Nb_2O_5 Johnson-Matthey Specpure. Nous avons utilisé une méthode de Czochralski automatisée permettant la croissance de cristaux de haute qualité optique et de diamètre constant. Les échantillons de $LiNbO_3$ dopés ont été réalisés en introduisant dans la solution fondue $2x10^{-4}$ à 10^{-3} mole/mole d' oxide de Cu, Cr, Ni et Er. Des parallélépipèdes, de dimensions minimales 2x4x10 mm^3, ont été taillés parallèlement aux axes cristallographiques déterminés par rayons X.

Les mesures de diffusion Raman ont été effectuées à T = 78 K en utilisant essentiellement la raie 647,1 nm d' un laser continu Kr^+. Dans le cas $LiNbO_3$:Er, nous avons employé la raie 514,5 nm d' un laser Ar^+, afin de s' affranchir de la fluorescence. L' intensité lumineuse n' a jamais excédé 500 mW. Nous avons travaillé en rétrodiffusion avec trois configurations du cristal et en diffusion à 90° avec deux configurations. Des polarisations différentes de la lumière incidente et diffusée, conduisent à onze géométries de symétries distinctes et un certain nombre de recoupements sont ainsi obtenus.

Nous avons utilisé un montage Raman classique équipé d' un double monochromateur à réseaux holographiques Jobin-Yvon (type Ramanor HG2S). La détection a été faite à l' aide d' une chaine de comptage comprenant un photomultiplicateur AsGa (RCA C31034) refroidi à -40 °C, un discriminateur et un compteur de photons (PAR 1109). Les données ont été stockées sur microordinateur Apple II permettant leur analyse et exploitation ultérieures.

Résultats expérimentaux Les résultats les plus marquants de notre étude

a) absence de modes supplémentaires (jusqu' à 2000 cm^{-1}),

b) aucune modification de fréquences et de largeurs de raies n' a été observée par l' introduction d' impuretés, dans la limite de la résolution expérimentale (< 2 cm^{-1}),

c) des changements d' intensité relatifs à certaines raies interviennent et peuvent être analysés comme suit:

i) dans le cas du Ni et de l' Er, les rapports d' intensité des modes permis par la symétrie présentent de faibles écarts (de l' ordre de 15%) avec ceux déterminés dans le cas des cristaux purs. Par contre, dans le cas du Cu et du Cr, ces écarts peuvent être un peu plus importants, allant même jusqu' à 50%,

ii) des dépolarisations apparaissent, de façon plus aléatoire, surtout dans les cristaux purs ou dopés au Cr et au Cu, et semblent dues à des effets secondaires.

Discussion Les variations d' intensité des raies Raman que nous venons de décrire peuvent être dues à un ensemble de causes:

a) dépolarisation de la lumière due à des imperfections de structure cristalline (tensions internes),

b) défauts d' orientations des échantillons (axes cristallographiques et positionnement du cristal, en tout état de cause, inférieurs à 2°),

c) variations de l' absorption de la lumière entre 600 et 700 nm dans le cas du Cr et du Ni,

d) variations du coefficient de réflexion des faces de l' échantillon,

e) ouverture du faisceau diffusé ($\approx 5^{o}$) défini par le système optique.

Pour effectuer une analyse détaillée de nos résultats, nous avons fait varier un certain nombre de ces paramètres comme l' orientation et le positionnement du cristal, la modification de l' ouverture du faisceau diffusé. Nous pouvons en déduire que les variations d' intensité des raies permises peuvent être imputables essentiellement à des tensions internes (cause a), tandis que les dépolarisations observées sont plutôt dues à l' ensemble des causes a, b, d et e.

Il est vraisemblable que les phénomènes observés par Semenov et Cherkasov /3/ dans le cas du $LiNbO_3$ dopé au Fe, soient attribuables à des causes identiques. Ceci démontre les difficultés rencontrées au cours de l' étude par diffusion Raman pour obtenir des résultats quantitatifs. Des mesures de biréfringence /9/ devraient pouvoir mettre en évidence l' influence des impuretés sur les tensions internes du cristal.

Par ailleurs, il apparaît que l'interprétation des propriétés de relaxation paramagnétique observées dans ces cristaux doit désormais être faite sans évoquer des modes localisés ou quasi-localisés dus à des impuretés ou des défauts.

Nous exprimons notre gratitude à M. le Professeur Lefrant pour l'intérêt qu'il a porté à cette étude.

Bibliographie

/1/ G. CORRADI, K. POLGÁR, I.M. ZARITSKII et V.S. VIKHNIN, Proc. Conf. Fransport Properties of Solids, Smolenice, Phys. and Appl. (Bratislava) 9, 109 (1982).

/2/ G. CORRADI, K. POLGÁR, V.S. VIKHNIN, L.G. DOVCHENKO et I.M. ZARITSKII, Soviet Phys. - Solid State, sous presse.

/3/ A.E. SEMENOV et E.V. CHESKASOV, Soviet Phys. - Solid State 22, 1105 (1980).

/4/ A.S. BARKER et R. LOUDON, Phys. Rev. 158, 433 (1967).

/5/ R. CLAUS et H.W. SCHRÖTER, 2ème Internat. Conf. Light Scattering in Solids, Flammarion, Paris 1971 (p. 244).

/6/ R. CLAUS, G. BORSTEL, E. WIESENDANGER et L. STEFFAN, Phys. Rev. B 6, 4878 (1972); Z..Naturf. 27a, 1187 (1972).

/7/ E. SCHULLER, R. CLAUS, H.J. FALGE et G. BORSTEL, Z. Naturf. 32a, 47 (1977).

/8/ I.P. KAMINOV et W.D. JOHNSTON, Jr., Phys. Rev. 160, 519 (1967).

/9/ W.D. JOHNSTON, Jr. et I.P. KAMINOV, Phys. Rev. 168, 1045 (1968).

(Received May 22, 1984)

phys. stat. sol. (b) 125, K17 (1984)

Subject classification: 6 and 20.1; 22.3

Sektion Physik der Karl-Marx-Universität Leipzig[1)] (a) and
Faculty of Physics, Polytechnic Institute of Kishinev (b)

Infrared Study of the Lattice Vibrations in Tetragonal CdP_2

By

H. SOBOTTA (a), H. NEUMANN (a), N.N. SYRBU (b), and V. RIEDE (a)

The tetragonal modification of CdP_2 crystallizes in one of the enantiomorphs D_4^4 or D_4^8 and contains eight molecules per primitive unit cell /1/. Therefore, we have 72 phonon branches the irreducible representation of which at the centre of the Brillouin zone is given by

$$\Gamma = 9\ A_1 \oplus 9\ B_1 \oplus 9\ A_2 \oplus 9\ B_2 \oplus 18\ E\ .$$

Of these modes one A_2 and one E are acoustic modes, the remaining 8 A_2 and 17 E modes are infrared active for incident radiation polarized parallel ($\vec{E} \parallel \vec{c}$) and perpendicular ($\vec{E} \perp \vec{c}$) to the tetragonal c axis of the crystals, respectively, and the A_1, B_1, B_2, and E modes are Raman active. In previous studies of the lattice vibrational properties of CdP_2 only Raman scattering measurements have been used to determine the frequencies of the fundamental long-wavelength optical modes /2 to 5/. In this manner, the frequencies of 9 A_1, 6 B_1, 8 B_2, and 13 E modes have been identified. With regard to the only infrared active A_2 modes some speculations have been made on the basis of infrared absorption and Raman scattering measurements in the two-phonon combination mode range /6 to 8/ which, however, must be considered as rather uncertain in view of the large number of possible two-phonon combination modes in this material. So far, no polarized infrared reflectivity measurements in the region of the fundamental lattice modes have been reported.

In the present note infrared reflectivity spectra of tetragonal CdP_2 are measured at 300 K in the wave number range $\bar{\nu}$ = 30 to 4000 cm^{-1} for both polarization directions $\vec{E} \perp \vec{c}$ and $\vec{E} \parallel \vec{c}$. The samples used in the experiments were vapour grown CdP_2 single crystals with the tetragonal c axis oriented parallel to the large surface plane of the specimens. To avoid imperfect polarization conditions due to the finite angle of incidence of 20° the spectra were always recorded with incident radiation polarized perpendicular to the plane

---

1) Linnéstr. 5, DDR-7010 Leipzig, GDR.

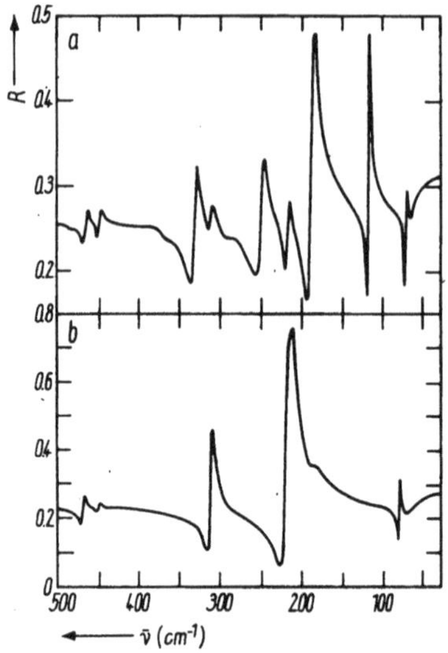

Fig. 1. Measured reflectivity spectra of tetragonal CdP_2 for a) $\vec{E} \perp \vec{c}$ and b) $\vec{E} \parallel \vec{c}$

of incidence /9/. The resulting reflectivity spectra for the wave number range $\bar{\nu}$ = 30 to 500 cm^{-1} are shown in Fig. 1. At $\bar{\nu}$ > 500 cm^{-1} no further structures could be found.

In order to determine the optical mode parameters the spectra were analysed in terms of a multiharmonic damped oscillator model which gives a complex dielectric function of the form

$$\varepsilon(\bar{\nu}) = \varepsilon_\infty + \sum_{j=1}^{n} \frac{s_j \bar{\nu}_j^2}{\bar{\nu}_j^2 - \bar{\nu}^2 - i\gamma_j \bar{\nu}_j \bar{\nu}}, \tag{1}$$

where ε_∞ is the high-frequency dielectric constant, n the number of lattice oscillators, and s_j, $\bar{\nu}_j$, and γ_j are the oscillator strength, the frequency, and the damping constant of the j-th mode. In evaluating the reflectivity spectra with (1) the non-vanishing angle of incidence was taken into account. The wave numbers $\bar{\nu}_{TO}$ and $\bar{\nu}_{LO}$ of the transverse and longitudinal optical modes were determined from the maxima in the imaginary part of the dielectric function $Im(\varepsilon(\bar{\nu}))$ and in the energy loss function $Im(-1/\varepsilon(\bar{\nu}))$, respectively. All the frequencies could be determined with an accuracy of $\Delta\bar{\nu} = \pm 0.5$ cm^{-1}.

The optimum fitting parameters resulting from the reflectivity spectra for $\vec{E} \perp \vec{c}$ (Fig. 1a) are compiled in Table 1 together with the corresponding E mode frequencies found in Raman scattering measurements /4/. We see that one additional E mode at about 250 cm^{-1} could be identified in our measurements. Obviously, the scattering efficiency of this mode is too low to be detected in the Raman spectra. On the other hand, it has been found in Raman scattering studies that there are further E modes with frequencies of 416, 299, 94.5, 88.5, and 77 cm^{-1} /4/ which are absent in the infrared reflectivity spectra and, therefore, should be characterized by very low oscillator strengths s_j.

In the case of $\vec{E} \parallel \vec{c}$ the reflectivity spectra exhibit five structures (Fig. 1b). The optimum fitting parameters for these lattice oscillators are given in Table 2.

Table 1

Parameters of the E modes in tetragonal CdP_2

$\bar{\nu}_{TO}(cm^{-1})$	$\bar{\nu}_{LO}(cm^{-1})$	s_j	γ_j	$\bar{\nu}_{Raman}$ /4/
466.2	466.7	0.013	0.0084	469
449.5	449.8	0.0094	0.0073	453
331.0	333.5	0.12	0.020	330
311.0	311.2	0.010	0.0080	313
248.9	252.3	0.21	0.033	-
216.9	217.9	0.066	0.020	215
184.6	190.6	0.60	0.036	191
117.4	118.8	0.23	0.014	117
70.4	70.7	0.061	0.085	72
$\varepsilon_{\infty\perp} = 8.1$	$\varepsilon_{0\perp} = \varepsilon_{\infty\perp} + \sum_{j=1}^{9} s_j = 9.4$			

Table 2

Parameters of the A_2 modes in tetragonal CdP_2

$\bar{\nu}_{TO}(cm^{-1})$	$\bar{\nu}_{LO}(cm^{-1})$	s_j	γ_j
469.7	470.4	0.023	0.0083
450.1	450.3	0.0074	0.0084
308.6	312.8	0.18	0.012
208.8	227.7	1.06	0.021
81.1	81.5	0.080	0.012
$\varepsilon_{\infty\parallel} = 7.7$	$\varepsilon_{0\parallel} = \varepsilon_{\infty\parallel} + \sum_{j=1}^{5} s_j = 9.0$		

On the basis of an analysis of two-phonon combination mode spectra it has been suggested that A_2 modes with frequencies of about 466 to 474, 449 to 454, and 309 to 316 cm^{-1} should be present in tetragonal CdP_2 /6 to 8/. As it can be seen from Table 2 these frequencies agree with the three high-energy A_2 modes found in our experiments.

In Table 1 and 2, we give also values for the static dielectric constants $\varepsilon_{0\perp}$ and $\varepsilon_{0\parallel}$, respectively. Since only 9 E and 5 A_2 modes of the 17 E and 8 A_2 modes expected from group theory could be identified in our measurements these values must be considered as lower bounds to the true $\varepsilon_{0\perp}$ and $\varepsilon_{0\parallel}$ values.

References

/1/ V.B. LAZAREV, V.YA. SHEVCHENKO, and S.F. MARENKIN, Izv. Akad. Nauk SSSR, Ser. neorg. Mater. 15, 1701 (1979).

/2/ M.YA. VALAKH, A.V. FEDOTOVSKII, and I.I. TICHINA, Fiz. tverd. Tela 15, 2225 (1973).

/3/ I.S. GORBAN, V.A. GORINYA, V.I. LUGOVOI, E.G. ULYANOVA, and I.I. TICHINA, Fiz. tverd. Tela 17, 2834 (1975).

/4/ I.S. GORBAN, V.A. GORINYA, R.A. DASHKOVSKAYA, V.I. LUGOVOI, A.P. MAKOVETSKAYA, and I.I. TICHINA, phys. stat. sol. (b) 86, 419 (1978).

/5/ J. WESZKA, I. MIOTKOWSKI, K. KLOC, and W. ŻDANOWICZ, phys. stat. sol. (b) 94, K97 (1979).

/6/ I.S. GORBAN, V.A. GORINYA, V.I. LUGOVOI, A.P. MAKOVETSKAYA, I.I. TICHINA, E.G. ULYANOVA, and M.A. ILIN, Zh. prikl. Spektroskopii 30, 890 (1979).

/7/ M.YA. VALAKH, J. WESZKA, and M.P. LISITSA, Ukr. fiz. Zh. 25, 1324 (1980).

/8/ N.N. SYRBU, in: Preparation and Investigation of New Materials for Semiconductor Technology, Ed. S.I. RADAUTSAN, Izd. Shtiintsa, Kishinev 1980 (p. 51)(in Russian).

/9/ L.P. MOSTELLER and F. WOOTEN, J. Opt. Soc. Amer. 58, 511 (1968).

(Received July 4, 1984)

phys. stat. sol. (b) 125, K21 (1984)
Subject classification: 4; 22.5.2

Physics Department, Agra College[1)]

Analysis of the Equation of State for LiF Crystal

By
J. SHANKER, W.N. BHENDE[2)], and P.S. BAKHSHI

Introduction Recently LiF crystal has been studied experimentally by Boehler and Kennedy /1/ who have measured the compression data for this crystal upto 30×10^8 Pa and 400 °C. These data provide new opportunity for studying the equation of state (ES) for alkali halide crystals. In the past, there have been two types of approaches for investigating ES. The first type of investigations makes use of phenomenological equations originally due to Murnaghan /2/, Birch /3/, and Keane /4/. The methods based on these equations and their modified forms are not capable of predicting reliable high pressure compression data. The extrapolation of low pressure ultrasonics is also difficult, because the results depend strongly on the choice of equation used to fit the measured data /5/. The phenomenological equations of state are thus of limited value.

The other type of approaches for studying ES has been based on interionic potentials /6 to 8/. They make use of pairwise or two-body potentials only and are therefore subject to the criticism that they do not explain the breakdown of the Cauchy relation $C_{12} = C_{44}$. A relevant criticism of two-body central potentials concerning their application to study ES has been made by Rouff and Chhabildas /9/.

In any case, the use of two-body potentials in LiF crystal is less justified because the magnitude of the Cauchy deviation for this crystal is largest and much higher than the corresponding values for other alkali halides /10/. In the present note we therefore include three-body interactions following the formalism originally due to Lundqvist /11/. These interactions have quantum-mechanical origin /12/. In the present note we follow a semi-empirical approach using the Hildebrand equation of state and Lundqvist's potential energy expression.

---

1) Agra 282002, India.

2) Permanent address: Dharampeth College, Nagpur, India.

Method of analysis We take the following expression for the crystal lattice energy:

$$W = -\frac{\alpha_M e^2(1+6f)^2}{r} + B(r) - \frac{C}{r^6} - \frac{D}{r^8}, \tag{1}$$

where α_M is Madelung's constant and f is the three-body charge transfer parameter as introduced by Lundqvist /11/ and analysed subsequently /10/. The last two terms are the van der Waals dipole-dipole and dipole-quadrupole interaction potentials. Magnitudes of these interactions have recently been estimated using the Slater-Kirkwood variational method /13/. Values of C and D thus obtained /13/ are used in the present note. The short-range overlap repulsive energy B(r) is considered effective between nearest neighbours and next-nearest neighbour ions. For an ionic crystal this energy is written as follows:

$$B(r) = 6b\beta_{+-}\exp\left(\frac{r_+ + r_- - r}{\rho_{+-}}\right) + 6b\beta_{++}\exp\left(\frac{2r_+ - r\sqrt{2}}{\rho_{++}}\right) + 6b\beta_{--}\exp\left(\frac{2r_- - r\sqrt{2}}{\rho_{--}}\right). \tag{2}$$

Equation (2) is similar to the one used in the Born model treatment /14/ except the hardness parameters ρ_{ij} which have different values for different pair interactions. This treatment is physically more justified /15/ than the assumption for a common hardness parameter taken in the Born model. Values of second neighbour hardness parameters ρ_{++} and ρ_{--} are evaluated directly from the values of overlap integrals /16/. Thus only two parameters b and ρ_{+-} are left to be determined in the repulsive energy expression (2). Before calculating them, it is essential to find the values of the three-body charge transfer parameter f and its derivative df/dr. For this purpose we make use of Lundqvist's expression for the Cauchy deviation

$$C_{12} - C_{44} = \frac{2.3301}{r^4} e^2 \left(r\frac{df}{dr}\right). \tag{3}$$

For evaluating df/dr we assume the exponential form of f /10/ originally suggested by Cochran /17/ viz.

$$f = f_o \exp\left(-\frac{r}{\varrho}\right) . \qquad (4)$$

For LiF crystal we take $\varrho = 0.0293$ nm corresponding to overlap integrals /16/. Equation (4) yields

$$\frac{df}{dr} = -\frac{f}{\varrho} \qquad (5)$$

and thus, f and df/dr, both are calculated with the help of (3) to (5). Now the repulsive parameters b and ϱ_{+-} are calculated using the Hildebrand equation of state:

$$P = -\left(\frac{\partial W}{\partial V}\right)_T + T\beta B_T , \qquad (6)$$

where β is the volume thermal expansion coefficient and B_T is the isothermal bulk modulus. For evaluating parameters we use (6) at T = 25 ^oC, and P = 0 and 3.09×10^8 Pa. The input data used and parameters thus calculated are reported in Table 1.

Table 1

Values of input data at room temperature and calculated potential parameters for LiF crystal

r_o (10^{-1} nm) 2.013	C_{12} (10^{10} Pa) 4.632	C_{44} (10^{10} Pa) 6.358		B_T (10^8 Pa) 665.1
β ($10^{-4}K^{-1}$) 0.999	r_+ (10^{-1} nm) 0.924	r_- (10^{-1} nm) 1.133		10^3 f 7.67
b (10^{-19} J) 0.780	ϱ_{+-} (10^{-1} nm) 0.451	ϱ_{++} (10^{-1} nm) 0.109		ϱ_{--} (10^{-1} nm) 0.430
C (10^{-25} J nm^6) 36	D (10^{-27} J nm^8) 9	β_{+-} 1.375	β_{++} 2	β_{--} 0.75

Table 2

Values of compression data for LiF crystal

T(^{o}C)	V/V_o	P(10^8 Pa)	
		calculated in the present study	experimental /1/
25	1	0	0
	0.9953	3.09	3.09
	0.9949	3.39	3.42
	0.9898	6.83	6.83
	0.9868	8.89	8.98
	0.9799	13.72	13.97
	0.9743	17.73	18.12
	0.9739	18.02	18.51
	0.9651	24.52	25.50
	0.9602	28.24	29.36
100	0.9993	5.45	5.61
	0.9947	8.50	8.60
	0.9892	12.22	12.44
	0.9863	14.22	14.88
	0.9793	19.13	19.46
	0.9737	23.15	23.64
	0.9733	23.44	24.05
	0.9637	30.59	31.00
200	0.9988	12.42	13.13
	0.9941	15.55	16.18
	0.9886	19.28	19.96
	0.9858	21.21	21.84
	0.9787	26.20	27.01
	0.9732	30.16	31.12
300	0.9984	19.33	20.65
	0.9937	22.46	23.80
	0.9882	26.20	27.52
	0.9853	28.20	29.46
400	0.9982	26.11	28.20
	0.9935	29.24	31.33

It should be mentioned here that (6) can be derived from the exact thermodynamic relation

$$P = -\left(\frac{\partial W}{\partial V}\right)_T + T\left(\frac{\partial P}{\partial T}\right)_V . \qquad (7)$$

We take $(\partial P/\partial T)_V$ equal to a constant, derived from an experimental value of βB_T. This is consistent with previous investigations /1/. $(\partial W/\partial V)_T$ has been taken to be a function of V only and independent of T. Following this semi-empirical approach based on (1) and (6) we find

$$P = \frac{X - Y}{6r_o^4} + T\beta B_T , \qquad (8)$$

where

$$X = -6r_o^4 \frac{d}{dV}\left[B(r)\right] \qquad (9)$$

and

$$Y = \alpha_M e^2\left[(1+6f)^2\left(\frac{V}{V_o}\right)^{-4/3} + 12\, r_o(1+6f)\frac{f}{\varrho}\left(\frac{V}{V_o}\right)^{-1}\right] + \\ + \frac{6C}{r_o^5}\left(\frac{V}{V_o}\right)^{-3} + \frac{8D}{r_o^7}\left(\frac{V}{V_o}\right)^{-11/3} ; \qquad (10)$$

r_o and V_o are the cation-anion nearest neighbour distance and volume, respectively, at zero pressure. Values of compression data for LiF crystal upto a pressure of about 30×10^8 Pa and temperature upto 400 ^{o}C or 673 K calculated from (8) to (10) are reported in Table 2. Calculated values are found to present good agreement with experimental data. This supports the validity of the semi-empirical approach used in the present work.

We are thankful to the referee for his valuable comments.

References

/1/ R. BOEHLER and G.C. KENNEDY, J. Phys. Chem. Solids 41, 1019 (1980).
/2/ F. MURNAGHAN, Proc. Nat. Acad. Sci. 30, 244 (1944).
/3/ F. BIRCH, Phys. Rev. 71, 809 (1947).
/4/ A. KEANE, Austral. J. Phys. 7, 3 (1954).
/5/ F. BIRCH, J. geophys. Res. 83, 1257 (1978).
/6/ M.P. VERMA and B. DAYAL, phys. stat. sol. 3, 901 (1963); 6, 545 (1964).
/7/ D.L. DECKER, J. appl. Phys. 36, 157 (1965); 37, 5012 (1966).

/8/ J. SHANKER and K. SINGH, phys. stat. sol. (b) 115, 381 (1983).
/9/ A.L. ROUFF and L.C. CHHABILDAS, J. appl. Phys. 47, 4867 (1976).
/10/ J. SHANKER, V.C. JAIN, and J.P. SINGH, Phys. Rev. B 22, 1083 (1980).
/11/ S.O. LUNDQVIST, Ark. Fys. 9, 435 (1955).
/12/ P.O. LÖWDIN, Adv. Phys. 5, 1 (1956).
/13/ J. SHANKER, G.G. AGRAWAL, and R.P. SINGH, J. chem. Phys. 69, 670 (1978).
/14/ M.P. TOSI, Solid State Phys. 16, 1 (1964).
/15/ J. SHANKER and D.P. AGRAWAL, J. Phys. Chem. Solids 41, 1003 (1980).
/16/ D.W. HAFEMEISTER and W.H. FLYGARE, J. chem. Phys. 43, 795 (1965).
/17/ W. COCHRAN, CRC Crit. Rev. Solid State Sci. 2, 1 (1971).

(Received March 20, 1984)

phys. stat. sol. (b) 125, K27 (1984)
Subject classification: 6 and 20; 13.5.2; 21.6; 22.2.2

Sektion Physik der Friedrich-Schiller-Universität Jena[1)]

Coupled Surface Plasmon-Phonon Polaritons in Bilayer Systems

By
L. WENDLER

In previous papers /1 to 3/ we have developed the theory of polaritons in bilayer systems (four media system). The importance of bilayer systems is given by the fact that they are often practically realized, moreover the bilayer system serves as a model for multilayer systems. In this note we look for surface polaritons existing in a bilayer system composed of an isotropic dielectric (polar semiconductor, ionic crystal) and a metal or n-type semiconductor. The dipole-active excitations of these media, phonons and plasmons, respectively, can couple with photons and the resulting elementary excitations are called phonon polaritons and plasmon polaritons, respectively. Surface polaritons owe their existence to the presence of surfaces. Their electromagnetic field is localized at the surface and decays exponentially in the direction perpendicular to the surface. Bringing two finite solids close to each other so that the surface of one lies in the region of the electromagnetic field of the surface polariton of the other, a coupling between the two surface polaritons takes place and a new one arises. Such a mixture of surface polaritons is most notable when their wave vectors as well as their frequencies are nearly equal. In the here discussed case we have, when the layers are separated, two different kinds of surface elementary excitations. Bringing the two layers close to each other, a coupling between the surface plasmon polaritons and the surface phonon polaritons takes place. The resulting coupled surface elementary excitations are coupled surface plasmon-phonon polaritons. The existence of such a type of coupled surface excitations was suggested by Nakayama /4/ and Eguiluz et al. /5/. Ikarashi /6/ has investigated the problem consisting of a semi-infinite metal separated by vacuum from a semi-infinite dielectric.

The configuration we are concerned with is a bilayer system in vacuum. The region $a > z > 0$ (z direction perpendicular to the surfaces) is filled by a dielectric medium and the region $0 > z > -b$ is filled by a metal. The wavelength of the in-

1) Max-Wien-Platz 1, DDR-6900 Jena, GDR.

vestigated polaritons is long enough to characterize the media by its macroscopic dielectric function. For the isotropic dielectric media with a single branch of infrared active optical phonons we will assume the following dielectric function neglecting damping and spatial dispersion:

$$\varepsilon_1(\omega) = \varepsilon_\infty \frac{\omega_L^2 - \omega^2}{\omega_T^2 - \omega^2} \tag{1}$$

with the optical dielectric constant ε_∞ and the longitudinal and transverse optical phonon frequency ω_L and ω_T, respectively. In metals the electron density oscillation can be treated in the polariton region in terms of the local and real dielectric function

$$\varepsilon_2(\omega) = 1 - \frac{\omega_p^2}{\omega^2} \tag{2}$$

with the electron plasma frequency ω_p. In the case of an n-type semiconductor whose electrons behave like those of a quasi-free electron gas (2) can be used as well when resolving the one by ε_∞. Since the two media are isotropic we can specialize on the waves that propagate in the x direction with the wave vector component $k_\parallel$ without loss of generality. As a condition that surface polaritons exist in the bilayer system we obtain the following dispersion relation /1, 2/:

$$\frac{(\varepsilon_1 + \varepsilon_2\alpha_1/\alpha_2)(\alpha_1/\alpha_0 - \varepsilon_1) + (\varepsilon_1 - \varepsilon_2\alpha_1/\alpha_2)(\alpha_1/\alpha_0 + \varepsilon_1)\exp(2\alpha_1 a)}{(\varepsilon_1 - \varepsilon_2\alpha_1/\alpha_2)(\alpha_1/\alpha_0 - \varepsilon_1) + (\varepsilon_1 + \varepsilon_2\alpha_1/\alpha_2)(\alpha_1/\alpha_0 + \varepsilon_1)\exp(2\alpha_1 a)} -$$

$$- \frac{\alpha_2/\alpha_0 + \varepsilon_2}{\alpha_2/\alpha_0 - \varepsilon_2}\exp(2\alpha_2 b) = 0 \ . \tag{3}$$

The decay constants α_0, α_1, and α_2 appearing in (3) are given by

$$\alpha_0 = \left(k_\parallel^2 - \frac{\omega^2}{c^2}\right)^{1/2} ,$$

$$\alpha_1 = \left(k_\parallel^2 - \varepsilon_1(\omega)\frac{\omega^2}{c^2}\right)^{1/2} ,$$

$$\alpha_2 = \left(k_\parallel^2 - \varepsilon_2(\omega)\frac{\omega^2}{c^2}\right)^{1/2} . \tag{4}$$

These constants must be real and positive in order to describe surface polaritons localized at the interface and the surfaces of the bilayer system. In the limiting case $k_{\parallel} \rightarrow \infty$ (3) gives

$$(\varepsilon_1 + 1)(\varepsilon_2 + 1)(\varepsilon_1 + \varepsilon_2) = 0$$

and

$$\varepsilon_1 = -1, \quad \varepsilon_2 = -1, \quad \varepsilon_1 = -\varepsilon_2 \ . \tag{5}$$

Equation (5) with dielectric functions ε_1 and ε_2 yields the asymptotic frequencies ω_{S1}, ω_{S2}, ω_{S3}, and ω_{S4} for large wave vectors,

$$\omega_{S1} = \left(\frac{\varepsilon_\infty \omega_L^2 + \omega_T^2}{\varepsilon_\infty + 1}\right)^{1/2} , \tag{6}$$

$$\omega_{S2} = \frac{\omega_p}{\sqrt{2}} , \tag{7}$$

$$\omega_{S3,4} = \left\{\frac{1}{2(\varepsilon_\infty + 1)}\left(\varepsilon_\infty \omega_L^2 + \omega_T^2 + \omega_p^2 \pm \left[(\varepsilon_\infty \omega_L^2 + \omega_T^2 + \omega_p^2)^2 - 4(\varepsilon_\infty + 1)\omega_T^2 \omega_p^2\right]^{1/2}\right)\right\}^{1/2} . \tag{8}$$

A bilayer system composed of two media with dielectric functions given by (1) and (2), always supports exactly four coupled surface plasmon-phonon polariton modes. The frequencies ω_{S1} and ω_{S2} (6, 7) are the asymptotic frequencies of single interface surface polaritons of a semi-infinite dielectric medium and a semi-infinite metal, respectively. The frequencies ω_{S3} and ω_{S4} are the asymptotic frequencies of interface surface polaritons of the half-space geometry dielectric medium / metal. That means mode 1 is mainly localized at the surface of the dielectric layer, mode 2 at the surface of the metallic layer, and modes 3 and 4 are mainly localized at the interface between the two layers /2/. For numerical work we have chosen GaP ($\varepsilon_\infty = 8.4$, $\hbar\omega_T = 0.04538$ eV, $\hbar\omega_L = 0.04996$ eV) and Ag ($\hbar\omega_p = 3.75$ eV) whose dielectric functions can be characterized by (1) and (2), respectively. They are shown in Fig. 1. The dispersion curves of the coupled surface plasmon-phonon polaritons are calculated from (3) and graphically shown in Fig. 2. As can be seen from Fig. 2, mode 1 and 4 start at the dispersion curve of light in vacuum and mode 2 and 3 start at

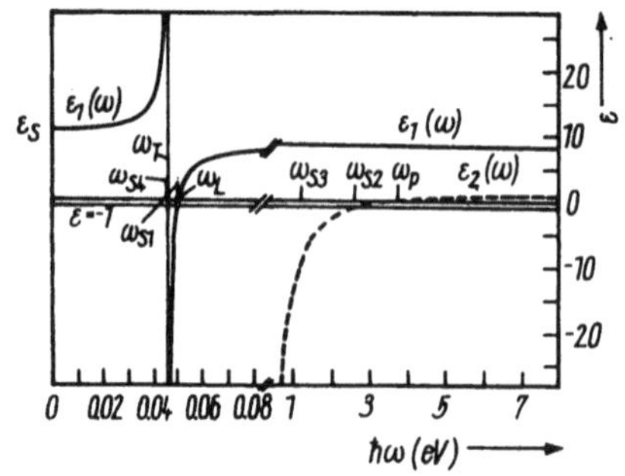

Fig. 1. Dielectric functions, $\varepsilon_1(\omega)$ (heavy lines) and $\varepsilon_2(\omega)$ (dashed lines) of GaP and Ag, respectively. The frequencies ω_{S1}, ω_{S2}, ω_{S3}, and ω_{S4} are the asymptotic frequencies for large values of the wave vector as described in the text

the upper part of the dispersion curve of bulk phonon-polaritons what has already approached its asymptotic value $\omega = ck_{\parallel}/\sqrt{\varepsilon_\infty}$. In supplement to it Fig. 3 shows the dispersion curves of the unretarded coupled surface plasmon-phonon polaritons of the bilayer system which are calculated from (3) with $c \to \infty$. For large values of the wave vector component $k_{\parallel}$ all surface modes approach the asymptotic frequencies ω_{S1}, ω_{S2}, ω_{S3}, and ω_{S4}, respectively.

The medium whose dielectric function is negative in the region where the surface polaritons exist is called the surface active medium and the other the surface inactive one. From Fig. 1 is to be seen that the dielectric layer supports only mode 1. The metallic layer is surface active to mode 2, 3, and 4. That means both interface modes 3 and 4 are supported by the metallic layer. Modes 1 and 4 exist in a small region in the infrared. They show only weak dispersion and their behaviour is phonon-like. Interesting is the phonon-like behaviour of mode 4 which is supported by the metallic layer. The reason for this is that the metallic layer is surface active in a wide frequency range ($\omega \approx 0$ to 10^{15} $s^{-1} < \omega_p$) but the dielectric layer is in the frequency region of mode 4 dominant to the behaviour of the coupled surface elementary excitation due to the presence of TO phonons.

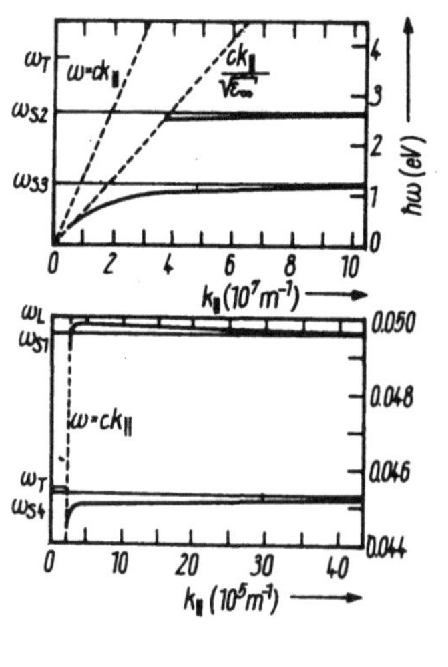

Fig. 2. Dispersion relation of coupled surface plasmon-phonon polaritons for a GaP/Ag bilayer system

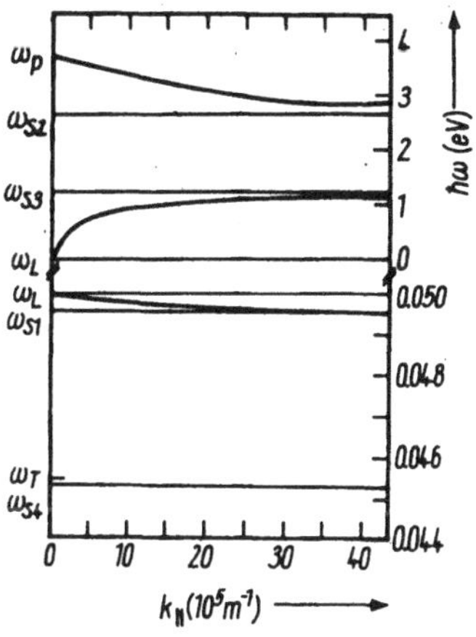

Fig. 3. Dispersion relation of unretarded coupled surface plasmon-phonon polaritons for a GaP/Ag bilayer system

Due to the dielectric properties of the metal in the infrared region the electric field of mode 4 is strong attenuated in z direction in the metallic layer. Consequently this mode shows only weak and in the unretarded case nearly no dispersion. Mode 2 lies in the ultraviolet frequency region, the typical region for phenomena connected with plasmons and corresponds in its behaviour with the surface plasmon polaritons discussed in /7/. The interface mode 3 is a very important excitation of the bilayer system consisting of a dielectric and a metallic layer while mostly representing the mixing character of the coupled surface plasmon-phonon polaritons. It starts at the phonon frequency ω_L and rises up to the visible frequencies. Since surface polaritons are proved to be sensitive tools to surface studies one has the main advantage to investigate the metal-semiconductor interface in the visible frequency region with surface mode 3.

References

/1/ L. WENDLER, Dissertation, Jena 1983.

/2/ L. WENDLER and E. JÄGER, phys. stat. sol. (b) 120, 235 (1983).

/3/ L. WENDLER, phys. stat. sol. (b) 123, 469 (1984).

/4/ M. NAKAYAMA, J. Phys. Soc. Japan 36, 393 (1974).

/5/ A. EGUILUZ, T.K. LEE, and J.J. QUINN, Phys. Rev. B 11, 4989 (1975).

/6/ T. IKARASHI, J. Phys. Soc. Japan 41, 1962 (1976).

/7/ L. WENDLER, submitted to phys. stat. sol. (b).

(Received June 13, 1984)

phys. stat. sol. (b) 125, K33 (1984)

Subject classification: 2 and 8; 21.2

Department of Physics (a) and Department of Physical Chemistry (b), Jadavpur University, Calcutta[1)]

Entropy and Heat Capacity of Liquid Metals

By

R.N. JOARDER (a), S. PALCHAUDHURI, and R.V. GOPALA RAO (b)

A combination of thermodynamic perturbation theory and pseudopotential formalism enables the determination of thermodynamics of liquid metals and alloys through the hard sphere model /1/. The calculations on simple, transition, and rare earth metals /2/ evidently show that this hard sphere description can predict even the temperature variation of several properties when a good temperature dependent packing density, η, is used for the calculation.

The entropy and heat capacities for such a system are given by

$$S = S_{hs} + S_{elec} , \tag{1}$$

where

$$\frac{S_{hs}}{k_B} = 2.5 + \ln\left[\Omega(Mk_B T/2\pi\hbar^2)^{3/2}\right] + \ln(1-\eta) + \frac{3}{2}\left\{1 - \frac{1}{(1-\eta)^2}\right\} ,$$

$$S_{elec} = \frac{\pi}{3} N(E_F)\, k_B^2 T ,$$

$$C_p = \frac{3}{2} R - RT\left\{\left(\frac{1}{1-\eta} + \frac{3}{(1-\eta)^3}\right)\frac{d\eta}{dT} - \frac{1}{\Omega}\left(\frac{d\Omega}{dT}\right)_p\right\} + \frac{z\pi^2 k_B T}{K_f^2}\left(1 + \frac{2}{3}\alpha_p T\right) , \tag{2}$$

$$C_v = C_p - T\alpha_p^2/\varphi\chi_T \tag{3}$$

and the terms have their usual meaning. The calculations at different temperatures are complete provided we have a good temperature dependent η at constant pressure. A simple consideration leads to a temperature dependent η /3/ given by

$$T(1 + 2\eta)^2/(1-\eta)^4 = A \exp\left[-\alpha(T - T_m)\right] , \tag{4}$$

where $A = T_m(1 + 2\eta_m)^2/(1-\eta_m)^4$, η_m is the packing density at the melting temperature T_m and α is a constant. A and α are obtainable by fitting the

1) Calcutta 700 032, India.

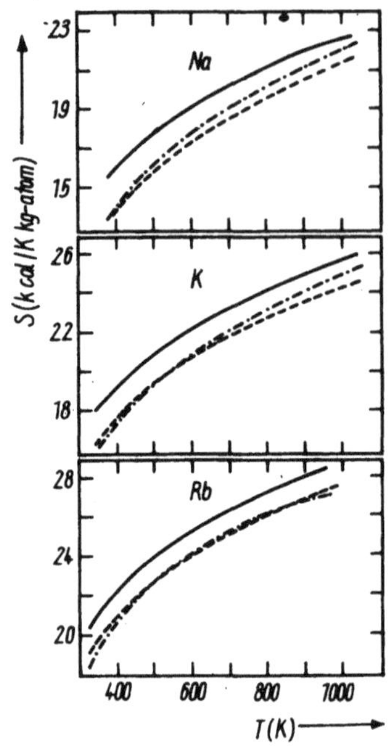

Fig. 1. Entropy of the alkali metals Na, K, Rb in dependence on temperature; —— exp, —·— /6/, --- present work

structure factor and its temperature derivative at the first peak at or near the melting point /3/.

The variation of the hard core volume, $w(\eta\Omega/N)$ with volume finds its expression in the pressure coefficient while that with temperature in the heat capacity /4/. Correct evaluation of the pressure coefficient and heat capacity surely means that this temperature dependent η accounts well the softness of the pseudo-atom. Thus

$$\left(\frac{\partial \ln w}{\partial \ln T}\right)_{\Omega} = \left(\frac{\partial \ln \eta}{\partial \ln T}\right)_{\Omega} = \frac{C_v}{C_p}\left(\frac{\partial \ln \eta}{\partial \ln T}\right)_p , \qquad (5a)$$

$$\left(\frac{\partial \ln w}{\partial \ln \Omega}\right)_T = 1 + \left(\frac{\partial \ln \eta}{\partial \ln \Omega}\right)_T = 1 + \left[\left(\frac{\partial \ln \eta}{\partial \ln T}\right)_p - \left(\frac{\partial \ln \eta}{\partial \ln T}\right)_{\Omega}\right]\frac{T}{\Omega}\left(\frac{\partial \Omega}{\partial T}\right)_p , \qquad (5b)$$

$$\left(\frac{\partial \ln \Omega}{\partial \ln T}\right)_{\eta} = -\left(\frac{\partial \ln \eta}{\partial \ln T}\right)_{\Omega} \Big/ \left(\frac{\partial \ln \eta}{\partial \ln \Omega}\right)_T . \qquad (5c)$$

We have taken three alkali metals for which the C_p variation at higher temperature is quite interesting, e.g. C_p goes through a minimum. The entropy and C_p variations with temperature are shown in Fig. 1 and 2. The densities at higher temperatures were taken from literature /5/. The agreement surely justifies our model. We have also given entropy and C_p variations based on Protopas et al.'s /6/ model. C_p calculation is a severe test of the correctness of the core variation. It is also evident that the electronic contribution is significant and has a role in determining the correct temperature variation. It is appropriate to mention here that Jones /7/ and Umar /8/ successfully calculated the temperature variation of C_p through ab initio calculations (e.g. core diameters obtained

Table 1

Thermodynamics of liquid alkali metals at the melting point

		Na	K	Rb
packing density η		0.475	0.465	0.459
$10^{-4}\ \alpha_p$		2.8	2.9	3.4
$-10^{-4}(\partial\eta/\partial T)_p$		3.154	3.426	3.897
$-\left(\frac{\partial \ln\eta}{\partial \ln\Omega}\right)_T$	present work	0.299	0.268	0.289
	Harder et al. /4/	0.407	0.462	0.512
$-\left(\frac{\partial \ln\eta}{\partial \ln T}\right)_\Omega$	present work	0.215	0.222	0.234
	Harder et al. /4/	0.275	0.262	0.280
$-\left(\frac{\partial \ln\Omega}{\partial \ln T}\right)_\eta$	present work	0.719	0.828	0.810
	Harder et al. /4/	0.680	0.570	0.550
$-\left(\frac{\partial \ln\Omega}{\partial \ln T}\right)_{\text{melting}}$	exper.	0.700	0.660	0.680
$10^{-7}\left(\frac{\partial p}{\partial T}\right)_\Omega$	present work	1.43	0.66	0.57
	exper.	1.33	0.72	0.70
$-\frac{S}{k_B}$	present work	4.53	4.33	4.21
	Hafner /9/	3.47	3.43	3.66
	exper.	3.45	3.45	3.63
$\frac{C_p}{k_B}$	present work	4.2	4.0	4.1
	Hafner /9/	4.0	4.6	5.0
	exper.	3.8	3.9	3.8
$\frac{C_v}{k_B}$	present work	3.7	3.6	3.6
	Hafner /9/	2.9	3.3	3.4
	exper.	3.4	3.4	3.3
$\gamma\left(=\frac{C_p}{C_v}\right)$	present work	1.1	1.1	1.1
	Hafner /9/	1.4	1.4	1.5
	exper.	1.1	1.1	1.2

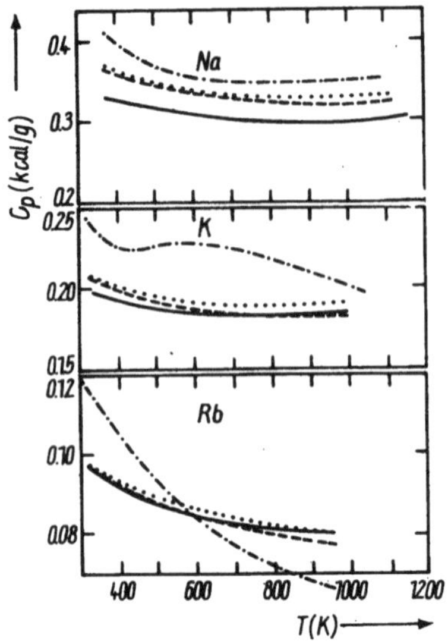

Fig. 2. Dependence of C_p on temperature for Na, K, Rb; —— exper. —·— /6/, --- calculated without electronic contribution, ··· calculated with electronic contribution

at different temperatures by the minimisation of free energy). The present method is simple, straightforward, and direct.

We have obtained C_v at the melting temperature from (3) using experimental values of expansivity and isothermal compressibility. The values of C_v, $\gamma (= C_p/C_v)$, $(\partial p/\partial T)_\Omega$ are quite reasonable (Table 1). This calculation enables us to calculate softness parameters of the core and these results are shown in Table 1 along with other results /4/. Comparison of these softness parameters with other polyvalent and noble metals shows that the alkali metals form a group of soft ions /4/. A comparison between $-(\partial \ln \Omega/\partial \ln T)_\eta$ and $-(\partial \ln \Omega/\partial \ln T)_{\text{melting}}$ evidently shows that η is almost constant along the melting line for these metals.

Concluding, the present method has the advantage over the ab initio calculations because of the lack of proper pseudopotential and correct screening functions.

References

/1/ I.H. UMAR, A. MEYER, M. WATEBE, and W.H. YOUNG, J. Phys. F 4, 1691 (1974).

/2/ W.H. YOUNG, Inst. Phys. Conf. Ser. 30, 1 (1977). Y. WASEDA, ibid. 20, 230 (1977). J.M. HARDER and W.H. YOUNG, Phys. Letters A 61, 468 (1977).

/3/ R.N. JOARDER and R.V. GOPALA RAO, phys. stat. sol. (b) 116, 299 (1983).

/4/ J.M. HARDER, M. SILBERT, I. YOKOYAMA, and W.H. YOUNG, J. Phys. F 9, 1005 (1979).

/5/ C.J. SMITHELLS, Metal Reference Book, Vol. 2, Butterworths Inc., 1967.

/6/ P. PROTOPAPAS, H.C. ANDERSON, and N.A.D. PARLEE, J. chem. Phys. 59, 15 (1973).

/7/ J. JONES, Phys. Rev. A 8, 3215 (1973).

/8/ I.H. UMAR, Ph.D. Thesis, Univ. of East Anglia, 1974.

/9/ J. HAFNER, Phys. Rev. A 16, 351 (1977).

(Received July 26, 1984)

phys. stat. sol. (b) 125, K37 (1984)
Subject classification: 10.2 and 20.1; 11; 22.5.2

Polytechnical Institute, Tomsk[1)]

Effect of Powerful Pulsed Electron Irradiation on Creation of Hole Colour Centres in KCl at 300 K

By
YU.M. ANNENKOV, V.F. STOLARENKO, and T.S. FRANGULIAN

The idea of this investigation was to study the properties of stable hole colour centre formation in alkali halides under powerful pulsed electron irradiation in comparison with the effects caused by low-density X-irradiation.

The investigations have been carried out using potassium chloride and potassium bromide crystals grown in the Institute of Physics of the Academy of Sciences of the Estonian SSR by sixtyfold zone melting. Measurements of the electrical conductivity showed the concentration of impurity-vacancy complexes in samples not to exceed about 2×10^{14} cm^{-3}.

The irradiation has been carried out using electrical pulses of nanosecond duration. Their parameters are: electron energy about 300 keV, current density in a pulse about 200 A/cm^2, pulse duration about 20 ns. The power of the dose absorbed in a pulse was up to about 10^{12} Gy/s. The time interval between single pulses was equal to 15 s. The X-raying has been carried out with the apparatus URS-70 (a tube with Mo anode, U = 50 kV, J = 20 mA), the power of the dose absorbed amounting to about 10 Gy/s.

Fig. 1 represents the additional optical absorption measured within the UV spectral range for KCl crystals under various doses of pulsed electron and X-ray irradiation. The additional absorption represents a wide non-elementary band with maximum shifting towards smaller wavelengths from 228 to 218 nm under increasing electron irradiation dose.

In the case of X-irradiation the maximum of the integral absorption occurs in the range of smaller wavelengths and undergoes a shift from 218 to 213 nm within the 3.6×10^4 to 7.3×10^5 Gy dose interval.

The analyses show that the radiation-induced optical absorption within the 200 to 400 nm wavelength range may be approximated by the sum of absorption bands with maxima at about 213, 233, 257, and 300 nm (Fig. 2). The bands at about 213, 233 nm are the most intensive ones.

---

1) Lenina 30, 634004 Tomsk, USSR.

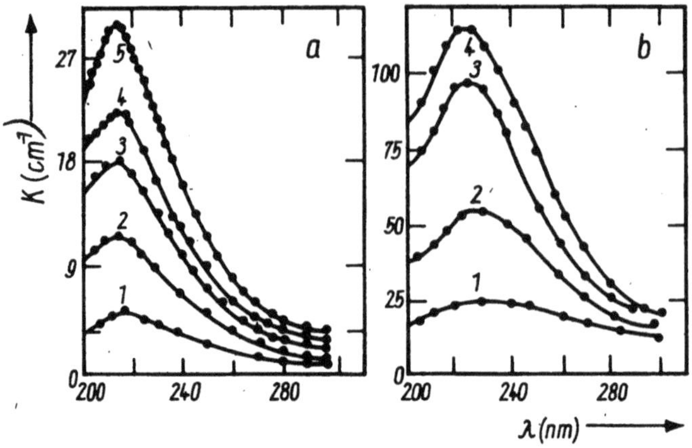

Fig. 1. Spectra of optical absorption for KCl crystals under a) various doses of X-ray irradiation (dose in Gy: (1) $3.6x10^4$, (2) $1.6x10^5$, (3) $3x10^5$, (4) $4.4x x10^5$, (5) $7.3x10^5$) and b) pulsed electron irradiation (dose in Gy: (1) $5x10^4$, (2) $5.5x10^5$, (3) $2.65x10^6$, (4) $3.65x10^6$)

Absorption bands with maxima at 233 and 213 nm are identified with V_2 and V_3 colour centres, respectively. The V_2 centre is modelled as a trihalide molecular ion (X_3^-) occupying two anion and one cation lattice sites /1/. The V_3 absorption is related to multihalide associates /2, 3/.

Thus, the relation between V_3 and V_2 colour centres characterizes the degree of aggregation of radiolysis hole products.

We have analysed the intensity ratios of V_2 and V_3 bands (K_{V_2}/K_{V_3}) under X-ray and electron irradiations. The data of the analysis are presented in Table 1.

Table 1

electron irradiation dose (Gy)	$\frac{K_{V_2}}{K_{V_3}}$	X-irradiation dose (Gy)	$\frac{K_{V_2}}{K_{V_3}}$
$5x10^5$	1.5	$3.6\ x10^4$	0.79
$2.15x10^6$	1.25	$1.56x10^5$	0.65
$8.65x10^6$	1.1	$8.4\ x10^5$	0.5

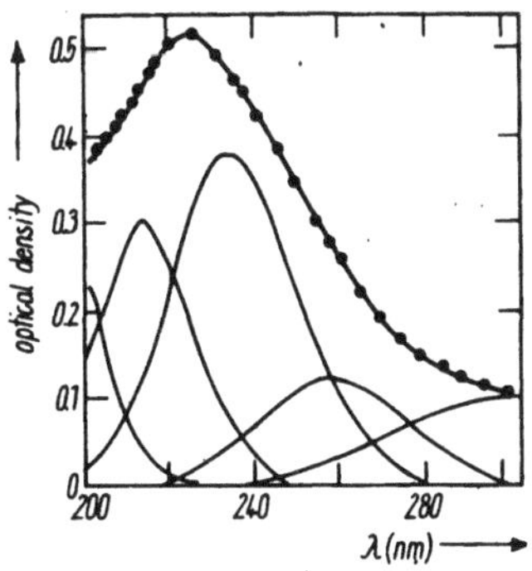

Fig. 2. Decomposition of optical absorption spectra for KCl into Gaussians

As follows from Table 1, the intensity of V_2 absorption within a wide range of electron irradiation doses is higher than that of V_3 absorption. The situation is different for X-irradiation, namely, one mostly observes V_3 centre accumulation.

The results presented indicate a substantial suppression of accumulation processes of multihalide associates under powerful electron irradiation. Qualitatively the analogous effect is characteristic of KBr, too.

To determine whether the stimulation of radiation annealing of halogen clusters is the cause of the effect observed we performed repeated electron irradiation of previously X-irradiated KCl samples with predominant V_3 centres. The repeated electron irradiation did not result in a decrease of the V_3 absorption. Hence, the reason mentioned above should be excluded.

In such a way, powerful pulsed irradiation itself suppresses the processes of H-defect aggregation.

Two reasons accounting for the effect are possible. One of them is based on the tunnel recombination on aggregates and the spatial distribution of colour centres.

With the simulation method used, the tunnel recombination being taken into account, the formation of uniform defect aggregates has been shown to depend on the irradiation intensity and duration /4, 5/. The short time intensive irradiation favours a uniform defect distribution over the crystal, the continuous low-intensity one resulting in the formation of areas of increased local concentration of identical defects.

On the other hand, the aggregation of radiolysis hole products substantially depends on the possibility of their capturing by traps which are cation vacancies. It is known that the presence of complexes, like divalent impurity metal ion-cation vacancy suppresses the formation of multi-halide associates. The small-radius cation impurity increases the radiational creation of cation Frenkel defects and so reduces the creation effectivity of V_3 centres /6/.

A high creation efficiency of cation sublattice defects may possibly be obtained under powerful pulsed electron irradiation. The process suppresses the aggregation of interstitial halogen and results in the creation of X_3^- centres and their small associates par excellence.

References

/1/ A. A. ELANGO and T.N. NURAKHMETOV, phys. stat. sol. (b) 78, 529 (1976).

/2/ A.E. RAUKHVARGER, YU.R. DZELME, and YU.E. TILIKS, Izv. Akad. Nauk Latv. SSR, Ser. Fizika i Tekhnika 6, 115 (1978).

/3/ V.E. ZIRAP, G.YA. ZARINYA, and V.YA. KRUMINSH, Abstr. IV Soviet Conf. Radiation Physics and Chemistry of Ionic Crystals, Salaspils 1978 (p. 121).

/4/ U.A. TALE, D.K. MILLERS and E.A. KOTOMIN, in: Electronic and Ionic Crystals, Vol. 2, Latvian State Univ., Riga 1974 (p. 43).

/5/ G.A. VALDATS and E.A. KOTOMIN, in: Electronic and Ionic Processes in Ionic Crystals, Vol. 6, Latvian State Univ., Riga 1977 (p. 57).

/6/ A.K. DAULETBEKOVA, Trudy Inst. Fiz. Akad. Nauk Eston. SSR 52, 121 (1981).

(Received March 1, 1984)

phys. stat. sol. (b) 125, K41 (1984)

Subject classification: 10.2 and 13.4; 20.3; 22.5.2

Institute of Applied Physics, Irkutsk State University[1] (a) and Institute of Physics, Academy of Sciences of the Belorussian SSR, Minsk[2] (b)

Influence of Oxygen Defects on the Position and Form of the F_2^+ Center Zero-Phonon Lines in LiF

By

V.A. CHEPURNOI (a), S.N. MYSOVSKII (a), I.A. PARFIANOVICH (a), V.N. SALOMATOV (a), and A.P. SHKADAREVICH (b)

F_2^+ center LiF crystals are widely used as active elements of tunable lasers /1, 2/ and as effective passive Q-switchers /3/. Crystals with OH^- impurity are most suitable in this respect due to the stability of the appropriate active centers at room temperature. The opinions concerning the mechanism of stabilization of the F_2^+ center in LiF:OH^- are contradictory. It has been suggested /4/ that the stabilization results from the fact that the impurity ion O^{2-} occurs very close to the F_2^+ center. It has also been supposed /5/ that impurity defects are involved in the stabilization of F_2^+ centers only as electron traps. Examination of details of optic spectra, in particular the position and form of zero-phonon lines may allow the question of the existence and the nature of impurity defects in the F_2^+ center environment in LiF to be answered.

In the present note LiF crystals of two types were used. The first type (LiF-I) was grown by the Stockbarger method from material purified from oxygen and hydroxyl impurities. The second type (LiF-II) was grown by the Kiropoulos method in air from raw material rich in oxygen-containing impurities, with an absorption coefficient in the OH^- band (3720 cm^{-1}) of about 1 cm^{-1}. High concentrations of unstable F_2^+ centers were created using LiF-I samples which were γ-irradiated at a dosage of 10^8 R from ^{60}Co at T = 196 K, then heated to 293 K and after annealing for 12 min placed into a He cryostat and cooled to 4.2 K. The absorption coefficient at the 645 nm maximum was 48 cm^{-1}. The stable $(F_2^+)^*$ centers were created using LiF-II crystals exposed to 2.5×10^7 R γ-irradiation at 360 K. The absorption coefficient at the band maximum (λ = 610.0 nm) was 1.5 cm^{-1}. Excitation and luminescence spectra

1) 20 Gagarin Boulevard, 664003 Irkutsk, USSR.

2) 70 Leninskii Prospekt, 220602 Minsk, USSR.

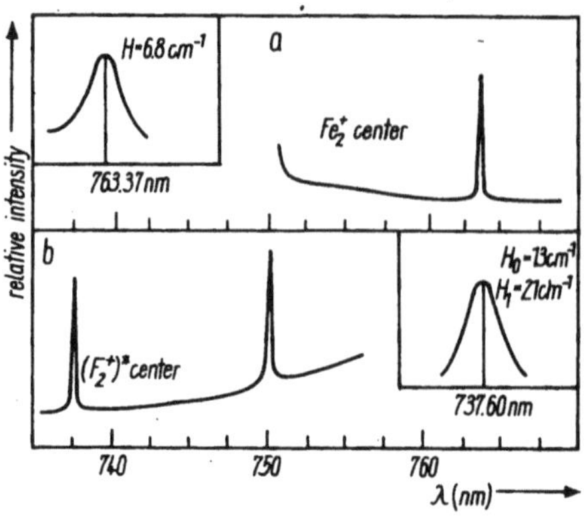

Fig. 1. a) Excitation and b) luminescence spectra fragments and zero-phonon line halfwidths of unstable and stable F_2^+ centres at 4.2 K

of the F_2^+ centers were measured at 4.2 K. A 3 kW xenon arc lamp served as an excitation source. The excitation spectra were measured by an MDP-23 monochromator with 0.13 nm resolution. The luminescence spectra were measured with a DFS-24 monochromator (0.04 to 0.12 nm resolution). FEU-125 was used as a receiver.

Fig. 1a shows a fragment of the F_2^+ center excitation spectrum in LiF-I. The position (λ = 763.37 nm) and halfwidth of this center zero-phonon line are in good agreement with absorption and luminescence data reported in /6, 7/.

For the $(F_2^+)^*$ centers (see Fig. 1b), a 26.8 nm displacement of the zero-phonon line to the shortwave region is observed. The zero-phonon line halfwidth also increases amounting to 13 cm^{-1}. The one-phonon line is displaced by 208 cm^{-1}, its halfwidth being 21 cm^{-1}. Comparison of the one-phonon line position with the density of phonon states in LiF displays a good correlation with the transverse acoustic mode, TA (L_3) is equal to 200 cm^{-1} in LiF /8/.

The experimental data obtained indicate the presence of impurity defects in the F_2^+ center environment in LiF. These defects are most likely to be OH^- radiolysis products. In this connection, we have calculated variations in the F_2^+ energy parameters in LiF, caused by perturbed impurity ions O^-. Defect configurations considered are illustrated in Fig. 2. The calculation has been carried out in the first order of perturbation theory. The calculation procedure is described in /9/. The wave functions for the F_2^+ ground and excited states in LiF are taken from /10/. The results are summarized in Table 1. ΔE_g, ΔE_u are changes in the ground and the excited state energies, respectively; ΔE =

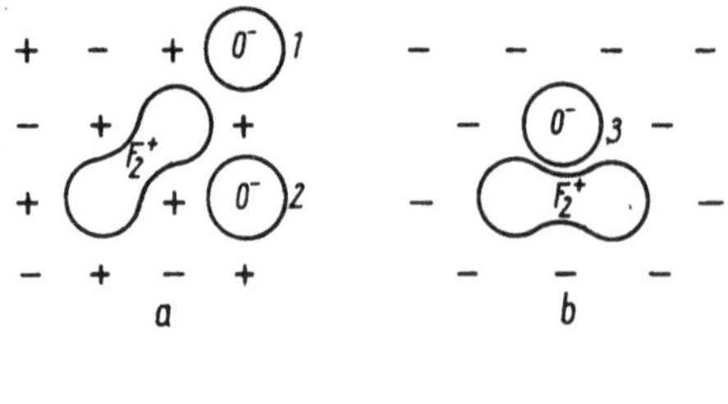

Fig. 2. Configurations of defects under investigation in a) the (100) and b) the (111) planes

Table 1

Energy parameters of stable (F_2^+) centers (eV)

method		ΔE_g	ΔE_u	ΔE
	1	-0.05	-0.07	-0.02
theory	2	-0.08	-0.07	+0.01
	3	-0.12	-0.08	+0.04
experiment				+0.058

$= \Delta E_u - \Delta E_g$ is the F_2^+ zero-phonon line displacement under the perturbing effect of the impurity ion.

As seen from a comparison of the calculated and experimental data, in the case of configuration 3, there is a fairly good agreement for the ΔE value and the difference between the F_2^+ and $(F_2^+)^*$ zero-phonon line positions observed experimentally.

Since the calculation has been performed for only one of the possible impurity defects, the above agreement with the experiment does not provide comprehensive information concerning the stabilized $(F_2^+)^*$ model. Besides, the presence of an impurity ion devoid of excess charge in the F_2^+ center environment cannot explain the mechanism of F_2^+ center stabilization. The mechanism suggested in /5/ is possible to be the case. The F_2^+ center perturbed by the O^- impurity ion (configuration 3) can be regarded as a preliminary model of the stabilized $(F_2^+)^*$ center in LiF.

References

/1/ L.F. MOLLENAUER, B.M. BLOOM, and A.M. DEL GAUDIO, Optics Letters 3, 48 (1978).

/2/ N.A. IVANOV, V.D. LOKHNYGIN, G.I. ONISHUKOV, A.A. FOMICHEV, V.M. KHULUGUROV, and V.A. CHEPURNOI, Zh. tekh. Fiz., Pisma 9, 321 (1983).

/3/ N.A. IVANOV, I.A. PARFIANOVICH, V.M, KHULUGUROV, and V.A. CHEPURNOI, Izv. Akad. Nauk SSSR, Ser. fiz. 46, 1985 (1982).

/4/ I.A. PARFIANOVICH, V.M. KHULUGUROV, V.D. LOBANOV, and N.T. MAKSIMOVA, Izv. Akad. Nauk SSSR, Ser. fiz. 43, 1126 (1979).

/5/ W. GELLERMAN, F. LÜTY, K.P. KOCH, and G. LITFIN, phys. stat. sol. (a) 57, 411 (1980).

/6/ Y. FARGE, G. TOULOUSE, and M. LAMBERT, C.R. Acad. Sci. (France) 262, 66 (1966).

/7/ Y. FARGE and M.P. FONTANA, Solid State Commun. 10, 333 (1972).

/8/ K.K. REBANE, Impurity Spectrum in Solids, Plenum Press, New York 1970.

/9/ V.N. SALOMATOV and I.A. PARFIANOVICH, phys. stat. sol. (b) 121, 401 (1984).

/10/ R.A. EVARESTOV, phys. stat. sol. 31, 401 (1969).

(Received June 26, 1984)

phys. stat. sol. (b) 125, K45 (1984)
Subject classification: 18.2

Institute of Physics, Faculty of Sciences, Novi Sad[1)]

Temperature Dependence of the Energy Gap of an Anisotropic Heisenberg Ferromagnetic Model

By
D.V. KAPOR, J.P. ŠETRAJČIĆ, and M.J. ŠKRINJAR

We have devoted several papers /1 to 4/ to the study of the low-temperature properties of the anisotropic Heisenberg model proposed by Buyers et al. /5/ and developed by Belorizky et al. /6 to 8/. The Hamiltonian of the model is given by

$$H = -I \sum_{\substack{\vec{n}<\vec{m} \\ n.n.}} \vec{S}_{\vec{n}} \vec{S}_{\vec{m}} + \delta I \sum_{\vec{n}} \sum_{r=x,y,z} S^r_{\vec{n}} S^r_{\vec{n}+\vec{\varepsilon}_r} , \tag{1}$$

where δ is the anisotropy parameter, while the complete notation was explained in the previous papers /1, 6/. Using the Green's function formalism for the bosonic representation of spins /9/, we have derived the system of self-consistent equations describing the properties of the system. The (dimensionless) energy is given by

$$E(\vec{k}) = \left[X^2(\vec{k}) + Y(\vec{k})R(\vec{k})\right]^{1/2} , \tag{2}$$

where the parameters $X(\vec{k})$, $Y(\vec{k})$, and $R(\vec{k})$ are determined from the system

$$X(\vec{k}) = \frac{1}{2}\left[\gamma(0) - \gamma(\vec{k})\right] - \frac{\delta}{2}\left[2 - \alpha(\vec{k})\right] + \frac{1}{N}\sum_{\vec{q}}\left[\frac{X(\vec{q})}{E(\vec{q})}\coth\frac{E(\vec{q})}{2\Theta'} - 1\right] \times$$
$$\times \left[M(\vec{k},\vec{q}) - \delta M'(\vec{k},\vec{q})\right] - \frac{\delta}{2N}\sum_{\vec{q}}\frac{Y(\vec{q})}{E(\vec{q})}(\cos q_x a - \cos q_y a)\coth\frac{E(\vec{q})}{2\Theta'} , \tag{3a}$$

$$Y(\vec{k}) = Y(\cos k_x a - \cos k_y a) ; \quad Y \equiv \left[\frac{2}{N}\sum_{\vec{q}}(\cos q_x a - \cos q_y a)\frac{\cos q_x a}{E(\vec{q})} \times \right.$$
$$\left. \times \coth\frac{E(\vec{q})}{2\Theta'} - 1\right]^{-1} , \tag{3b}$$

$$R(\vec{k}) = -Y(\vec{k})\left\{1 - \frac{2}{N}\sum_{\vec{q}}\left[\frac{X(\vec{q})}{E(\vec{q})}\coth\frac{E(\vec{q})}{2\Theta'} - 1\right]\right\} . \tag{3c}$$

---

1) Dr. I. Djuričića 4, Yu-21000 Novi Sad, Yugoslavia.

Here $\Theta' = k_B T/I$ is the dimensionless temperature and

$$\alpha(\vec{k}) = \cos k_x a + \cos k_y a; \quad \gamma(\vec{k}) = \sum_{r=x,y,z} \cos k_r a \quad , \tag{4a, b}$$

$$M(\vec{k}, \vec{q}) = \gamma(\vec{k}) + \gamma(\vec{q}) - \gamma(0) - \gamma(\vec{k}-\vec{q}) \quad , \tag{4c}$$

$$M'(\vec{k}, \vec{q}) = \alpha(\vec{k}) + \alpha(\vec{q}) - 2\cos(k_z - q_z)a - 2 \quad . \tag{4d}$$

We are interested in the behaviour of the energy gap at finite temperatures. It can be seen from (3a, b, c) and (4a, c) that

$$E(0) = \lim_{k \to 0} E(\vec{k}) = |X(0)| = \left| -\frac{\delta}{2N} \sum_{\vec{q}} \left[\frac{X(\vec{q})}{E(\vec{q})} \coth \frac{E(\vec{q})}{2\Theta'} - 1 \right] M'(\vec{k}, \vec{q}) - \right.$$
$$\left. - \frac{\delta}{2N} \sum_{\vec{q}} \frac{Y(\vec{q})}{E(\vec{q})} (\cos q_x a - \cos q_y a) \coth \frac{E(\vec{q})}{2\Theta'} \right| \quad . \tag{5}$$

We know that $E(0) \to 0$ for $\delta \to 0$ and that at T = 0 K, $E(0) \sim \delta^2$ /2/. We wish to examine the low temperature dependence of the gap in the approximation up to δ^2. This implies in fact that all relevant quantities should be expanded only up to linear terms in δ:

$$X(\vec{k}) = X_o(\vec{k}) + \delta X_1(\vec{k}); \quad E(\vec{k}) \approx X(\vec{k}); \quad Y(\vec{k}) \approx \delta Y_1(\vec{k}); \quad R(\vec{k}) \approx -\delta Y_1(\vec{k}) \; . \tag{6}$$

This gives the following result:

$$E(0) = \left| -\frac{\delta}{2N} \sum_{\vec{q}} \left[\coth \frac{X_o(\vec{q})}{2\Theta'} - 1 \right] M'(0, \vec{q}) - \frac{\delta^2}{2N} \sum_{\vec{q}} \frac{X_1(\vec{q})}{2\Theta'} M'(0, \vec{q}) \times \right.$$
$$\left. \times \left[1 - \coth^2 \frac{X_o(\vec{q})}{2\Theta'} \right] - \frac{\delta^2}{2N} \sum_{\vec{q}} \frac{Y_1(\vec{q})}{X_o(\vec{q})} (\cos q_x a - \cos q_y a) \coth \frac{X_o(\vec{q})}{2\Theta'} \right| \tag{7}$$

$X_o(\vec{k})$ is obviously the energy of the isotropic Heisenberg model ($\delta \to 0$) and according to our approximation it is precisely the Dyson result /9/. It must be isotropic, so bearing in mind the way in which q_x, q_y, and q_z enter into $M'(0, \vec{q})$, it can be seen that the first in the bracket must vanish. This means that at finite temperatures, there is no term in the gap proportional to δ.

The expression for E(0) can be modified in the following way:

$$E(0) = \left| \frac{\delta^2}{2N} \sum_{\vec{q}} \frac{X_1(\vec{q})}{2\Theta'} \frac{M'(0, \vec{q})}{\mathrm{sh}^2 \frac{X_o(\vec{q})}{2\Theta'}} - \right.$$

$$\left. - Y_1 \frac{\delta^2}{2N} \sum_{\vec{q}} \frac{(\cos q_x a - \cos q_y a)^2}{X_o(\vec{q})} \coth \frac{X_o(\vec{q})}{2\Theta'} \right| . \quad (8)$$

$X_o(\vec{k})$ and $X_1(\vec{k})$ are the solutions of the self-consistent equations

$$X_o(\vec{k}) = \frac{1}{2} \left[\gamma(0) - \gamma(\vec{k})\right] + \frac{1}{2N} \sum_{\vec{q}} \left[\coth \frac{X_o(\vec{q})}{2\Theta'} - 1\right] M(\vec{k}, \vec{q}) , \quad (9)$$

$$X_1(\vec{k}) = \frac{1}{2} \left[\alpha(\vec{k}) + - 2\right] + \frac{1}{2N} \sum_{\vec{q}} \frac{X_1(\vec{q})}{2\Theta'} \left[1 - \coth^2 \frac{X_o(\vec{q})}{2\Theta'}\right] M(\vec{k}, \vec{q}) -$$

$$- \frac{1}{2N} \sum_{\vec{q}} \left[\coth \frac{X_o(\vec{q})}{2\Theta'} - 1\right] M'(\vec{k}, \vec{q}) . \quad (10)$$

Our aim is to obtain the low-temperature behaviour, i.e. only the leading term in $\Theta \equiv k_B T$. For this purpose, we take the long wavelength approximation. It is not difficult to see (by inspection) that $X_o(\vec{k})$ and $X_1(\vec{k})$ both begin with $(ka)^2$, and that it is sufficient to keep only the temperature independent part of the coefficient in front of it, so

$$X_o(\vec{k}) \approx \frac{1}{2} (ka)^2 ; \qquad X_1(\vec{k}) \approx - \frac{1}{4} (ka)^2 \sin^2 \vartheta .$$

The subsequent procedure is a standard one. We proceed to the change of variables introducing $q^2 a^2 / 4\Theta' = x$ which means that $(qa)^2 d(qa) \sim \Theta'^{3/2}$ and each $(qa)^2$ introduces additional Θ' term. The only peculiarity of the treatment is the problem of the upper bound of the integral. It is customary to introduce the maximal wave number q_m, and then change it into $x_m = (q_m a)^2 / 4\Theta'$ which tends to infinity for low temperatures. In this work we must perform all calculations with finite x_m and only at the end set this limit (in the integral) to infinitiy, otherwise the expressions diverge.

Let us look at each term separately:

$$-\frac{1}{N}\sum_{\vec q}\frac{X_1(\vec q)}{2\Theta'}\frac{M(0,\vec q)}{\mathrm{sh}^2\frac{X_o(\vec q)}{2\Theta'}} = -\frac{1}{(2\pi)^3}\int_0^{q_m}\int_0^{\pi}\int_0^{2\pi}(qa)^2 d(qa)\sin\vartheta d\vartheta d\varphi \times$$

$$\times\frac{1}{2\Theta'}\left(-\frac{1}{4}q^2a^2\sin^2\vartheta\right)\mathrm{sh}^{-2}\frac{q^2a^2}{4\Theta'}(qa)^2(\cos^2\vartheta-\frac{1}{2}\sin^2\vartheta) =$$

$$= -\frac{8}{15}\Theta'^{5/2}\int_0^{x_m}dx\frac{x^{5/2}}{\mathrm{sh}^2 x}\ .$$

In the limit $x_m \to \infty$, it gives

$$-\frac{\pi}{2}\varrho^{5/2}\zeta_{5/2}\ .$$

ζ_n is Riemann's zeta function, and $\varrho = \Theta'/2\pi$, The second term is

$$\frac{Y_1}{N}\sum_{\vec q}\frac{(\cos q_x a-\cos q_y a)^2}{X_o(\vec q)}\coth\frac{X_o(\vec q)}{2\Theta'} = \frac{Y_1}{(2\pi)^3}\int_0^{q_m}\int_0^{\pi}\int_0^{2\pi}(qa)^2 d(qa)\sin\vartheta d\vartheta d\varphi \times$$

$$\times\frac{1}{4}(qa)^4\sin^4\vartheta(\sin^2\varphi-\cos^2\varphi)\coth\frac{(qa)^2}{4\Theta'} = Y_1\frac{8}{15\pi^2}\Theta'^{5/2}\int_0^{x_m}dx\, x^{3/2}\coth x.$$

In the limit of large x, $\coth x \approx 1 + e^{-2x}$ which gives two terms, one of them is temperature independent and the other one is proportional to $\Theta'^{5/2}$:

$$Y_1\left[\frac{(q_m a)^2}{75\pi}+\frac{3\pi}{5}\varrho^{5/2}\right].$$

q_m can be evaluated as the radius of the sphere having the same volume as the first Brillouin zone, while Y_1 was evaluated numerically to give finally

$$E(0) = \delta^2\left[0.256363+\frac{\pi}{4}(0.506228+\zeta_{5/2})\varrho^{5/2}\right]. \qquad (11)$$

Several conclusions can be made:

a) the gap remains proportional to δ^2 even at finite temperatures;

b) the leading term in the low-temperature expansion is $\Theta'^{5/2}$;

c) the temperature independent part of the gap is given by the proper general expression /2/, but as it was evaluated here only in the long wavelength limit, its value is overestimated /4/.

References

/1/ M.J. ŠKRINJAR, D.V. KAPOR, and J.P. ŠETRAJČIĆ, phys. stat. sol. (b) 103, 559 (1980).

/2/ M.J. ŠKRINJAR, D.V. KAPOR, and J.P. ŠETRAJČIĆ, phys. stat. sol. (b) 107, K91 (1981).

/3/ D.V. KAPOR, J.P. ŠETRAJČIĆ, and M.J. ŠKRINJAR, Rev. Res. Fac. Sci., Univ. Novi Sad, Phys. Ser. 12, 15 (1982).

/4/ D.V. KAPOR, K.R. SURLA, D.I. SURLA, J.P. ŠETRAJČIĆ, and M.J. ŠKRINJAR, Fizika 15, 401 (1983).

/5/ W.J.L. BUYERS, T.M. HOLDEN, E.C. SVENSSON, R.A. COWLEY, and M.T. HUTCHINGS, J. Phys. C 4, 2139 (1971).

/6/ E. BELORIZKY, R. CASALEGNO, and P. FRIES, phys. stat. sol. (b) 77, 495 (1976).

/7/ E. BELORIZKY, R. CASALEGNO, P. FRIES, and J.J. NIEZ, J. Physique 39, 776 (1978).

/8/ E. BELORIZKY, R. CASALEGNO, and J.J. NIEZ, phys. stat. sol. (b) 102, 365 (1980).

/9/ F.J. DYSON, Phys. Rev. 102, 1217, 1230 (1957).

(Received May 28, 1984)

phys. stat. sol. (b) 125, K51 (1984)

Subject classification: 18.4

Institute of Mathematics and Physics, Titograd[1] (a) and
Department of Physics and Meteorology,
Faculty of Natural and Mathematical Sciences, Belgrade[2] (b)

Ground-State Degeneracy of the Ising Antiferromagnet on the Hypercubic Lattice

By

D. HAJDUKOVIĆ (a) and I.M. SAVIĆ[3] (b)

In this paper we study the Ising model on the hypercubic lattice in d dimensions, described by the Hamiltonian

$$\mathcal{H} = J\sum_{\langle ij\rangle} S_i S_j - H\sum_i S_i \ , \tag{1}$$

where S_i is the z projection of the spin at site i ($S_i = S, S-1, \dots, -S$), J is the antiferromagnetic interaction ($J > 0$), the applied field is positive ($H > 0$) and parallel to the z axis, and $\langle ij\rangle$ stipulates that the summation is restricted over those pairs of lattice sites which are the nearest neighbours.

In a very strong field, at zero temperature, all spins should acquire their maximum projections ($S_i = S$). If a single spin reduces its projections to a smaller value δ, it would cause the following change in the energy of the system:

$$\Delta\mathcal{H} = (S - \delta)(H - 2dJS) \ , \tag{2}$$

where 2d is coordination number of the hypercubic lattice in d dimensions. It follows from (2) that for $H = 2dJS$, the equality $\Delta\mathcal{H} = 0$ stays valid for any number n of reduced spin projections, provided that the reduction does not occur at two neighbouring sites in the lattice. Thus, $H_c = 2dJS$ is the critical field value associated with the ground-state degeneracy

$$P = \sum_n p_n \ , \tag{3}$$

where p_n is the number of ground-state configurations for a given n. It is very difficult to obtain p_n for the entire lattice, but it is easy enough for a single

---

1) 81000 Titograd, Yugoslavia.

2) POB 550, 11001 Belgrad, Yugoslavia.

3) To whom the correspondence concerning this paper should be addressed.

hypercube only. So, for example, if we restrict ourselves to the case S = 1/2, we would have

$$p_0 = 1, \quad p_1 = 4, \quad p_2 = 2 \tag{4}$$

for a single square, and

$$p_0 = 1, \quad p_1 = 8, \quad p_2 = 16, \quad p_3 = 8, \quad p_4 = 2 \tag{5}$$

for a single cube.

Now, let us consider a single hypercube. There are $N \equiv 2^d$ spins located on such a hypercube and the possible values for n are n = 0, 1, 2, ..., N/2. Thus, the ground-state degeneracy of a single hypercube is

$$P = \sum_{n=0}^{N/2} p_n$$

and it is evident that

$$\sigma_u = \frac{1}{N} \ln \sum_{n=0}^{N/2} p_n \tag{6}$$

is an upper limit for the ground-state entropy of the entire lattice.

At this point we make a crucial assumption: information about ground-state entropy of the infinite system is retained in the values of p_n for a single hypercube. In accordance with this assumption and the expression (4) we shall try to find the ground-state entropy of the S = 1/2 model in the form

$$\sigma = \frac{1}{N} \ln \sum_n (-1)^{g(n)} p_n \ , \tag{7}$$

where the function g(n) is to be determined. Let us note that similar assumptions were very successful in the determination of the critical point of the Ising model on the hypercubic lattice /1/.

In order to find the function g(n) let us note that

$$\sigma_1 = \frac{1}{2} \ln(1 + 2S) \tag{8}$$

is a lower limit of the ground-state entropy of the Ising model on the hypercubic lattice /2/. The ground-state entropy (7) must lie between the limits (6) and (8). In the case of a square lattice in accordance with (4) this is possible only if

$$\sum_n (-1)^{g(n)} p_n = p_1 + p_2 - p_0 \ .$$

Thus, in this special case g(n) is of the form

$$g(n) = \begin{cases} 0, & n \geq \frac{N}{4} \\ 1, & n < \frac{N}{4} \end{cases} , \qquad (9)$$

where N is the number of spins on a single square. In the general case we do not know an actual rule how to obtain the function g(n). Because of that we shall follow paper /1/, and make a conjecture that (9) is valid for all d with $N = 2^d$. Thus, we expect that the ground-state entropy is determined by (7) and (9).

The numerical values obtained in such a way for square and simple cubic lattices are, respectively,

$$\sigma_s = \frac{1}{4} \ln 5 = 0.4024; \quad \sigma_{sc} = \frac{1}{8} \ln 17 = 0.3542$$

and these results are in good agreement with the results

$$\sigma_s = 0.4075; \quad \sigma_{sc} = 0.3563$$

which are obtained by the computer technique /2/.

At the end let us remark that in the case of square and simple cubic lattices (7) and (9), which are valid in the case S = 1/2, can be reduced to

$$\sigma = \frac{1}{N} \ln \left\{1 + 2^{N/2}\right\} = \frac{1}{2^d} \ln \left\{1 + (2S+1)^{2^{d-1}}\right\} . \qquad (10)$$

For very large S, (10) gives

$$\sigma = \frac{1}{2} \ln 2S$$

which is known to be exact /3/. Thus, there is some reason for the conjecture that (10) is a good approximation for <u>all</u> d and S.

References

/1/ D. HAJDUKOVIĆ, J. Phys. A 16, L193 (1983).

/2/ B.D. METCALF and C.P. YANG, Phys. Rev. B 18, 2304 (1978).

/3/ D. HAJDUKOVIĆ and S. MILOŠEVIĆ, Phys. Letters A 95, 57 (1983).

(Received June 25, 1984)

phys. stat. sol. (b) 125, K55 (1984)
Subject classification: 4 and 13; 22.6

V.A. Steklovh Mathematical Institute,
Academy of Sciences of the USSR, Leningrad[1)]

On the Effective Potential for the Anion in MgO and CaO Crystals

By
I.M. ANTONOVA

The method of calculation of the localized crystal orbitals of an ionic crystal has been proposed in /1/. In this method the crystal orbitals are considered as the solutions of the Hartree-Fock equations for the ion in an external field, simulating the influence of the crystal environment. The potential of this field must be chosen so as to provide the minimum value to the energy of the crystal. This effective potential does not enter the expression of the energy explicitly. It influences the energy indirectly by means of the crystal orbitals.

The direct variational procedure was used in /1/. A certain analytical expression with parameters for the potential was chosen and these parameters were considered as variational parameters. Rather satisfactory results were obtained by this method for MgO and CaO crystals in /1 to 3/.

It is evident that because of the variational nature of the method the results depend on the choice the particular form of the effective potential. In /1/ this potential was taken as

$$V_s(r) = V_{so}(1 - (1 - \exp(-\alpha_s r))^{n_s}) , \qquad (1)$$

where s is the type of the ion, V_{so} and α_s, n_s determine the depth and the effective radius of the potential well.

The form (1) of the potential is consistent with the general notion of the behaviour of the crystal field in ionic crystals: this field is almost constant in the core region of an ion and decreases with increasing distance from the ion. Besides, the form of the potential was chosen so that the integration of this potential with Slater-type orbitals (STO) is very simple.

In order to examine to what extent the method proposed is susceptible to the particular form of the effective potential, we carried out the whole calculations of the crystal orbitals and the cohesive properties of MgO and CaO, as in /1, 2/, for another form of the potential V_s, namely

---

1) Fontaka 27, 191011 Leningrad, USSR.

Table 1

MgO			CaO		
n_p	ζ_p	c_{2p}	n_p	ζ_p	c_{2p}
2	1.5566	0.63588	2	1.5566	0.57387
2	3.4300	0.29061	2	3.4300	0.30707
2	8.1268	0.01238	2	8.1268	0.00971
3	1.30	0.18109	3	1.15	0.26457

$$V_s(r) = V_{so} + \left(\frac{q_s}{r} - V_{so}\right)\left(1 - \exp(-\alpha_s r)\right)^{n_s} . \qquad (2)$$

Here q_s is the charge of the crystal ion s, the other parameters are the same as in (1). The potential (2) has Coulomb behaviour for large r in comparison with the exponential behaviour of the potential (1).

In the present note as in /1, 2/ only the anion O^{2-} crystal orbitals were calculated (the optimization of the STO basis was carried out), the free ion orbitals were taken for the cation crystal orbitals. The following values of variational parameters of the potential (2) for the ion O^{2-} (s = 2) were obtained:

$$V_{20} = -0.9 , \quad \alpha_2 = 1.40 , \quad n_2 = 10 \text{ for MgO} ,$$
$$V_{20} = -0.75, \quad \alpha_2 = 0.97 , \quad n_2 = 10 \text{ for CaO} . \qquad (3)$$

The STO basis and 2p orbital expansion coefficients for the O^{2-} ion in field (3) for MgO and CaO are given in Table 1.

The radial 2p orbitals of the O^{2-} ion in the effective potential (1) and (2) for MgO together with these potentials are shown in Fig. 1. The orbitals are essentially the same, the orbital in the potential (2) being slightly more diffuse. The difference between orbitals is less pronounced than the difference between potentials (1) and (2). This is because the effective potential (1) or (2) is only a part of the total potential for the electron in an ion. Another part is the self-consistent-field potential of the ion. The combined influence of both these parts of the potential results in the closeness of 2p orbitals, corresponding to the effective potentials (1) and (2).

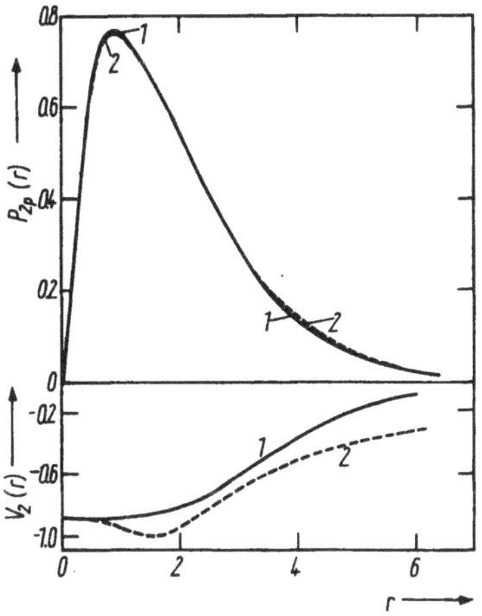

Fig. 1. Radial 2p orbitals of the O^{2-} ion in the effective potentials (1) and (2) together with both effective potentials

The values of the lattice constant d, cohesive energy E_{coh}, and the inverse compressibility $1/\beta$ of MgO and CaO calculated with use of the potential V_s (1) and (2) are shown in Table 2. The "fixed orbitals" version of /1/ (the crystal orbitals are calculated only for one interionic separation and used for the energy calculation for every interionic separation considered) was used. The nearest neighbours and nearest anion overlapping were taken into account and the van der Waals corrections were both included and omitted. For comparison the corresponding experimental values are also given in Table 2.

Table 2

		MgO			CaO		
	form of potential V_s	d (nm)	E_{coh} (kcal/mol)	$1/\beta$ (10^{11} dyn/cm^2)	d (nm)	E_{coh} (kcal/mol)	$1/\beta$ (10^{11} dyn/cm^2)
van der Waals corrections omitted	(1)	0.435	690	25.6	0.498	600	11.4
	(2)	0.437	690	24.5	0.496	600	12.4
van der Waals corrections included	(1)	0.428	700	26.2	0.476	630	14.4
	(2)	0.430	705	22.5	0.475	630	10
experiment		0.421[a)]	725[b)]	16.8[d)] 15.3[c)]	0.481[a)]	632[b)]	11.3[d)]

a) Obtained by Swanson (as quoted in /4/).
b) /5/.
c) Obtained by Bhagavantam (as quoted in /4/).
d) /6/.

The orbitals calculated with the use of the potential (2) produced a lower total energy of the crystal than those calculated with the potential (1). However, the difference of the total energies is rather small, about 2 kcal/mol. Therefore usually one can safely use either type of the effective potential. The calculated cohesive properties of MgO and CaO shown in Table 2 confirm this conclusion. Still, in some cases the effective potential (2) can be preferable. The polarizability of the crystal ion /3/ can be considered as an example.

References

/1/ I.V. ABARENKOV and I.M. ANTONOVA, phys. stat. sol. (b) 92, 389 (1979).
I.V. ABARENKOV and I.M. ANTONOVA, phys. stat. sol. (b) 93, 315 (1979).
/2/ I.M. ANTONOVA, phys. stat. sol. (b) 94, K165 (1979).
/3/ I.V. ABARENKOV, V.F. BRATZEV, and A.V. TULUB, Fiz. tverd. Tela 24, 272 (1982).
/4/ J.-L. CALAIS, K. MANSIKKA, G. PETTERSSON, and J. VALLIN, Ark. Fys. (Sweden) 34, 361 (1967).
/5/ A.J. COHEN and R.J. GORDON, Phys. Rev. B 14, 4593 (1976).
/6/ C.E. WEIR, J. Res. Nat. Bur. Standards 56, 187 (1956).

(Received May 28, 1984)

phys. stat. sol. (b) 125, K59 (1984)
Subject classification: 13.1; 22.5

Max-Planck-Institut für Festkörperforschung, Stuttgart[1)]

Deformation Potentials of the Valence Band Edge in CuBr

By
N.E. CHRISTENSEN

The present note is an addendum to the work /1/ describing first-principles calculations of deformation potentials in CuCl and CuI. The calculations presented here are, as in /1/, performed by means of self-consistent, relativistic linear-muffin-orbital (LMTO) band structure calculations, and the local-density approximation is used for exchange-correlation energies. The zincblende structure is open and therefore the basis used in the LMTO calculations includes two inequivalent "empty spheres" /1/. From total energy calculations for varying volume we find the theoretical equilibrium lattice constant 1.9% below the experimental value. The calculated bulk modulus is B = 392 kbar (at the observed volume) in agreement with experiment /2/, 39 GPa.

The results of the calculations of the deformation potentials for splitting of the state Γ^v_{15} (Γ^v_8 when spin-orbit coupling is included) induced by uniaxial strains are summarized in Table 1 (for details of notation, see /1/). The trigonal-shear deformation potentials d and d_1 are shown in Fig. 1 as functions of the internal-strain parameter ζ. If the value ζ = = 0.83 is chosen, then it follows from Fig. 1 that our calculated values of d as well as d_1 agree with the experiments of /5/. As for CuCl and CuI /1/ we also find for CuBr that the spin-orbit

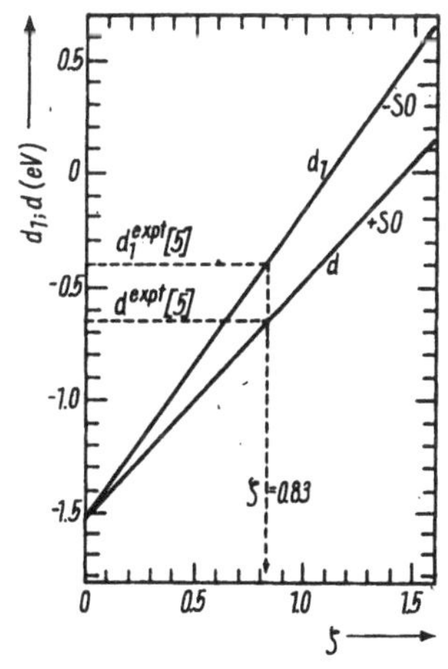

Fig. 1. Trigonal-shear deformation potentials for the valence-band top in CuBr as functions of internal-strain parameter ζ. The line labelled d_1 refers to calculations where the spin-orbit is omitted. Experimental results /5/ are indicated

1) Heisenbergstr. 1, D-7000 Stuttgart 80, FRG.

Table 1

Shear-deformation potentials of CuBr. All values are in eV

	present calculation	LCAO(-SO) /3/	experiments
b_1	-0.327	-0.45	
b	-0.338		-0.25 /4/ -0.4 /4/
b_2	-0.005		
d_1'	-1.529	-1.05	
d_1(0.83)	-0.400	-0.306	-0.4 /5/
d'	-1.538		
d(0.83)	-0.650		-0.65 /5/ -0.7 /5/
d_2(0.83)	-0.130		-0.13 /5/
d_o	-4.264	-3.5	-3.8 /6/
$d_{2,0}$	0.608		

coupling influences the trigonal-shear deformation potentials appreciably, whereas the tetragonal-shear values are almost the same in the calculations with and without SO; $b_2 \approx 0$.

The deformation potentials of the copper halides are anomalous when compared to the group IV and III-V isomorphic tetrahedral semiconductors. The optical deformation potentials, d_o, are all small in magnitude for the copper halides and the sign is for all three cases negative. Further d of CuCl is positive. These anomalies are due /3/ to the strong hybridization between the copper 3d states and the halogen p-states. This hybridization gives contributions which tend to compensate for the halogen p Cu-4p interactions /3/.

References

/1/ N.E. CHRISTENSEN, phys. stat. sol. (b) 123, 281 (1984).

/2/ R.C. HANSON, J.R. HALLBERG, and C. SCHWAB, Appl. Phys. Letters 21, 490 (1972).

/3/ A. BLACHA, H. PRESTING, and M. CARDONA, phys. stat. sol. (b) 126, 11 (1984).

/4/ J.L. DEISS, A. DAUNOIS, and O.V. GOGOLIN, EPS-Meeting, Lausanne 1983, unpublished.

/5/ A. BLACHA, S. VES, and M. CARDONA, Phys. Rev. B 27, 6346 (1983).

/6/ F. MESEGUER, J.C. MERLE, and M. CARDONA, Solid State Commun. 43, 511 (1982).

(Received June 12, 1984)

phys. stat. sol. (b) 125, K63 (1984)
Subject classification: 19; 22.5.3

Department of Physics, Tbilisi State University[1)]

Ligand Hyperfine Interaction of Er^{3+} in BaF_2 Single Crystals

By

A. DAVITULIANI, M. KIKNADZE, R. MIRIANASHVILI, and T. SANADZE

The ligand hyperfine interaction (HFI) of the trigonal fluorine centre of Er^{3+} in BaF_2 single crystals has been investigated by the method of radiofrequency discrete saturation (RFDS) /1/, which is the pulsed analogue of the ENDOR method.

The experiments have been carried out using an X-band superheterodyne spectrometer at liquid helium temperatures, on the EPR lines of the erbium isotopes with zero nuclear spin. The impurity concentration was about 0.003 to 0.05%. The angular dependence of the RFDS spectrum of eight fluorine nuclei in the nearest neighbourhood of the magnetic ion has been studied. Further we discuss the results of the investigation of the ligand HFI with nuclei 1 and 8, located along the trigonal Z-axis (Fig. 1).

For the Er^{3+} ions in BaF_2 the distance from the ground doublet up to the two nearest excited ones is merely 70 cm^{-1} and 76 cm^{-1} /2/. In this case the mixed parts of the Hamiltonian, containing products of the matrix elements of electron Zeeman and magnetic hyperfine interactions may give a significant contribution, having the form of the non-isotropic nuclear Zeeman interaction (pseudonuclear Zeeman effect) /3/. This effect may take place not only for the nucleus of the magnetic ion itself, but for the ligand nuclei, too /4/, though in the latter case the magnitude of the effect is significantly less.

As it came out, for the right interpretation of our experimental data it is necessary to take into account the ligand pseudonuclear Zeeman effect (contribution to resonance frequency ≈100 to 200 kHz).

The Hamiltonian of the magnetic ion and ligand nucleus can be written in the form

$$\mathcal{H} = g_{\parallel} \beta H_z S_z + g_{\perp}\beta(H_x S_x + H_y S_y) + h A_{\parallel} S_z I_z + h A_{\perp}(S_x I_x + S_y I_y) - \\ - g_{\parallel}^{(n)} \beta^{(n)} H_z I_z - g_{\perp}^{(n)} \beta^{(n)} (H_x I_x + H_y I_y) , \quad (1)$$

---

1) prospekt Chavchavadze 1, 380028 Tbilisi, USSR.

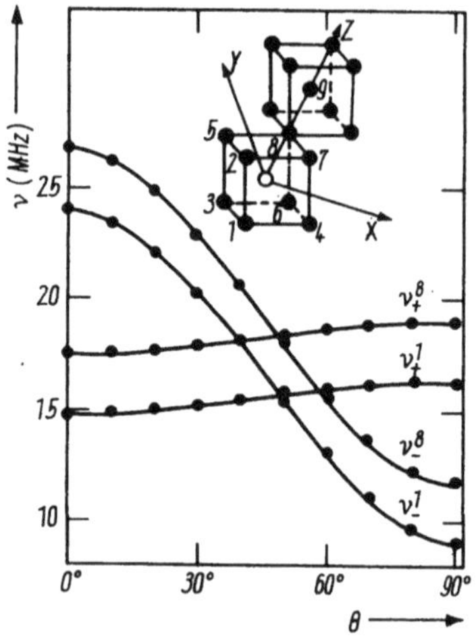

Fig. 1. Model of the immediate vicinity of Er^{3+} in BaF_2 and angular dependence of the RFDS spectrum of nuclei 1 and 8; in the inset ● F^-, o Er^{3+}

where $g_\parallel$, $g_\perp$ are the components of the electron g-tensor, $A_\parallel$, $A_\perp$ are components of the ligand HFI tensor, $g_\parallel^{(n)}$, $g_\perp^{(n)}$ are components of the nuclear g-tensor, β is the Bohr magneton, $\beta^{(n)}$ is the nuclear magneton. It is supposed that the symmetry of the nuclear g-tensor is the same as of the ligand HFI tensor.

For the RFDS frequencies, taking into account second order corrections in the ligand HFI (contribution ≈ 30 to 40 kHz), we have

$$\nu_\pm = \left\{ \left[\frac{g_\perp}{g} \left(\pm \frac{A_\perp}{2} - h \frac{A_\perp A_\parallel}{4g\beta H} \right) - \frac{1}{h} g_\perp^{(n)} \beta^{(n)} H \right]^2 \sin^2\Theta + \right.$$
$$\left. + \left[\frac{g_\parallel}{g} \left(\pm \frac{A_\parallel}{2} - h \frac{A_\perp^2}{4g\beta H} \right) - \frac{1}{h} g_\parallel^{(n)} \beta^{(n)} H \right]^2 \cos^2\Theta \right\}^{1/2} ; \qquad (2)$$

ν_+, ν_- are RFDS frequencies, corresponding to upper ($m = +1/2$) and lower ($m = -1/2$) electron states, Θ is the angle between the Z-axis and the magnetic field direction.

Table 1

Spin-Hamiltonian parameters

parameter \ nucleus	1	8
$g_\parallel$	5.938	5.938
$g_\perp$	7.150	7.150
$A_\parallel$ (MHz)	38.70	44.11
$A_\perp$ (MHz)	-25.53	-30.77
$g_\perp^{(n)}$	5.17	5.15
$g_\parallel^{(n)}$	5.46	5.47

The angular dependence of the RFDS spectrum has been fitted according to (2). The standard deviation of the experimental frequencies from their calculated values was about 10 kHz. The spin-Hamiltonian parameters are given in Table 1.

At the end we would like to mention that the investigation of the ligand HFI with nuclei 2-4 and 5-7 is a more complicated problem, as the ligand HFI tensor and the nuclear g-tensor have five independent components each. The results of these further investigations are to be published in nearest future.

References

/1/ B. BERULAVA, R. MIRIANASHVILI, D. DARASELIA, and T. SANADZE, Izv. Akad. Nauk SSSR, Ser. fiz. 47, 2314 (1983).

/2/ I. AIZENBERG, M. DAVIDOVA, B. MALKIN, A. SMIRNOV, and A. STOPOV, Fiz. tverd. Tela 15, 1345 (1973).

/3/ A. ABRAGAM and B. BLEANEY, Electron Paramagnetic Resonance of Transition Ions, Clarendon Press, Oxford 1970.

/4/ E. SECEMSKI and W. LOW, Phys. Rev. B 9, 4954 (1974).

(Received June 18, 1984)

phys. stat. sol. (b) 125, K67 (1984)

Subject classification: 5

Department of Applied Problems of Physics, Academy of Sciences of the Armenian SSR, Erevan[1)]

Time Dependence of Resonance γ-Radiation Modulated by Acoustic Excitations. Experiment

By

A.R. MKRTCHYAN, A.R. ARAKELYAN, R.G. GABRIELYAN, L.A. KOCHARYAN, G.R. GRIGORYAN, and M.M. SLAVINSKII

Investigations of the time dependence of the γ-resonance absorption line intensity in case of modulation by acoustic waves occupy a special place in modulation Mössbauer spectroscopy. The first works in that field performed by Perlow and Monahan and Perlow /1, 2/ correspond to the case when the acoustic oscillation frequency is greater than the absorption linewidth Γ_a. In those papers the counting rate of γ-quanta is shown to become dependent on time due to the transition of the γ-radiation frequency modulated by the source oscillations through the crystal. Harmonic composition and relative phase of the counting rate are sensitive to small shifts of energy and may be used to measure them.

Paper /3/ reports lineshapes in Mössbauer experiments with sinusoidal phase modulation at frequencies close to the linewidth of the Mössbauer state.

The present note suggests an experimental realization of the case considered in /4/ where the authors theoretically analysed the time-dependent Mössbauer spectroscopy when γ-radiation was modulated by low-frequency acoustic oscillations ($\Omega \ll \Gamma$).

In our experiment the source of the resonant γ-quanta was ^{57}Co in a chromium matrix and a stainless steel foil 10 μm thick was chosen as an absorber. The block diagram of the experimental device is presented in Fig. 1.

The γ-radiation passes through the oscillating absorber 1 and is registered by the multichannel analyser 4 operating in a multiscalar mode. The absorber is oscillated by the vibrosystem 2 supplied by a periodical voltage from the sound generator 5. The same generator simultaneously excites circuit 6 which forms the starting pulse from each period to start the internal sweep of the analyser channels.

---

1) Hr. Nersesyan str. 25, 375014 Erevan, USSR.

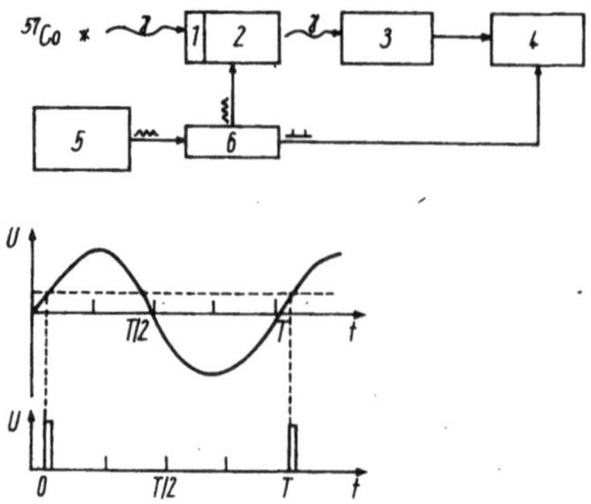

Fig. 1. Block diagram of the experiment. 1 Mössbauer absorber, 2 vibrosystem, 3 scintillation counter with photomultiplier, 4 multichannel analyser, 5 sound generator, 6 forming circuit, $T = 1/\Omega$

Fig. 2 and 3 show characteristic time dependences of the counting rate of the resonant γ-quanta corresponding to excitations with 3400 Hz and with different voltages at the vibrosystem transducer (U).

The experimental results in Fig. 2 and 3 may be explained in the following way. As it is shown in /4/, if the source or the absorber of γ-quanta is oscillating affected by an external periodical excitation, the energy of γ-quanta is periodically violated due to the Doppler effect, thus leading to oscillations of transmitted γ-quantum intensity N(t). In the approximation of a thin absorber it is described by

$$N(t) = N_o\left(1 - \frac{\frac{\Gamma_a}{\Gamma_{exp}}\beta}{\left(\Delta x + \frac{v(t)}{\lambda\Gamma_{exp}/2}\right)^2 + 1}\right) + N_b =$$

$$= N_\infty\left(1 - \frac{\frac{N_o}{N_\infty}\frac{\Gamma_a}{\Gamma_{exp}}\beta}{\left(\Delta x + \frac{v(t)}{\lambda\Gamma_{exp}/2}\right)^2 + 1}\right), \quad \Gamma_{exp} = \Gamma_a + \Gamma_s , \tag{1}$$

where N(t) is the counting rate at the moment of time t, N_o, N_∞ are the counting rates of the resonant and total quanta out of resonance, N_b is the background,

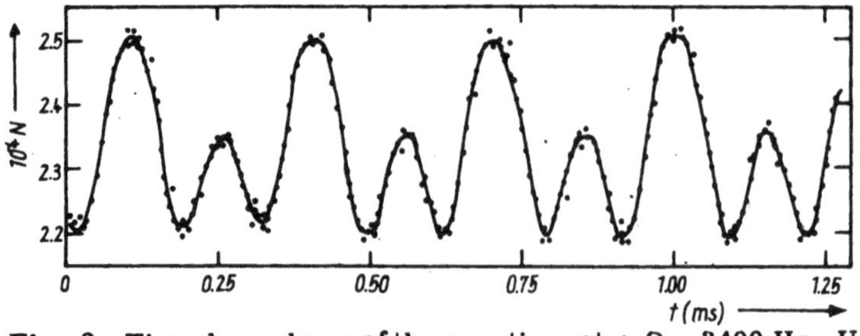

Fig. 2. Time dependence of the counting rate; Ω = 3400 Hz, U = 1.5 V

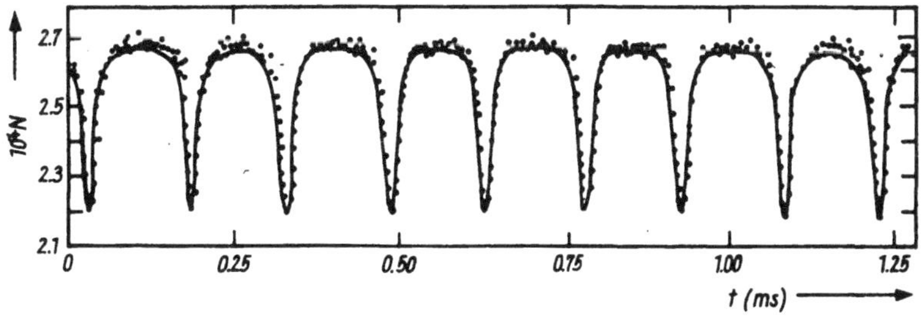

Fig. 3. Time dependence of the counting rate; Ω = 3400 Hz, U = 7 V

$\Delta x = (\omega_0 - \omega_0')/(\Gamma_{exp}/2)$ is the energy shift between the resonance frequencies of the radiation and absorption, v(t) is the velocity of the source or absorber motion, λ is the γ-quantum wavelength, β is the absorber effective thickness, and Γ_s is the radiation linewidth.

From comparison of experimental results in Fig. 2 and 3 with expression (1) for N(t) we find the following values of parameters: $\Delta\omega$ = 0.029mm/s, v = $= v_0 \sin(2\pi\Omega t+\varphi_0)$, v_0 = 0.124 mm/s, $\varphi_0 = -\pi/6$ (Fig. 2) and 2, v_0 = 0.57 mm/s, $\varphi_0 = -\pi/6$ (Fig. 3) (in our experiment we have $\lambda\Gamma_{exp}/2$ = 0.22 mm/s).

Thus, the time-dependent Mössbauer spectroscopy allows to measure the chemical shift $\Delta\omega$ as in /1/, but the experiment here is simpler and is easier realized. At the same time this method allows to determine the initial phase of oscillations φ_0 and the form of complex oscillations v(t) of the vibrosystem.

Concluding we should like to note the principal advantages of this method over the conventional Mössbauer spectroscopy. Firstly, time presentation acquires all advantages of the differential method in comparison with the integral one represented by the conventional Mössbauer spectroscopy. E.g., if the source is affected by complex oscillations, the form of the latter cannot be restored from the integral spectrum while it is not difficult to do so from the time spectrum (see /4/). Secondly, there is no need to scan the whole spectral line. Actually, the source velocity $v_0/(\lambda\Gamma_{exp}/2)) < 1$ in our experiment is rather small and it is impossible to obtain its conventional absorption spectrum. Thirdly, this method allows to measure the initial phase of oscillations $\varphi_0 = \varphi(t = 0)$ which makes it possible, in particular, to determine the propagation velocity and the attenuation coefficient of the acoustic waves in a non-resonant medium.

References

/1/ G.I. PERLOW, Phys. Rev. Letters 40, 896 (1978).
/2/ I.E. MONAHAN and G.I. PERLOW, Phys. Rev. A 20, 1499 (1979).
/3/ P. HELISTÖ, T. KATILA, W. POTZEL, and K. RISKI, Phys. Letters A 85, 177 (1981).
/4/ R.G. GABRIELYAN, L.A. KOCHARYAN, A.R. MKRTCHYAN, and H.A. UNANYAN, phys. stat. sol. (b) 120, K123 (1983).

(Received July 6, 1984)

phys. stat. sol. (b) 125, K71 (1984)
Subject classification: 13.4

Institute of Semiconductor Physics,
Academy of Sciences of the Lithuanian SSR, Vilnius[1)]

Field Tunneling from Impurities in Case of Anisotropic Effective Masses

By
A. DARGYS and A. MATULIS

In this note we will show that in case of anisotropic effective masses and strongly localized impurity wave functions the impurity-band tunneling probability depends on the inverse effective mass m_F^{-1} calculated in the direction of an electric field. The existence of anisotropic tunneling follows from /1, 2/, where tunneling probabilities were calculated in the directions of high symmetry with respect to constant energy ellipsoids. Experimentally, the pure field induced electron tunneling from shallow impurities to the conduction band of silicon was observed in /3, 4/.

The calculations will be performed in the effective mass approximation for a single ellipsoidal constant energy surface, Fig. 1. The impurity potential, in accordance with the Koster-Slater model /5/, will be described by a single-band wave-vector-independent matrix element. The paper /1/ will be closely followed, where one can find the necessary details. In /1/, using the Green function technique, it has been shown that in a weak field limit ($\hbar\Gamma \ll \Delta$) the impurity linewidth $\hbar\Gamma$ (or tunneling probability Γ) is

$$\hbar\Gamma = \operatorname{Im} G_0(\varepsilon)/\operatorname{Re}\left[dG_0(\varepsilon)/d\varepsilon\right]_{\varepsilon=-\Delta}, \tag{1}$$

where Δ is the level depth. $G_0(\varepsilon)$ is found from the conduction electron Green function $G_0(\vec{k}, \vec{k}'; \varepsilon)$ in the presence of an electric field F and in the absence of an impurity potential:

$$G_0(\varepsilon) = \int d\vec{k} \int d\vec{k}' \, G_0(\vec{k}, \vec{k}'; \varepsilon) \; . \tag{2}$$

The function $G_0(\vec{k}, \vec{k}'; \varepsilon)$ can be easily constructed from the electronic eigenfunctions written in the wave vector $\vec{k}$ representation. For weak electric fields and large level densities in the band ($eFa \ll E_m$, a is the lattice constant and E_m is the maximum energy of the band) the needed eigenfunctions, within good accuracy, may be calculated in the parabolic band approximation, with

1) 52 K. Pozhelos, 232000 Vilnius, USSR.

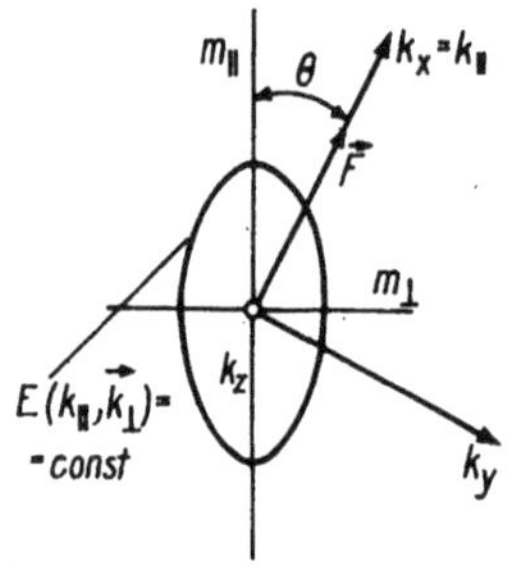

Fig. 1. The orientation of the reference axis and of the electric field with respect to the constant energy ellipsoid

finite bandwidth and Stark quantization neglected. In this approximation $G_0(\varepsilon)$ has the following form:

$$G_0(\varepsilon) = (2\pi)^2 (eF)^{-1} \int d\vec{k}_\perp \int_{-\infty}^{\infty} d\varepsilon_\parallel \frac{1}{\varepsilon - \varepsilon_\parallel} \times \times \left[\int_{-\infty}^{\infty} dk_\parallel \exp\left\{ -\frac{i}{eF} \int_0^{k_\parallel} dk'_\parallel \left[\varepsilon_\parallel - E(k'_\parallel, \vec{k}_\perp) \right] \right\} \right]^2 . \quad (3)$$

Here, $E(k_\parallel, \vec{k}_\perp)$ is the free electron (parabolic) dispersion law expressed in terms of wave vector components parallel, $k_\parallel \equiv k_z$, and perpendicular, $\vec{k}_\perp = (k_y, k_z)$, to the electric field, Fig. 1. In (3) the integration variable $\varepsilon_\parallel$ is, in fact, an energy eigenvalue.

In /1/ the calculations were restricted to the situation when the dispersion law can be written in the following form:

$$E(k_\parallel, \vec{k}_\perp) = E_1(k_\parallel) + E_2(\vec{k}_\perp) = \frac{\hbar^2 k_\parallel^2}{2m_\parallel} + E_2(\vec{k}_\perp) . \quad (4)$$

In this case the expression (3) factorizes and as a result a simple formula for Γ is obtained.

In general case, after the rotation of the electric field by an angle Θ, cf. Fig. 1, the dispersion law takes the rather complicated form

$$E(k_\parallel, \vec{k}_\perp) = \frac{\hbar^2}{2} \left[\frac{(k_\parallel + a k_y)^2}{m_F} + \frac{k_y^2}{m_\perp \cos^2\Theta + m_\parallel \sin^2\Theta} + \frac{k_z^2}{m_\perp} \right] , \quad (5)$$

$$a = m_F (m_\parallel^{-1} - m_\perp^{-1}) \sin\Theta \cos\Theta , \quad (6)$$

$$1/m_F = \cos^2\Theta / m_\parallel + \sin^2\Theta / m_\perp . \quad (7)$$

Here, $1/m_F$ is an inverse effective mass in the direction of the electric field. $m_\parallel$ and $m_\perp$ are effective masses parallel and perpendicular to the rotation axis of the energy ellipsoid.

The shape of (5) differs from that of (4). However, after the insertion of (5) into (3), the change of variables of integration, $k_{\parallel} + ak_y \rightarrow k_{\parallel}$, and the application of the property that the module is invariant with respect to addition to the argument of the exponent of any imaginary and of $k_{\parallel}$ independent function, one can rearrange the expression (3) so that the dispersion law figuring in (3) will assume exactly the same shape as (4), with the component $E_1(k_{\parallel})$, now, being $E_1(k_{\parallel}) = \hbar^2 k_{\parallel}^2/(2m_F)$. This means that the final expression for the tunneling probability obtained in /1/,

$$\Gamma = \left[eF/(4\sqrt{2m_{\parallel}\Delta})\right] \exp\left[-4\sqrt{2m_{\parallel}}\ \Delta^{3/2}/(3\hbar\, eF)\right] \qquad (8)$$

remains valid for a general case, i.e. for all directions of the electric field. The only difference is that, now, in (8) the effective mass $m_{\parallel}$ should be replaced by m_F.

It can be shown that this conclusion is true for a general case of constant energy ellipsoids, i.e. when there are no rotational axes. Of course, in this case in (8) a more general expression for m_F should be used.

References

/1/ E.N. KOROL, Fiz. tverd. Tela 19, 1266 (1977).
/2/ SH.M. KOGAN, Zh. eksper. teor. Fiz. 81, 2268 (1981).
/3/ S. ŽURAUSKAS and A. DARGYS, phys. stat. sol. (b) 121, 385 (1984).
/4/ A. DARGYS and S. ŽURAUSKAS, Fiz. Tekh. Poluprov. 18, 595 (1984).
/5/ G.F. KOSTER and J.C. SLATER, Phys. Rev. 95, 1167 (1954).

(Received June 7, 1984)

phys. stat. sol. (b) 125, K75 (1984)
Subject classification: 10.2 and 13.4; 22.1; 22.1.2

Belorussian Polytechnical Institute, Minsk[1)]

Noble Gas Atoms as Interstitials in Silicon and Diamond

By

A.V. MUDRYI, A.L. PUSHKARCHUK, V.D. TKACHEV, and A.G. ULYASHIN

The properties of noble gas interstitials in covalent crystals are important in many experimental applications, such as backscattering and channeling studies, sputtering, and others. Irradiation of silicon crystals by noble gas ions (He^+, Ne^+, Ar^+, Kr^+) and diamond crystals by He^+ and Ne^+ is known to produce defects which incorporate the noble gas atoms /1 to 4/. It was shown /1, 2, 4/ that noble gas atoms may exhibit chemical activity in silicon crystals. A theoretical investigation of interactions of He with the silicon crystal matrix was carried out in /5/.

Here we have studied the electron structure and the contributions from atomic orbitals (AO) of noble gas atoms to defect wave functions - molecular orbitals (MO) for interstitial complexes T + X (X = He, Ne, Ar, Kr, Xe) in silicon and diamond crystals. Atomic 1s orbitals on He atom; 2s and 2p on Ne atom; 3s and 3p on Ar atom; 4s, 4p, and 3d on Kr atom; 5s, 5p, and 4d on Xe atom with parameters given in /6/; 2s and 2p Slater orbitals on each carbon atom with the exponential parameters given in /7/ and ionisation potentials $I_{2s}(C) = 19.20$ eV, $I_{2p}(C) = 11.79$ eV /6/ were used. On silicon atoms 3s and 3p atomic orbitals with parameters given in /8/ were used. We used the extended Hückel theory with the approximations given in /9/. Calculations were made for a lattice consisting of 42 silicon (carbon) atoms of five nearest coordination spheres. Point groups D_2 and C_h were taken into account in the calculations by applying the method given in /10/. For this case symmetry-adapted linear combinations of the basis functions $|\varphi_{R\alpha, L}\rangle$ in notation /10/ are given as

$$\left|\Phi^{\Gamma_\lambda}_{\alpha, L}\right\rangle = \eta_g^{-1} \sum_{R \in g} \chi^{\Gamma_\lambda \otimes \Gamma_L}(R) \left|\varphi_{R\alpha, L}\right\rangle ,$$

where α is the number of atom; L is the number of AO from which $\left|\Phi^{\Gamma_\lambda}_{\alpha, L}\right\rangle$ is obtained; $\chi^{\Gamma_\lambda}(R)$ is the character of the irreducible representation Γ_λ for sym-

1) Leninskii prospekt 65, 220027 Minsk, USSR.

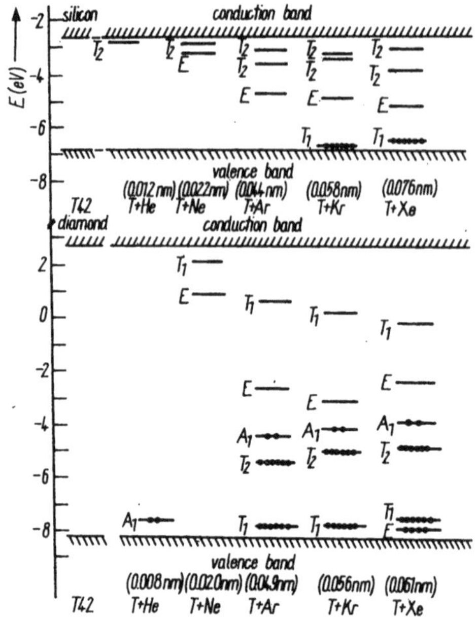

Fig. 1. Molecular orbital energy levels for T + X complexes in silicon and diamond. The values ΔR are shown in round brackets

metry operation R; η_g is the order of group g. The one-electron MO's $|\Psi_i^{\Gamma\lambda}\rangle$ of the system are constructed as a linear combination

$$|\Psi_i^{\Gamma\lambda}\rangle = \sum_{\alpha}{}' \sum_{L\in\alpha} C_i^{\alpha,L} |\Phi^{\Gamma\lambda}_{\alpha,L}\rangle ,$$

where summation $\sum'_{\alpha}$ is carried out over all atoms lying in the region from which the whole cluster may be constructed by applying the symmetry operation of the point group g. The total energy of the system is approximated by the sum of the energies of all occupied one-electron MO's

$$E = \sum_i n_i E_i ,$$

where E_i is the one-electron MO energy and n_i is the occupation number of the i-th MO. Equilibrium configuration of complexes has been obtained from the dependence of the total cluster energy on the lattice position of noble gas atoms. Different sites on the $(01\bar{1})$ plane of the silicon (diamond) lattice were studied. The calculations show that noble gas atoms occupy the tetrahedral interstitial positions, while the hexagonal interstitial site is the saddle point. The analysis of the interactions of noble gas atoms with the crystal lattice was carried out on MO coefficients $C_i^{\alpha,L}$ and on variations of atomic populations $\Delta N(\alpha)$ which describe the contribution from atom α to the i-th MO and the net total charge on this atom, respectively (Table 1). From Table 1 it is clear that the contribution from He, Ne, and Ar atoms to certain valent MO's is comparable to that from silicon (carbon) atoms of the crystal lattice (the maximum values of the $C_i^{\alpha,L}$ on the i-th MO are given). These contributions are increasing from Kr and Xe atoms. The energy structure of T + X complexes and the values of the lattice relaxations ΔR in a breathing mode around the tetrahedral position

Table 1

Some maximum values $C_i^{\alpha,L}$ for valent MO's of T + X (X = He, Ne, Ar) complexes in silicon and diamond

T + X, silicon				T + X, diamond			
E_i(eV)	- 8.482	-17.132	-25.384	E_i(eV)	-7.621	-25.793	-29.996
$C_i^{He,1s}$	0.124	0.245	0.928	$C_i^{He,1s}$	-0.760	- 0.430	0.445
$C_i^{Si,3s}$	- 0.123	- 0.423	0.150	$C_i^{C,2s}$	0.180	0.259	0.344
$C_i^{Si,3p}$	0.368	0.039	- 0.012	$C_i^{C,2p}$	-0.279	0.058	0.009
ΔN(He)		0.002e		ΔN(He)		0.0456e	
E_i(eV)	-11.136	-19.624	-23.346	E_i(eV)	-8.516	-21.754	-24.368
$C_i^{Ne,2s}$	0.092	0	0	$C_i^{Ne,2s}$	0.362	0	0
$C_i^{Ne,2p}$	0	- 0.195	0.959	$C_i^{Ne,2p}$	0	- 0.420	0.573
$C_i^{Si,3s}$	0.309	- 0.378	- 0.112	$C_i^{C,2s}$	-0.172	- 0.445	0.298
$C_i^{Si,3p}$	0.157	- 0.012	0	$C_i^{C,2p}$	0.293	0.166	- 0.085
ΔN(Ne)		0.0162e		ΔN(Ne)		0.0811e	
E_i(eV)	-10.593	-14.534	-16.093	E_i(eV)	-5.828	-20.240	-19.967
$C_i^{Ar,3s}$	0.173	0	0	$C_i^{Ar,3s}$	0	0	0.228
$C_i^{Ar,3p}$	0	- 0.5	- 0.573	$C_i^{Ar,3p}$	-0.522	0.249	0
$C_i^{Si,3s}$	0.268	0.498	- 0.523	$C_i^{C,2s}$	0.208	0.453	0.338
$C_i^{Si,3p}$	0.169	0.068	0.131	$C_i^{C,2p}$	-0.597	- 0.201	- 0.132
ΔN(Ar)		0.2111e		ΔN(Ar)		2.4433e	

are shown in Fig. 1. Thus the results of our calculations support the statement that noble gas atoms in silicon and diamond produce not only intrinsic defects but complexes involving noble gas atoms due to their chemical activity in the crystalline environment.

References

/1/ V.D. TKACHEV, N.S. MINAEV, A.V. MUDRYI, and C. SCHRODEL, Dokl. Akad. Nauk Beloruss. SSR 23, 315 (1979).

/2/ V.D. TKACHEV and A.V. MUDRYI, Proc. Internat. Conf. Radiation Physics in Semicond. and Related Materials, Tbilisi 1980 (p. 64).

/3/ A. GIPPINS, V.S. VAVILOV, A.M. ZAITSEV, and B.S. ZHAKUPBEKOV, Physica BC116, 187 (1983).

/4/ V.D. TKACHEV, A.V. MUDRYI, and N.S. MINAEV, phys. stat. sol. (a) 81, 313 (1984).

/5/ D.R. KAPLAN, C. WEIGEL, and J.W. CORBETT, phys. stat. sol. (b) 94, 359 (1979).

/6/ E. CLEMENTI and C. ROETTI, Atomic Data Nuclear Data Tables 14, 177 (1974).

/7/ E. CLEMENTI and D.L. RAIMONDI, J. chem. Phys. 38, 2686 (1963).

/8/ K.J. YIP, phys. stat. sol. (b) 66, 619 (1974).

/9/ M. WOLFSBERG and L. HELMHOLTZ, J. chem. Phys. 20, 837 (1952).

/10/ I.B. RIPS, Uchet simmetrii pri kvantovokhimicheskikh raschetakh, Zinatne, Riga 1978.

(Received June 18, 1984)

phys. stat. sol. (b) 125, K79 (1984)
Subject classification: 10.2 and 13.4; 22.6

Sektion Physik der Wilhelm-Pieck-Universität Rostock, Wissenschaftsbereich Isolatorphysik[1)]

Bethe-Lattice Approach to the Electronic Structure of Intrinsic Defects in v-SiO_2

By

V. BETHKENHAGEN and K. HÜBNER

The existence of intrinsic defects in vitreous (v) SiO_2 has been verified by EPR investigations of irradiated and annealed samples /1/. Furthermore, results of optical measurements (see, for example /2, 3/) give rise for the suggestion that intrinsic defects are one cause of localized states in the gap of this material. However, the calculations of the electronic states of intrinsic defects in v-SiO_2 performed until now (e.g. /4 to 7/) do not allow unique conclusions about the origin of observed optical transitions. An open question is the influence of the relaxation of the atomic network in the environment of a defect on the electronic structure of this defect. The present note involves calculations of this influence on a threefold coordinated Si atom (Si dangling bond) and a onefold coordinated oxygen atom (or non-bridging oxygen) within the v-SiO_2 network. The local electronic density of states (LDOS) of these defects is calculated using the Bethe-lattice method /8/ and a nearest-neighbour-tigt-binding Hamiltonian.

With the help of the Green-function technique /9/ the SiO_2 Bethe lattice with an Si-O-Si angle of 144° is transformed into an effective silicon Bethe lattice which contains two kinds of effective Si atoms (see Fig. 1). We use Si3s, Si3p, O2s, and O2p atomic wave functions as the basis set. The corresponding Slater-Koster parameters have been fitted to XPS and UPS data /10/ on the basis of recently published parameter sets /11, 12/. The parameter values used are $(pp\sigma) = 5.9$, $(pp\pi) = -1.6$, $(sp\sigma) = -5.4$, $(ps\sigma) = -7.0$, and $(ss\sigma) = -3.5$ (all in eV). The orbital energies in eV are $\langle Os |H| Os\rangle = -16.4$, $\langle Op |H| Op\rangle = -1.3$, $\langle Sis |H| Sis\rangle = 5.9$, $\langle Sip |H| Sip\rangle = 12.2$. The resulting electronic density of states of the SiO_2 Bethe lattice reveals a gap of about 10 eV (Fig. 3a, f)).

Fig. 2 involves schematic representations of the dangling-bond defects investigated and of the assumed bond-angle relaxations. The Si dangling bond leads

1) Universitätsplatz 3, DDR-2500 Rostock, GDR.

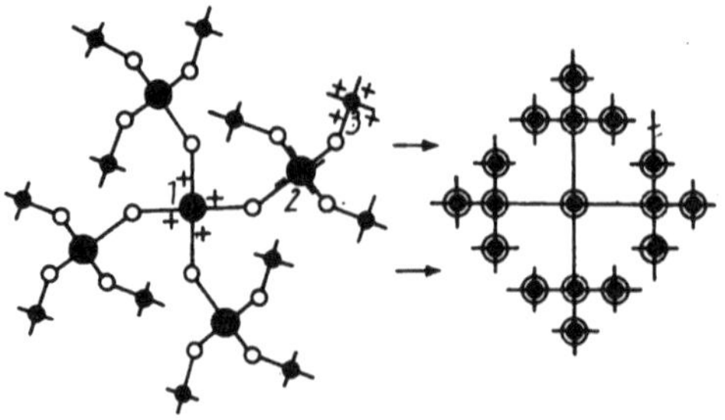

Fig. 1. The SiO_2 Bethe lattice and the related Bethe lattice of two different types of atoms resulting from SiO_4 tetrahedra with (111) and ($\bar{1}\bar{1}\bar{1}$) bond directions, which are marked by the + and - signs, respectively. ● silicon; o oxygen; ◉, two types of effective silicon

to a sharp peak below the bottom of the conduction band (Fig. 3b) at 9.8 eV. It has Si s and p character with dominance of s-like contributions.

Because of the neglect of orbital overlap in our tight-binding approximation it is necessary to modify the orbital energies of defect atoms from their bulk values (see /5/). So, the orbital energies of the Si defect are shifted downwards resulting in a downward motion of the defect state. A shift of the Si orbital energies by 2.4 eV leads to a gap state at 7.8 eV as shown in Fig. 3c.

The relaxation of the considered Si defect atom into the plane of the three oxygen neighbours causes a splitting of the gap peak into two distinct peaks, the upper one with p and the lower one with s character (Fig. 3e). This splitting occurs as a consequence of the loss of tetrahedral symmetry and indicates that it could be incorrect to assume an sp^3 hybridization at the threefold coordinated Si atom. It should be noted that a Si atom relaxed into the plane of its three oxygen neighbours is a part of the so-called E' center as discussed for example by Griscom /1/. As can be seen from Fig. 3e, another result of this relaxation is a shift of the gap states up to higher energies as described also by O' Reilly and Robertson /5/. Taking into account an additional change of the bond length we find a downward motion of the gap states by bond stretching and

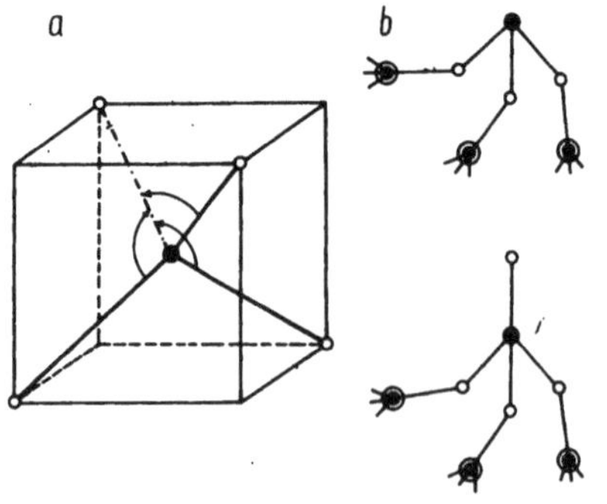

Fig. 2. a) Schematic representation of the relaxation of the three oxygen neighbours of the defects considered. The broken line describes the Si-O bond of the non-bridging oxygen being absent in the case that the Si atom is only threefold coordinated. b) Bethe-lattice representation of the defects investigated. The meaning of the circles is the same as in Fig. 1

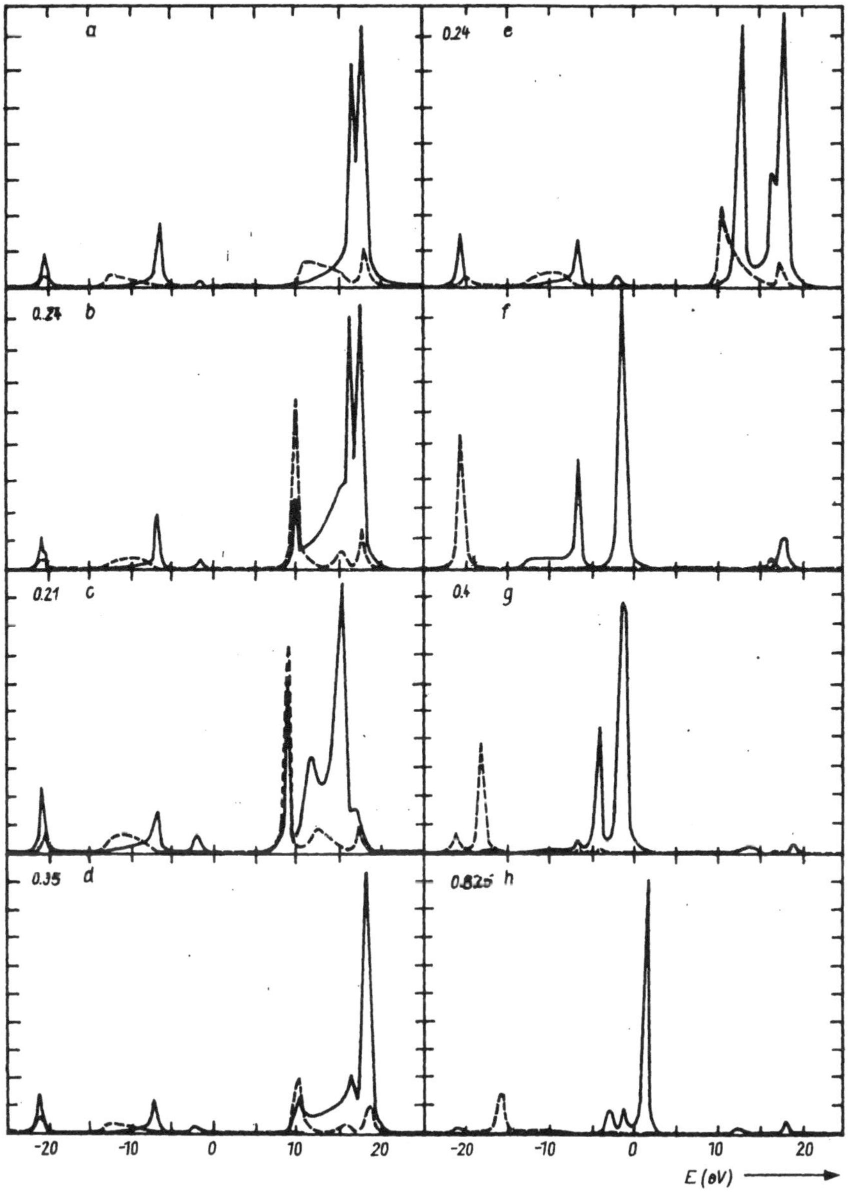

Fig. 3. Local electronic density of states (LDOS) of a a) normal Si atom in the SiO_2 Bethe lattice, b) threefold coordinated Si atom, c) threefold coordinated Si atom with orbital-energy shift, d) threefold coordinated Si atom with bond-length reduction, e) relaxed threefold coordinated Si atom, f) normal oxygen atom in the SiO_2 Bethe lattice, g) non-bridging oxygen, h) non-bridging oxygen with orbital-energy shift and relaxation; all in arbitrary units. The dashed lines show the s part of the LDOS and the solid lines the p part

an upward motion of the states by reduction of the bond length (Fig. 3d). The dependence of the tight-binding parameters on the bond length d is assumed to be of the form $p \sim d^{-2}$ /13/, where p is the corresponding Slater-Koster parameter.

Fig. 3 also contains the results for the non-bridging oxygen. The unmodified LDOS of an oxygen dangling bond does not reveal gap states (see Fig. 3b). However, in the valence band an oxygen p-like state appears at -4.2 eV. In correspondence with the shift of the orbital energies of the Si defect mentioned above the orbital energies of the oxygen defect must be shifted to higher energies. A shift of 2.7 eV gives a p-like peak at 1.4 eV in the gap (there hole traps were observed /7/) and a new p-like peak at -0.8 eV. Then, the defect peak in the valence band occurs at -3.1 eV. The corresponding LDOS is shown in Fig. 3h, where additionally a bond bending of 5° of the three Si-O bonds in the neighbourhood of the defect is taken into account, which shifts this valence-band peak to higher energies.

These results support the correlation of the experimentally observed red luminescence of an energy 1.9 eV with non-bridging oxygen defects as described by Skuja and Silin /3/, since the calculated energy difference between the two upper defect peaks near the valence-band edge is about 2 eV. However, this transition is related by our calculation to a broken O-Si bond and not to a splitting of the lone pair peak of oxygen by a dynamic Jahn-Teller effect as discussed in /3/.

To improve the significance of our results it seems to be necessary to consider not only a single defect atom, but a cluster containing the defect. Such calculations are in progress, and the results (also for other intrinsic defects of v-SiO_2 /14/) will be published in a forthcoming paper.

The authors are grateful to the Institute of Solid State Physics of the Latvian State University Riga, USSR, for the support of this work and to Dr. A. Silin and his colleagues for valuable discussions.

References

/1/ D.L. GRISCOM, J. non-crystall.Solids 40, 211 (1980).
/2/ S.W. MC KNIGHT and E.D. PALIC, J. non-crystall. Solids 40, 595 (1980).
/3/ L.N. SKUJA and A.R. SILIN, phys. stat. sol. (a) 70, 43 (1982).
/4/ F. HERMAN, D.J. HENDERSON, and R.V. KASOWSKI, in: The Physics of MOS Insulators, Ed. G. LUCOVSKY, S.T. PANTELIDES, and F.L.

GALEENER, Pergamon Press, 1980 (p. 107).

/5/ E.P. O'REILLY and J. ROBERTSON, Phys. Rev. B 27, 3780 (1983).

/6/ A. STERN and K. HÜBNER, phys. stat. sol. (b) 124, K171 (1984).

/7/ K. HÜBNER, J. Phys. C, submitted for publication.

/8/ J.D. JOANNOPOULOS, J. non-crystall. Solids 32, 241 (1979).

/9/ J.D. JOANNOPOULOS, Phys. Rev. B 16, 2764 (1977).

/10/ T.H. DISTEFANO and D.E. EASTMAN, Phys. Rev. Letters 27, 1560 (1971).

/11/ R.B. LAUGHLIN, J.D. JOANNOPOULOS, and D.J. CHADI, Phys. Rev. B 20, 5228 (1979).

/12/ S. CIRACI and S. ELLIALTROGLU, Solid State Commun. 40, 587 (1981).

/13/ W.A. HARRISON, Phys. Rev. B 24, 5835 (1981).

/14/ V. BETHKENHAGEN and K. HÜBNER, Physik der Halbleiteroberfläche, Vol. 15, Tagungsbericht Akademie der Wissenschaften der DDR, Zentralinstitut für Elektronenphysik, Berlin 1984, to be published.

(Received July 24, 1984)

phys. stat. sol. (b) 125, K85 (1984)
Subject classification: 13.3 and 14.3; 22.2.3

Sektion Physik der Humboldt-Universität zu Berlin[1)] (a) and International Laboratory of High Magnetic Fields and Low Temperatures, Wrocław[2)] (b)

Quantized Hall Effect in the n-Inversion Layer in InSb Grain Boundaries

By
R. HERRMANN (a), W. KRAAK (a), and M. GLIŃSKI (b)

The quasi two-dimensional electronic system at semiconductor interfaces has been investigated intensively in recent years (e.g. /1/). In the presence of a strong magnetic field perpendicular to the interface (surface) the electron motion is fully quantized due to Landau and surface quantization. Recently it was experimentally verified that the Hall conductivity in two-dimensional inversion layers in strong magnetic fields is quantized into integer multiples of e^2/h when the Fermi level lies in the localized states existing between the different Landau levels /2/. The Hall resistance ϱ_{xy} is given quite universally by the expression

$$\varrho_{xy} = \frac{h}{e^2 i} \quad (i = 1, 2, 3, \ldots) . \tag{1}$$

The quantized Hall effect manifests itself in the magnetotransport coefficients as plateaus in the Hall resistance ϱ_{xy} and as minima in the diagonal resistivity ϱ_{xx}. Recently fractional quantization of the Hall resistance was observed in the inversion layer of GaAs-$Al_xGa_{1-x}As$ heterostructures in the extreme quantum limit /3, 4/.

An analogous situation with similar electronic properties may exist in the grain boundary region in semiconductor bicrystals. Shubnikov-de Haas oscillations of the magnetoresistance and Hall effect have been observed in inversion layers adjacent to the grain boundary in Ge /5, 6/.

In previous publications /7, 8/ we have found from magnetotransport measurements that in p-InSb bicrystals with $p = N_A - N_D \approx 5 \times 10^{15}$ cm^{-3} in the vicinity of the grain boundary a degenerate n-type inversion layer exists. Pronounced quantum oscillations of the magnetoresistivity ϱ_{xx} and the Hall coefficient R_H

---

1) Hessische Str. 2, DDR-1040 Berlin, GDR.

2) Próchnik Str. 95, 53-529 Wrocław, Poland.

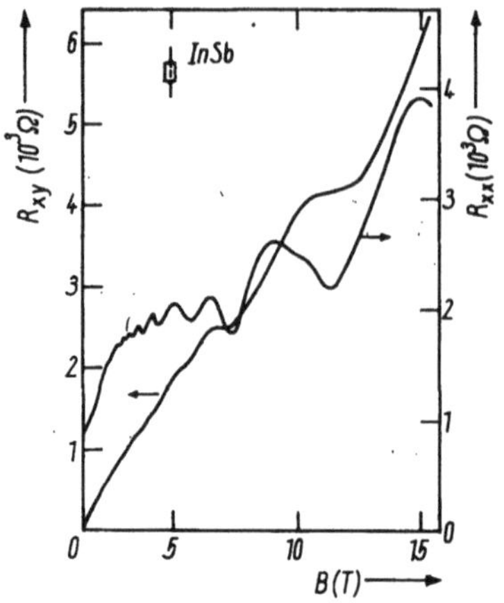

Fig. 1. Hall resistance R_{xy} and magnetoresistivity R_{xx} versus magnetic field at $T \approx 1.9$ K (the magnetic field is oriented normal to the grain boundary plane)

have been observed in magnetic fields above 0.45 T. The analysis of the quantum oscillations clearly suggests that two oscillations with different periods must be present and hence several subbands may be occupied. The electron densities of the subbands n_{si} ($n_{s1} = 3.5 \times 10^{11}$ cm^{-2}, $n_{so} = 1.1 \times 10^{12}$ cm^{-2}) and the effective cyclotron mass of the lower subband ($m_{co} = (0.023 \pm 0.002) m_0$) have been evaluated. From the Hall coefficient R_H and the boundary resistivity the total electron density in the inversion layer $n_s = 1.6 \times 10^{12}$ cm^{-2} and the carrier mobility $\mu \approx 1.1 \times 10^4$ cm^2/Vs have been determined.

In the present paper we report some striking new results on the magnetotransport properties in an n-inversion layer on p-InSb bicrystals in strong magnetic fields up to B = 17 T.

The InSb bicrystals have grown with a double-seed technique as described in /8/. Two [11$\bar{1}$] p-type InSb crystals ($p \approx 5 \times 10^{15}$ cm^{-3}) are tilted by an angle $\Theta = 13^0$ along the rotation axis [112]. A symmetrical β-grain boundary in [112] (1$\bar{1}$0) orientation with ($\bar{1}$10) symmetric plane was obtained. Fig. 1 shows typical data obtained at 1.9 K for the Hall resistance R_{xy} and the magnetoresistance R_{xx} as a function of magnetic field B.

The pronounced quantum oscillations of the diagonal component R_{xx} and the Hall component R_{xy} corresponding to Landau quantization of the lower subband ε_0 have been observed in a wide range of magnetic fields. In high magnetic fields (B>10 T) pronounced spin splitting is clearly evident. The quantum oscillations of the diagonal component R_{xx} and the strong angular dependence of R_{xx} and R_{xy} will be discussed in detail elsewhere /9/.

In high magnetic fields B > 10 T the Hall resistivity shows the characteristic features of the quantized Hall plateaus in finite regions of B; in addition, weaker plateaus are also apparent in lower fields (B = 5.3, 4.4, and 3.6 T). Simultaneously the diagonal resistivity R_{xx} drops to a minimum. These results resemble

Table 1

i	B(T)	$R_{xy}(10^3\Omega)$	$\frac{h/e^2}{R_{xy}} = n$	$\frac{h}{e^2 n}$ $(10^3\Omega)$
1	11.3	4.21 $\pm$ 0.03	6	4.302
2	7	2.508 $\pm$ 0.005	10	2.58
3	5.3	1.94 $\pm$ 0.04	14	1.84
4	4.4	1.46 $\pm$ 0.05	18	1.43
5	3.6	1.20 $\pm$ 0.05	22	1.173

in many respects the well-known quantized Hall effect at integral values of i. A summary of the experimental results is given in Table 1.

These results strongly suggest that the value of the Hall resistance at the plateaus is quantized in a multiple series. From the experimental results (Table 1) the following simple relation between the Hall resistance and the quotient of elementary constants h/e^2 can be derived

$$R_{xy} = \frac{1}{2}\,\frac{h/e^2}{2i+1}\quad (i = 0,\ 1,\ 2,\ \ldots)\,. \tag{2}$$

Thus the series is formed by multiples of the reciprocal of an odd integer. We would like to point out that the found quantum oscillations and the obtained series of plateaus R_{xy} differ in some respects from the properties accounted for the n-inversion layer on the surface structure. The finite value in the R_{xx} oscillations may be caused by the occupation of the second electric subband ε_1 which leads to a superposition of different Landau levels and probably to an increased level broadening due to intersubband scattering. On the other hand, the factor 1/2 in formulae (2) may be due to the existence of two different inversion layers adjacent to both sides of the grain boundary with the same Hall resistivity $(h/e^2)/(2i+1)$.

A first model for the inversion layer of the grain boundary may be a double layer of electrons separated by a localized positive charged layer. The degeneracy factor for each Landau level of the double layer is twice the factor of a single layer, thus the number of states per unit area of each level

$$N = 2\,\frac{eB}{h} \tag{3}$$

and therefore from the Hall resistivity $R_{xy} = B/n_s e$ follows (comparing with (2))

$$n_s = (2i + 1) N . \tag{4}$$

The steps between the Hall plateaus in our experiments are given by

$$\Delta R_{xy}^{-1} = 4 \frac{e^2}{h} , \tag{5}$$

e.g. for each Shubnikov-de Haas oscillation of the magnetoresistivity R_{xx} a twofold degererated Landau level crosses the Fermi energy following from the double layer degeneracy.

From the double layer degeneracy it can be supposed that pair coupling may be exist between the electrons on both sides of the grain boundary. Further experimental and theoretical investigations will be necessary to make clear this interesting problem.

The experiments have been performed in the International Laboratory for High Magnetic Fields and Low Temperatures in Wrocław (Poland).

The authors would like to thank the Academy of Science of the GDR and the coworkers of the International Laboratory for support in these experiments.

References

/1/ T. ANDO, A.B. FOWLER, and F. STERN, Rev. mod. Phys. 54, 437 (1982).

/2/ K.V. KLITZING, G. DORDA, and M. PEPPER, Phys. Rev. Letters 45, 494 (1980).

/3/ D.C. TSUI, H.L. STORMER, and A.C. GOSSARD, Phys. Rev. Letters 48, 1559 (1982).

/4/ H.L. STORMER, A. CHANG, D.C. TSUI, J.C.M. HWANG, A.C. GOSSARD, and W. WIEGMANN, Phys. Rev. Letters 50, 1953 (1983).

/5/ B.M. VUL and E.J. ZAVARITSKAYA, Zh. eksper. teor. Fiz. 76, 1089 (1979).

/6/ S. UCHIDA, G. LANDWEHR, and E. BANGART, Solid State Commun. 45, 869 (1983).

/7/ R. HERRMANN, G. NACHTWEI, and W. KRAKK, phys. stat. sol. (a) 83, K207 (1984).

/8/ R. HERRMANN, W. KRAAK, and G. NACHTWEI, Solid State Commun., in the press (1984).

/9/ R. HERRMANN, W. KRAAK, and G. NACHTWEI, phys. stat. sol. (b), to be published.

(Received July 16, 1984)

phys. stat. sol. (b) 125, K89 (1984)

Subject classification: 2 and 14.3.1; 14.3.4; 22.1.2

Institute of Physics, Czechoslovak Academy of Sciences, Prague[1)]

Drift Mobility of a-Si:H

By

J. KOČKA and K. ČERMÁK[2)]

The time of flight (TOF) technique is extensively used for the study of transport processes in amorphous hydrogenated silicon (a-Si:H), mainly for finding the electron (or hole) drift mobility. From the first TOF measurements done by Le Comber and Spear /1/ the electron drift mobility $\mu_e \approx 0.1\ cm^2/Vs$ has been found. After the continuous improvement of the undoped a-Si:H quality the same and other authors /2 to 4/ have found at room temperature now a typical value $\mu_e \approx (1\ to\ 2) cm^2/Vs$.

Recently Silver et al. (/5/ and references therein) have questioned the standard interpretation of TOF (mainly on the basis of the negative results of the reversal of the drift voltage during the drift of the photogenerated carriers and of the junction recovery) and concluded that the observed transient signals are related with the charge redistribution of the photoinduced space charge. They estimated the "true" drift mobility of a-Si:H $\mu_e \approx 800\ cm^2/Vs$, similar to crystalline Si.

Most recently Spear and Steemers /6/ and also Könenkamp et al. /7/ repeated under carefully optimised conditions the reversal of the drift voltage and found the reverse drift of photoinduced carriers. The main criticism of Silver et al.'s experiments concerned his long switching time and high excitation intensities.

We have done the standard "single shot" TOF measurements on Pt Schottky diodes from undoped a-Si:H (thickness 3.7 μm). A N_2 laser with the full or reduced intensity has been used, with less than 5 ns pulse duration. The drift field was applied ≈ 10 μs before the laser pulse. More experimental details and the raw experimental results are in /8/.

In Fig. 1 the room temperature non-dispersive transport of electrons photogenerated with the reduced laser intensity is seen. The voltage dependence of the corresponding reciprocal transit time is in Fig. 2. From the slope of this

---

1) Na Slovance 2, 180 40 Prague 8, Czechoslovakia.

2) VŠCHT, Leninovo nám. 565, 532 10 Pardubice, Czechoslovakia.

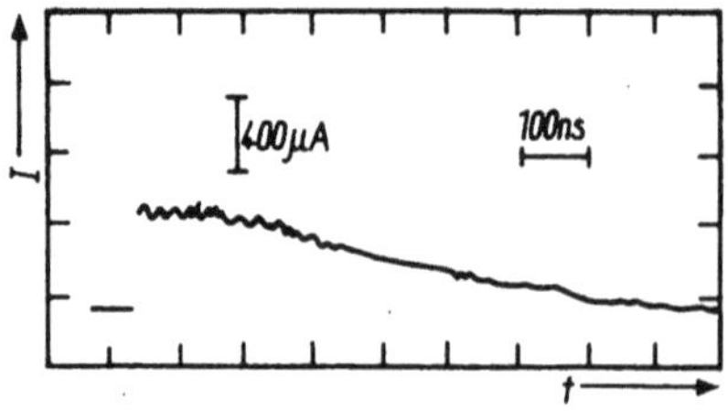

Fig. 1. Transient photocurrent for lower intensity of laser flash, measured on a-Si:H at 1 V (see text)

dependence the electron drift mobility $\mu_e = 0.6\ cm^2/Vs$ has been deduced.

The same experiment we have repeated on the same sample with the full laser intensity. The resulting current transient is in Fig. 3. A typical shape with a "cusp" /9, 10/ is observed which confirms that we have approached the space-charge limit. If we calculate /11/ the transit time t_T from the relation $t_c = 0.8\ t_T$ (where t_c is the "cusp" position) we can see from Fig. 2 that the resulting drift mobility (the slope) is also in this case $\mu_e = 0.6\ cm^2/Vs$.

We can speculate why Silver et al. /5/ did not observe the typical "cusp" shape, when they worked obviously near the space-charge limit. The reason may be, together with the high excitation level, the use of the relatively high repetition rate (although not specified we suppose 1 to 100 Hz - typical for used PRA lasers).

When the excitation level is low, both "single shot" and repetitive mode give the same results /10/. But the use of the repetitive mode near the space-charge limit can lead to the atypical results if no special precautions are used (for example "bleaching" laser pulses - see Spear /9/ or long waiting time ($\approx$10 s /4/) between following pulses).

We can conclude that even approaching the space-charge limit (high concentration of photoinduced carriers) still the correct drift mobility can be found when "single shot" TOF experiments (or repetitive one with the necessary precautions) are used. For the understanding of the reversal of the drifting carriers

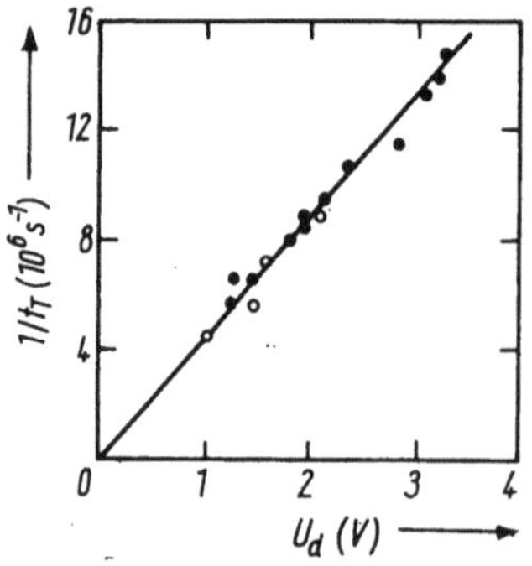

Fig. 2. Voltage dependence of the reciprocal transit time t_T (o taken from lower laser intensity results like in Fig. 1; ● taken from full laser intensitv results like in Fig. 3 and assuming $t_T = (1/0.8)t_c$)

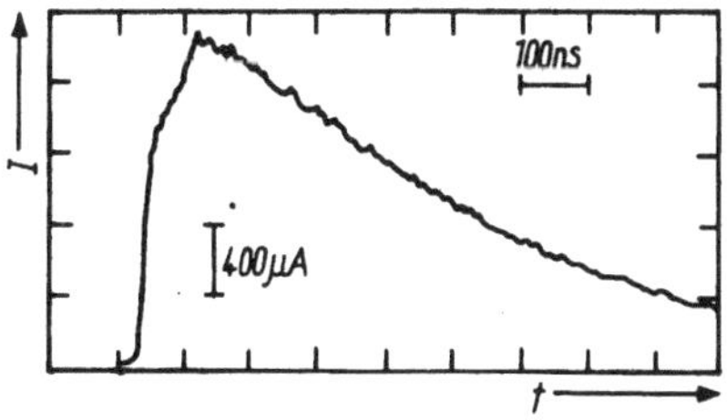

Fig. 3. Transient photocurrent measured on a-Si:H for full intensity of laser flash

under the high concentration of photoinduced charge more detailed experiments are necessary to explain the role of the "self-field", injection /7/ or repetition delay.

References

/1/ P.G. LE COMBER and W.E. SPEAR, Phys. Rev. Letters 25, 509 (1970).

/2/ T. TIEDJE, B. ABELES, D.L. MOREL, T.D. MOUSTAKAS, and C.R. WRONSKI, Appl. Phys. Letters 36, 695 (1980).

/3/ W.E. SPEAR and H. STEEMERS, Phil. Mag. B 47, L77 (1983).

/4/ R.A. STREET, Phys. Rev. B 27, 4924 (1983).

/5/ M. SILVER, E. SNOW, B. WRIGHT, M. AIGA, L. MOORE, V. CANNELLA, R. ROSS, S. PAYSON, M.P. SHAW, and D. ADLER, Phil. Mag. B 47, L39 (1983).

/6/ W.E. SPEAR and H.L. STEEMERS, Phil. Mag. B 47, L107 (1983).

/7/ R. KÖNENKAMP, A.M. HERMANN, and A. MADAN, Phil. Mag. B 48, L33 (1983).

/8/ K. ČERMÁK and J. KOČKA, Czech. J. Phys. B34, 216 (1984).

/9/ W.E. SPEAR, J. non-crystall. Solids 1, 197 (1969).

/10/ P.B. KIRBY, W. PAUL, P. JACQUES, and J.L. BREBNER, AIP Conf. Proc. 73, 207 (1981).

/11/ S.Z. WEISZ, A. COBAS, S. TRESTER, and A. MANY, J. appl. Phys. 39, 2296 (1968).

(Received July 3, 1984)

Pre-Printed Titles

Papers to be published
in "physica status solidi (b) - basic research"
Vol. 125, No. 2

Original Papers

B. A. VAID and K. C. SHARMA, Simla
Ultrasonic Attenuation in Liquid Metals

G. AMBRAZEVIČIUS, G. BABONAS, and YU. V. RUD, Vilnius
Excitation Spectra of Photoluminescence Bands in $Cd_{1-x}Mn_xTe$

C. PRASAD and J. D. DUBEY, Ranchi
Electronic Structure and Properties of Cubic Boron Nitride

M.I. VLADIMIR, V.A. MOSKALENKO, and V.V. SKAFARU, Kishinev
The Distribution of the Exchange Magnetic Fields in Spin Glasses

V. ČÁPEK, Prague
To the Interplay of Vibrational Relaxation and Exciton Motion

P. PETELENZ, Cracow
Optical Detrapping in Molecular Crystals (II)

V. SURIYANARAYANAN, S. MADHAVAN, and K. RAMACHANDRAN, Madurai
Isotope Diffusion in Cadmium Telluride

Y. MATSUOKA, S. TERAKADO, and S. TANIGAWA, Nagoya and Ibaraki
Positron Age-Momentum Correlation in Ultrafine Particles of Silver and Gold

S.W. BIERNACKI, Warsaw
Born-von Karman Cell Model for Deep Impurities

A. Z. MAKSYMOWICZ, T. STOBIECKI, E. JAROCKI, and W. KARAŚ, Kraków
Charge Transfer and Concentration Dependence of Magnetization in Amorphous Fe-B Alloys

M. A. ANDREEVA and R. N. KUZMIN, Moscow
Mössbauer Conversion Electron Spectra at Total Reflection

T. IWASHITA and N. URYÛ, Kumamoto and Fukuoka
The Curie Temperature of the Ising Ferromagnet on the Honeycomb Lattice with Mixed Spins

A.N. VASILEV and V. ČÁPEK, Moscow and Prague
Vibrational Relaxation in a Localized Excited Electronic State by the GME Method

K. ŽIČKUS, A. AUDZIJONIS, J. BATARŪNAS, and A. ŠILEIKA, Vilnius
The Fundamental Absorption Edge Tail of Ferroelectric SbSI

D. FRÖHLICH, A. NÖTHE, and K. REIMANN, Dortmund
Determination of Valence Band Parameters in ZnTe

M. KORKMAZ, B. AKTAŞ et M. DUPONT, Ankara et Aubière
Le spectre de R.P.E. de l'ion Mn^{2+} dans le tartrate de cadmium ($CdC_4H_4O_6 \cdot 5H_2O$)

S.K. PLACHKOVA and L.A. TERSIEV, Sofia
Phase Transitions and Intervalley Jahn-Teller Effect in the System $(Ge-Te)_{1-x}(AgSbTe_2)_x$

U.W. POHL, H.-E. GUMLICH, and W. BUSSE, Berlin
Zero-Phonon Lines in the Excitation Band Centred at 23 200 cm^{-1} Due to Mn^{2+} in Polymorphic ZnS

O.N. BORGES, L.A. AMARANTE RIBEIRO, and L.L. GONÇALVES, Belo Horizonte and Fortaleza
The XY Model with Nearest and Next-Nearest Neighbour Interaction in a Transverse Field

H. WERHEIT and R. FRANZ, Duisburg
States of Thermal Equilibrium in β-Rhombohedral Boron and Their Alteration by Optical Excitation

S.L. KAKANI and U.N. UPADHYAYA, Bhilwara and Udaipur
Superconducting and Magnetic Order Parameters in the Coexistence Region of Ternary Superconductors

F. ENGELKE, D. MICHEL, and F. PILLE, Leipzig
NMR Study of Dynamical Processes in Crystals of the TSCC Family

C. PISANI and E. MARCHESE, Torino
Critical Analysis and Simplified Formulation of the Travelling Cluster Approximation for Disordered Systems

V.A. KASIYAN, D.D. NEDEOGLO, A.V. SIMASHKEVICH, and I.N. TIMCHENKO, Kishinev
Hopping Conductivity and Electron Localization in Heavily Doped ZnSe

G. MILCHBERG, K. SAMINADAYAR, E. MOLVA, and H.R. ZELSMANN, Grenoble
Optical Determination of the Valence-Band Parameters in CdTe

A.A. ANDREEV, A.V. ZHERZDEV, A.I. KOSAREV, K.V. KOUGIA, and I.S. SHLIMAK, Leningrad
Photoluminescence and Recombination of Excess Carriers in Amorphous Hydrogenated Silicon

S.V. VOITIKOV and V.P. GRIBKOVSKII, Minsk
On the Theory of Crystal Excitation by Electric Discharges

I.G. RATISHVILI, Tbilissi
Annihilation of Frenkel Pairs in B.C.C. Metals (α-Iron) (I)

G. HEDER and J.M. SPAETH, Paderborn
ENDOR Investigation of H_i^0(Li) Centres in KCl

P. STŘEDA, Prague
Universal Transport Properties of Inversion Layers in Quantizing Magnetic Field

Papers to be published
in "physica status solidi (a) - applied research"
Vol. 85, No. 1

Review Article

V.A. PAVLOV, V.I. SHALAEV, and V.M. ALYABIEV, Sverdlovsk
Effect of Dislocation Structure and Creep and Fracture of Metals and Alloys (II)

Original Papers

V.E. SIDOROV, V.S. GUSHCHIN, and B.A. BAUM, Sverdlovsk
Magnetic Structure of γ-Iron Containing Oxygen

M. MESSAOUDI and J. DURAN, Paris
Conductivity in One-Dimensional Disordered Cs_2TCNQ_3 Salts

P.A. BESIRGANYAN, R.TS. GABRIELYAN, V.G. KOHN, and A.G. TONEYAN, Erevan and Moscow
Six-Beam X-Ray Diffraction in Ge Single Crystals

M. GABER and H.-J. FITTING, Rostock
Energy-Depth Relation of Electrons in Bulk Targets by Monte-Carlo Calculations

K. ŽĎÁNSKÝ and N.T. THUC HIEN, Prague
Analysis of DLTS Curves of Aggregated Deep Level Impurities

A.I. GUSEV, Sverdlovsk
Structural Vacancies in Nonstoichiometric Compounds at High Pressure

V.I. MALTSEV and V.G. VOLOGIN, Tomsk and Sverdlovsk
Neutron Diffraction Studies of the Magnetic Ground State in Li-Zn Ferrite

N. KURAMOCHI, H. MIZUBAYASHI, and S. OKUDA, Ibaraki
Dislocation Pinning at Low Temperatures in Nb

H. BERGER, YU.A. KULYUPIN, S.A. NEPIJKO, and N.N. SEDOV, Berlin and Kiev
Quantitative Interference Electron Microscopy for Spherical Objects

P. GAWORZEWSKI, E. HILD, F.-G. KIRSCHT, and L. VECSERNYÉS, Berlin, Frankfurt, and Budapest
Infrared Spectroscopical and TEM Investigations of Oxygen Precipitation in Silicon Crystals with Medium and High Oxygen Concentrations

J.J. ŻEBROWSKI and A. SUKIENNICKI, Warsaw
On the Inhibition of Dynamic Conversion in Bubble Garnet Domains by a Thin Midplane Layer

Y. KAMIURA and F. HASHIMOTO, Okayama
Origin of a Shallow Acceptor in Quenched Germanium

H.M. LEDBETTER, Boulder, Colorado
Monocrystal-Polycrystal Elastic Constants of a Stainless Steel

S.A. ALTEROVITZ, G.H. BU-ABBUD, J.A. WOOLLAM, and D.C. LIU, Lincoln, Nebraska, and Cleveland, Ohio
Structural and Chemical Analysis of a Silicon Nitride Film on GaAs by Null Ellipsometry

P.F. LUGAKOV and T.A. LUKASHEVICH, Minsk
Formation and Parameters of Boron-Divacancy Complexes in Irradiated p-Si

B.A. MEN and M.L. LEVITAN, Sverdlovsk
Analysis of the Ground State of a Solid Solution with Interactions up to k-th Co-ordination Spheres of the Crystal Lattice

P. LILLEY, M.R. CZERNIAK, and J.E. NICHOLLS, Manchester and Hull
Photoluminescence and Electrical Properties of Vapour Phase Epitaxial ZnSe Grown on GaAs

L. PŮST and Z. FRAIT, Prague
Method for Measurement of Magnetostriction Constants in (001) Thin Films Using FMR

D.S. MISRA, A. KUMAR, and S.C. AGARWAL, Kanpur
Intensity and Temperature Dependence of Steady-State Photoconductivity down to 8 K and DOS Distribution Obtained from These Measurements in a-Si:H

H. BOOYENS and J.H. BASSON, Pretoria
Piezoelectrically Induced Charge Distributions around Dislocations in CdTe and HgCdTe

19. Magnetic Resonance

20. Optical Properties
    20.1 Spectra. Optical Constants (X-Ray Spectra See 20)
    20.2 Lasers
    20.3 Luminescence (X-Ray Spectra See 20)

21. Metals. Alloys
    21.1 Transition Metals (Group Sc to Zn) and Their Alloys
        21.1.1 Fe and Fe Alloys
    21.2 Alkali Metals
    21.3 Alkali-Earth Metals
    21.4 Rare-Earth Metals
    21.5 Actinides
    21.6 Noble Metals
    21.7 Semi-Metals (Semiconducting Alloys 22.7)

22. Semiconductors and Ionic Crystals
    22.1 Elements
        22.1.1 Germanium
        22.1.2 Silicon
        22.1.3 Group VI Elements
    22.2 III–V Compounds
        22.2.1 Arsenides
        22.2.2 Phosphides
        22.2.3 Antimonides
        22.2.4 Mixed Crystals
    22.3 IV–IV, II–V, and III–VI Compounds
        22.3.1 Mixed Crystals
    22.4 II–VI Compounds
        22.4.1 Sulphides
        22.4.2 Selenides
        22.4.3 Tellurides
        22.4.4 Mixed Crystals
    22.5 Halides
        22.5.1 Silver Halides
        22.5.2 Alkali Halides
        22.5.3 Alkali-Earth Halides
        22.5.4 Mixed Crystals
    22.6 Simple Oxides (CdO, ZnO See 22.4)
        22.6.1 Mixed Crystals
    22.7 Semiconducting Intermetallic Compounds
        22.7.1 Mixed Crystals
    22.8 Tertiary and Higher Compounds
        22.8.1 Salts of Oxy-Acids (Carbonates, Phosphates, Silicates Including Mica etc.)
        22.8.2 Spinels, Garnets, Ferrites
    22.9 Organic Semiconductors

23. Solidified Gases

Attention

The preceding classification scheme is closely akin to the scheme used by Physics Abstracts to classify the various fields of solid state physics. A paper will either be classified under one or more of the subsections or, if none of these are relevant, it will just appear under the general section heading (i. e. sections and subsections are equivalent in this sense).

The *subject classification* given at the head of each paper begins with the main subject matter of the paper, and this is followed, if necessary, by figures referring to subsidiary subject matter and the type of solids investigated.

The papers published in this issue are listed in the order of their subject classification at the end of the table of contents.

Manuscripts and letters for physica status solidi (b) — basic research and physica status solidi (a) — applied research

should be sent to

DDR-1086 Berlin, Leipziger Straße 3–4, P. B. 1233

or

Internationaler Buchversand GmbH, 1000 Berlin 19, Eschenallee 21.

The object of "physica status solidi (b)" and "physica status solidi (a)" is to publish rapidly important and topical results concerning fundamental research in the field of solid state physics and aspects of application and experimental techniques of fundamental solid state research, respectively. The issues of "physica status solidi (b)" and "physica status solidi (a)" will be published on the 1st and the 16th day of each month, respectively. Either series will contain the following categories:

1. Review articles written by arrangement with the editors. They will be remunerated.
2. Original papers selected from those which have not been submitted for publication elsewhere, after being reviewed. (The latest date for acceptance of contributions is the 8th and the 23rd day of each month for "physica status solidi (b)" and "physica status solidi (a)", respectively, 50 days before the delivery of the journal.)
3. Short notes published in the off-set part of the journal. They are, if necessary after being reviewed, selected by editorial staff. (The latest date for acceptance of manuscripts for this part is the 12th and the 27th day of each month for "physica status solidi (b)" and "physica status solidi (a)", respectively, 16 days before the delivery of the journal.)
4. Pre-printed titles of papers to be published in the next issue of both series.

Editorial Information

Manuscripts should be submitted preferentially in English. Besides, papers in German, French, and Russian are admitted. The final decision on the incorporation of the papers in the series (b) or (a) is taken by the editorial staff after consultation of the referee. Authors are asked to observe the following rules:

1. Manuscripts have to be written as concisely as possible. As most, 20 and 4 pages (28 lines per page; abstracts, tables, references, figures, and captions of figures included, where 3 figures are counted as about one page) are admitted for original papers and short notes, respectively. All manuscripts should be type-written on one side only. Two copies are required. Badly arranged manuscripts are returned.
2. The author's name, initials, and institution should be given.
3. One abstract of about ten lines in English and one in German, French, or Russian are required for original papers.
4. Because of lack of space conclusions (if any) must not have the character of a summary (i. e. substantially a repetition of the contents of the abstract).
5. Formulae should not be written too small and not with pencil. Separate lines for formulae are desirable. Indistinct writing will lead to a return of the manuscript.
6. Drawings should be enclosed on separate sheets of paper, not exceeding the maximum and minimum sizes of 20×30 and 10×15 cm², respectively. Photographs should be submitted, on white paper, only for review articles and original papers. Duplicates are required.
7. Captions for the figures should be submitted on separate sheets of paper, and special instructions with regard to the size of illustrations should be given on these.
8. References should be numbered in order of appearance and should be given, on a separate sheet, in the form used in physica status solidi.
9. SI-units should be used.
10. Each paper should contain a proposal for the subject classification according to the classification scheme appearing on the second and third cover of each issue.

The authors will be sent galley proofs of review articles and original papers. Proof reading of short notes and pre-printed titles will be done by the editorial staff.

For each original paper and short note a total of 75 reprints is available free of charge. The number of reprints free of charge per review article is 25. Requests for additional reprints (against payment) should be given together with the submitted manuscripts. **The deadline for orders of additional reprints is the 15th and the 30th of each month** for "physica status solidi (b)" and "physica status solidi (a)", respectively, two weeks before the delivery of the journal.